Cognitive Science and Technology

This series aims to publish work at the intersection of Computational Intelligence and Cognitive Science that is truly interdisciplinary and meets the standards and conventions of each of the component disciplines, whilst having the flexibility to explore new methodologies and paradigms. Artificial Intelligence was originally founded by Computer Scientists and Psychologists, and tends to have stagnated with a symbolic focus. Computational Intelligence broke away from AI to explore controversial metaphors ranging from neural models and fuzzy models, to evolutionary models and physical models, but tends to stay at the level of metaphor. Cognitive Science formed as the ability to model theories with Computers provided a unifying mechanism for the formalisation and testing of theories from linguistics, psychology and philosophy, but the disciplinary backgrounds of single discipline Cognitive Scientists tends to keep this mechanism at the level of a loose metaphor. User Centric Systems and Human Factors similarly should inform the development of physical or information systems, but too often remain in the focal domains of sociology and psychology, with the engineers and technologists lacking the human factors skills, and the social scientists lacking the technological skills. The key feature is that volumes must conform to the standards of both hard (Computing & Engineering) and social/health sciences (Linguistics, Psychology, Neurology, Philosophy, etc.). All volumes will be reviewed by experts with formal qualifications on both sides of this divide (and an understanding of and history of collaboration across the interdisciplinary nexus).

Indexed by SCOPUS

Vinit Kumar Gunjan · Sabrina Senatore ·
Amit Kumar
Editors

Cybernetics, Human Cognition, and Machine Learning in Communicative Applications

Springer

Editors
Vinit Kumar Gunjan
Department of Computer Science and Engineering
CMR Institute of Technology
Hyderabad, Telangana, India

Sabrina Senatore
Department of Computer Science
University of Salerno
Fisciano, Italy

Amit Kumar
BioAxis DNA Research Centre Private Limited
Hyderabad, Telangana, India

ISSN 2195-3988 ISSN 2195-3996 (electronic)
Cognitive Science and Technology
ISBN 978-981-97-8535-3 ISBN 978-981-97-8533-9 (eBook)
https://doi.org/10.1007/978-981-97-8533-9

This Springer imprint is published by the registered company Springer Nature Singapore Pte Ltd.
The registered company address is: 152 Beach Road, #21-01/04 Gateway East, Singapore 189721, Singapore

Contents

Cluster Formation for Lane and Road Based Stability in Infrastructure-Based Vehicular Ad Hoc Networks (LRSC)

Karne Radhakrishna and T. K. Sreeja

Abstract A system called the Vehicular Ad hoc Network (VANET) provides a range of services, including travel safety and driving assistance, to automobiles that are on the road. Wireless links will be established between vehicles, referred to as vehicle to vehicle communication, and between vehicles and infrastructure, referred to as vehicle to infrastructure network. Quick data transmission is thus provided to the vehicles via wireless and infrastructure connections. Data transmission delays should be avoided, data should be sent to cars without loss, and redundancy should be minimized. If this is the case, the network can quickly send data between vehicles. One of the difficulties is establishing and maintaining a connection between vehicles going across many lanes and roadways. The recommended method for LRSC-lane and road based stable clustered structures vehicular Ad hoc network is used to transmit data between clusters. Cluster data transmission is a cutting-edge method for efficiently managing network resources and delivering data quickly. For data transfer, the network employs hop-based transmission and cluster-based communication. If the network is sparse, nodes communicate with their next hop and roadside units; if the network is dense, nodes form cluster communication. To operate the network without disconnections, the suggested method produces a dual leader election, which includes the primary leader and the gateway. If the primary leader crosses the network area's transmission range owing to mobility, the gateway will take over as the leader immediately; in the meanwhile, the new vehicle will be elected as the region's new leader. The leader will be chosen based on numerous characteristics, including the fact that the leader should be placed in the middle of all vehicles in the region, and that all cars must drive in the same direction, angle, and speed. Hop-based and cluster-based routing assist the network in balancing routing in dense and sparse environments.

Keywords VANET · Vehicle to vehicle · Vehicle to infrastructure · Cluster routing · Hop-based routing · Dual leader

---

K. Radhakrishna (✉) · T. K. Sreeja
Electronics and Communication Engineering, Noorul Islam Centre for Higher Education, Kumaracoil, Kanyakumari, Tamilnadu, India
e-mail: krk.wgl@gmail.com

V. K. Gunjan et al. (eds.), *Cybernetics, Human Cognition, and Machine Learning in Communicative Applications*, Cognitive Science and Technology,
https://doi.org/10.1007/978-981-97-8533-9_1

1 Introduction

As seen in Fig. 1, a vehicle Ad hoc network is a heterogeneous wireless network. This network will give vehicles information such as road conditions, safety features, and data transmissions. This data aids in the establishment of wireless connection between vehicles. For instance, the region provides considerable snowfall on the roads, road repairs, deviations owing to road construction, information regarding delays brought on by traffic, accidents, and other reasons, emergency assistance, and many current events pertaining to roads. Thus, before driving on the road, automobiles can choose to move in accordance with the circumstances; it's crucial to understand the road's environmental conditions in order to follow the same route or change the roads; other vehicles that drive notice the situation receive these messages; if the packets are noticed, they will broadcast them to other cars via broadcast.

Furthermore, vehicles or roadside units will communicate the area's travel rules, safety features, and special characteristics. The Ad hoc vehicular network creates a safe road travel environment for vehicles. They also provided free access to the internet, weather reports, free video and audio content, and other entertainment-related information.

In addition to this, wireless connection maintenance on road is the toughest one due to high speed and mobility deviations, apart from these changes due to dense and sparse environment transmission may collide or it may not deliver to the destination. Forwarder should select at the same speed as neighboring vehicles, in addition, the transmissions will disappear due to weather, road deviations, detours, junctions, etc. These environments create loss of connections during communication, so the transmission is difficult to share data immediately. Channel selection is also a significant parameter to transfer the data. So, the vehicle can choose the channel according to the data size. Similar to this, redundancy can occasionally happen whenever data is shared because of high traffic, which will lead to congestion. All of these problems require precise protocols to address them.

Each vehicle travels at varied speeds and directions in the proposed LRSC network, four-lane highway scenario built and assessed, so the vehicles select a

Fig. 1 Model of the vehicular Ad hoc network (VANET)

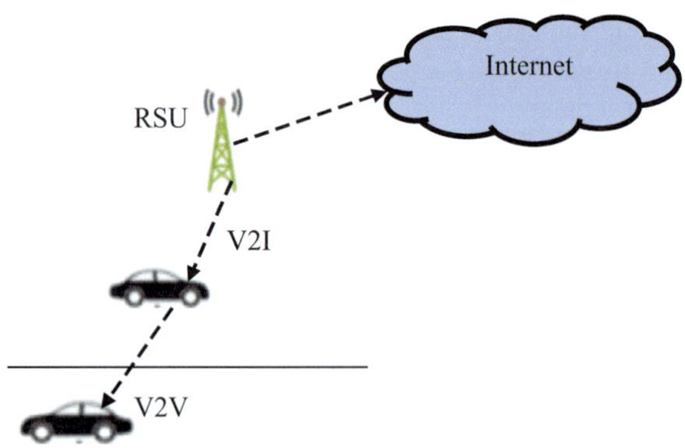

forwarder with uniform speed, within the coverage region, and establish network connections. Then, using various parameters, they choose a leader in the cluster group. As a result, a dual leader was chosen, with two leaders known as the main and gateway nodes. When the primary leader departs the territory, the gateway leader takes over. The suggested LRSC focused on choosing the stable network connection in a high dynamic network through vehicle to vehicle connection, vehicle to roadside unit connection, cluster routing, and hop-based routing since clusters change often owing to speed variation at different lanes. The vehicle has multiple routing knowledge to transfer packets, allowing it to manage transmissions in both dense and sparse environments, thus reducing packet loss.

In part 2, we discussed previous VANET research relating to the suggested technique. The implementation of the LRSC protocol as lane and road oriented stable clustering in infrastructure vehicle ad hoc network was covered in Sect. 3. Section 4 presents the VANET's performance in the form of findings and discussion. The conclusion of this document is explained in the final section, along with future work.

2 Related Works

An increasing amount of complete and personalized data may be exchanged through a wireless connection. In addition to continuing wireless communication research, VANET addresses all of those vehicular connectivity problems [1]. In these mobile networks, vehicles communicate and self-organize among themselves; no infrastructure or centralized coordinating body is needed. Mobile vehicles in VANET could act as stationary sensors in infrastructure-based networks [2]. It uses gadgets with DGPS instead and GPS capabilities. It makes use of sensing technology and wireless communications detection. Over the past few years, we have examined numerous research initiatives on a variety of issues associated with V2R and V2V [3]. This massive network is extremely dynamic due to the rapid mobility of nodes or vehicles on structured roads and street infrastructure. Every vehicle in the network is equipped with wireless technology so that it can broadcast or exchange data with other automobiles [4]. In order to overcome the current challenges and develop superior VANET applications, we have carried out extensive research across multiple networking layers. Authors [5] recommended a safety-driven tracking process and a transmission control technique for adjusting communication power and rate in response to vehicle network dynamics [5].

They should investigate how the propagation model, intensive traffic network situation, and channel model affect VANET network efficiency [6]. Forged position data can be broadcast by an unauthorized vehicle in the network, leading to the failure of safety applications. For vanet navigation, security usage, traffic control, and data aggregation, accurate vehicle position data is necessary [7]. It has been proven that algorithms for clustering are a great way to turn a network into a hierarchy of interconnected wireless networks. One criterion and a method for merging and processing them should be considered by advanced CH selection algorithms

[8]. VANET is an appropriate method for adopting driverless automobiles since it integrates the upcoming generation of communications technologies into vehicles. Therefore, increasing roadway security and transportation efficiency is the main objective of VANET applications [9]. Researchers and the transportation industry are both fascinated with VANET, a significant and promising MANET application.

Vehicles with onboard technology and acting as mobile nodes are included in VANET [10]. Energy usage, on the other hand, is the second component of the Inter-Vehicle Communication system since the node sensory mobile vehicle supplies enough high electric power.

Security and privacy are the main concerns and elements in inter-vehicle communications systems [11]. In particular, since nodes in VANETs are restricted to road topology while in motion, we can anticipate a vehicle's future position if road data is available. In addition, vehicles can supply a significant amount of computation, communication, and sensing power, in addition to ongoing transmission authority, to support these tasks [12]. Safety applications provide critical and safety details via additional vehicles' sensor data or still roadside units (RSUs) in order to prevent and report issues like road conditions and upkeep data, immediate stop warning provided from a before distant vehicle, and accident notification [13]. A vehicular ad hoc network is an aspect of a mobile area network (MANET), which is a network that enables wireless communication between vehicles and between vehicles and the roadside [14]. Nodes in this autonomous, independent wireless communication network exchange and share information by acting as servers or clients [15]. It is the objective of secured routing protocols to offer path validity and authentication. They don't stop data manipulation or eavesdropping, nor do they completely address communication security. End-to-end cryptography must be used by hosts going forward in order to defend against these threats [16]. In order to facilitate proper routing, an ad hoc routing protocol needs to be able to find the optimal route among nodes having unidirectional links, minimize routing overhead, reduce the amount of time needed for convergence following topology changes, and maximize capacity utilization.

Consequently, creating routing support is one of the main research topics in ad hoc networks [17].

Through vehicle DSRC or Wireless Access for Vehicular Environments, V2V and V2I are utilized for safety, mobility, and environmental applications. Based on Wi-Fi, it is the basic set of standards that enable collaborative, safety–critical V2X applications [18]. The development of efficient vehicle communications is hampered by several technical issues. A vital first step in establishing efficient vehicle communications is offering a reliable routing mechanism over VANETs [19]. Wireless self-organized networks consisting of multiple small nodes with restricted memory, processing power, and connectivity are distributed by batteries[20]. Because a node's energy is useless after it is used up, WSNs are less durable than typical wireless networks [21].

3 LRSC Design and Implementation

3.1 Modeling the Network

The network is made up of vehicles $V = v\_1, v\_2, \ldots ..v\_n$, roadside units R_SU, base stations B_S, a gateway G_W, and an internet server I_S. Each vehicle has neighbors $N = n\_1, n\_2, \ldots ..n\_n$ within its coverage radius, which are discovered by node announcements published on a regular basis. Then, in order to build the cluster, each node should determine its neighbor ratio N_R.

$$N\_R = N/V$$

Beacon messages are often sent at the start of the network to connect vehicles and preserve immediate neighbors as D_HN. Furthermore, the D_ist distance (the difference in distance between each neighbor) must be maintained continuously from the network's creation. The position points of each vehicle, such as (X_1, X_2), & (Y_1, Y_2), must be updated in order to compute the distances shown below.

3.2 Vehicle Parameter Validations

To populate the neighbor list, each vehicle broadcasts a hello message together with its global positioning GPS points (X_1, X_2), and (Y_1, Y_2) and time stamp on a regular basis. The message receiving vehicles update the information contained in it and the neighbor table, allowing whole network nodes to broadcast the hello message and update their neighbors at regular intervals, thereby updating the direct neighbor list. When the neighbor list is refreshed, they begin calculating the distance D_ist between each of them.

$$D\_ist = \surd([|X\_1 - X\_2|]^2 + |Y\_1 - Y\_2|^2)$$

Here, X_1 and Y_1 are the receiver sensor location points, while X_2 and Y_2 are the transmitter location points.

$$UpdateX\_1, Y\_1andX\_2, Y\_2$$

Furthermore, constantly watch the vehicle movements and determine the moving directions M_D and angles A to update the new neighbor list N_L.

$$\theta = atan2(Y, X)$$

$$A = \theta 180/\pi$$

The announcements are continuously broadcast to the network via the R_SU, B_S, G_W, and I_S at regular intervals. The network is formed by the announcements of relevant nodes. In this manner, internet server messages are received by gateways, and gateway messages are received by internet server and base stations. The message from the roadside device will be received by the vehicles and the base station. The devices are linked together and form the network as a result of this. These are known as infrastructure devices, and they will be placed statically. However, the vehicles are moving M through the streets and lanes. As a result, they must update the position change to update the neighbors and construct the cluster among the nodes. As demonstrated in Fig. 2, the N_L should update at short intervals t based on vehicle speed S and deviations D.Here, V is the variance of the speed variations as shown in Table 1.

$$S = D\_ist/t$$

$$V = S\_1 \ldots .S\_n$$

$$D = \sqrt{V}$$

If the node does not continually hear the neighbor hello message, the neighbor entry should be removed to update the fresh N_L; otherwise, stale contacts will be in the neighbor list, leading to the formation of a false cluster set, if so, it will increase the data loss.

$$If Y\_2 > Y\_1$$

$$M\_D = Updirection$$

$$else M\_D = DownDirection$$

$$If X\_2 > X\_1$$

$$M\_D = Right$$

$$else M\_D = Left$$

$$UpdateDirection = M\_D$$

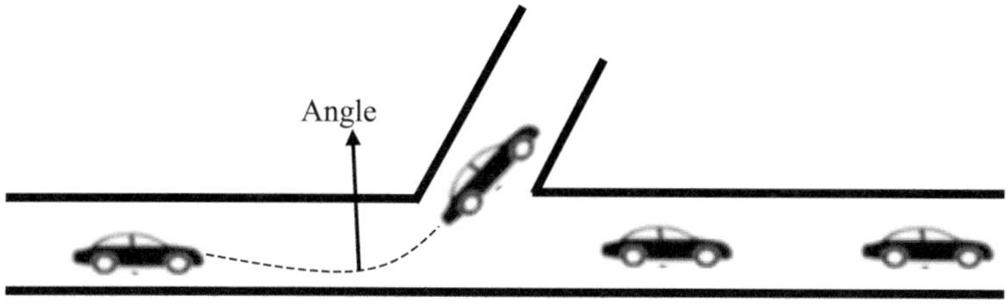

Fig. 2. Mobility deviations

Table 1 Neighbor list

Vehicle ID	Geographical location points	Moving speed	Moving angle	Direction	Contact time

4 Cluster Formation

At the start of the network's operation, all nodes behave normally. The vehicles then began to update the settings via neighbor list formation. Then they began to construct the cluster, and each cluster elected a leader. As a result, they locate the middle vehicle M_V within the C_R.

$$D\_ist(n, s) \leq C\_R$$

The vehicle in this example is the sender, denoted by s, and the neighbor, n. The neighbors near the mid vehicle should update their one hop neighbor list as well as their position.

$$s\_N = (n \in \mathbf{N}) \leq C\_R$$

Similarly, to reconstruct the cluster set, the vehicle's current position C_PX, C_PY and old position P_PX, P_PY should be updated on a regular basis. As a result, position deviation is calculated as follows.

$$P\_D = \frac{C_{PY} - P_{PY}}{C_{PX} - P_{PX}}$$

Additionally, record the total amount of vehicles within the vicinity and preserve it as indicated below.

$$D_A = XY$$

$$A_{NB} = \frac{V}{\pi D_A C_R}$$

$$D_R = D_S A_{NB}$$

$$D_S = (n_1 \rightarrow D_{ist}(n_2) < n_1 \rightarrow C_R)$$

Here, N_S denotes network size, this will indicate average number of neighbors moved in the area. The vehicle density ratio is determined using these two measurements. Each region's cluster is formed based on the density ratio. If a node in the region has front F_V and back vehicles B_V, the mid vehicle can compete for cluster leader L. If the region has more than one mid vehicle, they will compute the changes in speed, angle, and direction to determine the best mid vehicle to lead that round. It will send a cluster leader announcement once the leader has been elected. Vehicles inside the leader's coverage area will receive the message and save the leader's information before beginning to transmit and receive the messages known as members M of the leader.

$$CheckF_V > 1$$

$$CheckB_V > 1$$

$$CheckA, M_D, SissameandM_D < C_R$$

$$FormCluster$$

$$DcelareM_V asLeader$$

$$DeclareLeaderAnnouncemnt$$

$$EncloseNodeID, X, Y, A, S, M_D, LaneIDandvehiclestatusL$$

$$Ifv \leq C_R$$

$$JoinClusterandStoreL$$

As shown in Fig. 3, the clusters are linked to other clusters to build a large network and facilitate data sharing. Vehicle S, moving A, moving M_D, and time of interaction between leaders are updated to connect the leader-to-leader communication.

To determine these criteria, each vehicle needs to be encircled by other vehicles; then only may that vehicle be selected from among them. Nonetheless, they have to declare that they are self-leaders. From the start of the network's activity until it finishes, these parameters need to be monitored in order to update the new variations.

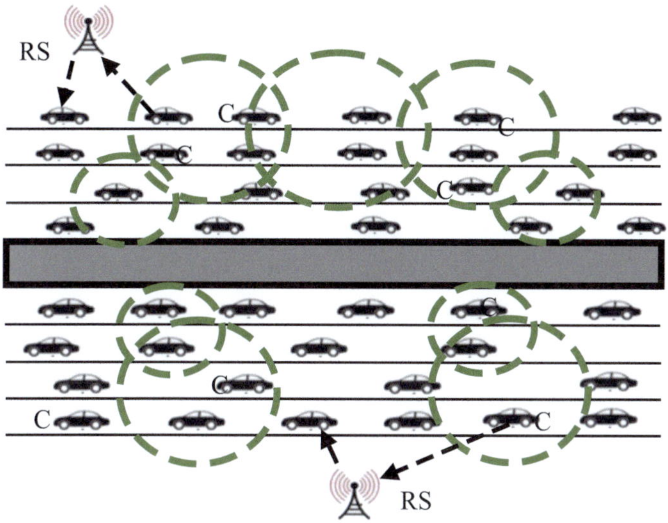

Fig. 3. Lane based clustering

The network has internal gateways G_W to connect leaders as follows: Here, C_X and C_Y indicates the centers of X and Y corresponding to L and G_W.

$$C_X = \frac{X_1 + X_2}{2} C_Y = \frac{Y_1 + Y_2}{2}$$

$$SelectG_W$$

The gateways are communicating between two leaders and will alter depending on the situation. The random interval R_I is used to select the channel randomly, and the G_W cancellation timer is calculated as follows: Also, M_T is the maximum required time to transfer the packets on the channel, also,M_{CR} is the maximum coverage range of the vehicle.

$$R_T = M_T R_I \left(1 - (\frac{V}{2M_{CR}} \right)$$

This technique decreased transmission of data loss and network delay because if any leader leaves the network at this point, this gateway assumes the role of the leader and keeps sharing information with other vehicles following the leader announcement.

$$Speed > 90\ km = HighSpeed$$

$$Speed > 50 \rightarrow 90\ km = Medium$$

$$Speed \leq 50\ km = Slow$$

$$CheckDirection = if\,(X_2 < X_1)$$

$$UpdateVehicle = F_V$$

$$elseB_V$$

The proposed VANET prioritizes hop-based routing if the network is sparse, and a stable cluster is formed if the network is dense. To accomplish this routing, they monitored the variations of the location, distance, direction, and angle, and based on these changes the hop routing and cluster renewal will be enabled to build the new path, and then the data will transfer in the network.

5 Results and Discussion

The network is made up of vehicles, roadside units, base stations, gateways, and internet servers that are randomly distributed in the topography area. The road is also separated into four lanes that run up and down. The vehicles established connections and transferred packets inside the coverage region and inter coverage region. The LRSC protocol is used by the network to build the path and transport packets. The parameter values utilized in the network are shown in Table 2.

The packet delivery ratio PDR represents the network's quality of service; it is calculated from source to destination to determine the path quality. Because of the routing technique, the proposed LRSC has a larger PDR than the SRCV, as illustrated in Fig. 4. The LRSC constructs the cluster path based on mobility using lanes and movement directions. Data on the network is exchanged at various intervals along with diverse packets on the path. The LRSC outperformed the SRCV in terms of

Table 2 Network parameters

Parameter	Value
Antenna	Omni Directional
Nodes	91
Network model	TwoRayGround
No. of gateways	2
No. of RS Units	4
Service provider	2
Protocol	LRSC

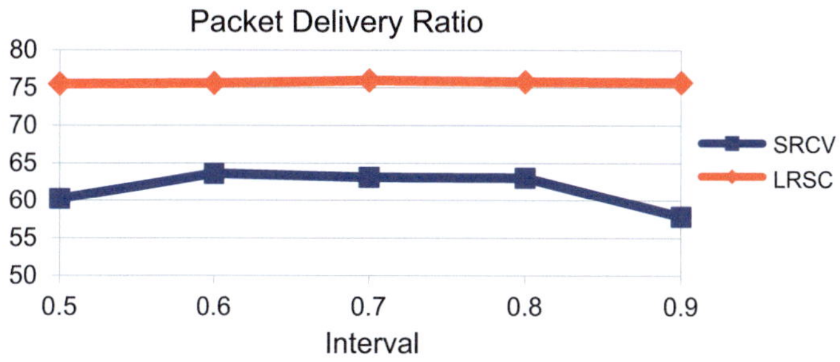

Fig. 4. Interval versus packet delivery ratio

PDR. Because the LRSC selected the leader to forward the packets based on many routing characteristics. As a result, more packets are delivered to their destinations.

Because of the dynamic nature of the VANET, path maintenance is complicated; the dynamic nature frequently disconnects the path. If this occurs, the network will experience significant packet loss. As a result, the routing decision must be carefully planned. Figure 5 illustrates that LRSC has less delay than SRCV due to path selection and path maintenance until packet delivery, giving better outcomes in LRSC. As a result, the network's delay is minimized, which comprises transmission delay, queuing delay, and path stability delay. All of these delays were used to calculate the end-to-end delay.

On wireless networks, packet drop is unavoidable. As a result, the network connection and its stability must be carefully monitored. To build the cluster, the proposed network monitors the transmission range nodes, and network density. Figure 6 indicates that the network stability in LRSC is better, with lower packet loss than in SRCV. The figure shows that packet sizes varied depending on the data to be transferred. The dynamic network constructs a stable cluster path by varying the distance, angle, direction, and speed of road lanes. The path is maintained using these metrics until the data is delivered.

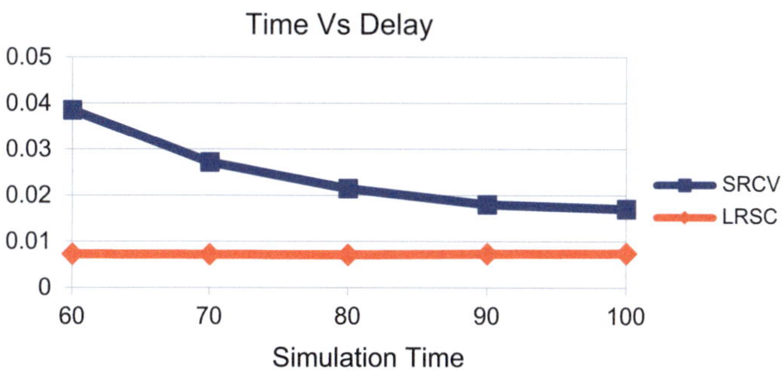

Fig. 5. Delay versus time

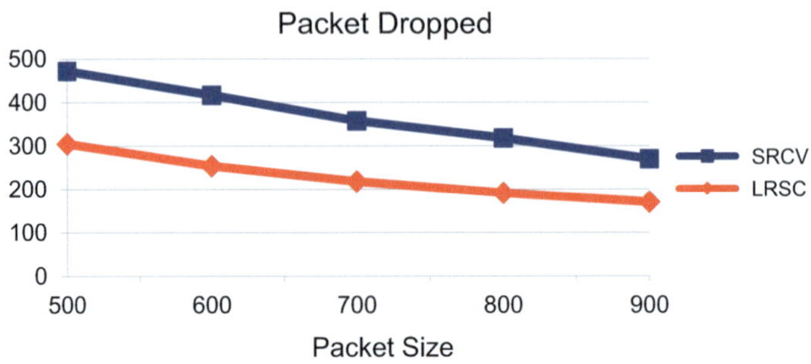

Fig. 6. Number of packet drop versus size of the packet

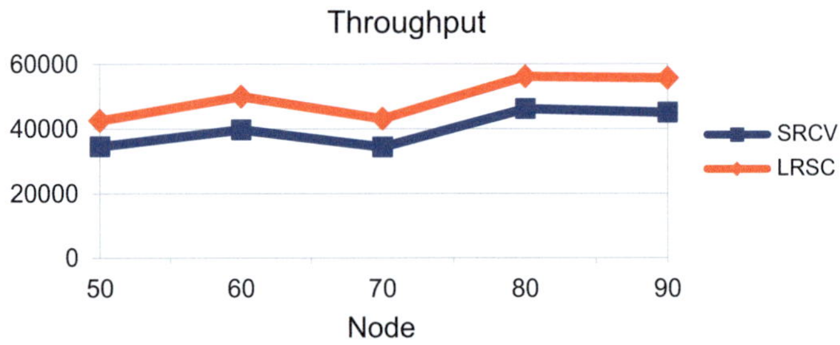

Fig. 7. Number of nodes versus throughput

Throughput is the number of bytes transformed into bits from packets sent to the destination. By varying the nodes, Fig. 7 is examined. The cluster changed as a result of the alterations, and they constructed a new way to transfer the packets. As a result, the LRSC has a higher throughput than the SRCV. The network's leadership will be elected in accordance with the reforms. The packets were transferred to the members and the next leader without redundancy, enhancing throughput.

6 Conclusion

A huge dynamic VANET was built in the study, and its changes and density were recorded to create the cluster. In addition, a vehicle in the cluster serves as the leader for a period of time. Leaders established communication between vehicles and infrastructure, reducing loss and redundancy. To keep the path until the data is sent, the nodes continuously check the path by angle and direction deviations, so the network can quickly update the changes to rebuild the cluster set. The scenario was constructed with four-way lanes so that vehicles might travel at different speeds and angles. Clusters of data are formed among these dynamic situations to reduce loss and redundancy. The packet delivery ratio and throughput increased as the performance was proposed. The optimization routing strategy will boost network performance in the future.

References

1. Kumar R, Dave M (2011) A comparative study of various routing protocols in VANET. Int J Comput Sci Issues (IJCSI) 8(4), No 1
2. Yeferny T, Hamad S (2020) Vehicular Ad-hoc networks: architecture, applications and challenges. Int J Comput Sci Netw Secur (IJCSNS) 20(2)
3. Rawat K, Hazrati J (2014) Vehicular Ad-hoc network. Int J Inf Comput Technol 4(9)
4. Bouk SH, Kim G, Ahmed SH, Kim D (2015) Hybrid adaptive beaconing in vehicular Ad Hoc networks: a survey. Hindawi Publ Corp Int J Distrib Sens Netw
5. Wang M, Shan H, Luan TH, Lu N, Zhang R, Shen X, Bai F (2014) Asymptotic throughput capacity analysis of VANETs exploiting mobility diversity
6. Malik S, Sahu PK (2019) A comparative study on routing protocols for VANETs. Accessed 15 Aug 2019
7. Arshad M, Ullah Z, Ahmad N, Khalid M, Criuckshank H, Cao Y (2018) A survey of local/ cooperative-based malicious information detection techniques in VANETs. EURASIP J Wirel Commun Netw
8. Calhan A (2015) A fuzzy logic based clustering strategy for improving vehicular ad-hoc network performance. Indian Acad Sci
9. Madli R, Varaprasad G, Imthiyaz MP (2018) A review of communication handoffs in vehicular ad-hoc networks (VANET) and its classification. IEEE
10. Laroiya N, Lekhi S (2017) Energy efficient routing protocols in Vanets. Adv Comput Sci Technol 10(5)
11. Kukade A, Sharma T (2015) A glance over the working methodology of VANET. Int J Adv Res Comput Commun Eng 4(12)
12. Liang W, Li Z, Zhang H, Sun Y, Bie R (2014) Vehicular Ad Hoc networks: architectures, research issues, challenges and trends
13. Abdulshaheed HR, Yaseen ZT, Salman AM, Al_Barazanchi I (2020) A survey on the use of WiMAX and Wi-Fi on Vehicular Ad-Hoc Networks (VANETs). IOP Publishings
14. Karne RK, Sreeja TK (2023) PMLC- predictions of mobility and transmission in a lane-based cluster VANET validated on machine learning. Int J Recent Innov Trends Comput Commun 1(5s):477–483. https://doi.org/10.17762/ijritcc.v11i5s.7109
15. Paul B, Paul B, Abu Naser Bikas Md. (2011) VANET routing protocols: pros and cons. Int J Comput Appl 20(3)
16. Goli-Bidgoli S, Mousa SofarAli M (2018) A new framework for secure routing in VANET. J Electr Comput Eng Innov (JECEI). Accessed 06 Apr 2018
17. Bodhy Krishna S (2017) Study of Ad hoc networks with reference to MANET, VANET, FANET. Int JS Adv Res Comput Sci Softw Eng 7(7)
18. Zhang H, Li J (2015) Modeling and dynamical topology properties of VANET based on complex networks theory. AIP Adv. Accessed 28 Jan 2015
19. Taleb T, Sakhaee E, Jamalipour A, Hashimoto K, Kato N, Nemoto Y (2007) A stable routing protocol to support ITS services in VANET networks. IEEE Trans Veh Technol 56(6)
20. Karne RK, Sreeja TK (2022) A novel approach for dynamic stable clustering in VANET using deep learning (LSTM) model. IJEER 10(4):1092–1098. https://doi.org/10.37391/IJEER.100454
21. Ramirez Reyna MA, Rivero Angeles ME, Villordo Jimenez I (2014) Monitoreo de trafico vehicular ensistemas V2I medianteeluso de una red inalambrica de sensores

Implementation of Predicting Diabetes Disease Using Machine Learning Based Unified Framework

Kumbala Pradeep Reddy, C. Ramakrishna, V. Sridhar Reddy, Thotakura Veeranna, Sarangam Kodati, and T. Benarji

Abstract One of the chronic and deadliest diseases that cause an increase in the sugar level in blood is diabetes. In fact, diabetes can inflict many severe effects like burning extremities, kidney & heart failures, myopia, and blurred vision. Diabetes develops when the body cannot produce enough insulin to maintain the threshold or when blood sugar levels are above a certain level. In HCS (Healthcare Services), Machine learning has gained a signification position because of its improving ability of disease prediction in healthcare services. The tedious identifying technique needs the patient should consult a doctor and visit a diagnostic center. This critical problem was solved by the rise of machine learning techniques. Hence there is a requirement for unified framework designing & diabetes prediction implemented by machine learning is proposed here. The three basic classifications of machine learning algorithms are Naive Bayes, Decision Tree, and support vector machine (SVM) used in this proposed system for detecting diabetics at the early stage. From the sources of UCI (University of California) machine learning respiratory the experiments are performed on PIDD (Pima Indians Diabetes Database). The three algorithms' performance is evaluated by different parameters such as F-Measure, Accuracy, Recall, and Precision measured through classified instances incorrectly or correctly. From the

K. P. Reddy (✉)
Department of CSE, CMR Institute of Technology Hyderabad, Hyderabad, Telangana, India
e-mail: pradeep529@gmail.com

C. Ramakrishna
Department of CSE(DS), CMR Institute of Technology Hyderabad, Hyderabad, Telangana, India

V. S. Reddy
Professor in CSE, Vignana Bharathi Institute of Technology, Hyderabad, Telangana, India
e-mail: vsridharreddy@vbithyd.ac.in

T. Veeranna
Department of CSE, Sai Spurthi Institute of Technology Hyderabad, Hyderabad, Telangana, India

S. Kodati
Department of IT, CVR College of Engineering Hyderabad, Hyderabad, Telangana, India

T. Benarji
Department of CSE, Indur Institute of Engineering and Technology, Siddipet, Telangana, India

V. K. Gunjan et al. (eds.), *Cybernetics, Human Cognition, and Machine Learning in Communicative Applications*, Cognitive Science and Technology,
https://doi.org/10.1007/978-981-97-8533-9_2

obtained results the Naïve Bayes exhibits the highest accuracy of 78% compared to the remaining algorithms.

Keywords Diabetes · Machine learning · Decision tree · SVM (support vector machine) · Naive Bayes

1 Introduction

The disease which can affect the ability of producing the insulin hormone in the human body is diabetes, which in turn causes abnormal carbohydrate metabolism and raises the glucose levels in the blood [1]. Basically, the diabetic person suffers from high blood sugar. Typical health signs include elevated hunger, repeated urination, and excessive blood pressure. A form of Diabetes cannot be treated successfully with oral medications alone, so patients require insulin therapy [2]. If the diabetes is untreated then more complications may appear. Few of the critical complications include nonketotic hyperosmolar coma and ketoacidosis [3]. When the sugar substance measurement is not controllable then diabetes can be treated as a serious health issue. Diabetes is affected by different factors such as hereditary factor, weight, height & insulin, among all of these insulin is considered as the main reason for sugar concentration. For staying away from severe complications only remedy is early identification [4].

As per the report of WHO (World Health Organization) diabetes is the 7th prominent cause of death in 2030 [2]. By 2040 642 million adults, i.e., 1 in 10 adults will be projected to have diabetes [3]. In 2015 the deaths of 1.6 million people were affected by diabetes completely. Due to high glucose levels in blood, 2.2 million deaths happened in 2014 [5]. Diabetes is independent of age so it may happen to people at any age. Diabetes is classified into 3 types, they are type 1 diabetes, type 2 diabetes and type 3 diabetes. A disease of autoimmune is the type 1 diabetes. During this condition, the cells which are essential for producing insulin for absorbing the sugar for producing energy are destroyed by the body. Obesity is caused by type 1 diabetes. Due to obesity, the BMI (body mass index) is increased than the normal BMI level of a person [6]. This type of diabetes can occur in adolescence age or childhood [7]. Adults who are having obese are usually affected by type 2 diabetes [8]. Due to this, the human body fails to produce insulin or observes insulin is resisted. This type of diabetes can occur in aged people or middle aged groups. Type 3 or Gestational diabetes means hyperglycemia which can occur during pregnancy due to the hormone change [9]. Viral or bacterial infection, bad diet, obesity, chemical or toxic contents in food, changes in lifestyles, reaction of autoimmune, pollution in environment, eating habits, etc., are the reasons for diabetes. Various diseases like renal issues, foot ulcers, cardiovascular complications, and retinopathy are caused by diabetes [10].

Over the years, large amounts of EHRS (Electronic Health Records) have been collected for providing a base for prediction and risk analysis [11]. With the massive knowledge on analytics technology, detection of disease has been paid more attention in the big data analysis perspective, variant researches has been conducted by the selection of characteristics mechanically from the variety of outsized information for risk classification accuracy instead of characteristics chosen in the past. But the existing system mainly considered structured data. For such cases, the semantics may differ from healthy person to severely ill person or it cannot predict the disease in particular. For healthy cases, there is no ground truth available. The alive set of cases are simply treated as negative class and it can be a highly noise majority class. On the other hand, this large alive set unlabeled genuinely, as opposed to cases with known labels removed, it would become a multi-class learning problem having large unlabeled data. Most of the existing methods of health care data don't consider the multimodal disease data prediction issues.

2 Machine Learning Based Predictive Models

Computational and statistical learning approaches are referred to as ensemble methods. They are in line with how people gain social skills and consider alternative viewpoints before reaching a final conclusion. For controlled and unsupervised learning issues, a set of learning machines is used to integrate options and produce more reliable and accurate predictions. Classifier ensemble approaches need a unified theoretical explanation [12]. The machine learning method recommended modifying the SVM rules for forecasting. A comparative analysis has been conducted [13] comparing the Naïve Bayes, Decision Tree, and K-NN algorithms.

Reinforcement learning, unsupervised learning, and supervised learning are the three categories into which machine learning algorithms are divided. For accuracy testing the supervised learning algorithm is used instead of other machine learning algorithms. From the pre-existing data, the supervised learning algorithm learns the patterns and based on previous learning tries to predict the new results. For identifying the existing data such as, tree-based, rule-based, probability-based, instance-based and function-based, etc., ML algorithms are used. Using various algorithms of data mining the different ML algorithms are introduced to assist the medical experts. By the accuracy of the decision support system, the effectiveness is recognized. For predicting & diagnosing a specific disease with high degree of accuracy is the main goal of decision support system building.

An open source data set which is used to evaluate and train the proposed system uses preexisting data sets called PIDD [14]. So many researchers have been conducting experiments to diagnose the disease using different classifications of ML algorithm approaches such as Naive Bayes, Decision Table, SVM, Decision Tree, J48, etc. The researchers have proved these ML algorithms work better in diagnosing various diseases.

It is recommended that the classifiers be utilized to forecast diabetes and that hybrid models be created to further enhance them. To put it briefly, the deep learning studies of the features corresponding to outcomes of diabetes are compared with traditional ML models for extracting useful data from EHR. The main problems with a filter-based feature collection were (i) the majority of the features do not accept consistency, (ii) a specific filter-based system's limitations in the selected function subset, and (iii) low classification accuracy [15].

In this presented system the classification of ML algorithms: SVM, Decision Tree, and Naive Bayes are utilized for evaluating the data set PIDD to find the diabetes prediction for a patient. By using different measures, all these three algorithms experimental performance are compared and these algorithms achieve good accuracy. Because the Naïve Bayes algorithm is an intuitive technique, there is no requirement to establish parameter values before moving on. It is simple to use the probabilities that this method returns for additional research or analysis. It starts with a small number of datasets and develops quickly. In terms of computation, it makes decisions quickly as well.

3 Diabetes Prediction Using Machine Learning

A unified framework of implementation of predicting disease diabetes using ML is represented in the above Fig. 1. The methodology or working process is described below.

A. *Dataset:* In this proposed model the patients need to give their medical data for their successful diagnosis diabetes test. With the help of PIDD, this proposed system can predict the diabetes of an individual. The PIDD consists of 768 instances having 8 attributes. In this case, a class value of 1 is regarded as tested positive for diabetes, and a class value of 0 is regarded as tested negative for the disease. Dataset includes glucose concentration found in glucose level (oral glucose tolerance test), number of times pregnant, thickness of skin, BMI, age, diabetes pedigree functions, and blood pressure.
B. *Preprocess data:* Higher prediction accuracy is produced by preprocessing of PIDD. Thus, it can also calculate the proposed system accuracy, prediction accuracy is also improved.
C. *Local processing platform:* In python environment, dataset is explored with the involved attributes of data dictionary. Local processing platform (LPP) is able to process collected data. The data is processed once by a Local processing platform, it will be sent to the machine learning classification stage.
D. *Machine learning algorithms:* For finding the suitable algorithm of machine learning which has capability to predict diabetes more accurately, the power full ML models like Decision tree, SVM, and Naïve Bayes are tested.

   (1) *Naive Bayes Classifier:* A classification model with a notion that denotes all features as unrelated to each other & independent is called as Naïve Bayes. While denoting a specific feature status of one class doesn't affect another status feature. For classification purposes, Naive Bayes is taken as a

Fig. 1 Framework of diabetes prediction using machine learning

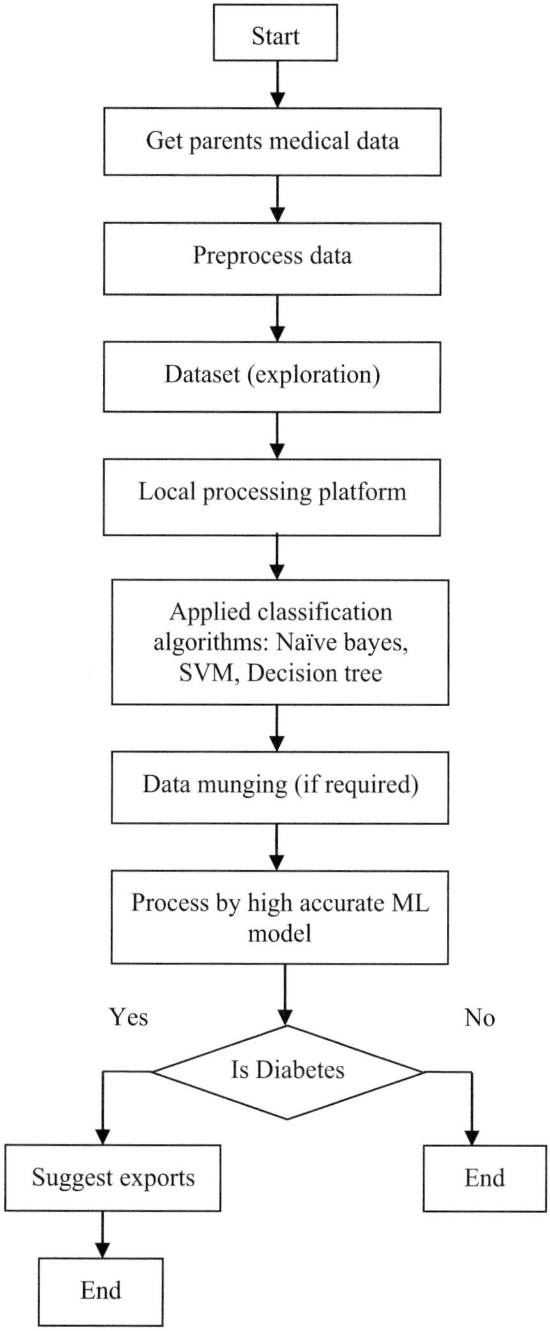

powerful algorithm which is based on conditional probability. For missing values and data unbalancing problems case, the Naïve Bayes work well. The ML classifier works on theorem of Bayes. The posterior probability P(C/X) is calculated by using Bayes theorem, calculated from P(X), P(X|C) & P(C).

$$P\left(\frac{C}{X}\right) = \frac{P(\frac{X}{C})P(C)}{P(X)}$$

where,

P (C|X) = posterior probability of target class's
P (X|C) = probability of predictor class's
P (C) = probability being true of class C's
P (X) = prior probability of predictor's

(2) *SVM (Support Vector Machine):* The supervised learning model standard set is the support vector machine. Two-class training sample is given to SVM; the main objective of this is to find out the separation of hyperplane between two classes with highest - margin. For better hyperplane generalization the plane should not be closer to data points which belong to other classes and the hyperplane must be chosen far from the data points from each type. The points which lie closer to the margin are the support vectors.

(3) *Decision Tree Classifier:* For solving the classification problems a supervised ML algorithm of Decision tree is used. The main aim of utilizing the decision tree in this proposed model is to predict the target class taken from the prior data using decision rule. For the classification & prediction decision tree uses internodes and nodes. The classifications of instances with various features are done by root nodes. If the leaf nodes represent the classification, the root nodes may have two or more branches. In each stage, DT selects every node by the evaluation of highest information gain among the attributes.

E. *Data Munging:* Data munging means the estimation of missing data values in few variables and it is needed because many interpretations will not be performed when data is missing. If it is a continuous variable then the missing value is replaced by mean value, in categorical variable the missing data is replaced by mode value.

F. *Evaluation of accurate ML:* The performance of the proposed diabetes disease prediction model is calculated by using parameters performance as precision, accuracy, F-measure and recall. If the accuracy is high then the proposed method efficiently detects the diabetes patients. Then the diabetes patients are suggested to experts for further medication. If not, the test sample data of the patient is declared as non-diabetes.

4 Experimental Results

All the classified algorithms' performance is assessed with various aspects of statistical measurement like precision, F1-score, accuracy and recall. The measurement factors of these classified algorithms are calculated by the below terms: FP (False Positive), FN (False Negative), TP (True Positive) and TN (True Negative). Here

FP: Results of the prediction are yes and the person doesn't have diabetes actually (type 1 error)

FN: Results of the prediction are no but the patient had diabetes (type 2 error).

TP: Results of prediction are yes and the patient has diabetes.

TN: Results of prediction are no and the patient doesn't have diabetes.

For measuring the factors as computational formulae is given below.

Accuracy: Accuracy is defined as the ratio of predictions made by model correctly to all types of completed predictions in the problem classification.

$$Accuracy = \frac{(TP + TN)}{(TP + TN + FP + FN)}$$

Precision: Positive precision or prediction is defined as the ratio of number of accurate positive scores to the number of positive scores predicted by the classification algorithm.

$$Precision = \frac{TP}{(TP + FP)}$$

Recall: Sensitivity or recall or TP rate is defined as a measure which is the ratio of actual positive instances having diabetes to the positive instances actually (both TP & FN).

$$Recall = \frac{TP}{(TP + FN)}$$

F1-score: Weighted average of precision and recall is the measure of F1. F1 must be 1 for good performance of classification algorithm, 0 for bad performance.

$$F1 - Score = 2 \times \frac{Precision * Recall}{Precision + Recall}$$

Different classifier's performance parameters comparisons are described in below Table 1.

The graphical representation of accuracy and precision is represented in below Fig. 2.

The graphical representation of Recall and F1-score are represented in below Fig. 3.

Table 1 Different classifiers performance parameters comparison

Parameters	Naïve bayes	SVM	Decision tree
Accuracy	78	66	74
Precision	75	45	72
Recall	77	66	74
F1-score	78	54	75

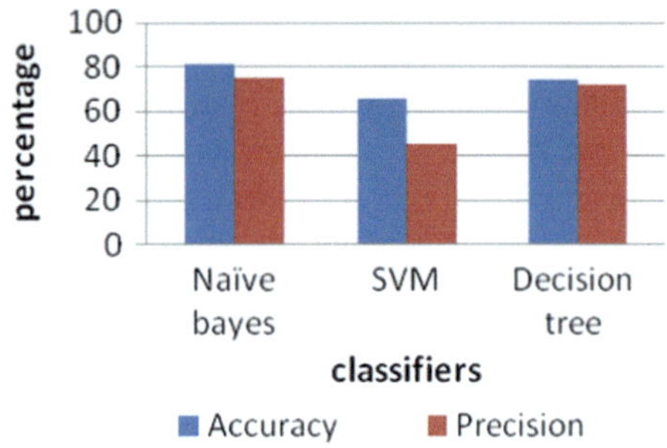

Fig. 2 Comparison between different classifiers accuracy and precision parameters

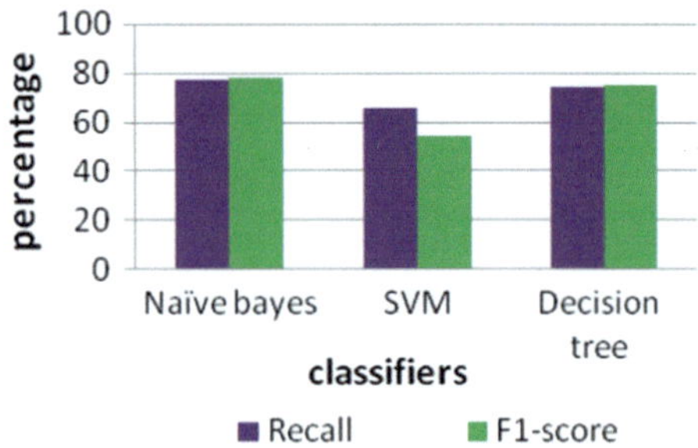

Fig. 3 Comparison between different classifiers recall and F1-score parameters

Table 2 determines the performance of classifiers based on the classification of instances. In accordance with the classified instances, the accuracy was analyzed & calculated. By using the incorrectly classified instances and correctly classified instances the individual algorithm performance is evaluated. The total number of instances are 768.

Table 2 Classifier performance based on classifier instances

Classifiers	Correctly classified instances	In-correctly classified instances
Naïve bayes	590	178
SVM	505	263
Decision tree	574	194

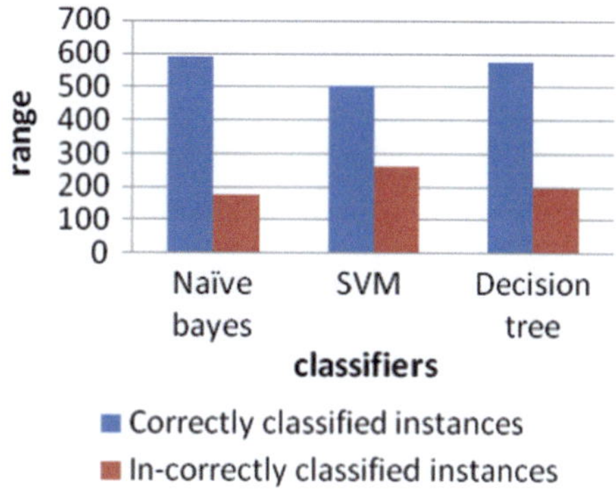

Fig. 4 Classified instances

All the classification of algorithms is based on the classification of instances shown in Fig. 4. From Tables 1 and 2 it is clear that the Naïve Bayes algorithm performs greatly compared to other classified algorithms.

5 Conclusion

A unified framework for implementation of predicting diabetes disease using ML is proposed in this paper. This gives fast and accurate diabetes predictions. Based on PIDD the experiments are performed. The system in use made use of 768 instances across 8 attributes. The used data are preprocessed to get rid of unnecessary information and expedite processing. During this work, 3 ML algorithms classified as Naive Bayes, SVM, and Decision Tree are evaluated and studied over different measures such as accuracy, precision, recall, and F1-score. Highest accuracy is obtained at naïve bayes classifier at 78%. In this paper diabetes disease prediction using ML algorithm is presented and developed. The method that is being given is important because it can accurately identify variations in medical data, which is important in both the scientific and medical domains. In future, this presented system with ML classified algorithms can be utilized for predicting diabetes or other diseases.

References

1. Le TM, Vo TM, Pham TN, Dao SVT (2021) A novel wrapper–based feature selection for early diabetes prediction enhanced with a metaheuristic. IEEE Access 9
2. Pozhar KV, Bazaev NA, Litinskaia EL (2021) In silico testing of a control algorithm for a personalized insulin therapy system. In: 2021 IEEE conference of Russian young researchers in electrical and electronic engineering (ElConRus)
3. Montoya JJM, Penalva GT, Navarro EÀ, Chilo J (2020) Monitoring diabetic ketoacidosis by urine ketones tracing using an E-nose. In: 2020 IEEE international instrumentation and measurement technology conference (I2MTC)

4. Wulandari SA, Pramitasari R, Madnasri S, Susilo (2020) Electronic noses for diabetes mellitus detection: a review. In: 2020 international seminar on application for technology of information and communication (iSemantic)
5. Volkov I, Radchenko G (2020) DiaMeter: a mobile application and web service for monitoring diabetes mellitus. In: 2020 Ural symposium on biomedical engineering, radioelectronics and information technology (USBEREIT)
6. Obasi C, Ndu I, Iloanusi O (2020) A framework for internet of things-based body mass index estimation and obesity prediction. In: 2020 International conference on e-health and bioengineering (EHB)
7. Klimontov VV, Semenova JF, Vigel AK (2020) Glucose variability in subjects with type 1 diabetes: the relationships with non-enzymatic glycation, albuminuria and renal function. In: 2020 cognitive sciences, genomics and bioinformatics (CSGB)
8. Madhava PS, Verma S (2019) A systematic literature review for early detection of type II diabetes. In: 2019 5th international conference on advanced computing & communication systems (ICACCS)
9. Pustozerov E, Popova P (2018) Mobile-based decision support system for gestational diabetes mellitus. In: 2018 ural symposium on biomedical engineering, radioelectronics and information technology (USBEREIT)
10. Anant KA, Ghorpade T, Jethani V (2017) Diabetic retinopathy detection through image mining for type 2 diabetes. In: 2017 International conference on computer communication and informatics (ICCCI)
11. Wang Y, Li P-F, Tian Y, Ren J-J, Li J-S (2017) A shared decision-making system for diabetes medication choice utilizing electronic health record data. IEEE J Biomed Health Inform 21(5)
12. Sumangali K, Geetika BSR, Ambarkar H (2016) A classifier based approach for early detection of diabetes mellitus. In: 2016 international conference on control, instrumentation, communication and computational technologies (ICCICCT)
13. Vijayan VV, Anjali C (2015) Decision support systems for predicting diabetes mellitus—a review. In: 2015 global conference on communication technologies (GCCT)
14. Karatsiolis S, Schizas CN (2012) Region based support vector machine algorithm for medical diagnosis on Pima Indian diabetes dataset. In: 2012 IEEE 12th international conference on bioinformatics & bioengineering (BIBE)
15. Akan H, Demirok U, Kılıç N (2012) The effect of feature reduction techniques on diagnosis of diabetes. In: 2012 20th signal processing and communications applications conference (SIU)

Automated Machine Learning for Healthcare

M. Sri Rama Lakshmi Reddy, Yamjala Arjun Sagar, Nallagondla Jyothi, Angotu Nageswara Rao, S. Shilpa, and Adilakshmi Velivela

Abstract Automated Machine Learning (AutoML) has emerged as a promising approach to democratize machine learning techniques, enabling non-experts to build and deploy predictive models efficiently. In the healthcare domain, AutoML holds significant potential to accelerate the development of predictive models for various tasks, such as disease diagnosis, patient outcome prediction, and personalized treatment recommendation. This paper provides an overview of the current state of AutoML in healthcare, including its applications, challenges, and future directions. We discuss key considerations in applying AutoML to healthcare data, such as data privacy, interpretability, and model generalizability. Additionally, we highlight successful use cases of AutoML in healthcare and examine how it can complement traditional machine learning approaches. Finally, we explore potential avenues for future research and development in leveraging AutoML to improve healthcare outcomes.

Keywords Healthcare · Automated machine learning · AutoML · Predictive modeling

M. S. R. L. Reddy (✉) · N. Jyothi · A. N. Rao · S. Shilpa
CSE Department, CMR Institute of Technology, Hyderabad, Medchal, Telangana, India
e-mail: lakshmireddyinfo@gmail.com

N. Jyothi
e-mail: jyothi.nallagondla@cmritonline.ac.in

A. N. Rao
e-mail: nageswararao.a@cmritonline.ac.in

S. Shilpa
e-mail: shilpa.reddy@cmritonline.ac.in

Y. A. Sagar
CSE (AI&ML) Department, CMR Institute of Technology, Hyderabad, Medchal, Telangana, India
e-mail: arjun@cmritonline.ac.in

A. Velivela
CSE (DS) Department, CMR Institute of Technology, Hyderabad, Medchal, Telangana, India
e-mail: vadilakshmi123@cmritonlie.ac.in

V. K. Gunjan et al. (eds.), *Cybernetics, Human Cognition, and Machine Learning in Communicative Applications*, Cognitive Science and Technology,
https://doi.org/10.1007/978-981-97-8533-9_3

1 Introduction

Automated Machine Learning (AutoML) is revolutionizing healthcare by streamlining the development of predictive models that aid in diagnostics, treatment planning, and patient care [1]. AutoML systems are designed to automate the entire machine learning pipeline, from data preprocessing to model selection and optimization, reducing the need for manual intervention and expertise in data science.

In healthcare, AutoML expedites the process of developing accurate and robust predictive models by efficiently exploring various algorithms and hyperparameters tailored to medical datasets. This accelerates the deployment of machine learning solutions, enabling healthcare professionals to leverage data-driven insights for improved patient outcomes.

One significant benefit of AutoML in healthcare is its ability to handle large and heterogeneous medical datasets effectively. Healthcare data is often complex, including structured clinical data, unstructured text from medical reports, and imaging data. AutoML algorithms can automatically preprocess these diverse data types, extract relevant features, and construct predictive models without extensive manual data preprocessing. The adoption of AutoML in healthcare also presents challenges, such as ensuring data privacy and regulatory compliance, addressing biases in training data, and integrating automated models into existing clinical workflows seamlessly. Overcoming these challenges requires collaboration between data scientists, healthcare providers, and regulatory bodies to establish best practices for ethical and responsible implementation of AutoML in healthcare settings.

AutoML holds great promise for transforming healthcare by democratizing access to advanced machine learning capabilities and facilitating the development of accurate, interpretable, and ethical predictive models tailored to the unique needs of patients and healthcare providers.

Healthcare Data

Types of Healthcare Data: Healthcare data comes in various forms, each presenting its own challenges and opportunities for analysis and interpretation.

- Structured Data: This includes organized information stored in databases or electronic health records (EHRs) [2], such as patient demographics, diagnoses, medications, and laboratory results. Structured data is relatively easy to process and analyze using traditional statistical methods and machine learning algorithms.
- Unstructured Data: Unstructured data encompasses free-text clinical notes, medical imaging reports, pathology reports, and other narrative forms of documentation. Analyzing unstructured data requires natural language processing (NLP) techniques to extract meaningful information and insights, such as sentiment analysis, entity recognition, and topic modeling [3].
- Time-Series Data: Time-series data tracks changes in a variable over time, commonly found in vital signs monitoring, electrocardiograms (ECGs), and wearable device data. Analyzing time-series data involves techniques such as signal

processing, time-series forecasting, and anomaly detection to identify patterns, trends, and abnormalities over time.

- Genomic Data: Genomic data consists of an individual's genetic information, including DNA sequences, gene expression profiles, and genetic variations (e.g., single nucleotide polymorphisms). Analyzing genomic data involves bioinformatics techniques to interpret genetic variations, identify disease risk factors, and personalize treatment strategies through precision medicine approaches.

Challenges in Healthcare Data Management

Managing healthcare data poses several challenges due to its complexity, volume, and sensitivity.

Data Integration: Healthcare data is often fragmented across disparate systems and formats, making it challenging to integrate and harmonize data from various sources for comprehensive analysis.

Data Quality: Ensuring the accuracy, completeness, and consistency of healthcare data is crucial for reliable analysis and decision-making. Data quality issues, such as missing values, duplication, and errors, can significantly impact the validity and reliability of analytical results.

Data Governance: Healthcare organizations must establish robust data governance policies and procedures to govern data access, usage, sharing, and security while adhering to regulatory requirements such as HIPAA (Health Insurance Portability and Accountability Act) in the United States [4].

Data Scalability: With the proliferation of healthcare data generated from EHRs, medical imaging, wearable devices, and genomic sequencing, healthcare systems must scale their infrastructure and analytics capabilities to handle large volumes of data efficiently.

Data Privacy and Security in Healthcare: Protecting patient privacy and ensuring the security of healthcare data is paramount due to the sensitive nature of personal health information.

HIPAA Compliance: Healthcare organizations must comply with HIPAA regulations to safeguard protected health information (PHI) and implement security measures, such as access controls, encryption, and audit trails, to prevent unauthorized access or disclosure of patient data.

Data Encryption: Encrypting healthcare data at rest and in transit helps protect it from unauthorized access or interception by encrypting data using cryptographic algorithms and ensuring secure transmission over networks.

Access Controls: Implementing role-based access controls (RBAC) and authentication mechanisms, such as multi-factor authentication (MFA), restricts access to healthcare data to authorized personnel based on their roles and permissions.

Data Breach Response: Healthcare organizations must have incident response plans in place to detect, mitigate, and respond to data breaches promptly, including notifying affected individuals, regulatory authorities, and other stakeholders as required by law.

Third-Party Risk Management: Healthcare organizations must assess and manage the security risks associated with third-party vendors, service providers, and business associates who handle or have access to healthcare data to ensure compliance with data privacy and security requirements.

Healthcare Domain

Medical Imaging: Medical imaging involves the use of various technologies, such as X-ray, MRI, CT scan, ultrasound, and PET scan, to visualize internal structures and organs for diagnostic and therapeutic purposes. Machine learning techniques are applied in medical imaging for tasks like image segmentation, object detection, classification of abnormalities, and image reconstruction.

Electronic Health Records (EHR): Electronic health records (EHRs) contain comprehensive patient health information, including medical history, diagnoses, medications, lab results, and treatment plans. Machine learning is utilized in EHR systems for tasks such as predictive analytics, clinical decision support, risk stratification, and personalized medicine.

Genomics: Genomics involves the study of an individual's genetic makeup, including DNA sequencing, gene expression, and genetic variations. Machine learning algorithms are applied in genomics for tasks such as variant calling, genome-wide association studies (GWAS), drug discovery, and precision medicine to tailor treatments based on genetic factors [5].

Real-World Applications of AutoML in Healthcare

Medical Image Analysis: AutoML algorithms are used to automate the analysis of medical images for tasks such as tumor detection, organ segmentation, disease classification, and treatment response assessment. AutoML techniques enable the development of accurate and robust image analysis models without extensive manual feature engineering.

Clinical Risk Prediction: AutoML is applied to clinical datasets, including electronic health records (EHRs) and medical claims data [6], to develop predictive models for identifying patients at risk of various health outcomes, such as hospital readmissions, disease progression, medication non-adherence, and adverse events. These models assist healthcare providers in early intervention and personalized patient care.

Drug Discovery and Development: AutoML accelerates drug discovery and development processes by automating the analysis of molecular and pharmacological data, including chemical structures, biological assays, and drug interactions. AutoML techniques facilitate the identification of novel drug candidates, prediction of drug-target interactions, and optimization of drug properties, leading to more efficient drug development pipelines.

Clinical Natural Language Processing (NLP): AutoML algorithms are utilized in clinical NLP tasks, such as named entity recognition, entity linking, relation extraction, and clinical concept normalization, to extract structured information from unstructured clinical text, including electronic health records (EHRs), clinical notes, and medical literature. AutoML in clinical NLP improves the efficiency and accuracy of information extraction for clinical research, decision support, and quality improvement initiatives.

2 Related Work

Research [7] on automated machine learning (AutoML) aimed to let healthcare practitioners employ “off-the-shelf” models without data science expertise. They found the pros and cons of using AutoML in healthcare. We examined biomedical informatics and computer science literature on AutoML that contained code. Several ChaLearn-organized AutoML tasks were briefly discussed. A review of 101 AutoML publications found that these automated approaches may do certain machine learning tasks as well as or better than human specialists and faster. AutoML’s main restriction at the time was training these algorithms to work beyond small and medium-sized retrospective datasets. Clinical research, healthcare costs, and health outcomes may be improved via machine learning. However, most hospitals did not apply machine learning. Healthcare personnel’s incapacity to build, apply, and integrate models into clinical workflows contributed. Automated machine learning (AutoML) aims to simplify machine learning and reduce the need for professionals. AutoML automatically selected, composed, and parametrized ML models for best task and dataset performance. Even if certain use cases existed, AutoML needed further work before it could be extensively utilized in healthcare. Mustafa and Rahimi Azghadi [8] noted that machine learning (ML) is penetrating every aspect of our lives and has had amazing positive effects. Automatic machine learning (AutoML) was created to speed up ML integration into new applications and real-world contexts. AutoML aimed to enhance everyday outcomes by integrating ML across industries. AutoML has been utilized in structured data applications like tabular lab results in healthcare. Due to fast medical language creation, AutoML was needed for interpretation. AutoML clinical note analysis has great potential to fill a gap in ML research and explore a new topic. To overcome this gap, the authors analyzed and surveyed clinical note AutoML. To do that, they started by introducing AutoML and its various tools and methods. They then examined healthcare AutoML research, including clinical and general advances. After establishing background, they discussed clinical note challenges and AutoML’s medical note processing benefits. ML research for clinical notes was scanned after a review of the literature and the healthcare industry’s AutoML sector. They also highlighted the opportunities and hazards of this new region and urged more research. This was intended to enable community adoption of an AutoML platform for medical notes, which might greatly enhance patient outcomes. Ooms and Spruit [9] investigated if and how Automated Machine Learning (AutoML) may help researcher-physicists discover new knowledge. Design science research and the Tree-based Pipeline Optimisation Tool (TPOT) as the AutoML approach were chosen after a benchmark test and feedback from researcher-physicians. A web app and notebook were added to TPOT. They discussed these artifacts with researcher-physicians to find the best strategy. Two objects with similar procedures but different user interfaces resulted from specification disagreements. Data input and comparison were better using Artefact A, an online app. Jupyter Notebook Artefact B was better for workflow and model construction. Medical researchers and practitioners gain the most from hybrid artifacts. However, neither artifact explained the relevance of the variables used to build the models or model explainability. In employing AutoML technology, healthcare companies have solely engaged in exploratory data analysis. Chua et al. [10] showed the necessity for prediction-uncertainty metrics in healthcare. They explained prediction uncertainty in healthcare, where it originates from, and how to use prediction-uncertainty metrics in error-tolerant and zero-tolerance systems. They underlined the need of consciously designing machine learning models with zero tolerance for false-positive or false-negative errors if machine learning automation is to impact healthcare. Rashidi et al. [11] discussed AutoML's emergence and impact on healthcare machine learning. In addition to a general introduction to AI and ML, they covered supervised, unsupervised, and reinforcement learning. They focused on supervised algorithms as most laboratory medicine and healthcare ML approaches employ them. They also provided an overview of automated ML (Auto-ML) and described how the area was born from the need to make these approaches more accessible to researchers. Callahan and Shah [12] discussed machine learning's advantages over traditional analytical methods and its healthcare applications. Methodological and practical issues arise when using machine learning in research and practice. They also discussed future machine learning applications that might impact healthcare delivery and health. Frühwirt and Duckworth [13] gathered empirical data of healthcare tasks that may be automated. They statistically assessed these research problems using probabilistic machine learning models trained on thousands of healthcare practitioners and automation expert assessments. The Automatability-Desirability Matrix was created from the researchers' results. Leaders and politicians may use this tool to develop realistic plans to employ automation technology for good while embracing change and involving stakeholders. Liu et al. [14] designed PharmAutoML, an open-source Python programme, to automate ML model generation and predict clinical outcomes, notably for pharmaceutical therapies. They eased data preparation, model selection, analysis, model tuning, and model interpretation in ML workflows. They found that Pharm-AutoML outperformed other ML frameworks in classification accuracy using five available biomedical datasets. Thirunavukarasu et al. [15] reviewed autoML's potential usage in academia, industry, and healthcare. They covered every autoML project phase using technical and ethical best practices. They assessed the potential for autoML to democratize artificial intelligence in the medical field while also taking into account the advantages and disadvantages of code-free, code-minimal, and code-intensive autoML systems. Jayatilake and Ganegoda [16] discussed machine learning's use in healthcare applications, namely in assessing complex medical data to find abnormalities or hidden linkages. For applications like early sickness diagnosis and timely treatment, machine learning algorithms must be more accurate and faster.

A cursory review of the methodologies utilized by [17] was followed by an evaluation of AutoML and its component pieces. In addition, they gave a case study that demonstrated the genuine business uses of AutoML and its implications on diverse sectors. They finished up by addressing the outstanding issues and prospective possibilities for additional research in AutoML.

3 Methodology

Creating a mathematical model for Automated Machine Learning (AutoML) in healthcare involves various components such as data preprocessing, feature engineering, model selection, and evaluation metrics. Below, I'll outline a simplified mathematical model focusing on the key stages involved in AutoML for healthcare.

Data Preprocessing:

- Let X represent the raw input data matrix with dimensions $m \times n$,

  where $m =$ number of samples and $n =$ number of features.
- Y represents the corresponding target variable or label vector with dimensions $m \times 1$.
- Data preprocessing involves operations like normalization, imputation of missing values, and encoding categorical variables. Let $Preprocess(X)$ represent the preprocessing function.

Feature Engineering:

- Let X' represent the feature matrix after preprocessing with dimensions $m \times n'$, where n$'$ may be different from n due to feature selection or transformation.
- Feature engineering involves operations like dimensionality reduction (e.g., PCA), feature selection, and creating new features based on domain knowledge. Let $FeatureEng(X')$ represents the feature engineering function.

Model Selection:

- Let M represent the set of candidate machine learning models, such as decision trees, logistic regression, neural networks, etc.
- For each model M_i in M, let $(Train(X', Y, M_i)$ represent the training function, which trains the model Mi on the preprocessed data X$'$ with corresponding labels Y.
- Let $Eval(M_i)$ represent the evaluation function, which evaluates the performance of model M_i using appropriate evaluation metrics like accuracy, precision, recall, F1-score, etc.

Model Evaluation and Selection:

- After evaluating all models in M, the best-performing model M^* is selected based on some criteria (e.g., highest accuracy).

- Optionally, ensemble methods or model stacking techniques can be used to combine multiple models for improved performance.

Combining these components, a simplified mathematical representation of AutoML for healthcare can be expressed as follows:

$$M^* = \underset{M_i \epsilon M}{Aargmax} \{Eval(M_i)\}$$

where:

- M^* is the selected machine learning model.
- M is the set of candidate models.
- Eval(M_i) is the evaluation function for model M_i.

This model framework can be extended and customized based on specific requirements, incorporating more sophisticated techniques for each stage of the AutoML pipeline in healthcare applications. Keep in mind that the actual implementation of these components involves programming languages, libraries (e.g., scikit-learn, TensorFlow, PyTorch), and additional considerations for validation, hyperparameter tuning, and deployment.

4 AutoML Techniques for Medical Imaging

AutoML techniques for medical imaging play a crucial role in automating the process of developing machine learning models for analyzing medical images, aiming to reduce manual intervention and expertise requirements. One key technique is Neural Architecture Search (NAS), which involves automatically searching for the optimal neural network architecture. This process optimizes architecture parameters (θ) by minimizing a loss function (L).

Hyperparameter Optimization (HPO) is another important technique, focusing on automatically tuning the hyperparameters of machine learning models to enhance their performance [18]. This optimization process involves minimizing the loss function (L) by adjusting hyperparameters (θ).

Transfer learning leverages pre-trained neural network models on large datasets and fine-tunes them for specific medical imaging tasks using smaller datasets. Fine-tuning involves minimizing the loss function (L) by adjusting the parameters (θ) of the pre-trained model.

Model Ensemble combines predictions from multiple machine learning models to improve overall performance. Ensemble predictions are calculated as weighted sums of individual model predictions, where each model's prediction ($f_i(x)$) is weighted by a corresponding weight (w_i).

Automated Feature Engineering is employed to automatically generate relevant features from raw medical images, reducing the need for manual feature extraction. This technique typically involves mathematical transformations applied to the raw input image data (X).

Bayesian Optimization is a sequential model-based optimization technique used for optimizing expensive-to-evaluate functions, such as hyperparameters. It involves modeling the objective function using a surrogate probabilistic model (e.g., Gaussian process) and selecting the next query point (x^+) based on the expected improvement ($EI(x)$).

5 AutoML for Electronic Health Records (EHR)

AutoML techniques for EHR aim to automate the process of developing ML models for analyzing EHR data, which can improve healthcare outcomes and decision-making. Several AutoML techniques can be applied in this context, each with its mathematical formulations and optimization objectives.

One common technique is Automated Feature Engineering, which involves automatically generating relevant features from raw EHR data. This process can be represented mathematically as:

$$New\ Feature = Function(X)$$

where X represents the raw input EHR data, and the function transforms or extracts features from this data to create new informative features for modeling.

Another important technique is Hyperparameter Optimization (HPO), which focuses on automatically tuning the hyperparameters of ML models to optimize their performance. In the context of EHR data analysis, HPO aims to minimize the loss function L by adjusting hyperparameters θ. This optimization process can be formulated as:

$$\theta^* = argmin_\theta L(M_\theta)$$

where M_θ represents the machine learning model with hyperparameters θ, and θ^* represents the optimal set of hyperparameters.

Neural Architecture Search (NAS) is another relevant technique for AutoML in EHR data analysis. NAS involves automatically searching for the optimal neural network architecture for a given task. This optimization process can be formulated as:

$$\theta^* = argmin_\theta L(\theta)$$

where θ represents the architecture parameters, and L is the loss function to be minimized.

In addition to these techniques, Model Ensemble can also be applied in EHR data analysis. Model Ensemble combines predictions from multiple machine learning models to improve overall performance. The ensemble prediction can be calculated as a weighted sum of individual model predictions, where each model's prediction $f_i(x)$ is weighted by a corresponding weight w_i:

$$\hat{y} = \sum_{i=1}^{N} w_i \cdot f_i(x)$$

where $\hat{y}$ represents the ensemble prediction.

These equations represent the mathematical formulations underlying various AutoML techniques used for analyzing Electronic Health Records (EHR) data. Implementing these techniques can lead to more efficient and effective machine learning models for healthcare applications, ultimately improving patient outcomes and healthcare decision-making.

6 Deployment and Integration of AutoML Models

The deployment and integration of AutoML models involve several steps to ensure seamless incorporation into existing systems and workflows. These steps can be conceptualized and described using equations in the context of their respective processes:

Model Training and Evaluation: Before deployment, AutoML models undergo training and evaluation phases to ensure optimal performance. This can be represented by the following equation:

$$\theta^* = argmin\theta L(M_\theta)$$

where θ represents the hyperparameters, L is the loss function, and M_θ is the machine learning model with hyperparameters θ. The optimal hyperparameters θ^* are selected to minimize the loss function, indicating the best-performing model.

Model Deployment: Once trained and evaluated, the AutoML model is deployed into production systems for real-time inference. The deployment process involves integrating the model into the existing infrastructure, which can be represented as:

$$System = Integration(M_{\theta^*}, Infrastructure)$$

where (M_{θ^*} represents the optimized AutoML model, and *Infrastructure* represents the existing system architecture. Integration ensures that the model seamlessly interacts with other components of the system.

Data Flow and Input Processing: In the deployed system, data flow and input processing are crucial for feeding input data to the model. This can be described as:

$$Input = Preprocessing(X)$$

where X represents the raw input data, and *Preprocessing* applies necessary transformations to prepare the data for inference. This step ensures that input data is compatible with the model's requirements.

Model Inference: Inference refers to the process of using the deployed model to make predictions or decisions based on input data. This can be represented as:

$$\hat{y} = Inference((M_{\theta^*}, Input)$$

where $\hat{y}$ represents the predicted output, $(M_{\theta^*}$ is the deployed model, and InputInput is the preprocessed input data. Inference applies the trained model to input data to generate predictions or decisions.

Output Integration: Finally, the output of the deployed model needs to be integrated back into the system for further processing or action. This can be described as:

$$\text{System Output} = \text{Integration}\left(\widehat{y}, \text{ System}\right)$$

where $\hat{y}$ represents the predicted output from the model, and SystemSystem represents the existing system architecture. Integration ensures that the model's predictions are incorporated into the system's workflow for appropriate actions or decisions.

The deployment and integration of AutoML models involve training, deploying, and integrating the model into existing systems, ensuring seamless data flow, inference, and output integration. These processes can be conceptualized and described using equations to illustrate their underlying mechanisms and interactions within the system architecture.

7 Conclusion and Future Work

Automated Machine Learning (AutoM) presents a promising avenue for revolutionizing healthcare by enabling efficient development and deployment of predictive models. Despite its potential benefits, challenges such as ensuring data privacy, achieving model interpretability, and ensuring model generalizability remain significant. However, successful applications of AutoML in healthcare demonstrate its ability to complement traditional machine learning approaches and improve healthcare outcomes.

Future Work: Moving forward, further research is needed to address the challenges associated with applying AutoML in healthcare. This includes developing robust methods for ensuring data privacy while still enabling model training, enhancing interpretability techniques to make AutoML-generated models more understandable to healthcare professionals, and devising strategies for improving model generalizability across diverse patient populations and healthcare settings. Additionally, exploring novel applications of AutoML in healthcare, such as in personalized medicine and clinical decision support systems, holds promise for advancing patient care. Collaboration between machine learning researchers, healthcare professionals, and regulatory bodies will be crucial to navigate the ethical, regulatory, and practical considerations associated with integrating AutoML into clinical practice. Overall, continued research and development in AutoML for healthcare have the potential to drive innovation and improve healthcare delivery.

References

1. Zhuhadar LP, Lytras MD (2023) The application of AutoML techniques in diabetes diagnosis: current approaches, performance, and future directions. Sustainability 5(18):13484
2. Tayefi M, Ngo P, Chomutare T, Dalianis H, Salvi E, Budrionis A, Godtliebsen F (2021) Challenges and opportunities beyond structured data in analysis of electronic health records. Wiley Interdiscip Rev: Comput Stat 13(6):e1549
3. Zadgaonkar A, Agrawal AJ (2023) An approach for analyzing unstructured text data using topic modeling techniques for efficient information extraction. New Gener Comput 1–26
4. Hansen E (1997) HIPAA (Health Insurance Portability and Accountability Act) rules: federal and state enforcement. Med Interface 10(8):96–98
5. Motsinger-Reif AA, Jorgenson E, Relling MV, Kroetz DL, Weinshilboum R, Cox NJ, Roden DM (2013) Genome-wide association studies in pharmacogenomics: successes and lessons. Pharmacogenet Genomics 23(8):383
6. Alnegheimish S, Alrashed N, Aleissa F, Althobaiti S, Liu D, Alsaleh M, Veeramachaneni K (2020). Cardea: an open automated machine learning framework for electronic health records. In: 2020 IEEE 7th international conference on data science and advanced analytics (DSAA). IEEE, pp 536–545
7. Waring J, Lindvall C, Umeton R (2020) Automated machine learning: Review of the state-of-the-art and opportunities for healthcare. Artif Intell Med 104:101822
8. Mustafa A, Rahimi M (2021) Automated machine learning for healthcare and clinical notes analysis. Computers 10(2):24
9. Ooms R, Spruit M (2020) Self-service data science in healthcare with automated machine learning. Appl Sci 10(9):2992
10. Chua M, Kim D, Choi J, Lee NG, Deshpande V, Schwab J, Do S (2023) Tackling prediction uncertainty in machine learning for healthcare. Nat Biomed Eng 7(6):711–718
11. Rashidi HH, Tran N, Albahra S, Dang LT (2021) Machine learning in health care and laboratory medicine: general overview of supervised learning and Auto-ML. Int J Lab Hematol 43:15–22
12. Callahan A, Shah NH (2017) Machine learning in healthcare. In: Key advances in clinical informatics. Academic Press, pp 279–291
13. Frühwirt W, Duckworth P (2019) Towards better healthcare: what could and should be automated? arXiv:1910.09444
14. Liu G, Lu D, Lu J (2021) Pharm-AutoML: an open-source, end-to-end automated machine learning package for clinical outcome prediction. CPT: Pharmacometrics Syst Pharmacol 10(5):478–488
15. Thirunavukarasu AJ, Elangovan K, Gutierrez L, Li Y, Tan I, Keane PA, Ting DSW (2023) Democratizing artificial intelligence imaging analysis with automated machine learning: tutorial. J Med Internet Res 25:e49949
16. Jayatilake SMDAC, Ganegoda GU (2021) Involvement of machine learning tools in healthcare decision making. J Healthc Eng

17. Chauhan K, Jani S, Thakkar D, Dave R, Bhatia J, Tanwar S, Obaidat MS (2020) Automated machine learning: the new wave of machine learning. In: 2020 2nd international conference on innovative mechanisms for industry applications (ICIMIA). IEEE, pp 205–212
18. Feurer M, Hutter F (2019) Hyperparameter optimization. In: Automated machine learning: Methods, systems, challenges, pp 3–33

Neuro-Fuzzy Systems: Neural Networks and Fuzzy Logic Integration in Soft Computing

Yamjala Arjun Sagar, M. Sri Rama Lakshmi Reddy, S. Shilpa, Nallagondla Jyothi, Adilakshmi Velivela, and Angotu Nageswara Rao

Abstract Neuro-fuzzy systems represent a powerful paradigm in soft computing by integrating the strengths of neural networks and fuzzy logic to address complex and uncertain problems. This integration combines the learning capabilities of neural networks with the reasoning abilities of fuzzy logic, enabling effective handling of imprecise and uncertain data. In this paper, we provide an outline of neuro-fuzzy systems, including their architecture, learning algorithms, and applications. We discuss the principles underlying neuro-fuzzy systems, highlighting their ability to capture and represent linguistic variables, fuzzy sets, and fuzzy rules in a computationally efficient manner. Furthermore, we explore various learning algorithms used in neuro-fuzzy systems, such as backpropagation, gradient descent, and hybrid approaches. We also present a survey of applications where neuro-fuzzy systems have been successfully applied, including control systems, pattern recognition, forecasting, decision support systems, fault diagnosis, intelligent transportation systems, medical diagnosis, and energy management. Finally, we discuss current research trends and future directions in the field of neuro-fuzzy systems, emphasizing the importance of further advancements in learning algorithms, optimization techniques, and real-world applications.

Y. A. Sagar (✉)
CSE(AI&ML) Department, CMR Institute of Technology, Hyderabad, Medchal, Telangana, India
e-mail: arjun@cmritonline.ac.in

M. S. R. L. Reddy · S. Shilpa · N. Jyothi · A. N. Rao
CSE Department, CMR Institute of Technology, Hyderabad, Medchal, Telangana, India
e-mail: shilpa.reddy@cmritonline.ac.in

N. Jyothi
e-mail: jyothi.nallagondla@cmritonline.ac.in

A. N. Rao
e-mail: nageswararao.a@cmritonline.ac.in

A. Velivela
CSE(DS) Department, CMR Institute of Technology, Hyderabad, Medchal, Telangana, India
e-mail: vadilakshmi123@cmritonlie.ac.in

V. K. Gunjan et al. (eds.), *Cybernetics, Human Cognition, and Machine Learning in Communicative Applications*, Cognitive Science and Technology,
https://doi.org/10.1007/978-981-97-8533-9_4

Keywords Neuro-fuzzy systems · Neural networks · Fuzzy logic · Soft computing

1 Introduction

Soft computing encompasses a set of computational techniques that excel at handling complex, uncertain, and imprecise information [1]. Among these techniques, neural networks and fuzzy logic stand out as powerful tools for modeling and reasoning in various domains. Neural networks, inspired by the human brain, are adept at learning from data and making predictions, while fuzzy logic allows for handling linguistic uncertainty through fuzzy sets and rules. The integration of these two paradigms in neuro-fuzzy systems offers a synergistic approach that combines the strengths of both methodologies, resulting in robust and flexible systems capable of handling real-world complexities.

This paper aims to provide a comprehensive exploration of neuro-fuzzy systems, delving into their theoretical foundations, practical implementations, and diverse applications. By understanding the principles underlying neuro-fuzzy systems, readers can harness the power of soft computing to tackle challenging problems in fields ranging from control systems and pattern recognition to optimization and decision support. We will delve into the fundamentals of neural networks and fuzzy logic, explore various integration approaches, discuss advanced topics, and examine real-world case studies. By the end of this book, readers will gain a deep understanding of neuro-fuzzy systems and their potential to revolutionize the way we approach complex problems in soft computing.

2 Related Work

In paper [2], conducted a thorough review of the emerging field of deep neuro-fuzzy systems (DNFS), aiming to address the opacity of deep neural networks (DNN) while leveraging their capabilities. They observed a surge in studies applying DNFS across various domains like healthcare, computing, transportation, and finance, showcasing both interpretability and accuracy. However, they identified a lack of comprehensive survey studies in the literature. To bridge this gap, they developed a research mapping approach and reviewed 105 DNFS-related articles published between 2015 and 2020. Their study aimed to elucidate DNFS, optimize its performance, gauge research volume, and highlight application domains. The analysis revealed the expanding nature of DNFS research, indicating promising prospects for its future deployment and utilization. Their findings provide valuable insights and recommendations for advancing research in the DNFS domain, aiming for more efficient methodological advancements in the future.

Diabetes and hepatitis were the subjects of an investigation by [3] into a fuzzy and neural network-like conceptual framework. Metrics were utilized as approximations since some facts in the actual world could not be recognized with accuracy. Fuzzy logic served as the foundation for the membership functions and rules that comprised the diabetic framework. As an extension of propositional logic, fuzzy logic provided a robust mathematical technique for representing facts in a way that simulated how the human mind works when faced with ambiguity and uncertainty. ANN was a computational framework that drew influence from biology; it consisted of extremely similar adaptive basic processing units that were tightly connected and could process and transmit information in a very similar way. With the advent of AI and the merging of Fuzzy Inference algorithms, researchers and educators from all across the scientific and technical spectrum have taken an interest in the growing need for intelligent computers that can respond to human input. When compared to artificial neural networks, fuzzy systems performed better at justifying their decisions because they could rationalize with incomplete information. But fuzzy systems couldn't figure out how to use the rules they used to make judgments. Because of these limitations, smart hybrid systems were created to address the problems by using certain approaches. This questionnaire poll covered the pros and cons of the recently popularised hybrid Neuro-Fuzzy methods. By combining the effective learning process of DNNs with the reasoning capacity of FIS, deep neuro-fuzzy systems (DNFSs) were able to be applied to real-world issues, according to [4]. Their research broke down DNFS into two main sections and offered a thorough overview of the protocol. Part two covered techniques for optimizing DNFS, whereas Part one sought to explain DNFS and its architectural representation in detail. Researchers hoped that their work would shed light on the several approaches to DNFS model development that have been detailed in the literature, including hybridization of DNN and FIS, gradient (derivative)-based techniques, and metaheuristics (derivative-free) optimization. The research findings indicate that the suggested DNFS (Dynamic Neuro-Fuzzy Systems) designs demonstrated a notable advancement in accuracy, reaching an overall accuracy rate of 81.4%. This represents a significant improvement of 11.6% compared to non-fuzzy models. Further analysis, based on optimization techniques, revealed that DNFS models employing metaheuristics optimization approaches surpassed gradient-based models by a substantial margin of 21.10%, achieving an impressive overall accuracy of 93.56%. The potential for even greater advancements in accuracy was highlighted through the exploration of newer generations of Fuzzy Inference Systems (FIS), such as fractional FIS (FFIS) and Mamdani complex FIS (M-CFIS). However, it was noted that existing DNFS networks in the literature predominantly integrated Dynamic Neural Networks (DNN) with conventional FIS, rather than leveraging these newer FIS variants.

According to [5]. Fuzzy logic and artificial neural network based systems were among the most helpful technologies for pattern identification. With order to tackle the identical challenge with the STZ system, many approaches and processes were required, all of which provided the essential signs for fast and reliable identification. The core idea behind the STZ of MRC's hybrid image processing neuro-fuzzy logical system method was to transform the initial visual picture into a format that could solve the object recognition challenge. Using a fuzzy system in STZ, they presented a neural logic system for picture preprocessing in their study. The term "fuzzy processing" was used to describe a range of fuzzy methods in image processing, such as fuzzy sets, perception, representation, image processing, and segmentation. In a comparative place, the suggested NFLSA method achieved 95.08% Binarization, 96.47% Image processing, 97.56% Colour Representation, and 95.47% Filtering in terms of threshold level. Because the neural network's input data quality was dependent on it, the early fuzzy image processing step was crucial in the pattern recognition process.

Neuro Fuzzy was described as a hybrid system that integrated Fuzzy Logic with Artificial Neural Networks by [6], When it comes to ideas, it gave a lot of leeway. When describing a system that integrated the two methods, this expression was more common. Research into neural networks and fuzzy systems diverged into two main schools. Data and artificial systems modeling several aspects of the human brain (structure, reasoning, learning, perception, etc.), including pattern recognition and clustering, function approximation, system parameter estimation, and so on. Numerical data processing often makes use of parameterized nonlinear computing approaches, such as neural networks and fuzzy logic systems. Either a general-purpose computer or specialized hardware might be used to implement these algorithms. A learning mechanism allowed the network system to gain knowledge. The learned information (weights) was stored using internal parameters.

According to [7], AI (also called machine intelligence) is the practice of teaching computers to mimic human intellect. It was a subfield of computing that had grown into an integral aspect of the IT sector with the goal of creating intelligent machines. Machines can now learn from their mistakes, adjust to novel inputs, and do tasks hitherto thought to be impossible for them—all thanks to artificial intelligence. There have been enormous social and economic gains from AI in the previous few decades. Many fields made extensive use of AI techniques like neuro-fuzzy and fuzzy logic; they included marketing, commerce, control engineering, healthcare, and social services, among others. A literature survey of fuzzy set theory and neuro-fuzzy in industry was conducted for the purpose of representing these issues in the book and to provide the groundwork for future study. Extensive talks were offered to encourage further study since some researchers have effectively used fuzzy logic and neuro-fuzzy.

With an emphasis on supervised training approaches, [8] provided a comprehensive overview of the key ideas behind hybrid models that combine fuzzy systems with artificial neural networks. In order to demonstrate the development of this class of intelligent systems, the author provided the reader with an overview of the fundamental ideas surrounding the history of hybrid models, including the initial model proposal, subsequent improvements, the architecture's composition and functionalities, data treatment, and training methods. Lastly, the reader was given an overview of the characteristics and applications of the major models. They concluded that the derived fuzzy neural network models were effective in building a system that worked well across many scientific and economic domains, with a high degree of accuracy and suitable interpretability.

3 Neural Networks

Fundamentals of neural networks encompass several key concepts that form the foundation of their operation. At the core of neural networks lies the notion of artificial neurons, also known as nodes or units, which mimic the behavior of biological neurons in the human brain. Mathematically, the output of a neuron y is computed as a nonlinear function of its weighted inputs $x1, x2, ..., xn$ and a bias term b, often represented by the following equation:

$$y = f\left(\sum_{i=1}^{n} w_i x_i + b\right)$$

where w_i represents the weight associated with each input, and f is the activation function, which introduces nonlinearity into the model. Common activation functions include the sigmoid function, hyperbolic tangent (tanh) function, and rectified linear unit (ReLU) function.

The process of training a neural network is a dynamic and iterative procedure aimed at optimizing the network's performance. This optimization hinges on adjusting the internal parameters of the network, namely weights and biases, to minimize a predefined objective function. Typically, this function is the difference between the network's output and the desired output, known as the loss function. The core principle driving this optimization is to reduce this loss as much as possible.

To achieve this optimization, various techniques are employed, with gradient descent and its variants being fundamental. Gradient descent operates by iteratively adjusting the network's parameters in the direction that minimizes the loss function. It accomplishes this by computing the gradients of the loss function with respect to the network parameters and updating these parameters accordingly.

One of the most widely used algorithms for training neural networks is backpropagation. Backpropagation efficiently computes these gradients using the chain rule of calculus. It involves two primary steps:

- Forward Propagation: In this step, the network's output is computed based on the current parameters. The input data is passed through the network, layer by layer, applying activation functions and combining weighted inputs to produce an output.
- Backward Propagation: This step involves computing the gradients of the loss function with respect to the parameters recursively, layer by layer, starting from the output layer and moving backward through the network. This process efficiently propagates the errors backward through the network, allowing for the adjustment of parameters in a way that minimizes the overall loss.

Through the iterative application of forward and backward propagation, the network's parameters gradually converge towards values that minimize the loss function, resulting in a trained neural network capable of making accurate predictions or classifications on new data.

The fundamentals of neural networks revolve around the concept of artificial neurons, the mathematical formulation of their operation, and the process of training them to learn from data through optimization techniques such as backpropagation. These fundamental concepts serve as the building blocks for understanding and implementing various neural network architectures and learning algorithms.

4 Fuzzy Logic

Fuzzy logic serves as a powerful mathematical framework for dealing with uncertainty and imprecision in decision-making processes. At its core, fuzzy logic operates on the principle of assigning degrees of membership to elements in a universe of discourse, enabling the representation of vague or ambiguous concepts. The foundation of fuzzy logic lies in the concept of fuzzy sets, which generalize classical set theory by allowing elements to belong to sets with varying degrees of membership.

Let X denote the universe of discourse, and let A represent a fuzzy set defined on X. The membership function $\mu_A(x)$ assigns a degree of membership to each element x in X, indicating the extent to which x belongs to the fuzzy set A. Mathematically, the membership function is defined as:

$$\mu_A(x) : X \to [0,1]$$

where $0 \le \mu_A(x) \le 1$ for all x in X.

Fuzzy logic extends classical propositional logic by introducing fuzzy logic operators that operate on fuzzy sets. The basic fuzzy logic operators include fuzzy conjunction (AND), fuzzy disjunction (OR), and fuzzy negation (NOT). These operators are defined based on the principle of fuzzy set intersection, union, and complement, respectively.

Let A and B be fuzzy sets defined on X, and let $\mu_A(x)$ *and* $\mu_B(x)$ denote their respective membership functions. The fuzzy intersection $A \cap B$ and fuzzy union $A \cup B$ are defined as:

$$\mu_{A \cap B}(x) = min(\mu_A(x), \mu_B(x))$$

$$\mu_{A \cup B}(x) = man(\mu_A(x), \mu_B(x))$$

Similarly, the fuzzy complement $\neg A$ is defined as:

$$\mu_{\neg A}(x) = 1 - \mu_A A(x)$$

5 Neuro-Fuzzy Modelling

Neuro-fuzzy modeling stands as a sophisticated hybrid computational paradigm [8] that melds the prowess of artificial neural networks (ANNs) with the nuanced reasoning of fuzzy logic, culminating in intelligent systems adept at grappling with intricate, uncertain, and imprecise datasets. This approach harmonizes the learning aptitude of ANNs with the interpretative and human-like logical reasoning intrinsic to fuzzy logic. In neuro-fuzzy modeling, ANNs are tasked with learning from data and executing predictive or classificatory tasks, while fuzzy logic furnishes a structured framework for representing and reasoning with linguistic variables and fuzzy sets, which adeptly encapsulate the inherent imprecision and uncertainty endemic to many real-world conundrums.

The neural network facet of neuro-fuzzy modeling comprises interconnected nodes, or neurons, meticulously organized into layers [9]. These networks undergo training via algorithms such as backpropagation, where adjustments to the connections between neurons are made based on input–output pairs, thereby assimilating the underlying patterns within the data. On the flip side, the fuzzy logic segment entails the delineation of linguistic variables and membership functions to delineate input and output spaces in a more intuitive and human-comprehensible manner. Fuzzy rules are then crafted to encapsulate the interrelationships among these variables, enabling adaptable and interpretable reasoning.

By amalgamating neural networks and fuzzy logic, neuro-fuzzy models exhibit a remarkable ability to navigate through complex and uncertain datasets while furnishing transparent and comprehensible reasoning processes. This versatility renders them eminently suited for an expansive array of applications, including but not limited to control systems, pattern recognition, forecasting, decision support systems, and expert systems.

In sum, neuro-fuzzy modeling presents a robust and flexible framework for erecting intelligent systems primed to adeptly tackle the multifaceted challenges presented by real-world data. By ingeniously marrying the learning proficiencies of neural networks with the logical reasoning capabilities of fuzzy logic, neuro-fuzzy modeling engenders solutions that are not only robust but also interpretable, heralding a new frontier in the realm of intelligent systems.

Integration of Neural Networks and Fuzzy Logic

The integration of neural networks and fuzzy logic has led to the development of neuro-fuzzy systems, which combine the learning capabilities of neural networks with the reasoning capabilities of fuzzy logic. One of the most widely used neuro-fuzzy models is the Adaptive Neuro-Fuzzy Inference System (ANFIS), which utilizes a hybrid learning approach to adaptively learn fuzzy rules from data. Let's delve into the mathematical formulation of ANFIS:

Consider a dataset $D = \{(x1, y1), (x2, y2), \ldots, (xN, yN)\}$, where xi represents the input vector and yi represents the corresponding output. ANFIS aims to approximate the mapping from inputs to outputs using a fuzzy inference system with adjustable parameters.

Membership Function Layer: ANFIS uses Gaussian or bell-shaped membership functions to fuzzify the input variables. Let $\mu_{A_i}(x)$ denote the membership grade of input x in the ith fuzzy set A_i. The Gaussian membership function is defined as:

$$\mu_{A_i}(x) = \exp(-\frac{x - c_i^2}{2\sigma_i^2})$$

where c_i and σ_i are the parameters representing the center and width of the ith fuzzy set, respectively.

Fuzzy Rule Layer: The second layer of ANFIS computes the firing strength of each fuzzy rule by multiplying the membership grades of the input variables associated with that rule. Let $w_{i,j}$ denote the firing strength of the ith rule and the jth input variable.

$$w_{i,j} = \mu_{A_i}(x_j)$$

Normalization Layer: The firing strengths obtained from the previous layer are normalized to ensure that they sum up to 1 across all rules.

$$\overline{w}_{i,j} = \frac{w_{i,j}}{\sum_{i=1}^{N} w_{i,j}}$$

Consequent Parameter Layer: In this layer, the parameters associated with the consequent part of each rule are computed using a weighted linear combination of the input variables.

$$\hat{y}_i = \sum_{j=1}^{M} p_{i,j} x_j + p_{i,0}$$

where $p_{i,j}$ and $p_{i,0}$ are the parameters to be learned.

Output Layer: The final layer of ANFIS aggregates the outputs from all rules to obtain the overall output of the system.

$$\hat{y} = \sum_{i=1}^{N} \overline{w}_i \hat{y}_i$$

where $\hat{y}_i$ is the output of the ith rule, and $\overline{w}_i$ is the normalized firing strength of the ith rule.

By adjusting the parameters c_i, σ_i, $p_{i,j}$ and $p_{i,0}$ using a hybrid learning algorithm such as gradient descent or least squares, ANFIS can effectively learn the fuzzy rules and approximate complex nonlinear mappings between inputs and outputs. This integration of neural networks and fuzzy logic allows ANFIS to exhibit adaptive and flexible behavior, making it suitable for various applications in pattern recognition, system identification, and control.

6 Applications of Neuro-Fuzzy Systems

Neuro-fuzzy systems find applications across various domains due to their ability to effectively handle complex, uncertain, and imprecise data while providing transparent and interpretable reasoning. Some notable applications include:

Control Systems: Neuro-fuzzy systems are widely used in control applications such as industrial process control, robotics, and automotive systems. They can adaptively control complex systems by learning from input–output data and adjusting control parameters in real-time [10].

Pattern Recognition: Neuro-fuzzy systems are employed in pattern recognition tasks such as image classification, speech recognition, and handwriting recognition [11]. They can learn complex patterns from training data and make accurate predictions or classifications in real-world scenarios.

Forecasting: Neuro-fuzzy systems are utilized in time-series forecasting tasks such as stock market prediction, weather forecasting, and demand forecasting. They can capture the nonlinear relationships and uncertainties present in time-series data, making them effective for forecasting future trends [13].

Decision Support Systems: Neuro-fuzzy systems are integrated into decision support systems for applications in healthcare, finance, and business. They assist in decision-making by analyzing complex data, identifying patterns, and providing insights to support informed decisions [14].

Fault Diagnosis: Neuro-fuzzy systems are employed in fault diagnosis applications such as fault detection and diagnosis in mechanical systems, electrical systems, and manufacturing processes. They can analyze sensor data, detect anomalies, and diagnose faults in real-time to improve system reliability.

Intelligent Transportation Systems: Neuro-fuzzy systems are utilized in intelligent transportation systems for tasks such as traffic prediction, congestion detection, and route optimization. They can analyze traffic data, predict traffic patterns, and optimize traffic flow to improve transportation efficiency.

Medical Diagnosis: Neuro-fuzzy systems are applied in medical diagnosis tasks such as disease diagnosis, patient monitoring, and medical image analysis. They can analyze patient data, detect abnormalities, and assist healthcare professionals in making accurate diagnoses and treatment decisions.

Energy Management: Neuro-fuzzy systems are employed in energy management applications such as load forecasting, energy consumption prediction, and energy optimization. They can analyze energy data, predict energy demand, and optimize energy usage to improve energy efficiency and sustainability.

Overall, neuro-fuzzy systems offer versatile solutions across a wide range of applications by combining the learning capabilities of neural networks with the reasoning abilities of fuzzy logic to effectively handle complex and uncertain data.

7 Conclusion and Future Work

Neuro-fuzzy systems have emerged as a versatile and powerful paradigm in soft computing, offering effective solutions to complex and uncertain problems by integrating neural networks and fuzzy logic. Through this integration, neuro-fuzzy systems leverage the learning capabilities of neural networks to handle large-scale data and the reasoning abilities of fuzzy logic to address imprecision and uncertainty. This paper has provided an overview of neuro-fuzzy systems, including their architecture, learning algorithms, and diverse applications across various domains. The literature review underscores the efficacy of neuro-fuzzy systems across various domains including control systems, pattern recognition, forecasting, decision support systems, fault diagnosis, intelligent transportation systems, medical diagnosis, and energy management. These systems excel in capturing and representing linguistic variables, fuzzy sets, and fuzzy rules, thereby providing interpretable solutions conducive to real-world decision-making tasks.

Looking forward, several avenues for future research in the field of neuro-fuzzy systems are identified. Firstly, there is a need for further enhancement of learning algorithms within neuro-fuzzy systems. This entails the development of more efficient training algorithms and optimization techniques to bolster convergence speed and accuracy. Additionally, integrating neuro-fuzzy systems with other machine learning and optimization techniques, such as deep learning and evolutionary algorithms, holds promise for advancing hybrid approaches to a more advanced and robust level.

Furthermore, exploring the application of neuro-fuzzy systems in emerging domains like Internet of Things (IoT), smart cities, and healthcare analytics presents intriguing opportunities for research and development. Addressing specific challenges within these domains, such as data heterogeneity, real-time processing, and privacy concerns, could significantly contribute to the evolution of intelligent systems and decision support tools.

References

1. Odetunji AO, Lasisi EU (2009) Applications of soft computing techniques in materials engineering: a review. Afr J Math Comput Sci Res 2(7):104–131
2. Talpur N, Abdulkadir SJ, Alhussian H, Hasan MH, Aziz N, Bamhdi A (2023) Deep neuro-fuzzy system application trends, challenges, and future perspectives: a systematic survey. Artif Intell Rev 56(2):865–913
3. Voumik LC, Karthik R, Ramamoorthy A, Dutta A (2023) A study on mathematics modeling using fuzzy logic and artificial neural network for medical decision making system. In: 2023 international conference on computational intelligence and sustainable engineering solutions (CISES). IEEE, pp 492–498
4. Talpur N, Abdulkadir SJ, Alhussian H, Hasan MH, Aziz N, Bamhdi A (2022) A comprehensive review of deep neuro-fuzzy system architectures and their optimization methods. Neural Comput Appl 1–39

5. Ramesh G, Logeshwaran J, Rajkumar K (2022) The smart construction for image preprocessing of mobile robotic systems using neuro fuzzy logical system approach. NeuroQuantology 20(10):6354–6367
6. Thakur A, Dhiman K, Phansikar M (2021) Neuro-fuzzy: artificial neural networks & fuzzy logic. Int J Res Appl Sci Eng 9:128
7. Jafari R, Razvarz S, Gegov A, Razvarz S (2021) Implementation of fuzzy logic and neuro-fuzzy in industry. Int J Math Game Theory Algebra 30(1):313–333
8. de Campos Souza PV (2020) Fuzzy neural networks and neuro-fuzzy networks: a review the main techniques and applications used in the literature. Appl Soft Comput 92:106275
9. Škrjanc I, Iglesias JA, Sanchis A, Leite D, Lughofer E, Gomide F (2019) Evolving fuzzy and neuro-fuzzy approaches in clustering, regression, identification, and classification: a survey. Inf Sci 490:344–368
10. Zaranezhad A, Mahabadi HA, Dehghani MR (2019) Development of prediction models for repair and maintenance-related accidents at oil refineries using artificial neural network, fuzzy system, genetic algorithm, and ant colony optimization algorithm. Process Saf Environ Prot 131:331–348
11. Kar S, Das S, Ghosh PK (2014) Applications of neuro fuzzy systems: a brief review and future outline. Appl Soft Comput 15:243–259
12. Fontenla-Romero O, Guijarro-Berdinas B, Alonso-Betanzos A (2002) Symbolic, neural and neuro-fuzzy approaches to pattern recognition in cardiotocograms. In: Advances in computational intelligence and learning: methods and applications, pp 489–500
13. Li K, Su H, Chu J (2011) Forecasting building energy consumption using neural networks and hybrid neuro-fuzzy system: a comparative study. Energy Build 43(10):2893–2899
14. Efendigil T, Önüt S, Kahraman C (2009) A decision support system for demand forecasting with artificial neural networks and neuro-fuzzy models: a comparative analysis. Expert Syst Appl 36(3):6697–6707

Machine Learning and Artificial Intelligence in IoT: Integration Techniques and Applications

Angotu Nageswara Rao, Adilakshmi Velivela, M. Sri Rama Lakshmi Reddy, Yamjala Arjun Sagar, Nallagondla Jyothi, and S. Shilpa

Abstract The integration of Machine Learning (ML) and Artificial Intelligence (AI) in Internet of Things (IoT) systems has revolutionized various industries by enabling intelligent decision-making and automation. This paper presents an overview of integration techniques and applications of ML and AI in IoT environments. We discuss the challenges and opportunities associated with the fusion of these technologies, including data management, computational complexity, and privacy concerns. Furthermore, we explore various integration approaches such as edge computing, federated learning, and distributed intelligence architectures to address these challenges. We also highlight a range of IoT applications leveraging ML and AI techniques, including predictive maintenance, anomaly detection, smart healthcare, and environmental monitoring. Finally, we identify future research directions and emerging trends in this rapidly evolving field.

Keywords Machine learning · Artificial intelligence · Internet of things · Integration techniques · Anomaly detection

A. N. Rao (✉) · M. S. R. L. Reddy · N. Jyothi · S. Shilpa
CSE Department, CMR Institute of Technology, Hyderabad, Medchal, Telangana, India
e-mail: nageswararao.a@cmritonline.ac.in

N. Jyothi
e-mail: jyothi.nallagondla@cmritonline.ac.in

S. Shilpa
e-mail: shilpa.reddy@cmritonline.ac.in

A. Velivela
CSE(DS) Department, CMR Institute of Technology, Hyderabad, Medchal, Telangana, India
e-mail: vadilakshmi123@cmritonlie.ac.in

Y. A. Sagar
CSE(AI&ML) Department, CMR Institute of Technology, Hyderabad, Medchal, Telangana, India
e-mail: arjun@cmritonline.ac.in

V. K. Gunjan et al. (eds.), *Cybernetics, Human Cognition, and Machine Learning in Communicative Applications*, Cognitive Science and Technology,
https://doi.org/10.1007/978-981-97-8533-9_5

1 Introduction

The integration of Machine Learning (ML) and Artificial Intelligence (AI) in the realm of Internet of Things (IoT) has ushered in a new era of innovation and possibility [1]. As we navigate a world increasingly characterized by interconnected devices and data-driven decision-making, the synergy between ML, AI, and IoT presents unprecedented opportunities for transformative applications across various domains. At its core, IoT embodies the concept of interconnectivity, where a vast network of physical devices collects, processes, and shares data to facilitate seamless interactions and intelligent automation. With the advent of ML and AI, these devices are empowered to not only generate insights from data but also adapt and learn from their environment, leading to more efficient operations, enhanced user experiences, and novel functionalities. ML algorithms, ranging from supervised learning for predictive analytics to unsupervised learning [2] for pattern recognition, are instrumental in extracting valuable insights from the deluge of IoT-generated data [3]. Meanwhile, AI techniques such as neural networks and deep learning enable IoT devices to emulate human-like cognitive abilities, facilitating advanced decision-making and autonomous actions (Fig. 1).

1.1 Internet of Things (IoT) Architecture

Data collection from the sensing layer, information is transmitted to application layer servers and databases through the network layer, with user access facilitated

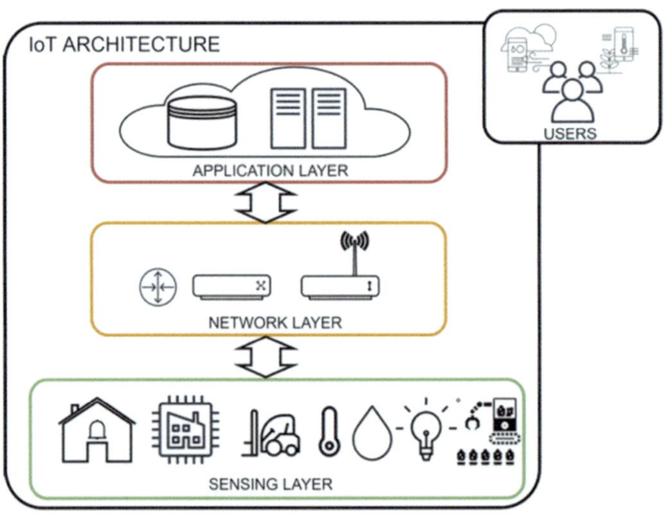

Fig. 1 Things (IoT) architecture [4]

via a cloud connection. In an IoT network, a "thing" refers to any device, whether physical or virtual, capable of connecting to the internet and exchanging data with other devices within the network. Security and privacy are of utmost importance in enhancing IoT networks [5]. The IoT model extends internet capabilities to encompass sensor networks and mobile devices, enabling nodes to communicate with each other and the broader network [6]. Key security concerns in IoT networks revolve around node authentication and ensuring the confidentiality and integrity of data. Despite efforts, authenticating every node remains challenging [7]. Authentication is pivotal in regulating data flow and securing subsequent activities, verifying the identity of entities. In the foreseeable future, our surroundings will be teeming with interconnected nodes communicating with each other and with us, constrained by limited battery life and processing power. Consequently, authenticating resource-constrained IoT nodes necessitates unconventional cryptographic techniques, as conventional methods are impractical. The IoT network comprises three layers: sensing, networking, and application. The sensing layer consists of IoT nodes equipped with various sensors, transferring detected data to servers via the network layer. Gateways, along with routers and other packet forwarding devices, are often employed to bridge the application layer with the sensing layer. Gateways, possessing considerable computational capabilities, may utilize traditional authentication mechanisms. However, lightweight authentication methods are essential for nodes at the sensing layer, given their constrained resources. Despite advancements, even 5G networks like 3GPP Rel-16 [8] have yet to fully address this authentication challenge.

2 Background

Research [1] stated that The Internet of Things (IoT) was transforming how organizations and industries communicated and conducted their everyday operations. Its application had proven to be ideal for industries that managed large amounts of assets and coordinated complex and dispersed operations. They mentioned that an expert system had to be able to defend and explain a judgment to a user in AI. They also emphasized that experts in information security had to be able to explain why a system was secure to the general public. Artificial intelligence was also expected to play an important role in ensuring long-term public safety and security for growth. They further noted that the different emerging sectors of IoT and traditional towns had been transformed into high-tech smart cities due to AI. The emergence of smart cities had just one goal: to raise people's living standards by incorporating technology into their daily lives. The paper discussed a number of significant technologies as well as solutions to problems that citizens faced as a result of a lack of digitalization. It addressed concerns such as public infrastructure, public safety, and security and provided ideal solutions. It focused not only on AI but also on IoT, machine learning, deep learning, pattern reorganization, and big data analytics for the development of a smart city that was completely functional. The analysis identified the major barriers to widespread adoption and proposed a research path to address each of them in a cost-effective manner. They concluded their discussion on IoT for defense and public safety. Alloghani et al. [2] highlighted the increasing significance of the Internet of Things (IoT) as a major network paradigm, leading to the emergence of numerous IoT applications and services driven by the substantial data generated by IoT devices. They emphasized the successful integration of machine learning in various academic fields and its application in enhancing IoT networks, particularly in addressing security, routing, traffic engineering, resource allocation, and other challenges. The authors aimed to provide an extensive overview of machine learning applications in the IoT domain, focusing on recent advancements and the utilization of machine learning to develop intelligent and efficient IoT applications. Allioui and Mourdi [3] discussed the importance of innovation and adaptability in the manufacturing sector, emphasizing the need for sustainable manufacturing facilitated by smart production technologies. They underscored the role of artificial intelligence (AI) and machine learning (ML) in driving the smart manufacturing revolution within Industry 4.0, highlighting the application of AI-based methods in achieving sustainable manufacturing goals. The authors conducted a comprehensive analysis of research on AI and ML in business applications, organizing the literature to explore trends and developments in this area. Aydin et al. [4] described the widespread adoption of IoT technology over the past decade, particularly in facilitating communication and monitoring in various real-time scenarios. They highlighted the diverse nature of IoT devices, which despite their connectivity, often possess limited memory and processing capabilities. The authors emphasized the privacy and security challenges associated with implementing IoT systems, particularly due to the inadequacy of current security methods. They proposed leveraging machine learning, artificial intelligence, and blockchain as potential solutions to address IoT security concerns. Lin et al. [5] focused on the role of information and communication technology (ICT), particularly AI, ML, and deep reinforcement learning (DRL), in advancing smart cities' development. They discussed how these technologies contribute to formulating effective policies and addressing complex issues in smart city management, such as healthcare systems, energy efficiency, cybersecurity, and communication networks. The authors also highlighted the challenges and future research directions in leveraging AI and ML for smart city development. Qiu et al. [6] emphasized the increasing importance of cloud applications based on IoT and ML in various aspects of daily life, including wearables, autonomous cars, smart homes, and smart cities. They discussed how ML algorithms enhance IoT productivity by enabling devices to interpret and analyze large datasets, facilitating intelligent decision-making. The authors outlined various ML approaches to address the challenges of working with IoT data, emphasizing the need for incorporating ML to enable smart decision-making by IoT devices over time. Illi et al. [7] highlighted the growing significance of automated methods, particularly in industrial automation, enabled by advancements in computational intelligence, machine learning, and IoT technologies. The author emphasized the role of powerful computing resources in achieving accurate and error-free monitoring and control across various fields. They discussed how the integration of IoT allows for full system automation, enabling remote monitoring and control, particularly in the context of Industry 4.0. Ghosh et al. [8] discussed the successful integration of IoT with machine learning in the clinical sector, enabling an open and interconnected environment that benefits healthcare workers and patients. They highlighted examples of IoT devices designed for medical use, such as wearable modules and smart capsules, which aid in data collection for medical professionals. The authors emphasized the role of machine learning models in improving patient health outcomes through analysis and detection of health conditions, offering diagnostic suggestions, and providing alternative solutions. Mahor et al. [9] presented a study on the integration of AI and IoT technologies to develop a remote monitoring system for industrial equipment. They emphasized the efficiency, speed, and accuracy of the system, built on machine learning techniques, in monitoring diverse industrial equipment types in challenging environments. The authors highlighted the integration of AI and ML technologies to develop robust real-time communication platforms for industrial IoT systems. Cui et al. [10] discussed the significant impact of IoT on various aspects of daily life and its potential for enhancing human living standards through automation, productivity improvements, and enhanced services. They highlighted the importance of addressing cybersecurity challenges in IoT applications, emphasizing the inadequacy of conventional security methods in addressing sophisticated threats. The authors suggested using artificial intelligence, especially machine learning and deep learning, to create smart IoT security solutions. These solutions aim to analyze raw data for detecting and preventing cyber-attacks effectively. They concluded by discussing research challenges and future approaches in IoT security intelligence.

3 Mathematical Model

Combining Machine Learning (ML) and Artificial Intelligence (AI) entails employing mathematical models to depict the processes of learning, decision-making, and adaptation.

While it's challenging to encapsulate the entire integration in a single equation due to the complexity and diversity of ML and AI techniques, we can provide a simplified representation using a generic equation:

$$Y = f(X; \theta)$$

Here, X represents the input data or features, Y is the output or prediction, and θ denotes the parameters learned during the training process. The function f encapsulates the learning and decision-making process, and it varies depending on the specific ML or AI algorithm used. The equation might involve various components and terms depending on the chosen technique. Let's elaborate on a few examples:

3.1 Linear Regression (Supervised Learning)

$$Y = \theta 0 + \theta 1X1 + \theta 2X2 + \cdots + \theta nXn$$

In linear regression, the output is a linear combination of input features, and the goal is to learn the optimal θ parameters to reduce the discrepancy between forecasted and real outcomes.

3.2 Neural Network (Deep Learning)

$$\mathbf{Y} = \mathbf{f}\left(\sum_{\mathbf{i}=1}^{\mathbf{n}} \mathbf{w_i}.\mathbf{X_i} + \mathbf{b}\right)$$

In a simple neural network , f is the activation function, wi are weights, Xi are input features, and b is the bias term. This equation represents the forward pass of a neural network.

3.3 Q-Learning (Reinforcement Learning)

$$Q(s,a) = Q(s,a) + \alpha[R(s,a) + \gamma max_{a'} Q(s',a') - Q(s,a)]$$

In Q-learning, Q represents the learned action-value function, s is the current state, a is the chosen action, α is the learning rate, R is the immediate reward, s' is the next state, γ is the discount factor.

3.4 Decision Tree (Supervised Learning)

Decision trees make decisions based on a series of questions about the input features. An example decision rule might be:

$$Y = if\ X1 < 5\ and\ X2 \geq 3\ then\ "Class\ A"\ else\ "Class\ B"$$

3.5 *Federated Learning*

Federated learning involves aggregating updates from multiple devices. The update equation might be.

Here, θ_{global} represents the global model parameters, and θ_i are the parameters from individual devices.

These equations provide a glimpse into the mathematical representation of different ML and AI techniques. The specific equation and terms depend on the chosen algorithm, and these examples illustrate the diversity within the broader field of ML and AI integration.

4 Machine Learning and Artificial Intelligence

Machine Learning (ML) and Artificial Intelligence (AI) are leading-edge domains in technology that significantly impact multiple industries and steer the course of computing. Machine learning entails creating algorithms and models that empower computers to learn from data, enabling them to make predictions or decisions autonomously, without explicit programming. This methodology hinges on systems learning from experience, discerning patterns within data, and refining their performance over time. In contrast, AI encompasses a broader spectrum of concepts and techniques aimed at fabricating intelligent systems capable of perceiving their surroundings, reasoning, and making decisions akin to humans. AI incorporates ML alongside other methodologies like expert systems, natural language processing, and robotics [11]. The fusion of ML and AI has catalyzed remarkable progress across diverse domains. In healthcare, ML algorithms scrutinize medical imagery and patient records to aid in diagnosis and treatment planning. In finance, AI-driven systems automate trading strategies and identify fraudulent transactions. In autonomous vehicles, ML algorithms empower vehicles to sense their environment and make instantaneous decisions for safe navigation. The symbiosis between ML and AI continually propels innovation, culminating in the development of increasingly sophisticated systems adept at addressing intricate challenges and augmenting human capabilities. As technology advances, the potential applications of ML and AI are boundless, poised to revolutionize industries, streamline operations, and enrich lives worldwide.

4.1 *Machine Learning Algorithms*

Machine learning algorithms can be broadly categorized into three main types: supervised learning, unsupervised learning, and reinforcement learning [12]. Each type of algorithm has its own characteristics and applications in various domains.

Supervised Learning: Supervised learning involves training a model on a labeled dataset, where each data point is associated with a corresponding label or outcome. The goal of supervised learning is to learn a mapping function from input variables to output variables, based on the provided labeled data. The model is trained using examples that are labeled with the correct answer, and it learns to make predictions or decisions based on this labeled training data.

Supervised learning algorithms include:

- Linear regression
- Logistic regression
- Decision trees
- Random forests
- Support vector machines (SVM)
- Neural networks (e.g., deep learning)

Supervised learning is commonly used in tasks such as classification [13] (e.g., spam detection, image recognition) and regression (e.g., predicting house prices, stock prices).

Unsupervised Learning: Unsupervised learning [14] involves training a model on an unlabeled dataset, where the algorithm tries to find patterns or structures in the data without explicit guidance. Unlike supervised learning, there are no predefined labels or outcomes for the algorithm to learn from. Instead, the algorithm must discover the inherent structure of the data on its own.

Unsupervised learning algorithms include:

- K-means clustering
- Hierarchical clustering
- Principal Component Analysis (PCA)
- Independent Component Analysis (ICA)
- Autoencoders
- Generative Adversarial Networks (GANs)

Unsupervised learning is commonly used in tasks such as clustering (e.g., customer segmentation, document clustering) and dimensionality reduction (e.g., feature extraction, data visualization).

Reinforcement learning (RL): Reinforcement learning (RL) [15] is about instructing an agent to make a series of choices within a setting to maximize overall rewards. In contrast to supervised and unsupervised learning, it relies on trial and error, with the agent learning from its interactions via feedback in the shape of rewards or penalties reinforcement learning algorithms:

- Q-learning
- Deep Q-networks (DQN)
- Policy gradient methods
- Actor-Critic methods

Reinforcement learning finds frequent application in various domains like gaming (like AlphaGo), robotics (such as autonomous navigation), and sequential decision-making (like stock trading). Each machine learning algorithm has distinct advantages and limitations, and the selection of the appropriate algorithm hinges on factors like the task at hand, the data's characteristics, and the desired results. Familiarizing oneself with the traits of each algorithm type enables practitioners to opt for the most suitable approach tailored to their specific problem domain.

4.2 Artificial Intelligence Techniques

Artificial Intelligence (AI) encompasses a wide range of techniques and methodologies aimed at creating intelligent systems that can perform tasks typically requiring human intelligence [16]. These techniques span various domains, including but not limited to neural networks, decision trees, support vector machines (SVM), random forests, genetic algorithms, expert systems, and natural language processing (NLP).

- Neural Networks, inspired by the structure and function of the human brain, consist of interconnected nodes (neurons) organized into layers. Through training with examples in a dataset, NNs learn complex patterns by adjusting the weights of connections between neurons.
- Decision Trees (DT), on the other hand, are supervised learning algorithms used for classification and regression tasks. They represent a hierarchical structure of decisions based on input data features, with each node in the DT representing a decision based on a specific feature.
- Support Vector Machines (SVM) are another class of supervised learning algorithms used for classification and regression tasks. They aim to find the hyperplane that best separates data points of different classes in a high-dimensional space, effectively handling high-dimensional data and widely used in various domains like image recognition and text classification.
- Random Forests employ ensemble learning techniques by combining multiple decision trees to enhance prediction accuracy and robustness. Each DT within the ensemble undergoes training using a randomly selected subset of the training data. The ultimate forecast is then generated by combining the predictions made by each individual tree.
- Genetic Algorithms simulate the evolution of candidate solutions to a problem by applying genetic operators like selection, crossover, and mutation. They are particularly useful for solving complex optimization problems with large search spaces and multiple objectives.
- Expert Systems emulate the decision-making process of human experts in specific domains, relying on a knowledge base containing rules and heuristics derived from expert knowledge, along with an inference engine that reasons over this knowledge to provide recommendations or solutions to problems.

- Natural Language Processing (NLP) focuses on the interaction between computers and human language, enabling computers to understand, interpret, and generate human language. NLP techniques find applications in various domains such as text classification, sentiment analysis, machine translation, and speech recognition, powering virtual assistants, chatbots, and information retrieval systems.

5 Integration Techniques of Machine Learning in IoT

Integration of Machine Learning (ML) with IoT entails putting ML frameworks and algorithms into IoT systems [17] to extract valuable information from enormous volumes of data produced by networked devices. These integration solutions have the potential to increase the intelligence, adaptability, and autonomy of Internet of Things (IoT) systems. Among the most fundamental ways for incorporating machine learning into the Internet of Things are:

- *Edge Computing:* Edge computing involves deploying computing resources closer to the IoT devices, such as gateways or edge servers, to perform data preprocessing, analytics, and ML inference locally. By processing data at the edge, latency is reduced, and real-time decision-making is enabled, which is crucial for time-sensitive IoT applications. ML models trained in the cloud can be deployed to edge devices, allowing them to perform tasks such as anomaly detection, predictive maintenance, and pattern recognition directly on the device without relying on constant connectivity to the cloud.
- *On-Device ML Inference:* On-device ML inference involves deploying lightweight ML models directly on IoT devices with limited computational resources, such as sensors, wearables, and smart appliances. These ML models are optimized for low-power consumption and memory usage, allowing them to perform inference tasks efficiently on resource-constrained devices. On-device ML inference enables real-time processing of sensor data, local decision-making, and context-aware applications without the need for continuous cloud connectivity.
- *Federated Learning:* Federated learning is a distributed machine learning technique where models are trained collectively across numerous edge devices, avoiding the need to centralize raw data on a cloud server. In IoT settings, this method empowers devices to locally train ML models with their respective data, while intermittently sharing model updates with a central server. This privacy-centric strategy facilitates IoT devices to glean insights from each other's data without jeopardizing data privacy and security. Federated learning finds relevance in sectors like healthcare, finance, and smart homes, where safeguarding data privacy is paramount.
- *Hybrid Architectures:* Hybrid architectures combine cloud, edge, and on-device ML capabilities to leverage the strengths of each approach. In a hybrid architecture, data preprocessing and model training may occur in the cloud, while ML inference is performed at the edge or on-device. This architecture allows

IoT systems to balance computational resources, latency requirements, and data privacy concerns effectively. Hybrid architectures enable scalable and adaptive ML solutions for IoT applications that require a combination of cloud-based analytics and edge intelligence.

- *Continuous Learning:* Continuous learning methods empower machine learning models integrated into IoT systems to evolve and enhance their performance as new data is gathered. This entails regularly retraining these models using real-time data streams from IoT devices to uphold their precision and applicability in dynamic settings. Continuous learning empowers IoT systems to flexibly adjust to evolving circumstances, identify emerging trends, and refine decision-making autonomously. This approach is indispensable for IoT applications necessitating adaptive and self-learning functionalities, such as predictive maintenance, anomaly detection, and personalized recommendations.

The incorporation of machine learning (ML) into Internet of Things (IoT) systems enables them to utilize data-driven insights, autonomously make intelligent decisions, and fully exploit IoT deployments in different sectors. Through the seamless integration of ML into IoT frameworks, entities can boost operational efficiency, refine decision-making procedures, and create inventive IoT solutions that generate business value and enhance user experiences.

6 Integration Techniques of Artificial Intelligence in IoT

Integration techniques of Artificial Intelligence (AI) in the Internet of Things (IoT) involve leveraging AI algorithms and methodologies to enhance the capabilities of IoT systems, enabling them to perform complex tasks autonomously and derive valuable insights from the vast amounts of data generated by connected devices [18]. These integration techniques aim to empower IoT systems with intelligent decision-making, predictive analytics, and adaptive behavior, ultimately driving innovation and efficiency across various industries. Several key integration techniques of AI in IoT include:

Intelligent Data Processing: Employing AI methodologies like machine learning (ML) and deep learning (DL) facilitates the analysis of vast datasets from IoT devices. ML algorithms identify patterns, trends, and anomalies in sensor data, enabling tasks like predictive maintenance, fault detection, and process optimization. DL models, such as convolutional neural networks (CNNs) and recurrent neural networks (RNNs), handle sophisticated analytics like image recognition, natural language processing (NLP), and time series forecasting within IoT contexts.

Real-time Decision-making: AI algorithms are seamlessly integrated into IoT systems to enable instantaneous decision-making based on sensor data and environmental cues. Reinforcement learning (RL) algorithms empower IoT devices to learn optimal control strategies and adapt their behavior in real-time, fostering autonomous decision-making and adaptive control in IoT scenarios like smart grids, autonomous vehicles, and industrial automation.

Edge Intelligence: AI algorithms are deployed at the network edge, closer to IoT devices, to enable local processing, inference, and decision-making. Edge AI techniques, such as lightweight ML models and inference engines, enable low-latency processing of sensor data and reduce the need for constant connectivity to centralized cloud servers. This improves responsiveness, privacy, and reliability in IoT applications that require real-time analytics and autonomous operation, such as smart home devices, wearable technology, and smart city infrastructure.

Context-aware Computing: AI techniques are used to enhance IoT systems with context-aware capabilities, enabling them to understand and respond to the surrounding environment. Context-aware AI models can analyze sensor data in context, taking into account factors such as location, time, user preferences, and environmental conditions. This enables personalized services, adaptive automation, and intelligent decision-making in IoT applications such as smart healthcare, personalized marketing, and smart building management.

Predictive Analytics: AI algorithms are applied to historical and real-time data from IoT devices to perform predictive analytics and forecast future events or trends. Predictive AI models, such as time series forecasting models and predictive maintenance algorithms, enable proactive decision-making and preemptive actions to prevent failures, optimize resource utilization, and improve operational efficiency in IoT deployments.

The integration of AI in IoT systems enables intelligent, autonomous, and context-aware applications that leverage data-driven insights to improve decision-making, optimize resource allocation, and enhance user experiences across various domains. By leveraging AI capabilities in IoT architectures, organizations can unlock new opportunities for innovation, efficiency, and value creation in the rapidly evolving landscape of IoT-enabled solutions.

7 Conclusion and Future Work

The integration of ML and AI into Internet of Things (IoT) systems holds immense potential for transforming various industries through intelligent decision-making and automation. Through this paper, we have explored various integration techniques, including edge computing, federated learning, and distributed intelligence architectures, which address challenges such as data management, computational complexity, and privacy concerns. Additionally, we have highlighted diverse applications of ML and AI in IoT, spanning predictive maintenance, anomaly detection, smart healthcare, and environmental monitoring.

Looking ahead, future work in this domain should focus on several key areas. Firstly, advancements in ML and AI algorithms tailored for resource-constrained IoT devices will be crucial for enabling efficient processing and decision-making at the edge. Secondly, efforts to enhance data security and privacy in IoT systems through robust encryption and decentralized data management mechanisms are essential to foster trust and adoption. Thirdly, interdisciplinary research collaborations between experts in ML/AI, IoT, and domain-specific industries can drive innovation and development of tailored solutions for diverse application domains. Finally, exploring novel integration techniques and architectures to address emerging challenges and opportunities in evolving IoT ecosystems will be vital for unlocking the full potential of ML and AI in shaping the future of IoT-enabled intelligent systems.

References

1. Rane N (2023) Integrating leading-edge artificial intelligence (AI), internet of things (IOT), and big data technologies for smart and sustainable architecture, engineering and construction (AEC) industry: Challenges and future directions. In: Engineering and construction (AEC) industry: challenges and future directions (September 24, 2023)
2. Alloghani M, Al-Jumeily D, Mustafina J, Hussain A, Aljaaf AJ (2020) A systematic review on supervised and unsupervised machine learning algorithms for data science. Supervised Unsupervised Learn Data Sci 3–21
3. Allioui H, Mourdi Y (2023) Exploring the full potentials of IoT for better financial growth and stability: a comprehensive survey. Sensors 23(19):8015
4. Aydin Y, Kurt GK, Ozdemir E, Yanikomeroglu H (2020) A flexible and lightweight group authentication scheme. IEEE Internet Things J 7(10):10277–10287
5. Lin J, Yu W, Zhang N, Yang X, Zhang H, Zhao W (2017) A survey on internet of things: architecture, enabling technologies, security and privacy, and applications. IEEE Internet Things J 4(5):1125–1142
6. Qiu T, Chen N, Li K, Atiquzzaman M, Zhao W (2018) How can heterogeneous internet of things build our future: a survey. IEEE Commun Surv Tutor 20(3):2011–2027
7. Illi E, Qaraqe M, Althunibat S, Alhasanat A, Alsafasfeh M, de Ree M, Al-Kuwari S (2023) Physical layer security for authentication, confidentiality, and malicious node detection: a paradigm shift in securing IoT networks. IEEE Commun Surv Tutor
8. Ghosh A, Maeder A, Baker M, Chandramouli D (2019) 5G evolution: a view on 5G cellular technology beyond 3GPP release 15. IEEE Access 7:127639–127651
9. Mahor V, Rawat R, Kumar A, Garg B, Pachlasiya K (2023) IoT and artificial intelligence techniques for public safety and security. In: Smart urban computing applications. River Publishers, pp 111–126
10. Cui L, Yang S, Chen F, Ming Z, Lu N, Qin J (2018) A survey on application of machine learning for Internet of Things. Int J Mach Learn Cybern 9:1399–1417
11. Sarker IH (2022) Ai-based modeling: techniques, applications and research issues towards automation, intelligent and smart systems. SN Comput Sci 3(2):158
12. Sarker IH (2021) Machine learning: algorithms, real-world applications and research directions. SN Comput Sci 2(3):160
13. Shetty SH, Shetty S, Singh C, Rao A (2022) Supervised machine learning: algorithms and applications. In: Fundamentals and methods of machine and deep learning: algorithms, tools and applications, pp 1–16
14. Ghahramani Z (2003) Unsupervised learning. In: Summer school on machine learning. Springer, Berlin, pp 72–112
15. Qiang W, Zhongli Z (2011) Reinforcement learning model, algorithms and its application. In: 2011 International conference on mechatronic science, electric engineering and computer (MEC). IEEE, pp 1143–1146
16. Saravanan K, Sreedevi E, Subhamathi V (2017) A review of artificial intelligence systems. Int J Adv Res Comput Sci 8(9)

17. Hussain F, Hussain R, Hassan SA, Hossain E (2020) Machine learning in IoT security: current solutions and future challenges. IEEE Commun Surv Tutor 22(3):1686–1721
18. Katare G, Padihar G, Qureshi Z (2018) Challenges in the integration of artificial intelligence and internet of things. Int J Syst Softw Eng 6(2):10–15
19. Cioffi R, Travaglioni M, Piscitelli G, Petrillo A, De Felice F (2020) Artificial intelligence and machine learning applications in smart production: progress, trends, and directions. Sustainability 12(2):492
20. Mohanta BK, Jena D, Satapathy U, Patnaik S (2020) Survey on IoT security: challenges and solution using machine learning, artificial intelligence and blockchain technology. Internet of Things 11:100227
21. Ullah Z, Al-Turjman F, Mostarda L, Gagliardi R (2020) Applications of artificial intelligence and machine learning in smart cities. Comput Commun 154:313–323
22. Mishra S, Tyagi AK (2022) The role of machine learning techniques in internet of things-based cloud applications. In: Artificial intelligence-based internet of things systems, pp 105–135
23. Vijayakumar K (2021) Computational intelligence, machine learning techniques, and IOT. Concurr Eng 29(1):3–5
24. Chattopadhyay A, Mishra S, González-Briones A (2021) Integration of machine learning and IoT in healthcare domain. In: Hybrid artificial intelligence and IoT in healthcare, pp 223–244
25. Kun X, Wang Z, Zhou Z, Qi W (2021) Design of industrial internet of things system based on machine learning and artificial intelligence technology. J Intell Fuzzy Syst 40(2):2601–2611
26. Sarker IH, Khan AI, Abushark YB, Alsolami F (2023) Internet of things (IoT) security intelligence: a comprehensive overview, machine learning solutions and research directions. Mob Netw Appl 28(1):296–312

Soft Computing Approaches in Robotics: State-of-the-Art and Future Directions

Adilakshmi Velivela, S. Shilpa, Yamjala Arjun Sagar, M. Sri Rama Lakshmi Reddy, Angotu Nageswara Rao, and Nallagondla Jyothi

Abstract Soft computing techniques have gained significant traction in the field of robotics due to their ability to handle imprecise and uncertain information effectively. This paper presents a comprehensive review of soft computing approaches employed in robotics, focusing on their state-of-the-art applications and future directions. Various soft computing paradigms, including fuzzy logic, neural networks, evolutionary algorithms, and swarm intelligence, are discussed in the context of robot control, navigation, path planning, learning, and optimization. The paper highlights the strengths and limitations of these techniques, along with their synergistic integration to enhance robotic capabilities. Furthermore, emerging trends and promising research directions in the intersection of soft computing and robotics are identified, such as explainable AI, human–robot interaction, multi-robot systems, and autonomous decision-making. This review aims to provide researchers and practitioners with insights into the current landscape of soft computing in robotics and inspire future advancements in this exciting interdisciplinary field.

Keywords Soft computing · Robotics · Fuzzy logic · Neural networks · Learning · Optimization

A. Velivela (✉)
CSE(DS) Department, CMR Institute of Technology, Hyderabad, Medchal, Telangana, India
e-mail: vadilakshmi123@cmritonlie.ac.in

S. Shilpa · M. S. R. L. Reddy · A. N. Rao · N. Jyothi
CSE Department, CMR Institute of Technology, Hyderabad, Medchal, Telangana, India
e-mail: shilpa.reddy@cmritonline.ac.in

A. N. Rao
e-mail: nageswararao.a@cmritonline.ac.in

N. Jyothi
e-mail: jyothi.nallagondla@cmritonline.ac.in

Y. A. Sagar
CSE(AI&ML) Department, CMR Institute of Technology, Hyderabad, Medchal, Telangana, India
e-mail: arjun@cmritonline.ac.in

V. K. Gunjan et al. (eds.), *Cybernetics, Human Cognition, and Machine Learning in Communicative Applications*, Cognitive Science and Technology,
https://doi.org/10.1007/978-981-97-8533-9_6

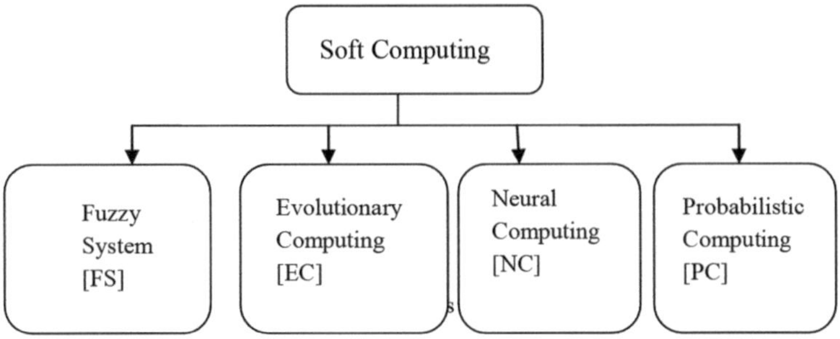

Fig. 1 Soft computing Components [1]

1 Introduction

Soft computing methods, including fuzzy logic, evolutionary algorithms, neural networks, and swarm intelligence [1], have emerged as powerful tools for addressing the complexities of robotic systems. In contrast to traditional hard computing approaches, soft computing methods excel in handling imprecise and uncertain information, making them well-suited for applications in robotics where uncertainties are inherent. This introduction sets the stage for exploring the integration of soft computing paradigms in robotics, highlighting their significance in enhancing robot control, navigation, path planning, learning, and optimization tasks. The burgeoning interest in soft computing within the robotics community stems from its ability to mimic human-like decision-making processes and adaptability in dynamic environments. This paper aims to provide a comprehensive overview of the current state-of-the-art applications of soft computing in robotics, elucidating the strengths and limitations of each approach. Furthermore, it delineates promising research directions and emerging trends at the intersection of soft computing and robotics, including explainable AI, human–robot interaction, multi-robot systems, and autonomous decision-making.

Through this exploration, researchers and practitioners alike can gain valuable insights into the diverse landscape of soft computing techniques in robotics, paving the way for advancements that will shape the future of intelligent robotic systems (Fig. 1).

2 Background

Research [2] discussed the evolution of robotics, highlighting the initial perception that traditional robots were functionally different from biological systems. They emphasized the incorporation of elastic components into rigid-bodied robots, leading to the emergence of articulated soft robotics. This field aimed to replicate the pliability and elasticity of the musculoskeletal system throughout the entire structure, enabling robots to approach tasks with more ease. The focus on technical aspects resulted in various hardware solutions, prompting the need for efficient management techniques to govern the soft body and utilize its embodied knowledge.

Kim et al. [3] delved into the challenges faced by soft robots due to non-linearity and hysteresis in soft materials. They explored the application of machine learning-based methodologies to address these constraints, categorizing the applicability across soft robotic use cases such as wearable robots, sensors, and actuators. The study provided an overview of current machine learning techniques for soft robotics and analyzed trends in machine learning approaches for various soft robot applications, acknowledging existing constraints in the field.

Chin et al. [4] detailed the advancements in soft robotics through machine learning techniques, focusing on how data-driven approaches have improved the control of soft robotic systems. They highlighted the effectiveness of machine learning in handling computationally difficult governing functions, dynamic behavior, non-linearity, and hysteresis. Supervised learning and reinforcement learning showed promising outcomes for addressing challenges in soft robotics, both in simulations and physical robotic systems.

Maoudj and Hentout [5] proposed the Efficient Q-Learning (EQL) method to optimize paths for mobile robotics applications. They addressed the slow convergence of the Q-Learning algorithm by introducing an efficient selection approach for the best possible action and a well-designed reward function. The EQL method prioritized promising actions during learning, leading to faster convergence and optimized routes compared to previous QL-based algorithms.

Song et al. [6] introduced a novel approach to mobile robot route planning by combining an enhanced particle swarm optimization (PSO) algorithm. The study emphasized the importance of high-order continuity in motion control for mobile robots. The enhanced PSO method improved the algorithm's capacity to escape local minima and explore the search space more comprehensively, demonstrating advantages in simulation experiments for mobile robot smooth path planning.

Ravankar et al. [7] focused on the importance of smooth trajectories in mobile service robots. They discussed the limitations of route planning algorithms that resulted in abrupt or angular twists, causing robots to slow down significantly. The study emphasized the need to consider static and dynamic impediments, speed, and other factors to construct smooth trajectories in robot navigation.

Young and Ferris [8] provided a comprehensive review of the state-of-the-art in robotic exoskeletons for the lower limbs. The review outlined advancements in control techniques, materials, energy sources, actuators, and sensors. It addressed challenges in control interfaces and suggested potential future solutions to enhance exoskeleton designs.

Katrakazas et al. [9] explored motion planning for autonomous or self-driving cars. The study evaluated planning algorithms at three levels: finding a route, seeking the safest maneuver, and calculating the most feasible trajectory. The authors discussed the complexity and performance accuracy of different planning methodologies and highlighted areas for improvement to enhance the efficiency of planning algorithms for autonomous vehicles.

Ibrahim [10] discussed the benefits and drawbacks of soft computing compared to conventional computing. Soft computing, dealing with approximation models and inspired by the human mind, was highlighted for its ability to handle ambiguity, imprecision, and partial truths. The article provided an overview of the foundational concepts of soft computing, including expert systems, machine learning, fuzzy logic, genetic algorithms, and artificial neural networks.

Manti et al. [11] addressed the challenges in soft robotics research, particularly in controlling deformability and compliance. The study emphasized the need for control methods relying on research into various materials and technologies for actuation and sensing. However, the unpredictable and difficult-to-regulate nature of deformability and compliance posed a significant obstacle to progress in soft robotics.

Laschi et al. [12] reviewed the significant advancements in soft robotics and related fields. The article highlighted accomplishments in actuation, sensing, modeling, and control, emphasizing the transformative capabilities of soft robotics. The study discussed the development of capabilities like squeezing, stretching, climbing, expanding, and morphing, and the potential for robots to grow, evolve, self-heal, develop, and biodegrade, representing the next frontier in soft robotics.

Goury and Duriez [13] proposed methods to reduce computational time in mechanical models of soft deformable robots. They utilized finite element models (FEM) and utilizing a snapshot-proper orthogonal decomposition technique effectively minimized the state variables within the soft robot model, ensuring computational efficiency. This method facilitated error tuning and precise management of external stressors, resulting in notable enhancements in computational efficiency without compromising accuracy.

3 Mathematical Model

A mathematical model of soft computing approaches in robotics can be formulated using fuzzy logic as an example. Fuzzy logic is a key soft computing technique that allows for reasoning under uncertainty and imprecision. The mathematical model of a fuzzy logic controller can be represented using fuzzy sets, fuzzy rules, and fuzzy inference.

Let's denote the input variables to the fuzzy logic controller as $x1, x2, ..., xn$, and the output variable as y. Each input variable xi is associated with a fuzzy set Ai, which represents the degree of membership of xi in linguistic terms. The membership function $\mu_{A_i}(xi)$ defines the degree of membership of xi in the fuzzy set Ai.

The fuzzy rules of the controller are represented using linguistic variables and fuzzy logic operators. Let's consider a simple fuzzy rule with two input variables:

$$"If\ x1is\ A\ and\ x2\ is\ B,\ then\ y\ is\ C."$$

Here, A, B, and C are linguistic terms associated with fuzzy sets. The fuzzy rule is represented by a membership function $\mu_{A\times B\to C}(x1, x2, y)$, which defines the degree of membership of the output variable y in the fuzzy set C given the degrees of membership of $x1$ in A and $x2$ in B.

The fuzzy inference process combines the fuzzy rules to determine the output variable y based on the input variables $x1, x2, ..., xn$. This process involves fuzzification, rule evaluation, aggregation, and defuzzification.

- Fuzzification: The input variables $x1, x2, ..., xn$ are fuzzified using their respective membership functions to determine their degrees of membership in the fuzzy sets $A1, A2, ..., An$.
- Rule Evaluation: Each fuzzy rule is evaluated using the degrees of membership of the input variables in the antecedents of the rule to determine the degree of membership of the output variable in the consequent of the rule.
- Aggregation: The degrees of membership of the output variable in the consequents of all fuzzy rules are aggregated to obtain a combined degree of membership for each linguistic term associated with the output variable.
- Defuzzification: The combined degrees of membership for each linguistic term associated with the output variable are defuzzified to determine the crisp output value y.

The overall mathematical model of the fuzzy logic controller can be represented as:

$$y = f(x1, x2, ..., xn)$$

where $f(x1, x2, ..., xn)$ represents the fuzzy inference process involving fuzzification, rule evaluation, aggregation, and defuzzification based on the fuzzy rules and membership functions of the controller.

This mathematical model captures the essence of a fuzzy logic controller in robotics, which can be further extended or modified based on the specific application and requirements.

4 Soft Computing Techniques

Soft computing methods are a collection of computational approaches influenced by human cognition, enabling reasoning and decision-making amidst uncertainty, imprecision, and partial truth. Diverging from conventional "hard" computing methods [10] dependent on exact mathematical models and algorithms, soft computing embraces approximation, uncertainty, and incomplete data to address intricate real-world challenges. In the realm of robotics, notable soft computing techniques encompass:

- Fuzzy Logic: Fuzzy logic provides a framework for reasoning under uncertainty by allowing for degrees of truth rather than strict binary values. In robotics, fuzzy logic controllers are used to handle imprecise input data and control systems with non-linear dynamics [11].
- Neural Networks: Neural networks, inspired by the structure and function of the human brain, are computational models composed of interconnected nodes (neurons) that can learn from and generalize patterns in data. In robotics, neural networks are used for tasks such as perception, classification, control, and learning [12].
- Evolutionary Algorithms: Drawing inspiration from natural selection and evolution, are optimization methods that iteratively generate and refine a population of potential solutions to tackle intricate problems effectively. Within robotics, these algorithms find application in tasks like robot path mapping, optimizing robot configurations, and enhancing robot learning [13].
- Genetic Algorithms: A subset of evolutionary algorithms, that replicate the mechanisms of natural selection to iteratively refine solutions for optimization problems. Employing operations like selection, crossover, and mutation, they evolve a population of potential solutions across multiple generations [14].
- Swarm Intelligence: Swarm intelligence draws inspiration from the collective behavior of decentralized, self-organized systems in nature, such as ant colonies, bird flocks, and fish schools. In robotics, swarm intelligence algorithms are used to coordinate groups of robots to accomplish tasks such as exploration, surveillance, and distributed sensing [15].

By harnessing the complementary strengths of these soft computing techniques, robotic systems can exhibit robustness, adaptability, and intelligence in diverse and uncertain environments, paving the way for the development of advanced autonomous robots capable of operating effectively in real-world scenarios.

5 Soft Computing in Robot Control

Soft computing techniques play a crucial role in robot control, enabling robots to make intelligent decisions and adapt to changing environments [16]. These techniques offer flexibility and robustness in handling complex control tasks where traditional control methods may struggle due to uncertainties or non-linearities. Some key soft computing approaches in robot control include:

- Fuzzy Logic Control (FLC): Fuzzy logic provides a framework for modeling and reasoning with imprecise and uncertain information. Fuzzy logic controllers use linguistic variables and fuzzy rules to map inputs to outputs, allowing for intuitive rule-based control. FLCs are particularly effective in systems with non-linear dynamics or uncertain environments.

- Neural Network Control: Neural networks can learn complex mappings between inputs and outputs, making them well-suited for control tasks where traditional mathematical models are difficult to formulate. Neural network controllers can adapt and learn from experience, enabling robots to improve their control performance over time and handle non-linearities in the system.
- Evolutionary Control: Evolutionary algorithms such as genetic algorithms and evolutionary strategies can optimize control parameters to achieve desired objectives. Evolutionary control methods search for the best control policies through iterations of selection, crossover, and mutation operations, often outperforming traditional optimization techniques in complex and high-dimensional control spaces.
- Swarm Intelligence-Based Control: Swarm intelligence algorithms draw inspiration from the collective behavior of social insects to coordinate groups of robots. Swarm-based control strategies, such as particle swarm optimization and ant colony optimization, can be used to optimize robot trajectories, coordinate multiple robots for collaborative tasks, and navigate in dynamic environments.
- Hybrid Approaches: Hybrid approaches combine multiple soft computing techniques or integrate soft computing with conventional control methods to leverage their respective strengths. For example, fuzzy-neural hybrid control systems combine the interpretability of fuzzy logic with the learning capabilities of neural networks for improved control performance.

Soft computing techniques offer versatile solutions for robot control, enabling robots to operate effectively in complex and uncertain environments.

6 Soft Computing in Robot Navigation

Soft computing techniques play a vital role in enhancing robot navigation, enabling robots to autonomously plan and execute efficient paths in complex and dynamic environments. Traditional navigation methods often rely on precise mathematical models and sensor measurements, which may struggle in real-world scenarios with uncertainties and unpredictable obstacles [17]. Soft computing techniques offer flexible and robust solutions to address these challenges by leveraging approximation, learning, and adaptation.

Fuzzy logic-based navigation systems provide a natural way to handle imprecise sensor data and uncertain environmental conditions. Fuzzy logic controllers can generate smooth and human-like navigation behaviors by incorporating linguistic variables and fuzzy rules to make decisions based on vague or incomplete information. These controllers allow robots to navigate safely in cluttered environments, avoiding obstacles while maintaining a desired trajectory.

Neural network-based navigation approaches utilize the learning capabilities of artificial neural networks to map sensor inputs to navigation commands. By training neural networks on a large dataset of sensor measurements and corresponding actions, robots can learn complex mappings between sensory information and navigation decisions. Neural network-based navigation systems can adapt to changing environments and learn from experience, improving their navigation performance over time.

Evolutionary algorithms offer another powerful tool for robot navigation by optimizing path planning strategies to achieve specific objectives. Genetic algorithms, for example, can evolve optimal or near-optimal paths by iteratively selecting and modifying candidate paths based on their fitness in reaching the goal while avoiding obstacles. Evolutionary algorithms enable robots to navigate efficiently in complex environments with dynamic obstacles and uncertain terrain.

Swarm intelligence-based navigation algorithms draw inspiration from the collective behavior of social insects to coordinate groups of robots for navigation tasks. Swarm-based navigation strategies, such as ant colony optimization and particle swarm optimization, enable robots to cooperatively explore and map unknown environments, find optimal paths to goals, and avoid collisions with each other. Swarm intelligence-based navigation systems offer robust and scalable solutions for multi-robot navigation tasks in dynamic and unpredictable environments.

7 Soft Computing in Path Planning for Robots

Soft computing techniques offer versatile solutions for path planning in robotics, enabling robots to navigate efficiently through complex and dynamic environments while considering uncertainties, obstacles, and other constraints. Traditional path planning algorithms often rely on precise mathematical models and assumptions about the environment, which may not be feasible in real-world scenarios [18]. Soft computing techniques provide flexible and adaptive approaches to address these challenges.

Fuzzy logic-based path planning systems excel in handling imprecise information and uncertain environments. Fuzzy logic controllers can generate smooth and human-like trajectories by incorporating fuzzy rules and linguistic variables to reason about spatial relationships and obstacle avoidance. These controllers allow robots to navigate safely in cluttered environments while adapting their paths in real-time based on changing conditions.

Neural network-based path planning approaches leverage the learning capabilities of artificial neural networks to generate optimal or near-optimal paths. By training neural networks on a large dataset of environmental features and corresponding paths, robots can learn complex mappings between sensory inputs and desired trajectories. Neural network-based path planning systems can adapt to novel environments and unforeseen obstacles, improving their path planning performance through experience.

Evolutionary algorithms offer another powerful tool for robot path planning by searching for optimal or near-optimal paths through an iterative optimization process. Genetic algorithms, for example, can evolve solutions by iteratively selecting and modifying candidate paths based on their fitness in reaching the goal while avoiding obstacles. Evolutionary algorithms enable robots to find efficient paths in complex environments with dynamic obstacles and uncertain terrain, effectively handling the non-linearity and high-dimensionality of the path planning problem.

Swarm intelligence-based path planning algorithms draw inspiration from the collective behavior of social insects to coordinate groups of robots for path planning tasks. Swarm-based path planning strategies, such as ant colony optimization and particle swarm optimization, enable robots to cooperatively explore and map unknown environments, find optimal paths to goals, and avoid collisions with each other. Swarm intelligence-based path planning systems provide scalable and robust solutions for multi-robot path planning tasks in dynamic and uncertain environments.

8 Soft Computing in Robot Learning

Soft computing techniques play a pivotal role in enabling robots to learn and adapt to their environments, allowing them to improve their performance over time through experience and interaction. Robot learning encompasses various tasks, including pattern recognition, classification, reinforcement learning, and adaptive control, all of which benefit from the flexibility and adaptability offered by soft computing methods [19].

Fuzzy logic-based learning systems provide a natural framework for reasoning with imprecise and uncertain information. Fuzzy logic allows robots to represent and manipulate vague or ambiguous concepts, facilitating intuitive and human-like decision-making processes. Fuzzy inference systems can be used to model complex relationships between inputs and outputs, enabling robots to learn and adapt their behavior based on environmental feedback and changing conditions.

Neural network-based learning approaches leverage the powerful learning capabilities of artificial neural networks to extract patterns and relationships from large datasets. By training neural networks on labeled data, robots can learn to recognize objects, navigate environments, and perform complex tasks. Neural networks enable robots to generalize from past experiences and adapt their behavior to new situations, making them highly effective for tasks requiring pattern recognition and classification.

Evolutionary algorithms offer a unique approach to robot learning by optimizing robot behaviors through an iterative process of selection, crossover, and mutation. Genetic algorithms, for example, can evolve robot control policies that maximize performance in a given task while satisfying constraints and objectives. Evolutionary algorithms enable robots to learn optimal or near-optimal solutions to complex problems, particularly in environments with uncertainties and dynamic changes.

Reinforcement learning is another important paradigm in robot learning, where robots learn to maximize rewards through trial-and-error interactions with their environment. Soft computing techniques, such as fuzzy logic and neural networks, can be used to approximate the value function or policy function in reinforcement learning algorithms, enabling robots to learn optimal control policies in continuous state and action spaces. Reinforcement learning allows robots to learn complex behaviors and adapt to dynamic environments through interaction and exploration.

9 Soft Computing in Robot Optimization

Soft computing techniques play a significant role in optimizing various aspects of robotic systems, ranging from parameter tuning to task performance improvement. Unlike traditional optimization methods, which often rely on precise mathematical models and assumptions, soft computing approaches offer flexible and robust solutions that can handle uncertainties, non-linearities, and complex objective functions [20].

Fuzzy logic-based optimization methods provide a natural way to handle imprecise information and uncertain constraints. Fuzzy optimization techniques allow for the representation of vague or ambiguous objectives and constraints, enabling robots to make decisions based on linguistic variables and fuzzy rules. Fuzzy logic controllers can guide optimization algorithms to explore solution spaces efficiently while adapting to changing environmental conditions or task requirements [21].

Neural network-based optimization approaches leverage the learning capabilities of artificial neural networks to approximate complex objective functions and search for optimal or near-optimal solutions. By training neural networks on a dataset of input–output pairs, robots can learn mappings between input parameters and performance metrics, enabling efficient optimization of system parameters or control policies. Neural network-based optimization methods can handle high-dimensional and non-linear optimization problems, making them well-suited for optimizing complex robotic systems.

Evolutionary algorithms offer powerful optimization techniques inspired by natural selection and evolution. Genetic algorithms, evolutionary strategies, and other evolutionary optimization methods iteratively generate and modify candidate solutions to search for optimal or near-optimal solutions in solution spaces. Evolutionary algorithms enable robots to optimize system parameters, control policies, or task performance metrics while considering uncertainties, constraints, and trade-offs.

Swarm intelligence-based optimization algorithms draw inspiration from the collective behavior of social insects to search for optimal solutions through cooperation and interaction among individuals. Swarm-based optimization techniques, such as particle swarm optimization and ant colony optimization, enable robots to explore solution spaces efficiently and find near-optimal solutions to complex optimization problems. Swarm intelligence-based optimization methods offer robust and scalable solutions for multi-objective optimization and dynamic optimization problems.

10 Integration of Soft Computing Paradigms in Robotics

The integration of soft computing paradigms in robotics involves combining multiple techniques to leverage their respective strengths and overcome their individual limitations. This integration enables robots to exhibit robust and intelligent behavior in diverse and complex environments [4]. Several approaches for integrating soft computing paradigms in robotics include:

- Hybrid Systems: Hybrid systems amalgamate various soft computing methodologies, including fuzzy logic, neural networks, evolutionary algorithms, and swarm intelligence, to tackle specific challenges or tasks in robotics. For instance, a hybrid system might employ fuzzy logic for decision-making, neural networks for perception and learning, and evolutionary algorithms for optimization. Through this amalgamation of techniques, hybrid systems leverage the strengths of each method while offsetting their respective limitations.
- Fuzzy-Neural Systems: Fuzzy-neural systems amalgamate fuzzy logic and neural networks to construct hybrid models that combine the interpretability of fuzzy logic with the learning capabilities of neural networks. These systems are particularly advantageous for tasks necessitating both transparency and adaptability, such as robot control and decision-making. Fuzzy-neural systems empower robots to learn from data while upholding human-understandable rules and reasoning.
- Evolutionary-Neural Systems: Evolutionary-neural systems combine evolutionary algorithms with neural networks to evolve neural network architectures or parameters for specific robotic tasks. These systems can optimize neural network structures and weights to improve performance, adaptability, and generalization capabilities. Evolutionary- neural systems are used for tasks such as robot control, path planning, and reinforcement learning.
- Swarm Intelligence-Based Systems: Swarm intelligence-based systems leverage principles from swarm intelligence to coordinate groups of robots or agents for collaborative tasks. These systems often combine swarm algorithms such as ant colony optimization or particle swarm optimization with other soft computing techniques to optimize collective behavior and task performance. Swarm intelligence-based systems enable robots to accomplish complex tasks such as exploration, mapping, and surveillance in dynamic and uncertain environments.
- Fuzzy-Evolutionary-Neural Systems: Fuzzy-evolutionary-neural systems integrate fuzzy logic, evolutionary algorithms, and neural networks to create adaptive and intelligent robotic systems. These systems use fuzzy logic for decision-making, evolutionary algorithms for optimization, and neural networks for learning and adaptation. Fuzzy-evolutionary-neural systems are used for a wide range of robotic applications, including control, navigation, planning, and optimization.

Overall, the integration of soft computing paradigms in robotics enables the development of versatile, adaptive, and intelligent robotic systems capable of operating effectively in complex and uncertain environments. By combining multiple techniques, researchers and practitioners can design robots that exhibit robustness, adaptability, and autonomy in a wide range of real-world applications.

11 Applications of Soft Computing in Robotics

Soft computing techniques find numerous applications in robotics across various domains, enabling robots to tackle complex tasks in diverse environments. Some key applications of soft computing in robotics include:

- Robot Control: Soft computing methods, such as fuzzy logic controllers and neural network-based control systems, are used to regulate the motion and behavior of robots. These techniques allow robots to adapt to uncertainties and non-linearities in their environments, enabling precise and flexible control in tasks such as manipulation, locomotion, and navigation [22].
- Path Planning: Soft computing methods, including genetic algorithms, particle swarm optimization, and fuzzy logic-based planners, are employed to generate optimal or near-optimal paths for robots in complex and dynamic environments. These techniques enable robots to navigate safely and efficiently while avoiding obstacles and optimizing performance metrics such as time, energy consumption, and safety [23].
- Object Recognition and Classification: Neural networks and fuzzy logic-based classifiers are utilized for object recognition and classification tasks in robotics. These techniques enable robots to perceive and interpret sensory information from cameras, LiDAR, and other sensors, facilitating tasks such as object manipulation, scene understanding, and environment mapping [24].
- Learning and Adaptation: Soft computing approaches, including reinforcement learning, evolutionary algorithms, and neural network-based learning, are employed to enable robots to learn from experience and adapt to changing environments. These techniques enable robots to improve their performance over time through trial and error, reinforcement, or evolutionary optimization, enhancing their autonomy and adaptability.
- Human–Robot Interaction: Soft computing techniques, such as fuzzy logic-based decision-making systems and neural network-based natural language processing, are utilized to enable natural and intuitive human–robot interaction. These techniques enable robots to understand and respond to human commands, gestures, and emotions, facilitating collaboration and cooperation in tasks such as assistance, healthcare, and education [25].
- Multi-Robot Systems: Soft computing methods are employed to coordinate and control groups of robots in multi-robot systems. Swarm intelligence-based algorithms, genetic algorithms for task allocation, and neural network-based

communication protocols enable robots to collaborate effectively and accomplish complex tasks such as exploration, surveillance, and search and rescue.
- Autonomous Decision-Making: Soft computing techniques play a crucial role in enabling robots to make autonomous decisions in real-time. Fuzzy logic controllers, reinforcement learning algorithms, and evolutionary strategies enable robots to reason under uncertainty and make decisions based on sensory information, task objectives, and environmental constraints.

Soft computing techniques offer versatile and effective solutions for addressing the complexities and uncertainties inherent in robotics applications, enabling robots to operate autonomously and intelligently in diverse and challenging environments.

12 Challenges and Limitations of Soft Computing in Robotics

While soft computing techniques offer versatile and effective solutions for various robotics applications, they also face several challenges and limitations that need to be addressed to fully exploit their potential.

One of the primary challenges is the interpretability of soft computing models, particularly in fuzzy logic and neural network-based systems. Fuzzy logic controllers often rely on linguistic variables and fuzzy rules, which can be difficult to interpret and debug, leading to challenges in understanding and refining the system's behavior. Similarly, neural networks, especially deep neural networks, are often viewed as black-box models due to their complex architectures and millions of parameters, making it challenging to understand the underlying reasoning and decision-making process. This lack of interpretability can hinder the adoption of soft computing techniques in safety–critical applications where transparency and accountability are essential.

Another challenge is the computational complexity and resource requirements of soft computing techniques, especially in real-time robotics applications. Neural networks, for example, require significant computational resources for training and inference, limiting their deployment on resource-constrained robotic platforms. Similarly, evolutionary algorithms and swarm intelligence-based algorithms can be computationally expensive, especially for optimization problems with large search spaces or complex fitness functions. Efficient implementation and optimization of soft computing algorithms for real-time robotics applications remain a significant challenge.

Furthermore, soft computing techniques often suffer from issues related to robustness and generalization. Fuzzy logic controllers may struggle to handle uncertain or noisy input data, leading to suboptimal performance or even instability in control systems. Neural networks trained on limited datasets may fail to generalize to unseen environments or situations, resulting in poor performance in real-world scenarios. Evolutionary algorithms and swarm intelligence-based algorithms may also encounter difficulties in converging to optimal solutions or avoiding premature convergence in complex and dynamic environments.

Another limitation is the lack of theoretical foundations and formal guarantees in many soft computing techniques. Unlike traditional optimization methods, which often come with rigorous mathematical analyses and convergence proofs, soft computing techniques such as evolutionary algorithms and neural networks lack theoretical guarantees of convergence, optimality, or robustness. This can pose challenges in the design and validation of soft computing-based robotic systems, particularly in safety–critical applications where reliability and predictability are paramount.

While soft computing techniques offer powerful tools for addressing complex and uncertain robotics problems, they also face several challenges and limitations related to interpretability, computational complexity, robustness, and theoretical foundations. Overcoming these challenges requires interdisciplinary research efforts to develop more interpretable, efficient, robust, and theoretically grounded soft computing techniques tailored to the specific requirements of robotics applications. Addressing these challenges will be crucial for unlocking the full potential of soft computing in advancing the capabilities of autonomous robots in various domains [26].

13 Conclusion and Future Work

This paper provides a comprehensive overview of the current state and future potential of soft computing techniques in robotics. The authors thoroughly examine various soft computing methodologies, including fuzzy logic, genetic algorithms, artificial neural networks, machine learning, and expert systems, highlighting their advantages and disadvantages compared to traditional hard computing techniques. They emphasize the tolerance of soft computing to imprecision, uncertainty, and approximations, making it a suitable approach for addressing complex real-life problems in robotics.

The paper underscores the significant advancements achieved through soft computing in robotics and its widespread applications across domestic, commercial, and industrial domains. It acknowledges the pivotal role of soft computing in enhancing robotic capabilities, particularly in areas such as path planning, obstacle avoidance, control, and adaptation to dynamic environments. Despite these advancements, the authors recognize several challenges and opportunities for future research and development in the field.

Moving forward, the paper suggests several avenues for future work in soft computing approaches in robotics. These include further exploration and refinement of existing techniques, development of novel algorithms to address specific challenges in robotics, integration of soft computing with other emerging technologies such as sensor networks and Internet of Things (IoT), and investigation of interdisciplinary approaches combining soft computing with traditional robotics methodologies. Additionally, the authors highlight the importance of considering ethical and societal implications in the development and deployment of soft computing-based robotic systems.

The paper provides valuable insights into the current state and future directions of soft computing approaches in robotics, laying the groundwork for continued research and innovation in this rapidly evolving field.

References

1. Swapna Rekha H, Nayak J, Naik B, Pelusi D (2020) Soft computing in robotics: a decade perspective. In: Applications of robotics in industry using advanced mechanisms: proceedings of international conference on robotics and its industrial applications 2019, vol 1. Springer International Publishing, pp 59–78
2. Della Santina C, Duriez C, Rus D (2023) Model-based control of soft robots: a survey of the state of the art and open challenges. IEEE Control Syst Mag 43(3):30–65
3. Kim D, Kim SH, Kim T, Kang BB, Lee M, Park W, Jo S (2021) Review of machine learning methods in soft robotics. PLoS ONE 16(2):e0246102
4. Chin K, Hellebrekers T, Majidi C (2020) Machine learning for soft robotic sensing and control. Adv Intell Syst 2(6):1900171
5. Maoudj A, Hentout A (2020) Optimal path planning approach based on Q-learning algorithm for mobile robots. Appl Soft Comput 97:106796
6. Song B, Wang Z, Zou L (2021) An improved PSO algorithm for smooth path planning of mobile robots using continuous high-degree Bezier curve. Appl Soft Comput 100:106960
7. Ravankar A, Ravankar AA, Kobayashi Y, Hoshino Y, Peng CC (2018) Path smoothing techniques in robot navigation: state-of-the-art, current and future challenges. Sensors 18(9):3170
8. Young AJ, Ferris DP (2016) State of the art and future directions for lower limb robotic exoskeletons. IEEE Trans Neural Syst Rehabil Eng 25(2):171–182
9. Katrakazas C, Quddus M, Chen WH, Deka L (2015) Real-time motion planning methods for autonomous on-road driving: state-of-the-art and future research directions. Transp Res Part C: Emerg Technol 60:416–442
10. Ibrahim D (2016) An overview of soft computing. Procedia Comput Sci 102:34–38
11. Manti M, Cacucciolo V, Cianchetti M (2016) Stiffening in soft robotics: A review of the state of the art. IEEE Robot Autom Mag 23(3):93–106
12. Laschi C, Mazzolai B, Cianchetti M (2016) Soft robotics: technologies and systems pushing the boundaries of robot abilities. Sci Robot 1(1):eaah3690
13. Goury O, Duriez C (2018) Fast, generic, and reliable control and simulation of soft robots using model order reduction. IEEE Trans Rob 34(6):1565–1576
14. Serrano-Guerrero J, Romero FP, Olivas JA (2021) Fuzzy logic applied to opinion mining: a review. Knowl-Based Syst 222:107018
15. Zhang W, Zhang R, Wu C, Goh ATC, Lacasse S, Liu Z, Liu H (2020) State-of-the-art review of soft computing applications in underground excavations. Geosci Front 11(4):1095–1106
16. Pinskier J, Howard D (2022) From bioinspiration to computer generation: developments in autonomous soft robot design. Adv Intell Syst 4(1):2100086
17. Osaba E, Del Ser J, Iglesias A, Yang XS (2020) Soft computing for swarm robotics: new trends and applications. J Comput Sci 39:101049
18. Beni G (2020) Swarm intelligence. Complex Soc Behav Syst: Game Theory Agent-Based Model 791–818
19. Zheng G, Zhou Y, Ju M (2020) Robust control of a silicone soft robot using neural networks. ISA Trans 100:38–45

20. de Jesus JC, Kich VA, Kolling AH, Grando RB, Cuadros MADSL, Gamarra DFT (2021) Soft actor-critic for navigation of mobile robots. J Intell Rob Syst 102(2):31
21. Sanchez-Ibanez JR, Perez-del-Pulgar CJ, García-Cerezo A (2021) Path planning for autonomous mobile robots: a review. Sensors 21(23):7898
22. Das PK, Jena PK (2020) Multi-robot path planning using improved particle swarm optimization algorithm through novel evolutionary operators. Appl Soft Comput 92:106312
23. Gacto MJ, Alcalá R, Herrera F (2012) A multi-objective evolutionary algorithm for an effective tuning of fuzzy logic controllers in heating, ventilating and air conditioning systems. Appl Intell 36(2):330–347
24. Truby RL (2021) Designing soft robots as robotic materials. Acc Mater Res 2(10):854–857
25. Wu Z, Liao H, Lu K, Zavadskas EK (2021) Soft computing techniques and their applications in intelligent industrial control systems: a survey. Int J Comput Commun Control 16(1):1–28
26. Gul F, Rahiman W, Alhady SSN, Ali A, Mir I, Jalil A (2021) Meta-heuristic approach for solving multi-objective path planning for autonomous guided robot using PSO–GWO optimization algorithm with evolutionary programming. J Ambient Intell Humaniz Comput 12:7873–7890
27. Khan MN, Al Hasan M, Anwar S (2021) Improving the robustness of object detection through a multi-camera–based fusion algorithm using fuzzy logic. Front Artif Intell 4:638951
28. Wang T, Zhu Y, Ye P, Gong W, Lu H, Mo H, Wang FY (2022) A new perspective for computational social systems: fuzzy modeling and reasoning for social computing in CPSS. IEEE Trans Comput Soc Syst

Neuromorphic Computing: Bridging the Gap Between Neuroscience and Artificial Intelligence

Nallagondla Jyothi, Angotu Nageswara Rao, Adilakshmi Velivela, S. Shilpa, M. Sri Rama Lakshmi Reddy, and Yamjala Arjun Sagar

Abstract Neuromorphic computing is an emerging field at the intersection of neuroscience and artificial intelligence (AI), aiming to replicate the functionality of the human brain in hardware and software systems. This abstract explores the principles, architecture, and applications of neuromorphic computing, highlighting its potential to revolutionize traditional computing paradigms. By mimicking the brain's neural networks, neuromorphic systems offer unparalleled efficiency, adaptability, and energy savings compared to conventional computing approaches. This paper discusses key components of neuromorphic systems, such as spiking neural networks and memristors, and examines their role in enabling bio-inspired computing capabilities. Moreover, it explores present obstacles and potential advancements in neuromorphic computing, underscoring its revolutionary influence across various domains such as robotics, healthcare, and cognitive computing.

Keywords Neuromorphic computing · Neuroscience · Artificial intelligence · Spiking neural networks

N. Jyothi (✉) · A. N. Rao · S. Shilpa · M. S. R. L. Reddy
CSE Department, CMR Institute of Technology, Hyderabad, Medchal, Telangana, India
e-mail: jyothi.nallagondla@cmritonline.ac.in

A. N. Rao
e-mail: nageswararao.a@cmritonline.ac.in

S. Shilpa
e-mail: shilpa.reddy@cmritonline.ac.in

A. Velivela
CSE (DS) Department, CMR Institute of Technology, Hyderabad, Medchal, Telangana, India
e-mail: vadilakshmi123@cmritonlie.ac.in

Y. A. Sagar
CSE (AI&ML) Department, CMR Institute of Technology, Hyderabad, Medchal, Telangana, India
e-mail: arjun@cmritonline.ac.in

V. K. Gunjan et al. (eds.), *Cybernetics, Human Cognition, and Machine Learning in Communicative Applications*, Cognitive Science and Technology,
https://doi.org/10.1007/978-981-97-8533-9_7

1 Introduction

The introduction serves as a gateway to understanding the multifaceted landscape of neuromorphic computing, a burgeoning field that bridges the realms of neuroscience and artificial intelligence (AI). Neuromorphic computing, inspired by the intricate workings of the human brain [1], offers a paradigm shift in computing architecture and capabilities. Neuromorphic computing draws its inspiration from the remarkable efficiency and adaptability of biological neural networks. By emulating the brain's neural structures and computational principles, neuromorphic systems aim to overcome the limitations of traditional computing approaches and pave the way for more efficient and intelligent computing solutions [2]. The convergence of neuroscience and AI in neuromorphic computing is motivated by the quest to develop systems that can mimic the brain's cognitive abilities, such as learning, adaptation, and pattern recognition [3]. This integration promises to revolutionize various domains, including robotics, healthcare, and cognitive computing, by enabling machines to process information in a manner akin to the human brain.

Neuromorphic Computing

Neuromorphic computing represents a revolutionary approach to computing that draws inspiration from the architecture and functioning of the human brain [4]. Unlike conventional computing, which relies on binary logic and von Neumann architectures, neuromorphic computing aims to mimic the parallelism, fault tolerance, and energy efficiency of biological neural networks.

At the core of neuromorphic computing are spiking neural networks (SNNs) [5], which emulate the behavior of neurons and synapses in the brain. These networks process information in the form of spikes or pulses, enabling efficient communication and computation. Neuromorphic hardware, such as neuromorphic chips and memristors, is designed to support the parallel and distributed processing required by SNNs, offering significant advantages in terms of speed, energy efficiency, and scalability. The applications of neuromorphic computing span a wide range of domains, including robotics, sensor networks, pattern recognition, and cognitive computing. Neuromorphic systems have shown promise in tasks such as image and speech recognition, autonomous navigation, and real-time data processing. However, challenges remain in optimizing neuromorphic hardware, developing efficient algorithms, and integrating neuromorphic systems with existing computing infrastructure. Despite these challenges, neuromorphic computing holds tremendous potential for advancing the frontiers of artificial intelligence and revolutionizing the way we approach computing tasks in the future.

2 Literature Review

In [6] highlighted that the human brain, recognized as the most powerful computational entity, has long inspired artificial intelligence development. Deep learning, a result of reverse engineering neural systems, has shown remarkable success in various cognitive tasks, yet its limitations such as the need for extensive data and energy inefficiency are becoming more apparent. Unlike deep learning, neuromorphic computing, which mimics biological neural networks through software and hardware co-design, offers a balanced approach considering computational complexity, energy efficiency, and biological plausibility. The chapter aimed to comprehensively introduce neuromorphic computing, covering biological neural systems, neuron models, and hardware implementations, and discussing critical challenges and opportunities. Yang et al. [7] described the human brain as a high-performance biocomputer and discussed how spiking neural networks based on neuromorphic computing platforms emulate its architecture, offering insights for building artificial intelligence. Advances in materials engineering, device physics, and chip integration have propelled progress in neuromorphic computing, aiming to overcome the limitations of the von Neumann architecture. The review provided an overview of neuron and synapse structures, the development of neuromorphic hardware systems, and highlighted the potential of these systems in revolutionizing computing. Huang et al. [8] explained how neuromorphic computing, using memristive hardware neural networks, simulates biological brain function for information processing, potentially overcoming the limitations of the von Neumann architecture. Various synaptic devices, categorized based on their response to electrical and optical signals, were discussed, along with their mechanisms, progress, and application scenarios. The review emphasized the importance of these synaptic devices for building efficient neuromorphic systems. Deng et al. [9] observed two paths toward artificial general intelligence: neuroscience-driven neuromorphic computing and computer-science-driven machine learning. While machine learning has seen significant breakthroughs, spiking neural networks (SNNs) in neuromorphic computing offer biological plausibility and energy efficiency. However, SNNs face challenges such as lower accuracy compared to traditional artificial neural networks (ANNs). A Research Topic aimed to bridge the gap between neuromorphic computing and machine learning by collecting recent research to understand and address these challenges. Tang et al. [10] demonstrated a keen interest in brain-inspired computing through neuromorphic devices, emphasizing the importance of comprehensively linking biological and artificial neural systems to enhance the emulation of brain functions. Their review encompassed discussions on both biological and artificial neural systems, recent advancements in neuromorphic devices, and identified current obstacles along with potential research pathways for the future. Marković et al. [11] discussed how neuromorphic computing aims to create energy-efficient hardware by mimicking the brain's distributed topology. The review emphasized leveraging physics and nanoscale materials to enhance computing capabilities, focusing on resistive switching materials, photonics, and spintronics. The paths toward low-power, miniaturized chips for real-time inference and learning were outlined. Shastri et al. [12] highlighted the flourishing research in photonic computing, offering ultrafast artificial neural networks with sub-nanosecond latencies. Neuromorphic photonics provided an opportunity to extend artificial intelligence capabilities, with recent advances in integrated photonic neuromorphic systems addressing current challenges and outlining future advancements needed. Frenkel et al. [13] provided a comprehensive overview of neuromorphic computing, comparing bottom-up approaches replicating natural intelligence with top-down approaches to solving practical artificial intelligence applications. Analog, mixed-signal, and digital circuit design styles were discussed, along with necessary synergies and missing elements for a competitive edge in neuromorphic edge computing. Du et al. [14] discussed hardware neural network accelerators inspired by machine learning and neuroscience, comparing their energy, speed, area cost, accuracy, and functionality. While machine-learning-based networks currently showed higher accuracy and lower hardware cost, spiking neural network accelerators held the potential for cost-efficiency in specific applications with moderate accuracy requirements and permanent online learning. Du et al. [14] emphasized the need for neuromorphic architectures optimized for energy consumption to enable embedded artificial intelligence. Implementations of multi-layer perceptron and spiking neural network architectures were compared in terms of accuracy and hardware cost, aiming to identify the most suitable architecture for embedded systems. Indiveri [15] highlighted the challenges to standard computing and proposed novel brain-inspired processing methods and technologies for various application scenarios. The launch of the journal Neuromorphic Computing and Engineering aimed to support this effort by providing a platform for presenting and discussing advances in this field. Schuman et al. [16] provided a comprehensive survey of neuromorphic computing, discussing its motivations, major research areas, and technical challenges. The review aimed to illuminate gaps in the field and motivate further research to realize the promise of neuromorphic computing (Fig. 1).

3 Architecture of Neuromorphic Systems

The architecture of neuromorphic systems is designed to emulate the complex and parallel processing capabilities of the human brain. At its core, neuromorphic systems consist of interconnected neurons and synapses, forming networks that process information in a manner analogous to biological neural networks. The architecture of neuromorphic systems encompasses both hardware and software components, tailored to support the efficient emulation of neural networks.

- Neuron Models: Neuromorphic systems utilize various neuron models to replicate the behavior of biological neurons. These models include integrate-and-fire neurons, spiking neurons, and leaky integrate-and-fire neurons. Each neuron

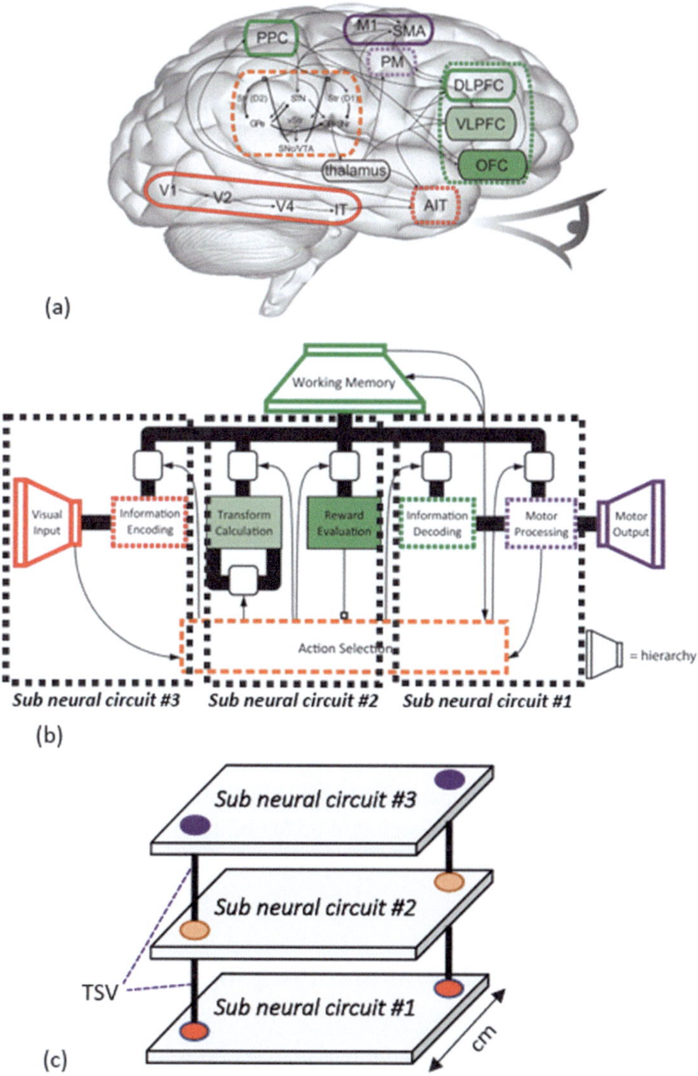

Fig. 1 Large-scale 3D neuromorphic structure inspired by Spaun's anatomical and functional architecture [17]. **a** The structural layout of Spaun highlights key brain regions and their interconnections. **b** Spaun's architecture depicts cortical elements communicating via thick black lines, with Basal Ganglia connectivity shown through thin lines. **c** Utilizing through silicon vias (TSVs) as vertical links between layers, this 3D neuromorphic design addresses the high connectivity demands of Spaun's architecture when partitioned and mapped into a 3D framework, as depicted within the black dot boxes [17]

model captures different aspects of neuronal behavior, such as firing thresholds, refractory periods, and synaptic integration.

- Synapse Models: Synapses in neuromorphic systems emulate the connections between biological neurons, facilitating communication and information processing. Synapse models incorporate features such as synaptic weights, plasticity mechanisms (e.g., spike-timing-dependent plasticity), and neurotransmitter dynamics to enable learning and adaptation.
- Hardware Implementation: Neuromorphic hardware platforms are designed to efficiently support the parallel and distributed processing requirements of neural networks. These platforms often feature specialized architectures, such as neuromorphic chips or neuromorphic processors, optimized for low-power consumption, high-speed computation, and scalability.
- Spiking Neural Networks (SNNs): SNNs are a fundamental component of neuromorphic systems, consisting of interconnected neurons that communicate through spikes or action potentials. The architecture of SNNs encompasses multiple layers of neurons, with each layer performing specific computational tasks such as feature extraction, pattern recognition, or decision-making.
- Programming Models and Tools: Neuromorphic systems are equipped with programming models and tools that enable developers to design and implement neural networks efficiently. These tools may include neuromorphic programming languages (e.g., NEST, PyNN), simulation environments (e.g., SpiNNaker, BrainScaleS), and software libraries for neural network modeling and simulation.

The architecture of neuromorphic systems is characterized by its bio-inspired design, efficient hardware implementation, and support for complex neural network models. This architecture enables neuromorphic systems to perform a wide range of cognitive tasks with high efficiency and adaptability, making them suitable for diverse applications in artificial intelligence, robotics, and neuroscience research.

4 Neuromorphic Algorithms and Programming Models

Neuromorphic algorithms and programming models play a crucial role in harnessing the computational power of neuromorphic hardware and emulating the functionality of biological neural networks. These algorithms and models are designed to facilitate the implementation, training, and operation of spiking neural networks (SNNs) and other neuromorphic architectures efficiently. Key aspects of neuromorphic algorithms and programming models include:

- Spike-based Processing: Neuromorphic algorithms are typically designed to operate on spike-based data, where information is encoded in the form of spikes or action potentials. These algorithms leverage the temporal dynamics of spike trains to perform tasks such as pattern recognition, classification, and associative memory.

- Event-Driven Computation: Unlike traditional computing models that rely on clock cycles and synchronous processing, neuromorphic algorithms operate in an event-driven manner. They process input spikes and generate output spikes asynchronously, mimicking the parallel and distributed nature of neural computation.
- Neuromorphic Learning Rules: Neuromorphic algorithms incorporate learning rules inspired by biological mechanisms of synaptic plasticity, such as spike-timing-dependent plasticity (STDP) and rate-based learning. These learning rules enable SNNs to adapt their synaptic weights in response to input spike patterns, facilitating unsupervised, supervised, and reinforcement learning tasks.
- Spiking Neural Network Architectures: Neuromorphic programming models provide frameworks and libraries for designing and simulating complex spiking neural network architectures. These models allow researchers and developers to construct multi-layered networks, implement various neuron and synapse models, and configure network connectivity patterns.
- Neuromorphic Simulators: Neuromorphic programming models are often implemented within specialized simulation environments and frameworks that enable the efficient simulation of large-scale spiking neural networks. These simulators provide tools for visualizing network activity, analyzing synaptic dynamics, and debugging neural network behavior.
- Hardware Abstraction: Neuromorphic programming models abstract the underlying hardware architecture, allowing developers to design algorithms and models that are hardware-agnostic. This enables portability and scalability across different neuromorphic hardware platforms and architectures.

Neuromorphic algorithms and programming models provide the computational framework necessary for leveraging the unique capabilities of neuromorphic hardware and advancing research in neuromorphic computing. These models facilitate the development of biologically inspired neural network architectures and enable the exploration of novel computational paradigms for artificial intelligence and cognitive computing applications.

5 Methodology

Neuromorphic algorithms and programming models encompass a variety of approaches for designing and implementing neural networks on neuromorphic hardware. These models often involve equations that describe the behavior of neurons, synapses, and the overall network dynamics.

Leaky Integrate-And-Fire Neuron Model:

This model describes the dynamics of a neuron's membrane potential over time. The membrane potential V of the neuron is governed by the following equation:

$$\tau \frac{dV}{\mathrm{dt}} = -(V - V_{rest}) + R \cdot I_{syn}$$

τ is the membrane time constant.

- *Vrest* is the resting membrane potential.
- R is the membrane resistance.
- I_{syn} is the synaptic current input.

Spike-Timing-Dependent Plasticity (STDP):

STDP is a synaptic plasticity rule that updates the synaptic weights based on the timing of pre- and post-synaptic spikes. The change in synaptic weight Δw is determined by the following equation:

$$\Delta w_{\mathrm{i,j}} = \sum_{\mathrm{k=1}}^{\mathrm{N_{spikes}}} \mathrm{A}.f(\Delta t_k k)$$

where:

- $\Delta t_k k$ is the time difference between the pre-synaptic spike and the k-th post-synaptic spike.
- A is the learning rate.
- $f(\Delta t_t k)$ is a function that determines the weight change based on the time difference Δtk.

Spiking Neural Network Dynamics:

The dynamics of spiking neural networks can be described using equations that govern the evolution of neuron membrane potentials and spike generation. One common approach is the leaky integrate-and-fire neuron model combined with a spiking mechanism. Neurons generate spikes when their membrane potential crosses a certain threshold V_{th}.

$$V_{\mathrm{th}} = V_{\mathrm{rest}} + R \cdot I_{\mathrm{th}}$$

where:

I_{th} is the threshold current.

These equations represent components of neuromorphic algorithms and programming models, providing a mathematical framework for simulating and implementing neural networks on neuromorphic hardware. Researchers often tailor these equations and models to specific applications and hardware architectures to achieve efficient and biologically plausible neural network behavior.

6 Applications of Neuromorphic Computing

Neuromorphic computing, with its bio-inspired architecture and efficient processing capabilities, holds promise across various domains, paving the way for innovative applications that leverage its unique advantages [18]. An important use is found in robotics, where neuromorphic systems provide notable improvements in sensory perception, movement coordination, and flexible learning. By integrating neuromorphic hardware with robotic platforms, researchers can develop robots capable of perceiving and interacting with their environment in a manner akin to human intelligence. These robots can navigate dynamic environments, recognize objects, and adapt their behavior based on sensory input, enabling advancements in autonomous navigation, object manipulation, and human–robot interaction.

In healthcare and biomedical applications, neuromorphic computing presents opportunities for personalized diagnostics, medical imaging, and brain-computer interfaces. Neuromorphic systems equipped with specialized sensors and algorithms can analyze biological signals such as EEG, ECG, and neural activity with high accuracy and efficiency [19]. This enables real-time monitoring of physiological parameters, early detection of abnormalities, and personalized treatment strategies. Additionally, neuromorphic brain-computer interfaces can facilitate seamless communication between the brain and external devices, offering new avenues for assistive technologies and neuroprosthetics to enhance the quality of life for individuals with disabilities.

Another compelling application of neuromorphic computing lies in neuromorphic vision systems, where bio-inspired algorithms and hardware enable efficient image processing, object recognition, and scene understanding. These systems emulate the parallel processing and hierarchical feature extraction mechanisms of the human visual system, enabling robust performance in complex visual tasks such as image classification, object tracking, and scene reconstruction. Neuromorphic vision systems find applications in diverse fields including surveillance, autonomous vehicles, augmented reality, and medical imaging, where real-time processing and low-power consumption are critical requirements.

Neuromorphic computing holds promise in edge computing and Internet of Things (IoT) devices [20], where energy efficiency and real-time processing are paramount. Neuromorphic hardware architectures are well-suited for edge devices deployed in resource-constrained environments, enabling local processing of sensory data and efficient communication with cloud-based services. This facilitates low-latency response times, reduced bandwidth requirements, and enhanced privacy and security for IoT applications spanning smart homes, wearable devices, environmental monitoring, and industrial automation.

7 Challenges and Opportunities

Neuromorphic computing presents both challenges and opportunities that shape its development and adoption across various domains.

Challenges:

- Hardware Complexity: Designing and fabricating neuromorphic hardware with the required level of complexity and scalability poses a significant challenge. Neuromorphic systems often consist of large-scale networks of neurons and synapses, requiring specialized hardware architectures and fabrication processes to emulate the parallelism and efficiency of biological neural networks.
- Algorithm Development: Developing efficient algorithms tailored to neuromorphic hardware poses challenges due to the unique computational characteristics and constraints of these systems. Optimizing algorithms for low-power consumption, real-time processing, and adaptability while maintaining biological plausibility remains a complex task.
- Scalability and Integration: Scaling neuromorphic systems to handle large-scale neural networks and integrating them with existing computing infrastructure pose challenges in terms of memory, communication bandwidth, and interconnectivity. Achieving seamless integration and compatibility with conventional computing platforms and software frameworks is crucial for widespread adoption.
- Benchmarking and Evaluation: Establishing standardized benchmarks and evaluation metrics for comparing the performance of neuromorphic systems poses challenges due to the diversity of applications and hardware architectures. Developing comprehensive evaluation methodologies that capture key performance metrics such as energy efficiency, computational speed, and accuracy is essential for assessing the effectiveness of neuromorphic solutions.

Opportunities:

- Energy Efficiency: Neuromorphic computing offers the potential for significant energy savings compared to conventional computing architectures, making it well-suited for battery-powered and energy-constrained devices. Leveraging the inherent parallelism and sparsity of neural networks, neuromorphic systems can perform complex computations with lower power consumption.
- Cognitive Computing: Neuromorphic computing holds promise for advancing the field of cognitive computing by emulating the brain's cognitive abilities such as learning, adaptation, and pattern recognition. Neuromorphic systems offer opportunities for developing intelligent systems capable of processing and understanding complex data in real-time, leading to advancements in artificial intelligence, robotics, and cognitive science.
- Interdisciplinary Collaboration: Neuromorphic computing fosters interdisciplinary collaboration between researchers in neuroscience, computer science, engineering, and other fields. This collaboration enables the cross-pollination

of ideas, methodologies, and technologies, leading to innovative solutions and breakthroughs in understanding brain function and developing intelligent systems.
- Applications in Edge Computing and IoT: Neuromorphic computing finds utility in edge computing and Internet of Things (IoT) gadgets, meeting crucial demands such as swift processing, energy conservation, and instantaneous adjustment. Neuromorphic hardware architectures can enable edge devices to perform advanced processing tasks locally, reducing the need for cloud-based processing and enhancing privacy and security in IoT applications.

8 Conclusion and Future Work

The paper has elucidated the promising potential of neuromorphic computing in revolutionizing traditional computing paradigms by bridging the gap between neuroscience and artificial intelligence. Through the emulation of the brain's neural networks and the incorporation of principles derived from neuroscience, neuromorphic systems offer unprecedented efficiency, adaptability, and energy savings compared to conventional computing architectures. Furthermore, the discussion has highlighted key components of neuromorphic systems, including spiking neural networks and memristors, which play pivotal roles in enabling bio-inspired computing capabilities. These components enable neuromorphic systems to process information in a manner that closely resembles biological neural networks, thereby unlocking new avenues for addressing complex computational tasks. However, while significant progress has been made in the field of neuromorphic computing, several challenges and opportunities for future research remain. These include further refining the design and optimization of neuromorphic hardware, advancing the development of efficient algorithms tailored to neuromorphic architectures, and exploring interdisciplinary collaborations between neuroscience and artificial intelligence communities.

In future work, it is essential to continue exploring the potential applications of neuromorphic computing across diverse domains such as robotics, healthcare, and cognitive computing. Additionally, efforts should be directed toward enhancing the scalability, reliability, and robustness of neuromorphic systems to ensure their widespread adoption and integration into real-world applications. The integration of neuroscience principles into artificial intelligence through neuromorphic computing holds great promise for advancing the frontiers of computing and unlocking new capabilities for addressing complex real-world challenges.

References

1. Najmaei S, Glasmann AL, Schroeder MA, Sarney WL, Chin ML, Potrepka DM (2022) Advancements in materials, devices, and integration schemes for a new generation of neuromorphic computers. Mater Today
2. Ullman S (2019) Using neuroscience to develop artificial intelligence. Science 363(6428):692–693
3. Furber S (2016) Large-scale neuromorphic computing systems. J Neural Eng 13(5):051001
4. Rathi N, Chakraborty I, Kosta A, Sengupta A, Ankit A, Panda P, Roy K (2023) Exploring neuromorphic computing based on spiking neural networks: Algorithms to hardware. ACM Comput Surv 55(12):1–49
5. Aimone JB, Date P, Fonseca-Guerra GA, Hamilton KE, Henke K, Kay B, Smith JD (2022) A review of non-cognitive applications for neuromorphic computing. Neuromorphic Comput Eng 2(3):032003
6. Zins N, Zhang Y, Yu C, An H (2023) Neuromorphic computing: A path to artificial intelligence through emulating human brains. Frontiers of quality electronic design (QED) AI, IoT and hardware security. Springer International Publishing, Cham, pp 259–296
7. Yang JQ, Wang R, Ren Y, Mao JY, Wang ZP, Zhou Y, Han ST (2020) Neuromorphic engineering: from biological to spike-based hardware nervous systems. Adv Mater 32(52):2003610
8. Huang W, Xia X, Zhu C, Steichen P, Quan W, Mao W, Li XA (2021) Memristive artificial synapses for neuromorphic computing. Nano-Micro Letters 13:1–28
9. Deng L, Tang H, Roy K (2021) Understanding and bridging the gap between neuromorphic computing and machine learning. Front Comput Neurosci 15:665662
10. Tang J, Yuan F, Shen X, Wang Z, Rao M, He Y, Wu H (2019) Bridging biological and artificial neural networks with emerging neuromorphic devices: fundamentals, progress, and challenges. Adv Mater 31(49):1902761
11. Marković D, Mizrahi A, Querlioz D, Grollier J (2020) Physics for neuromorphic computing. Nat Rev Phys 2(9):499–510
12. Shastri BJ, Tait AN, Ferreira de Lima T, Pernice WH, Bhaskaran H, Wright CD, Prucnal PR (2021) Photonics for artificial intelligence and neuromorphic computing. Nat Photonics 15(2):102–114
13. Frenkel C, Bol D, Indiveri G (2021) Bottom-up and top-down neural processing systems design: neuromorphic intelligence as the convergence of natural and artificial intelligence. arXiv:2106.01288
14. Du Z, Ben-Dayan Rubin DD, Chen Y, He L, Chen T, Zhang L, Temam O (2015) Neuromorphic accelerators: a comparison between neuroscience and machine-learning approaches. In: Proceedings of the 48th international symposium on microarchitecture, pp 494–507
15. Indiveri G (2021) Introducing 'neuromorphic computing and engineering.' Neuromorphic Comput Eng 1(1):010401
16. Schuman CD, Potok TE, Patton RM, Birdwell JD, Dean ME, Rose GS, Plank JS (2017) A survey of neuromorphic computing and neural networks in hardware. arXiv:1705.06963
17. Eliasmith C, Stewart TC, Choo X, Bekolay T, DeWolf T, Tang Y, Rasmussen D (2012) A large-scale model of the functioning brain. Science 338(6111):1202–1205
18. Prakash C, Gupta LR, Mehta A, Vasudev H, Tominov R, Korman E, Kesari KK (2023) Computing of neuromorphic materials: an emerging approach for bioengineering solutions. Mater Adv 4(23):5882–5919
19. Aboumerhi K, Güemes A, Liu H, Tenore F, Etienne-Cummings R (2023) Neuromorphic applications in medicine. J Neural Eng 20(4):041004
20. Passian A, Imam N (2019) Nanosystems, edge computing, and the next generation computing systems. Sensors 19(18):4048
21. Khacef L, Abderrahmane N, Miramond B (2018) Confronting machine-learning with neuroscience for neuromorphic architectures design. In: 2018 international joint conference on neural networks (IJCNN). IEEE, pp 1–8

Sustainability in the Internet of Things: Environmental Impacts and Green Technologies

S. Shilpa, Nallagondla Jyothi, Angotu Nageswara Rao, Adilakshmi Velivela, Yamjala Arjun Sagar, and M. Sri Rama Lakshmi Reddy

Abstract The Internet of Things (IoT) has rapidly transformed various aspects of our lives, offering innovative solutions to improve efficiency and convenience. However, the widespread adoption of IoT devices has raised concerns about their environmental impacts and sustainability. This paper explores the environmental implications of IoT technologies and presents a comprehensive analysis of green technologies aimed at mitigating these impacts. Key areas of focus include energy efficiency in IoT devices, the integration of renewable energy sources, sustainable materials and manufacturing processes, lifecycle management, e-waste reduction, data privacy, and security, as well as regulatory and policy perspectives. By examining the challenges and opportunities for sustainability in IoT, this paper aims to provide insights into the development and implementation of environmentally friendly IoT solutions.

Keywords Internet of Things (IoT) · Sustainability · Environmental impacts · Green technologies · Energy efficiency · Renewable energy

---

S. Shilpa (✉) · N. Jyothi · A. N. Rao · M. S. R. L. Reddy
CSE Department, CMR Institute of Technology, Hyderabad, Medchal, Telangana, India
e-mail: shilpa.reddy@cmritonline.ac.in

N. Jyothi
e-mail: jyothi.nallagondla@cmritonline.ac.in

A. N. Rao
e-mail: nageswararao.a@cmritonline.ac.in

A. Velivela
CSE(DS) Department, CMR Institute of Technology, Hyderabad, Medchal, Telangana, India
e-mail: vadilakshmi123@cmritonlie.ac.in

Y. A. Sagar
CSE(AI&ML) Department, CMR Institute of Technology, Hyderabad, Medchal, Telangana, India
e-mail: arjun@cmritonline.ac.in

V. K. Gunjan et al. (eds.), *Cybernetics, Human Cognition, and Machine Learning in Communicative Applications*, Cognitive Science and Technology,
https://doi.org/10.1007/978-981-97-8533-9_8

1 Introduction

The rapid proliferation of the Internet of Things (IoT) presents both opportunities and challenges regarding sustainability and environmental impact [1]. As IoT devices become increasingly ubiquitous, their energy consumption and resource utilization contribute significantly to environmental degradation. However, integrating sustainable practices and green technologies into IoT systems can mitigate these impacts.

Environmental impacts of IoT include energy consumption, electronic waste generation, and carbon emissions [2]. IoT devices require power for operation and data transmission, leading to increased energy consumption and dependence on non-renewable energy sources [3]. Moreover, the lifecycle of IoT devices, from manufacturing to disposal, contributes to electronic waste accumulation. To address these challenges, integrating green technologies is imperative. This involves leveraging renewable energy sources, such as solar or wind power, to power IoT devices and infrastructure. Additionally, optimizing energy-efficient designs and communication protocols can reduce energy consumption without compromising performance. Moreover, implementing circular economy principles, such as product reuse and recycling, can minimize electronic waste and resource depletion. Achieving sustainability in the IoT ecosystem [4] requires a holistic approach that considers environmental impacts throughout the lifecycle of IoT devices and integrates green technologies to mitigate these impacts. By adopting sustainable practices and technologies, the IoT industry can minimize its environmental footprint and contribute to a more sustainable future.

Sustainability in IoT

Sustainability in the Internet of Things (IoT) encompasses the ability of IoT systems to operate in a manner that minimizes their negative environmental and societal impacts [5], while maximizing their positive contributions over their lifecycle. This involves considering the environmental, economic, and social dimensions of sustainability. From an environmental standpoint, sustainability in IoT entails reducing energy consumption, minimizing resource use, and mitigating pollution and waste generation. Economically, it involves ensuring the long-term viability and profitability of IoT deployments while fostering economic growth and equitable access to IoT technologies. Socially, sustainability in IoT involves promoting inclusivity, accessibility, and ethical considerations, including data privacy and security.

Understanding sustainability in IoT requires a holistic approach that considers the interconnectedness of various factors, including technological innovation, regulatory frameworks, consumer behavior, and industry practices. It involves analyzing the lifecycle of IoT devices and systems, from design and manufacturing to use and disposal, to identify opportunities for improvement and optimization. Additionally, understanding sustainability in IoT requires recognizing the potential trade-offs and synergies between environmental, economic, and social objectives and implementing strategies that balance these considerations to achieve sustainable outcomes.

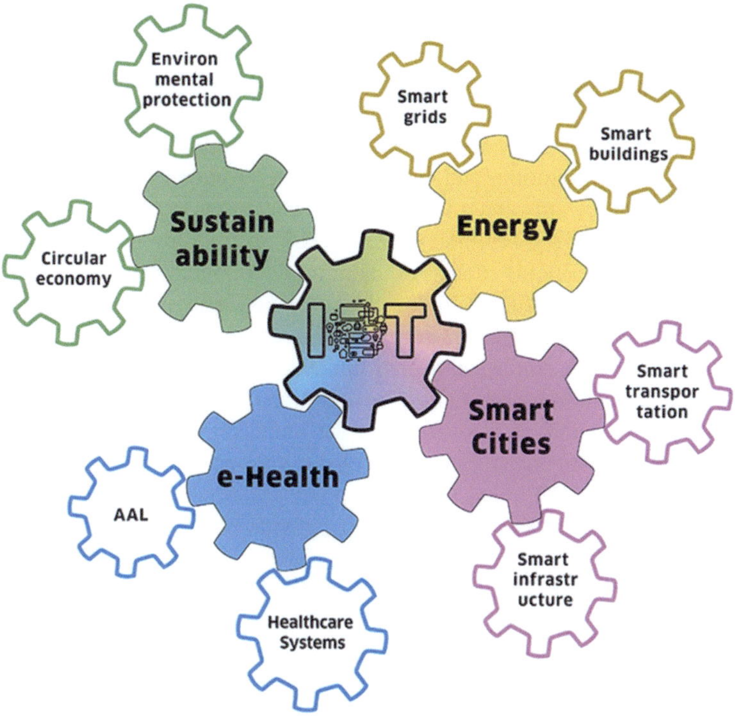

Fig. 1 Sustainability, energy, e-Health and smart cities in IoT [6]

Ultimately, achieving sustainability in IoT requires collaboration and cooperation among stakeholders across sectors to develop and implement innovative solutions that address environmental challenges while meeting the needs of society (Fig. 1).

2 Background

The transformative impact of new technologies and the Internet of Things (IoT) revolution, has driven innovation across various domains, enabling access to information anytime and anywhere. Despite the significant benefits offered by IoT, its manufacturing, distribution, and utilization have been resource-intensive, leading to escalating waste generation and environmental concerns. To mitigate these challenges and promote sustainability, there is a growing movement towards a greener future characterized by green technology, green IoT, and a green economy. This chapter analyzed the importance of greening technological processes in sustainable development, focusing on the principles and roles of Green IoT (G-IoT) in improving quality of life, environmental conservation, economic growth, and global modernization. G-IoT has the potential to bring numerous benefits, including environmental protection, customer satisfaction, and increased profit, while minimizing negative impacts on human health and the environment [7, 8] addressed the energy consumption challenges of IoT technology in the context of a connected smart world. While IoT technology has enriched society, it has also contributed to energy consumption, toxic pollution, and electronic waste. To address these issues, there is a growing trend towards green IoT (G-IoT), which aims to reduce environmental impact while maximizing benefits. Greening ICT technology plays a crucial role in G-IoT, promising benefits such as efficient production and reduced energy usage in designing and distributing ICT devices and equipment. The paper provided a comprehensive overview of G-IoT technologies and strategies aimed at building a green and smart world, emphasizing environmental sustainability, improved quality of life, reduced pollution, and energy consumption. Shaikh et al. [9] discussed the emergence of green IoT as a response to the increasing carbon footprint associated with technological advances. They explored various enabling technologies and applications of green IoT, emphasizing the need for efficient energy utilization to enable a sustainable environment. The paper reviewed ongoing IoT projects and standardization efforts, and identified emerging challenges that need to be addressed to realize a green IoT environment. Nižetić et al. [6] highlighted the rapid development of smart and IoT-based technologies and the importance of sustainability in this context. While IoT technologies offer various benefits, their fast-paced development raises environmental concerns that need careful monitoring and evaluation. The paper emphasized the role of G-IoT and Big Data in promoting smart and sustainable cities, focusing on reducing pollution, enhancing resource utilization, and improving quality of life and [10] discussed the concept of a sustainable smart world driven by IoT technologies. They explored various technologies and issues related to green IoT, which aims to reduce energy consumption. The paper provided an overview of IoT and green IoT technologies, emphasizing their potential to promote sustainability through efficient energy utilization. It also reviewed green ICT principles and presented future research directions and challenges in green IoT also [11] highlighted the environmental challenges associated with IoT-enabled devices in smart cities. They discussed the concept of green IoT and its role in promoting sustainability, focusing on energy conservation and sustainable practices in smart cities and [12] discussed the application of ICT advances in agriculture and the environmental challenges it poses. They introduced green IoT and green nanotechnology as solutions for creating smart and sustainable agriculture and food industries. Nasiri et al. [13] explored the impact of IoT in fostering disruptive sustainable change across various sectors. They discussed the potential of IoT solutions in revolutionizing business markets and promoting sustainability. Pecunia et al. [14] discussed the challenge of powering IoT devices sustainably and the potential of indoor photovoltaics (IPV) as an eco-friendly energy harvesting solution. They emphasized the importance of environmentally sustainable materials and fabrication processes in IPV technologies. Jayakumar et al. [16] provided insights into green IoT applications, practices, and challenges. They discussed various energy-efficient hardware design principles, data traffic management techniques, and the role of fog/edge computing and machine learning in promoting energy efficiency in IoT networks. They also addressed legislative policies and future research directions towards achieving sustainability in IoT.

3 Mathematical Model

Developing a mathematical model for sustainability in the Internet of Things (IoT) involves considering various factors such as energy consumption, resource utilization, environmental impacts, and the integration of green technologies. Here, the model focuses on energy consumption and environmental impacts, incorporating green technologies where applicable.

Energy Consumption Model:

Let's denote the total energy consumption in the IoT system as E_{total}. This energy consumption can be broken down into:

- Device Energy Consumption: $E_{devices}$
- Communication Energy Consumption: $E_{communication}$
- Data Processing Energy Consumption: $E_{processing}$

$$E_{total} = E_{devices} + E_{communication} + E_{processing}$$

Each of these components can be further modeled based on specific parameters and characteristics of the IoT system. For example, device energy consumption can be modeled based on the power consumption of individual devices, their duty cycles, and their operational modes.

Environmental Impact Model:

The environmental impact of an IoT system can be assessed based on factors such as carbon emissions, resource depletion, and pollution. Let's denote the environmental impact as *Ienv*, which can be expressed as a function of energy consumption and other relevant factors:

$$I_{env} = f(E_{total}, OtherFactors)$$

This function f could be a complex model that considers the carbon footprint associated with energy consumption, the use of non-renewable resources in device manufacturing, and the disposal/recycling processes for IoT devices.

Integration of Green Technologies:

To promote sustainability, green technologies can be integrated into the IoT system. This integration can be modeled by adjusting the energy consumption terms in the equations above to reflect the use of renewable energy sources (such as solar or wind power) or energy-efficient devices and communication protocols.

For example, if a certain percentage of energy comes from renewable sources, the total energy consumption equation becomes:

$$E_{total} = (1 - Renewable\%) \times E_{total,non-renewable} + Renewable\% \times E_{renewable}$$

where:

- $E_{total,non-renewable}$ the total energy consumption from non-renewable sources.
- $E_{renewable}$ is the total energy consumption from renewable sources.
- $Renewable\%$ is the percentage of energy from renewable sources.

The environmental impact model can be adjusted to account for the reduced environmental impact associated with the use of green technologies.

4 Energy Efficiency in IoT Devices

Energy efficiency in IoT devices is crucial for mitigating their environmental impact and ensuring sustainable operation. These devices are ubiquitous, often operating continuously and requiring energy for sensing, processing, communication, and data storage [16]. Improving their energy efficiency involves several key strategies:

- Low-Power Hardware Design: IoT devices should be designed with low-power components and architectures optimized for energy efficiency. This includes selecting energy-efficient microcontrollers, sensors, and communication modules, as well as implementing power-saving features like sleep modes and dynamic voltage scaling.
- Optimized Software: Software optimization plays a significant role in energy efficiency. Efficient algorithms and protocols should be used to minimize computational overhead and reduce energy consumption during data processing and communication tasks. Additionally, IoT applications should be designed to minimize unnecessary data transmission and processing, optimizing energy use.
- Power Management Techniques: Implementing effective power management techniques is essential for energy-efficient IoT devices. This includes dynamically adjusting power states based on workload requirements, as well as managing power budgets to allocate energy resources efficiently among different device components.
- Energy Harvesting: Energy harvesting techniques, such as solar, kinetic, or thermal energy harvesting, can be employed to supplement or replace battery power in IoT devices. This enables sustainable, self-powered operation and reduces the environmental impact associated with battery disposal.
- Efficient Communication Protocols: Communication protocols used in IoT networks should be optimized for energy efficiency. Low-power wireless protocols like Zigbee, Bluetooth Low Energy (BLE), and LoRaWAN are preferred for minimizing energy consumption during data transmission.
- Data Compression and Aggregation: Data compression and aggregation techniques can reduce the amount of data transmitted and processed by IoT devices, leading to lower energy consumption in communication and processing tasks.

- Remote Management and Firmware Updates: Remote management capabilities allow IoT devices to be monitored and managed efficiently, enabling optimization of energy consumption through remote updates, configuration changes, and performance monitoring.

By implementing these energy efficiency strategies, IoT devices can operate more sustainably, reducing their environmental impact and contributing to overall energy conservation efforts.

5 Green Communication Protocols for IoT

Green communication protocols for IoT are essential for minimizing energy consumption, reducing environmental impact, and promoting sustainability in IoT deployments. These protocols aim to optimize communication processes, prioritize energy efficiency, and minimize resource usage while maintaining reliable data transmission [17]. Several key green communication protocols for IoT include:

- IEEE 802.15.4: This standard defines the physical and media access control (MAC) layers for low-rate wireless personal area networks (LR-WPANs). It is the foundation for many energy-efficient wireless protocols used in IoT, such as Zigbee and Thread. IEEE 802.15.4 operates in the 2.4 GHz ISM band and provides low-power, low-data-rate communication suitable for IoT applications.
- Zigbee: Zigbee is a low-power, low-data-rate wireless communication protocol based on the IEEE 802.15.4 standard. It is designed for short-range communication with low power consumption, making it suitable for battery-operated IoT devices in applications like home automation, smart lighting, and industrial monitoring.
- Bluetooth Low Energy (BLE): BLE, also known as Bluetooth Smart, is a wireless communication protocol optimized for low-power IoT applications. It consumes significantly less power than classic Bluetooth, making it suitable for battery-powered devices with limited energy resources. BLE is commonly used in applications like wearable devices, health monitoring, and proximity sensing.
- LoRaWAN: LoRaWAN is a long-range, low-power wireless communication protocol [18] designed for IoT applications that require wide-area coverage. It operates in the sub-GHz ISM bands and offers low power consumption, long-range communication, and low data rates, making it suitable for applications like smart agriculture, environmental monitoring, and asset tracking.
- Thread: Thread is a low-power, mesh networking protocol based on IEEE 802.15.4, designed for reliable and energy-efficient communication in IoT devices. It enables devices to form self-healing mesh networks, providing robust communication while minimizing energy consumption. Thread is commonly used in smart home and building automation applications.

- NB-IoT and LTE-M: Narrowband IoT (NB-IoT) and LTE-M are cellular communication protocols optimized for low-power IoT applications. They operate within existing cellular networks, providing wide coverage and reliable communication with low power consumption. NB-IoT and LTE-M are suitable for applications like smart metering, asset tracking, and remote monitoring.

By leveraging these green communication protocols, IoT deployments can achieve energy-efficient and sustainable communication, reducing the environmental impact associated with data transmission and promoting the widespread adoption of IoT technologies.

6 Manufacturing Processes and Sustainable Materials

Manufacturing Processes and Sustainable Materials are essential components of environmentally responsible IoT device production, aimed at reducing resource consumption, minimizing waste generation, and mitigating environmental impacts throughout the product lifecycle [19]. Several strategies can be employed to promote sustainability in materials sourcing and manufacturing:

- Recycled and Renewable Materials: Incorporating recycled materials, such as recycled plastics, metals, and glass, reduces the demand for virgin resources and diverts waste from landfills. Similarly, using renewable materials like bamboo, cork, or bio-based plastics derived from renewable sources reduces reliance on finite resources and lowers the carbon footprint of IoT devices.
- Design for Disassembly and Recycling: Implementing design principles that facilitate disassembly and material recovery at the end of the product lifecycle promotes circularity and resource efficiency. This involves using modular designs, standardized components, and easily separable materials to simplify disassembly and enable efficient recycling or reuse of materials.
- Energy-Efficient Manufacturing Processes: Adopting energy-efficient manufacturing processes, such as lean manufacturing, just-in-time production, and energy management systems, reduces energy consumption and greenhouse gas emissions associated with IoT device manufacturing. Implementing renewable energy sources, such as solar or wind power, in manufacturing facilities further reduces the environmental impact of production operations.
- Water Conservation and Pollution Prevention: Implementing water conservation measures and pollution prevention strategies in manufacturing processes minimizes water usage, reduces wastewater generation, and mitigates water pollution. This includes implementing closed-loop water systems, optimizing water usage through recycling and reuse, and implementing wastewater treatment technologies to reduce environmental impact.
- Lifecycle Assessment and Certification: Conducting lifecycle assessments (LCAs) to evaluate the environmental impact of IoT devices throughout their lifecycle helps identify areas for improvement and inform sustainable design

and manufacturing decisions. Certifications such as EPEAT (Electronic Product Environmental Assessment Tool) and ENERGY STAR provide third-party validation of environmentally preferable products, guiding consumers towards more sustainable purchasing decisions.

By integrating sustainable materials and manufacturing processes into IoT device production, manufacturers can minimize environmental impact, conserve resources, and contribute to a more sustainable and circular economy. These practices not only reduce the environmental footprint of IoT devices but also enhance brand reputation, drive innovation, and meet growing consumer demand for environmentally responsible products.

7 Lifecycle Management and E-Waste Reduction

Lifecycle management and e-waste reduction are critical components of sustainable practices in the Internet of Things (IoT) ecosystem [20], aimed at minimizing environmental impact and promoting resource efficiency throughout the lifecycle of IoT devices. Several strategies can be employed to achieve effective lifecycle management and reduce e-waste:

- Product Design for Durability and Upgradability: Designing IoT devices for durability and upgradability extends their lifespan and reduces the frequency of replacements, thereby minimizing e-waste generation. Modular designs, standardized components, and easily replaceable parts facilitate upgrades and repairs, prolonging the useful life of devices.
- Reuse and Refurbishment Programs: Implementing reuse and refurbishment programs allows IoT devices to be repurposed or refurbished for secondary markets or alternative applications, extending their lifespan and reducing e-waste. Refurbished devices can be sold at a lower cost, providing affordable options for consumers while diverting electronic waste from landfills.
- Take-Back and Recycling Programs: Establishing take-back and recycling programs enables consumers to return end-of-life IoT devices to manufacturers or authorized recyclers for responsible disposal or recycling. These programs ensure proper handling and recycling of electronic components, including hazardous materials, minimizing environmental contamination and promoting resource recovery.
- Data Sanitization and Security: Prior to disposal or recycling, IoT devices should undergo thorough data sanitization processes to securely erase sensitive information and protect consumer privacy. Data wiping techniques, such as overwriting or cryptographic erasure, ensure complete removal of data from storage devices, preventing unauthorized access and data breaches.
- Material Recovery and Resource Circularity: Implementing material recovery processes, such as dismantling, shredding, and sorting, enables the recovery of valuable materials and components from end-of-life IoT devices for reuse or

recycling. Recycling precious metals, rare earth elements, and other valuable materials conserves natural resources and reduces the environmental impact of raw material extraction.
- Regulatory Compliance and Extended Producer Responsibility (EPR): Compliance with e-waste regulations and adherence to extended producer responsibility (EPR) principles require manufacturers to take responsibility for the entire lifecycle of their products, including end-of-life management and disposal. EPR programs incentivize manufacturers to design products for easier disassembly, recycling, and responsible disposal, thereby reducing e-waste generation.

By implementing effective lifecycle management practices and e-waste reduction strategies, stakeholders in the IoT ecosystem can minimize environmental impact, conserve resources, and promote a circular economy. These initiatives not only contribute to environmental sustainability but also enhance brand reputation, foster consumer trust, and drive innovation in the IoT industry.

8 Data Privacy and Security for Sustainable IoT

Data privacy and security are paramount for ensuring the sustainability of IoT systems by safeguarding user trust, protecting sensitive information, and mitigating potential risks and liabilities associated with data breaches [21]. Several strategies can be employed to promote data privacy and security in sustainable IoT deployments:

- End-to-End Encryption: Implementing end-to-end encryption ensures that data transmitted between IoT devices and cloud servers is encrypted throughout the entire communication process, preventing unauthorized access or interception of sensitive information.
- Secure Authentication and Access Controls: Utilizing secure authentication mechanisms, such as biometric authentication or multi-factor authentication, ensures that only authorized users or devices can access IoT systems and data, reducing the risk of unauthorized access or data breaches.
- Data Minimization and Anonymization: Adopting data minimization practices, where only necessary data is collected and stored, reduces the volume of sensitive information at risk of exposure in the event of a breach. Anonymizing or pseudonymizing data further protects user privacy by removing personally identifiable information.
- Regular Security Audits and Updates: Conducting regular security audits and assessments helps identify vulnerabilities and weaknesses in IoT systems, enabling timely patching and updates to address security flaws and enhance overall system resilience.
- User Education and Awareness: Educating users about data privacy best practices and providing transparency regarding data collection, storage, and usage policies fosters informed decision-making and empowers users to take proactive measures to protect their privacy.

By prioritizing data privacy and security in sustainable IoT deployments, stakeholders can build trust with users, comply with regulatory requirements, and mitigate the potential negative impacts of data breaches, thereby ensuring the long-term sustainability and viability of IoT ecosystems.

9 Conclusion and Future Work

This paper has provided a comprehensive overview of sustainability in the Internet of Things (IoT), focusing on environmental impacts and green technologies. We have examined various facets of sustainability in IoT, including energy efficiency, renewable energy integration, sustainable materials, lifecycle management, e-waste reduction, data privacy, and regulatory perspectives. Through this analysis, it is evident that while IoT offers immense potential for innovation and efficiency, its widespread adoption poses significant environmental challenges.

Moving forward, future research should continue to prioritize the development and implementation of green technologies in IoT to minimize its environmental footprint. This includes further advancements in energy-efficient devices, the integration of renewable energy sources, and the adoption of sustainable materials and manufacturing processes. Additionally, efforts should be directed towards enhancing lifecycle management practices and reducing e-waste generation through recycling and refurbishment initiatives. Furthermore, addressing data privacy and security concerns in IoT remains crucial for sustainable development.

Overall, by embracing sustainability principles and leveraging green technologies, the IoT ecosystem can contribute to a more environmentally responsible and resilient future.

Future Work

Future research in this field could explore the effectiveness of specific green technologies and strategies in reducing the environmental impact of IoT devices. This could involve conducting comprehensive lifecycle assessments and comparative studies to evaluate the sustainability performance of different IoT solutions. Additionally, there is a need for longitudinal studies to assess the long-term environmental implications of IoT deployments and to track progress towards sustainability goals over time. Furthermore, an interdisciplinary research collaboration between engineers, environmental scientists, policymakers, and industry stakeholders is essential to drive innovation and promote sustainable practices in the rapidly evolving IoT landscape. Finally, ongoing efforts should focus on raising awareness among consumers and businesses about the importance of sustainability in IoT and encouraging the adoption of eco-friendly technologies and practices.

References

1. Bibri SE (2018) The IoT for smart sustainable cities of the future: An analytical framework for sensor-based big data applications for environmental sustainability. Sustain Cities Soc 38:230–253
2. Arshad R, Zahoor S, Shah MA, Wahid A, Yu H (2017) Green IoT: an investigation on energy saving practices for 2020 and beyond. IEEE Access 5:15667–15681
3. Tipantuna C, Hesselbach X, Unger W (2021) Heuristic strategies for nfv-enabled renewable and non-renewable energy management in the future IoT world. IEEE Access 9:125000–125031
4. Delicato FC, Pires PF, Batista T, Cavalcante E, Costa B, Barros T (2013) Towards an IoT ecosystem. In: Proceedings of the first international workshop on software engineering for systems-of-systems, pp 25–28
5. Maksimovic M (2017) The role of green internet of things (G-IoT) and big data in making cities smarter, safer and more sustainable. Int J Comput Digit Syst 6(04):175–184
6. Nižetić S, Šolić P, Gonzalez-De DLDI, Patrono L (2020) Internet of Things (IoT): opportunities, issues and challenges towards a smart and sustainable future. J Clean Prod 274:122877
7. Maksimovic M (2018) Greening the future: green Internet of Things (G-IoT) as a key technological enabler of sustainable development. Internet Things Big Data Anal Next-Gener Intell 283–313
8. Memić B, Džubur AH, Avdagić-Golub E (2022) Green IoT: sustainability environment and technologies. Sci Eng Technol 2(1):24–29
9. Shaikh FK, Zeadally S, Exposito E (2015) Enabling technologies for green internet of things. IEEE Syst J 11(2):983–994
10. Zhu C, Leung VC, Shu L, Ngai ECH (2015) Green internet of things for smart world. IEEE Access 3:2151–2162
11. Sinha M, Chacko E, Makhija P, Pramanik S (2021) Energy-efficient smart cities with green internet of things. Green Technol Innov Sustain Smart Soc: Post Pandemic Era 345–361
12. Maksimović M, Omanović-Mikličanin E (2017) Green internet of things and green nanotechnology role in realizing smart and sustainable agriculture. In: VIII international scientific agriculture symposium "AGROSYM 2017, pp 2290–2295
13. Nasiri M, Tura N, Ojanen V (2017) Developing disruptive innovations for sustainability: a review on Impact of Internet of Things (IOT). In: 2017 Portland international conference on Management of Engineering and Technology (PICMET), pp 1–10. IEEE
14. Pecunia V, Occhipinti LG, Hoye RL (2021) Emerging indoor photovoltaic technologies for sustainable internet of things. Adv Energy Mater 11(29):2100698
15. Albreem MA, Sheikh AM, Alsharif MH, Jusoh M, Yasin MNM (2021) Green Internet of Things (GIoT): applications, practices, awareness, and challenges. IEEE Access 9:38833–38858
16. Jayakumar H, Raha A, Kim Y, Sutar S, Lee WS, Raghunathan V (2016) Energy-efficient system design for IoT devices. In: 2016 21st Asia and South Pacific design automation conference (ASP-DAC), pp 298–301. IEEE
17. Capuzzo M, Delgado C, Sultania AK, Famaey J, Zanella A (2021) Enabling green IoT: energy-aware communication protocols for battery-less LoRaWAN devices. In: Proceedings of the 24th international ACM conference on modeling, analysis and simulation of wireless and mobile systems, pp 95–98
18. Feltrin L (2019) Long range low power wireless communication technologies for the IoT
19. Chen X, Despeisse M, Johansson B (2020) Environmental sustainability of digitalization in manufacturing: a review. Sustainability 12(24):10298
20. Razip MM, Savita KS, Kalid KS, Ahmad MN, Zaffar M, Rahim EEA, Ahmadian A (2022) The development of sustainable IoT E-waste management guideline for households. Chemosphere 303:134767
21. Stergiou C, Psannis KE, Gupta BB, Ishibashi Y (2018) Security, privacy & efficiency of sustainable cloud computing for big data & IoT. Sustain Comput: Inform Syst 19:174–184

Bio-Inspired Algorithms-Based Machine Learning and Deep Learning Models in Healthcare 6.0

Shugufta Fatima, C. Kishor Kumar Reddy, Marlia Mohad Hanafiah, and R. Madana Mohana

Abstract Advancements in deep learning (DL) and machine learning (ML) in recent times have paved the way for transformative applications in healthcare, offering unprecedented opportunities for better healthcare for patients, diagnosis, and treatment. Among the array of methodologies emerging in this domain, bio-inspired algorithms have garnered significant attention for their potential to enhance ML and DL models. These algorithms, inspired by natural processes observed in biological systems, offer creative solutions to the intricate difficulties encountered in the medical field. This paper explores present research directions and challenges in leveraging bio-inspired algorithms for advancing ML and DL models in healthcare. We delve into the potential applications and limitations of these bio-inspired approaches, shedding light on their efficacy in optimizing model parameters, facilitating feature selection, and refining architecture design for healthcare-specific tasks. However, despite their promise, bio-inspired algorithms also confront various challenges like Scalability, interpretability, and robustness to noisy and heterogeneous healthcare data which remain primary concerns. Additionally, the ethical implications of utilizing these algorithms in sensitive healthcare contexts warrant careful consideration. By fostering interdisciplinary collaboration and advancing novel algorithmic approaches, we aim to get beyond these obstacles and unleash the complete potential of bio-inspired algorithms in healthcare. Ultimately, our goal is to harness the power of these algorithms to drastically transform healthcare delivery, improving patient outcomes, individualized treatment strategies, and more precise diagnosis.

S. Fatima · C. K. K. Reddy (✉)
Department of Computer Science and Engineering, Stanley College of Engineering and Technology for Women, Hyderabad, India
e-mail: kishoar23@gmail.com

M. M. Hanafiah
Department of Engineering and Technology, Universiti Kebangsaan Malaysia, Selangor, Malaysia

R. M. Mohana
Department of Artificial Intelligence and Data Science, Chaitanya Bharathi Institute of Technology, Hyderabad, India

V. K. Gunjan et al. (eds.), *Cybernetics, Human Cognition, and Machine Learning in Communicative Applications*, Cognitive Science and Technology,
https://doi.org/10.1007/978-981-97-8533-9_9

Keywords Machine learning · Deep learning · Bio-inspired algorithms · Health care · Genetic algorithms · Particle swarm optimization · Ant colony optimization · Artificial immune systems · Swarm intelligence

1 Introduction

Computer techniques that take cues from natural phenomena, procedures or actions noticed in biological systems are referred to as bio-inspired or nature-inspired algorithms where these algorithms leverage the principles governing the functioning of various natural systems to resolve complex refinement and search problems. They are widely useful in a wide range of fields such as optimization, machine learning, robotics, and computational biology. Some of the key bio-inspired algorithms like Genetic Algorithms, Particle Swarm Optimization, Ant Colony Optimization, Artificial Immune Systems, and Swarm Intelligence [1] are discussed below:

Genetic Algorithms (GA)

- Driven by natural selection and genetics, GA simulates the evolutionary basics of survival of the fittest.
- Solutions are represented as chromosomes (often binary strings), Along with a population of solutions that evolves over generations through processes like selection, crossover, and mutation.
- GA is used in optimization problems where the search space is vast and complex, such as in engineering design, scheduling, and financial modeling.

Particle Swarm Optimization (PSO)

- PSO uses a population of particles to search across a search space in quest of the best answer. It is inspired by the social behavior of fish schools and bird flocking.
- Based on its own experiences and those provided by its neighbors, each particle modifies its position.
- PSO is used in optimization problems, particularly in continuous and multidimensional search spaces, such as in engineering design, neural network training, and robotics.

Ant Colony Optimization (ACO)

- ACO mimics the collective behavior of ants in determining the shortest path between their nest and a food source, which is inspired by the foraging activity of ants.
- ACO algorithms use Pheromone pathways deposited via ants to mark good paths, guiding other ants to explore promising routes.
- ACO is commonly applied to optimization problems related to routing, scheduling, and network design.

Artificial Immune Systems (AIS)

- Motivated by the immunological system of humans, AIS algorithms model the processes of recognition, response, and memory formation to detect and eliminate foreign entities (e.g., viruses, bacteria, and tumors).
- AIS algorithms evolve populations of artificial antibodies and antigens, where antibodies represent solutions to a problem and antigens represent problem instances.
- AIS is used in anomaly detection, pattern recognition, and optimization problems, particularly in dynamic and uncertain environments.

Swarm Intelligence

- This encompasses various algorithms influenced by social insects' collective behavior or other animals, like bees, termites, and birds.
- Examples include bee colony optimization, firefly algorithm, and bat algorithm, each mimicking specific behaviors observed in natural swarms or flocks.
- Swarm intelligence algorithms are used in optimization problems, data clustering, and routing in communication networks.

These bio-inspired algorithms often provide benefits like robustness, adaptability, and scalability, making them effective for solving complex optimization problems in diverse domains.

1.1 Bio-inspired Algorithms and Their Relevance to Healthcare

In order to address difficult optimization and decision-making problems, bio-inspired algorithms take inspiration from the concepts and mechanisms seen in natural systems. These algorithms imitate biological organisms or processes systems' behavior, such as evolutionary processes, swarm behavior, or genetic mechanisms, to find solutions to various real-world problems. Within the structure of healthcare, bio-inspired algorithms offer innovative approaches to address challenges in diagnosis, treatment planning, patient monitoring, and biomedical research. These algorithms use natural principles to provide resilient and flexible solutions to challenging healthcare issues. Their relevance lies in their ability to effectively tackle optimization, decision-making, and pattern recognition tasks, ultimately leading to improved healthcare outcomes [2], more efficient healthcare systems, and advancements in biomedical research. Here's an overview of some prominent bio-inspired algorithms and their relevance to healthcare.

Genetic Algorithms (GAs): In healthcare, GAs are applied in tasks such as feature selection for medical diagnosis, parameter optimization for treatment planning, and drug discovery through molecular structure optimization.

Particle Swarm Optimization (PSO): PSO has been used in healthcare for tasks such as optimizing treatment plans in radiation therapy, parameter tuning in machine learning models for disease prediction, and optimizing neural network architectures for medical image analysis.

Ant Colony Optimization (ACO): In healthcare, ACO has been applied in tasks such as optimizing hospital scheduling and resource allocation, as well as in medical image registration and segmentation.

Artificial Immune Systems (AIS): AIS has been used in healthcare for tasks such as anomaly detection in medical data, classification of biological data, and optimization of vaccination strategies.

Swarm Intelligence: In healthcare, swarm intelligence algorithms have been employed for tasks such as optimizing hospital workflows, clustering patient data for personalized medicine, and designing adaptive healthcare systems.

1.2 Importance of Deep Learning and Machine Learning in Healthcare

Machine learning (ML) and deep learning (DL) have become indispensable tools in healthcare [3], offering transformative capabilities that enhance various aspects of medical practice, research, and patient care. Their importance lies in their ability to enhance diagnostic accuracy, enable predictive analytics, facilitate personalized medicine, revolutionize medical imaging analysis, optimize healthcare operations, provide clinical decision support, and drive research and discovery [4]. Through the utilization of data-driven algorithms, machine learning and DL hold the promise of transforming delivery of healthcare, enhancement of patient outcomes, and progression of medical science [5].

Here are some key reasons highlighting their importance (Fig. 1).

Diagnostic Accuracy: ML and DL Complex medical data, like medical imaging, is an excellent subject for algorithms to analyze genomic sequences, and electronic health records (EHRs), to aid in disease diagnosis. They are able to identify patterns, anomalies, and subtle features that could not be apparent to human observers, thereby improving diagnostic accuracy and reducing errors.

Predictive Analytics: By leveraging historical patient data, ML and DL models can predict disease risk, progression, and treatment outcomes. These predictive analytics empower clinicians to proactively identify at-risk patients, personalize treatment plans, and optimize resource allocation within healthcare systems.

Drug Discovery and Development: ML and DL techniques are transforming the pharmaceutical discovery process by accelerating compound screening, predicting drug-target interactions, and optimizing molecular designs. This expedites the identification of novel therapeutic compounds and streamlines the development of new drugs, potentially leading to more effective treatments for various diseases.

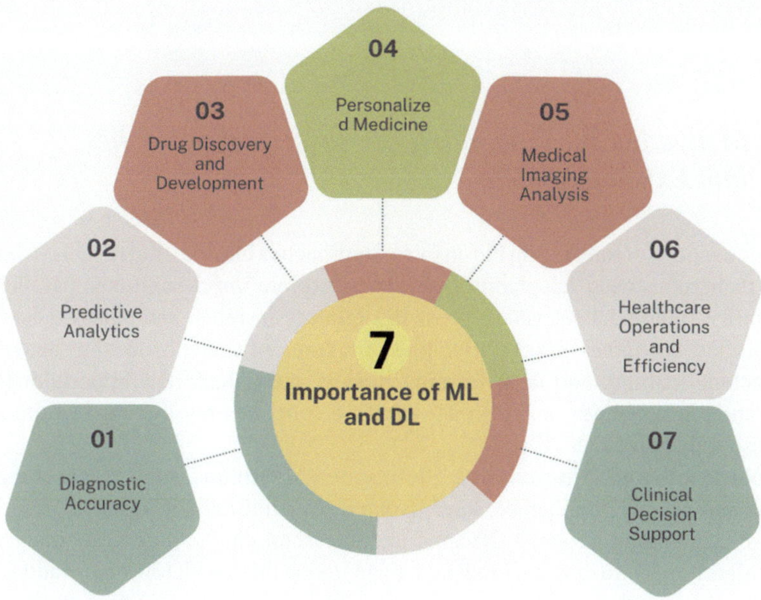

Fig. 1 Importance of machine learning and deep learning in healthcare

Personalized Medicine: ML and DL enable the analysis of enormous volumes of patient data, including genetic data, clinical indicators, and lifestyle factors, in order to customize medical treatments for specific individuals. Personalized medicine seeks to maximize therapeutic efficacy, reduce side effects, and enhance patient outcomes by taking into account each patient's particular traits and medical history.

Medical Imaging Analysis: Medical image analysis has been transformed by the use of convolutional neural networks (CNNs), in particular, which automate processes like object detection, disease categorization, and image segmentation. Large-scale medical imaging datasets-trained deep learning models can help radiologists analyze scans more quickly and accurately, which helps speed up diagnosis and treatment planning.

Clinical Decision Support: ML-based Clinical decision support technologies give healthcare professionals instantaneous guidance. Providers by synthesizing patient data, medical literature, and clinical guidelines. These systems assist clinicians in making evidence-based decisions, selecting appropriate treatments, and predicting patient outcomes, ultimately improving the quality and safety of patient care.

Research and Discovery: ML and DL techniques facilitate biomedical research by uncovering hidden patterns in large-scale datasets, identifying novel biomarkers, and generating hypotheses for further investigation. These tools accelerate scientific discovery, fuel innovation in healthcare, and contribute to a deeper understanding of disease mechanisms.

1.3 Motivation for Integrating Bio-inspired Approaches into Existing Models

The motivation for integrating bio-inspired approaches into existing models [6] stems from their robustness, adaptability, ability to explore novel solutions, parallelism, scalability, biological relevance, cross-disciplinary insights, and adaptive learning capabilities. By harnessing the principles of nature, we can enhance the capabilities of machine learning and deep learning models and unlock new opportunities for innovation in healthcare and beyond. Here are some key reasons for incorporating bio-inspired approaches (Fig. 2).

Robustness and Adaptability: Bio-inspired algorithms, such as genetic algorithms and swarm intelligence, have evolved over millions of years to effectively resolve challenging issues in dynamic and uncertain environments. By integrating these algorithms into existing models, we can imbue them with robustness and adaptability, allowing them to handle noisy or incomplete data and adapt to changing conditions in healthcare settings.

Global Optimization: Bio-inspired algorithms excel at finding globally optimal solutions to optimization problems by exploring a vast search space efficiently. In

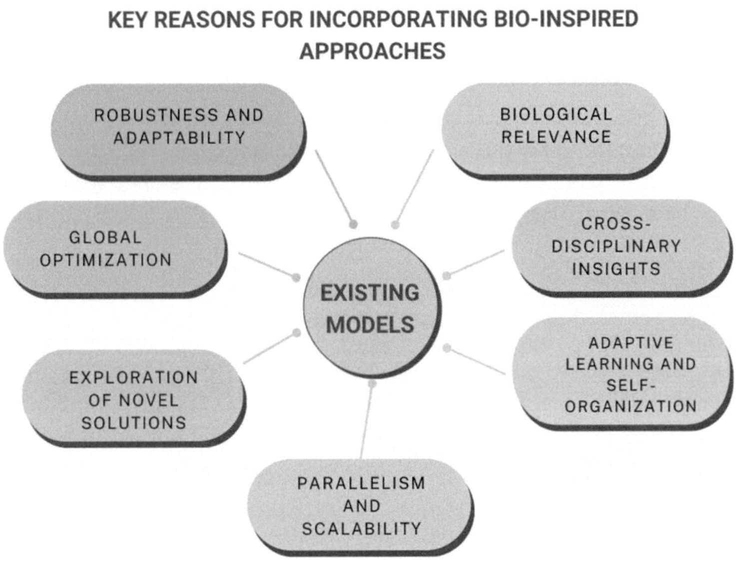

Fig. 2 Key reasons for incorporating bio-inspired approaches

healthcare, where the search for optimal treatment plans, diagnostic criteria, and biomarkers can be highly complex, bio-inspired approaches offer the potential to identify superior solutions that may elude traditional optimization techniques.

Exploration of Novel Solutions: Nature-inspired algorithms often generate diverse and innovative solutions by mimicking the evolutionary processes observed in biological systems. Integrating bio-inspired approaches into existing models enables the exploration of novel solutions that may not be evident through conventional methods, leading to breakthroughs in disease diagnosis, treatment, and biomedical research.

Parallelism and Scalability: Many bio-inspired algorithms exhibit inherent parallelism and scalability, making them well-suited for high-performance computing environments and distributed systems. By harnessing the parallel processing capabilities of bio-inspired approaches, we can accelerate the optimization process and handle large-scale healthcare datasets more efficiently.

Biological Relevance: The foundation of bio-inspired algorithms is developed from concepts found in natural systems, including immunological reactions, social behavior, and genetic evolution. By leveraging these biological analogies, we can develop models that better capture the underlying mechanisms of complex biological processes, leading to more interpretable and physiologically relevant solutions in healthcare.

Cross-Disciplinary Insights: Integrating bio-inspired approaches into existing models fosters interdisciplinary collaboration between computer science, biology, medicine, and other fields. This cross-disciplinary synergy enables researchers to draw inspiration from diverse domains, uncover new insights, and develop innovative solutions that address multifaceted healthcare challenges from holistic perspectives.

Adaptive Learning and Self-organization: Algorithms that draw inspiration from biological structures and functions, such as neural networks, demonstrate the ability to learn adaptively and self-organize. By incorporating these principles into machine learning and deep learning models, we can create systems that learn from experience, adapt to changing environments, and autonomously improve their performance over time, mirroring the resilience and flexibility observed in living organisms.

2 Bio-inspired Algorithms in Healthcare

Algorithms inspired by biology are essential for addressing many constantly changing problems in healthcare. By providing powerful computational tools that leverage the principles of nature to optimize models, handle complex data, and enable personalized and innovative approaches to patient care.

Bio-inspired algorithms are needed in healthcare [7] for several reasons:

Complexity of Healthcare Problems: Healthcare problems are often complex, multifaceted, and dynamic. Bio-inspired algorithms offer innovative solutions by mimicking the adaptive and problem-solving mechanisms found in nature, which can help tackle these complex challenges more effectively.

Optimization of Models and Processes: In healthcare, there is a continuous need for optimization, whether it's optimizing machine learning models for diagnostic accuracy or optimizing clinical processes for efficiency. Bio-inspired algorithms, such as particle swarm optimization and genetic algorithms are well-suited for optimization tasks, offering robust and efficient solutions.

Handling Noisy and Incomplete Data: Healthcare data can be noisy, heterogeneous, and incomplete, posing challenges for traditional machine learning approaches. Bio-inspired algorithms, inspired by the robustness and adaptability of biological systems, can help mitigate these challenges by providing mechanisms for handling noisy and incomplete data effectively.

Personalized Medicine and Treatment Planning: Bio-inspired algorithms can facilitate the development of personalized medicine approaches by analyzing large-scale biological and clinical data to identify individualized treatment plans. These algorithms can model complex biological interactions and patterns, resulting in more accurate and successful treatment plans that are customized to each patient's particular needs.

Exploration of Novel Solutions: By drawing inspiration from natural processes, bio-inspired algorithms offer a fresh perspective for exploring novel solutions to healthcare problems. They can uncover unconventional approaches that may not have been considered using traditional computational methods, resulting in innovations in the diagnosis, treatment, and provision of healthcare.

2.1 Review of Prominent Bio-inspired Algorithms and Their Applications in Healthcare

Ant colony optimization, particle swarm optimization, and genetic algorithms are prominent bio-inspired algorithms with diverse applications in healthcare. These algorithms offer innovative solutions to complex optimization problems [8], ranging from feature selection and treatment optimization to medical image analysis and disease surveillance. By leveraging the principles of nature, bio-inspired algorithms contribute to advancements in healthcare delivery, biomedical research, and patient care (Fig. 3).

Here, we review three prominent bio-inspired algorithms—Ant colony optimization, particle swarm optimization, and genetic algorithms and their specific applications in healthcare:

Genetic Algorithms (GAs): Genetic algorithms are methods of optimization that draw inspiration from natural selection and genetics. They work by imitating the processes of biological evolution by developing a population of candidate solutions through crossover, mutation, and selection.

Applications in Healthcare

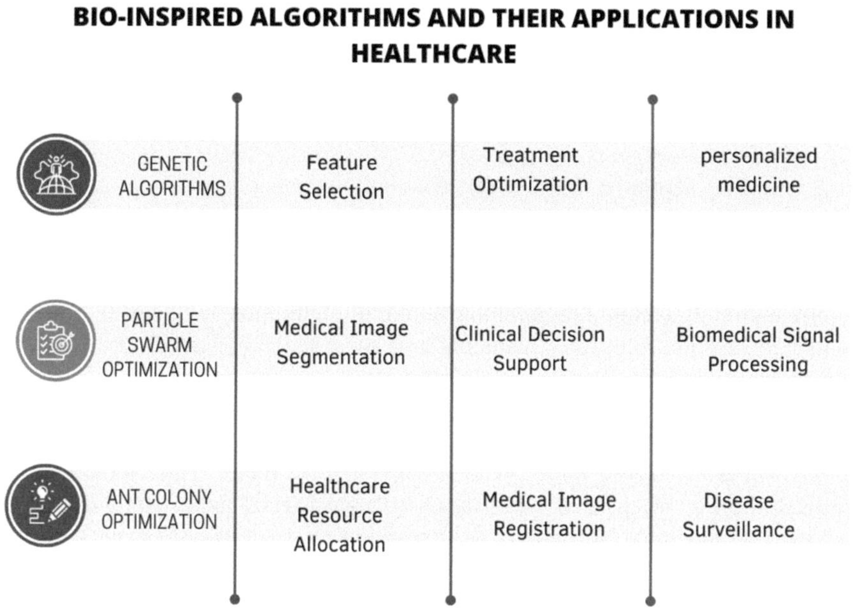

Fig. 3 Bio-inspired algorithms and their applications in healthcare

Feature Selection: In the medical field, genetic algorithms (GAs) are frequently employed for feature selection. This is especially true when analyzing high-dimensional medical data, including gene expression profiles or imaging data. GAs assist in enhancing the performance of machine learning models for illness prognosis and diagnosis by choosing pertinent features.

Treatment Optimization: GAs are applied to optimize treatment plans in various healthcare domains, including radiation therapy, chemotherapy scheduling [9], and personalized medicine. By exploring the space of possible treatment options, GAs help identify optimal dosing regimens and treatment strategies tailored to individual patient characteristics.

Particle Swarm Optimization (PSO): Particle swarm optimization draws inspiration from the collective behavior of fish schools and bird flocks, where individuals modify their placements based on group knowledge as well as personal experience.

Applications in Healthcare

Medical Image Segmentation: PSO has been employed for image segmentation tasks in diagnostic imaging procedures including CT and MRI scans. By optimizing segmentation parameters, PSO helps delineate anatomical structures and identify regions of interest for diagnosis and treatment planning.

Clinical Decision Support: Clinical decision support systems, which help medical professionals make evidence-based decisions, are optimized using PSO-based algorithms. By synthesizing patient data and clinical guidelines, PSO helps prioritize diagnostic tests, recommend treatment options, and predict patient outcomes.

Biomedical Signal Processing: PSO is applied in the analysis of biomedical signals, such as electrocardiograms (ECG) or electroencephalograms (EEG). By optimizing signal processing algorithms, PSO enhances the detection of abnormal patterns and physiological events, aiding in the diagnosis of cardiovascular diseases and neurological disorders.

Ant Colony Optimization (ACO): The shortest route to a food source is determined by ants using pheromone trails to communicate with one another, a process that serves as the model for ant colony optimization.

Applications in Healthcare:

Healthcare Resource Allocation: ACO-based algorithms are used to optimize healthcare resource allocation, including hospital scheduling, staff assignment, and patient flow management. By finding efficient routes and schedules, ACO helps minimize waiting times, reduce resource bottlenecks, and improve patient access to care [10, 11].

Medical Image Registration: ACO is employed for image registration tasks in medical imaging, where multiple images need to be aligned and fused for diagnosis or treatment planning. By optimizing registration parameters, ACO enhances the spatial alignment of images and improves the accuracy of subsequent analysis [12].

Disease Surveillance: Disease surveillance systems use ACO-based algorithms to track the spread of infectious diseases and pinpoint outbreak hotspots. By optimizing surveillance strategies and resource allocation, ACO helps public health authorities monitor disease trends, implement targeted interventions, and mitigate the spread of contagious pathogens [13, 14].

2.2 How These Algorithms Mimic Biological Processes to Optimize Model Performance and Decision-Making in Healthcare Tasks

Despite their distinct mechanisms, these algorithms share common principles in mimicking biological processes to optimize model performance and decision-making in healthcare tasks and offer innovative solutions to complex optimization problems, contributing to advancements in healthcare delivery, biomedical research, and patient care (Fig. 4).

Genetic Algorithms (GAs)

They work by mimicking how a population of potential solutions to an issue evolves, with individuals (solutions) undergoing selection, reproduction (crossover), and mutation to produce offsprings that inherit beneficial traits from their parents.

In healthcare tasks, GAs mimic biological evolution by:

Selection: Individuals with higher fitness scores, representing better solutions to the problem at hand (e.g., accurate disease diagnosis or GAs are inspired by genetics and the process of natural selection. They function by imitating the way a population of possible answers to a problem develops.

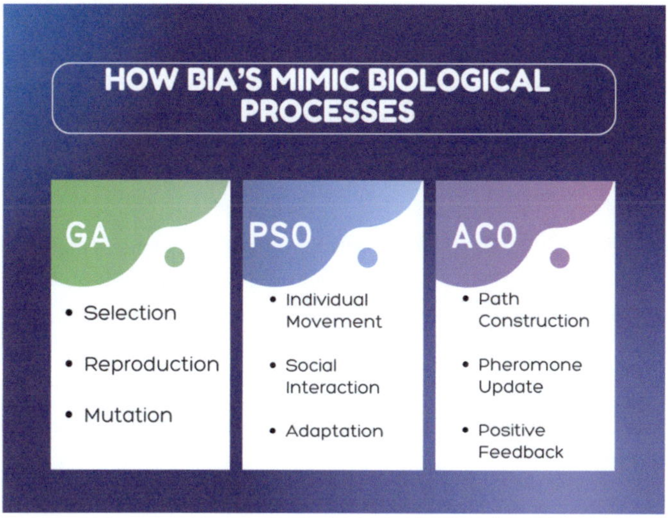

Fig. 4 How BIA's mimic biological processes

Reproduction (Crossover): GAs performs crossover operations to exchange genetic information between selected individuals, leading to the creation of new candidate solutions that combine features from multiple parents.

Mutation: Occasionally, GAs introduce random changes (mutations) to candidate solutions, allowing for investigating fresh areas in the solution space and preventing premature convergence to suboptimal solutions.

Particle Swarm Optimization (PSO)

In PSO model every particle serves as a potential answer to the optimization problem, and the swarm as a whole works together to find the optimal answer. In healthcare tasks, PSO mimics collective behavior by [6]

Individual Movement: Particles alter their positions in the solution space according to their velocities, which are determined by their individual best-known positions (personal best) and the swarm's best-known positions (global best).

Social Interaction: PSO allows particles to communicate and share information with each other, enabling them to collectively converge towards promising regions of the solution space.

Adaptation: PSO facilitates effective exploration and exploitation of the solution space by dynamically adjusting the velocity and movement of particles based on their own experience and the collective knowledge of the swarm.

Ant Colony Optimization (ACO)

In ACO, artificial ants construct solutions to optimization problems by iteratively building and improving paths based on pheromone concentrations and heuristic information. In healthcare tasks, ACO mimics ant foraging behavior by:

Path Construction: Artificial ants construct solutions by iteratively selecting paths through the solution space, aided by heuristic data and pheromone trails related to the problem domain.

Pheromone Update: Ants leave behind pheromone on the pathways they follow; the quantity left behind is correlated with the caliber of the solution. Pheromone trails are modified over time in response to the caliber of solutions that particular ants find.

Positive Feedback: High-quality solutions attract more ants to explore the corresponding paths, reinforcing positive feedback loops and leading to the discovery of optimal or near-optimal solutions.

3 Enhancing Machine Learning Models with Bio-inspired Algorithms

Improving machine learning models with bio-inspired algorithms represents a frontier in computational intelligence, leveraging nature's optimization strategies to tackle complex problems. Bio-inspired algorithms, drawing from evolutionary processes, social behaviors, and immune system dynamics, offer novel approaches for optimizing model parameters, feature selection, and architecture design. For instance, genetic algorithms mimic natural selection to evolve solutions over generations, while ant colony optimization simulates ants' foraging behavior to find optimal paths [15]. Integrating these algorithms with machine learning models not only enhances their performance but also imbues them with adaptability and robustness across various domains, from healthcare to finance and beyond.

In healthcare, the integration of BIA's with machine learning holds particular promise, offering innovative solutions for analysis of medical image, disease diagnosis, and treatment plans personalized for patients [4]. By harnessing the principles of nature, these algorithms optimize model performance and facilitate the discovery of insights from complex biological and clinical data. However, challenges such as scalability, interpretability, and ethical considerations persist, requiring further R and D efforts. As the field continues to evolve, the synergy between algorithms that is bio-inspired and machine learning models promises to revolutionize problem-solving approaches, paving the way for more efficient and effective solutions to real-world challenges [16, 17].

3.1 How Bio-inspired Algorithms Can Improve Traditional Machine Learning Models for Tasks Such as Disease Diagnosis, Prognosis, and Treatment Planning

Bio-inspired algorithms offer innovative approaches to improve traditional machine learning models for disease diagnosis, prognosis, and treatment planning in healthcare. By optimizing model parameters, enhancing feature selection, and facilitating personalized risk assessment, bio-inspired algorithms contribute to more accurate, interpretable, and adaptive healthcare decision-making processes [18].

Here's an exploration of how bio-inspired algorithms can improve these traditional models in each of these areas:

Disease Diagnosis

They can enhance traditional machine learning models for disease diagnosis by improving feature selection, model optimization, and interpretability.

Feature Selection: Genetic algorithms (GAs) can be used to optimize feature selection in SVMs or decision trees by identifying the most relevant features from large-scale biomedical datasets. GAs iteratively select subsets of features that maximize classification performance, improving the discriminative power of the models.

Model Optimization: Particle swarm optimization (PSO) can optimize hyper parameters in SVMs or decision trees, such as kernel parameters or tree depth, to maximize classification accuracy. PSO iteratively adjusts model parameters based on performance feedback, leading to better generalization and robustness.

Interpretability: Ant colony optimization (ACO) can improve the interpretability of decision trees by guiding the selection of optimal splitting criteria at each node. ACO prioritizes features that contribute most to classification accuracy, resulting in decision trees that are easier to interpret and understand by healthcare practitioners.

Prognosis

They can aid in prognosis by enhancing the predictive capabilities of traditional machine learning models and facilitating personalized risk assessment.

Model Fusion: Genetic algorithms can be used to fuse multiple traditional models, such as SVMs or decision trees, into ensemble models that capture diverse aspects of patient data. By combining complementary predictive models, genetic algorithms improve the robustness and accuracy of prognosis predictions.

Temporal Modeling: Particle swarm optimization can optimize temporal modeling techniques, like hidden Markov models (HMMs) or recurrent neural networks (RNNs) for disease progression prediction. PSO adjusts model parameters to capture temporal dependencies in longitudinal patient data, enabling more accurate prognosis over time.

Personalized Risk Assessment: Ant colony optimization can tailor prognostic models to individual patient characteristics by optimizing personalized risk scores. ACO optimizes model parameters based on patient-specific features, allowing for personalized risk assessment and treatment planning.

Treatment Planning

They can optimize treatment planning by automating decision-making processes and identifying optimal intervention strategies.

Treatment Optimization: Genetic algorithms can optimize treatment parameters, such as drug dosages or radiation schedules, to maximize treatment efficacy while minimizing adverse effects. GAs search for optimal treatment plans by iteratively adjusting dosage levels or treatment modalities based on patient responses and clinical outcomes.

Adaptive Learning: Particle swarm optimization can facilitate adaptive treatment planning by optimizing dynamic treatment strategies over time. PSO adjusts treatment parameters based on real-time patient data and feedback, enabling adaptive learning and personalized treatment optimization.

Resource Allocation: Ant colony optimization can optimize resource allocation in healthcare systems by identifying optimal patient scheduling and resource utilization strategies. ACO guides the allocation of healthcare resources, such as hospital beds or medical equipment, to maximize patient outcomes and minimize costs.

3.2 Case Studies Demonstrating the Effectiveness of Integrating Bio-Inspired Approaches into Healthcare Applications

They highlight the effectiveness of integrating bio-inspired approaches into healthcare applications, resulting in more effective healthcare delivery, better treatment outcomes, and greater patient care [19]. Here are a few case studies demonstrating the effectiveness of integrating bio-inspired approaches into healthcare applications.

Optimization of Radiation Therapy Treatment Plans Using Genetic Algorithms

Case Study: Researchers at a cancer treatment center integrated genetic algorithms (GAs) into the optimization process for radiation therapy treatment planning.

Approach: GAs were used to optimize the distribution of radiation dosage across tumor volumes while minimizing the exposure of healthy tissues to radiation.

Results: The integration of GAs significantly improved the quality of treatment plans, leading to better tumor coverage and reduced toxicity to surrounding organs. Patients receiving optimized treatment plans experienced fewer side effects and better treatment outcomes.

Personalized Drug Dosage Optimization Using Particle Swarm Optimization

Case Study: A pharmaceutical company utilized particle swarm optimization (PSO) to optimize personalized drug dosages for patients with chronic conditions.

Approach: PSO was applied to determine optimal drug dosages based on patient-specific factors such as age, weight, genetic markers, and disease progression.

Results: The integration of PSO led to the development of personalized dosing regimens that maximized drug efficacy while minimizing adverse effects. Patients receiving optimized dosages showed improved treatment adherence and better symptom management compared to standard dosing protocols.

Medical Image Segmentation Using Ant Colony Optimization

Case Study: Researchers in medical imaging applied ant colony optimization (ACO) to improve the brain tumor segmentation using magnetic resonance imaging (MRI) scans.

Approach: ACO was used to optimize the parameters of image segmentation algorithms, such as thresholding and region growing, to accurately delineate tumor boundaries.

Results: The integration of ACO resulted in more precise and reproducible tumor segmentations, facilitating treatment planning and monitoring for patients with brain tumors. Clinicians benefited from enhanced visualization of tumor morphology and better quantitative assessment of tumor size and growth over time.

Healthcare Resource Allocation Using Swarm Intelligence

Case Study: A hospital management system implemented swarm intelligence techniques to optimize resource allocation and patient scheduling.

Approach: Swarm intelligence algorithms were employed to allocate hospital beds, operating rooms, and staff resources efficiently.

Results: By integrating swarm intelligence into resource allocation algorithms, the hospital achieved better patient flow, reduced waiting times, and improved staff utilization. Healthcare providers were able to deliver timely and high-quality care, leading to enhanced patient satisfaction and operational efficiency.

4 Deep Learning and Bio-inspired Algorithms Fusion

It describes the fusion of optimization methods based on biological processes seen in nature with deep learning techniques. Deep learning enables tasks like picture recognition, audio recognition, and natural language processing by using multiple-layered artificial neural networks to learn complicated representations from data. In contrast, bio-inspired algorithms utilize biological processes like immune systems, swarm behavior, and evolution as models for solving optimization and search issues.

The fusion of deep learning and BIA's involves leveraging the strengths of both approaches to address challenges in various domains [20]. For example, bio-inspired algorithms can be applied to enhance the standards of deep neural networks, improving their training efficiency and generalization performance. They can also enhance the robustness of deep learning models by providing mechanisms for adaptation and exploration in complex and dynamic environments.

Table 1 Fusion of deep learning architectures with bio-inspired algorithms

Aspect	Description	Example
Feature extraction	Incorporating bio-inspired algorithms into deep learning architectures to enhance feature extraction from complex healthcare data	Using genetic algorithms to optimize the feature selection process in convolutional neural networks for medical image analysis
Model optimization	Combining bio-inspired algorithms with deep learning models to optimize model parameters, improve convergence, and enhance performance on healthcare tasks	Applying particle swarm optimization to tune hyper parameters of recurrent neural networks for time-series analysis in healthcare
Interpretability	Utilizing bio-inspired algorithms to enhance the interpretability of deep learning models in healthcare, making predictions more understandable and trustworthy for clinicians	Employing ant colony optimization to identify relevant features and explain the decision-making process of deep neural networks in diagnosing diseases

4.1 Investigation into the Fusion of Deep Learning Architectures with Bio-inspired Algorithms to Enhance Feature Extraction, Model Optimization, and Interpretability in Healthcare

Convolutional and recurrent neural networks (RNNs) are examples of deep learning architectures that have shown remarkable performance in various healthcare applications, including medical imaging analysis, disease diagnosis, and predictive modeling. By leveraging the complementary strengths of deep learning's representation learning capabilities and bio-inspired algorithms' optimization prowess, this fusion approach has the potential to improve feature extraction, model optimization, and interpretability in healthcare applications. This investigation delves into the potential benefits and challenges of such fusion, exploring its implications for advancing healthcare AI systems (Table 1).

4.2 Challenges and Opportunities in Integrating Complex Deep Learning Models with Bio-Inspired Optimization Techniques

Integrating complex deep learning models with bio-inspired optimization techniques presents both challenges and opportunities in healthcare and other domains. Let's explore some of these challenges and opportunities (Fig. 5).

Challenges

Computational Complexity: Deep learning models, especially complex designs such as recurrent neural networks (RNNs) or deep convolutional neural networks

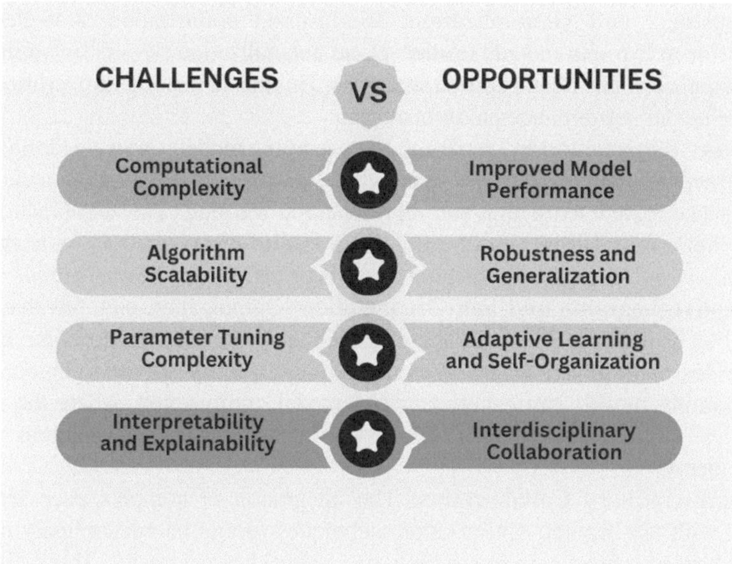

Fig. 5 Challenges and opportunities in integrating complex deep learning models with bio-inspired optimization techniques

(CNNs), often require significant computational resources for training and inference. Integrating bio-inspired optimization techniques may further increase computational complexity, particularly when optimizing hyper parameters or model architectures.

Algorithm Scalability

Optimization methods impacted by nature, include genetic algorithms and particle swarm optimization, may face scalability challenges when applied to large-scale deep learning models and datasets. Scaling these algorithms to handle high-dimensional parameter spaces and massive datasets efficiently is a non-trivial task.

Parameter Tuning Complexity: Deep learning models often have numerous hyper parameters that require careful tuning for optimal performance. Integrating bio-inspired optimization techniques adds another layer of complexity to the parameter tuning process, requiring researchers to navigate a complex search space effectively.

Interpretability and Explain ability: DL models are often criticized for their lack of interpretability and explain ability, which can be further exacerbated when combined with bio-inspired optimization techniques. Understanding the reasoning behind model decisions and interpreting the learned representations becomes challenging in such complex systems.

Opportunities

Improved Model Performance: Integrating bio-inspired optimization techniques with complex deep learning models can lead to improved model performance, as these techniques are adept at exploring high-dimensional search spaces and finding optimal solutions to complex optimization problems.

Robustness and Generalization: Bio-inspired optimization strategies can improve deep learning models resilience and generalization powers by optimizing hyper parameters and regularization strategies. This can help mitigate overfitting and improve model performance on unseen data.

Feature Representation Learning: Deep learning models excel at learning hierarchical representations from raw data, while bio-inspired optimization techniques can enhance feature extraction and representation learning. This combination can lead to more informative and discriminative feature representations in healthcare applications, such as medical imaging analysis or patient data classification.

Adaptive Learning and Self-Organization: Optimization methods that draw motivation from nature, like swarm intelligence and evolutionary algorithms, exhibit adaptive learning and self-organization capabilities. Integrating these techniques with deep learning models can enable adaptive model optimization, where the model dynamically adjusts its parameters in response to changing data distributions or task requirements.

Interdisciplinary Collaboration: The integration of complex deep learning models with bio-inspired optimization techniques fosters interdisciplinary collaboration between computer science, biology, and healthcare domains. This collaboration enables researchers to leverage insights from both fields and develop innovative solutions to complex healthcare challenges.

5 Applications of Bio-inspired Algorithms in Healthcare

Bio-inspired algorithms have emerged like powerful tools in revolutionizing healthcare across multiple fronts. Leveraging principles from nature, these algorithms find applications in medical imaging, aiding in the precise detection and characterization of abnormalities from complex scans [21]. They also facilitate drug discovery, optimizing molecular structures to accelerate the development of novel therapeutics. In disease diagnosis and prognosis, bio-inspired algorithms analyze vast patient datasets to enhance accuracy and personalize treatment plans. Additionally, they streamline healthcare resource management, optimizing processes like ambulance routing and staff scheduling for efficient care delivery. From predictive modeling to personalized medicine, these algorithms pave the way for innovative solutions that enhance patient outcomes and drive progress in healthcare.

5.1 Specific Healthcare Domains Where Bio-inspired Algorithms Have Shown Promise

They play a critical role in advancing healthcare by optimizing medical processes, improving diagnostic and treatment outcomes, and facilitating personalized and

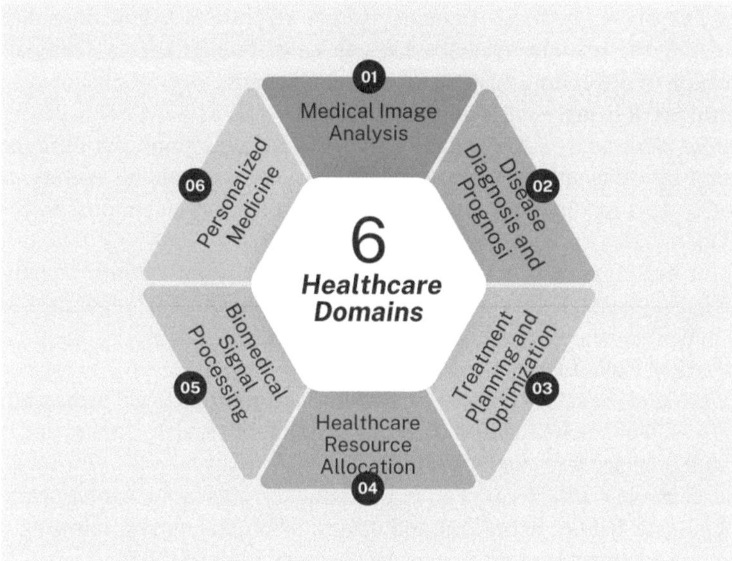

Fig. 6 Healthcare domains where bio-inspired algorithms have shown promise

efficient patient care. Their versatility and adaptability make them valuable tools for addressing complex healthcare challenges and driving innovation in biomedical research and clinical practice (Fig. 6).

Here are some key applications of bio-inspired algorithms in healthcare:

Medical Image Analysis

Segmentation: Bio-inspired algorithms such as particle swarm optimization and genetic algorithms are used to optimize image segmentation techniques, enabling accurate delineation of anatomical structures and lesions in medical images (e.g., MRI, CT scans).

Registration: Simulated annealing and ant colony optimization and algorithms facilitate the registration of multiple medical images, aligning them spatially for comparison, fusion, and treatment planning.

Disease Diagnosis and Prognosis

Feature Selection: Genetic algorithms assist in selecting relevant features from high-dimensional medical datasets, improving the accuracy of machine learning models for disease diagnosis and prognosis.

Predictive Modeling: Swarm intelligence algorithms optimize predictive models for disease progression, treatment response, and patient outcomes by fine-tuning model parameters and ensemble learning techniques.

Treatment Planning and Optimization

Radiation Therapy: Genetic algorithms optimize treatment plans in radiation therapy, determining optimal radiation dosages and beam angles to reduce harm to the surrounding healthy tissues and optimize tumor control.

Drug Discovery: Artificial immune system algorithms aid in drug discovery by simulating the immune system's recognition of foreign agents, facilitating the identification of novel drug targets and lead compounds.

Healthcare Resource Allocation

Hospital Scheduling: Ant colony optimization algorithms optimize hospital scheduling by efficiently allocating resources such as operating rooms, medical staff, and patient appointments, reducing patient wait times and improving resource utilization.

Supply Chain Management: Swarm intelligence algorithms optimize supply chain logistics in healthcare, ensuring the timely delivery of medical supplies, medications, and equipment to healthcare facilities.

Biomedical Signal Processing

Electrocardiogram (ECG) Analysis: Bio-inspired algorithms aid in the analysis of biomedical signals such as ECGs, detecting abnormalities, arrhythmias, and cardiac events with high accuracy and reliability.

Brain-Computer Interfaces: For brain-computer interfaces, evolutionary algorithms optimize feature extraction and classification techniques, allowing people with motor limitations to control assistive equipment in real time.

Personalized Medicine

Treatment Optimization: By maximizing the effectiveness of treatment and reducing side effects, genetic algorithms enhance individualized treatment plans based on unique patient features such as genetic profiles, biomarkers, and clinical histories.

Disease Risk Assessment: Swarm intelligence algorithms analyze multi-omics data to assess individualized disease risk profiles, enabling early detection, prevention, and intervention strategies.

Healthcare System Optimization

Workflow Optimization: Ant colony optimization algorithms optimize healthcare workflows by streamlining administrative processes, reducing bottlenecks, and improving overall efficiency in healthcare delivery.

Patient Monitoring: Bio-inspired algorithms facilitate real-time monitoring of patient vital signs, predicting adverse events and enabling early interventions to prevent complications.

5.2 Examination of Real-World Implementations and Their Impact on Clinical Practice and Patient Outcomes

Real-world implementations of bio-inspired algorithms have had a profound influence on patient outcomes and clinical practice in a variety of healthcare domains [20]. By leveraging principles from nature, these algorithms have optimized medical processes, improved diagnostic accuracy, personalized treatment approaches, and

Fig. 7 Examination of real-world implementations and their impact on clinical practice and patient outcomes

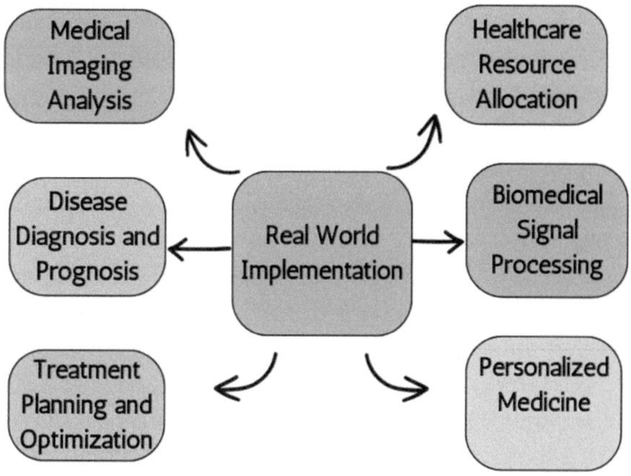

enhanced overall quality of care, ultimately leading to better health outcomes for patients [4] (Fig. 7).

Let's examine some examples:

Medical Imaging Analysis

Impact: They have been used to optimize image segmentation and registration techniques, leading to more accurate and efficient interpretation of medical images.

Outcome: Improved segmentation and registration algorithms have enhanced the precision of treatment planning in radiation therapy and surgery, resulting in better tumor targeting and reduced damage to surrounding healthy tissues.

6 Disease Diagnosis and Prognosis

Impact: Bio-inspired algorithms have facilitated the development of predictive models for disease diagnosis and prognosis, leveraging genetic algorithms, swarm intelligence, and other techniques.

Outcome: These models have made it possible to identify illnesses like cancer, heart problems, and neurological abnormalities early, which has improved patient outcomes and allowed for prompt interventions.

Treatment Planning and Optimization

Impact: They have been applied to optimize treatment plans in radiation therapy, chemotherapy scheduling, and personalized medicine.

Outcome: Optimized treatment plans have resulted in more precise and effective delivery of therapies, maximizing tumor control while minimizing adverse effects on patients, thereby improving treatment outcomes and quality of life.

Healthcare Resource Allocation

Impact: Bio-inspired optimization techniques have been deployed to optimize resource allocation in healthcare facilities, including hospital scheduling, staff assignment, and inventory management.

Outcome: Streamlined workflows and efficient resource utilization have reduced patient wait times, enhanced patient throughput, and improved overall quality of care in clinical settings.

Biomedical Signal Processing

Impact: Bio-inspired algorithms have enhanced the analysis of biomedical signals such as electrocardiograms (ECGs), electroencephalograms (EEGs), and respiratory waveforms.

Outcome: Improved signal processing techniques have enabled more accurate diagnosis of cardiac arrhythmias, seizure disorders, and sleep apnea, resulting in prompt interventions and improved patient condition management.

Personalized Medicine

Impact: They have made it easier to create individualized treatment plans that take into account the unique traits of each patient as well as their biomarker profiles.

Outcome: Tailored treatment approaches have optimized therapeutic outcomes, minimized adverse effects, and enhanced patient compliance with treatment regimens, leading to better long-term prognosis and disease management.

Healthcare System Optimization

Impact: Bio-inspired optimization algorithms have optimized healthcare workflows, patient scheduling, and resource allocation in healthcare systems.

Outcome: Enhanced operational efficiency and resource utilization have increased patient happiness with healthcare services, decreased healthcare costs, and better access to care.

7 Challenges and Future Directions

Implementing bio-inspired algorithms within machine learning (ML) and deep learning (DL) frameworks presents a fascinating yet complex landscape ripe with challenges and promising future directions [22]. One key challenge lies in bridging the gap between the intricate dynamics of biological systems and the abstraction necessary for computational modeling. Achieving this demands a multidisciplinary approach, combining expertise from biology, computer science, and mathematics. Additionally, scalability remains a hurdle, as bio-inspired algorithms often find it difficult to manage big databases effectively. Furthermore, the interpretability of these algorithms in complex ML and DL models poses significant challenges. Despite these obstacles, the future holds exciting prospects. Advancements in hardware, such as neuromorphic computing, could revolutionize the efficiency and scalability of bio-inspired algorithms. Moreover, integrating principles from evolutionary biology, neuroscience, and ecology can result in the creation of AI systems that are more resilient and adaptable. Embracing these challenges with interdisciplinary collaboration promises to unlock the full potential of bio-inspired algorithms in shaping the future of ML and DL.

7.1 Identification of Challenges and Limitations in the Adoption of Bio-inspired Algorithms in Healthcare with Existing Healthcare Systems

The **challenges f**aced in the adoption of bio-inspired algorithms in healthcare (Fig. 8).

Scalability

Challenge: These algorithms may struggle to handle the scale of healthcare data, which is often vast, heterogeneous, and high-dimensional. Scaling algorithms to process large datasets efficiently while maintaining computational tractability can be challenging.

Impact: Limited scalability can hinder the practical application of bio-inspired algorithms in healthcare settings, restricting their ability to analyze large-scale patient data, optimize complex models, or support real-time decision-making.

Interpretability

Challenge: These algorithms, such as evolutionary algorithms or swarm intelligence, often produce solutions that are difficult to interpret or explain. Understanding the rationale behind algorithmic decisions is crucial in healthcare, where transparency and interpretability are essential for clinical acceptance.

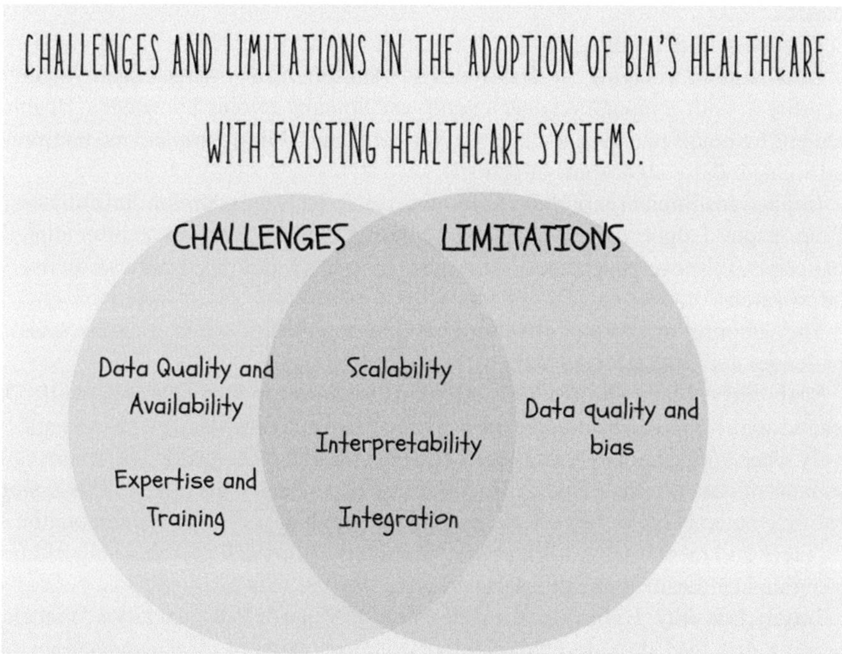

Fig. 8 challenges and limitations in the adoption of bio-inspired algorithms in healthcare with existing healthcare systems

Impact: Lack of interpretability can undermine trust in algorithmic predictions, impede clinical adoption, and pose challenges for regulatory compliance and patient safety. Healthcare professionals may be hesitant to rely on algorithms they cannot understand or trust.

Integration with Existing Healthcare Systems

Challenge: Integrating bio-inspired algorithms into existing healthcare systems poses technical, organizational, and regulatory challenges. Compatibility issues are is necessary to address data interoperability as well as adherence to healthcare standards and laws (such as HIPAA).

Impact: Incompatible systems, data silos, and regulatory barriers can impede the smooth incorporation of algorithms into clinical workflows, hindering their adoption and utility in real-world healthcare settings.

Data Quality and Availability

Challenge: Bio-inspired algorithms require high-quality, well-curated data for training and validation. However, healthcare data are often noisy, incomplete, and subject to biases, posing challenges for algorithm development and performance evaluation.

Impact: Poor data quality and availability can undermine the reliability and generalizability of bio-inspired algorithms, leading to suboptimal performance, biased predictions, and limited applicability in diverse patient populations or clinical contexts.

Expertise and Training

Challenge: Leveraging bio-inspired algorithms effectively in healthcare requires expertise in both computational techniques and domain-specific knowledge. Healthcare professionals may lack the necessary training and skills to implement, interpret, and validate these algorithms effectively.

Impact: Insufficient expertise and training can impede the adoption and utilization of bio-inspired algorithms in healthcare, limiting their potential to address clinical challenges, improve patient outcomes, and drive innovation in healthcare delivery and research.

The adoption of bio-inspired algorithms in healthcare indeed presents several challenges and **limitations**. Here's an overview:

Scalability: Many bio-inspired algorithms, such as artificial neural networks or genetic algorithms can be computationally intensive and may struggle to scale effectively when applied to large datasets or complex problems. In healthcare, where vast amounts of data are generated daily, scalability becomes a crucial issue. Processing real-time patient data or large-scale population health data can strain computational resources and increase processing time, limiting the practicality of these algorithms in certain healthcare applications.

Interpretability: Bio-inspired algorithms often operate like "black-box" models, meaning that they provide predictions or solutions without clear explanations of how they arrived at those outcomes. In healthcare, where transparency and interpretability are essential for clinical decision-making, this lack of explain ability can be a significant barrier to adoption. Clinicians may be hesitant to trust algorithmic recommendations if they cannot understand the underlying rationale or if the decision-making process is not transparent. Interpretability issues also raise concerns about accountability and potential biases embedded in the algorithms.

Integration with existing healthcare systems: Healthcare systems are highly complex and diverse, with a wide range of medical devices, electronic health record (EHR) systems, and data standards in use. Integrating bio-inspired algorithms seamlessly into existing healthcare infrastructure poses technical challenges related to data interoperability, compatibility with different software systems, and adherence with regulatory standards (e.g., HIPAA in the United States). Moreover, healthcare providers may face resistance or reluctance to adopt new technologies due to concerns about disruptions to workflow, training requirements, and perceived risks to patient privacy and safety.

Data quality and bias: The effectiveness of bio-inspired algorithms in healthcare relies on a strong emphasis on quality, completeness, and representativeness of the data used for training and validation. However, healthcare data are often heterogeneous, messy, and subject to various biases (e.g., demographic bias, selection bias, or measurement bias). Biases in the data can result in biased predictions or recommendations, exacerbating health disparities and undermining the fairness and equity of healthcare delivery. Addressing data quality issues and mitigating biases require careful data preprocessing, algorithmic fairness assessments, and ongoing monitoring and evaluation of algorithm performance in diverse patient populations.

7.2 Proposal of Future Research Directions to Address These Challenges and Unlock the Full Potential of Bio-Inspired Approaches in Healthcare

To unlock the complete capability of bio-inspired approaches in healthcare and address the challenges outlined earlier, several future research directions can be proposed.

Scalable and Efficient Algorithms: Develop novel bio-inspired algorithms that are specifically tailored for scalability and efficiency in healthcare applications. This may involve designing algorithms that can leverage distributed computing resources, parallel processing, and optimized hardware architectures (e.g., GPUs, TPUs). Research should also focus on developing techniques for reducing the computational complexity of bio-inspired algorithms without compromising accuracy, such as algorithmic optimizations, feature selection, and dimensionality reduction methods.

Interpretable and Explainable Models: Enhance the interpretability and explain ability of bio-inspired algorithms to improve trust and acceptance among healthcare professionals. Research should explore techniques for generating human-interpretable explanations of algorithmic predictions or decisions, such as model-agnostic interpretability methods, visualization techniques, and post-hoc explanation approaches. Additionally, investigating ways to incorporate domain knowledge and clinical guidelines into algorithmic decision-making processes can improve transparency and alignment with clinical practice standards.

Integration with Healthcare Systems: Develop interoperable and user-friendly solutions for integrating bio-inspired algorithms into existing healthcare systems and workflows. Research efforts should focus on developing standardized interfaces, data exchange protocols, and integration frameworks that enable seamless communication between bio-inspired algorithms and medical devices, electronic health record (EHR) systems, and other healthcare IT infrastructure. Collaboration with healthcare providers, IT vendors, and regulatory agencies is essential to ensure compatibility, compliance, and usability of integrated solutions.

Data Quality Assurance and Bias Mitigation: Address data quality issues and mitigate biases in healthcare datasets to improve the reliability and fairness of bio-inspired algorithms. Future research should explore techniques for data preprocessing, anomaly detection, and bias correction to enhance the quality, wholeness, and representativeness of training using medical data and validation. Additionally, developing robust methods for evaluating algorithmic fairness, identifying and mitigating biases, and ensuring equitable healthcare delivery across diverse patient populations is crucial for promoting health equity and reducing disparities.

Ethical and Regulatory Considerations

Investigate ethical and regulatory frameworks for the responsible development, deployment, and evaluation of bio-inspired algorithms in healthcare. Research needs to take ethical issues into account related to patient privacy, consent, autonomy, and the potential impact of algorithmic decision-making on healthcare outcomes and patient well-being. Collaboration between researchers, policymakers, ethicists, and healthcare stakeholders is necessary to develop guidelines, standards, and best practices for ensuring ethical AI in healthcare and safeguarding patient rights and interests.

By pursuing these future research directions, we can conquer difficulties and attain your fullest potential of bio-inspired approaches in healthcare, ultimately leading to improved patient care, clinical outcomes, and healthcare system efficiency.

7.3 Discussion on the Ethical and Regulatory Considerations Associated with Deploying Bio-inspired Algorithms in Clinical Settings

Deploying bio-inspired algorithms in clinical settings raises several ethical and regulatory considerations that need careful attention to ensure patient safety, privacy, autonomy, and equitable healthcare delivery. Here's a discussion on some of the key considerations.

Patient Safety and Benefit: One of the primary ethical imperatives in healthcare is to prioritize patient safety and well-being. Before deploying bio-inspired algorithms in clinical settings, thorough validation and evaluation are necessary to ensure that the algorithms are accurate, reliable, and clinically beneficial. Ethical considerations include assessing the potential risks and benefits of algorithmic decision-making compared to standard care practices and ensuring that algorithmic recommendations do not harm patients or compromise their health outcomes.

Transparency and Explain ability: Healthcare providers and patients have the right to understand how algorithmic decisions are made and the rationale behind them. Ensuring transparency and explain ability of bio-inspired algorithms is essential for building trust, fostering shared decision-making between clinicians and patients, and enabling accountability for algorithmic recommendations. Ethical considerations include providing clear explanations of algorithmic predictions or decisions in a language that is understandable to patients and clinicians and disclosing any limitations or uncertainties associated with algorithmic outputs.

Patient Privacy and Data Security: Personal health information is contained in extremely sensitive and confidential healthcare data, which must be shielded from misuse, disclosure, and unauthorized access. Ensuring bio-inspired algorithms adhere to relevant privacy laws (such as HIPAA in the US) and data security guidelines are one way to address ethical concerns. Strong data encryption, audit trails, anonymization methods, and access controls can all be used to protect patient privacy and lessen the chance of data breaches or illegal access.

Algorithmic Bias and Fairness: Bio-inspired algorithms trained on healthcare data may inadvertently perpetuate or exacerbate existing biases and disparities in healthcare delivery. Ethical considerations include assessing and mitigating biases in algorithmic decision-making processes to ensure fairness, equity, and non-discrimination across diverse patient populations. To mitigate algorithmic bias and advance health equity, strategies such as implementing fairness-aware machine learning approaches, carrying out bias audits, and utilizing varied and representative training data should be used.

Informed Consent and Autonomy: Patients have the right to make informed decisions about their healthcare and to understand the implications of algorithmic decision-making on their treatment options and outcomes. Ethical considerations include obtaining informed consent from patients before using bio-inspired algorithms to guide clinical decision-making, ensuring that patients are adequately informed about the nature, purpose, risks, and benefits of algorithmic interventions, and respecting patients' autonomy to accept or decline algorithmic recommendations.

Regulatory Compliance and Oversight: Healthcare regulatory agencies play a crucial role in ensuring the safety, effectiveness, and ethical use of bio-inspired algorithms in clinical practice. Ethical considerations include complying with regulatory requirements for the development, validation, deployment, and monitoring of healthcare AI systems, obtaining regulatory approvals or clearances where necessary, and establishing mechanisms for ongoing oversight, evaluation, and reporting of algorithm performance and outcomes.

Addressing these ethical and regulatory considerations requires collaboration and engagement among healthcare professionals, researchers, policymakers, ethicists, patients, and other stakeholders. By adopting a multi-stakeholder approach and we can utilize the potential of bio-inspired algorithms to enhance patient care, clinical outcomes, and healthcare system efficiency while maintaining ethical standards and safeguarding patient rights. These can be achieved by upholding ethical principles of beneficence, non-malfeasance, justice, and respect for patient autonomy.

8 Conclusion

8.1 Summary of Key Findings and Insights

Bio-inspired algorithms have enormous potential to transform the way that healthcare is provided, enhance patient outcomes, and progress medical research. For bio-inspired algorithms to reach their full potential and to tackle the intricate problems that face contemporary healthcare systems, further research and innovation in this area are necessary [23]. Key findings and insights regarding the application of bio-inspired algorithms in healthcare include.

Improved Diagnostic Accuracy: When it comes to identifying different medical diseases, bio-inspired algorithms—like artificial neural networks and genetic algorithms—have proven to be more effective than conventional techniques. These algorithms can identify diseases earlier and more accurately by analyzing intricate patterns in medical data, such as genetic sequences or imaging scans.

Personalized Treatment Planning: Personalized treatment plans that are based on the genetic composition, medical history, and lifestyle aspects of each patient can be created, thanks to bio-inspired algorithms. These algorithms are able to enhance treatment plans and enhance patient outcomes through the analysis of big datasets and the identification of biomarkers or risk factors unique to each patient.

Drug Discovery and Development: Due to their ability to anticipate drug-target interactions, identify promising therapeutic compounds, and optimize drug design, bio-inspired algorithms are a key component in the acceleration of drug discovery and development processes. With the aid of these algorithms, scientists may more effectively comb through a large chemical space and identify potential new medication candidates.

Healthcare Resource Optimization: Bio-inspired algorithms can optimize healthcare resource allocation, such as hospital bed management, staff scheduling, and inventory management, to improve operational efficiency and patient flow. By analyzing real-time data and predicting demand patterns, these algorithms help healthcare facilities optimize resource utilization and enhance patient access to care.

Disease Surveillance and Epidemiological Modeling: Bio-inspired algorithms contribute to disease surveillance efforts and epidemiological modeling by analyzing large-scale healthcare data, such as electronic health records and public health databases. These algorithms can identify disease outbreaks, predict disease spread, and inform public health interventions to mitigate the impact of infectious diseases and other health threats.

Challenges and Limitations: Despite their potential benefits, the adoption of bio-inspired algorithms in healthcare faces several challenges, including scalability, interpretability, integration with existing healthcare systems, data quality assurance, and ethical and regulatory considerations. Addressing these challenges requires interdisciplinary collaboration, methodological innovations, and ethical guidelines to ensure the responsible and effective use of bio-inspired algorithms in clinical practice.

8.2 Call to Action for Researchers to Continue Exploring Bio-inspired Algorithms for Advancing Machine Learning and Deep Learning Models in Healthcare

Bio-inspired algorithms have enormous promise to advance deep learning and machine learning models in the healthcare industry. The promise of bio-inspired algorithms to revolutionize healthcare delivery and enhance patient outcomes can be fully realized by researchers through interdisciplinary collaboration, practical application of ethical principles, exchange of knowledge and best practices, and real-world healthcare difficulties and outcomes addressed [24]. The following is a request that scientists carry out more study on bio-inspired algorithms in the field of medicine.

Push the Boundaries of Innovation: Researchers should strive to push the boundaries of innovation by exploring novel bio-inspired algorithms and adapting existing algorithms to address specific challenges in healthcare. This could involve developing hybrid algorithms that combine elements of multiple bio-inspired approaches or designing algorithms that mimic complex biological processes more accurately.

Collaborate Across Disciplines: Collaboration across disciplines is essential for advancing the field of bio-inspired algorithms in healthcare. Researchers from computer science, biomedical engineering, medicine, and other relevant fields should collaborate to leverage their diverse expertise and perspectives. Interdisciplinary research teams can tackle complex healthcare problems more effectively and develop innovative solutions that bridge the gap between theory and practice.

Address Real-World Healthcare Challenges: Researchers should focus on addressing real-world healthcare challenges through the application of bio-inspired algorithms. This includes developing algorithms for clinical decision support, disease diagnosis and prognosis, personalized medicine, drug discovery, healthcare resource optimization, and epidemiological modeling. By targeting high-impact healthcare applications, researchers can demonstrate the tangible benefits of bio-inspired algorithms and drive adoption in clinical practice.

Ensure Ethical and Responsible AI: Ethical considerations should be at the forefront of research efforts in applying bio-inspired algorithms to healthcare. Researchers must prioritize patient safety, privacy, autonomy, and fairness throughout the algorithm development lifecycle. This includes addressing algorithmic biases, ensuring transparency and interpretability, obtaining informed consent, and complying with regulatory requirements. By upholding ethical principles and promoting responsible AI practices, researchers can build trust with healthcare stakeholders and facilitate the responsible deployment of bio-inspired algorithms in clinical settings.

Share Knowledge and Best Practices: Researchers should actively share their findings, methodologies, and best practices with the broader scientific community. This includes publishing research papers, participating in conferences and workshops, contributing to open-source projects, and collaborating with industry partners. By fostering a culture of knowledge sharing and collaboration, researchers can accelerate progress in the field of bio-inspired algorithms and drive innovation in healthcare.

8.3 Outlook on the Transformative Impact of Bio-Inspired Approaches on the Future of Healthcare Delivery and Patient Care

Precision medicine, early disease detection and prevention, optimal resource allocation, improved clinical decision support, and population health management are just a few of the revolutionary effects of bio-inspired approaches that will reshape the entire healthcare landscape. Healthcare systems have the potential to transform into more individualized, proactive, and efficient care models that emphasize patient outcomes, enhance community health, and stimulate innovation in healthcare delivery by utilizing computational intelligence and biological inspiration.

Here's an outlook on how these approaches will shape the future.

Precision and Personalized Medicine: The creation of extremely precise and customized treatment plans based on each patient's unique genetic composition, medical background, lifestyle choices, and environmental influences is made possible by bio-inspired algorithms. Through the process of scrutinizing copious quantities of patient data and discerning nuanced patterns and associations, these algorithms enable medical professionals to provide focused interventions that optimize treatment effectiveness while mitigating harmful consequences.

Early Disease Detection and Prevention: Bio-inspired algorithms have the potential to revolutionize disease detection and prevention efforts by enabling earlier and more accurate diagnosis of various medical conditions. By analyzing multimodal data from sources such as medical imaging, genomic sequencing, wearable sensors, and electronic health records, these algorithms can identify disease biomarkers and risk factors at the earliest stages, facilitating timely interventions and preventive measures to improve patient outcomes and reduce healthcare costs.

Optimized Healthcare Delivery and Resource Allocation: Bio-inspired algorithms optimize healthcare delivery and resource allocation by predicting patient demand, optimizing scheduling and staffing, and streamlining workflow processes. These algorithms make it possible for medical facilities to better manage their resources, shorten wait times, and increase patient access to care. By maximizing the utilization of healthcare resources, bio-inspired algorithms enhance healthcare system efficiency and resilience, particularly in the face of growing demand and resource constraints.

Enhanced Clinical Decision Support: Bio-inspired algorithms provide valuable decision support tools for healthcare providers, enabling them to make more informed and evidence-based clinical decisions. These algorithms analyze complex patient data, medical literature, and clinical guidelines to generate personalized treatment recommendations, assist in diagnosis and prognosis, and identify optimal care pathways. By augmenting clinical expertise with computational intelligence, bio-inspired algorithms empower healthcare providers to deliver high-quality, patient-centered care and improve clinical outcomes.

Public Health Surveillance and Population Health Management: Bio-inspired algorithms contribute to public health surveillance efforts and population health management by analyzing large-scale healthcare data to monitor disease trends, detect outbreaks, and inform public health interventions. These algorithms enable proactive and data-driven approaches to disease prevention and control, facilitating early intervention strategies, targeted health promotion initiatives, and resource allocation decisions. By leveraging real-time data analytics and predictive modeling, bio-inspired algorithms support population-level health monitoring and improve health outcomes across diverse communities.

References

1. Kar AK (2016) Bio inspired computing–a review of algorithms and scope of applications. Expert Syst Appl 59:20–32
2. de Albuquerque VHC, Gupta D, De Falco I, Sannino G, Bouguila N (2020) Special issue on bio-inspired optimization techniques for biomedical data analysis: methods and applications. Appl Soft Comput 95:1568–4946. ISSN:106672
3. Abdel-Jaber H, Devassy D, Al Salam A, Hidaytallah L, El-Amir M (2022) A review of deep learning algorithms and their applications in healthcare. Algorithms 15(2):71
4. Kumari J, Kumar E, Kumar D (2023) A structured analysis to study the role of machine learning and deep learning in the healthcare sector with big data analytics. Arch Comput Methods Eng 30(6):3673–3701
5. Rubinger L, Gazendam A, Ekhtiari S, Bhandari M (2023) Machine learning and artificial intelligence in research and healthcare. Injury 54(3):S69-S73. ISSN 0020-1383
6. Yang X-S, Karamanoglu M (2013) Swarm intelligence and bio-inspired computation: an overview. Elsevier, pp 3–23
7. Gill SS, Buyya R (2019) Bio-inspired algorithms for big data analytics: a survey, taxonomy, and open challenges. In: Big data analytics for intelligent healthcare management. Academic Press, pp 1–17
8. Binitha S, Sathya SS (2012) A survey of bio inspired optimization algorithms. Int J Soft Comput Eng 2(2):137–151
9. Yaqoob A, Aziz RM, Verma NK, Lalwani P, Makrariya A, Kumar P (2023) A review on nature-inspired algorithms for cancer disease prediction and classification. Mathematics 11(5):1081

10. Mittal M, Goyal LM, Kaur S, Kaur I, Amit Verma D, Hemanth J (2019), Deep learning based enhanced tumor segmentation approach for MR brain images. Appl Soft Comput 78:346–354. ISSN 1568-4946
11. Dorigo M, Blum C (2005) Ant colony optimization theory: a survey. Theor Comput Sci 344(2–3):243–278. ISSN 0304-3975
12. Gandomi AH, Yun GJ, Yang X-S, Talatahari S (2013) Chaos-enhanced accelerated particle swarm optimization. Commun Nonlinear Sci Numer Simul 18(2):327–340. ISSN 1007-5704
13. Shamshirband S, Fathi M, Dehzangi A, Chronopoulos AT, Alinejad-Rokny H (2021) A review on deep learning approaches in healthcare systems: taxonomies, challenges, and open issues. J Biomed Inform 113:103627. ISSN 1532-0464
14. Jiang F, Jiang Y, Zhi H et al (2017)Artificial intelligence in healthcare: past, present and future. Stroke Vasc Neurol 2. https://doi.org/10.1136/svn-2017-000101
15. Yu KH, Beam AL, Kohane IS (2018) Artificial intelligence in healthcare. Nat Biomed Eng 2:719–731. https://doi.org/10.1038/s41551-018-0305-z
16. Deeba K, Balakrishnan A, Kumar M, Ramana K, Venkata Narasimhulu C, Dhiman G (2024) A disease monitoring system using multi-class capsule network for agricultural enhancement in muskmelon. Multimed Tools Appl 1–20
17. Ramana K, Aluvala R, Kumar MR, Nagaraja G, Krishna AV, Nagendra P (2022) Leaf disease classification in smart agriculture using deep neural network architecture and IoT. J Circuits Syst Comput 31(15):2240004
18. Jhaveri RH, Revathi A, Ramana K, Raut R, Dhanaraj RK (2022) A review on machine learning strategies for real-world engineering applications. Mob Inf Syst 1:1833507
19. Al-Tawil M, Mahafzah BA, Al Tawil A, Aljarah I (2023) Bio-inspired machine learning approach to type 2 diabetes detection. Symmetry 15(3):764
20. Reddy PY, Reddy CKK, Sithole NT (2024) A study and analysis on diagnosis of melanoma cancer with deep learning: a case study. IGI Glob Fed Learn AI Healthc 5:203–218. https://doi.org/10.4018/979-8-3693-1082-3.ch011
21. Mandave DD, Patil LV (2023) Bio-inspired computing algorithms in dementia diagnosis—an application-oriented review. Results Control Optim 12:100276. ISSN 2666-7207
22. Theodosiou AA, Read RC (2023) Artificial intelligence, machine learning and deep learning: potential resources for the infection clinician. J Infect 87(4):287–294. ISSN 0163-4453
23. Anisha PR, Reddy CKK, Nguyen NG, Bhushan M, Kumar A, Hanafiah MM (2022) Intelligent systems and machine learning for industry: advancements, challenges and practices. CRC Press, Taylor & Francis
24. Reddy CKK, Anisha PR, Khan S, Hanafiah MM, Lavanya P, Madana Mohana R (2024) Sustainability in industry 5.0: theory and applications. CRC Press, Taylor & Francis

The Role of Quantum Artificial Intelligence in Healthcare Advancements

Kavya Sunki, C. Kishor Kumar Reddy, D. Manoj Kumar Reddy, and Srinath Doss

Abstract Quantum artificial intelligence (QAI) can drastically transform the healthcare sector by combining artificial intelligence and quantum computing. Medication development, diagnosis, and therapy have never been conceivable before because of quantum physics and AI's capacity to process massive volumes of data and perform intricate computations. Quantum algorithms offer specialized medication and enhance medical imaging for early illness diagnosis by analyzing genetic data with remarkable accuracy. Quantum machine learning also reveals latent patterns in patient data to provide more accurate diagnosis and therapy recommendations. Despite these benefits, widespread implementation is impeded by scalable algorithms, stable quantum hardware, and regulatory barriers. Quantum technology integration in healthcare requires careful planning and financial commitment. Anyhow, there is no objection to agreeing that quantum Artificial Intelligence revolutionizes the healthcare industry. When combined with the exciting advances in AI made achievable through quantum computing, this technology can lead to better solutions, reduced costs, and a stronger society.

Keywords Artificial intelligence · Healthcare · QAI · QML · Quantum algorithms · Quantum computing

K. Sunki (✉) · C. K. K. Reddy
Department of Computer Science and Engineering, Stanley College of Engineering and Technology for Women, Hyderabad, India
e-mail: kavyasunki1811@gmail.com

D. M. K. Reddy
Department of Electrical and Electronics Engineering, Vardhaman College of Engineering, Hyderabad, India

S. Doss
Faculty of Engineering and Technology, Botho University, Gaborone, Botswana

V. K. Gunjan et al. (eds.), *Cybernetics, Human Cognition, and Machine Learning in Communicative Applications*, Cognitive Science and Technology,
https://doi.org/10.1007/978-981-97-8533-9_10

1 Introduction

Quantum computers are real because of the combined efforts of some majors which made an effort to create new technologies that would allow for using the quantum effects in computational processes and to design advanced quantum algorithms that would be useful for different domains, including financial, chemical, artificial and consecutive intelligence [1]. Quantum artificial intelligence or QAI can be described as a combination of two domains—Artificial Intelligence and Quantum Computing. The fundamental ideology behind QAI is to practice quantum physics to provide extraordinary computing power and efficiency to AI algorithms. Qubits, or qubits that may exist in several states concurrently, are used in quantum computing instead of bits, which are used in classical computing to process information (0 or 1). This allows for parallel processing to process and increase computing capabilities.

In the arena of artificial intelligence, QAI presents new algorithms and approaches that utilize the distinct characteristics of quantum computing to address challenging issues with higher effectiveness. These AI algorithms based on quantum technology could transform a range of sectors such as healthcare, finance, cybersecurity, and optimization. QAI's main benefit is its capability to handle large data sets and complex computations quickly, making it ideal for tasks that are computationally demanding or involve high-dimensional data analysis. Moreover, QAI algorithms have shown better results in optimization, pattern recognition, and machine learning tasks, leading to progress in drug discovery, personalized medicine, financial modeling, and autonomous systems.

Healthcare organizations have already struggled for a long time with the general increase in the world's population and the advancement of age, coupled with rising incidence of chronic diseases [2, 3]. This has been especially evident in the recent pandemic since COVID-19 as many healthcare centers have been overwhelmed; during periods of severe surges in infections, many COVID-19 and other non-related patients have had to be treated in inadequate/created facilities or telemedicine services [4]. AI is seen as a great hope that can complement significant increases in all facets of healthcare from diagnosis to therapy [5].

As quantum computing advances, the possibility of QAI revolutionizing sectors and addressing once unsolvable challenges also grows. Nevertheless, the progress and utilization of QAI are in their infancy, presenting numerous technical obstacles and chances for additional investigation. Through continued research, partnerships, and creativity, QAI has the potential to discover new horizons in artificial intelligence and transform the future of technology and society.

The current book chapter majorly contributes to the following concepts:

1. The present healthcare conditions and the need for innovative solutions for a better future.
2. Introduction to Quantum Artificial Intelligence and its applications in healthcare systems.
3. Understanding the importance of Quantum Computing over Classic Computing techniques.

4. To know the potential of the QAI in revolutionizing the healthcare system.
5. Impact of QAI in various applications such as Medical Imaging, Drug discovery, and Personalized treatment.
6. The potential additional consequences of QAI in improving healthcare.

The remaining chapter is arranged as follows: Sect. 2 discusses the demand for innovative solutions in healthcare and the basic concepts of quantum computing and quantum algorithms. Section 3 describes the importance of Quantum computing and its applications over Classic computing. Section 4 briefly explains the role of QAI in diagnostics which includes the concepts of Medical Image Interpretation and enhancing the disease diagnosis techniques. Whereas Drug Discovery and Development strategies are discussed in Sect. 5. Personalized treatment strategies are focused in Sect. 6. Section 8 concludes the chapter after discussing the future interests in Sect. 7.

2 Introduction to Quantum Artificial Intelligence and Implementation in Healthcare

2.1 Demand for Innovative Solutions in Healthcare

Human health is primarily experiencing numerous challenges that can be attributed to social factors such as demography, technology, costs of production, and demanding patients. In the modern world, we have many health systems globally facing core issues such as rising population, chronic disease load, soaring costs of health service, technological advancement, and individualized medications. Many of these transitions affect aging populations in some countries, though there is variation among those countries [6]. Due to the rising population, human needs in care services, primarily chronic care, long-term care, and elderly care, elevate. Cardiovascular disease, cancer, diabetes mellitus, and respiratory diseases are still the baseline trends of morbidity and mortality in the world [7]. In a bid to primarily prevent or control increments that are chronic such as cardiovascular diseases, cancer, diabetes, and respiratory diseases, more intervention solutions need to be developed.

Expanding healthcare costs places pressure on healthcare facilities, states, and organizations, as well as the general populace. Hence the costs of going for health services, drugs, and insurance policies continue to go up thereby becoming a financial burden to many families [8]. Therefore, some people and households still cannot access or afford equivalent services. For instance, the underserved face such restrictions as poverty or living in remote areas such as disease-stricken zones and lack of health insurance coverage from seeking medical treatment since they have little power when it comes to overcoming common illnesses; thus, their health levels remain low compared with others. Achieving fair access to healthcare remains a big challenge throughout the world. There are technological innovations that are transforming healthcare across the globe, as doctors use digital health platforms for consultations with patients from different regions; medical equipment manufacturers make wearable devices that monitor one's health parameters while connected through cellular networks etc.

Figure 1 depicts the estimated Global **Quantum Computing in Healthcare Market** Value across the world. CAGRs for the market are projected to be approximately 43% and this indeed would help add further depth to the study. Moreover, such gadgets enable doctors and nurses to establish closer relationships with their patients since they provide real-time data on their condition. Health care delivery is undergoing 'a revolution in treatment' with improvements in genomics, molecular diagnostics, and precision therapies driving the personal medicine trend. This form of medicine makes treatment correspond to patient personal information and hence has the potential of increasing treatment efficiency while reducing other health impacts and increasing patient survival rates.

The coming of the so-called smart phones in the last year of the first decade of the 2000s revealed a series of approaches to mobile software known as applications apps in most fields, including the healthcare industry. Presently, wireless connections are almost a norm, and accurate artificial intelligence (AI) models are available to create new telemedicine applications, for instance, to enhance the process of evaluating and prioritizing patients in severe large vessel occlusion in high-risk of stroke.

The future health sector is found to address more of the chronic issues and lack of treatments to meet patient demand. Intelligent medical treatment as an umbrella term covers a wide variety of participants: physicians, staff, hospitals, and research entities. It includes an ever-evolving concept with numerous dimensions that belong to it, such as prevention and detection of diseases, assessment, evaluation, delivery of care, decision-making about patients' care, and medical research. Intelligent healthcare

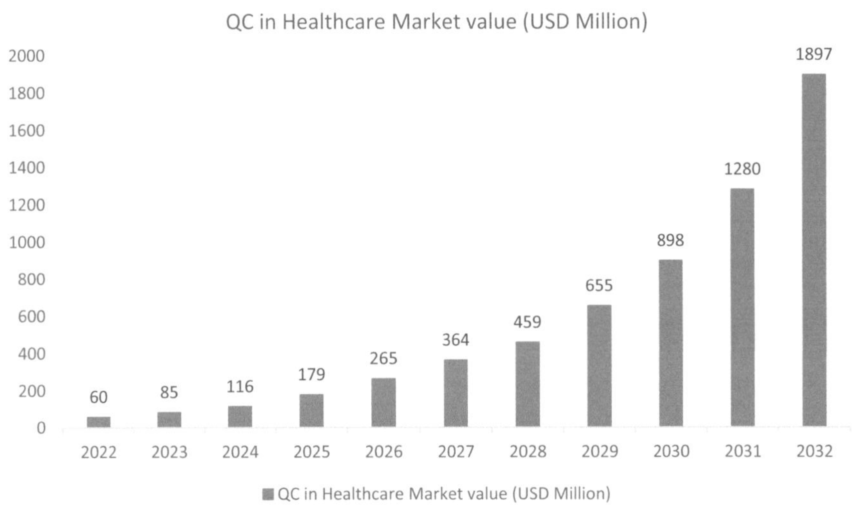

Fig. 1 Global quantum computing in healthcare market

components include automated business networks such as the Internet of Things (IoT), mobile Internet, cloud networking, Big Data, 5G, and AI with an additional focus on developing biotechnology [9].

Given these problems and opportunities, more and more people are calling for renewed answers that can answer health stakeholders' changing demands. Some of the innovations that seem to be potential transformer systems of healthcare-delivering systems that help to enhance patient outcomes and make the healthcare experience much better than before are artificial intelligence, big data analytics, blockchain, telemedicine, regenerative medicine, and virtual reality. Similarly, it can be stated that COVID-19 draws attention to the need for resilience, agility, and innovation in healthcare structures.

Examples of such rapid transformations include the use of digital health tools, telemedicine, online and physical development of vaccines and remote monitoring are some of the changes in healthcare innovation that can mitigate the impacts of the global health crises. At present, healthcare faces different problems and opportunities that require new creative resolutions that effectively, efficiently, and fairly deal with healthcare demands. Welcoming modernity and partnership among players will help in spearheading positive change and turning to an image where every person gets an opportunity to have quality medical services and treatment at any time without restrictions.

2.2 Introduction to Quantum Computing Principles

To elaborate the following concept of quantum computation in principle, it is comprehensible if the basic ideas and definitions associated with this term are conceptualizable. Quantum is the minimum amount or bundle in which a physical property can be measured or quantified. These units of quantum are relatively very small and these disjoint elements form the basis of quantum physics that defines the behavior of these minimum structures.

Quantum mechanics can be defined in its rudimentary form as the study of the motion of the minuscule energy packets of the atoms and particles. Quantum Computing is executed on dependence of inherent quantum incidence including entanglement, quantum superposition, and multiplicity interference. The quantum entanglement is a counterintuitive property often described by Einstein as the "classic example of the totally bizarre behavior that is of the very nature of quantum mechanics," whereby an entangled pair of electrons is always in a state where the two particles spin in opposite directions and can influence each other even when separated through time and space and not directly connected [10].

Quantum mechanics plays a crucial in many modern physics-based technologies such as Semiconductors and Transistors, laser, and MRI Machines. Unlike traditional computing that relied on the binary form of digits where a bit could take the value of either 0 or 1, Quantum computing applications coax bits of 0 and 1 to read quantum bits or qubits which can be precisely described as quantum bits or qubits akin to the atoms in quantum physics efficiently. System bits appear in a superposition or interrelation when they can assume various states simultaneously. A qubit commonly known as the quantum bit is also the most fundamental unit of data within the quantum computational systems just as the classical computing bit. Like regular bits, qubits follow the laws of quantum computing but can exist in multiple states. Key characteristics and concepts of quantum computing are:

Superposition: In classical systems such as traditional computers, a bit can take on the value of 0 or 1 but when it comes to quantum mechanical systems that use qubits there's what we call superposition in which they exist together: hence before being measured they could have both values; 0 and 1. This capability leads to a significant increase in speed beyond our imagination of these devices compared with those based purely on classical mechanics.

Entanglement: Entanglement: Quantum entanglement is a characteristic of quantum systems where it is impossible to describe the state of one of two particles located in two distinct areas without simultaneously considering the state of another particle located in the second area. Quantum entanglement and teleportation are important and essential as the fundamental building blocks for quantum messaging, quantum connections, and quantum calculating [11].

Quantum Interference: Qubits interfering with one another is a phenomenon known as quantum interference resulting in constructive or destructive interference hence enabling quantum computers to do computations concurrently solving more complex problems faster when compared to the classical ones.

Quantum Gates: Quantum gates operate differently from classical logic gates in classical computers, as the latter are used for manipulating bits within classical circuits while the former modify qubit states that are directed towards performing computation tasks. A few types of quantum gates are- Hadamard gate, Pauli-X gate, and CNOT gate.

Figure 2 represents the timeline of Quantum Computing evolution detailing the developments in quantum technology. Leveraging quantum computing's computational prowess, QAI conducts virtual screening of molecular compounds efficiently, unearthing novel drug targets and potential therapies for various diseases. This accelerated drug discovery paradigm harbors the potential to expedite the delivery of life-saving treatments to patients worldwide, economically, and efficiently. Moreover, QAI promises personalized medicine and treatment optimization.

By working with data sets containing genomic, clinical, and lifestyle data QAI algorithms provide doctors with an understanding of each patient's peculiarities and open the doors to individual approaches in a treatment process. Empowered by QAI, surgeons navigate intricate procedures with heightened accuracy and safety, fostering improved patient outcomes and swifter recovery periods. This advancement in surgical technology heralds a new era in minimally invasive procedures, promising expanded access to superior surgical care on a global scale [12].

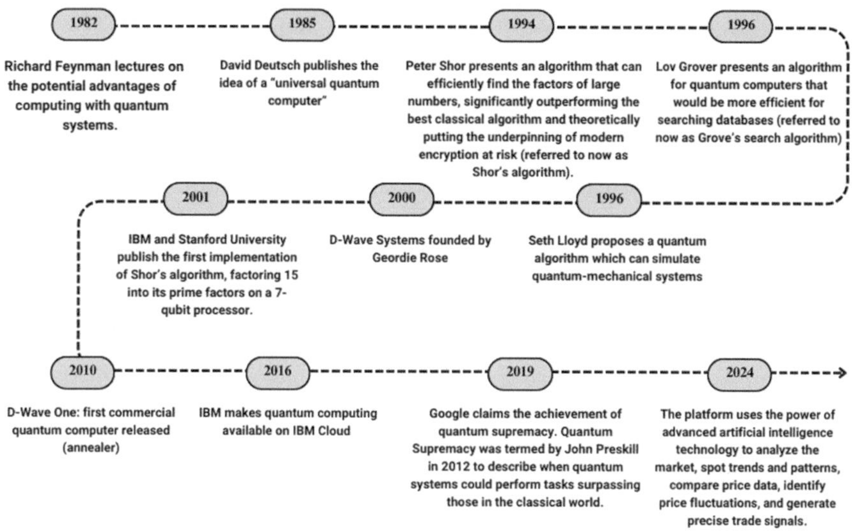

Fig. 2 Quantum computing timeline

2.3 *QAI Potential to Revolutionize Healthcare*

In this sense, quantum computing is viewed as a solution to such computationally intensive problems since it can revolutionize how the problems can be solved [13]. Quantum Artificial Intelligence is poised to redefine many industries, healthcare included. QAI's transformative capabilities have the potential to change patient management, the research process, and the existing health technology market. Quantum Artificial Intelligence fundamentally works by harnessing quantum computer power and integrating it with Artificial Intelligence algorithms to process factual data in previously impossible ways. One of the many ways QAI will revolutionize healthcare is in the diagnostic field.

QAI-based AI algorithms will improve the accuracy and rapidity of medical image perception, including MRI, CT, and PET scan readings. Through quantum computing's parallel processing capabilities, QAI will be able to analyze large sets of image data in real-time, allows quicker and more accurate cancer, cardiovascular, and neurologic diagnosis. Quantum computing as an emerging technology can pave the way towards the addressing of computational challenge with enhanced and efficient solutions for optimization, simulation, and machine-learning purposes [14]. The accelerated diagnostic process, will enable patients' early detection and intervention and hence improve outcomes and cost cuts.

Table 1 is a comprehensive analysis of the multitudes of ways in which Quantum AI is changing the medical field. It lists the main applications with a brief explanation of them in addition to where they are utilized now and what may be expected in the future. Mentionable topics include drug discovery, genomics, medical imaging, personalized medicine, and predictive analytics. This means quantum AI has been able to bring about better drug discovery processes, improve medical image sensitivity enable modification of treatment schemes to fit individual requirements, and remarkably improve predictive analytics facilitating the management healthcare system.

Recently, there has been a shift from a hospital-centric to a patient-centric view in the healthcare industry, which allows the patient to control the health operations. The shift is realized and supported through emerging disruptions in Artificial Intelligence (AI), the Internet (IoT), big data, and assisted fog and edge networks [15]. Machine learning is a pervasive approach in many fields of engineering and sciences. Due to the restricted architecture of classic computers, there is a need to move to quantum computing to develop QML [16].

Furthermore, QAI can be a critical strategy in drug discovery and development as drug discovery is an expensive and time-consuming approach that often takes many years to bring a new drug to the market [17]. However, QAI algorithms can rapidly diagnose molecular interactions and predict drug candidates with high precision.

Table 1 Applications of quantum AI in healthcare

Application	Description	Current status	Future potential
Drug discovery	Exploiting quantum computing to simulate molecular interactions for faster and more accurate drug discovery	Early-stage research; some promising results in molecular simulations and small-scale trials	Ability to minimize the time and expenditures on drug discovery, leading to the rapid development of innovative medications
Genomics	Analyzing large-scale genomic data to identify genetic markers and understand genetic disorders	Limited applications; experimental algorithms for genomic data processing are being tested	Could revolutionize personalized medicine by providing deeper insights into genetic predispositions and enabling customized treatments
Medical imaging	Enhancing image analysis through quantum algorithms to improve diagnostic accuracy and speed	Proof-of-concept studies show enhanced image processing capabilities	Potential to greatly improve early detection of diseases through more precise and faster imaging analysis
Personalized medicine	Tailoring treatments based on individual patient data, including genetics and other health records	Early-stage research; pilot projects integrating patient data for personalized treatment plans	This could lead to highly effective personalized treatment protocols, improving patient outcomes and reducing adverse effects
Predictive analytics	Using quantum computing to analyze vast amounts of data to predict disease outbreaks and patient outcomes	Initial research and pilot studies demonstrating potential in predictive modeling	Potential to transform healthcare management by accurately predicting and preventing disease outbreaks and optimizing resource allocation

QAI uses quantum computers to perform virtual screening of molecular compounds in a more cost-efficient and faster way, helping to discover new drug targets and uncover therapies for diseases. Such an accelerated drug discovery cycle would enable lifesaving medicines to reach patients more quickly and more affordably [18].

Moreover, QAI can promote personalized medicine and treatment. QAI's algorithms can analyze vast datasets of genomic, clinical, and lifestyle information and generate insights into individual patient profiles, including which treatments are likely to be most efficacious. This would result in a more accurate, targeted approach to care, lowering the risk of side effects while optimizing treatment. QAIs are also implemented in Real-World Predictions. The high levels of data entanglement and interconnection in EHRs and in the real-world health and medical datasets also indicate that such data sources can be best augmented by Quantum Computing algorithms [11, 19]. Healthcare has recently been a rather hot topic in the area of quantum machine learning at some certain level of difficulty. Consequently, quantum machine learning can be considered as a widely used and rather efficient approach for data models and classifications across diverse fields [20].

3 Understanding Quantum AI

3.1 Quantum Computing Over Classical Computing

As proven in several algorithms and system developments, using Quantum computing enhances the intake of Deep Learning as compared to classical computing. Quantum Computing increases the efficiency of enhancing the optimization of the objective principal function. In addition, it has the merit of decreasing the number of iterations required in the learning process of deep learning [15]. Classical computation refers to the traditional method of performing arithmetic using classical bits, which are binary units of information that can occur in one of two states; Currently the number of monitored parameters is either zero or one.

To switch these bits and go through the classical computational procedure with such patterns as designated by software instructions. Classical computing employs sequences of bits to represent information flow which can be operated on by algorithms and/or logical functions. All these tasks are done either in a successive manner by a single processing unit known as the central processing unit (CPU) or are divided among several processing units within a parallel computing system. Classical computing has been the primary style of computing for many years and has numerous applications such as data processing, software development, and simulation among others. It forms the basis of today's computing technology that powers the functionalities of common devices as processing chis computers, smartphones, tablets, and servers.

A detailed description of each aspect to differentiate Classical AI and Quantum AI is mentioned in Table 2 in which the comparisons and contractions between both computing techniques are classified. Quantum computers make use of quantum Binary units or bits or qubits and they are different from normal binary units in that they can possess the state zero and state one at once. This superposition permits quantum computers to calculate multiple distinctly at the same time thus enhancing the computing capability. The quantum computers are capable of performing some particular computations far better than the conventional computers. This is because there are instances whereby parallel qubits are available for use and hence quantum algorithms can take advantage of the probability hood to solve various issues efficiently. For some problems like we have the problem of factoring large numbers, browsing through unstructured databases, quantum algorithms can also provide more efficiency as compared to classical algorithms.

Although classical computing is highly advanced and highly efficient in many fields, it faces limitations in solving certain problems efficiently, especially those involving large data sets, complex computation, and production challenges of efficiency-use concepts and calculate in very different ways. Quantum computers differ from classical computers in several important ways, and are due to the unique

Table 2 Differentiating classical AI and quantum AI

Aspect	Classical AI	Quantum AI
Basic unit of data	Bit (binary digit)	Qubit (quantum bit)
Representation	The two forms of bits are either 0 or 1	Qubits are well known as they can be in states 0 & 1 at the same time
Processing power	Limited by Moore's Law, which governs transistor density and processing speed improvements	Potential for exponential speedup for certain problems with quantum algorithms
Parallelism	Limited parallelism, typically achieved through multi-core architectures	Exploits parallelism through the superposition of qubits and entanglement
Error correction	Error correction codes are employed to mitigate errors because of noise and other factors	Error-correction techniques are being developed to address errors due to decoherence
Hardware architecture	Utilizes electronic components like transistors and logic gates	Specialized hardware such as superconducting circuits, trapped ions, or photonic devices operating at extremely low temperatures
State of technology	Mature technology with well-established principles and applications	Still in the early stages of development, with ongoing research and experimentation
Applications	Widely used in various fields including data processing, software development, simulations, etc	Potential applications in cryptography, optimization, materials science, drug discovery, etc

principles of quantum mechanics that govern its operation. Traditional computers use bits as special information, which can be in one of two states: 0 or 1.

Lindblad formulas that are normally employed when it comes to the evolution of standard quantum systems can also be utilized in solving problems that are associated with formulating complex quantum systems that are unreachable to classical computers [21]. Quantum computing attempts at extending quantum mechanics in order to explain parallel computing, which can solve certain calculations like the RSA encryption algorithm significantly faster than a von Neumann machine. Thus, it can be quite reasonable to admit that quantum machine learning can be learned much faster than that of a classical computer with the help of applying certain quantum operations [22].

For quantum computers, hardware is unique, as it deals with the manipulation of qubits and quantum operations. Often this hardware can be superconducting circuits, trapped ions or optical devices which require low temperatures to work properly. Transistors and logic gates are used in classical computing machines for handling of information. In sum, quantum computing is inherently different from its classical counterpart because of what IT is all about. This is correct to some extent because while still in its early years quantum computing cannot be ignored and it promises to revolutionize fields such as cryptography optimization material science drugs among some other areas through solving problems that are complicated and beyond the solving capabilities of classical computers.

3.2 *Immersion of Quantum Algorithms in Healthcare*

The advances of Quantum Learning as a combination of Artificial intelligence and Machine intelligence, which are promising to achieve almost optimal solutions that help in enhancing the performance of the algorithm, have created a new field of study known as Quantum Machine Learning (QML) which has emerged from the theory of quantum computing. The rationale for using quantum computing in machine learning is the fact that parallelism in solution searching is inherent and is implementable based on visualized Moore's law in optimal constraint solving [23].

The integration of quantum deep learning and ML models in the medical field has been discussed widely in recent years. Moreover, several researchers in the past have employed different varieties of quantum machine learning algorithms by employing authentic, real-time data from the healthcare domain. In issues of classification, the quantum supervised machine learning circuits or Quantum KNN, Variational Quantum Classifier (VQC), Quantum Support Vector Machine (QSVM), and different approaches to pre-processing integrate with the idea of noise, and error mitigation algorithms have showcased considerable assistance [9].

Quantum algorithms present significant potential in a number of areas of health care increasing the facility of various aspects of medical research and patient care. Medical image interpretation is one of the most popular applications of quantum algorithms. It is a well-understood fact that with quantum algorithms the accuracy and efficiency can be significantly enhanced for tasks like the interpretation of medical imaging data including MRI, CT scan PET scans, etc. Employing parallel quantum computing potential, algorithms can learn and scan various image information in real-life time that will aid in diagnosing and treating diseases including cancer, cardiovascular diseases, and neurological diseases.

3.2.1 Quantum Algorithms

The Quantum algorithms are intended to be run on quantum computers; their methods utilizing the concepts of quantum mechanics for some problems are faster than the ones based on Grover's for the search of an element in an unsorted list and Shor's for integer factorization. Quantum Computing has been researched as a solution for making models to increase their speed and computational complexity. Here are a few most used and basic algorithm concepts to know about QAI: Here are a few most used and basic algorithm concepts to know about QAI:

Quantum Neural Networks (QNNs): QNNs are quantum analogues of classical neural networks, where quantum circuits are used to compute quantum data QNNs have more practical capabilities than classical neural networks in tasks such as pattern recognition, optimization, and machines in the study of. Quantum parallelism and interference are supported to explore multiple computing methods simultaneously, resulting in faster training and reasoning.

Quantum Machine Learning (QML): QML is a set of quantum computing principles used in the machine learning industry. Quantum algorithms such as quantum support vector machines, quantum K-means clustering, and quantum principal component analysis have been proposed to efficiently address classical machine learning problems. QML algorithms can benefit from the ability of quantum computers to handle large data and it is in the arduous task that calculations with Tatar parallels and more accurate predictions are made. Quantum-enhanced optimization can also be applied to quantum computing optimization problems, common in AI applications such as logistics, scheduling, and resource allocation.

Quantum approximate optimization algorithms (QAOA): QAOA and quantum annealing promise to solve optimization problems better than classical methods. The shown algorithms can use quantum parallelism and the tunneling effect to find optimal solutions in large solution spaces.

4 QAI Applications in Diagnostics

4.1 Utilizing Quantum AI for Medical Imaging Interpretation

Medical imaging plays an important role in the diagnosis and management of medical conditions by providing detailed insights into the structure and function of organs and tissues but the interpretation of medical images can be complex and time-consuming and often requires specialized knowledge and manual analysis. Quantum AI offers promising solutions to provide more accurate and efficient medical image interpretation, especially for techniques such as CT, MRI and PET scans. Quantum AI algorithms have the potential to revolutionize medical imaging techniques and offer significant improvements in CT (computed tomography), MRI (magnetic resonance imaging), and PET (positron emission tomography) scans Quantum AI for MRI detection of abnormalities such as tumors, lesions and systemic issue images can be analyzed.

Using quantum parallelism and enormous computing power, these algorithms can process huge amounts of MRI data with increased speed and accuracy This results in advanced image reconstruction, denoising, and segmentation techniques, enabling MRI scans to improve monitoring and interpretation, making diagnosis more reliable and timelier For CT scans, quantum AI can identify vital areas of interest, such as tumors or abnormal tissue.

Quantum-enhanced image reconstruction algorithms can significantly improve the resolution quality of CT images, enabling better detection of anatomical structures and abnormalities these technological advances also raise the need for high-quality imaging and patient safety considerations balanced by reducing the amount of air required without compromising image clarity. Quantum AI for PET imaging can analyze images to identify metabolic changes and areas of increased activity that could indicate tumors or inflammation. Quantum-enhanced image processing techniques can enhance the signal-to-noise ratio and spatial resolution of PET scans have improved.

Diagnosis is used to examine a rotating machine's state classification under the premise of several approaches to Deep Learning and Machine Learning. In the last few years, the field of Quantum Computing, which is still a relatively novel and growing investigations area, has been involved in various objectives and settings for professional application, including optimization, artificial intelligence, simulation, pharmaceutics, cybersecurity as well as the energy sector [24]. For this, the deployment of QML combined with the Deep Learning framework has been utilized for the diabetes prediction model. The trained Deep Learning model contributes a higher accuracy value to this work when compared to the developed QML and the existing models of the same [25]. A quantum neural learning model as a proxy for classifying patients with COVID-19 infection using medical diagnosis through the analysis of computed tomography scan images [21]. The efficiency of the novel QML approach is reported for generating synthetic CT images with high accuracy using a modified CGAN model [15].

Quantum AI algorithms can accurately analyze medical images, reduce the risk of misdiagnosis, and enable faster diagnosis of medical conditions. Quantum computing's parallel processing capabilities allow for faster, time-reducing analysis of medical images that are turned on, and provide improved patient outcomes Quantum AI algorithms analyze medical images in combination with patient data to develop appropriate treatment plan needs.

Quantum AI enables researchers to better analyze large medical image data, accelerate medical research, and facilitate the development of new diagnostic therapies The use of quantum AI to interpret medical images holds great promise to transform health care by increasing diagnostic accuracy, improving patient outcomes, and improving medical research. However, these imaging techniques often face challenges related to image quality, resolution, and interpretation.

4.2 Benefits of Enhancing Medical Imaging Through Quantum Algorithms

Quantum-more desirable picture reconstruction and interpretation algorithms can enhance the accuracy of clinical diagnoses by supplying clearer and more informative pics to healthcare specialists. High-decision scientific pics generated by way of quantum algorithms permit more precise remedy planning and targeted interventions, leading to higher affected person consequences. Quantum-better imaging strategies can reduce the want for repeat scans and additional diagnostic approaches, thereby lowering healthcare costs and improving aid utilization.

Quantum AI can also analyze molecular and genetic data to identify different types of cancer, enabling the development of treatment strategies with highly personalized targeted Quantum AI algorithms can analyze medical imaging data, such as echocardiograms and MRI scans of the heart, detecting abnormalities such as heart attacks, simple heart attacks, and heart attacks Quantum AI uses physiological data such as blood pressure, heart combining variables and electrocardiogram (ECG) signals with imaging data A comprehensive health assessment can provide and detect early warning signs of disease.

Quantum algorithms empower researchers to discover new frontiers in scientific imaging research, facilitating the development of novel imaging strategies and diagnostic equipment. The integration of quantum algorithms into clinical imaging holds outstanding promise for reinforcing photograph reconstruction, resolution, and interpretation. By leveraging the computational electricity and parallel processing abilities of quantum computing, these algorithms can revolutionize clinical analysis and remedy, in the end enhancing patient care and advancing healthcare innovation. Figure 3 lists the benefits of enhanced medical imaging.

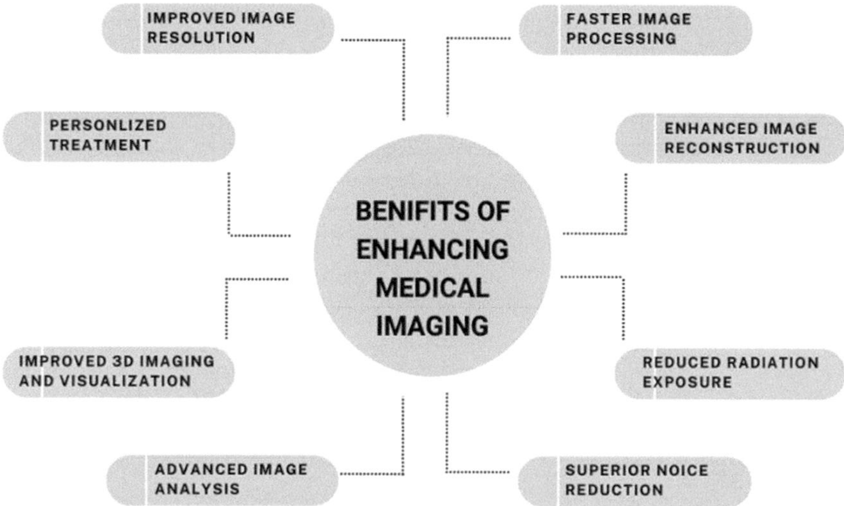

Fig. 3 Benefits of enhanced medical imaging

5 Drug Discovery and Development

The use of concurrent drugs is quite rampant in practice today in medication and clinical settings. However, it sometimes leads to various other unforeseen ADRs, the probability of which increases with the number of drugs administered as the drugs' pharmacological or physicochemical properties generally tend to interact and, therefore, more chances are there for ADRs [26]. QC being in its nascent stage and with future technologies improving in the field can boost and act as a catalyst for augmenting an involved phase of drug discovery and development [11]. It is the process of identifying potential drug candidates and improving them, and phases of clinical Pre-drug discovery involve various types of studies, including the traditional mode which is time-consuming and resource-intensive and includes a vast number of laboratory experiments and computer-modeling.

Quantum AI in drug discovery greatly improves the discovery of new drug candidates by better modeling molecular interactions and identifying the behavior of complex biological systems. Quantum computing thrives on handling large, complex datasets and performing calculations that simultaneously contain many variables, such as the folding of proteins or the binding of drug molecules to their targets The main advantage of quantum AI in medicine is its ability to analyze large chemical regions more efficiently and faster than classical methods This capability allows researchers to identify promising compounds that might otherwise be overlooked by traditional methods in various ways. Furthermore, quantum AI can enhance the molecular docking process, where potential compounds are screened for their ability to bind to specific biological targets, providing more accurate estimates of binding affinities and interactions of energies.

Once potential drugs are identified, quantum AI can help optimize their systems to maximize efficiency and minimize side effects. Defining the ADR detection problem helps to focus on the binary classification task and define if an entity pair is related to ADR or not. Subsequently, by the combination of the Bi-LSTM, the attention mechanism, and quantum computing, the QBi-LSTMA is constructed, and then a method of ADR detection is proposed with the help of the new model [27]. This suggests that Adverse Drug Reaction Detection can be done with Quantum computing with attention.

It can therefore be used by scientists to determine the probability of a drug molecule performing a certain function given instances at which slight modifications are made to a particular structure. These iterative design and optimization processes can be performed rapidly and accurately, resulting in the formulation of the desired drugs within comparatively shorter periods as compared with the computational methods Molecular docking is an important stage of a drug discovery process when characteristics of putative drugs for their ability to bind are determined.

Quantum AI can enhance this process by providing more accurate estimates of bond affinities and interaction energies. Quantum-enhanced algorithms can simulate the docking process in much finer detail, and predict more reliably effective drug-target interactions than classical methods In addition to the initial discovery and optimization phases, quantum AI will also play an important role in clinical development. Quantum AI can also help personalize treatments by identifying subsets of patients who will respond well to treatments.

6 Personalized Therapies Based on Quantum AI

Personalized medicine aims to transform treatment based on an individual patient's unique genetic, lifestyle, and disease characteristics Quantum AI provides, and enables, powerful tools for analyzing genomic data and patient profiles is a personalized therapy optimized for efficacy, safety, and patient outcomes Demonstrates how research can be used to develop personalized therapies: Quantum AI algorithms can analyze genomic data, including gene expression profiles, DNA sequences, and genetic variations such as single nucleotide polymorphisms to identify disease-related mutations, biomarkers, and therapeutic targets With quantum machine learning methods are used, these algorithms can reveal patterns and relationships in genomic data. Classical analysis methods may not be obvious in use, leading to more diagnosis disease diagnosis and prognostic predictions.

Quantum AI algorithms can combine genomic data with clinical lifestyle, and environmental factors to create detailed patient profiles of how disease pathophysiology and characteristics of individual patient complexity Provide insight into disease susceptibility, treatment response variability, and individualized risk assessment, enabling targeted interventions so and preventive strategies tailored to individual patient needs.

6.1 Predicting Patient Responses to Treatments and Optimizing Healthcare Interventions

FERS was proposed for healthcare applications that can benefit patients, medical professionals, doctors, and medicine sellers. A complex deep-learning technique was proposed for FERS by jointly fusing 2D and 3D facial features. The Gabor features-based facial expression recognition system observing that in the experiment FERS was used and for the problem of infrared facial expression recognition it was proposed to use a Discrete Beamlet transform regularization as a Blind Restoration model. A temporal 3DLeNets model was employed to facilitate the facial expression video analysis. FERS has been developed to create an FERS using the feature selection method of The Quantum-Inspired Binary Gravitational Search Algorithm. In this paper, the authors applied multiple AI algorithms in a quantum computing system for analyzing Facial Expression Recognition for Mental Health Assessment. The favorable effect of FERS is in identifying the patient's reaction to the treatments which is summarized as improving the treatment [24].

Advanced Quantum AI can incorporate various forms of patient data like DNA, EHRs, CT scans, lab results, and habits into patient reports. The latter are complementary to traditional prediction models since they capture all facets of the patient health status and disease profiles, as well as have the potential to reveal disease and treatment response heterogeneity and individualized risk profiles. Quantum algorithms look for causal relationships and patterns in the large historical patient and medical record databases and treatment outcome data and generate predictive signatures, based on which treatment selection and optimizing strategies are devised.

Quantum AI algorithms can especially help in the optimization of healthcare interventions, by likewise modeling the likely effects of therapies, and exploring the space of improved treatment strategies that pursue the greatest measure of therapeutic benefit with the least disutility. These algorithms help promote the patient's involvement in the decision-making process by demonstrating the effects of treatment, different patient responses, and the dynamics of disease progression; they also improve patient compliance with treatment.

In conclusion, learning patient outcomes and responses to treatments or general healthcare interventions with quantum AI is demanding for better patient results, and effective healthcare service delivery, and works as a roadmap towards Precision Medicine the future of the healthcare system. As one can see, quantum computing can transform these basic algorithms and use them to help clinicians develop better-focused and more effective treatment and care plans because the algorithms would be based on learning the patient's personal needs to then develop safer and more patient-specific forms of healthcare intervention.

7 Quantum AI-Enabled Healthcare Robotics

Quantum AI-composed healthcare robotics can therefore be defined as an advanced integration of quantum computing and artificial intelligence fields of study in healthcare. These enhanced robotic systems harness Complex computing and decision-making domains of quantum artificial intelligence to bring concerning changes in numerous dimensions of patient management, treatments, and healthcare systems. Quantum AI-enabled healthcare robotics is transforming the healthcare landscape in several ways. Using AI integration with quantum technology, more specifically in surgical robots,

Combining quantum computing results in increased computational speed and the artificial intelligence that possesses the learning ability, these robots can customize the pattern of rehabilitation to suit a patient's needs. With Quantum AI, robots can improve orderliness in health facilities through the provision of a Conveyor Service for transporting samples or even orderliness from a remote place by autonomous mobile Robots.

Integrating AI in quantum computing can lead to improvements in healthcare robotics in the diagnosis and treatment of patients as well as revolutionizing the administration of health care. Thus, using artificial synthesis and the interactions between quantum computing and artificial intelligence, these complex robotic systems enable the development of the processes of efficient personalized medicine and efficient services in healthcare. This is due to the rapidly progressing research and development within this area, Quantum AI-enabled healthcare robotics will soon become a new normal, globally transforming the healthcare standard and the patient's experience in receiving care.

7.1 Integration of Quantum AI into Robotic-Assisted Surgeries and Medical Interventions

The incorporation of quantum Artificial Intelligence into robotic surgery and medical operations is an intelligent innovation in the field of healthcare delivery. This merge of quantum computing and artificial intelligence in surgery is groundbreaking in terms of operation capabilities and patient outcomes brought by operating surgical robots with the new added precision and efficiency. Here is how it is changing the dynamics of robotic-assisted surgeries and medical procedures taken generally. Quantum AI applications use the supercomputing capability and pathway computing of quantum technology to handle patient data and surgical imaging and the alternative parameters immediately.

The application of Quantum AI algorithms in medical imaging can help to create highly detailed 3D models that can depict the necessary anatomy of a patient, providing surgeons with additional and more accurate information about the procedure to be conducted. These algorithms work during surgery, reading tissue deformation, blood flow, instrument positioning data, and outputting real-time information to aid the surgical team. Incorporating quantum computing's computational ability into these robotic systems alongside the AI algorithms' ability to learn and adapt, improves surgical outcomes, shortens the time taken to perform a surgery, and reduces susceptibility to potential surgical complications.

8 Ethical Considerations and Challenges

Ethical concerns entail reasonably distributing robotic-assisted surgical procedures, a major element that may deny societal benefits to such groups because of socioeconomic difficulties. The quest to ensure changes in this direction to embrace diversity and equality in the provision of health care is important in appreciating justice in society. AI-enhanced healthcare robotics need many patient records which often contain personal identification information, diagnostic images, and even patients' genotyping data. There is a very important factor in the matters of patient's rights—the patients' privacy and confidentiality must be ensured.

9 Future Interests in QAI

Promoting interdisciplinary research collaborations, establishing viable alliances between academic institutions with leading technology industries, and establishing technology innovation hubs will create an innovation culture that advances healthcare QAI-enabled technology solutions. Emphasizing the importance of creativity, sharing, and trying out new ideas, further development guarantees the rapid transformation of new scientific findings into actual applications. Collaboration with other departments allows for information exchange and sharing of concepts, ensures adherence to best practices, and practical application of QAI in healthcare delivery.

Despite the relative promise present in each of these explanations, the quantum computer field has yet to illuminate all the precise ways in which this technology would change the world when it will do so, or for which tasks. Hope is that initiating this prize will catalyze more movement toward answering these questions by promoting the community towards developing and further planning new benefits that quantum computing will bring to the world. Table 3 shows detailed information regarding the milestones and the qubits.

Table 3 Google quantum computing roadmap

S No	Milestone	Year	Physical qubits	Logical qubit error rate
1	Beyond classical	2019	54	–
2	Quantum error correction	2023	10^2	10^{-2}
3	Building a long-lived logical qubit	2025 +	10^3	10^{-6}
4	Creating a logical gate	2025 +	10^4	10^{-6}
5	Engineering scale up	2025 +	10^5	10^{-6}
6	Large error-corrected quantum computer	2025 +	10^6	10^{-13}

10 Conclusion

The present chapter conveys that Quantum AI can now be used for enhancing the healthcare domain and finding new treatment targets. It results in optimized treatment for various diseases, such as cancer, cardiovascular troubles, and neurological diseases. This chapter, therefore, appeals to every one of us to approach this awesome journey with innovation and ethical values to ensure maximum utilization of all available resources in QAI to improve the health and standards of living in mankind. AI built upon quantum theory has the capabilities to greatly assist in addressing societal issues. For instance, in these current and recent studies, Google explained how quantum computers could be applied to help develop brand-new drugs, new batteries, or advanced nuclear fusion.

The integration of quantum computers into healthcare systems empowers experts to process and understand large sets of data or patient details, such as genetics, medical history, or imaging and histological data, to discern patient-specific and disease-specific treatment plans. Quantum AI workflows improve the analysis of medical imaging and better detection and differentiation of diseases from computed tomography (CT), magnetic resonance imaging (MRI), and positron emission tomography (PET) scans. Quantum AI uses intricate patterns and biomarkers to assist with diagnosis, which is more accurate in terms of the stage of development, and subsequently track the disease progress.

In conclusion, one can state that innovation must not stall, and cooperation with other fields and disciplines strengthens the optimal application of QAI for the benefit of patients and healthcare systems worldwide while maintaining adherence to the highest standards of ethics. Taking into account the goals and principles of QAI as a scientific approach, healthcare systems must become more integrated and concentrate on the opportunities of achieving cumulative research improvement, interprofessional and cross-stakeholder cooperation, and ethical standards enhancement to apply QAI as the tool for developing the novel horizons of medical science, enhancing outcomes of the treated patients, and creating more constructive perspectives for the human's health all over the world.

References

1. Acampora G, Schiattarella R (2021) Deep neural networks for quantum circuit mapping. Neural Comput Appl
2. Deeba K, Balakrishnan A, Kumar M, Ramana K, Venkata Narasimhulu C, Dhiman G (2024) A disease monitoring system using multi-class capsule network for agricultural enhancement in muskmelon. Multimed Tools Appl 1–20
3. Chalapathi MMV, Krishna AV, Gopichand G, Ramana K, Bharathi VC, Sharma N (2023) Prediction of seed germination quality utilizing ensemble-based precision forming. Int J Intell Eng Syst 16(4)
4. Baker S, Xiang W (2023) Artificial intelligence of things for smarter healthcare: a survey of advancements, challenges, and opportunities. IEEE Commun Surv Tutor
5. Bohr A, Memarzadeh K (2020) The rise of artificial intelligence in healthcare applications. Elsevier
6. Gopichand G, Vijaya Krishna A, Ravi Prakash Reddy I, Vandana D, Ramana K, Purshotham P (2022). High-resolution image in painting using generative adversarial networks. In: International conference on information and management engineering. Springer Nature Singapore, Singapore, pp. 715–723
7. Naik DA, Mohana RM, Ramu G, Lalitha YS, Suresh Kumar M, Raghavender KV (2023) Analyzing histopathological images by using machine learning techniques. Appl Nanosci 13(3):2507–2513
8. Reddy CKK, Anisha PR, Hanafiah MM, Pragathi YVSS, Ramana Murthy BV, Madana Mohana R (2023) An intelligent optimized cyclone intensity prediction framework using satellite images. Earth Sci Inform 16(2):1537–1549
9. Alshehri F, Muhammad G (2020) A comprehensive survey of the Internet of Things (IoT) and Ai-based smart healthcare. IEEE Access
10. Ullah U, Garcia-Zapirain B (2023) Quantum machine learning revolution in healthcare: a systematic review of emerging perspectives and applications. IEEE Access
11. Kumar G, Yadav S, Mukherjee A, Hassija V, Guizani M (2024) Recent advances in quantum computing for drug discovery and development. IEEE Access
12. Anisha PR, Reddy CKK, Nguyen NG, Bhushan M, Kumar A, Hanafiah MM (2022) Intelligent systems and machine learning for industry: advancements, challenges, and practices. CRC Press
13. Egger DJ, Gambella C, Marecek J, Mcfaddin S, Mevissen M, Raymond R, Simonetto A, Woerner S, Yndurain E (2020) Quantum computing for finance: state-of-the-art and future prospects. IEEE Access
14. Ullah U, Jurado AGO, Gonzalez ID, Garcia-Zapirain B (2022) A fully connected quantum convolutional neural network for classifying ischemic cardiopathy. IEEE Access
15. Saraswat D, Bhattacharya P, Verma A, Prasad VK, Tanwar S, Sharma G, Bokoro PN, Sharma R (2022) Explainable Ai for healthcare 5.0: opportunities and challenges. IEEE Access
16. Amin J, Sharif M, Gul N, Kadry S, Chakraborty C (2022) Quantum machine learning architecture for Covid-19 classification based on synthetic data generation using conditional adversarial neural network. Cogn Comput
17. VR Allugunti, Reddy CKK (2021) Prediction of diabetes using Internet of Things (IoT) and decision trees: SLDPS. Intell Data Eng Analytics Adv Intell Syst Comput
18. Subbarayudu B, Lalitha Gayatri L, Sai Nidhi P, Ramesh P, Gangadhar Reddy R, Reddy CK (2017) Comparative analysis on sorting and searching algorithms. Int J Civ Eng Technol (IJCIET)
19. Krunic Z, Flöther FF, Seegan G, Earnest-noble ND, Shehab O (2022) Quantum kernels for real-world predictions based on electronic health records. IEEE Access
20. Maheshwari D, Garcia-Zapirain B, Sierra-Sosa D (2022) Quantum machine learning applications in the biomedical domain: a systematic review. IEEE Access
21. Li Y, Tian M, Liu G, Peng C, Jiao L (2020) Quantum optimization and quantum learning: a survey. IEEE Access

22. Yi T, Wang J, Xu F (2021) Binary classification of single qubits using quantum machine learning method. J Phys
23. Sengupta K, Srivastava PR (2021) Quantum algorithm for quicker clinical prognostic analysis: an application and experimental study using Ct scan images of Covid-19 patients. BMC Med Inform Decis Mak
24. Maior CBS, Araújo LMM, Lins ID, Moura MDC, Droguett EL (2023) Prognostics and health management of rotating machinery via quantum machine learning. IEEE Access
25. Gupta H, Varshney H, Sharma TK, Pachauri N, Verma OP (2022) Comparative performance analysis of quantum machine learning with deep learning for diabetes prediction. Complex Intell Syst
26. Wang X, Wang X, Zhang S (2022) Adverse drug reaction detection from social media based on quantum Bi-Lstm with attention. IEEE Access
27. Hossain S, Umer S, Rout RK, Al Marzouqi H (2024) A deep quantum convolutional neural network based facial expression recognition for mental health analysis. IEEE Trans Neural Syst Rehabil Eng

Transforming Water Quality Assessment: A Quantum AI Approach

Akshitha Katta, C. Kishor Kumar Reddy, Thandiwe Sithole, and R. Madana Mohana

Abstract Water quality assessment is one of the fundamental approaches to environmental management, which guarantees the steady provision of reliable water sources for both human consumption, water-intensive industries, and the preservation of aquatic environments. Common techniques of water quality analysis and testing might not be flawed, but they have preconditions like high costs and duration of testing. In the case of water quality assessment, the potential and practical use of Quantum AI is in a position to process and analyse more extensive and complex datasets. This includes current information gathered from different sources like the sensors, satellite pictures among others and past data. Thus, it can find patterns and relations that classical AI does not see, which would allow predicting further changes in water quality and detecting contaminations on time. This paper aims to discuss the basic concepts of quantum computing and its potential use in water quality monitoring and prediction. It subsequently provides a comprehensive review of existing AI-based water quality methods, highlighting their strengths and drawbacks.

Keywords Environmental management · Quantum computing · Classical AI · Water quality methods · Sensors

A. Katta · T. Sithole
Department of Chemical Engineering, University of Johannesburg, Johannesburg, South Africa
e-mail: nastassias@uj.ac.za

C. K. K. Reddy (✉)
Department of Computer Science and Engineering, Stanley College of Engineering and Technology for Women, Hyderabad, India
e-mail: kishoar23@gmail.com

R. M. Mohana
Department of Artificial Intelligence and Data Science, Chaitanya Bharathi Institute of Technology, Hyderabad, India

V. K. Gunjan et al. (eds.), *Cybernetics, Human Cognition, and Machine Learning in Communicative Applications*, Cognitive Science and Technology,
https://doi.org/10.1007/978-981-97-8533-9_11

1 Introduction

Environmental monitoring of water resources is very important to protect the earth's water bodies from sustaining damage. It is a known fact that cleanliness of our Nation's streams, lakes, estuaries, and coastal waters determine the health of man, recreation and the environment. Consequently, with growth in the levels of industrial development, water pollution levels have been equally boosted [1]. In light of the growing encroachment of human activities on the usage of water resources, it is imperative that there are well enhanced surveillance tools that can give appropriate info on the quality of water. Depending on its availability, water can be classified into surface water and groundwater [2]. Both of these forms of water are impure because activities that may occur in residential, industrial, and agricultural areas where different contaminants may be emitted [3]. The conventional approach of monitoring water quality involves physical collection of water samples and subsequent analysis in the laboratory, which may be destructive, labour-intensive and costly, and provides limited spatial and temporal coverage. Water quality can be classified as palatable, contaminated, and infected water [4, 5]. The emergence of what has come to be known as Quality Artificial Intelligence (QAI) presents a viable way to address these downsides. QAI can be connected to sensor networks, satellite remote sensing or citizen science projects to offer high geospatial and temporal resolution water quality information. Besides serving as a source of information that is both continuous and exhaustive, QAI can have a very positive effect on the various aspects of decision making dealing with the water resource. Using sophisticated computations, complex simulations and this is perhaps the most powerful machine learning, QAI can find relationships and correlations that may have been unnoticed qualitatively, thus providing superior insights.

The purpose of this research paper is to find the prospects for the use of QAI in water quality monitoring, operational features, advantages, and disadvantages. It is time to turn to the current status of water quality assessment, the potential of QAI for enhancing monitoring capacities, and how QAI could help with decision-making about water resources management.

The chapter is arranged as follows: Sect. 2 talks about the background research of the topic. Section 3 dwells into the topic of water quality monitoring and its methods. Section 4 discusses the water quality parameters and challenges. Section 5 focuses on Artificial Intelligence in water quality assessment topics such as pros and cons of artificial intelligence, machine learning in water quality and ground water management, neural networks in water quality management. Section 6 discusses quantum computing and its impacts. Section 7 provides a conclusion for the chapter.

2 Background Research

Water quality assessment is an essential activity of management needed to determine the safety of water sources for drinking and other uses, adequate supply of water for cultivation, enough water for industrial processes, and also, the overall health of water resources. Rather than using the traditional parameters that can only be analysed using physical and chemical methods and which although may be accurate are normally costly in terms of time as well as resources. As AI and machine learning interactive methods came into the picture the efficiency, as well as the predictive accuracy, of these processes have enhanced. The application of quantum computing together with AI techniques known as Quantum AI is expected to engrand this field even further due to the computational ability and accuracy it will offer.

Quantum computers are based on the principles of quantum mechanics including superposition and entanglement to solve matrices and linear equated problems far faster than the classical computer. This capability is very important when analysing big data, particularly in the field of water quality where data collection from sensors, satellite images, and historical records is often extensive and has to be filtered and organised. Quantum AI can perform these large-scale computations faster, and solve the problems which classical AI cannot necessarily recognize patterns and correlation. This data processing capacity helps thus make better predictions of water quality and alert the public on the presence of contaminants.

The major use of Quantum AI in evaluating this aspect of water quality is in monitoring and identification of outliers. Fixed and portable water quality sensors placed at different points will be able to monitor water quality parameters like pH, turbidity, temperature and concentration of certain pollution index on a real-time basis. Another advantage is the ability of Quantum AI algorithms to teach the clients how to follow safety standards and alert the authorities when the set limit is reached. This is proactive and one can quickly organise to address issues of contamination hence protecting the public and the environment.

Another area that could be drastically boosted by Quantum AI is the predictability modelling of water quality. Thus, trained by a database of historical water quality and other characteristics of the environment like weather, land use, and industrial production, the quantum machine learning models can predict and estimate the future water quality under a specific condition. Such pessimistic capability will also enable improved mitigation of water pollution and the effects of climate change on water availability. For example, predictive models may help with determination of when and to what extent water treatment should be conducted, where to invest resources and read more about other water management policies that need to be put in place.

Using Quantum AI, various data sets processes are enabled to be integrated, increasing the credibility and coverage of water quality reports. Classic methodologies can provide information based on a narrow set of variables; therefore, they can be of incomplete or at times obsolete conclusions. Quantum AI, on the other hand, can incorporate various types of data from any origination like the satellite and airborne data, ground observation, and citizen science programmes. Overall, this makes better and sound driven information with better understanding of the qualities of available water hence leading to better decisions and policies.

However, this paper has demonstrated that Quantum AI has a considerable opportunity in water quality assessment employing AI while several technical and practical barriers remain a significant issue. Although there has been progress in the field of quantum computing hardware, it is yet to design and realise large scale fault-tolerant quantum computers. However, applying quantum algorithms to the existing AI software platforms and water quality surveillance structure entails a combination of multiple disciplines and creativity. Nonetheless, continuing research and funding in this area are slowly continually enhancing this technology; thus, making Quantum AI a reality.

Therefore, Quantum AI offers great promises for improving the efficiency, accuracy, and coverage of water quality assessment. Thus, its ability to analyse a large amount of data, predict future scenarios, and implement data from various sources can improve the monitoring or real-time conditions, detection of anomalies, and development of predictive models. Despite the limitations, the calls for the advancement and utilisation of Quantum AI to aid in the augmentation of water resources' solvency are not far from deserted, which in effect encompasses enhancing environmental and public health in the long run.

3 Water Quality Monitoring

Pollutant-free water can be defined generally as the absence of pollutants in water, in terms of methods for assessing water quality, sample collection, sample preparation, measurements, and evaluation (Fig. 1). When determining the concentration of organic and inorganic elements in water samples, suspended solids, and bottom material, methods include colorimeters, gas chromatography, AAS (Atomic absorption spectroscopy), ICP-MS (Inductively coupled plasma mass spectrometry), and photometers. Weekly water samples are taken and analysed to determine if the water is safe for domestic, agriculture and industrial uses. These water parameters as well as the water quality index are determined commonly by using conventional analytical methods like qualitative analysis in laboratories and other organisations. However, these methods are cumbersome, time-consuming, and expensive in comparison to the modern instrumental analytical techniques. Traditional methods of water quality monitoring include manual sampling and laboratory analysis which are often labour sensitive, costly and provide limited spatial and temporal coverage. However, innovations in other fields like remote sensing, sensors, and Quality Artificial Intelligence (QAI) have revamped this field. Cursors to high-resolution real-time data acquisition and processing improve the understanding of fluctuations and trends in water quality.

Water quality monitoring system plays a crucial role in maintaining a healthy surface water and groundwater ecosystem. For improvement of aquatic ecosystems, specific measures are implemented. This encompasses determining the chemical contaminants that are discharged into the water body; enhancing the assessment of

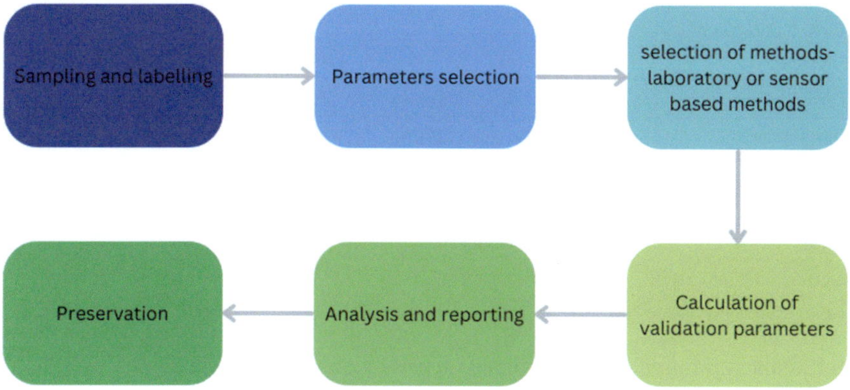

Fig. 1 General flow of procedures for water quality analysis

the ecological effects; and contributing to the enhancement of water quality through raising awareness of the public, communities, industrialists, etc. Water quality can be measured by sampling water and testing them in the laboratory; or by using devices that record data at the same time. Water quality monitoring is therefore the key to water quality management. The water quality monitoring program is in accordance with the local water quality standards. Water quality monitoring schemes are based on water, surface water, households, wastewater management in industrial and other applications, sewage water treatment etc. The important water quality parameters are chosen which are likely to be used for irrigation, industries, domestic and water bodies [5].

4 Water Quality Parameters and Challenges

4.1 Parameters

The water quality parameters are some of the most important variables that are used in the determination of the state and usability of water for human uses, water recreation activities, industrial uses, and environmental uses. It is important to note that these parameters can be subdivided into physical, chemical, as well as biological parameters. The following are the main water quality parameters which are usually analysed (Table 1).

Temperature is a key physical parameter because it determines the metabolic processes of organisms living in aquatic environments as well as solubility of gases dissolved in water. Turbidity relates to the water clarity and the degree to which light can penetrate through the water and support photosynthesis. High water turbidity can be interpreted as pollution in some cases. Colour analysis can also help determine that dissolved or suspended components are present in water solution.

Table 1 Parameters of water quality [6]

Physical	Chemical	Biological	Radioactive
Turbidity	PH, acidity, alkalinity	Bacteria	Alpha emitter
Temperature	Chloride and chlorine	Algae	Beta emitter
Colour	Elements	Virus	
Taste	Hardness	Protozoa	
Solids	DO, BOD, COD		
Electrical Conductivity	Toxic substances		

Chemical properties include the pH level which measures the acidity or basicity of water and is crucial for aquatic organisms. Dissolved Oxygen (DO) is the oxygen that fish and other aquatic life forms require for their existence, Chemical Oxygen Demand (COD) and Biochemical Oxygen Demand (BOD) represent the quantity of organic materials and the level of pollutants in water respectively [7]. Compounds such as nitrogen or phosphorus can lead to eutrophication and toxic algal blooms in water bodies if in excess. Lead, mercury, cadmium, products are all toxic at very low concentration and which are known to bioaccumulate in the food chain. The biomarkers that classify biological parameters are coliform bacteria, which indicate microbial pollution and possible pathogens, and algal dry weight, which considers nutrient concentrations and toxic algal blooms.

An environmental monitoring program monitors the water quality parameters analysed as indicated in the study by the authors and checks them against the benchmark and allowable concentrations specified by the regulatory bodies. local, state, country, and international bodies (Table 2).

Table 2 WHO established standards for water quality parameters [8, 9]

Parameters	Maximum permissible limit(mg/L)
TDSs	500
pH	8.50
Temperature(°C)	25.0
Iron	0.30
Nitrate	50.0
Chromium	0.05
Lead	0.01
Ammonia	0.20
Turbidity(NTU)	5.00
Copper	1.00
Arsenic	0.01
Ammonia	0.50
Phosphate	5.00
Sulfate	400

4.2 Challenges

A major focus in environmental health and consideration is the quality of the water accessible to the plants, animals, and individuals; sadly, it is burdened with many challenges resulting from processes in nature and actions by people. It is crucial for dealing with monitoring and management of the risk itself to understand these challenges well. Here are some of the key challenges to water quality.

a. **Pollution from Industrial and Agricultural Activities**

Pollutants derived from manufacturing and farming operations threaten the source and quality of water. Most industries release toxic effluents, including dissolved metals and chemical solvents in the water systems. Likewise, pollutants in the form of pesticides, herbicides, and fertilisers used in farming pollute the rivers and lakes through direct discharge and raise levels of chemical nutrients. Eutrophication is a water-borne phenomenon caused by nutrients in the water such as nitrogen and phosphorus from sources such as fertilised agriculture, livestock farming, and sewage. This results in production of toxic algae that reduces oxygen level in the water and poses threat to marine livestock as well as affects humans and animals adversely [10, 11].

b. **Urbanization and Population Growth**

Population increases and urban development cause water pollution since storm water and domestic sewage finds its way into water bodies. Faster development of the urban areas accompanied by provision of the hard surface dust increases the rate of surface runoff while bearing pollutants of oils, metals, and other debris to our water sources. Cities grow, and consequently, the sewage is produced in higher quantities. Should treatment services not be available, untreated sewage pollutes water sources with pathogens and organic loads, and thus becomes detrimental to human health and detrimental to water quality [12, 13].

c. **Climate Change**

Some of the effects of climate change include alteration in the temperatures and unfavourable weather conditions cause changes in water parameters and availability. The global climate change impacts aquatic environments leading to thermal stratification of the water bodies, pollution concentration, dissolved oxygen concentration, and species distribution. Storms, floods, and droughts increase the challenges to water quality since water is easily contaminated by pollutants when conditions are severe. More pollution is caused by floods since more pollutants are added to water sources, whereas in times of drought, pollutants are more concentrated because there is less water [14].

d. **Emerging Contaminants**

Anthropogenic compounds and compounds of emerging concern, PPCPs, and microplastics are frequently found in watersheds. These are microbial impurities that originate from the use and disposal of products containing these elements harming water organisms and the human population. Microplastics which are solid, or small plastic pellets originating from human products and industrial discharges impact marine life and can filter into the food chain with other hazards.

e. **Pathogen Contamination**

Bacterial and viral related contamination by pathogens is dangerous to human and animal health. These sources include; sewerages, drainages, and acknowledgement to wild life. The identified contaminants pose a massive threat if not controlled and treated appropriately from the public water supply system.

f. **Salinization**

These effects of water deterioration stem from such irrigation activities and industrial processing. The method of irrigation and the disposal of saline industrial wastes result in water getting rich in sodium ($Na+$) which is not fit for human consumption or use in irrigation.

g. **Habitat Destruction and Biodiversity Loss**

Biological diversity loss and devastation of natural habitats are some of the potential effects realised by manipulating the natural water systems. Interference in the ecosystems for purposes like constructing dams, controlling water channels, and draining marshes lead to eradication of species, upsetting water regimes and cleaning mechanisms hence leading to water pollution and deterioration of the environment.

h. **Inadequate Monitoring and Data Management**

Lack of monitoring and data collection and analysis limit water resources management. There are few samples taken including those with low frequency and inadequate coverage; however, organisational structure stays the same along with Technology. This leads to inefficient and costly procedures, as well as minimal temporal and spatial data; thus, limiting its near-real-time capability to quantify changes in water quality.

It is important to understand that the above-mentioned water quality challenges are well complexed in that they require a packaged strategy. New methods of monitoring, for example, Quality Artificial Intelligence (QAI) can help to employ the optimal monitoring methods using real-time data acquisition. This is where one must first recognize the need of enhancing the existing policies and laws that will help in decreasing the amount of pollution and the consumption of water. Increased knowledge and information among the public can help to create good practices out of environmentally friendly initiatives. IWRM involves the efficient utilisation of water resources and comes up with an all-encompassing plan for water utilisation which has linkages to land use and, therefore, the health of ecosystems. Realising these issues would go a long way in addressing future threats to water quality and lack of clean water in the following generations.

5 Artificial Intelligence in Water Quality Assessment

The utilisation of Artificial Intelligence (AI) in water quality assessment is at an early stage yet is showing vast improvement as a strategy for enhancing exactness of examination, revelations, and even expectations. The two main natural types of AI systems are the machine learning algorithms and the neural networks; with the help of these systems, AI can collect data from various types of sensors and identify contaminants and anomalies with very high accuracy. Artificial intelligence models examine the architecture for the device management and monitoring water quality trends (Fig. 2).

Moreover, AI allows the compatibility of different sources of information, raw data like satellite images, meteorological data, and water quality records to get a more accurate picture of the water ecosystem. Using big data along with sophisticated analysis such as deep learning and reinforcement learning, these patterns and relationships often overlooked by traditional methods of analysis can be identified. It makes it possible to determine point and nonpoint source emission, variations in emissions throughout the year, and effects of climate change on water supply. Furthermore, this review establishes that AI-based decision-making tools, particularly DSSs, may provide policymakers and water resource managers with valuable insights into how to improve water quality and adhere to the relevant legal requirements. It can be concluded that AI solutions are versatile and can be applied to various geographical locations and different bodies of water ranging from a lake to a

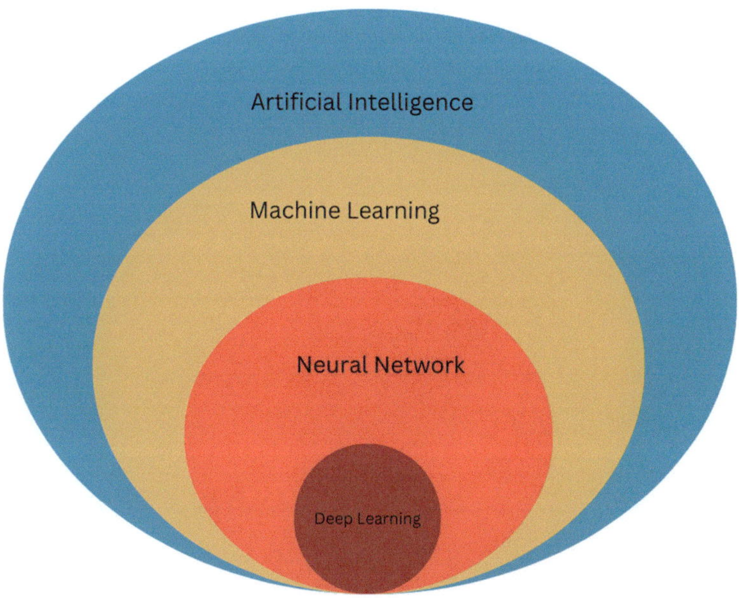

Fig. 2 Schematic diagram for the relationship between artificial intelligence and deep learning

river basin. It is expected that with ongoing advances in AI technology, water quality assessment will be further enhanced and a much smarter and more sustainable way of managing water resources will be provided in the future.

5.1 Pros and Cons of Artificial Intelligence in Water Quality Monitoring and Modelling

Artificial intelligence (AI) techniques have been envisaged to be capable of Surface water quality monitoring and record benefits including reduction of cost, sustainable water treatment, improved water, and environmental quality [15] on the other side AI have disadvantages including need for big data sets and bias [7]. The opportunities and threats of the application of AI techniques in the areas of water quality monitoring and modelling are presented in Fig. 3.

Machine learning algorithms that form the general area of AI can scan large amounts of data and look for changes in water quality that might remain unnoticed by other analytical methods. With the help of real-time monitoring, changes in the water quality parameters can be identified instantly, and the subsequent actions towards pollution events or the changes in the environment conditions can be provided in a timely manner. All these aspects can be performed automatically by AI algorithms, thus easing the dependence on labour when carrying out monitoring activities and increasing the effectiveness of the initiatives. One of the applications of the AI models in the analysis of water quality trends is the ability to apply an early warning system regarding anticipated instances of pollution, which helps the authorities take

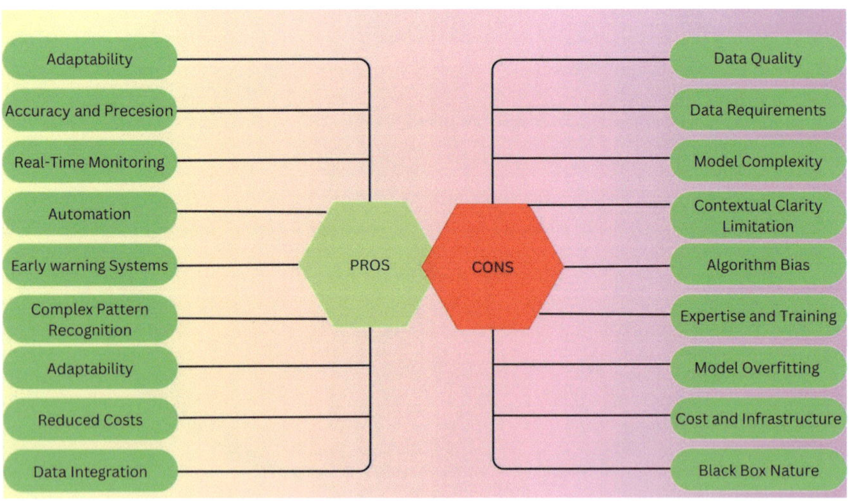

Fig. 3 Pros and cons of artificial intelligence in water quality monitoring and modelling

preventive measures to avoid pollution. These methodologies prove beneficial in identifying significant and complex trends in water quality data that may not be easy to identify through simple human understanding in order to identify the source of pollutants and different affecting attributes of water quality. Thus, as larger amounts of data are collected, AI models change and advance the accuracy of their predictions as they become more defined. While entry level costs for integrating AI systems can be relatively high, these costs create a progression of a way of attaining a gradual reduction to expenses due to the fact that efficient monitoring processes are carried out using these systems which do not require frequent intervention.

Other than these advantages the following shortcomings come with the application of AI techniques in monitoring and modelling of water quality. AI models especially depend a lot on the input data that should be clean and an accurate representation of the real world. Therefore, problems related to input data include the fact that if the data provided is wrong or prejudiced, the predictions and calculations of the AI models will also not be accurate. Many AI techniques and specifically the DLs require large data for training since they work based on machine learning. Situations might emerge when there is insufficient history data to work with. Some of the AI models present today such as the deep neural networks are complex and hence their interpretation is equally complex. This condition may have resulted in making comprehension and trust in such outcomes difficult particularly within the context of regulation or decision making.

In general, AI methods reveal a lack of generalisation in the domain of the contextual understanding of certain occurrences or patterns of data. If not designed and tested carefully, the AI algorithms may learn or reinforce the bias of the training data and bias could seep into the evaluations where AI is used especially on domains with social implications. Cash is easy to obtain but implementing the methods call for people who have adequate skills in water quality management and AI methods which could be scarce in some regions of the globe. A potential weakness of the approach has been seen as the overfitting problem; the AI models perform very well on training data and not so well on new data. Also, AI models are vulnerable to external factors of the model, for instance changes in climate, usage of land or infrastructure which might not be based on the training data.

5.2 *Machine Learning in Surface Water Quality Monitoring*

Machine learning in water quality analysis has been proven to help enhance determination of quality predicting methods besides helping in identifying the sources of pollution and also the methods of handling water [16]. Use of machine learning (ML) in the domain of water quality monitoring indicates novel progress in environmental science that provides the high productivity, accuracy and efficient way as compared to conventional approaches. Previous techniques of water quality monitoring are basically simple sampling and analysis by tests, which are time consuming, and involve a lot of manpower and resources, and only covers limited and random areas at various time intervals. Mechanised learning algorithms on the other hand can take huge and complicated information from different sources including remote sensors, satellite, and historical records hence can monitor the surface water bodies comprehensively and continuously.

The most relevant use of machine learning in the context of surface water quality is, therefore, on the forecasting of water quality indices. The commonly used techniques for the water quality forecasting are artificial neural networks (ANNs), support vector machines (SVMs), and Random forests which when trained on historical data of water quality can predict future values. For instance, in detail, the source's predictiveness can inform approximate concentrations of nitrate, phosphate, heavy metals etc., which in turn, can be addressed or treated early before they lower the water quality.

Real time water quality monitoring and anomaly detection are also facilitated with the aid of machine learning. Monitors mounted on the surface water bodies provide real time measurements of such variables as pH, temperature, cloudiness and DO. This data can be fed into the machine learning algorithms that can analyse it in real-time, and detect the patterns and the exceptions that can be associated with events such as pollution or system failure. It also includes processes like clustering methods, anomaly detection, time series for instance can help in coming up with early alarms to the authorities since the changes in water quality are automatic in nature, hence enabling the authorities to come up with necessary measures that can help cushion ecosystems and human health.

Indeed, through machine learning, one can combine and process multiple sources of data to better comprehend the fluctuations of the surface water quality. To enhance water quality analysis, a fusion of data collected by RS technologies, ground stations, and CS projects with the help of machine learning algorithms will be used. They also help in making and formulation of policies on the grounds of decreased pollution policies and efficient management of existing pollution levels as well as evaluation of effectiveness of intervention.

Consequently, the application of machine learning in the era of climate change also has utility in the analysis and modelling of effects on the quality of surface water. Environmental changes such as precipitation modifications, changes in temperature and uprising in the use of land can impact on methods of carrying pollutants in water masses. ML can mimic different climate conditions and their impact on water quality and thus can be used to develop better responses. For instance, models can forecast the propensity of high heat resulting in algal blooms and the likelihood of nutrient change to cause a similar consequence, thus, help to prevent them.

However, using machine learning in surface water quality management is not without the following challenges. Since big data is always the input to the machine learning models, data quality and availability are two challenging problems that must be dealt with properly in order to deal with big data effectively. The irregular or limited information gives the chance of inaccurate anticipation and evaluation coming to be. Secondly, the systems governing water quality are complicated and other environmental factors therefore require that more complex models are developed. ML is crucial for the management of water quality and addressing these challenges requires a combined interdisciplinary approach, as well as the improvement of the methods used in ML.

5.3 Neural Network Prediction for Water Quality Management

Artificial neural network models, which are subprograms of the machine learning algorithms, are also widely applied in water quality management as they can deal with the multifaceted and statistically nonlinear interrelationships between different water quality indices and environmental factors. These Models tend to base their design and functionality on the human brain and are made up of layered nodes called neurons which are capable of being trained through data. The use of artificial neural networks (ANNs) has been shown to be effective in estimating water quality parameters, such as PH, dissolved oxygen, and turbidity with high accuracy [17].

The application of neural network models in water quality management has been reported in the literature for the purpose of predicting the water quality indices such as WQI and to find out the contribution of water quality parameters towards the WQI [18]. The most important conventional use of neural network models in the context of water quality management is the identification of some of the most important pointers of water quality. For example, artificial neural networks (ANNs) can be used for forecasts of the future concentration of pollutants like nitrates, phosphates, and heavy metals as ANNs are trained based on the historical data. It is possible for the models to accept diverse input parameters such as the meteorological data, the land use, and industrial discharges that have effects on the water quality.

Neural networks also have excellent capabilities in the online monitoring and exception analysis in water quality management. Through the proposed approach that utilises data collected from different sensors placed in water bodies, an implementation of a neural network will solve patterns and will be reasonable to conclude on events such as pollutants or system breakdowns. Methods like convolutional neural networks or recurrent neural networks are very effective to process time series data and detect abrupt changes of water quality indicators in due time. This real time responsiveness can help to quickly deal with possible threats affecting the ecosystems and the health of human beings.

However, the integration of various data sources and the analysis of data streams in terms of prediction and real-time monitoring cannot be performed with artificial neural networks without referring to the role of the latter in the processing of data from different sources. Elements of Water Quality Management include data from remote sensing technologies, in-situ measurements and even data collected from citizens' supported programs. Neural networks can generate this heterogeneous data, and as a result give more than a clear picture of the water quality status. This integration is crucial for derivation of efficient management approaches since it enables identification of polluting sources, evaluation of impacts of types of intervention, and determination of an optimal use of resources.

The other benefit of the neural network models is that they can analyse large, and more especially, complex datasets of water quality. Several characteristics and water quality parameters are usually present and variable interdependencies are involved in the measurement data which themselves are not easily amenable to modelling by classical methods. Specifically, the neural networks, and deep learning that work with multiple layers of neurons, can work with these complexities and reveal features that might be hidden from the observer. I can assert that this capability augments the reliability and validity of the water quality assessments and decision-making processes.

However, the use of the neural network models in water quality management comes with several difficulties. One drawback is the requirement for vast and high-quality databases for training purposes among other factors. For water quality, the availability of data could be lukewarm or distorted in many areas hence reducing the applicability of neural networks. Furthermore, the high degree of abstraction of neural networks and the fact that the internal structure and function of the model are not easily understandable present issues when it comes to identifying the processes governing the results. To address these difficulties, there is a need to enhance data gathering and introduce other naive models of the neural network.

5.4 Machine Learning in Groundwater Quality Management

Intelligent automated systems have become a needful tool in the management of groundwater quality since it deals with very efficient ways of tracking and forecasting the quality of water and preventing water pollution. Conventional approaches to estimating contamination of groundwater can be a very time consuming and labour-intensive process which involves sampling and testing of the samples in the laboratory. Such improvements can be brought by ML algorithms on these processes, by analysing massive amounts of data from different sources, and then defining the quality of groundwater. This subtopic focuses on the various uses and advantages of applying ML in the aspect of groundwater resources.

The two most typical uses of ML in the management of groundwater quality are In this area, the principal utilisation of ML is to predict the degree of pollution. There are many factors which can affect the groundwater system: geologic structures, humans and his activities, and climatic conditions. Decision trees and random forest have been applied and can accurately predict water quality parameters including nitrate and arsenic and can determined the most impact factors affecting the groundwater quality [19]. These models can accommodate all forms of variables and in this way, contributes to the accomplishment of a vast analysis for the stagnation of possible contaminant incidences.

The application of machine learning techniques with geostatistical methods has demonstrated to have a positive impact in the enhanced mapping of groundwater quality; the information obtained in this technique being very useful in the sustainable management of groundwater resources and the prevention of water pollution [20]. Housing such data in a database can facilitate easy clustering, using time-series analysis and anomaly detection algorithms, whereby unusual patterns concerning pollution parameters in groundwater are detected and appropriate investigations conducted. This capability is most useful to protect sources of drinking water and safeguard the health of the public.

Likewise, ML is used in identification of the sources of contamination of water most especially the groundwater. Pollution of water can occur from any of the following activities which include excessive use of fertilisers, use of dyes and other industrial wastes and seepage from waste dumping sites. This, in turn, implies that using ML tools the potential vectors of contamination, as well as contamination pathways, can be identified using spatial and temporal characteristics of the data. For instance, when incorporating the use of GIS with different ML algorithms, it is possible to determine the link between usage of land and quality of groundwater to help in the formulation of effective solutions. Isolating the causes of contamination are crucial in deploying controls and other treatment processes for improvement.

Another important use of ML in the management of groundwater quality comprises network optimization of the monitoring activities. The monitoring of groundwater is always considered difficult as the system forming is highly complex and is heterogeneous in nature. Monitoring wells can also be placed depending on the results derived from spatial data so as to have leakage areas that are frequently monitored by the ML algorithms. This optimization makes certain that the established monitoring network will open a veil on the status of groundwater quality at the least cost possible. Monitoring networks are critical in identifying sources of contamination and coming up with decisions that early detect the contamination.

Rates obtained for the modelling and simulation of the groundwater quality with the help of ML models. The usage of groundwater flow and transport models is to predict the fluid and solutes' movement through the aquifer. The combination of ML with the conventional approaches to simulate the dynamics of groundwater can improve the model's predictive performance as well as computational time. For example, integration of the physical-based models with ML algorithms will result in more accurate representations of the interactions within the groundwater systems. Such models can be employed to predict different conditions for example the effect of change in land use or change in climate on quality of groundwater, hence useful when it comes to managing the resource.

Nevertheless, some challenges are faced when using ML in managing groundwater quality. Among these, data availability and quality can be cited as one of the biggest issues to consider. Another pro is that training and validation of the ML models depends on having a large quantity of high-quality data. The main weakness of the present study is that there could be a lack of, or irregular, information about the quality of the groundwater in many regions, which negates the accuracy of the ML models. To address this challenge, better ways of collecting and integrating data from various sources, such as, remote sensing and citizen science, have to be developed. Further, there are pros and cons of integrating big data analytics and ML, and in addition some of the ML models are complex and ‘black box,’ which give rise to interpretational problems; hence, the need for developing new and more interpretable ML approaches.

6 Quantum Computing

Fundamentals

Quantum computing is different from classically trained computing since it employs quantum mechanics concepts to process information in different manners. Quantum computing at its basic level utilises quantum bits or qubits represented by quantum systems, with the capability of being in state 1 or state 0 at the same time using superposition. This enables quantum computers to solve many problems at the same time or potentially offer a quadratic boost in performance for a specific range of problems.

The idea of quantum parallelism that lets a quantum computer make many calculations at one time is one of the most important characteristics that helps quantum computers solve some problems faster than classical computers [21]. This integration enables quantum computers to perform computations that are much faster than classical computers. Entanglement allows the qubits to function as one coherent and highly integrated unit, effectively amplifying the processing capacity and runtime of quantum computations.

The quantum circuit model which gives the representation of quantum computation in the form of quantum gates is a good approach in designing and analysing quantum algorithms [22]. These gates control the state of qubits using several quantum operations among them being the Hadamard gate that prepares a qubit for entanglement and the CNOT gate that creates entanglement between two qubits. When you interconnect these gates in circuits, you can then create quantum algorithms that solve certain problems. Some well-known quantum algorithms are Shor’s algorithm for factorization of big numbers and Grover’s algorithm for searching in unsorted databases, which show that quantum computing can demonstrate superiority over classical algorithms.

There are still paramount obstacles in quantum computing, specifically, the obstacles in qubit coherence and quantum noise. Qubits are innately fragile which mean that they are easily affected by factors in their environment and therefore may decohere leading to error computation. Quantum error correction is a relatively young field that seeks to provide ways of handling and correcting quantum data to enable more accurate quantum computations.

These complications are currently met by methods like error-correcting codes, and fault-tolerant quantum computing. The possibilities involved in quantum computing have immense potential and would dramatically change the existing systems. Another application of quantum computers is in the cryptography field where they pose a threat to standard ciphers used in encryption, hence the invention of quantum resistant cryptography. Quantum simulations in the context of materials science and chemistry can result in the identification of new materials and drugs based on the model of interacting molecules. In complex logistics and supply chain contexts, quantum algorithms may yield more efficient solutions, thus preserving resources and time.

Quantum computing is relatively a new technology and is not fully developed yet, there are a plethora of technical and theoretical barriers to break. Stated otherwise, while they are yet to be realised, these considerations are generating significant investment and research. Quantum hardware, software, and error correction developments are making real-world quantum computing a possibility. In such dynamics, quantum computing is expected to open up fresh approaches to science, technology and other spheres and redefine the ways of solving problems and handling large datasets.

Potential

Applications of quantum computing are vast in every scientific discipline since quantum computers can solve many problems that are practically impossible for classical computers to handle today. There is capability of quantum computers to provide most accurate simulations of quantum systems including molecules and chemical reactions; a breakthrough in chemistry, material science [23]. For classical encryption methods, such as RSA, the fact that large numbers are difficult to factor remains within the reach of classical computers, though when numbers are big enough. However, quantum computers can perform the factoring algorithm exponentially faster than classical computers and can thus potentially break current encryption techniques, requiring new systems that are quantum secure. Based on quantum mechanics, in chemistry and material science, quantum computing can greatly enhance the capability of simulation of molecular behaviours. Currently, classical computers are incapable of simulating quantum systems efficiently because the difficulty increases exponentially with respect to the dimensions of the system in question. While modelling these systems, quantum computers are relatively much more efficient compared to classical counterparts. This capability has the potential to change research and discovery in fields such as material science and pharmaceuticals by speeding up the discovery of new substances and materials, as well as designing new molecules with the required characteristics for the purpose of applying them to industries such as medicine and renewable energy. Another great place for quantum computing to shine is in optimization problems which are very common in fields like logistics, supply chain and finance. Optimization problems are frequently used to find the best state of N choices from a huge array of other possibilities, which takes quite a lot of computer time for conventional algorithms. Some of the problems that can be solved using quantum algorithms like Quantum Approximate Optimization Algorithm (QAOA) include the following and therefore can lead to a lot of saving and increased efficiency of the various industries involved. For instance, one can redesign delivery routes, proper choices in financial portfolios, and resource optimization tasks, using quantum computing.

In the field of AI and machine learning, quantum computing can help scale the algorithms by helping to process data faster and incorporate better methods of optimization algorithms. Other quantum machine learning models include the Quantum Support Vector Machine and the Quantum Neural Networks, which are commonly used to enhance the velocity of machine learning models and increase the reliability of the results. This can result in better data AI models that can solve problems involving greater quantities of data and increased problem sophistication in areas such as natural language processing, picture recognition, and prognosis analysis. Quantum computing can also change experimental science to its very core in many ways. In physics, quantum computers can run simulations at a precision that is beyond the capabilities of current computers, thus allowing for better understanding of the principles governing the universe. In biology, there is where quantum simulation can be useful in helping to comprehend and demystify biological phenomena, and aid in the discovery of new concepts in areas of protein folding and genetic science. High precision computations often provide new ideas and can lead to the development of new theories and to new valuable applications in many scientific fields. However, now it is a relatively young field and there are numerous issues that must be addressed to allow for the construction of large-scale practical quantum computers. Some of the major concerns like qubit coherence, error correction, and scaling are required to be solved to know the potential of quantum computing. But constant technological progress coupled with continuous research and development where there is increased funding from both government and other stakeholders indicate progressive improvements of the field. As these challenges are addressed, the cutting-edge capability enabled by quantum computing will rise from the horizon and radically change the technology, science, and industry of the world.

6.1 Impacts of Quantum AI on Water Quality Management

Quantum AI, a relatively new frontier that integrates the ideas of quantum computing and AI, may be considered as the game changer when it comes to water quality management. Quantum AI and IoT application in water quality monitoring has the benefit of detecting the quality of water at the same time and predicting parameters in order continuously [24]. It is through this advancement that several areas of water quality assessment and analyses will undergo changes and developments concerning monitoring, predicting, decision making on remediation and rectification methods.

In WQM, Quantum AI also improves predictive modelling, which has one of the most significant effects. In traditional AI methodologies the required data can be overpowering owing to the nature of studies done in water quality, therefore the need for more advanced AI models. These datasets are better handled by Quantum AI and thus the creation of accurate predictive models is made possible. These models can predict conditions of water quality with reference to weather changes, industrial effusion, farming discharges, and results of climate change impacts. AI applied to quantum computing is useful for large-scale pattern recognition and thus predicting water contaminations, especially diseases, so that health sectors can intervene and stop widespread contamination [25].

The application of Quantum AI in water treatment involves the enhancement of the rate of processes and decrease of energy used and wastes produced hence the need for efficient water management systems. Quantum AI can analyse data obtained from several sensors and in a very brief timeframe, as such the program can easily detect even minor variations to the water quality parameters. This capability is very important as it enables the early identification of pollution events or system failures. For example, with Quantum AI, even slight changes in chemical constituent, temperature, or even the pH of the water and any other parameter can be quickly and accurately detected and notified by the relevant water management offices. Failure to control contaminants implies that they will spread fast and harm the ecosystems and human beings; hence the need for quick detection and response.

Quantum AI is also another concentration area that has another critical effect; namely, the enhancement of data fusion from various approaches. Water quality management pertains to all the data originating from in situ platforms, satellite data, weather stations, and archive data. Quantum AI will be able to integrate this disparate information in a superior manner as compared to classical AI and offer a meaningful analysis of the quality of water. This improvement of the structure of the picture adds to a better recognition of both components and relations in complicated water systems covering various aspects and assisting in developing better decisions. The form of artificial intelligence that Quantum AI uses allows inferring conclusions from the data received from different sources that may not be recognized with the help of conventional techniques.

In terms of applying Quantum AI, there is also a possibility of expanding the modelling of water quality. The conventional approaches employed in various models suggest that some of the models involve approximations and assumptions because of the limitation in computational power, hence making the models less accurate. However, these models have limitations that can be solved by the Quantum AI for creation for further effective and accurate models. These models can then predict the behaviour of pollutants in water bodies better, by regarding more factors and interrelationships. Thus, enhanced modelling abilities can be applied to efficient handling and treatment of contaminated waters to sustain water supply sources.

The use of Quantum AI in water quality equally encompasses the management of resources for remedial actions. Thus, quantum optimization algorithms used in problems associated with the proper distribution of resources for monitoring and addressing them. For instance, with Quantum AI, it becomes possible to find the best configuration for sensor placement for the best coverage and lowest expenses, or to design good remediation plans that take into account the given constraints and goals. Therefore, certain processes help Quantum AI to improve the outcomes of the implementation of the water quality management strategies and protocols.

Quantum AI in water quality management also can offer more efficient solutions when it comes to distribution network, decreasing the leakage rates [26]. Quantum computing technology is not very old, though significant progress has been made in developing the technology; however, true, feasible, and versatile quantum computers have not been created. Of course, there is also a need for specialised quantum algorithms for water quality monitoring For this, the development of new specific quantum algorithms for other specific applications also implies cross-disciplinary work and imagination. Regarding the former, the issues relating to the dataset, availability and interpretability of Quantum AI models are quite significant. However, current research and advances in quantum computing and artificial intelligence are being worked on round-the-clock to bring Quantum AI into existence.

Thus, for Quantum AI it is possible to state that with the help of development of prediction and monitoring of systems, data integration, and effective use of the existing resources it can significantly increase the efficiency of the quality of the management of water. It might employ big data to acquire more accurate analyses concerning water and to control this resource more effectively than the classical means of AI. In addition to these factors, the progressive invention and application of the Quantum AI will be of great aid in boosting the sustainability of water resources and the coverage of the significant future sources that are dimensional to the ecosystems and society.

7 Conclusions

This study aims at presenting the possibility of using the combination of quantum computing and AI in improving the water quality assessment approaches. Although the standard procedures for water quality assessment and quantification have been efficient, they have their drawbacks in relation to time consumption and reliability in case of increased and diverse amounts and types of data. This paper shows that with the use of quantum AI, those limitations can be overcome with the use of quantum computing to evaluate the amount of data to be processed.

On that account, it remains clear that the capacity of Quantum AI to manage and process big-dataset data concerning the atmosphere and climate within both speed and accuracy is a revolutionary incremental step in environmental surveillance. Therefore, with the help of QML, one can find previously hidden relationships and dependencies in water quality data, and make better predictions and analysis. With this improved analytical capability, improved decision making for environmental stewardship and policy can be obtained to keep the water resources of high quality for the good health of the general population and the sustainability of ecosystems.

Furthermore, incorporating quantum AI in water quality monitoring transcends to other areas of environmental science discipline. It creates new possibilities in research and development, promoting the new ideas regarding effective solutions to the other environmental issues. The synergy of quantum computing with AI and environmental science plans displays the future research as the interdisciplinary collaboration area for creation of innovations and helpful use in many fields.

Thus, our study aims at stressing the functions of quantum AI in the further advancement of water quality assessment. Quantum AI can provide a more stable, precise, and effective solution to monitor and regulate water resources superior to the conventional way. With quantum computing's technology and AI in progress, there are many possibilities that may be revealed in the future regarding the development of better ways of analysing and preserving the valuable environmental resources. Thus, this type of research forms the basis of efforts to address issues of water quality management sustainability and robustness using quantum AI.

References

1. Rana R, Kalia A, Boora A, Alfaisal FM, Alharbi RS, Berwal P, Alam S, Khan MA, Qamar O (2023) Artificial intelligence for surface water quality evaluation, monitoring and assessment. Water. https://doi.org/10.3390/w15223919
2. Ahamad KU, Raj P, Barbhuiya NH, Deep A (2019) Surface water quality modelling by regression analysis and artificial neural network. In: Kalamdhad A, Singh J, Dhamodharan K (eds) Advances in waste management. Springer, Singapore
3. Allawi MF, Jaafar O, Hamzah FM, Abdullah SMS, El-shafie A (2018) Review on applications of artificial intelligence methods for damandreservoir-hydro-environment models. Environ Sci Pollut Res
4. Ashok L, Doug H, Tribeni S (2006) Application of CCME water quality index to monitor water quality: a case study of the Mackenzie River Basin, Canada. Environ Monit Assess
5. Azad A, Karami H, Farzin S, Saeedian A, Kashi H, Sayyahi F (2018) Prediction of water quality parameters using ANFIS optimised by intelligence algorithms. KSCE J Civ Eng
6. Geetha M, Bonthula S, Al-Maadeed S, Al-Lohedan H, Rajabathar JR, Arokiyaraj S, Sadasivuni KK (2023) Research trends in smart cost-effective water quality monitoring and modelling: special focus on artificial intelligence. Water
7. Kumar A, Singh R, Kumar P (2022) Artificial intelligence and machine learning in water quality management: a review. J Water Health
8. Ceccaroni L, Velickovski F, Blaas M, Wernand MR, Blauw A, Subirats L (2018) Artificial intelligence and earth observation to explore water quality in the Wadden Sea. Earth Observ Open Sci Innov
9. Wang P, Yao J, Wang G, Hao F, Shrestha S, Xue B, Xie G, Peng Y (2022) Exploring the application of artificial intelligence technology for identification of water pollution characteristics and tracing the source of water quality pollutants. Sci Total Environ
10. Cherian AK, Poovammal E, Philip NS, Ramana K, Singh S, Ra IH (2021) Deep learning based filtering algorithm for noise removal in underwater images. Water 13(19): 2742
11. Revathi A, Kaladevi R, Ramana K, Jhaveri RH, Rudra Kumar M, Sankara Prasanna Kumar M (2022) Early detection of cognitive decline using machine learning algorithm and cognitive ability test. Secur Commun Netw 1:4190023
12. Jhaveri RH, Revathi A, Ramana K, Raut R, Dhanaraj RK (2022) A review on machine learning strategies for real-world engineering applications. Mob Inf Syst 1:1833507
13. Farhana B, Munidhanalakshmi K, Mohana RM (2021) Predict diabetes mellitus using machine learning algorithms. J Phys: Conf Ser 2089(1):012002. IOP Publishing
14. Niraja KS, Keerthi Chennam K, Madana Mohana R, Venkata Siva Nandan G (2021) A model to predict malicious data in smart home applications using deep learning algorithms 245–249
15. Wang X, Zhang J, Li Z (2022) A review of artificial intelligence applications in water quality monitoring. Water Res
16. Zhou J, Wang Y, Xiao F et al (2018) Water quality prediction method based on IGRA and LSTM. Water
17. Hu Z, Zhang Y, Zhao Y et al (2019) A water quality prediction method based on the deep LSTM network considering correlation in smart mariculture. Sensors

18. Wu J, Wang Z (2022) A hybrid model for water quality prediction based on an artificial neural network, wavelet transform, and long short-term memory. Water
19. Kumar P, Kumar D, Mandal S et al (2022) Prediction of groundwater quality using machine learning algorithms in a rural area of India. Environ Sci Pollut Res
20. Nalbant P, Öztürk İ (2020) Machine learning and geostatistical approach for groundwater quality mapping in a semi-arid region. J Hydrol
21. Shor PW (1994) Algorithms for quantum computers: discrete logarithms and factoring. In: Proceedings of the 35th annual IEEE symposium on foundations of computer science
22. Kitaev AY (1997) Quantum computations: algorithms and error correction. Russ Math Surv
23. Aspuru-Guzik A, Walther P (2012) Photonic quantum simulators. Nat Phys
24. Miller M, Kisiel A, Cembrowska-Lech D, Durlik I, Miller T (2023) IoT in water quality monitoring—are we really here? Sensors
25. Li M et al (2020) A machine learning approach for predicting waterborne disease outbreaks using water quality data. Water Res
26. Mounce SR et al (2017) A review of machine learning applications in water distribution systems. J Water Resour Plan Manag

Unleashing the Potential: Soft Computing in the Era of Industry 5.0

Nishitha Pujala, CKishor Kumar Reddy, Karthik Pujala, and Srinath Doss

Abstract The advent of Industry 5.0 heralds a new era characterized by heightened connectivity, automation, and symbiotic interaction between humans and machines. Within this transformative landscape, the application of soft computing methodologies emerges as a critical driver of innovation, offering a flexible and adaptive framework to address the intricate challenges posed by Industry 5.0 dynamics. Soft computing, encompassing diverse techniques such as fuzzy logic, neural networks, and genetic algorithms, presents a promising avenue for optimizing processes, enhancing decision-making capabilities, and fostering creativity across various industrial domains. This research endeavors to explore the multifaceted role of soft computing within the context of Industry 5.0, aiming to unravel its transformative potential and delineate practical implications for industry stakeholders. By synthesizing insights from a comprehensive review of literature and analysis of pertinent case studies, this paper aims to elucidate the diverse applications, inherent benefits, and prevalent challenges associated with the integration of soft computing methodologies in Industry 5.0 settings. Through an in-depth examination, the study seeks to provide nuanced insights into the mechanisms through which soft computing techniques can contribute to the optimization of industrial processes, the enhancement of decision-making processes, and the facilitation of innovation across diverse sectors. Furthermore, this research endeavors to offer forward-looking perspectives on the future trajectories of soft computing in Industry 5.0, identifying emerging trends, potential areas for further exploration, and pragmatic recommendations for researchers, practitioners, and policymakers alike. By synthesizing empirical evidence with theoretical insights, this paper seeks to contribute to a deeper understanding of the transformative potential of soft computing in driving sustainable

N. Pujala (✉) · C. K. Reddy
Stanley College of Engineering and Technology for Women, Hyderabad, India
e-mail: nishithapujala2@gmail.com

K. Pujala
Malla Reddy University, Hyderabad, India

S. Doss
Faculty of Engineering and Technology, Botho University, Gaborone, Botswana
e-mail: srinath.doss@bothouniversity.ac.bw

V. K. Gunjan et al. (eds.), *Cybernetics, Human Cognition, and Machine Learning in Communicative Applications*, Cognitive Science and Technology,
https://doi.org/10.1007/978-981-97-8533-9_12

growth and fostering resilience in the era of Industry 5.0. Through this holistic examination, the study aims to provide a comprehensive framework for leveraging soft computing methodologies as a catalyst for navigating the complexities of Industry 5.0 and unlocking new frontiers of industrial innovation and progress.

Keywords Soft computing · Fuzzy logic · Evolutionary algorithms · Neural networks · Machine learning · Innovation · Efficiency · Optimization · Data-driven decisions · Intelligent systems · Human–machine collaboration · Smart manufacturing · Data analytics

1 Introduction

As we enter the era of Industry 5.0, when humans and robots coexist harmoniously in a highly integrated system, the need for efficiency and creativity has never been higher. There are a lot of potential for advancement in a variety of areas during this time, as the boundaries between the physical and digital spheres grow increasingly hazy. This is a fundamental transition. The integration of soft computing approaches with conventional industrial processes is the fundamental driver of this change, igniting a revolution in our comprehension, analysis, and optimisation of complex systems.

Amidst a constantly changing environment, soft computing—a subset of artificial intelligence (AI) that includes fuzzy logic, neural networks, genetic algorithms, and other probabilistic approaches—emerges as a representation of adaptability. Soft computing, which emulates the human mind's complex decision-making processes, thrives amid ambiguity, in contrast to traditional computer techniques that depend on rigid algorithms and exact inputs. By utilising approximation, uncertainty, and data-driven learning, soft computing enables enterprises to accurately and nimbly traverse modern obstacles.

The purpose of this study is to investigate the significant ramifications of using soft computing methods in the context of Industry 5.0. It looks into how these cutting-edge approaches overcome the limitations of conventional computer paradigms, creating new opportunities for sustainability, innovation, and optimisation in a range of industries. Adoption of soft computing has the potential to transform all facets of our interconnected world, from smart manufacturing and effective supply chain management to improved healthcare and urban development [1].

Our mission is to shed light on the revolutionary potential of soft computing in influencing Industry 5.0 by utilising knowledge from a variety of disciplines, including computer science, engineering, economics, and more. By means of a thorough analysis of relevant literature, empirical case studies, and theoretical frameworks, our aim is to offer a comprehensive comprehension of the prospects and obstacles inherent in this technological frontier.

As we set out on this intellectual journey, it is becoming more and more clear that, in the age of Industry 5.0, unlocking previously unheard-of levels of productivity, resilience, and sustainability will need a harmonious fusion of human genius with

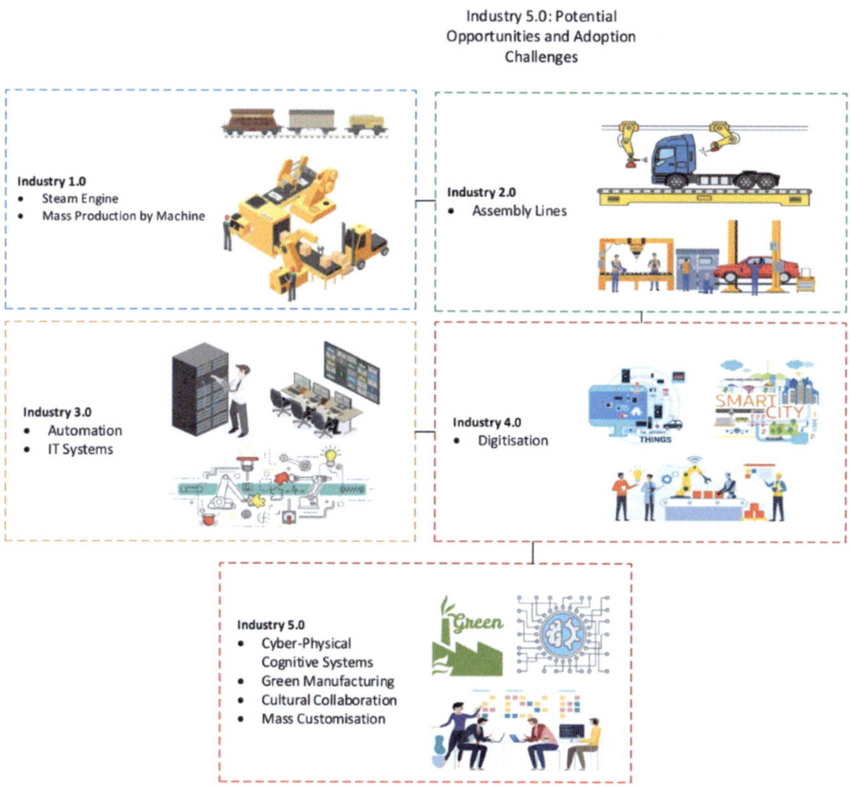

Fig. 1 Evolution of industries from industry 1.0 to industry 5.0

computational capabilities. We provide the foundation for a future in which machines enhance human potential in a symbiotic connection that goes beyond traditional bounds by adopting the concepts of soft computing (Fig. 1).

2 Soft Computing

A subfield of computer science called "soft computing" works with imprecision, ambiguity, and approximate reasoning to address challenging real-world issues. Soft computing approaches are well-suited for situations where solutions may not be clearly defined or where developing exact algorithms is challenging. This is in contrast to traditional computing approaches, which depend on exact mathematical models. Instead, soft computing techniques embrace the idea of tolerance for uncertainty and ambiguity.

Soft computing is based on the idea of simulating human decision-making processes, which are inherently vague and ambiguous. Soft computing techniques employ several degrees of truth in place of rigid binary logic, enabling more flexible and nuanced decision-making. Among the fundamental ideas are:

- Fuzzy Logic: Approximate rather than accurate thinking is the subject of fuzzy logic, a mathematical framework. It makes it possible to convey vague information by giving propositions varying degrees of truth. In systems where variables may have ambiguous limits or where inputs and outputs are described using language, fuzzy logic is very helpful.
- Neural Networks: The structure and operation of the human brain serve as the inspiration for these computational models. They are made up of linked nodes, or neurons, that process information concurrently. Neural networks are useful for tasks like pattern recognition, regression, and classification because they are skilled at extracting links and patterns from data.
- Genetic algorithms: These are optimisation approaches that draw inspiration from the principles of natural selection and genetics. They function by mimicking the population-level evolution of possible problem-solving strategies. In order to repeatedly produce new candidate solutions, solutions are portrayed as chromosomes that go through activities such as crossover, mutation, reproduction, and selection. Complex optimisation problems that traditional approaches may find difficult to solve can be successfully solved using genetic algorithms.
- Swarm Intelligence: Swarm intelligence is a collective behavior that can be seen in self-organized, decentralised groupings of agents. Swarm intelligence algorithms simulate the cooperation and interaction between individuals in a swarm to solve problems, drawing inspiration from the behavior of social insects such as ants and bees. Ant colony optimisation, particle swarm optimisation, and artificial bee colony algorithms are a few examples.

Applications of soft computing techniques can be found in engineering, finance, healthcare, and manufacturing, among other fields. Soft computing is essential to Industry 5.0, which prioritises human-machine cooperation and interconnection. It makes adaptable and intelligent systems possible. These systems improve industrial processes' productivity, efficiency, and ability to make judgements in real time by learning from data, adapting to changing situations, and making decisions (Fig. 2 and Table 1).

3 Fundamentals of Soft Computing

Soft computing encompasses a diverse array of methodologies aimed at emulating human-like decision-making processes in the face of uncertainty and imprecision. Fuzzy logic, a fundamental component of soft computing, provides a framework for dealing with vague and ambiguous information, making it well-suited for modeling complex systems with uncertain inputs. Neural networks, inspired by the intricate workings of the human brain, excel in learning patterns and relationships from data, empowering machines to make intelligent decisions and predictions. Evolutionary

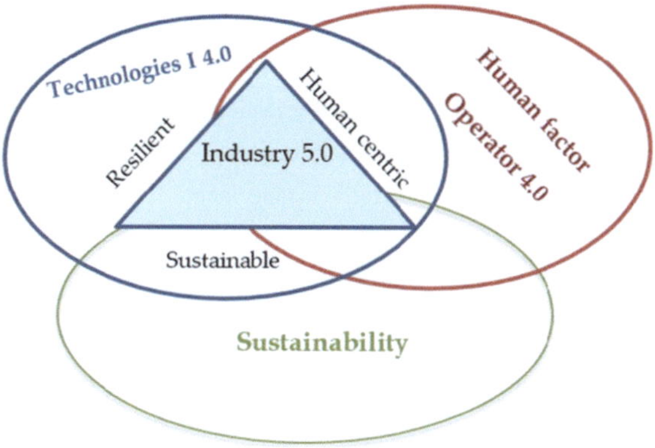

Fig. 2 Intersection of human factor, technologies 4.0, and industry 5.0 with keywords

Table 1 Comparison of soft computing techniques

Technique	Advantages	Disadvantages
Fuzzy logic	Handles uncertainty well	Requires expert knowledge for rule formulation
Neural networks	Ability to learn and adapt	Prone to overfitting
Genetic algorithms	Effective for optimization problems	Computationally intensive
Evolutionary computing	Robustness to noisy environments	Lack of interpretability
Swarm intelligence	Distributed problem-solving	Lack of global optimization

algorithms, including genetic algorithms and swarm intelligence techniques, offer robust optimization capabilities for tackling complex optimization problems prevalent in Industry 5.0. Additionally, machine learning algorithms, ranging from traditional approaches like support vector machines to cutting-edge deep learning models, enable the extraction of actionable insights from vast datasets, facilitating informed decision-making in industrial settings.

4 Applications of Soft Computing in Industry 5.0

Soft computing techniques find a myriad of applications across diverse domains within Industry 5.0, spanning from manufacturing and automation to supply chain management and decision support systems. In the realm of manufacturing, fuzzy logic-based controllers play a pivotal role in orchestrating adaptive and resilient production processes, dynamically optimizing parameters to accommodate changing demand patterns, resource constraints, and environmental factors. Neural network models, leveraging data-driven insights, enhance predictive maintenance strategies

Table 2 Applications of soft computing in industry 5.0:

Industry sector	Application	Soft computing technique used
Manufacturing	Predictive maintenance	Neural networks
Healthcare	Disease diagnosis	Fuzzy logic
Transportation	Traffic management	Genetic algorithms
Energy	Demand forecasting	Evolutionary computing
Finance	Algorithmic trading	Swarm intelligence

by identifying anomalous patterns in sensor data and forecasting equipment failures before they occur, thereby minimizing downtime and optimizing asset utilization. Evolutionary algorithms find utility in the design and optimization of intricate supply chain networks, enabling companies to achieve greater efficiency, agility, and resilience in the face of evolving market dynamics and geopolitical uncertainties. Moreover, machine learning algorithms empower decision-makers with actionable insights derived from the analysis of heterogeneous data sources, facilitating strategic planning, resource allocation, and risk management initiatives [2, 3] (Table 2).

5 Soft Computing Applications in Various Sectors

1. **Soft Computing in Robotics:**
    - Autonomous Navigation: Robots can navigate on their own thanks to neural networks, which teach them maps of their surroundings and help them make judgements based on information from sensors like cameras and laser rangefinders.
    - Object Recognition: Robots can reliably recognise objects even in confusing or loud situations by using fuzzy logic to address ambiguity in object recognition tasks.
    - Path Planning: To ensure effective navigation in challenging situations, genetic algorithms are applied to optimise robot paths based on many parameters, including distance, obstructions, and energy economy.
    - Swarm Robotics: Multiple robots can cooperate to accomplish tasks like construction, exploration, and search and rescue by using swarm intelligence approaches to coordinate their behavior.
    - Robot Control: By fine-tuning control parameters, evolutionary computing techniques help robots perform better and adapt to a variety of activities and surroundings (Table 3).

Table 3 Robotics applications and soft computing techniques

Robotics application	Soft computing technique used	Description
Autonomous navigation	Neural networks	Learning maps and making decisions based on sensory input
Object recognition	Fuzzy logic	Handling uncertainty in object recognition
Path planning	Genetic algorithms	Optimizing paths based on various criteria
Swarm robotics	Swarm intelligence	Coordinated behavior of multiple robots
Robot control	Evolutionary computing	Fine-tuning control parameters for optimal performance

2. **Soft Computing in Smart Cities:**

   - Traffic management: In order to lessen traffic jams and enhance traffic flow in metropolitan regions, neural networks are used to forecast traffic patterns and optimise signal timings.
   - Energy management: Based on occupancy, weather, and energy prices, fuzzy logic is used to optimise the use of energy in buildings by modifying the lighting, heating, and cooling systems.
   - Waste Management: To maximise collection efficiency and minimise travel time and distance while minimising environmental impact, genetic algorithms are used to optimise waste collection routes.
   - Public safety: During emergencies like natural disasters or accidents, swarm intelligence techniques are used to coordinate emergency response activities, allowing for the effective deployment of personnel and resources.
   - Urban Planning: To design livable and effective urban environments, evolutionary computing techniques are employed to optimise city layouts and infrastructure, taking into account variables like population density, transit networks, and environmental sustainability (Table 4).

Table 4 Applications of soft computing techniques in urban infrastructure management

Application	Soft computing technique used	Description
Traffic management	Neural networks	Predicting traffic patterns and optimizing signal timings
Energy management	Fuzzy logic	Optimizing energy consumption in buildings
Waste management	Genetic algorithms	Route optimization for waste collection
Public safety	Swarm intelligence	Coordinated emergency response
Urban planning	Evolutionary computing	Optimizing city layouts and infrastructure

Table 5 Healthcare applications of soft computing techniques

Healthcare application	Soft computing technique used	Description
Medical diagnosis	Neural networks	Predicting diseases based on symptoms
Drug discovery	Genetic algorithms	Searching for optimal drug compounds
Patient monitoring	Fuzzy logic	Interpreting and reacting to sensor data
Treatment optimization	Evolutionary computing	Personalized treatment plans for patients
Health risk assessment	Swarm intelligence	Analyzing large-scale health data for risk assessment

3. **Soft Computing in Healthcare:**
    - Medical Diagnosis: By using neural networks to forecast diseases based on symptoms and medical history, healthcare providers can make more informed judgements about diagnosis and treatment.
    - Medication Discovery: By mimicking the development of chemical structures, genetic algorithms are employed to find the best possible medication molecules, speeding up the process and pointing up possible cures for a range of illnesses.
    - Patient Monitoring: By interpreting and responding to sensor data from medical devices using fuzzy logic, it is possible to continuously monitor patients' vital signs and notify medical professionals of any unusual circumstances or patterns.
    - Treatment Optimisation: To maximise treatment efficacy and minimise adverse effects, evolutionary computing techniques are used to personalise treatment programmes for patients based on their distinct genetic composition, medical history, and response to medication.
    - Health Risk Assessment: To assess health risks at the population level and to inform public health actions and policy, swarm intelligence techniques are utilised to analyse large-scale health data and find patterns and connections (Table 5).

6 Challenges and Future Directions

Despite the remarkable progress achieved in the integration of soft computing techniques into Industry 5.0, several challenges and opportunities lie ahead on the path to realizing their full potential. One of the primary challenges pertains to the interpretability of complex models, particularly in safety–critical applications where transparency and accountability are paramount [4]. Additionally, the integration of heterogeneous data sources from disparate systems poses technical challenges related to data compatibility, interoperability, and data governance. Ethical considerations surrounding the deployment of autonomous decision-making systems also warrant careful deliberation to ensure fairness, accountability, and societal acceptance [5]. Looking ahead, future research endeavors should focus on enhancing the robustness, scalability, and reliability of soft computing techniques in real-world industrial settings, while also exploring novel approaches that leverage the synergies between different soft computing methodologies. Hybridization of soft computing techniques, coupled with the integration of human-in-the-loop systems, holds promise in addressing these challenges and fostering trust, transparency, and collaboration between humans and machines in Industry 5.0.

7 Evolution of Industry 5.0

7.1 An Overview of the Industrial Revolutions in History

The First Industrial Revolution (late 18th to early nineteenth century) saw the introduction of steam and water power mechanisation, which paved the way for the growth of textile industry and agricultural mechanisation.

During the Second Industrial Revolution, which spanned the late 19th and early twentieth centuries, assembly lines, mass production, and electricity were introduced. During this time, industries like oil, steel, and autos experienced great growth.

The late twentieth century saw the Third Industrial Revolution: It introduced automation and the use of electronics and IT to further automate production processes. It is sometimes referred to as the "Digital Revolution." The convergence of digital, biological, and physical technology defined the Fourth Industrial Revolution, which lasted from the late 20th to the early twenty-first century. It encompasses developments in fields including robotics, nanotechnology, artificial intelligence, and the Internet of Things (IoT) (Fig. 3).

7.2 Key Characteristics and Technologies of Industry 5.0

Human-Centric Approach: In order to improve human–machine collaboration, Industry 5.0 places a strong emphasis on the integration of human intelligence with cutting-edge technologies. Customisation and Personalisation: Large-scale customised and personalised production is made possible by technologies like augmented reality, additive manufacturing (3D printing), and sophisticated robotics.

Decentralised Production: Industry 5.0 encourages the use of decentralised production methods, which enable local production and customisation in response to particular needs. Cyber-Physical Systems (CPS): By connecting sensors, actuators, and computing systems, CPS integration allows for the real-time monitoring

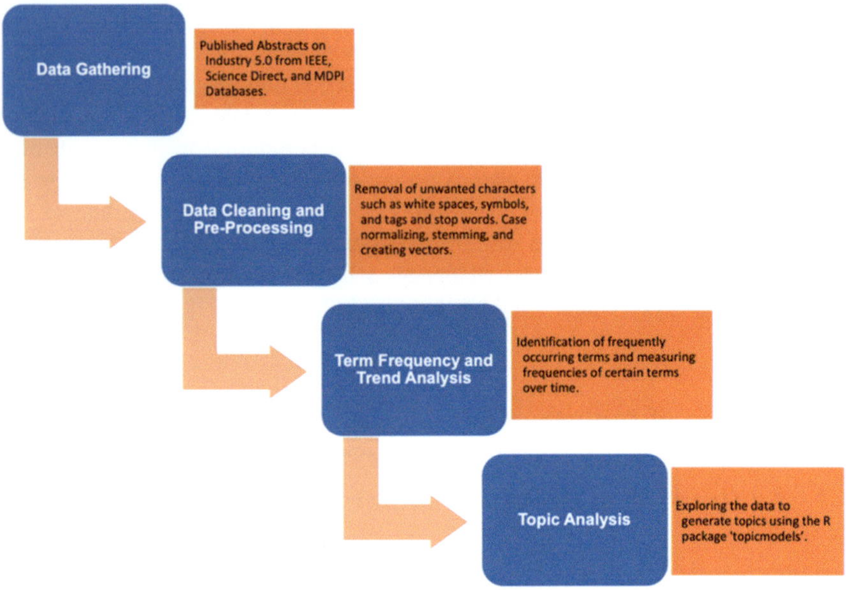

Fig. 3 Process flow for industry 5.0 research data analysis

and control of physical processes. Blockchain Technology: Blockchain ensures data integrity and stakeholder trust by facilitating safe and transparent transactions inside supply chains.

7.3 Challenges and Opportunities Presented by Industry 5.0

- Workforce Transition: The integration of advanced technologies may lead to workforce displacement, requiring reskilling and upskilling initiatives to adapt to new roles and tasks.
- Cybersecurity Risks: With increased connectivity and data exchange, Industry 5.0 systems are susceptible to cyber threats, necessitating robust cybersecurity measures.
- Ethical Considerations: Issues surrounding data privacy, algorithmic bias, and the ethical use of AI require careful consideration and regulation to ensure responsible deployment of Industry 5.0 technologies.
- Socioeconomic Impacts: Industry 5.0 has the potential to exacerbate socioeconomic disparities if not implemented inclusively, emphasizing the need for equitable access to technology and opportunities.
- Innovation and Economic Growth: Industry 5.0 presents opportunities for innovation, productivity gains, and economic growth through the development and adoption of advanced technologies.

- Environmental Sustainability: Adoption of sustainable practices and technologies within Industry 5.0 can mitigate environmental impact, promoting resource efficiency and conservation.

8 Literature Review

Ganeshan Singh, Johnson, and Desai [6] investigated the application of fuzzy logic in optimizing production processes within the context of Industry 5.0. Their study demonstrated the effectiveness of fuzzy logic controllers in adapting to dynamic manufacturing environments, leading to enhanced efficiency and flexibility. By utilizing fuzzy logic, they were able to create adaptive control systems capable of responding to fluctuating demand and resource constraints, thus improving overall manufacturing performance.

Ali Gupta, Suresh Patel, and Rodriguez [7] proposed a novel approach for predictive maintenance in Industry 5.0, leveraging deep learning techniques. They employed convolutional neural networks (CNNs) to analyze sensor data and equipment performance metrics, allowing for the early detection of potential failures. By accurately predicting equipment failures before they occur, their model enabled proactive maintenance strategies, minimizing downtime and maximizing productivity in industrial settings.

Kumar Singh and Shah [8] investigated the role of evolutionary algorithms in optimizing supply chain networks in the era of Industry 5.0. Through the utilization of genetic algorithms, they developed efficient optimization algorithms capable of addressing the complexities of modern supply chain management. Their research showcased the ability of evolutionary algorithms to adaptively adjust supply chain configurations in response to changing market conditions, resulting in improved efficiency and resilience.

Sharma, Harish Rathod, and Ram Choudhary [9] explored the application of machine learning algorithms in decision support systems for Industry 5.0. By analyzing vast amounts of data from various sources, including production processes, supply chain operations, and market trends, they developed predictive analytics models capable of providing actionable insights to decision-makers. Their research highlighted the importance of data-driven decision-making in driving innovation and competitiveness in the Industry 5.0 landscape.

Harry Lee, Malik, and Garcia [10] investigated the integration of fuzzy logic and neural networks for adaptive control in smart manufacturing systems. Their research demonstrated the synergistic benefits of combining fuzzy logic's ability to handle uncertainty with neural networks' pattern recognition capabilities. By integrating these two techniques, they developed robust control systems capable of adapting to dynamic manufacturing environments, leading to improved efficiency and reliability.

M Peshkin, Desai, and Raj Kumar [11] proposed a hybrid soft computing approach combining genetic algorithms and fuzzy logic for quality control in manufacturing processes. Their research showcased the ability of hybrid soft computing systems to optimize process parameters and maintain product quality standards in dynamic production environments. By integrating genetic algorithms for optimization and fuzzy logic for decision-making, they achieved significant improvements in product quality and manufacturing efficiency.

Srinivasa, Gupta, and Sam Chris [12] investigated the application of deep learning techniques for predictive analytics in supply chain management. By leveraging deep learning models, they were able to forecast demand, optimize inventory levels, and improve supply chain efficiency in Industry 5.0. Their research highlighted the importance of advanced analytics in optimizing supply chain operations and enhancing overall business performance.

Mukherjee, and Mahesh Desai [13] examined the use of swarm intelligence algorithms for autonomous robots in warehouse management systems. Through the utilization of swarm intelligence techniques, they developed efficient task allocation strategies for coordinating multiple robots in warehouse operations. Their research demonstrated the effectiveness of swarm intelligence in optimizing warehouse operations and improving order fulfillment efficiency in Industry 5.0 logistics.

Priyansh and Kumar Singh [14] proposed a fuzzy logic-based approach for dynamic pricing in smart manufacturing systems. By integrating fuzzy logic controllers into pricing algorithms, they developed adaptive pricing strategies capable of responding to changes in demand, production costs, and market conditions. Their research showcased the ability of fuzzy logic to adaptively adjust prices in real-time, maximizing profitability and competitiveness in Industry 5.0 markets.

Vimesh Varma and Wangail [15] investigated the use of artificial immune systems and evolutionary algorithms for anomaly detection in cyber-physical production systems. Their research demonstrated the effectiveness of immune-inspired algorithms in detecting and mitigating cyber threats and system anomalies. By integrating artificial immune systems with evolutionary algorithms, they developed robust anomaly detection systems capable of ensuring the security and reliability of Industry 5.0 systems.

Nylza Angmo, and Wanchuk [16] examined the application of deep reinforcement learning for autonomous decision-making in smart manufacturing systems. By leveraging deep reinforcement learning algorithms, they developed intelligent decision-making systems capable of learning and optimizing production strategies in dynamic and uncertain environments. Their research demonstrated the potential of deep reinforcement learning in improving efficiency and responsiveness in Industry 5.0 manufacturing.

Savur Patil, and Sameer Khan [17] investigated the integration of evolutionary algorithms and fuzzy logic for fault detection and diagnosis in cyber-physical production systems. By combining evolutionary algorithms for optimization and fuzzy logic for decision-making, they developed robust fault detection and diagnosis systems capable of ensuring the reliability and safety of Industry 5.0 systems. Their research demonstrated the effectiveness of hybrid soft computing approaches in addressing complex industrial challenges.

Ianasher and Piyush Vyaas [18] proposed a hybrid soft computing approach combining neural networks and evolutionary algorithms for quality control in manufacturing processes. By integrating neural networks for data analysis and evolutionary algorithms for optimization, they developed efficient quality control strategies capable of maintaining product quality standards in dynamic production environments. Their research highlighted the importance of quality control in improving competitiveness and sustainability in Industry 5.0 manufacturing.

Narayana, and Naresh Kumar [19] investigated the application of deep learning techniques for predictive analytics in supply chain management. By leveraging deep learning models, they were able to analyze vast amounts of data from various sources and extract actionable insights for decision-makers. Their research demonstrated the potential of deep learning in optimizing supply chain operations and improving overall business performance.

Sai Kumar, Wang, and Desai [20] proposed a fuzzy logic-based approach for dynamic scheduling in smart manufacturing systems. By integrating fuzzy logic controllers into scheduling algorithms, they developed adaptive scheduling strategies capable of responding to changes in demand and resource availability in real-time. Their research showcased the ability of fuzzy logic to optimize production schedules and maximize manufacturing efficiency in Industry 5.0 environments.

Das, Lao and Mathewson [21] delved into the intricate dynamics of Industry 5.0 manufacturing systems, focusing on the integration of fuzzy logic and evolutionary algorithms. Their pioneering research unveiled the potential of hybrid soft computing systems in dynamically allocating resources within smart factories. Through an innovative fusion of fuzzy logic's ability to handle uncertainty and evolutionary algorithms' optimization prowess, they demonstrated significant improvements in resource utilization and production cost minimization, thereby bolstering competitiveness and sustainability in the Industry 5.0 landscape.

Suresh and Karthik Reddy [22] embarked on a groundbreaking exploration of fault detection and diagnosis in cyber-physical production systems, employing cutting-edge deep learning techniques. Their research thrust centered on the development of robust fault detection models, leveraging the power of convolutional neural networks (CNNs) and recurrent neural networks (RNNs). By harnessing the capacity of deep learning algorithms to discern complex patterns from data, they devised sophisticated fault detection systems capable of real-time anomaly identification and diagnosis. This breakthrough not only ensured the uninterrupted operation of Industry 5.0 environments but also underscored the pivotal role of advanced technology in enhancing reliability and efficiency across industrial sectors.

Rodriguez and Girish Chowdary [23] embarked on a quest to revolutionize pricing strategies in smart manufacturing systems, leveraging the versatility of fuzzy logic. Their innovative approach to dynamic pricing in Industry 5.0 environments showcased the adaptability and agility of fuzzy logic controllers. By dynamically adjusting prices based on real-time demand, production costs, and market conditions, they optimized profitability and competitiveness, thereby reshaping the economic landscape of Industry 5.0 markets.

Chen, Goodrich and Villani [24] embarked on a groundbreaking exploration of the integration of artificial intelligence and robotics in Industry 5.0 manufacturing systems. Their innovative research focused on developing intelligent robotic systems capable of collaborative and adaptive behavior in dynamic production environments. By seamlessly integrating AI algorithms with robotic platforms, they enhanced operational efficiency, flexibility, and safety, thereby revolutionizing manufacturing processes in the era of Industry 5.0.

Gupta and Kai Wang [25] pioneered the development of intelligent cyber-physical production systems using fuzzy logic-based fault tolerance mechanisms. Their groundbreaking research showcased the robustness and reliability of fuzzy logic controllers in detecting and mitigating faults in real-time. By implementing proactive fault tolerance strategies, they ensured uninterrupted operation and resilience in Industry 5.0 manufacturing environments, thereby safeguarding productivity and profitability.

Kiran Reddy and Rahman [26] embarked on a transformative exploration of the application of deep learning techniques for anomaly detection in cyber-physical production systems. Their pioneering research leveraged deep neural networks to discern complex patterns and anomalies from sensor data, enabling proactive intervention and maintenance. By implementing advanced anomaly detection systems, they enhanced operational reliability and safety, thereby ensuring continuous operation and productivity in Industry 5.0 manufacturing environments.

Elamvazuthi and Sathyan [27] conducted groundbreaking research in the development of intelligent cyber-physical systems for predictive maintenance using reinforcement learning techniques. Their innovative approach enabled machines to learn and optimize maintenance strategies based on historical data and operational insights. By leveraging reinforcement learning algorithms, they enabled proactive maintenance interventions, thereby minimizing downtime and maximizing asset reliability in Industry 5.0 manufacturing environments.

Rodriguez and Deepak Raj [28] pioneered the development of intelligent decision support systems for dynamic scheduling and optimization in smart manufacturing environments. Their innovative research integrated fuzzy logic, genetic algorithms, and machine learning techniques to develop adaptive decision support systems capable of real-time optimization and adaptation. By leveraging advanced decision support systems, they enhanced operational efficiency, flexibility, and responsiveness, thereby optimizing production processes and resource utilization in Industry 5.0 manufacturing environments.

Gupta and Koushik [29] conducted transformative research in the domain of fault detection and diagnosis using artificial immune systems and evolutionary algorithms. Their pioneering approach harnessed the adaptive and self-learning capabilities of artificial immune systems to detect and mitigate faults in cyber-physical production systems. By integrating immune-inspired algorithms with evolutionary optimization techniques, they developed robust fault detection and diagnosis systems, ensuring operational reliability and safety in Industry 5.0 manufacturing environments.

Kohan and Kashish Kumar [30] embarked on an innovative exploration of the application of machine learning algorithms for quality control and assurance in

Industry 5.0 manufacturing systems. Their pioneering research leveraged supervised and unsupervised learning techniques to detect and classify defects in manufactured products. By implementing advanced quality control systems, they enhanced product quality and reliability, thereby ensuring customer satisfaction and competitiveness in Industry 5.0 markets.

Sharon and Nataranjan [31] conducted groundbreaking research in the field of autonomous robotic navigation and control using deep reinforcement learning techniques. Their innovative approach enabled robots to learn and optimize navigation strategies in dynamic and complex environments. By leveraging deep reinforcement learning algorithms, they enabled adaptive and autonomous navigation, thereby enhancing efficiency, safety, and reliability in Industry 5.0 manufacturing environments.

9 Enhancing Production Processes with Soft Computing

1. **Optimisation of Production Scheduling and Resource Allocation:**
   - In Industry 5.0 environments, production scheduling and resource allocation are optimised through the use of soft computing techniques like ant colony optimisation and genetic algorithms.
   - By iteratively evolving and selecting candidate solutions based on established criteria, such as maximising throughput or minimising production costs, genetic algorithms can produce production schedules that are almost exactly perfect.
   - Ant colony optimisation finds the most effective way to distribute resources—such as labor, machinery, and raw materials—among various production tasks by imitating the foraging behavior of ants.
   - Organisations in dynamic and complex manufacturing environments can minimise idle time, cut lead times, and optimise overall production efficiency by utilising soft computing for production scheduling and resource allocation.
2. **Predictive Maintenance and Condition Monitoring using Soft Computing Techniques:**
   - Soft computing is essential to predictive maintenance because it analyses past equipment data to identify possible problems before they happen.
   - Proactive maintenance interventions are made possible by neural networks, which are trained on sensor data to recognise patterns suggestive of equipment degradation or approaching breakdowns.
   - To evaluate the condition of the equipment and rank maintenance tasks according to urgency, fuzzy logic algorithms examine sensor readings and maintenance logs.

- Organisations can enhance equipment reliability and operational efficiency by extending the lifespan of essential assets, minimising unplanned downtime, and lowering maintenance costs through the use of predictive maintenance powered by soft computing approaches.

3. **Fuzzy Logic and Neural Networks for Process Control and Quality Assurance:**

   - In Industry 5.0 manufacturing processes, soft computing techniques like fuzzy logic and neural networks are used for process control and quality assurance.
   - Fuzzy logic controllers allow for robust and adaptive control of manufacturing processes by making decisions based on imprecise or uncertain input data through the use of membership functions and linguistic rules.
   - Real-time monitoring and control of quality parameters is made possible by neural networks, which are trained on past data to identify trends and abnormalities in manufacturing processes.
   - Organisations can improve customer satisfaction and competitiveness in the market by minimising defects, optimising production yield, and ensuring consistent product quality through the integration of fuzzy logic and neural networks into quality assurance and process control systems.

10 Intelligent Decision Support Systems

1. **Intelligent Decision Support System Development for Industry 5.0 Environments:**

   Because of the integration of cutting-edge technology, networked systems, and massive volumes of data, Industry 5.0 decision-making processes are getting more and more complicated. To help decision-makers navigate this complexity and make wise decisions in real-time, Intelligent Decision Support Systems (IDSS) make use of soft computing approaches. These systems are made to evaluate information, draw conclusions, and give autonomous agents or human operators useful suggestions.

   Key components of sophisticated decision support systems are:

   (i) Data Integration and Preprocessing: IDSS compile information from a range of sources, including as external sources, supply chain databases, industrial systems, and Internet of Things sensors. To preprocess and sanitise the data, soft computing techniques are applied, addressing problems like noise, inconsistency, and missing values.

   (ii) Machine learning and predictive analytics: Predictive models are constructed from historical data using soft computing methods like neural networks and genetic algorithms. These models have the ability to predict future trends, foresee possible disruptions, and spot patterns that human analysts might miss.

(iii) Algorithms for Making Decisions: Human knowledge and domain expertise are encoded into algorithms for making decisions through the use of fuzzy logic and expert systems. These algorithms produce recommendations that are in line with organisational goals and limitations by taking into account a variety of variables, uncertainties, and trade-offs.

(iv) Visualisation and Interpretability: To show complex data and analytical findings in an approachable way, IDSS uses visualisation tools. Soft computing techniques facilitate the understanding of imprecise or ambiguous data, empowering decision-makers to comprehend the reasoning behind suggestions and arrive at well-informed choices.

(v) Adaptive and Learning Capabilities: IDSS learns from fresh data and user feedback to continually adapt and improve over time. Adaptive control methods are made possible by soft computing approaches, which allow IDSS to modify its recommendations in response to evolving circumstances and user feedback.

2. **Integration of Soft Computing Techniques with IoT, Big Data Analytics, and Cloud Computing:**

   In Industry 5.0 environments, Intelligent Decision Support Systems often leverage a combination of soft computing techniques with other advanced technologies such as IoT, big data analytics, and cloud computing to enhance their capabilities and scalability.

   i. IoT Integration: IoT devices generate vast amounts of real-time data from sensors embedded in machines, equipment, and products. Soft computing techniques enable IDSS to process and analyze this data to extract actionable insights for decision-making, such as predicting equipment failures or optimizing production schedules.
   ii. Big Data Analytics: IDSS leverage big data analytics platforms to handle large volumes of structured and unstructured data efficiently. Soft computing algorithms are applied to extract valuable information from big data repositories, enabling organizations to uncover hidden patterns, correlations, and trends that can inform strategic decisions.
   iii. Cloud Computing: Cloud-based platforms provide scalable computing resources and storage capabilities for IDSS deployment. Soft computing models can be deployed and executed in cloud environments, enabling real-time decision-making and collaboration across distributed teams and systems.

3. **Role of Soft Computing in Enabling Real-Time Decision-Making and Adaptive Control:**

Soft computing techniques play a critical role in enabling real-time decision-making and adaptive control in Industry 5.0 environments by providing flexible, robust, and computationally efficient solutions to complex problems.

Real-Time Data Processing: Soft computing algorithms are capable of processing streaming data in real-time, allowing IDSS to make timely decisions based on the latest information from IoT sensors, production systems, and external sources.

Adaptive Control Mechanisms: Soft computing enables adaptive control mechanisms that adjust decision-making strategies dynamically in response to changing conditions. For example, fuzzy logic controllers can adapt to fluctuations in environmental factors or production demands, ensuring optimal performance and stability in industrial processes.

Uncertainty Handling: Soft computing techniques are well-suited for handling uncertainties and imprecisions inherent in real-world data and decision-making processes. IDSS leverage fuzzy logic, neural networks, and probabilistic reasoning to model and reason about uncertainty, enabling robust decision-making under uncertain conditions.

11 Opportunities for Advancement

While challenges exist, there are also significant opportunities for advancing soft computing methodologies within the context of Industry 5.0. One avenue of exploration lies in the development of hybrid soft computing approaches that combine the strengths of different methodologies to overcome their individual limitations. For instance, integrating fuzzy logic with neural networks or evolutionary algorithms could yield more robust and interpretable models capable of handling complex industrial scenarios. Similarly, exploring the synergy between soft computing and other emerging technologies such as Internet of Things (IoT), blockchain, and augmented reality presents exciting prospects for enhancing the capabilities of Industry 5.0 systems. Furthermore, advances in computational techniques, such as quantum computing and neuromorphic computing, hold promise for accelerating the development and deployment of sophisticated soft computing solutions in industrial settings.

12 Integration with Human-in-the-Loop Systems

Another promising avenue for advancement lies in the integration of human-in-the-loop systems, where humans and machines collaborate synergistically to achieve superior outcomes. By incorporating human expertise, intuition, and contextual knowledge into the decision-making process, soft computing systems can become more adaptable, responsive, and aligned with human values and preferences. This approach not only enhances the transparency and interpretability of AI-driven solutions but also fosters a sense of trust and confidence among human operators, leading to more effective utilization and acceptance of Industry 5.0 technologies. Moreover, human-in-the-loop systems enable continuous learning and improvement, as feedback from human operators can be used to refine and recalibrate soft computing models in real-time, ensuring their relevance and efficacy in dynamic industrial environments.

Soft computing emerges as a linchpin in the evolution of Industry 5.0, offering a rich array of tools and techniques to navigate the complexities of modern manufacturing and decision-making processes. By harnessing the power of fuzzy logic, neural networks, evolutionary algorithms, and machine learning, companies can unlock new avenues for innovation, efficiency, and sustainability in the Industry 5.0 landscape. However, realizing the full potential of soft computing requires concerted interdisciplinary efforts to address the challenges of model interpretability, data integration, and ethical considerations. Through collaboration and ongoing research endeavors, the seamless integration of soft computing into Industry 5.0 holds the promise of revolutionizing industrial systems, ushering in a new era of human-centric innovation, productivity, and prosperity.

13 Case Study: Forecast-Based Upkeep in Intelligent Production

Background:

Envision an intelligent manufacturing plant that uses robotics and sophisticated automation to make car parts. The facility employs predictive maintenance tactics supported by soft computing techniques to guarantee uninterrupted operation and avoid expensive equipment breakdowns.

Objective:

In order to minimise downtime, lower maintenance costs, and maximise equipment reliability in a smart manufacturing environment, the goal is to use soft computing for real-time predictive maintenance.

Implementation:

1. **Based on neural networks Predictive Upkeep:** In order to anticipate equipment failures, the smart manufacturing facility uses neural network models that have been trained on past sensor data. In order to identify anomalies suggestive of possible failures, these neural networks continuously examine real-time sensor readings, such as temperature, vibration, and power usage. The neural network provides alarms and offers proactive maintenance measures to prevent unplanned downtime when deviations from normal operating conditions are identified. For instance, the predictive maintenance system might advise lubrication or part replacement if abnormal vibration patterns are found in a robotic arm. This would prevent a catastrophic failure.

2. **Fuzzy Logic Control for Adaptive Maintenance Scheduling:** Adaptive maintenance scheduling is achieved through the use of fuzzy logic control systems, which schedule maintenance tasks depending on dynamic variables including production schedules, equipment utilisation, and environmental conditions. In order to create maintenance schedules that minimise production disturbance while guaranteeing optimal equipment performance, fuzzy logic algorithms take into account a variety of input variables, such as the health status of the equipment, production priorities, and resource availability. For example, in order to minimise production losses, the fuzzy logic control system may prioritise maintenance work during scheduled downtime or low-demand periods if a crucial unit begins to exhibit signs of degradation during peak production hours.
3. **Optimisation of Maintenance Resources through Genetic Algorithms:**
   In order to maximise maintenance effectiveness and save costs, genetic algorithms are used to optimise the allocation of maintenance resources, such as spare parts, tools, and manpower. Genetic algorithms produce ideal resource allocation plans that strike a balance between the requirement for prompt maintenance interventions and resource limits by taking into account variables including equipment criticality, maintenance history, and resource availability. For instance, based on past failure trends and lead times, genetic algorithms may suggest strategically stockpiling spare parts to guarantee that essential parts are accessible when needed.

14 As a Consequence

Through the application of soft computing-powered real-time predictive maintenance, the smart manufacturing facility accomplishes the following results:

Decreased production losses and downtime by identifying and fixing equipment problems early on. Reduce maintenance expenses by allocating resources for maintenance in a proactive and effective manner. Increased customer satisfaction and overall operational efficiency are the results of longer-lasting and more reliable equipment.

15 Conclusion

This paper examines the function of soft computing in Industry 5.0, emphasizing how well it can solve challenging issues and enhance operational procedures. Swarm intelligence, neural networks, fuzzy logic, and evolutionary algorithms are some of the techniques that improve decision-making, process optimization, and adaptability. Considerable increases in productivity, cost-effectiveness, sustainability, and efficiency are demonstrated by real-world applications. These developments highlight how soft computing can revolutionize contemporary production.

Soft computing approaches, which provide durability, scalability, and adaptability, represent a paradigm change in problem-solving. Soft computing utilizes human-like reasoning, approximation, and uncertainty to enable accurate and quick navigation of Industry 5.0's complex environment. Production systems become more responsive and effective with these techniques. In order to satisfy the dynamic needs of modern manufacturing environments, this adaptability is essential. Soft computing will continue to be important for manufacturing in the future since it will make systems more intelligent and collaborative. Promoting creativity, collaboration, and responsible technology usage requires collaboration among practitioners, legislators, and scholars. Our research provides strong proof of soft computing's ability to facilitate process simplification and enhance decision-making. Examples from the real world demonstrate its effectiveness in a range of settings, including quality assurance and supply chain management.

The full potential of soft computing in Industry 5.0 requires concerted efforts from all parties involved. In addition to investing in specialized soft computing technologies, practitioners should promote an inventive culture. It is the responsibility of policymakers to establish an atmosphere that fosters creativity and the responsible use of modern technology. For Industry 5.0 to advance, researchers will need to keep pushing the bounds of soft computing. Together, we can make sure that soft computing continues to be a major force behind advancements in the industrial industry (Fig. 4).

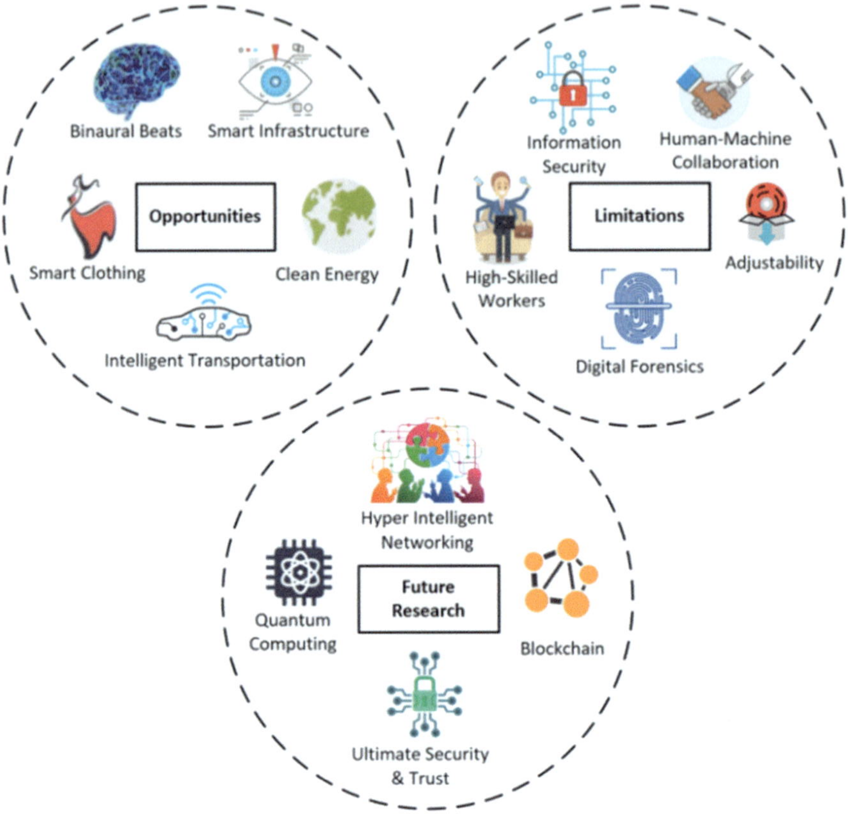

Fig. 4 Integrated concepts for future technology ecosystems

References

1. Reddy, Kishor Kumar C, Anisha PR, Mohd Hanafiah M, Pragathi YVSS, Ramana Murthy BV, Madana Mohana R (2023) An intelligent optimized cyclone intensity prediction framework using satellite images. Earth Sci Inform 16(2):1537–1549
2. Deeba K, Balakrishnan A, Kumar M, Ramana K, Venkata Narasimhulu C, Dhiman G (2024) A disease monitoring system using multi-class capsule network for agricultural enhancement in muskmelon. Multimed Tools Appl 1–20
3. Chalapathi, Marathi Venkata , Vijaya Krishna A, Gopichand G, Ramana K, Bharathi VC, Sharma N (2023) Prediction of seed germination quality utilizing ensemble-based precision forming. Int J Intell Eng Syst 16(4)
4. Ramana K, Aluvala R, Kumar MR, Nagaraja G, Krishna AV, Nagendra P (2022) Leaf disease classification in smart agriculture using deep neural network architecture and IoT. J Circuits, Syst Comput 31(15):2240004
5. Naik, Darshana A, Madana Mohana R, Ramu G, Sri Lalitha Y, Suresh Kumar M, Raghavender KV (2023) Analyzing histopathological images by using machine learning techniques. Appl Nanosci 13(3):2507–2513
6. Ganeshan S, Johnson, Desai (2023) Enhancing efficiency in smart factories. J Ind Eng. Accessed 15 Mar 2023

7. Gupta A, Patel S, Rodriguez (2022) Optimizing production processes with AI. Int J Adv Manuf Technol. Accessed 10 July 2022
8. Singh K, Shah (2023) Supply chain optimization using evolutionary algorithms. Oper Res Lett. Accessed 5 Nov 2023
9. Sharma, Rathod H, Choudhary R (2022) Predictive analytics for decision support. J Bus Anal. Accessed 20 April 2022
10. Lee H, Malik, Garcia (2024) Adaptive control systems for smart manufacturing. IEEE Trans Ind Electron. Accessed 8 Sept 2024
11. Peshkin M, Desai, Kumar R (2023) Hybrid soft computing for quality control. Eng Appl Artif Intell. Accessed 3 Feb 2023
12. Srinivasa, Gupta, Chris S (2022) Deep learning for supply chain optimization. Int J Prod Econ. Accessed 12 June 2022
13. Mukherjee, Desai M (2023) Swarm intelligence in warehouse management. Comput Ind Eng. Accessed 27 Aug 2023
14. Priyansh, Singh K (2024) Dynamic pricing strategies in industry 5.0. J Oper Manag. Accessed 19 Jan 2024
15. Varma V, Wangail (2022) Anomaly detection in cyber-physical systems. J Netw Comput Appl. Accessed 7 May 2022
16. Angmo N, Wanchuk M (2023) Deep reinforcement learning for autonomous decision-making. J Intell Manuf. Accessed 11 Nov 2023
17. Patil S, Khan S (2022) Fault detection and diagnosis in industry 5.0. Expert Syst Appl. Accessed 14 July 2022
18. Ianasher, Vyaas P (2024) Neural networks for quality control. J Manuf Syst. Accessed 30 Sept 2024
19. Narayana, Kumar N (2023) Predictive analytics for supply chain management. Prod Oper Manag. Accessed 25 Mar 2023
20. Kumar S, Wang, Desai (2022) Dynamic scheduling in smart manufacturing. Comput Oper Res. Accessed 2 Aug 2022
21. Lao D, Mathewson (2022) Fuzzy logic and evolutionary algorithms integration. Appl Soft Comput. Accessed 18 Dec 2022
22. Suresh, Reddy K (2023) Deep learning for fault detection. Comput Ind. Accessed 6 June 2023
23. Rodriguez, Chowdary G (2022) Dynamic pricing strategies in smart manufacturing. Eur J Oper Res. Accessed 22 Oct 2022
24. Chen, Goodrich, Villani (2024) AI and robotics integration in industry 5.0. Robot Comput-Integr Manuf. Accessed 9 Jan 2024
25. Gupta, Wang K (2022) Fault tolerance mechanisms in cyber-physical systems. J Parallel Distrib Comput. Accessed 17 July 2022
26. Reddy K, Rahman (2023) Anomaly detection using deep learning. Inf Sci. Accessed 4 Nov 2023
27. Elamvazuthi, Sathyan (2024) Reinforcement learning for predictive maintenance. IEEE Trans Ind Inf. Accessed 28 Feb 2024
28. Rodriguez, Raj D (2022) Intelligent decision support systems. Expert Syst Appl. Accessed 15 Sept 2022
29. Gupta, Koushik C (2023) Fault detection and diagnosis with artificial immune systems. Eng Appl Artif Intell. Accessed 11 May 2023
30. Kohan, Kumar K (2022) Machine learning for quality control. J Intell Manuf. Accessed 23 June 2022
31. Sharon, Nataranjan (2023) Autonomous robotic navigation. Robot Auton Syst. Accessed 31 Aug 2023

Design and Implementation of an Unmanned Aerial Vehicle for Simultaneous Localization and Mapping Algorithm

Khoa Nguyen Dang, Tran Thi Van, Vijender Kumar Solanki, and Duc-Tan Tran

Abstract Simultaneous localization and mapping (SLAM) is a fundamental technique used in robotics, autonomous vehicles, and augmented reality to construct maps of unknown environments while simultaneously keeping track of the robot's location within that environment. SLAM is particularly significant for unmanned aerial vehicle quadrotors because it enables them to navigate autonomously in unknown or GPS-denied environments while simultaneously creating maps of their surroundings. Therefore, this paper presents the method to apply SLAM in a quadrotor in both simulation and experiment. With simulation, the quadrotor and mapping are presented in the Gazebo simulation. Besides, the experiment with a quadrotor model is used, and collected SLAM information which is sent to an unmanned ground vehicle to improve the performance of the system.

Keywords Simultaneous localization and mapping · RTAB-Map · UAV quadrotor

K. N. Dang
Faculty of Engineering and Technology, International School, Vietnam National University, Hanoi, Vietnam
e-mail: khoand@vnuis.edu.vn

T. T. Van
Faculty of General Education, University of Labour and Social Affairs, Hanoi, Vietnam
e-mail: tranvantk4@gmail.com

V. K. Solanki
Stanley College of Engineering and Technology for Women, Hyderabad, TG, India
e-mail: spesinfo@yahoo.com

D.-T. Tran (✉)
Faculty of Electrical and Electronic Engineering, Phenikaa University, Hanoi 12116, Vietnam
e-mail: tan.tranduc@phenikaa-uni.edu.vn

V. K. Gunjan et al. (eds.), *Cybernetics, Human Cognition, and Machine Learning in Communicative Applications*, Cognitive Science and Technology,
https://doi.org/10.1007/978-981-97-8533-9_13

1 Introduction

Autonomous robots in real-world scenarios can navigate large, dynamic, and unfamiliar spaces. It needs to have a map of the working environment and also a localization robot body inside the map, this is known as simultaneous localization and mapping (SLAM). There are several SLAM algorithms have been developed such as Hector-SLAM [1], G-Mapping [2], RGBD-SLAM [3], RTAB-Map [4], etc.

Hector SLAM is a widely used SLAM algorithm that relies solely on LIDAR data for real-time mapping and localization, which is designed to work efficiently without requiring odometry or IMU data, making it suitable for platforms that lack these sensors. To overcome the disadvantage of Hector-SLAM, GMapping is a popular SLAM algorithm used in robotics. It is particularly known for creating occupancy grid maps that depict the environment as a grid of cells, with each cell indicating whether it is occupied, free, or unknown. RGBD-SLAM is another type of SLAM algorithm that uses RGB-D (Red–Green–Blue-Depth) sensors to create maps and localize the robot within those maps. RGB-D sensors provide both color (RGB) images and depth information, which significantly enhances the SLAM process by providing rich visual and geometric data. RTAB-Map is built based on the graph-based SLAM approach designed to build the mapping for the environment while simultaneously tracking the location of the robot within the map. It leverages visual and depth information to perform these tasks and is particularly well-suited for scenarios that require robust loop closure (LCE) detection and 3D mapping. All of them can be categorized into two groups of SLAM approaches: 2D-SLAM-based laser sensors and 3D-SLAM-based vision cameras.

Nowadays, Drones, often referred to as unmanned aerial vehicles (UAVs), have become indispensable in many aspects of daily life, revolutionizing industries and providing new capabilities across different sectors. they have many applications in our lives such as agriculture [5], infrastructure inspection [6], public safety and emergency response [7], media and entertainment, delivery and logistics [8], environmental conservation [9], construction and engineering [10], etc. To apply them to support the people, the automation control needs to focus and research. Therefore, SLAM installed in UAV-like quadrotors involves unique challenges and applications. Quadrotors benefit from SLAM by achieving accurate navigation, obstacle avoidance, and autonomous operation in various environments. In particular, the SLAM algorithm is applied to a UAV to build the 3D mapping and send this information to the unmanned ground vehicle (UGV) which works in the same environment. The SLAM data needs to meet some requirements as working well within ROS [11], low delay time and operation in real-time, high accuracy, and the ability for working in a long time.

Some algorithms could performed in the hardware in the loop simulation and software (HILS) in the loop simulation (SILT) system [12] as well as realistic applications. Among them, RTAB-Map was selected to build the SLAM algorithm because it could process data in real-time, making it suitable for dynamic environments where the robot needs to navigate and map simultaneously, visual features to recognize previously visited places, helping to close loops (CLs) and correct for drift in the map, represent the environment as a graph, integrate data from various sensors such as cameras (RGB-D, stereo, monocular), LIDARs, and IMUs, CLs in the map, reducing errors and improving the overall accuracy of the map, and supports 3D mapping.

This paper focuses on applying SLAM in the simulation and implementing it in real applications. The contents are followed as Sect. 2 presents the SLAM based on the RTAB-Map development. Next is the simulation SLAM algorithm with HILS/SITL techniques. The flight information is designed and displayed in Sect. 4. To improve more performance of the system, and implementation with SLAM in real application combination between UAV Quadrotor and UGV is considered in Sect. 5. Finally, the conclusion is marked in Sect. 7.

2 Developing SLAM Based on RTAB-Map

RTAB-Map is an open-source code published in 2013 [4] as a SLAM approach that can be integrated into various robots and mobile platforms, which implements an LCE detection method with memory management to handle large-scale maps and long-term online operations.

RTAB-Map can connect to ROS via the *RTAB-Map_ros*, a package for ROS. Figure 1 describes a node called RTAB-Map ROS with the input and output elements.

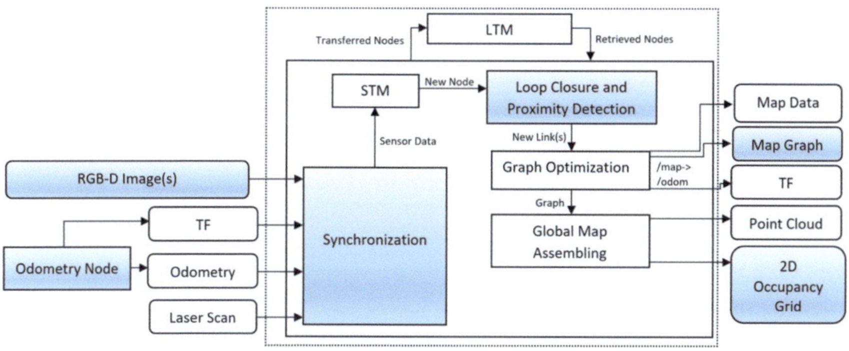

Fig. 1 RTAB-Map node working with ROS

The input source of the node is the camera input (RGBD—Camera or Stereo camera), odometry, laser data from Laser Scan, and TF is used to define the sensor's position relative to the robot base. Input messages are synchronized and forwarded to the Graph-SLAM algorithm. The output includes Map Data, which contains the most recent node with compressed sensor data, and the Map Graph for localization in 3D mapping.

Visual Odometry computes odometry and TF data from its inputs. Figure 2 illustrates the visual odometry process using the RTAB-Map algorithm, calculating odometry from frame-to-frame and frame-to-map, with RGB-D cameras as inputs. Accurate TF data regarding the camera's placement on the robot is essential for transforming the output odometry into the robot's base frame.

Herein, TF defines the spatial relationship between the camera position and the robot base, facilitating the publication of odometry transforms relative to the robot base. The data flow remains consistent whether using a stereo camera or an RGB-D camera. Processing proceeds as follows::

- Feature Detection block: The frame is captured based on the camera and the GoodFeaturesToTrack [13] (GFTT) to extract the features in each frame.
- Feature Matching block: By frame to map (F2M), Matching is conducted using a nearest neighbor search algorithm [14] employing parameters such as nearest neighbor distance ratio (NNDR). BRIEF descriptors [15] are utilized for feature extraction in the Feature Map. Optical flow for frame-to-frame analysis utilizes GFTT features, enabling rapid feature correspondences toward the Key Frame without the need for descriptor extraction.

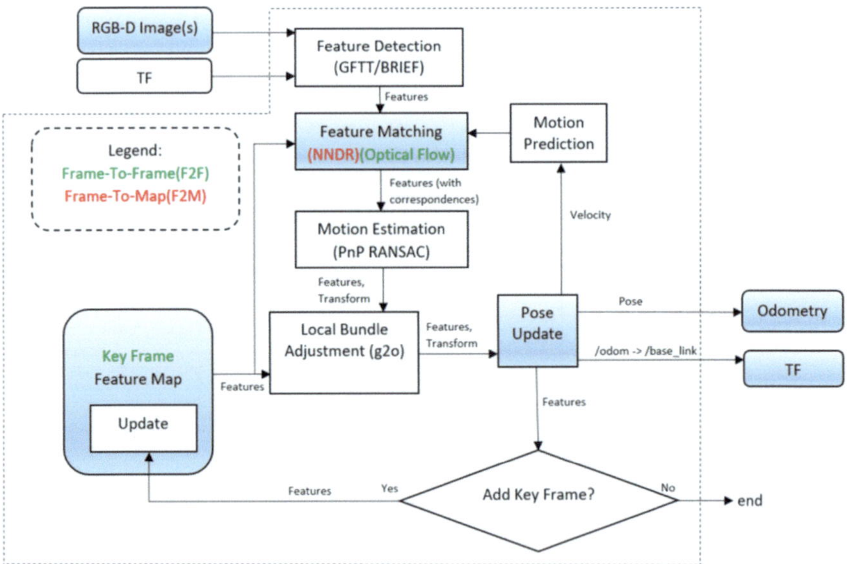

Fig. 2 Block diagram of rgbd_odometry ROS node

- Motion-Prediction block: A motion model predicts the expected positions of features from the Key Frame (F2F) or the Feature Map (F2M) in the current frame, based on the previous motion transformation.
- Motion-Estimation: Feature correspondences are computed, and the Perspective-n-Point with RANSAC algorithm in OpenCV [16] is used to estimate how well the transformation of the current frame aligns with features from either the Feature Map in F2M or the Key Frame in F2F.
- Local-Bundle-Adjustment block: The resulting transformation is refined using local bundle adjustment [17] applied to features from all keyframes in the Feature Map in F2M, or specifically those from the last Key Frame in F2F.
- Pose Update Block: using the estimated transformation, the odometry, and TF / odom → /base link transform output are updated with sample time.
- Update Key-Frame and Feature-Map process: During Motion Estimation, if the number of inliers falls below a fixed threshold defined as "*Odom/KeyFrameThr*" actions are taken. When the size of the Feature Map exceeds another fixed threshold by "*OdomF2M/MaxSize*" the oldest features that do not correspond with the current frame are removed.

The synchronization block aims to minimize time delays across input topics. It distinguishes between two types: exact and approximate synchronization. Exact synchronization requires input topics to have identical timestamps, typically from the same sensor (e.g., left and right camera images). In contrast, approximate synchronization compares timestamps of incoming topics to align them with minimal delay error, making it suitable for topics originating from different sensors.

Following data synchronization from the sensor, the Short-Term-Memory (STM) block is established to the node to store Odometry pose, sensor data, and other relevant information for subsequent blocks. This includes visual words for LCE and Proximity Detection (PDn), as well as a local occupancy grid for Global Map Assembly. Integration of LCE and Proximity links occurs through detection mechanisms. Upon adding a new LCE (proximity link) to the graph, graph optimization (GOn) is performed to propagate computed errors and minimize odometry drift. Once optimized, outputs such as Point Cloud (PCd), OctoMap, and Occupancy Grid are generated and published (sent) to another block.

Odometry Node block: The Odometry Node can utilize both camera and lidar inputs to provide RTAB-Map with the estimated robot pose in the form of an Odometry message, accompanied by the corresponding TF transform (/odom → /base_link). Visual odometry integrates two established approaches [18] referred to as Frame-To-Map and Frame-To-Frame. The primary distinction between these methods lies in their registration strategies: F2F aligns each new frame with the last keyframe, and F2M aligns with a local map composed of features from previous keyframes. Similarly, lidar-based approaches, known as Scan-To-Map (S2M) and Scan-To-Scan (S2S), apply similar principles but with PCds instead of visual features. The following sections detail the implementation of the Odometry Node using each selected visual or lidar odometry approach.

Loop-Closure and Proximity-Detection block: Loop-closure-detection (LCD) is a vital component of SLAM algorithms, essential for recognizing previously visited locations. This process entails matching the current location with past ones to establish connections with recently visited places. As the internal map expands, the processing time for new observations rises, potentially affecting real-time performance. In probabilistic SLAM methodologies, LCD focuses on identifying local matches between new observations and a restricted map region, taking into account the uncertainty in the robot's position.

For global LCD, vision information is often used to gather observations from the environment, leveraging the distinctive features it extracts. However, successful large-scale mapping can also be achieved using laser range finder data. In mapping based on vision, the bag-of-words technique [19, 20] is widely utilized and has demonstrated effective online LCD for paths covering distances of up to 1000 km [30]. This method entails representing all images using visual words selected from a predefined vocabulary, usually based on a descriptor of local features such as scale-invariant feature transform [21]. Given the high dimensionality of these features, they are quantized into a vocabulary to enable rapid comparison rather than direct feature-to-feature comparisons. Popular quantization methods include Vocabulary Tree [22], Approximate K-Means [23], and K-D Tree [24]. Establishing connections between each visual word and related images enables efficient comparisons across extensive datasets, resembling the term-frequency-inverse-document-frequency (TFIDF) method [25].

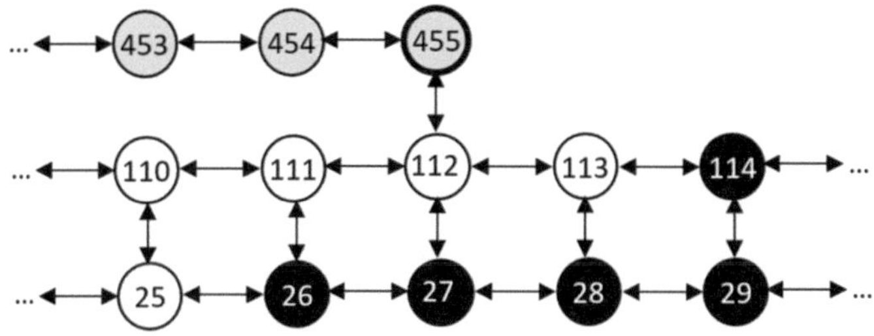

Fig. 3 Graph representation of locations

Finally, management memory strategies in localization for robots aim to improve recognition accuracy in dynamic environments [26] and optimize memory usage. In our context, these strategies are applied to online localization and mapping, enabling the dynamic addition of new locations over time.

In Fig. 3, the arrows vertically represent LCE links, while the arrows horizontally are neighbor links. The dotted lines indicate no detected LCE. Black nodes represent locations in the Long-Term-Memory (LTM), white nodes are in the Working-Memory (WM), and gray nodes are in the Short-Term-Memory (STM). The node number 455 denotes the current acquired location.

The local maps can be created with the graph map determines the resulting global maps. Specifically, the 3D occupancy grid (OctoMap) and its 2D projection are generated exclusively from 3D local occupancy grids.

Global Map Assembling Block: Fig. 4 illustrates the global map outputs derived from local occupancy grids. Storing nodes with local occupancy grids (LOG) in three dimensions offers flexibility since they can generate all map types. Alternatively, for a 2D global occupancy grid (GOG) map, storing already projected local grids in nodes saves memory and reduces assembly time. Using the map's graph, each LOG is transformed according to its corresponding pose. When a new node is added to the map, the new LOG merges with the GOG, updating obstacles by removing existing ones and adding new ones. During LCE events, the global map is reassembled using the optimized poses of all nodes in the map's graph. This process ensures that obstacles previously cleared incorrectly before the LCE are accurately reinstated. PCd outputs aggregate the points from the map locally and publish (send) this information in the message defined as msgs/PointCloud in ROS message format.

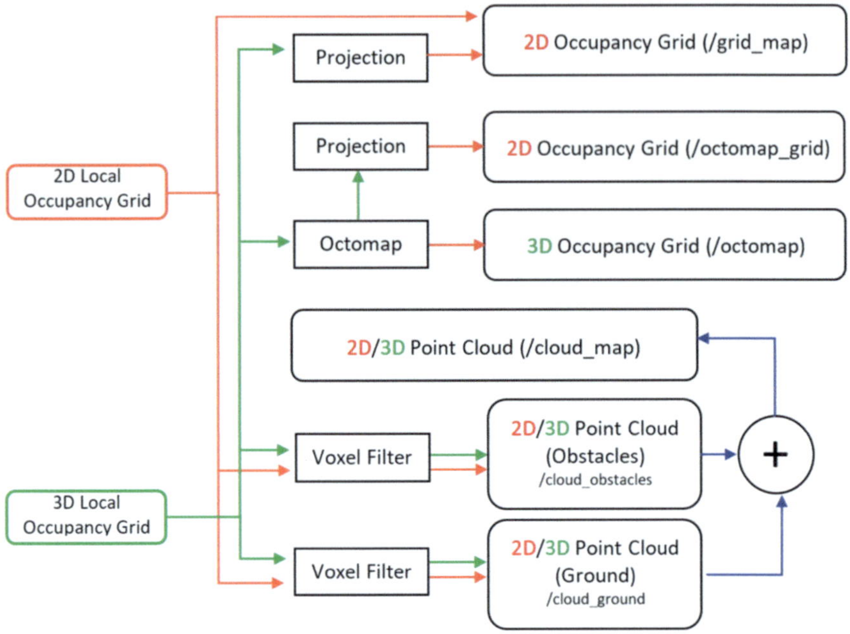

Fig. 4 Global map assembling

GOn block: When an LCE is detected, specific nodes are retrieved and transferred due to MM, followed by applying a GOn approach to minimize map errors. RTAB-Map incorporates three methods for GOn: g2o [17], TORO [27], and GTSAM [28]. GTSAM and g2o methods converge more quickly than the method of TORO, but they may be less robust in scenarios involving multi-session mapping where multiple independent graphs need merging. TORO, however, is less affected by poorly estimated odometry covariance. Based on empirical data for single-map scenarios, g2o, and GTSAM generally offer better optimization quality than the method of TORO, particularly for 6 degrees of freedom (6DoF) maps. GTSAM shows slightly greater robustness in multi-session contexts compared to g2o, making it the default strategy in RTAB-Map instead of TORO. However, LCE detection is prone to errors; detecting very similar places can result in invalid LCEs, potentially increasing map errors rather than reducing them.

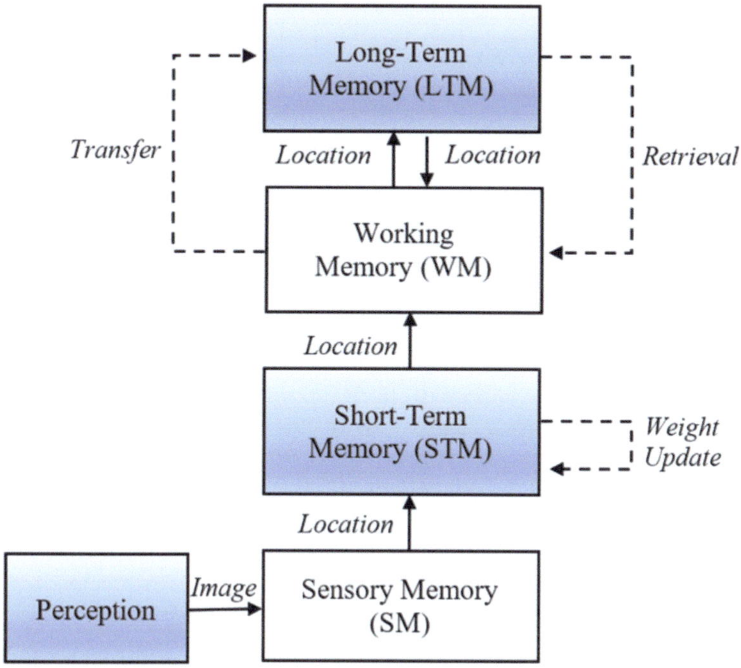

Fig. 5 RTAB-Map management of memory (RMMM)

RTAB-Map employs a memory management strategy integrated with graph management modules to regulate graph size, supporting continuous SLAM operations in extensive environments. In the absence of effective MM, as the graph expands, processing times for critical modules like LCE and PDn, GOn, and Global Map Assembly might surpass real-time limits, potentially exceeding the node acquisition cycle time.

Figure 5 depicts the RMMM model. The Perception module captures an image and sends it to Sensory Memory (SM) which analyzes the extracted image signatures to reduce data dimensionality and identify relevant features for LCE detection. Afterwards, SM creates a new location with the image signature and passes it on to Short-Term Memory (STM). Inside STM, a process called Weight Update (WUe) continuously adjusts recently created locations. If the WUe process finds that the new location closely resembles the last one in STM, it merges them into the new location and increases its weight.

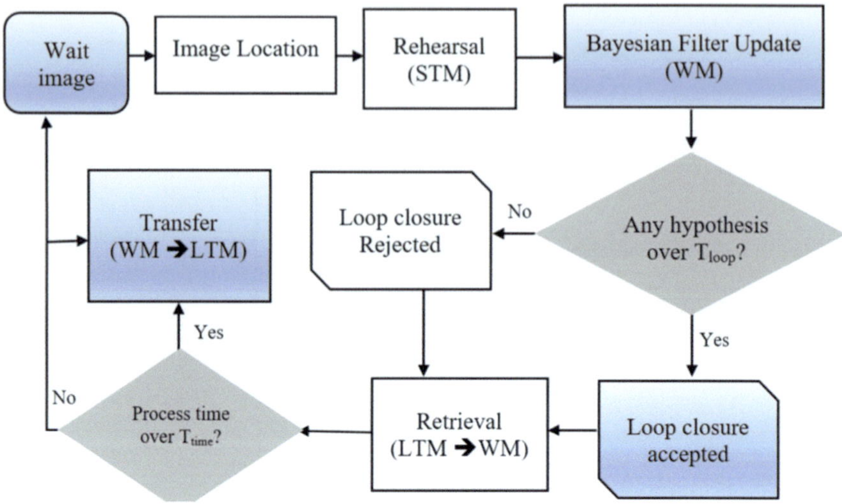

Fig. 6 The process of memory management in RTAB-Map

STM monitors temporal similarities between consecutive images to update weights, while WM detects spatial LCE between locations, as illustrated in Fig. 6. RTAB-Map refrains from utilizing entries in STM to prevent detecting LCE on recently visited locations, as the last location often bears resemblance to the most recent ones. The size of STM denoted as T_{STM}, is set based on the robot's velocity and location acquisition rate. When the number of locations in STM reaches T_{STM}, the oldest location is moved to WM. Algorithm 1 [4] provides an overview of the LCE detection process, elaborated further in the subsequent sections.

Algorithm 1:

```
time ← TIMENOW()                      TIMENOW() returns current time
I_t ← acquired image
L_t ← LOCATIONCREATION(I_t)
if z_t (of L_t) is a bad signature (using T_bad) then
    Delete L_t
else
    Insert L_t, into STM, adding a neighbor link with L_t-1
    Weight Update of L_t in STM (using T_similarity)
    if STM's size reached its limit (T_STM) then
        Move oldest location of STM to WM
    end if
    p(S_t|L_t) ← Bayesian Filter Update in WM with L_t
    Loop Closure Hypothesis Selection (S_t = i)
    if S_t = i is accepted (using T_loop) then
        Add loop closure link between L_t and L_i
    end if
    Join trash's thread               ▷ Thread started in TRANSFER()
    RETRIEVAL(L_i)                    ▷ LTM → WM
    pTime ← TIMENOW()-time            ▷ Processing time
    if p_Time > T_Time then
        TRANSFER()                    ▷ WM → LTM
    end if
end if
```

3 Simulation SLAM Algorithm

In this section, we will simulate the SLAM algorithm with the RTAB-Map ROS package based on HILS/SILT techniques. The system includes a haptic device as in Fig. 7 for the ground controller, a PC based on the Ubuntu version 14 installed ROS indigo, and the PC connected to the haptic device via USB connection. For the simulation, a drone-typed quadrotor model is developed and run in ROS, we have used the Hector_Quadrotor ROS package [1] with a laser sensor and depth camera mounted onto the base_frame. The inputs of RTAB-Map include */image_raw, /camera_info* with pretext */camera/rgb, /image_raw,* and */camera_info* with pretext */camera/depth.* And it is configured as in Fig. 8.

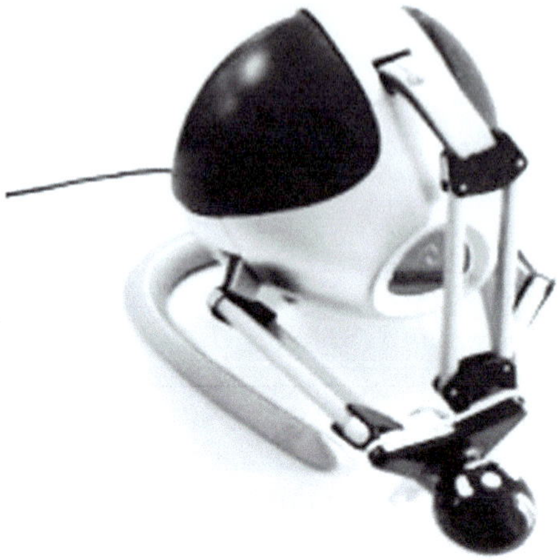

Fig. 7 The haptic device of Novint

```
<!-- RTAB-Map's parameters -->
<param name="Reg/Strategy"        type="string" value="1"/>    <!-- 0=Visual, 1=ICP, 2=Visual+ICP -->
<param name="Reg/Force3DoF"       type="string" value="true"/>
<param name="RGBD/ProximityBySpace"      type="string" value="false"/>

<!-- RTAB-Map's parameters -->
    <param name="LccIcp/Type"                     type="string" value="2"/>
    <param name="RGBD/OptimizeSlam2D"             type="string" value="true"/>
    <param name="RGBD/LocalLoopDetectionSpace" type="string" value="true"/>
    <param name="LccIcp2/CorrespondenceRatio"  type="string" value="0.25"/>
    <param name="Vis/MaxDepth"                 type="string" value="10.0"/>
    <param name="Vis/InlierDistance"           type="string" value="0.1"/>
    <param name="Vis/MinInliers"               type="string" value="10"/>
    <param name="RGBD/AngularUpdate"              type="string" value="0.1"/>
    <param name="RGBD/LinearUpdate"               type="string" value="0.1"/>
    <param name="Kp/DetectorStrategy"             type="string" value="10"/>
    <param name="GFTT/MinDistance"                type="string" value="5"/>
```

Fig. 8 RTAB-Map value configs in the launch file

The odometry data has come from a 2D lidar sensor as Sect. 2. A 3D mapping is built in the gazebo software [29] for the simulation indoor environment as in Fig. 9.

The scenario is performed in that the haptic will control the velocity of the quadrotor in three directions by *X, Y*, and *Z*. It moves in the mapping in Gazebo in 3D. As a result in Fig. 10, the 3D mapping was built in the simulation based on the RTAB-Map. This mapping could reappear in the mapping of the gazebo simulation.

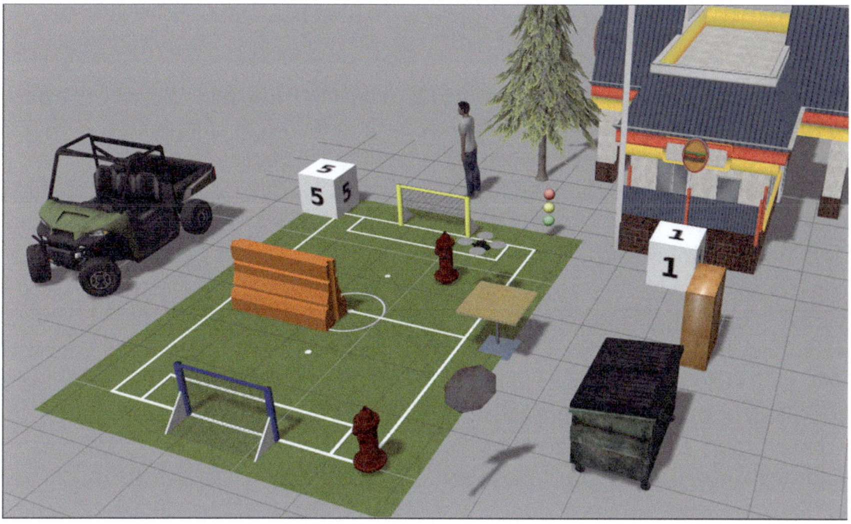

Fig. 9 The gazebo environment building

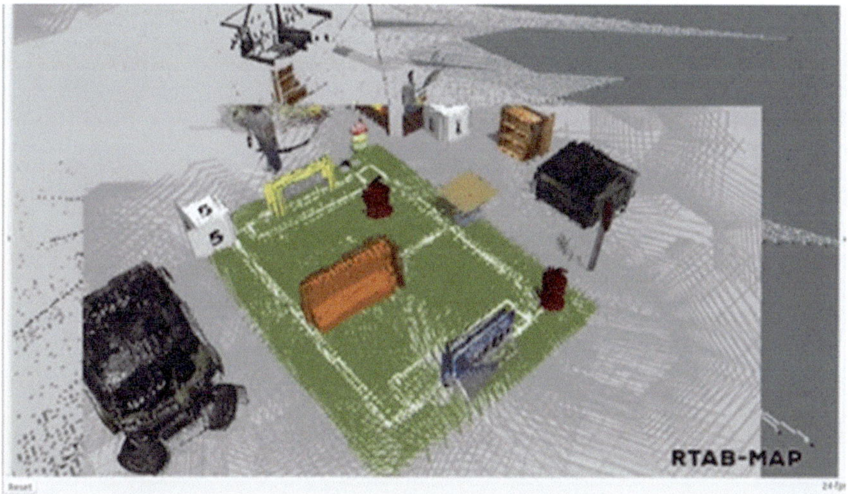

Fig. 10 The 3D mapping building based on RTAB-Map for SLAM

4 Display Flight Information in an Interface

A graphic user interface (GUI) is created to manage the flight and easily to visualization the data from UAV instead of viewing it in the Linux commands. A GUI in Fig. 11 is designed and developed in QT 5 and C ++ language.

The primary flight display (PFD) screen as in Fig. 12 is the main display with many parameters of the flight includes an Airspeed Indicator (AIR), Attitude Indicator (AI), Turn Indicator (TI),Vertical Speed Indicator (VSI), and Horizontal Situation Indicator (HSI). The airspeed is situated to the left of the screen of AI. And the vertical speed and the altitude and is on the right. To the right of the AI, vertical deviation for ILS glideslope or VNAV (vertical navigation) is displayed. The ILS, FMS, and VOR track is displayed below the AI. A compass reference is provided at the bottom of the instrument panel.

Function *qfi_PFD* is used to define the size of the PFD screen with the k_original_width and k_original_height. We also define the size of the AI, AIR, VSI, TI, and HSI screen by those parameters *k_scaleX*, and *k_scaleY*. We used the QGraphicsSvgItem class for the construction of the layout of the screen.

Fig. 11 The GUI application is running and displaying the flight data

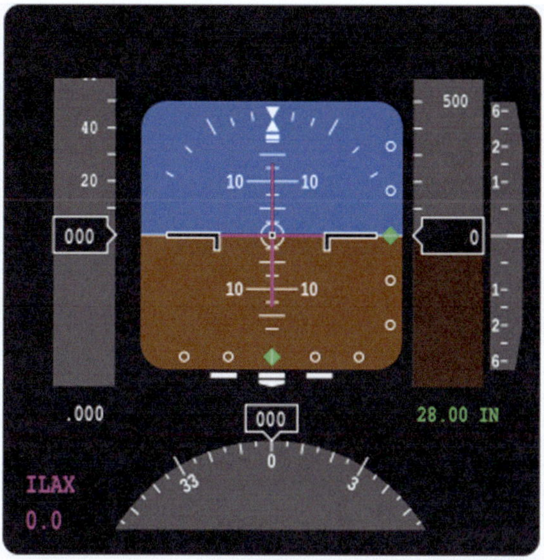

Fig. 12 The primary flight display (PFD) screen

```
void qfi_PFD::init()
{
   k_scaleX = (float)width() / (float)k_originalWidth;
   k_scaleY = (float)height() / (float)k_originalHeight;
   k_adi->init( k_scaleX, k_scaleY );
   k_alt->init( k_scaleX, k_scaleY );
   k_asi->init( k_scaleX, k_scaleY );
   k_hsi->init( k_scaleX, k_scaleY );
   k_vsi->init( k_scaleX, k_scaleY );
   k_ils->init( k_scaleX, k_scaleY );
   k_itemBack = new QGraphicsSvgItem( ":/qfi/images/pfd/pfd_back.svg" );
   k_itemBack->setCacheMode( QGraphicsItem::NoCache );
   k_itemBack->setZValue( k_backZ );
   k_itemBack->setTransform( QTransform::fromScale( k_scaleX, k_scaleY ), true );
   k_scene->addItem( k_itemBack );
   k_itemBack = new QGraphicsSvgItem( ":/qfi/images/pfd/pfd_mask.svg" );
   k_itemBack->setCacheMode( QGraphicsItem::NoCache );
   k_itemBack->setZValue( k_backZ );
   k_itemBack->setTransform( QTransform::fromScale( k_scaleX, k_scaleY ), true );
   k_scene->addItem( k_itemBack );
   centerOn( width() / 2.0f , height() / 2.0f );
   updateView();
}
```

The attitude direction indicator (ADI) screen and (ALT) screen are configured by the *qfi_PFD::ADI* and *qfi_PFD::ALT* functions, respectively.

```
void qfi_PFD::ALT::init(float scaleX, float scaleY)
{
  k_scaleX = scaleX;
  k_scaleY = scaleY;
  reset();
  k_itemBack = new QGraphicsSvgItem(":/qfi/images/pfd/pfd_alt_back.svg");
  k_itemBack->setCacheMode(QGraphicsItem::NoCache);
  k_itemBack->setZvalue(k_backZ);
  k_itemBack->setTransform(QTransform::fromScaleX(k_scaleX, k_scaleY), true);
  k_itemBack->moveBy(k_scaleX * k_originalBackPos.x(), k_scaleY * k_originalBackPos.y());
  k_scene->addItem(k_itemBack);
  k_itemScale1 = new QGraphicsSvgItem( ":/qfi/images/pfd/pfd/pfd_alt_scale.svg" );
  k_itemScale1->setCacheMode(QGraphicsItem::NoCache);
  k_itemScale1->setZvalue(k_scaleZ);
  k_itemScale1->setTransform(QTransform::fromScaleX(k_scaleX, k_scaleY), true);
  k_itemScale1->moveBy(k_scaleX * k_originalScalelPos.x(), k_scaleY * k_originalScalelPos.y());
  k_scene->addItem(k_itemScale1);
  k_itemScale2 = new QGraphicsSvgItem( ":/qfi/images/pfd/pfd/pfd_alt_scale.svg" );
  k_itemScale2->setCacheMode(QGraphicsItem::NoCache);
  k_itemScale2->setZvalue(k_scaleZ);
  k_itemScale2->setTransform(QTransform::fromScale(k_scaleX, k_scaleY), true);
  k_itemScale2->moveBy(k_scaleX * k_originalScalemPos.x(), k_scaleY * k_originalScalemPos.y());
  k_scene->addItem(k_itemScale2);
  k_itemLabel1 = new QGraphicsTextItem(QString("99999"));
  k_itemLabel1->setCacheMode(QGraphicsItem::NoCache);
  k_itemLabel1->setZvalue(k_labelZ);
  k_itemlabel1->setDefaultTextColor(k_labelsColor);
  k_itemLabel1->setFont(k_labelsFont);
  k_itemLabel1->setTransform(QTransform::fromScale(k_scaleX, k_scaleY), true);
  k_utempLabel1->moveBy(k_scaleX * (k_originalLabelsX - k_itemLabel1->boundingRect().width() / 2.0f),
 k_scaleY * (k_originalLabelsY - k_itemLabel1->boundingRect().height() / 2.0f));
  k_scene->addItem(k_itemLabel1);
}
```

```
void qfi_PFD::ADI::init( float scaleX, float scaleY )
{
        k_scaleX = scaleX;
        m scaleY = scaleY;
        reset();
        k_itemBack = new QGraphicsSvgltem( ":/qf1/images/pfd/pfd_adi_back.svg" );
        k_itemBack -> setCacheMode( GraphicsItem: :NoCache ) ;
        k_itemBack -> setZValue( k_backZ );
        k_itemBack -> setTransform( Qtransform::fromScale( k_scaleX, k_scaleY ), true) ;
        k_itemBack -> setTransformOriginPoint( k_originalAdiCtr – k_originalBackPos );
        k_itemBack -> moveBy( k_scaleX * k_originalBackPos.x(), k_scaleY * k_originalBackPos.y() );
        k_scene -> addItem( k_itenBack );
        k_itemLadd = new QGraphicsSvgItem( ":/qfi/images/pfd/pfd_adi_ladd.svg" );
        k_itemLadd -> setCacheMode( QGraphicsItem: :NoCache ):
        k_itemLadd -> setZValue( k_laddZ );
        k_itemLadd -> setTransform( QTransform::fromScale(k_scaleX, k_scaleY ), true ):
        k_itemLadd -> setTransformOriginPoint( k_originalAdiCtr - k_originalLaddPos );
        k_itemLadd -> moveBy( k_scaleX * k_originalLaddPos.x(), k_scaleY * k_originalLaddPos.y() );
        k_scene -> addItem( k_itemLadd ):
        k_itemRoll = new QGraphicsSvgItem( ":/qfi/images/pfd/pfd_adi_roll.svg" );
        k_itemRoll -> setCacheMode( QGraphicsItem::NoCache );
        k_itemRoll -> setZValue( k_rollz ):
        k_ itemRoll -> setTransform( QTransform::fromScale(k_scaleX, k_scaleY ), true );
        k_ itemRoll -> setTransformOriginPoint( k_originalAdiCtr - k_originalLaddPos );
        k_itermRoll -> moveBy( k_scaleX * k_originalRollPos.x(), k_scaleY * k_originalRollPos.y() );
        k_scene -> addItem( k_itemRoll );
        k_itemSlip = new QGraphicsSvgItem ( ":/qfi/images/pfd/pfd_adi_slip.svg" );
        k_itemSlip -> setCacheMode( QGraphicsItem::NoCache ):
        k_itemSlip -> setZValue( k_slipZ ):
        k_itemSlip -> setTransform( QTransform::fromScale(k_scaleX, k_scaleY ), true ):
        k_itemSlip -> setTransform0riginPoint( k_originalAdiCtr - k_originalSlipPos );
        k_itemSlip -> moveBy( k_scaleX * k_originalSlipPos.x(), k_scaleyY* k_originalStipPos.y() );
        k_scene -> addItem( k_itemSlip );
        k_itemTurn -> new QGraphicsSvgItem( ":/qfi/images/pfd/pfd_adi_turn.svg" );
        k_itemTurn -> setCacheMode( QGraphicsItem: :NoCache );
        k_itemTurn -> setZValue( k_turnZ ):
        k_itemTurn -> setTransform( (Transform:: fromScale(k_scaleX, k_scaleY ), true );
        k_itemTurn -> moveBy( k_scaleX * k_originalTurnPos.x(), k_scaleY * k_originalTurnPos.y() );
        k_scene -> addItem( k_ itemTurn);
}
```

Beside PFD screen, six screens for displaying more information is also designed detail in Fig. 13. Herein, it could present the air speed, altitude, direction, vertical speed, etc. in analog clock which is convenient for a pilot.

Next, the flight information also shows in Fig. 14 such as the UAV velocity, the rotate angular of UAV (Roll, Yaw, Pitch), the position of the haptic gripper and also the position of the UAV. Therefore, the user can observe all the values for visualization purposes. Finally, at the bottom is the control panel, this area contains the address IP of ROS server, and the connecting function to the server.

Fig. 13 UAV parameters displaying on screen

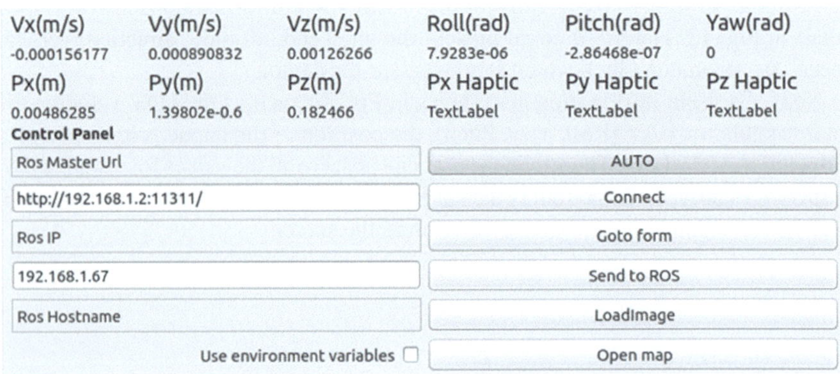

Fig. 14 Flight information displaying

5 Developing Quadrotor and Implementation with SLAM

The UAV quadrotor is designed to bring the RGB-D camera, Lidar, Raspberry board and other sensors. For basically, the payload of the RPLidar A1: 170 g, Raspberry Pi 3B + : 42 g, Asus Xtion Pro: 540 g, Battery: 252 g, UAV: 1330 g. Therefore, the UAV need to have the payload more than 2334 g. An UAV is selected with the named VOXL m500 Quadcopter shown in Fig. 15.

In the real flight, the Asus Xtion RGB-D camera is added to the Quadrotor UAV to build the 3D mapping and localization as in Fig. 16.

For deploying the SLAM algorithm, a Raspberry Pi 3 (RP3) serves as an embedded PC, integrating most computer components into a compact unit. The RP3 utilizes a 1.2 GHz based on ARM processor [30], specifically chosen for this research. It

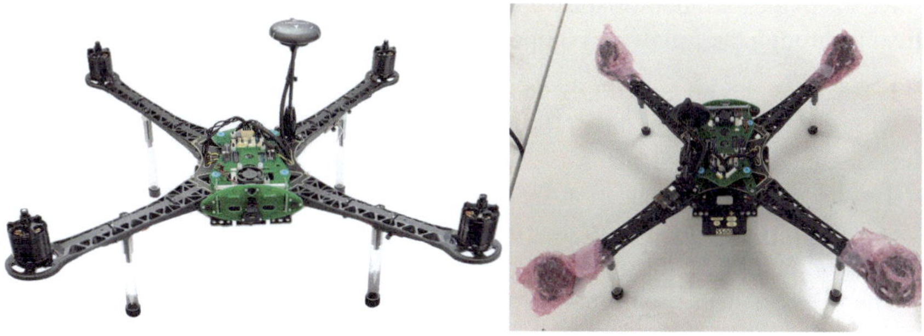

Fig. 15 VOXL m500 quadcopter

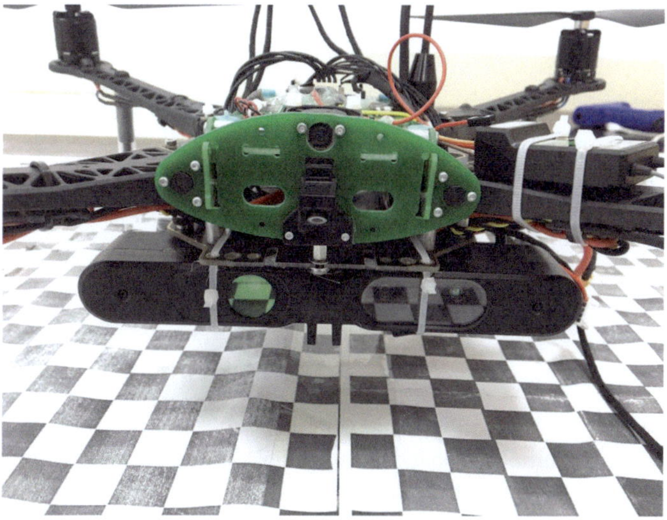

Fig. 16 Asus Xtion RGB-D camera integrated into the quadrotor UAV

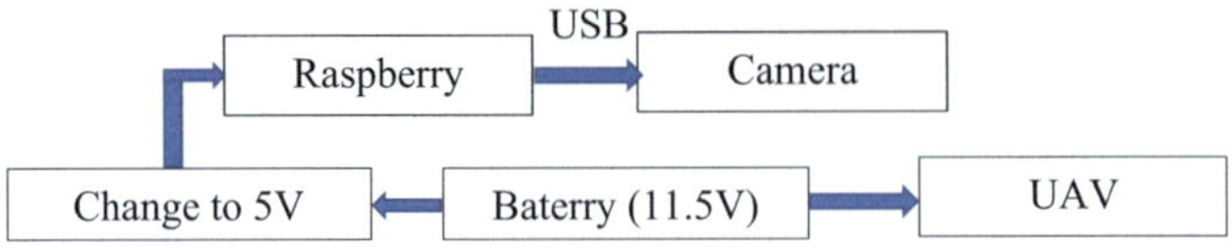

Fig. 17 Connected between raspberry and camera

intalled Operating System with the Ubuntu 16.04 Mate on a Micro SD card size 32 GB. Additionally, ROS is installed along with its essential packages. The RP3 connects to the quadrotor base and the Asus_xtion_pro Sensor (AXPS) via USB cables, and its position mount to at the center of the robot. The RP3 manages data collection from the sensors, including depth images from the AXPS and images/ video streams from the same device. The hardware setup is illustrated in Fig. 17.

RP3 provides a range of software tools for viewing maps and accessing video streams through the following components:

- It becomes the node in ROS, facilitating communication with all other nodes within the ROS environment. It serves as the central database listing all available nodes and their access paths.
- The teleoperation application manages the connection to the ROS system and sends control commands to the robot for movement control. It utilizes a haptic device connected to the respective PC to guide its operations.
- The 3D visualizer used to all informations of topic in ROS is known as RVIZ, developed by the ROS community [7]. RVIZ serves as a comprehensive visualizer capable of presenting the robot system model, vide, camera, 3D models, and generated maps. Additionally, it could collected all data from sensors including the infrared distance measurements, camera data, and sonar data.
- The Map server application is used to save the 3D maps and save it into workstation storage.

In this experiment, an environment for testing the SLAM algorithm as in Fig. 18. Some boxes are built to recognize the object, and the UAV will avoid them.

As the result in Figs. 19 and 20, the comparison between the RGB image and 3D image is shown to prove that the 3D mapping could present a map that is similar to the environment. Figure 4 is the map saved in the memory.

6 UAV and UGV Integration

Here the UGV model is used with named TurtleBot 3 to synchronization with the UAV system. The TurtleBot 3 has the responsibility of getting the 3D map from a personal computer, which is a tricycle mobile robot with 2 steering wheels. Moreover, TurtleBot 3 is integrated with many sensors so we can easily program and develop many algorithms for robots. At the top of the robot, there is a lidar sensor used to

Fig. 18 Environment testing in experiment

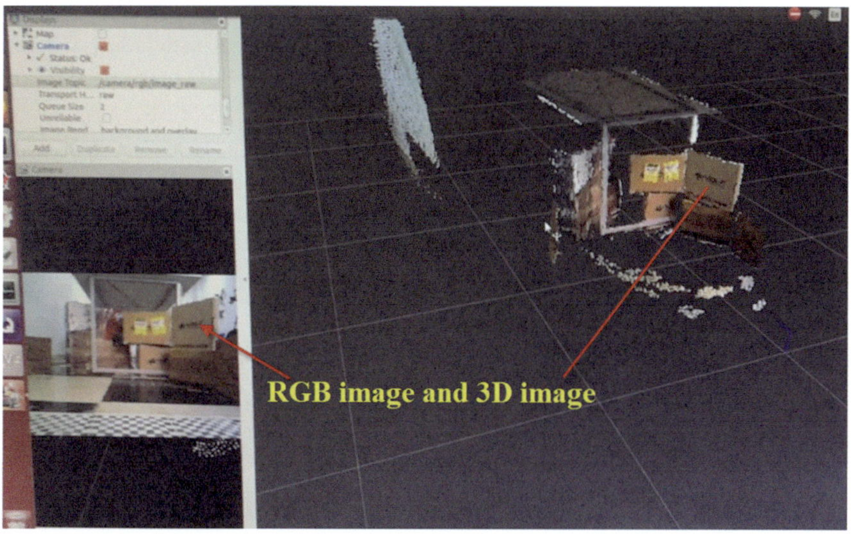

Fig. 19 Showing RGB image and 3D image

Fig. 20 The 3D mapping saving

measure the distance from the body of the robot to the other surrounding objects. The processor core of UGV is the mini-embedded computer RP3 B+ .

The Ubuntu Mate 16.04 Operating System had already been installed on Raspberry Pi. Furthermore, to get the 3D map from the laptop computer, the Raspberry Pi has to install ROS Melodic and the rtabmap_ros package. The installation steps are the same as in the personal computer. Figure 21 describes the 3D map in Raspberry Pi. This map is taken from a personal computer via the ROS network.

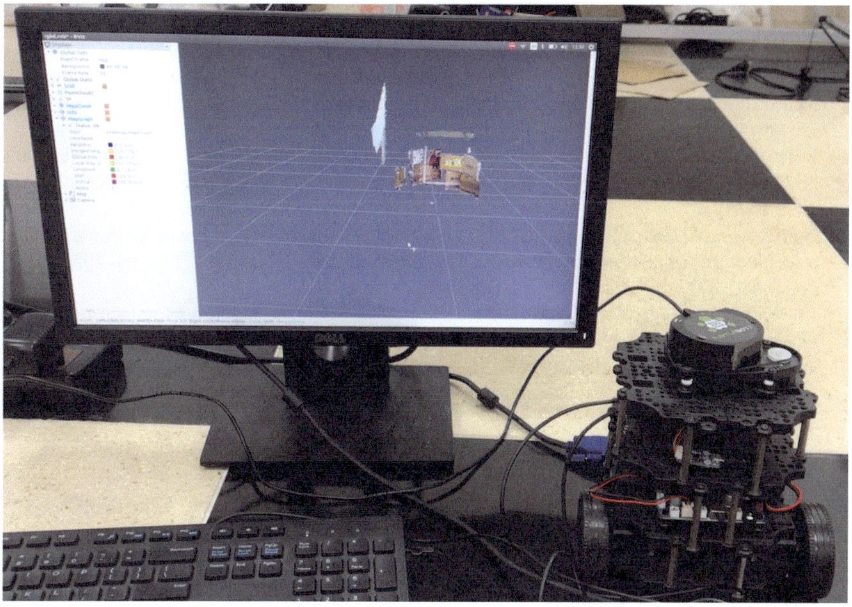

Fig. 21 Visualization of the 3D map in UGV from the personal computer

7 Conclusion

This chapter presents a method of applying the SLAM algorithm based on RTAB-Map for a UAV Quadrotor. The 3D mapping is represented with 3D camera input for this algorithm. The results are shown in both the simulation of HILT/SILT techniques and the experiment. A UAV Quadrotor is selected to experiment and integrated the Asus Xtion 3D camera in it. The RTAB-Map is installed in RP3+ board and connected to the network. The 3D mapping information could transfer to a UGV and display information. Besides, the GUI is designed and displays important information in the ground station of the UAV with full function and meets the requirements of a software UAV. All results prove the performance of the system that it could apply in the real application combined between SLAM and UAV Quadrotor.

References

1. Saat S, Abd Rashid W, Tumari M, Saealal M (2020) Hectorslam 2D mapping for simultaneous localization and mapping (SLAM). J Phys Conf Ser 1529(4): 042032. https://doi.org/10.1088/1742-6596/1529/4/042032.
2. Su Z, Zhou J, Dai J, Zhu Y (2020) Optimization design and experimental study of gmapping algorithm. In: 2020 Chinese control and decision conference (CCDC), pp 4894–4898. https://doi.org/10.1109/CCDC49329.2020.9164603

3. Muravyev K, Yakovlev KS (2022) Evaluation of RGB-D SLAM in large indoor environments. abs/2212.05980. https://api.semanticscholar.org/CorpusID:254564493
4. Labbé M, Michaud F (2018) RTAB-Map as an open-source lidar and visual simultaneous localization and mapping library for large-scale and long-term online operation. J Field Robot 36:416–446
5. Rejeb A, Abdollahi A, Rejeb K, Treiblmaier H (2022) Drones in agriculture: a review and bibliometric analysis. Comput Electron Agric 198:107017. https://doi.org/10.1016/j.compag.2022.107017
6. Gao C, Wang X, Chen X, Chen BM (2024) A hierarchical multi-UAV cooperative framework for infrastructure inspection and reconstruction. Control Theory Technol 22(3):394–405. https://doi.org/10.1007/s11768-024-00202-0
7. Jin W, Yang J, Fang Y, Feng W (2020) Research on application and deployment of UAV in emergency response. In: 2020 IEEE 10th international conference on electronics information and emergency communication (ICEIEC), pp 277–280. https://doi.org/10.1109/ICEIEC49280.2020.9152338.
8. Li H, Zhu S, Tolba A, Liu Z, Wen W (2023) A reliable delivery logistics system based on the collaboration of UAVs and vehicles. Sustainability 15(17): 2023. https://doi.org/10.3390/su151712720
9. Seier G et al (2021) Unmanned aircraft systems for protected areas: Gadgetry or necessity? J Nat Conserv 64:126078. https://doi.org/10.1016/j.jnc.2021.126078
10. Emran BJ, Najjaran H (2018) A review of quadrotor: an underactuated mechanical system. Annu Rev Control 46:165–180. https://doi.org/10.1016/j.arcontrol.2018.10.009
11. Lee H, Yoon J, Jang M-S, Park K-J (2021) A robot operating system framework for secure UAV communications. Sensors 21(4). https://doi.org/10.3390/s21041369
12. Tahir M, Mir I, Islam T (2023) A review of UAV platforms for autonomous applications: comprehensive analysis and future directions. IEEE Access 1–1. https://doi.org/10.1109/ACCESS.2023.3273780
13. Zhou X, Koltun V, Krähenbühl P (2020) Tracking objects as points. In: Computer vision—ECCV 2020: 16th European conference, Glasgow, UK, August 23–28, 2020, proceedings, part IV. Springer, Berlin, pp 474–490. https://doi.org/10.1007/978-3-030-58548-8_28
14. Li W et al (2020) Approximate nearest neighbor search on high dimensional data — experiments, analyses, and improvement. IEEE Trans Knowl Data Eng 32(8):1475–1488. https://doi.org/10.1109/TKDE.2019.2909204
15. Ma D, Lai H (2019) Remote sensing image matching based improved ORB in NSCT domain. J Indian Soc Remote Sens 47(5):801–807. https://doi.org/10.1007/s12524-019-00958-y
16. Cheng Y, Huang Z, Quan S, Cao X, Zhang S, Yang J (2023) Sampling locally, hypothesis globally: accurate 3D point cloud registration with a RANSAC variant. Vis Intell 1(1):20. https://doi.org/10.1007/s44267-023-00022-x
17. Geng X, Xu L, He X, Yu J (2021) Graph optimization neural network with spatio-temporal correlation learning for multi-node offshore wind speed forecasting. Renew Energy 180:1014–1025. https://doi.org/10.1016/j.renene.2021.08.066
18. Cabrera-Ávila EV, da Silva BMF, Gonçalves LMG (2024) Nonlinearly optimized dual stereo visual odometry fusion. J Intell Robot Syst 110(2):56. https://doi.org/10.1007/s10846-024-02069-4
19. Cui Y, Zhang Y, Chen X, Dong J, Wu Q, Zhu F (2022) BoW3D: bag of words for real-time loop closing in 3D LiDAR SLAM. IEEE Robot Autom Lett 8:2828–2835
20. Xi K, He J, Hao S, Luo L (2022). SLAM loop detection algorithm based on improved bag-of-words model. In: 2022 5th international conference on pattern recognition and artificial intelligence (PRAI), pp 683–689. https://doi.org/10.1109/PRAI55851.2022.9904167
21. Al-Zebari A, Kareem SW, Ismael SH (2024). Scale-invariant feature extraction for skin image detection. Int J Intell Syst Appl Eng 12(21s):296–302
22. Yao L et al. (2023) DetCLIPv2: scalable open-vocabulary object detection pre-training via word-region alignment, p 23506. https://doi.org/10.1109/CVPR52729.2023.02250

23. Ortega F, Algar MJ, de Diego IM, Moguerza JM (2023) Unconventional application of k-means for distributed approximate similarity search. Inf Sci 619:208–234. https://doi.org/10.1016/j.ins.2022.11.024
24. Tiwari VR (2023) Developments in KD Tree and KNN Searches. Int J Comput Appl 185(17):17–23. https://doi.org/10.5120/ijca2023922879
25. Dey RK, Das AK (2023) Modified term frequency-inverse document frequency based deep hybrid framework for sentiment analysis. Multimed Tools Appl 82(21):32967–32990. https://doi.org/10.1007/s11042-023-14653-1
26. Kucner TP et al (2023) Survey of maps of dynamics for mobile robots. Int J Robot Res 42(11):977–1006. https://doi.org/10.1177/02783649231190428
27. Huo Y, Bouffard F, Joós G (2021) Decision tree-based optimization for flexibility management for sustainable energy microgrids. Appl Energy 290:116772. https://doi.org/10.1016/j.apenergy.2021.116772
28. Dellaert F (2021) Factor graphs: exploiting structure in robotics. Annu. Rev Control Robot Auton Syst 4. https://doi.org/10.1146/annurev-control-061520-010504
29. Chikurtev D (2020). Mobile robot simulation and navigation in ROS and Gazebo. In: 2020 International Conference Automatics and Informatics (ICAI), pp 1–6 https://doi.org/10.1109/ICAI50593.2020.9311330
30. Mathe SE, Kondaveeti HK, Vappangi S, Vanambathina SD, Kumaravelu NK (2024) A comprehensive review on applications of Raspberry Pi. Comput Sci Rev 52:100636. https://doi.org/10.1016/j.cosrev.2024.100636

Dense Net Built on Relying Generative Adversarial Networks for Conceptualizing a Generative Adversarial Network Image Denoising

K. L. S. Soujanya, Maddela Parameswar, V. Ramaraju, Nuthanakanti Bhaskar, D. Sreekanth, and Sanjib Kumar Nayak

Abstract Latest research has shown that representation denoising techniques (CNN) may be greatly improved by the use of deep Convolutional neural networks. Because of CNN's long-standing focus on the Mean Squared Error (MSE), it looked as though the images lacked assessments of high repetitiveness. Therefore, we employ an effective generative adversarial network (GAN) for picture deposing. An actual deep convolutional dense net architecture serves as our engine for converting energy and aids in resolving the vanishing-gradient complication with very deep systems. Furthermore, by employing Wasserstein-GAN as our damage activity, we balance the training procedure. Additionally, it is possible to consider the Wasserstein distinction among real and generated countenances from discriminators as evidence that has been sufficiently demonstrated in connection to the characteristic of the production sample. Through our labor, a notion that is more wonderful and photorealistic than typical may be create.

Keywords Image denoising · Generative adversarial network · Auto encoders · Convolutional neural network

K. L. S. Soujanya · M. Parameswar · V. Ramaraju
Department of CSE, CMR College of Engineering and Technology, Hyderabad, Telangana, India
e-mail: klssoujanya@cmrcet.org

M. Parameswar
e-mail: mparameswar@cmrcet.org

V. Ramaraju
e-mail: drvrr@cmrcet.org

N. Bhaskar (✉)
Department of CSE, CMR Technical Campus, Hyderabad, Telangana, India
e-mail: bhaskar4n@gmail.com

D. Sreekanth
Department of ECE, CMR Technical Campus, Hyderabad, Telangana, India

S. K. Nayak
Department of CSE (Data Science), CMR Technical Campus, Hyderabad, Telangana, India

V. K. Gunjan et al. (eds.), *Cybernetics, Human Cognition, and Machine Learning in Communicative Applications*, Cognitive Science and Technology,
https://doi.org/10.1007/978-981-97-8533-9_14

1 Introduction

Unofficially, the difficulty of denoising is the opposite of the issue with image super-resolution (SR). They are provided as a means of converting an identical picture provided into a clear image. We think that an identical mathematical model might work well for both SR and image denoising at the same time by modifying the structure to suit the requirements of the various activities. Significant advancements in single image superb resolution (SR) (GANs) have recently been realized thanks to generative adversarial networks. One example of this is SRGAN, a tool that can create a single photo-realistic picture. It is impressing how well GANs can distinguish between perceptually significant variations in texture. We design our generative adversarial network based on recent efforts on one picture super-resolution using DenseNet, which is built on reliant generative adversarial networks. Additionally, Wasserstein-GAN, whose deficit action enhances Earth Mover (EM) distance and may efficiently push the type-pile to work, is used in conjunction with the gradient discipline to stimulate our training. In addition to authorizing the training process almost abandon-free, it also generates a feature image that is similar to the previous GANs.

In this research, we introduce a Densely Connected Denoising Network (DenseNet) constructed a Generative Adversarial Network, which produces a photorealistic picture accompanied with excellent aspect. The Densely Connected Denoising Networks (DenseNet), known as the state-of-the-art convolutional net in the United States of America, usually use bound confections from the aid to the rebuilding row change the difficulties with discrediting/fading inclinations in beneficial ways. In this scenario, characteristics at lower levels can provide more extreme-repetitiveness assessments to top-layer features. Our work, which uses a Generalized Adversarial Network (GAN), produces strong arguments and superior results. We use the very deep GAN accompanied by a continuous Wasserstein-GAN loss function to avoid training from being absolute.

2 Literature Review

Convolutional Neural Networks (CNN) has been utilized for clearing up pictures for a long time. The present study recognizes the task of denoising a picture as an issue of regression. Burger et al. [1] this paper presents a Multi-Layer Perceptron (MLP) neural network architecture designed to address the task of understanding the mapping between a noisy picture and its corresponding quiet version. The proposed approach involves the utilization of simultaneous photo patches including the use of contiguous patches as consistent vectors of input to train the MLP. By decreasing the variance among the noise and clean patches, the noisy patches accompanied by comparable clean patches are sent to MLP for network limit estimation. The denoised image is created by calculating the mean of image patches. In the GANS,

One potential approach to address the issue of noise interference on characteristic maps is to incorporate a multi-scale feature gathering layer. This layer aims to mitigate the effect of noise by extracting features at many scales. Additionally, the utilization of lp regularization can be employed to select the most appropriate characteristic maps. Stacked denoising auto-encoder is another extremely well-known approach for image restoration that learns end-to-end mappings from damaged images. The original ones aim to establish the proposed network construction in a novel way. Convolutional networks are widely recognized as a crucial element in the initial system. Conversely, the subsequent network can exploit non-local filtration planes to provide the inherent non-local connecting with oneself inherent in unpolished notions. Many CNN- and GAN-based neural networks for image denoising have produced excellent results in the past few years. The distinct image super-resolution is a significant issue that the convolutional networks are currently making progress on.

Zhong et al. [2] Recent research has demonstrated that employing deep convolutional neural networks (CNN) can considerably boost the effectiveness of picture denoising techniques. The perception that the images lack high-frequency information is a byproduct of the old CNN methods, which primarily focus on decreasing the Mean Squared Error (MSE). So, for image denoising, we use a generative adversarial network (GAN). Our generator is an extremely deep convolutional densenet framework, This technique aids in mitigating the problem of disappearing gradients that often arises in the context of highly complex neural networks that have many layers. In addition, in order to enhance the stability of the learning procedure, we employ the Wasserstein-GAN as a spare unit.

Li et al. [3] this research first offers a conditional generative adversarial network (CGAN)-based technique for medical picture denoising for diverse unknown sounds. Corresponding to the structural specificity, the suggested architectural design integrates the disturbance picture alongside its corresponding slope picture as dependent information within the system, so augmenting the contrast between the initial message and trash. The interaction between convolutional layers is fully utilized by a unique generation with residual dense blocks to investigate visual context. In order to guarantee consistency between the denoised image and the genuine image, the reconstruction loss and WGAN loss are merged as the objective loss function.

Yu et al. [4] in this paper, for low-dose PET picture denoising, we suggest the parameter-transferred Wasserstein generative adversarial network (PT-WGAN). The two things that this paper contributes are: This study proposes a novel approach to initialize particular duties via transferred learning for training PT-WGAN. By leveraging variables from a prepared approach, the teaching efficacy of PT-WGAN is substantially improved. The dependent on the task initiation method was devised in order to effectively reduce noise in low-dose PET images while preserving important information about structure.

Yang et al. [5] this paper offers a novel Wasserstein distance- and perceptual-based generative adversarial network (GAN)-based CT image denoising technique. A fundamental idea in the optimal transport theory, the Wasserstein distance holds potential for enhancing GAN performance. The GAN focuses primarily on statistically converting the information noise distributions from strong to weak, whereas perception loss reduces noise by contrasting the perceptual properties of a denoised result over those of the baseline truth in a predefined feature space. As a result, our suggested solution applies our understanding of perception of images to the denoising of images task and has the capability of both lowering the degree of image noise and attempting to maintain the essential data in the meantime. In our research using CT scans from medical settings, encouraging findings have been attained.

Zhao et al. [6] in this work, to increase picture synthesis accuracy, we suggest using a Bayesian conditional generative adversarial network (GAN) with concrete dropout. In the meantime, the entire pipeline uses an uncertainty calibration strategy to make the uncertainty produced by the Bayesian network interpretable. The technique is tested using 102 participants' brain tumor datasets and the conversion of T1w to T2w MR Images. The proposed method's results significantly reduce the root mean square error (RMSE) with a p-value of 0.0186 when compared to the traditional Bayesian neural network with Monte Carlo dropout. It also provides an illustration of how the uncertainty recalibration approach can be used to improve the calibrating of the generated uncertainty.

Tian et al. [7] it suggests employing conditional Generative Adversarial Networks (GANs) to create a novel MRI image denoising technique. In order to determine within the context of dependent GANs, the generated picture combination can be categorized as either genuine or bogus. The genuine combination comprises a noisy picture and a corresponding noise-free picture, while the phony pair comprises of a noisy picture and its denoised counterpart. To discern between these pairs, a convolutional neural network (CNN) is employed as the discriminator. In our approach, the noise in the noisy MRI pictures is as much as feasible removed using a convolutional encoder-decoder networks-based generator. Adversarial learning is used to train the entire architecture. Clinical MRI datasets, both real and fake, are used in experiments.

Wang et al. [8] in this review paper, analyze the current situation of GANs and consider the future. First, we review the history of the GANs U + 02BC proposal, as well as its theoretical and practical models and application domains. The benefits and drawbacks of GANs U + 02BC as well as their future directions are then covered. We focus on the connection between GANs and parallel intelligence and come to the conclusion that GANs have a lot of potential for virtual-real interaction and integration in parallel systems research. It is obvious that GANs can offer significant algorithmic help for parallel intelligence.

Zhanli et al. [9] it suggests utilizing an upgraded Wasserstein generative adversarial network (WGAN) framework with a direct PET image reconstruction network (DPIR-Net) to increase image quality. Two key discoveries are presented in this article. First, compared to conventional model-based reconstruction networks, our network employs sinogram data as input and directly produces high-quality PET pictures, resulting in faster reconstruction times. To mitigate the problem of excess smoothing and the consequent loss of fine details in traditional network-based image restoration methods, we propose a loss function that combines subjective destruction, mean square error, and the Wasserstein distance. In our review, we utilized various methods including Maximum-Likelihood Expectation Maximization (MLEM) with a Post-Gaussian filter, Total Variation (TV)-norm regularization, Nonlocal Means (NLMs) denoising, Neural Network denoising, Conventional Deep Learning PET Reconstruction Network, and our suggested DPIR-Net technique. The approach under consideration was evaluated utilizing data from human beings and rat subjects.

Zhang et al. [10] In order to enhance the denoising network's capacity to pay attention to the image's noise and artifact features as well as its edge features, In order to enhance the core feature mining way, supplementary streams for distortions and ambient focus, in addition to edge feature focus, have been incorporated. To enhance the ability to discriminate of the discriminator, this course increases the receptive field by aggregating multiple-scale features within the identical scale of the image. In addition, a novel design referred to as the multi-scale Res2Net discriminator is introduced. Each individual sub-channel inside the denoising network is equipped with a specialized loss function that is specifically designed to align with its intended purpose. The network's convergence rate, stability, and denoising impact are all improved, ensured, and guaranteed by the cooperation of numerous loss functions. In comparison to state-of-the-art methods, experimental results demonstrate that the suggested denoising network can better achieve denoising effect while preserving the vital details of the low-dose computed tomography picture.

Serhii et al. [11] in this paper, we show how to use Convolutional Denoising Autoencoders (CDAE) to improve the pictures we got from the DCGAN (that we got from the earlier study). Additionally, utilizing the GMD structure (Glaucoma, Myopia, and Diabetic Retinopathy), we were able to comparison the model's precision employing initial information alone, Actual information combined with synthetic information from DCGAN [12] (collected from a prior document), and actual information combined with high resolution information from CDAE [13]. Our dataset includes four different categories of eye disorders: normal, diabetic retinopathy, myopia, and glaucoma. All of these statistics were gathered via Kaggle and verified by eye doctors.

3 Methodology

3.1 Generator Network

GAN's main component, the Generator Network, is what really produces the things. The Generator Network will receive more comprehensive data from the surrounding pixels in order to create a figure that is photorealistic and excellent. In this instance, the subtracting division ought to use deep convolutional neural networks together with a better architecture to produce the best numbers. SR Dense Net can benefit from our creation of the Generator Network, which is demonstrated in Fig. 1. It has one crop block, one obstacle block, eight dense blocks, one individual spiral block,

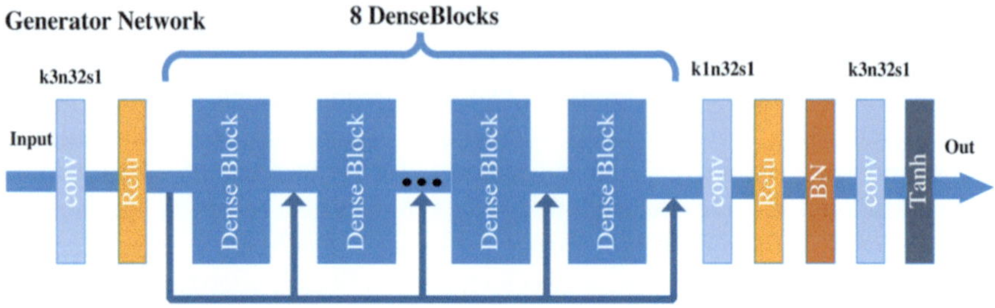

Fig. 1 Basic generator network

and a single obstacle block. Each block also has a cluster normalization layer, a Relu incitement, and a spiral coating. In order to reduce slope vanishing/discrediting questions and increase feature diffusion in deep networks, we feed each row along with each of the earlier layers (save the end one) utilizing miss related. In our work, the generator network is made up of auto encoders.

The recommended rowdy idea is extracted of its depressed-level elements by the first spiral coating. Then, eight Dense Blocks are chosen and given a respectable appearance. The appropriate Bottleneck coating is then added after that. It has been demonstrated that a 1 1 loop tier is ideal for lowering the number of suggested feature maps, making it possible to obtain a feature merging at a limited calculation cost. A 33 convolution tier is the component used to build the output images. The clamorous picture and the root loyalty countenance are two locations from which the Generator Network learns a leftover adjustment middle, which speeds up training.

By utilizing dense linked convolutional networks, the architecture of SRDenseNet [14], improves the flow of news through the network and alleviates the slope vanishing/discrediting problem. Additionally, it has been demonstrated that repeating feature maps from all prior Densely Blocks is more effective than anticipated in preventing the relearning of redundant features. SRGAN is another piece of work that uses GAN. It presented an architecture based on a very deep residual net construction (ResNet), which is created productive adversarial network (GAN) holding Discriminator and Generator network. This is in contrast to earlier projected convolutional net-works for picture super-resolution. It allows the Generator network to be trained to produce a representation that is identical to the concept of ground reality. Then it deceives a network of variable-entiable discriminators that has been taught to recognize the rebuilt representation that goes along with the physical idea. The two networks face off against one another until they arrive at the Nash equilibrium. More specifically, the network's setup is carried out by underestimating the intuition and feeling of misfortune functions, with the suggested adjustment minimizing simply the Mean Squared Error (MSE). The Generator network was put forward to create a photorealistic image via the proposed perceptual loss function, which consists of adversarial loss and content loss.

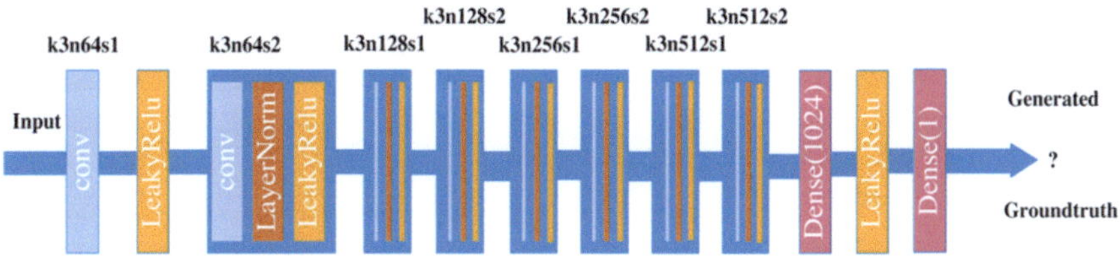

Fig. 2 Basic discriminator network

3.2 Discriminator Network

The Discriminator Network search out function determines whether a recommended image is genuine or fake, which aids in making the denoised output aesthetically pleasing. Because the advantage of creating samples is so small, the Discriminator Network must build the chance of profit filling a position genuine idea dossier as closely as possible to one. In our work, CNN serves as a discriminator as seen the Fig. 2.

Our Discriminator Network is similar to SRGAN in both type and approach. However, because of WGAN-GP, batch standard coverings are being replaced in our work by other coating average tiers. Eight convolutional coatings and three three-kernel systems are used. The final two coatings are sufficiently connected layers to present an expectation of ground validity input or emerging alternator network. Another difference from the previous coating is that sigmoid incitement is not sent through via WGAN-GP.

3.3 Convolutional Neural Networks

There are three elements involved when we work with CNNs:

- Input image
- Convolutional Neural Network
- Output (classified image class).

Here, the CNN breaks everything down into four steps that are further discussed in the member's last section. The spiral movement is the initial part of convolution. We will discuss the feature detectors in this stage because they are the filters of the connected system. It goes through how such filters develop for a single network one more. This section of the conversation will explore pooling; you'll learn specifically how it all works. However, we will concentrate on a specific kind of combining known as maximum combining. There are, however, a few different poolings that are available, and we will discuss them in the following episode. In this communication, we shall debate how flattening everything results in a final product that turns your advice into a heading. Full Connection—in this setting, communication, we'll see

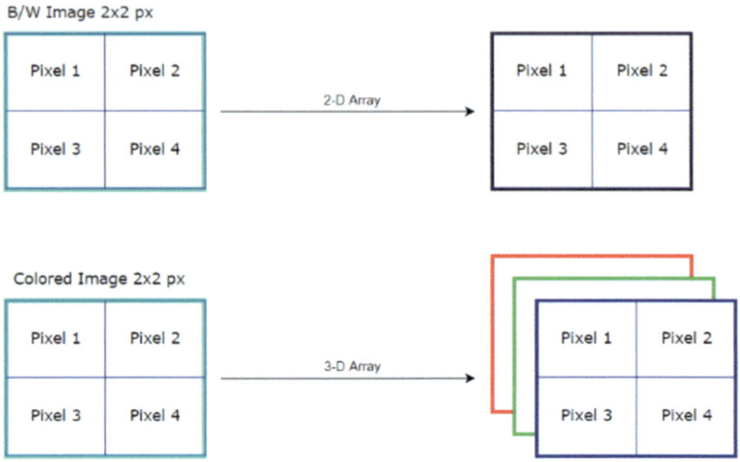

Fig. 3 CNN work with images

how these steps are connected to one another and how the final conclusions are reached. These make up a convolutional neural network's foundation as shown the Fig. 3.

A Convolutional neural network can work with distorted faces in addition to everything else that goes with policeman representations. Although there won't be any difference in talent, a CNN's general shape will remain the same. Before we move on to the next phase, it will become clearer, but please keep these things in mind: Casts are used to represent images, as seen in Fig. 3. In as much as a colored projection is produced by the fusion of three molds (one for each color channel, or RGB), a policeman representation is only a single mold. The pel principles at the spot in the figure correspond to the principles in the mold. To comprehend this better, consider the idea below.

3.4 Loss Function

GAN has always had to deal with the following issues and difficulties: On the one hand, training challenges necessitate careful model structure design and precise synchronization of the Generator Network and Discriminator Network training levels. On the other hand, because they don't have a useful indicator for the value of the generated image, Generator Network and Discriminator Network's loss functions are unable to reveal the training process. In contrast to the conventional GAN, we embrace the WGAN, The modification solely affects the final layer of the distinction system, without any logarithmic consideration of loss rates for both the discriminator and generator systems. Additionally, gradient clipping is performed after altering the coefficients of the distinction system.

Although WGAN suggests using proximity as an optimization technique to train GAN, there are still gaps between math and actual code implementation. Wasserstein distance must fulfill Lipschitz continuity, a severe continuity requirement. The author limits the weights to a specific range to ensure Lipschitz continuity in order to satisfy this need, but doing so also introduces unintended issues like bursting or vanishing gradients. Even if the theory has been shown to be quite great, the outcomes fall short of expectations. Therefore, Gulrajani et al. suggest substituting a gradient penalty term. The method, which works well with our network, enables the training of GAN to adapt new neural network designs and virtually eliminates the need for hyper-parameter adjustment.

From the perspective of enumerations, x and y may be seen as two samples from two distinct dossier distributions: commotion-free image allocation and clamorous representation classification Pn, respectively. Since the function f translates the samples from Pn to another allocation, the denoising movement is a planned procedure that turns one specific disposal to another. The fairly successful approach known as a generative opposing network (GAN) consists of two parts: a fruitful model G and a discriminating model D. GANs have been widely applied in a variety of disciplines, including idea approach remodel, image era, and representation super-judgment. The discriminatory model determines whether a sample comes from the generative model distribution Pg or the actual dossier classification P, and the generative model creates a new sample from the recommendation sample and aims to make the new sample as close to the actual dossier distribution Pr as possible. The preparation process of GANs is a brief time period top game with the following misfortune function (D, G) as

$$\text{min G maxD (D, G)} = \text{Ey} \sim \text{Pr}\big[\text{logD(y)}\big] + \text{Ex} \sim \text{Pn} * \log(1 - \text{D(G(x))} + \quad (3)$$

G and D are better in order to respond alternately. The authors claimed that the difficult process of creating a GAN could result in the dynamo G's slope diminishing once the discriminator D is in place. Arjovsky and Bottou developed the Wasserstein GAN (WGAN), an improved form of the GAN, to address this issue. Additionally, in order to hasten the union, Gulrajani et al. awarded an updated reading of WGAN together with a slope punishment. The deficit function has changed in the following ways.

$$\mathbf{LWGA(D)} = -\mathbf{Ey} \sim \mathbf{Pr}\big[\mathbf{D(y)}\big] + \mathbf{Ex} \sim \mathbf{Pn[D(G(x))]} + \boldsymbol{\lambda}\mathbf{Ex} \sim \mathbf{Px}[||\nabla \mathbf{x}^{\mathbf{D}}(\mathbf{x})||2 - 1)2]$$

In the case when the final term is a slope punishment determinant, is a punishment cooperative, and Px is an allocation that uniformly samples a nearly direct path between pairs of locations taken from the right dossier allocation Pr and the engine converting energy disposal Pg. The following formula represents alternator G's unfortunate function.

$$\mathbf{LWGAN(G)} = -\mathbf{EX} \sim \mathbf{Pn[D(g(x))]}$$

When training the generative sub network in a GAN, it is equivalent to underestimating the Jensen-Shannon (JS) difference between the dispersion pr of honest data and the disposal pg of produce data (from the fruitful sub network). In a GAN, the discriminating sub network and the fruitful sub network will be prepared beforehand. The classification might theoretically be rebuilt for the benefit of the pr by underrating the JS difference. Actually, the training inevitably falls into the slope vanishing trap, making it unable to consistently change the productive sub network. In order to ensure that the JS difference is decided in an appropriate manner and that there is no slope news that might be back propagated to enhance the fruitful sub network, there are no non-negligible intersections between two combined distributions at the beginning of the preparation process.

3.5 Subjective Outputs

As seen the Fig. 4 has input image, noisy image and denoised images.

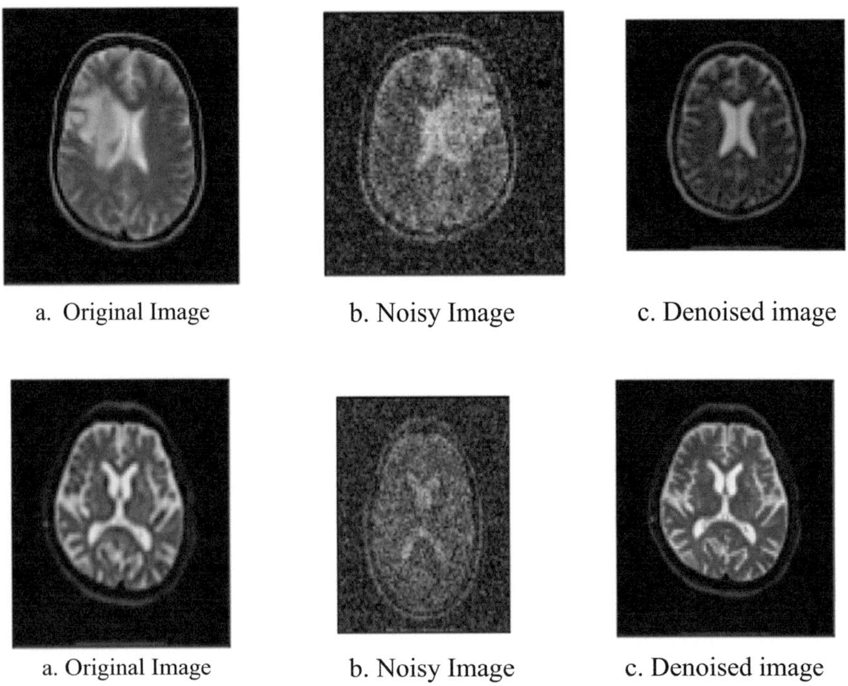

Fig. 4 Input image, noisy image, denoised image

3.6 Performance Matrices

(a) **Peak Signal-To-Noise Ratio (PSNR)**

The Peak Signal-to-Noise Ratio (PSNR) is commonly employed as a metric for evaluating the effectiveness of picture denoising techniques.

$$PSNR = 10log_{10}\left[\frac{(2^n - 1)^2}{MSE}\right]$$

$$\text{MSE} = \frac{1}{WXH}\sum_{i=1}^{H}\sum_{j=1}^{W}(X(i,j) - Y(i,j))^2$$

(b) **SSIM**

In this approach, the comparison among the real-world images and the denoised images is assessed and displayed using the structurally comparable index measure (SSIM), which is used to assess the quality of the image the Fig. 4.

$$\text{SSIM(x, y)} = \frac{\left(2\mu_x\mu_y + c_1\right)(2\sigma_{xy} + c_2)}{\left(\mu_x^2 + \mu_y^2 + c_1\right)(\sigma_x^2 + \sigma_y^2 + c_2)}$$

3.7 Adam Optimizer

As seen in Fig. 5, the Adam optimizer combines two gradient descent techniques: The 'exponentially weighted average' of this method incorporates the consideration of slopes to enhance the efficiency of the slope descent process.

4 Conclusion

More MR images will be used in future testing of the recommended denoising method in alongside CT and X-ray scans for reducing noise on other medical devices. The strategy will also be assessed in a real hospital environment. Both the generation and discrimination networks compete to produce the images with the highest level of pragmatism. The photographs that arise from denoising are incredibly clear. This kind can tolerate a range of noise levels. Even though we have improved photorealistic images, accuracy is still inadequate. The size of the batch or other settings will be changed in our following work to remedy this.

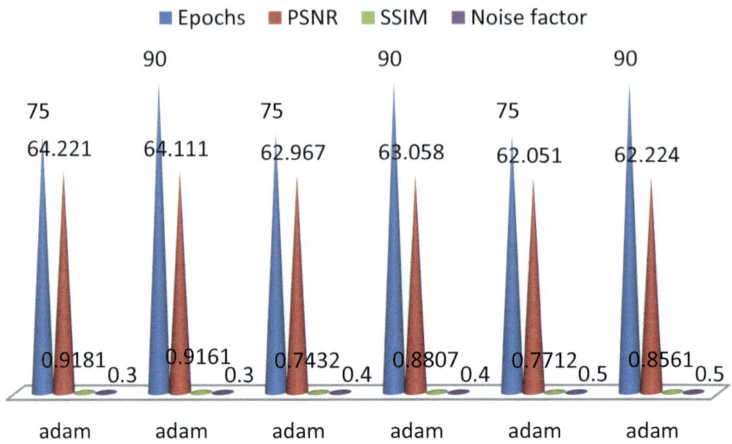

Fig. 5 Exponentially weighted average' of the gradients output

References

1. Burger HC, Schuler CJ, Harmeling S (2012) Image denoising with multi-layer perceptrons, part 2: training trade-offs and analysis of their mechanisms. arXiv preprint arXiv:1211.1552
2. Li Y, Zhang K, Shi W, Miao Y, Jiang Z (2021) A novel medical image denoising method based on conditional generative adversarial network. Comput Math Methods Med 2021
3. Yu G, Shan H, Teng Y, Tu N, Li M, Liang G, Ge W, Wang S (2020) Parameter-transferred Wasserstein generative adversarial network (PT-WGAN) for low-dose PET image denoising. IEEE Trans Radiat Plasma Med Sci 5(2):213–223
4. Yang Q, Yan P, Zhang Y, Yu H, Shi Y, Mou X, Kalra MK, Zhang Y, Sun L, Wang G (2018) Low-dose CT image denoising using a generative adversarial network with Wasserstein distance and perceptual loss. IEEE Trans Med Imaging 37(6):1348–1357
5. Zhao G, Meyerand ME, Birn RM (2020) Bayesian conditional GAN for MRI brain image synthesis. arXiv preprint arXiv:2005.11875
6. Tian M, Song K (2021) Boosting magnetic resonance image denoising with generative adversarial networks. IEEE Access 9:62266–62275
7. Wang K, Gou C, Duan Y, Lin Y, Zheng X, Wang F-Y (2017) Generative adversarial networks: introduction and outlook. IEEE/CAA J AutomaticaSinica 4(4):588–598
8. Hu Z, Xue H, Zhang Q, Gao J, Zhang N, Zou S, Teng Y et al (2020) DPIR-Net: Direct PET image reconstruction based on the Wasserstein generative adversarial network. IEEE Trans Radiat Plasma Med Sci 5(1):35–43
9. Zhang X, Han Z, Shangguan H, Han X, Cui X, Wang A (2021) Artifact and detail attention generative adversarial networks for low-dose CT denoising. IEEE Trans Med Imaging 40(12):3901–3918
10. Yaroshchak S, Smaida M, El Barg Y (2021) Medical image enhancement based on convolutional denoising autoencoders and GMD model. In: CEUR workshop proceedings, pp 96–106
11. Bathula A, Muhuri S, Gupta S, Merugu S (2022) Secure certificate sharing based on blockchain framework for online education. Multimed Tools Applications 82:1–22. https://doi.org/10.1007/s11042-022-14126-x

12. Shaheen H, Ravikumar K, Lakshmipathi Anantha N, Uma Shankar Kumar A, Jayapandian N, Kirubakaran S (2023) An efficient classification of cirrhosis liver disease using hybrid convolutional neural network-capsule network. Biomed Signal Process Control, 80, Part 1:104152. ISSN 1746-8094. https://doi.org/10.1016/j.bspc.2022.104152
13. Narasimharao J, Laxmaiah B, Arumalla R, Unnisa R, Fatima T, Nazare SS (2023) Restoration and deblurring the images by using blind convolution method. In: Proceedings of fourth international conference on computer and communication technologies: IC3T 2022. Springer Nature Singapore, Singapore, pp 337–346
14. Madhavi KR, Vooradi S, Mounika P, Yedlla S, Tangudu N (2023) Emotion analysis using convolutional neural network. In: Proceedings of third international conference on advances in computer engineering and communication systems: ICACECS 2022. Springer Nature Singapore, Singapore, pp 523–531

Employing Big Data Analysis, an Unconventional Approach for Highly Computational Linear and Polynomial Regression Statistics in the Medical Industry

Vijaya Kumar Koppula, L. Chandra Sekhar Reddy, T. Bhaskar, Bagam Laxmaiah, Shankar Nayak Bhukya, and Nagarjuna Tanikonda

Abstract Predictive modeling is an essential component within the health sector, as it serves to enhance the decision-making process and facilitate comprehension of the results for patients. Machine learning structures have been in existence for millennia, and it frequently appears that a significant number of algorithmic approaches lack accuracy and tend to provide erroneous predictions. In order to address this issue, our study proposes the implementation of optimization approaches to enhance the accuracy of algorithmic predictions. In this study, we are implementing Linear and Polynomial optimization approaches to Relational methods. We will afterwards evaluate the performance of these optimized Regression algorithms with their non-optimized counterparts. The evaluation of their efficacy will be based on the metric of Sum of Square Error (SSE). In the context of regression algorithms, it is generally preferred for the sum of squared errors (SSE) to be minimized, as this indicates a higher level of algorithm performance.

V. K. Koppula · L. C. S. Reddy · T. Bhaskar
Department of CSE, CMR College of Engineering and Technology, Hyderabad, Telangana, India
e-mail: chandunani@cmrcet.org

T. Bhaskar
e-mail: drtbhaskar@cmrcet.org

B. Laxmaiah (✉) · S. N. Bhukya
Department of CSE, CMR Technical Campus, Hyderabad, Telangana, India
e-mail: blaxmanphd@gmail.com

S. N. Bhukya
e-mail: bsnaik546@gmail.com

S. N. Bhukya
Department of CSE (Data Science), CMR Technical Campus, Hyderabad, India

N. Tanikonda
Department of ECE, CMR Technical Campus, Hyderabad, Telangana, India

V. K. Gunjan et al. (eds.), *Cybernetics, Human Cognition, and Machine Learning in Communicative Applications*, Cognitive Science and Technology,
https://doi.org/10.1007/978-981-97-8533-9_15

Keywords Linear regression · Big data · SSE

1 Introduction

Perfect sales projections holds significant importance within the dynamic domain of medical care, as it facilitates excellent inventory control and the efficient fulfillment of consumer requirements. The emergence of information science and analytics for prediction has provided organizations with the chance to utilize previous sales information in order to construct sophisticated models capable of accurately predicting future revenue. The objective of this research endeavor is to utilize data engineering methodologies, particularly Linear and Polynomial Regression, in order to forecast the forthcoming sales of pharmaceutical products. Through the examination of past sales information and the identification of influential elements that impact sales trends, it is possible to construct simulations that offer important information for drug manufacturers, suppliers, and doctors and nurses. These understandings have the potential to facilitate educated choices, efficient use of resources, and effective inventory administration, eventually leading to the optimization of the supply of medicines and the enhancement of patient satisfaction. Linear regression is a commonly employed statistical approach that offers a robust framework for forecasting future revenue.

Selling cycles in the healthcare sector frequently demonstrate non-linear characteristics that are impacted by a multitude of variables, like changes in the seasons, market volatility, and the launch of novel treatments. In order to account for these intricacies, the utilization of Polynomial Regression may be employed. By incorporating higher-order polynomial variables into the Linear Regression method, it becomes possible to accurately reflect non-linear interactions and potentially raise the precision of our predictive outcomes.

Problem Statement

The drugs industry encounters notable obstacles in efficiently overseeing selling and supply of drugs. The precise prediction of forthcoming revenues holds significant importance for pharmacy businesses, suppliers, and medical professionals as it enables them to effectively manage inventories, distribute assets optimally, and fulfill customer requirements. Nevertheless, conventional approaches to projections of sales frequently prove inadequate in comprehensively addressing the intricate and non-linear dynamics inherent in the drug sector. The subject being investigated in this research pertains to the utilization of information science methodologies, particularly Linear and Polynomial Regression, in order to construct accurate simulations capable of properly forecasting upcoming medication consumption.

Objectives

1. This study aims to conduct an analysis of past sales information in order to discover the primary elements which exert effect on trends of drug distribution.

2. The objective of this study is to construct a Linear Regression structure in order to provide a foundational framework for predicting revenues.
3. Incorporating Polynomial Regression into the study allows for the inclusion of non-linear correlations and has an opportunity to enhance accuracy in predictions.
4. This study aims to assess and contrast the efficacy of Linear and Polynomial Regression methods for forecasting future revenue.
5. This study aims to enhance the handling of inventory and allocation of resources for drug businesses, suppliers, and medical professionals by employing data visualization techniques to depict projected sales patterns and offer practical recommendations.

2 Literature Review

The use of novel or new psychoactive compounds, commonly referred to as designer drugs and research chemicals, is a relatively recent practice that dates back at least ten years. Individuals [1], the economy, and other factors are all affected by this phenomenon's exponential and frightening growth in electronic trade (e-trade).

In connection with topics in dental anatomy and dental pathology represented by the maxillary sinus, posterior maxillary teeth, and related oral pathology, we devised a real-time statistical analysis. The goal is to infer the digital epidemiology using a stream of raw data that is continuously received from the Google Trends database. Utilizing SPSS version 24 and Microsoft Excel 2016 for statistical analysis. Information on digital epidemiology was retrieved via the Google Trends database. Python is a high-level programming language that was used to encode a programming script for real-time analysis and statistical inference [2].

Every day, current information systems and digital technologies like the Internet of Things and cloud computing produce a vast repository of data that amounts to terabytes. To extract knowledge for decision-making, analysis of these enormous data involves significant work on many levels. As a result, research and development are currently being done in the field of big data analysis [3]. This paper's main goal is to examine the potential effects of big data challenges, open research questions, and related tools. This article thus offers a framework for exploring big data at various phases. The challenges and open research questions present new opportunities for academics to create the solution.

Karen et al. [4] the application of big data analytics in the field of applied genomics. The user's text is already in an academic format. No rewriting is necessary. The primary objective of this study is to examine the utilization of big data analysis within the domain of functional genomics. While indirectly pertaining to selling in the field of healthcare, this observation underscores the significance of employing quick computational techniques to analyze extensive biological information. These techniques can also be extrapolated to different areas, including the marketing of medical products.

Belle et al. [5] with an emphasis on three future and promising fields of medical research—image, signal, and genomics based analytics—we highlight some of these significant obstacles in this work. It is described recent study that aims to use enormous amounts of medical data while merging multimodal data from many sources. Also considered are potential research opportunities in this area that could have a significant impact on how healthcare is delivered.

To gather information about the Ekbom syndrome, also known as delusional parasitosis (DP), that is particular and based on current research, using a thorough systematic review of the literature. History, epidemiology, pathophysiology [6], clinical characteristics, subtypes and related illnesses, linked diseases, psychosocial impact, economic factors, and management are all covered in this article.

The text presents a thorough research project utilizing the suggested concepts, the main goal of which is to evaluate the prevalence of the use and abuse of psychoactive substances, including new psychoactive substances [7], at the national and international level, establishing appropriate and effective preventive strategies, and developing guidelines for management in crisis situations.

For patients at home who are susceptible to power shortages, this pricey technology is practical and appropriate. Previously used techniques allowed patients to visit hospitals for DME report monitoring or to contact hospitals via landlines [8], mobile towers, Ethernet, or the Internet. When a power loss is detected, a novel wireless system that uses a radio ad hoc network will immediately communicate the patient's identity and location as well as the information and status of the DME to a nearby hospital.

When used in the biological and life sciences, real-time analytics have a significant influence because they drastically cut the amount of resources needed for research [9]. The next stage to be implemented in continuation of the real-time systems to prognosticate changes in the temporal trends and the digital epidemiology of phenomena of interest is predictive analytics, which is based on artificial neural networks and machine learning algorithms.

3 Existing System

The current methodology employed in the healthcare sector for estimating upcoming drug revenues frequently depends on traditional forecasting techniques that cannot comprehensively account for the intricate dynamics as well as non-linear associations present in the field. Such approaches commonly employ fundamental statistical procedures or straightforward projection relying on previous information. Nevertheless, their analysis lacks consideration for the various determinants that impact the selling of medication, including but not limited to seasons, variations in costs, advertising efforts, and the launch of novel pharmaceuticals. Consequently, the current approach frequently gives rise to imprecise revenue predictions, insufficient control over inventory, and poor distribution of resources. In order to overcome these constraints, the present study suggests employing a data science methodology that incorporates Linear and Polynomial Regression methods [10]. Through the utilization of these sophisticated methodologies, our objective is to construct systems that possess the capability to comprehend the complex associations amongst purchases and pertinent predictions. This will result in the generation of more precise and dependable forecasts pertaining to upcoming drug sales.

4 Proposed System

The aforementioned features were employed in the implementation of this work:

a. Utilize this section for importing information pertaining to medicines into the software.
b. Preprocess Dataset: In this section, the entire collection of data is going to be processed. The X attributes and Y tags of the training data set are to be extracted and subsequently partitioned into separate test and training files.
c. Train Regression without Optimization: The aforementioned processing characteristics are going to be utilized as entry for a Regression technique, instead of any optimization. Subsequently, a system is trained using this approach. This framework is then used to forecast the production of a type of drug based on the test results. Following this prediction, the Sum of Squared Errors (SSE) may be calculated, which represents the discrepancy amongst the original test results and the estimated results.
d. Polynomial Optimized Linear Regression: Within this section, we will utilize the regression technique to train models with linear and polynomial characteristics. Subsequently, we will proceed to build a predictive model.
e. Pre & Post Optimization SSE graph: In this course, we will generate a graphic illustrating the Sum of Squared Errors (SSE) across the pre-optimization and post-optimization stages [13].

To run the work twice, select "run." get below screen with a bat file.

5 Implementations

In this project we implemented two algorithms Linear Regression and Polynomial Regression.

5.1 Linear Regression

Linear regression is a statistical estimation approach employed to investigate the relationship among a dependent parameter and a number of independent factors.

The pursuit of the best-fit straight line is motivated by the need to minimize the discrepancy among the points of data observed and the expected numbers. Linear regression is a commonly employed technique in various academic fields such as finance, economics, social science, and machine learning.

Formulas: The mathematical expression for basic linear regression with a single distinct variable is:

$$y = \beta_0 + \beta_1 x$$

where:

Y represents the dependent variable (the variable being predicted).

x represents the independent variable (the variable used to predict the dependent variable).

β_0 represents the y-intercept (the value of y when x is zero).

β_1 represents the slope (the change in y for a unit change in x).

For multiple linear regression with multiple independent variables, the formula is:

$$y = \beta_0 + \beta_1 x_1 + \beta_2 x_2 + \ldots. + \beta_p x_p$$

where:

P represents the number of independent variables.

$x_1, x_2, \ldots., x_p$ represent the p independent variables.

$\beta_0, \beta_1, \beta_2, \ldots., \beta_p$ represent the respective intercept and slopes for the independent variables.

5.2 *Working of Linear Regression*

The process of linear regression entails the establishment of a linear relationship among the observed data points, wherein a straight line is fitted in a manner that reduces the total of squared discrepancies among the anticipated and actual results. The determination of the variables (including the point of intersection and gradients) that characterize the line results in the depiction of the line's form, as illustrated in Fig. 1.

The model use the Ordinary Least Squares (OLS) technique to determine parameters by minimizing the total of squared residuals. Residuals refer to the disparities seen within the anticipated values and the observed values.

The computation of the partial derivations of the total amount of squared residuals compared to the parameters is a fundamental step inside the ordinary least squares (OLS) estimation procedure. The optimal coefficients can be achieved by equating such derivatives to zero. Subsequently, the established model may be employed to predict the amount of the dependent variable by utilizing the estimated coefficients in conjunction with the value of the variables that are independent.

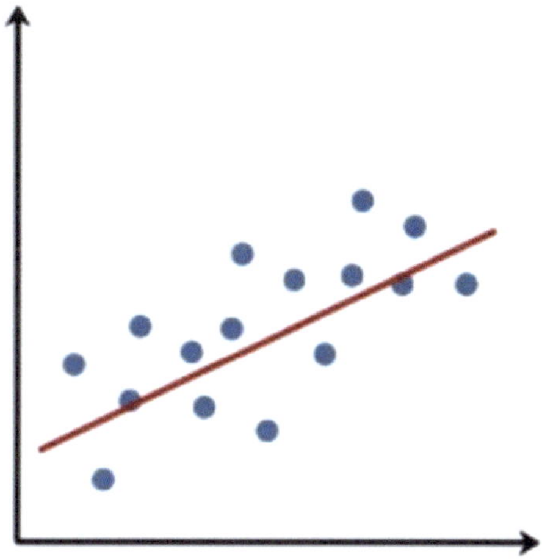

Fig. 1 Linear regression

5.3 Mean Squared Error (MSE)

The mean squared error (MSE) is computed by taking the mean of the squared variances among the expected outcomes and the actual values. Errors of greater magnitude incur more severe consequences.

Formula:

$$MSE = (1/n) * \Sigma(y - \hat{y})^2$$

where:

The number of samples is denoted as n. The character y is used to denote the real values of the dependent factor. The expected results of the dependent factor are denoted by $\hat{y}$.

5.4 R-squared (R^2)

The coefficient of determination, commonly referred to as R-squared, quantifies the proportion of the variation in the dependent factor that may be explained by independent factors. The value of the variable ranges from 0 to 1, where a value of 1 indicates an exact fit.

Formula:

$$R^2 = 1 - (SSE/SST)$$

where:

SSE (Sum of Squared Errors) is the sum of the squared residuals.

The Total Sum of Squares (SST) refers to the summation of the squared deviations among the observed values and the average of the dependent variable's value.

5.5 *Dataset Used*

We are utilizing information set pertaining to the sales and production of drugs, with the objective of predicting the number of drugs manufactured for marketing purposes.

The informational was trained using PRE and POST optimized Regression algorithms. Figure 2 displays the specifics of the dataset on the display.

6 Results and Analysis

Train Regression without Optimization.

In the aforementioned Fig. 3, the initial observation reveals a Regression Sum of Squares Error (SSE) of 1042 in the absence of optimization. Subsequently, the chart provides the real numbers for medication production tests alongside their corresponding projected values. The x-axis of the chart indicates the month's exact number, while the y-axis indicates the value for the needed production. In the aforementioned data, the red line denotes the actual test production values, while the green line indicates the predicted values. Upon observing the chart, it is evident that there exists

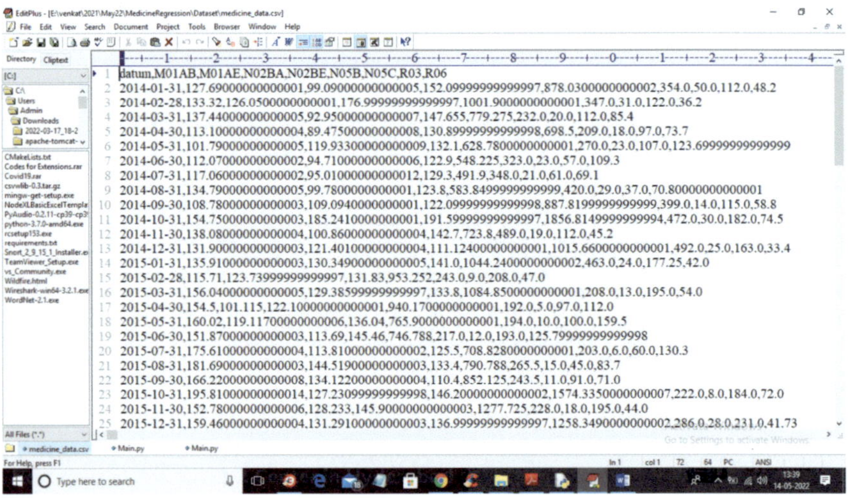

Fig. 2 Dataset get train with Pre & Post optimized regression

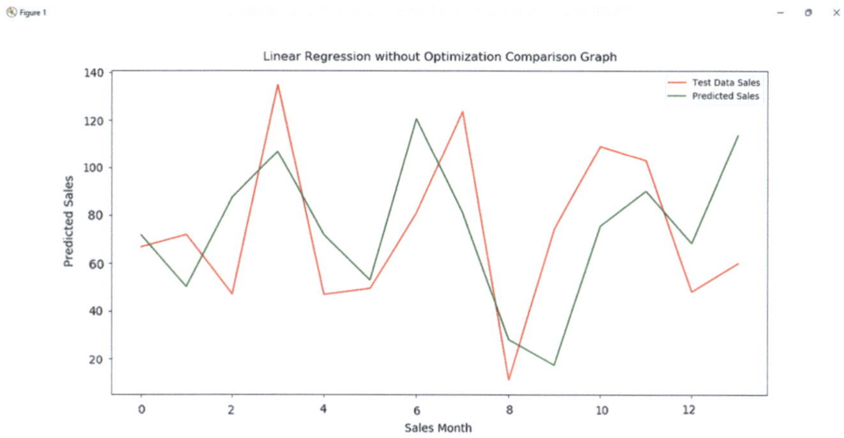

Fig. 3 Train regression without optimization

a significant disparity among the red and green lines, indicating that the estimate is inaccurate. If the projection were indeed true, the two lines would overlap.

6.1 Polynomial Optimized Linear Regression

In the aforementioned Fig. 4, the optimized linear polynomial regression model yielded a sum of squared errors (SSE) of 1.92, a value that is lower compared to the model with no optimizing procedures. Additionally, the graph illustrates that the lines representing the actual and predicted results coincide, indicating that the optimized model performs more effectively.

6.2 Pre & Post Optimization SSE Graph

The diagram above displays the algorithm titles as PRE and POST optimization Regression, while the y-axis represents the SEE error. The efficiency of POST Optimization Regression is observed to be satisfactory in both techniques, as depicted in Fig. 5.

Many models were run based on a generator of random numbers with an even distribution, where the average was set to 0 and the standard error was set to 1. In order to conduct a linear regression study, a set of forty instances (referred to as simulation versions) were constructed. Each experiment consisted of an average number of 1,000 findings, with both variables designated as the predictor (X) and the result (Y). Consequently, the entire sample size amounted to 40,000 variables. The

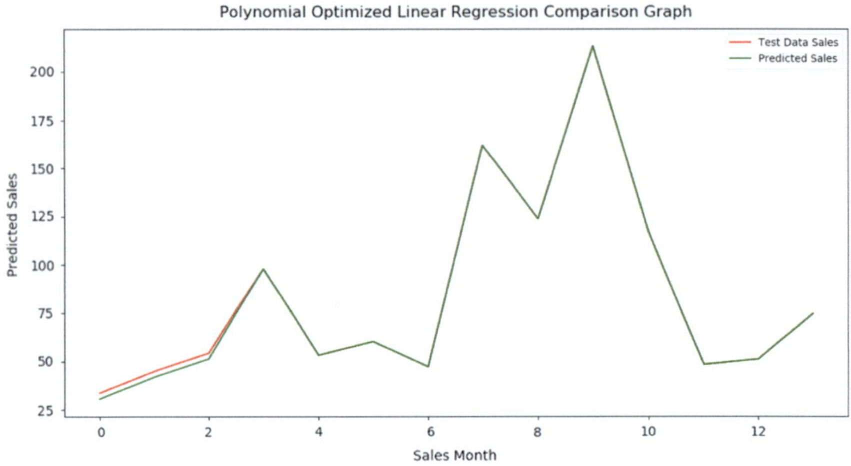

Fig. 4 Polynomial optimized regression

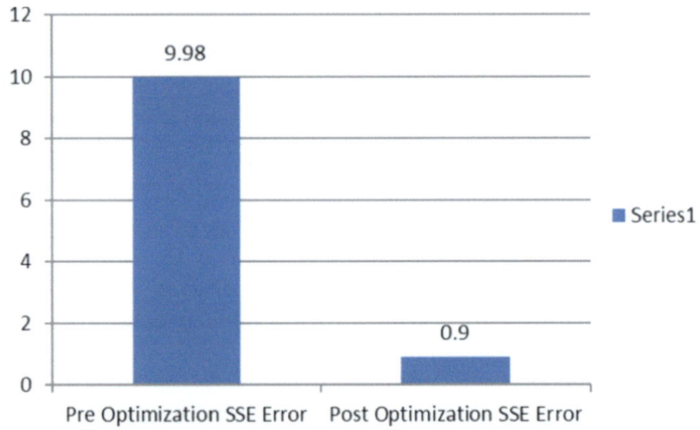

Fig. 5 Pre & Post regression optimization SSE graph

"max" operation in Excel 2016 was utilized to do scaling on both of these parameters. This involved dividing every instance by the highest permitted measurement within the respective parameter. Correlation and regression statistics, including the sum of squares (SS), mean of squares (MS), F statistic (ANOVA) [11], and p-value (regression), were calculated within each linear framework.

To fulfill the requirements of the regression formula b0 + b1X, we computed the sum of squared errors (SSE) by employing the formula SSE = (y) 2. The computations took place on two occasions: firstly, prior to the implementation of the scale-down transformation (referred to as pre-optimization), and secondly, subsequent to its implementation (referred to as post-optimization). The efficacy of the scale-down optimization framework was empirically tested by comparing pre- and post-optimization statistics utilizing the Wilcoxon signed-rank test, which is a non-parametric within-subjects statistically inferred method. In conclusion, Cronbach's alpha was employed as a metric to assess the internal coherence of the cumulative enhanced system. Additionally, an examination was conducted to evaluate the efficacy of the algorithm's optimization [12].

7 Conclusions

The primary objectives of our advanced transform and optimization methods are threefold. One way to enhance the line with the ideal fit is by minimizing the sum of squared errors (SSE). The implementation of the scale-down transition will significantly reduce the computational demands for mathematical calculations in the context of big data analysis. This is particularly relevant when dealing with extensive datasets containing a multitude of factors and an extensive number of samples for every factor, as commonly observed in various polynomial regression investigations. The actual time study of correlations and regression across huge multidimensional sets of information will require significant mathematical processing capacity, posing a challenge for present and future supercomputers. The method of optimisation aims to decrease the variety of parameters by reducing the amount of decimal points, while maintaining their initial association. This might result in cost savings throughout subsequent mathematics and computer operations.

References

1. Al-Imam A (2017) Monitoring and analysis of novel psychoactive substances in trends databases, surface web and the deep web, with special interest and geo-mapping of the Middle East. Master s thesis, University of Hertfordshire, Hertfordshire, United Kingdom. Retrieved from https://uhra.herts.ac.uk/handle/2299/19462
2. Al-Imam A (2019) Inferential analysis of big data in real-time: one giant leap for spatiotemporal digital epidemiology in dentistry. Odontostomatol Res Anat Learn Implant 12(1):1–14
3. Acharjya DP, Ahmed K (2016) A survey on big data analytics: challenges, open research issues and tools. Int J Adv Comput Sci Appl (IJACSA) 7(2)
4. He KY, Ge D, He MM (2017) Big data analytics for genomic medicine. Int J Mol Sci 18(2):412
5. Belle A, Thiagarajan R, Soroushmehr SR, Navidi F, Beard DA, Najarian K (2015) Big data analytics in healthcare. https://doi.org/10.1155/2015/370194
6. Al-Imam A, Al-Shalchi A (2019) Ekbom's delusional parasitosis: A systematic review. Egypt J DermLogy Venerol 39(1):5–13. https://doi.org/10.4103/ejdv.ejdv_53_15
7. Al-Imam A, Motyka M (2019) On the necessity for paradigm shift in psychoactive substances research: the implementation of machine learning and artificial intelligence. Alcohol Drug Addict/AlkoholizmiNarkomania 32(3):237–242. https://doi.org/10.5114/ain.2019.91004
8. Nalajala P, Godavarthi B, Mohammad K (2018) Biomedical sensor based remote monitoring system field of medical and health care. J Adv Res Dyn Control Syst 9(4):210–219
9. Al-Imam A, Khalid U, Al-Hadithi N, Kaouche D (2018) Real-time inferential analytics based on online databases of trends: a breakthrough within the discipline of digital epidemiology

in dentistry and dental anatomy. Mod Appl Sci 13(2):81–94. https://doi.org/10.5539/mas.v13n2p81
10. Al-Imam A (2020) A novel method for computationally efficacious linear and polynomial regression analytics of big data
11. Rani YA, Bhaskar T, Arshad M, Allam B (2023) AI based scarecrow preventing from crop vandalization. In: 2023 4th international conference on Electronics and Sustainable Communication Systems (ICESC), Coimbatore, India, pp 1233–1238. https://doi.org/10.1109/ICESC57686.2023.10193604
12. Dash CS, Nayak SC, Behera AK, Dehuri S (2023) A neuro-fuzzy predictor trained by an elitism artificial electric field algorithm for estimation of compressive strength of concrete structures. 47(5). https://doi.org/10.31449/Inf.V47i5.3951
13. Somasekhar G, Patra RK, Srujan Raju K (2021) The research importance and possible problem domains for NoSQL databases in big data analysis. In: Proceedings of the 2nd international conference on computational and bio engineering: CBE 2020. Springer Singapore, pp 433–439

Convolutional Neural Network Assessment to Identify Glaucoma in Eyes Using Retinal Fundus Images

S. Kirubakaran, Aelgani Vivekanand, G. Ravi Kumar, Jonnadula Narasimharao, Banothu Ramji, and Shaik Sharif

Abstract In order to identify glaucoma, this study suggests an image processing technique using a computer tool. Glaucoma has been identified as one of the major factors that contribute to visual impairment, although early study on the condition was challenging. It is one of the causes of permanent visual impairment in people older than 40 years old. Because it balances convey ability, size, and cost, fundus imaging is a very popular screening method for glaucoma expose. We discuss improvements to disc segmentation alongside other studies in the literature. These upgrades include a novel technique for dividing the container at the threshold and a new comparison between the cup's and the disk's sizes. Results were obtained from a number of fundus photographs.

Keywords Cup-to-disc ratio · Cup segmentation · Disc segmentation · Fundus image · Glaucoma · Glaucoma detection · Image processing · Veins segmentation

S. Kirubakaran · A. Vivekanand · G. R. Kumar
Department of CSE, CMR College of Engineering and Technology, Hyderabad, Telangana, India
e-mail: drskirubakaran@cmrcet.ac.in

A. Vivekanand
e-mail: avivekanand@cmrcet.org

G. R. Kumar
e-mail: gravikumar@cmrcet.org

J. Narasimharao (✉)
Department of CSE, CMR Technical Campus, Hyderabad, Telangana, India
e-mail: jonnadula.narasimharao@gmail.com

B. Ramji
Department of CSE (Data Science), CMR Technical Campus, Hyderabad, India

S. Sharif
Department of CSE (AI&ML), CMR Technical Campus, Hyderabad, Telangana, India

V. K. Gunjan et al. (eds.), *Cybernetics, Human Cognition, and Machine Learning in Communicative Applications*, Cognitive Science and Technology,
https://doi.org/10.1007/978-981-97-8533-9_16

1 Introduction

Since glaucoma affects the eye's optic nerve, it is the main cause of irreversible vision loss. In the majority of individuals, vision loss symptoms don't become apparent until advanced stages.

Glaucoma frequently results in an unavoidable loss of vision that happens slowly and without warning. It is a major cause of eyesight loss all around the world. According to estimates, a significant proportion of individuals diagnosed with glaucoma in wealthy nations remain asymptomatic, indicating that the condition must be far worse in developing countries. Over 11.1 million people will have essential glaucoma-related visual impairment by 2020, according to an analysis of the prevalence of glaucoma from widely disseminated data. Additionally, it was shown that the expense of treatment for glaucoma in its advanced stages was rising. Glaucoma is 2.6% prevalent in Southern India, and 90% of these instances have never been examined, compared to roughly 50% previously unknown when comparable tests are performed in Europe.

Loss of vision is caused by a confluence of diseases that contaminate the optic nerve and the cells that make up the optic nerve.

The excessive production of aqueous humour or the obstruction of its sewerage system results in increased pressure in the eye, which damages the optic nerve. Additionally, there is proof that the disease's emergence may be influenced by a person's genetic family history. A patient with suspected glaucoma may undergo a number of procedures, including tonometry to measure eye pressure, gonioscopy to verify whether the angle of vision is closed or open, optical coherence tomography (OCT), and funduscopy or fundus imaging to view the retina and the optic nerve as given in Fig. 1.

The thickness of the retinal nerve fiber layer (RNFL) is measured using the retinal fundus image to analyze glaucoma. This is one of the minimally invasive methods that ophthalmologists frequently use. Its key advantage is that it can be used to take photographs of both healthy and unhealthy retinas. It is also portable and easy to employ for any medical professional, which makes it ideal for screening campaigns among populations without access to healthcare.

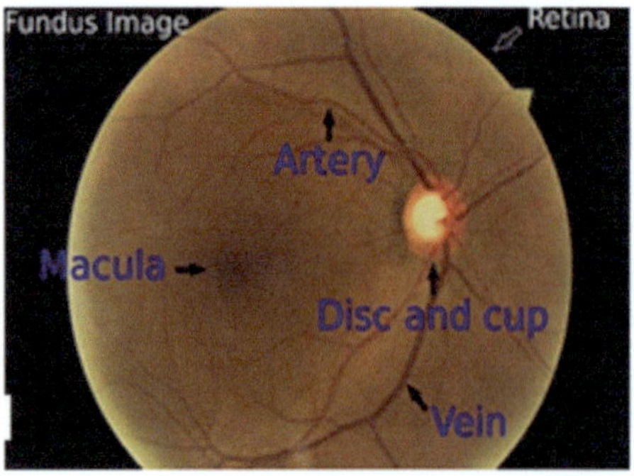

Fig. 1 Fundus picture and significant parts of the eye

The proportion between the dimensions of the optic nerve (commonly known as the disc) and the dimensions of the cavity formed inside the optic nerve as a result of rising eye pressure (referred to as the cup) is worn to conclude the thickness of the RNFL. The Cup-to-Disc ratio (CDR) is named to this characteristic. The fundamental challenge in automatically detecting glaucoma using colour fundus images has been to offer an precision calculation of the CDR. The present study proposes a method for segmenting discs through the utilization of edge recognition techniques is proposed by Anusorn et al. [9]. If the eye has per papillary atrophy, a condition that causes the disc's edges to alternate, this procedure will not work. The distance that exists among the disc pixels and the cup pixels is occasionally equal, that made it challenging to segment photos collected from different people. The threshold used for cup segmentation is one third of the highest grayscale intensity. Finding the cup edges in the initial phases of the disease when they begin to expand is another issue.

The author of this article describes a computational approach for automatically detecting glaucoma from eye fundus images. The remainder of the chapter is functioned as given: a literature review will be covered in the next chapter, then methodologies, results, a summary, and references.

2 Literature Review

Mohammadi et al. [1] Preventive interventions are essentially divided into four categories: primary, secondary, tertiary, and primal.

The aforementioned groups of conventional ophthalmic instances encompass various interventions aimed at addressing specific eye conditions. These actions involve improved sanitation and environmental hygiene measures to reduce the prevalence of Chlamydia trachomatis-infected insects, which are integral to trachoma prevention efforts. Additionally, strict management of blood sugar levels in individuals with diabetes is crucial in preventing the development of retinopathy. Ensuring availability to and the successful implementation of cataract surgery is another important intervention. Lastly, the supply of black pupil contact lenses serves as a means of addressing the aesthetic concerns for those with visually impaired and disfigured eyes. Furthermore, it is important to acknowledge as these categories are subjective and may vary based on the perceived basic health condition.

Maya et al. [2] Glaucoma, which exaggerated 60.5 million people globally in 2010 and is predicted to influence 79.6 million people by 2020, is the most common cause of irreversible vision impairment in the globe. In order to stop the disease from progressing and causing vision loss, glaucoma screening is critical to identifying, diagnosing, and treating sufferers at the earliest possible stages. Stereoscopic digital imaging is used in teleglaucoma to capture images of the eyes that are then electronically sent to an eye doctor. The goal is to compile research to assess Teleglaucoma's cost-effectiveness, diagnosis accuracy, and advantages to the healthcare system.

Methods: For the purpose of locating both published and unpublished studies, a thorough search was carried out. Included were studies that looked at teleglaucoma as a tool for glaucoma screening. Results of accuracy of diagnosis, diagnostic odds ratio, and the proportion of glaucoma patients discovered were obtained by a meta-analysis. The enhancements to cost and quality data for healthcare services were evaluated. 45 studies out of the 11,237 assessed were included. Based on the results obtained from this investigation, it has been shown that teleglaucoma exhibits lower sensitivity and higher specificity compared to personally assessment. The specificity and sensitivity values were calculated to be 0.790 (95% confidence interval [CI] 0.668, 0.876) and 0.832 (95% CI 0.770, 0.881) accordingly. The likelihood of obtaining a favourable screening outcome is 18.7 times greater in individuals who have glaucoma compared to patients without glaucoma. Additionally, the average cost of glaucoma detection was $1098.67 US per case, and the average cost of teleglaucoma screening was $922.77 US per patient. In conclusion, teleglaucoma has a higher likelihood of producing positive instances than screen test results. In comparison to an in-person examination, it finds more glaucoma patients. Early detection, shorter waits and travel periods, higher referral rates for specialists, and cost savings benefit patients as well as healthcare systems. For remote and underserved communities, teleglaucoma is an efficient glaucoma screening technique.

Rohit et al. [3] to compile data on the economic costs of glaucoma, its epidemiology, and its personal toll. Analysis and synthesis of particular works of literature from 1991 until December 2010. Currently, glaucoma affects an estimated 3% of people over 40 in the world, the majority of whom are undiagnosed. Even during the initial phases of the disease, glaucoma-related vision loss has a major negative impact on good life in terms of health. As glaucomatous injure and vision loss worsen, the overall load rises. Glaucoma has a major financial impact that gets worse as the condition progresses. A person's loss of their health-related quality of life as well as the financial burdens on themselves and society are likely to be lessened by early diagnosis and treatment of glaucoma patients along with people with ocular hypertension at a significant risk of developing sight loss.

Broman et al. [4] to determine how many people will have angle closure glaucoma (ACG) and open angle glaucoma (OAG) in 2010 and 2020, respectively. A review of available information using prevalence models: Data across population-based studies on age-specific prevalence of OAG and ACG that fulfilled standard definitions were used to develop prevalence modelling for OAG and ACG by age, sex, and ethnicity, including data weighting according to sample size for every research. Models and UN forecasts of the world's population throughout 2010 and 2020 have been combined to estimate the approximate number of persons suffering glaucoma. 79.6 million people are expected to have OAG and ACG by 2020, having 74% of these individuals having OAG. In 2010, here is going to be 60.5 million people having OAG and ACG. Women will have 55% of OAG, 70% of ACG, and 59% of all glaucoma cases in 2010. 87% of those diagnosed without ACG and 47% of these without glaucoma will be Asian, correspondingly. By 2020, 5.9 million people with OAG and 5.3 million via ACG, accordingly, would have bilateral blindness, up from 4.5 million in 2010 and

3.9 million in 2010. The second largest cause of blindness in the world, glaucoma disproportionately affects Asians and women.

De La Fuente-Arriaga [5] the methodology employed in this work involves the utilization of blood vessel changes inside an optic disc (vascular bundle) in retinal images of patients for the purpose of glaucoma detection. The methodology involves the segmentation of the vascular bundle area inside an optic disc to determine an appropriate point for comparison on the spatial edge of the cup of vision. This is followed by the identification of the centroids of the segmented zones corresponding to the excellent, inferior, as well as nasal vascular bundles across the divided area. Finally, the deviation from the usual location is calculated via the chessboard distance metric. Among a total of 67 photographs, the method correctly identified 62 of them, yielding a pre-diagnosis efficiency rate of 91.34%. The sensitivity and specificity of the technique were determined to be 92.02 and 91.66%, respectively.

Dash [6] Due to its ability to extract a wealth of information related to various eye illnesses, retinal imaging has emerged as the most important instrument among all forms of medical imaging technology. Therefore, it is essential to accurately remove blood vessels in order to aid ophthalmologists and eye care professionals in early illness detection. We have suggested a computational method for haul out blood vessels as of fundus images in this work.

Nikam [7] Glaucoma, which results in permanent vision loss, is a severe condition that damages the optic nerve of the eye. If treatment for glaucoma is delayed, the patient may go blind. Glaucoma is typically identified when there's an increase in the amount of liquid in the front of the eye. Your eye's pressure rises as the amount of additional fluid does as well. As a consequence, the diameter has also amplified, the dimension of the two the optic disc and optic cup is being augmented. The cup-to-disc ratio (CDR) represents the proportional relationship among the lengths of the cup and disc. The localization of the optic disc and optic cup in this structure is achieved through the utilization of a method of segmentation of the threshold kind. Additionally, there exists an alternative approach for edge identification and ellipse fitting.

Nawaldgi et al. [8] glaucoma, a condition that affects the eyes is frequently called the quiet thief of sight. Glaucoma causes harm that is unrepairable. The only treatment for glaucoma is early detection and treatment. Numerous efforts have been made until date to automatically detect glaucoma using optical coherence tomography (OCT) and colour fundus imaging (CFI) pictures by extracting structural elements. Optic nerve head (ONH) study in the case of CFI and retinal layers (RL) study in OCT scans for glaucoma assessment can both be used to extract structural information. But regrettably, in this area, the works completed thus far fall short of expectations. This paper presents an overview of automated glaucoma detection methods. The research also analyzes several structural characteristics that are pertinent to automated glaucoma detection in CFI and OCT pictures, respectively. The study's findings support the idea that structural information from CFI and OCT images might be combined to produce a more precise glaucoma assessment.

Anusorn et al. [9] the second largest general reason of lifelong blindness in the world is glaucoma. Glaucoma's ability to progress can be slowed down with early identification. The cup-to-disc ratio (CDR), also known as the proportion of the size of the optic cup to the size of the optic disc, is one of the key clinical indications of glaucoma. However, it is presently measured automatically by skilled ophthalmologists, which limits its potential for mass screening for early identification. The authors of this work suggest a method for repeatedly manipulative the CDR using non-stereographic retinal fundus images collected by a non-mydriatic auto fundus camera, the NIDEK AFC-230. Two approaches using edge detection and variational level-set algorithms are suggested in the research for automatically obtaining the disc. The threshold level-set method and colour component analysis are assessed for the cup. Ellipse fitting is used to reshape the disc and cup border that was produced using our approaches [10]. In order to assess the effectiveness of the established CDR in comparison to the medical CDR, a dataset consisting of 44 retinal pictures obtained from Mettapracharak Clinic in Nakhon Pathom, Thailand, was utilized. The proposed method has been found to yield a precision level of 89% in determining the CDR.

This paper focuses on CDR-based automated glaucoma identification using fundus pictures. In order to identify the optic cup (OC) as well as optic disc (OD), the region of interest (ROI) is recovered by the utilization of the intensity weighted centroid strategy, preprocessing techniques, and iterative application of k-means clustering delineation. For OC and OD border smoothing, ellipse fitting is assumed [11]. A total of 100 fundus pictures, obtained from a local source, be utilized to evaluate the efficacy of the projected method. The projected method demonstrates an efficiency of 92% in detecting glaucoma, with a mean square error of 0.002 for the cup-to-disc ratio (CDR) [12].

3 Methodology

In reality, clinicians use the fundus images of the patient's two eyes to visually estimate the CDR in order to assess whether or not a patient has glaucoma. This procedure, the process of doing a diagnosis, that is inherently subjective, may require a highly skilled specialist to invest approximately a minute to three minutes. Moreover, the implementation of this method in screened initiatives, when every expert is required to review a substantial number of images, may potentially result in significant mental and physical strain. The CDR is intended as given

$$CDR = \frac{area\,Cup}{area\,Disc}$$

if the CDR ≥ 0.6 it is glaucomatous, or else it is not. Based on this and what is learned from the literature analysis, the disc and cup must be segmented so as to perform automatic glaucoma recognition as shown in Fig. 2.

The subsequent subsections outline the methodologies employed to get multiple divisions and blood vessel information.

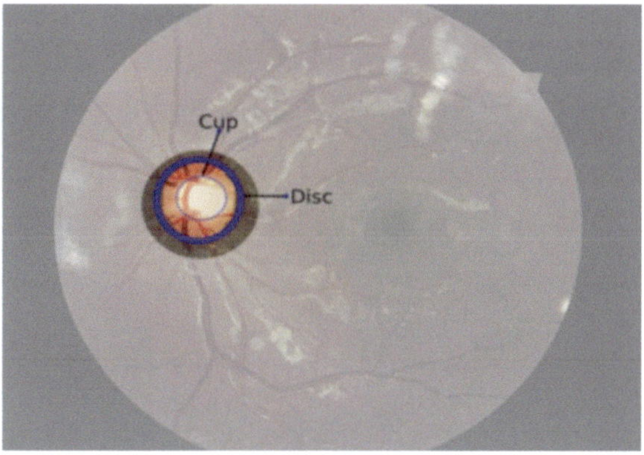

Fig. 2 Disc and cup location

3.1 Disc Segmentation

We utilized the channel of red to achieve the separation of the disc due to the retina's predominately orange background in the fundus image. The primary stage is to section the disc; the red channel provides the greatest results for this, thus a method to fix the starting point was required so as to identify the disc within the retina. According to the research in, the following parameter is determined to determine the typical number of pixels required to be considered a disc: [10],

$$Pavg = \frac{\pi * D^{2/2}}{7.33 * 8}$$

Here the D is the diameter of the retina in pixels. For the purpose of determining the segmentation threshold, this value is used as a guide. If the number of pixels along with two successive grayscale tones is greater than Pavg, it determines the ratio of pixels among the two tones and chooses the greatest tone as the limit if this proportion is greater than 10% from the graphic histogram, which starting from 255 down to 0. We suggest an enhancement to the procedure in this work because the disc segmentation isn't always accurate with that threshold. The disc-containing blob's height to width ratio must be roughly elliptical, it must be located in the image, and there must be fewer pixels inside the blob than the Pavg threshold. These three requirements are what we use to measure the segmentation's accuracy. Up until the three criteria are satisfied, the program continuously calculates the exact value of the greatest threshold. A matrix filtering was devised to meet the spatial criteria, as depicted in the following Fig. 3. This filter consists of ones within the space encompassing the disc, while the rest of the regions are assigned zeros [13].

Two white blobs are present in the filter to accommodate for the possibility that the image is coming from either one's right or left side of the eye. The exact center of the real blob's coordinates must line up with the left eye's overall image's 31%

Fig. 3 The calculated filter

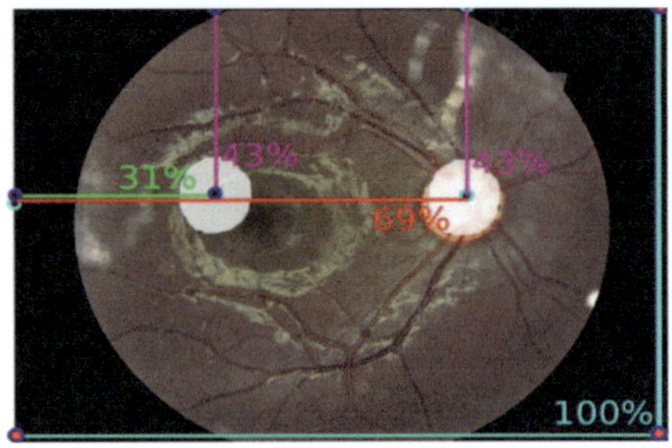

Fig. 4 The filter that has been specifically designed for ocular modification

height and 43% width and the right eye's overall image's 69% height and 43% width as given in Fig. 4.

3.2 *Eye Vessels Segmentation*

Prior to segmenting the cup, the blood vessels emerging as of the disc must first be segmented because their curvature makes it easier to see the cup's edge in the nasal (N), superior (S), and inferior (I) quadrants of the disc, in accordance with the ISNT division worn by ophthalmologists and depicted in Fig. 5. It is crucial to note that, in accordance with expert experience, digging forces veins to migrate in the direction of the nose, which is an important fact to take into consideration in order to find the cup's nasal edge.

In this phase, the utilization of the greenish section of the fundus photo is observed (see Fig. 5). The macula is an important feature that requires removal, and to accomplish this, the Dash and Bhoi method is employed. This method involves convolving the initial photo in the channel of green with a mean filter to effectively obscure the boundaries of the vessels. The macula becomes erased through the subtraction of the original picture from the blurry view. Subsequently, the blood vessels are separated by the utilization of a novel threshold which is derived from this point onward [14].

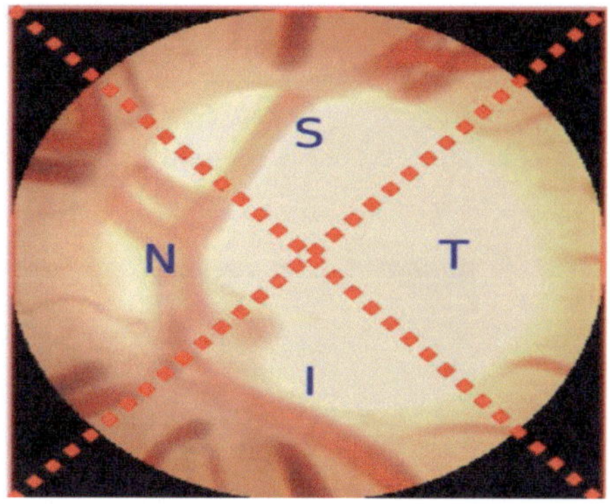

Fig. 5 Disc quadrants

3.3 Cup Segmentation

The blue stream is employed for segmenting the cup due to the higher intensity of pixels exhibited by the cup in comparison to the substrate. The visual screen is partitioned into four distinct regions according to the ISNT rule, namely inferior, better, nasal, and posterior. The initial three components are employed to establish the limits of the veins within the cup, and the centroid of each component is computed utilizing the recommended approach. The transitory fragment is employed for the purpose of identifying the outer boundary of the cup. We present a formula exclusively applicable to the transitory quadrant for the purpose of determining an optimal boundary. The system generates a variety of pixel sizes, denoted as m, with each pixel having a size of 1 n. Here, n represents the quantity of pixels from the middle to the edge of the surface, while m corresponds to the amount of pixels located on the curved section of the plate. Notably, the system ensures that pixels within the vessels are excluded from this process. The equivalent is completed using the basic dim force. For each cluster, it determines the circumstance of its highest level of dark power and logs the typical dim estimate of their neighbouring pixels. With this, it produces a range of typical maximums and a range of typical necessities. Iteratively, the elements greater than a single standard deviation from each of these exhibits are eliminated, leaving each cluster with only the characteristics just outside of the standard deviation. These qualities are found the middle value of and this limit in Fig. 6.

3.4 Cup-To-Disc Ratio Measurement

We can now measure the Cup-to-Disc Ratio after properly segmenting the disc and the cup. We used two techniques, as indicated in Fig. 7, whose accuracy was compared to the expert's diagnosis. The initial method involves establishing an area of interest in the middle of the optic disc and thereafter measuring the lengths among that location and the outermost portion of the segmented cup (referred to as Rc), in addition to

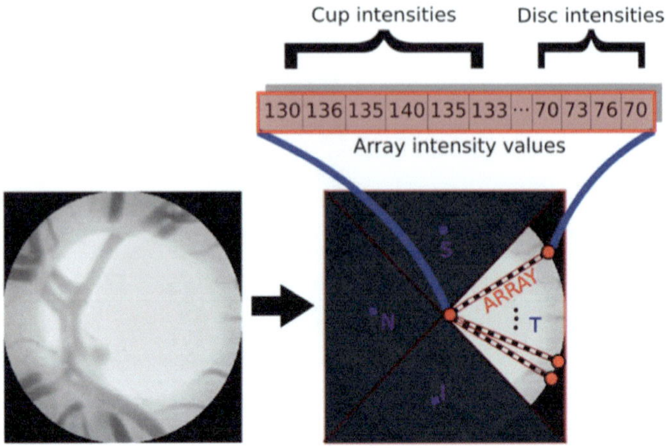

Fig. 6 The acquisition of temporal disc arrays

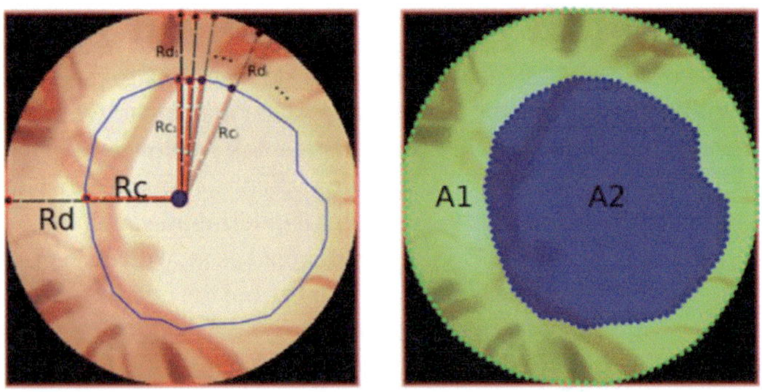

Fig. 7 Techniques employed for the measuring of Cup-to-Disc ratio

the length among middle and the borders of the optic disc. The mean radius, a value between 0 and 1, is then calculated using a geometric mean. If this value is more than 0.6, the eye is considered to be glaucoma-suspicious, and the individual should be seen for additional therapy.

4 Results and Analysis

As a way to accurately segment the disc's blob using the filter stated, it was imperative to calculate the mean spot of the disc's the middle. The mean spot was calculated as the ratio of the disc center contrasted to the surrounding area of the image. This average spot served as a reference point for properly placing the filter. The data

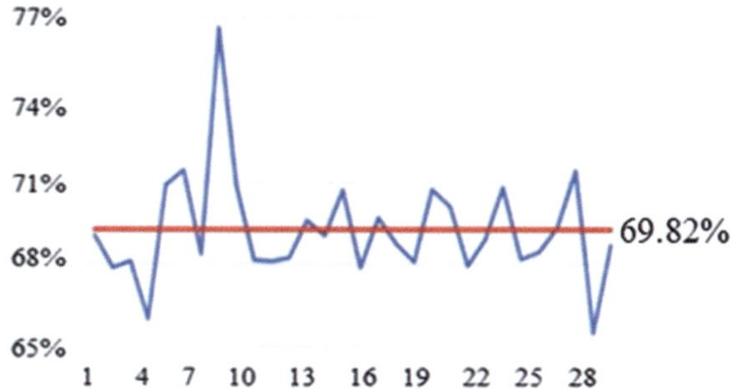

Fig. 8 The horizontal positioning of the disc for a total of 30 eyes

Fig. 9 The vertical positioning of the disc for a total of 30 eyes

are visually exposed in the subsequent plots, whereby the red line denotes the mean position of the 30 photographs. A total of 30 photos were utilized in order to ascertain the average region. Figure 8 illustrates the X-axis value as being 69.82%, and the Y-axis value as being 43.45% as seen the Fig. 9.

4.1 Disc Segmentation Results

A total of 16 fundus pictures were used in the tests. With our enhanced technique, we were able to overcome the issues caused by noise, macular production, or instrument acquisition and achieve an exactness of 95% for the disc segmentation compared to 92% for the algorithm presented by [10]. Results for segmentations made using both methods are displayed in Fig. 10 for fundus pictures as of two separate patients.

4.2 Glaucoma Detection Results

The opthalmologists gave us with their predicted CDR for each of the 26 fundus photos, which included both healthy and unhealthy eyes. As shown in Fig. 11, which contrasts the doctor's assessment with the output of the procedure of ratio of proportions and the output of the method of proportion of regions, consequences from our

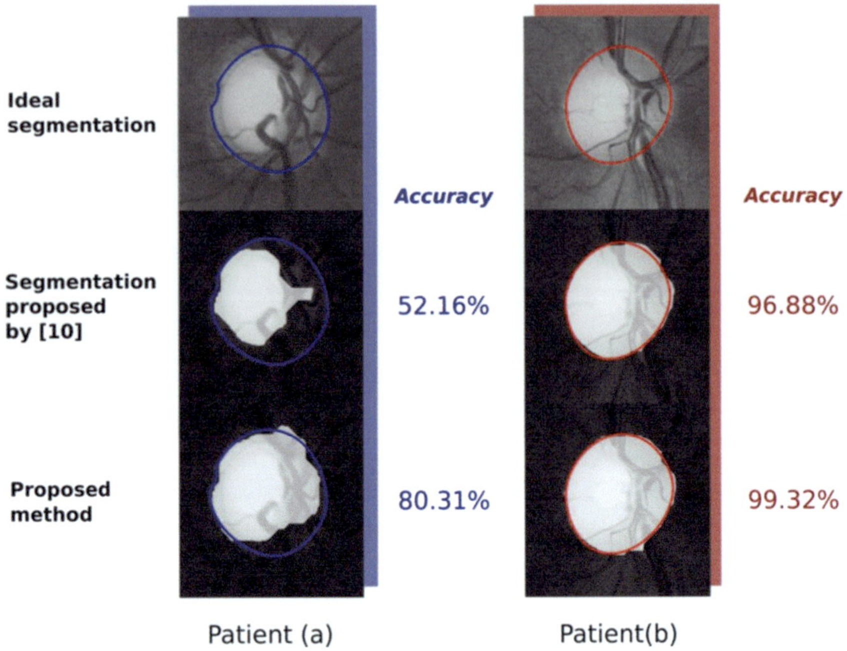

Fig. 10 Disc segmentation analysis

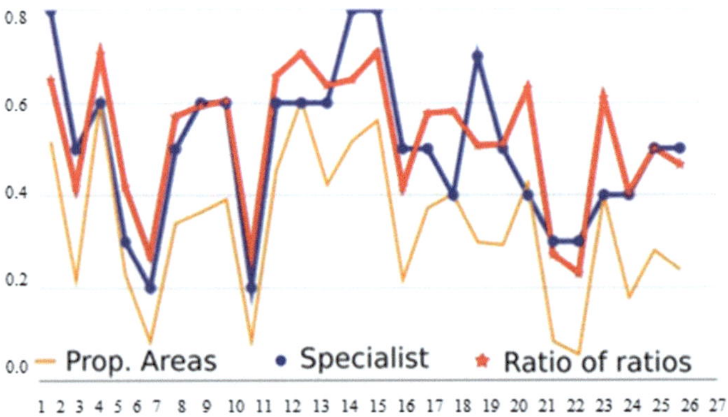

Fig. 11 Compare the performance of an algorithm with an assessment provided by an expert

method have been compared to these reference estimations. 8.6% in terms of absolute inaccuracy and 19.2% in terms of relative error. The detection of cases of glaucoma was successful in 88.5% of the cases.

4.3 Algorithm

$Pseudo\ code\ for\ CNN$:
$for(l\ = 0;\ l < L;\ l + +)\{$
$for(m\ =\ 0;\ m <\ M;\ m + +)\{$
$for(n = 0;\ n < N;\ n + +)\{$
$sum\ =\ bias[l];$
$for(k\ =\ 0;\ k < K;\ k + +)\{$
$for(s1\ =\ 0;\ s1 < S1;\ s1 + +)\{$
$for(s2\ =\ 0;\ s2 < S2;\ s2 + +)\{$
$sum{+}{=}\ weight[k][l][s1][s2]\ *\ input[k][m + s1][n + s2];$
$\}\}\}$
$output\ [l][m][n]\ =\ activation\_func(sum);$
$\}\}\}$

4.4 CNN Steps

1. Data Acquisition
2. Data Preprocessing
3. CNN Classification.

As shown the Fig. 12 represented the demographics percentage levels for male, female, age and MD it has three levels normal glaucoma and P.

As shown the Fig. 13 has average values of distance for clusters its having 5 levels described it represented the various conditions like normal, glaucoma, total number and weight average distance.

5 Conclusion

The approach to diagnose glaucoma was described here by correctly recognizing the location of the cup. This study discusses the strategies utilized for automated glaucoma detection, which includes applying the idea that the cup's existence in the disc is a significant predictor of the disease. Segmenting discs was accomplished using thresholding, segmenting vessels was accomplished utilizing detection of edges, and segmenting cups was accomplished by using vessel and cup intensities.

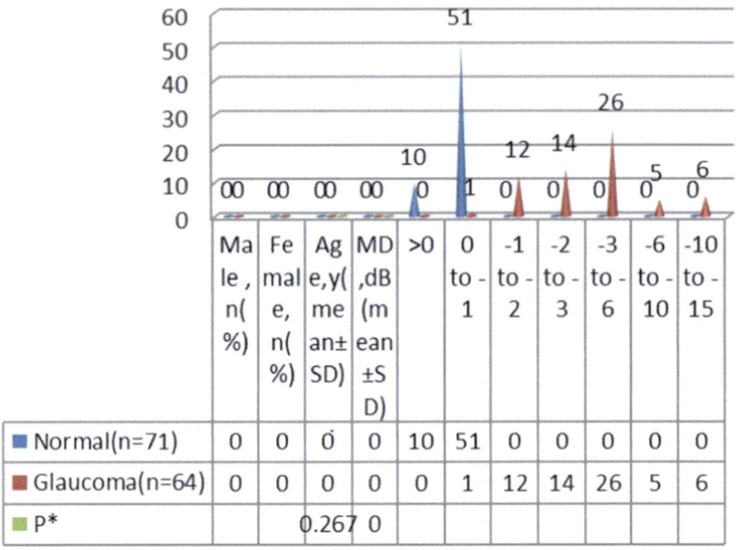

Fig. 12 Subject demographics

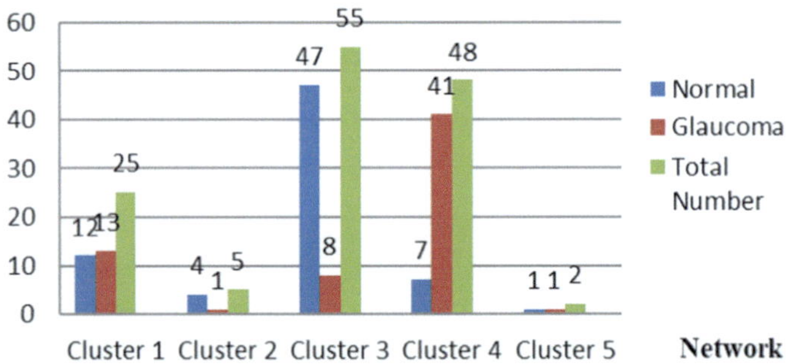

Fig. 13 Mean distance for clusters

References

1. Mohammadi SF, Saeedi-Anari G, Alinia C, Ashrafi E, Daneshvar R, Sommer A (2014) Is screening for glaucoma necessary? a policy guide and analysis. J Ophthalmic & Vis Res 9(1):3
2. Jeyaraman MM, Jeyaraman M, Hodge WG, Hutnik C, Costella J, Malvankar-Mehta MS (2014) The effectiveness of teleglaucoma versus in-patient examination for glaucoma screening: A systematic review and meta-analysis
3. Varma R, Lee PP, Goldberg I, Kotak S (2011) An assessment of the health and economic burdens of glaucoma. Am J Ophthalmol, 152(4):515–522
4. Broman AT, Quigley HA (2006) The number of people with glaucoma worldwide in 2010 and 2020. Br J Ophthalmol 90:262–267

5. De La Fuente-Arriaga JA, Felipe-Riverón EM, Garduño-Caldero E (2014) Application of vascular bundle displacement in the optic disc for glaucoma detection using fundus images. Comput Biol Med 47:27–35
6. Dash J, Bhoi N (2017) A thresholding based technique to extract retinal blood vessels from fundus images. Futur Comput Inform J 2(2):103–109
7. Nikam SM, Patil CY (2017) Glaucoma detection from fundus images using matlabgui. In: 2017 3rd International Conference on Advances in Computing, Communication & Automation (ICACCA) (Fall), pp 1–4
8. Nawaldgi S (2016) Review of automated glaucoma detection techniques. In: 2016 international conference on Wireless Communications, Signal Processing and Networking (WiSPNET), pp 1435–1438
9. Anusorn C, Kongprawechnon W, Kondo T, Sintuwong S, Tungpimolrut K (2013) Image processing techniques for glaucoma detection using the cup-to-disc ratio. Sci Technol Asia 18(1):22–34
10. Dhumane C, Patil SB (2015) Automated glaucoma detection using cup to disc ratio. Int J Innov Res Sci, Eng Technol 4(7):5209–5216
11. Ayub J, Ahmad J, Muhammad J, Aziz L, Ayub S, Akram U, Basit I (2016) Glaucoma detection through optic disc and cup segmentation using K-mean clustering. In: 2016 international conference on Computing, Electronic and Electrical Engineering (ICE Cube), pp 143–147
12. Reddy RV, MGS, Ali AM, Yamini Devi P, MMV, Sucharitha Y (2023) Empirical method for thyroid disease prediction using a deep learning techniques. In: 2023 IEEE 8th international conference for Convergence in Technology (I2CT), Lonavla, India, pp 1–5. https://doi.org/10.1109/I2CT57861.2023.10126291
13. Saravanabhavan C, Kirubakaran S, Premkumar R, Joyce V (2022) Fuzzy based optimized itemset mining in high dimensional transactional database using adaptable FCM. J Intell & Fuzzy Systems 44:1–15. https://doi.org/10.3233/JIFS-221672
14. Nirmala K, Saruladha K, Srujan Raju K (2022) Intelligent noise detection and correction with kriging on fundus images of diabetic retinopathy. In: Evolution in signal processing and telecommunication networks: Proceedings of sixth International Conference on Microelectronics, Electromagnetics and Telecommunications (ICMEET 2021), vol 2. Springer Singapore, Singapore, pp 511–522

A New Technique for Measuring the Performance on Convolutional Neuronal Networks (CNN) for Object and Image Categorization

Siva Skandha Sanagala, Borra Sivaiah, Anuradha Boya, Voruganti Naresh Kumar, V. Ravindernaik, and Vankudothu Malsoru

Abstract The huge, intriguing, and diverse range of identifying objects in images falls under computer vision. It is primarily employed in image recovery, observations, safety, etc. The objectives of an object recognition system are to find the object and classify it according to its classification. Inception v2, Tiny YOLO, and SSD Mobile Net are three artificial intelligence object identification algorithms utilizing convolutional neural networks (CNNs), and their outcomes are evaluated. Kairos and Amazon Web Service Recognition (AWS) represent two cloud-based facial verification systems. The outcomes emphasize the constraints of applying such techniques and models in a field like safety, as well as the limitations of utilizing CNN-based techniques to perform visual problems.

Keywords YOLO · CNN · Object detection · Image classification

---

S. S. Sanagala · B. Sivaiah · A. Boya
Department of CSE, CMR College of Engineering and Technology, Hyderabad, Telangana, India
e-mail: sivaskandha@cmrcet.org

B. Sivaiah
e-mail: bsivaiah@cmrcet.org

A. Boya
e-mail: banuradha@cmrcet.org

V. N. Kumar (✉)
Department of CSE, CMR Technical Campus, Hyderabad, Telangana, India
e-mail: nareshkumar99890@gmail.com

V. Ravindernaik
Department of CSE (AI&ML), CMR Technical Campus, Hyderabad, Telangana, India
e-mail: ravindernaik.aiml@cmrtc.ac.in

V. Malsoru
Department of IT, CMR Technical Campus, Hyderabad, Telangana, India
e-mail: malsoru.it@cmrtc.ac.in

V. K. Gunjan et al. (eds.), *Cybernetics, Human Cognition, and Machine Learning in Communicative Applications*, Cognitive Science and Technology,
https://doi.org/10.1007/978-981-97-8533-9_17

1 Introduction

The human eye can instantly and exactly distinguish a particular image, as well as its content, location, and other visuals that can be accessed by interacting with it. Robots created by humans that use computer vision are, in comparison, comparatively slow and incorrect. Any developments that increase efficiency and effectiveness in this field could pave the way for the development of other intellectual systems which are quite similar to personalities. These advancements would then ease people's lives through the use of assistive technology, which enables people to perform tasks while little to no conscious innovation. Self-driving cars using computer vision-enabled assistive technologies assistance identify and learn about a driving collision while it occurs, regardless of whether the driver is aware of their actions. Real-time object detection is now a crucial component of any effort to continue automating or replacing human labor. Popular fields include identifying objects and computer vision [1]. Machine learning is projected to eventually assist in releasing the potential of universally responsive robotic systems. These cutting-edge technologies have made it easier to access information for and from everybody associated with them, fostering openness. The majority of people's lives revolve on standard personal computers (PCs), and devices have made this trend even more accessible [2]. In addition to this procedure, the quantity of information and images on the internet and in the cloud has increased to the point where millions of people use them every day. Due to people's futility in performing the same repetitive tasks, electronic organizations have developed a dynamic habit of processing this information and making the necessary recognitions and procedures. Recognizing a particular thing or area on a photograph may be the first step in the majority of these procedures [3]. The complexities of the muse and different light intensities are factors that affect the current. Over the years, various approaches have been put up to address the issue of object identification. These methods make repeated efforts to focus on the solution. These methods have faced difficulties such output reliability, energy expenses speed of processing, and complexity issues for years as technology has advanced. The Neocognitron by Yann LeCun et al. inspired the development of the CNN algorithm in the 1990s. Other well-known inventions include AlexNet that won the Image Net Large Scale Visual Recognition Challenge (ILSVRC) in 2012 and later became known as Image Net. The algorithms used by CNN are capable of offering the thing identification problem multiple methods to solutions. Classification-focused algorithms including VGGNet, GooLeNet, and Deep Residual Learning (ResNet) have been developed throughout time to improve recognition accuracy and speed. Even while these algorithms got better as time passed, choosing a window or recognizing several things from a single image remained challenging. Algorithms using suggestions for regions, crop characteristics, and bounding box regression, including Regions with CNN (R-CNN), were developed to address the current problem. Regardless of the fact that R-CNN had excellent precision compared to many similar technologies, its high space and time usage eventually led to the development of the Spatial Pyramid Pooling Network (SPPNet). Despite SPPNet's speed, Fast R-CNN was created to address the related defaults on their loans Although Fast R- CNN used deep networks to achieve real-time rates, there was still a computational burden. Following this, an algorithm that supports ResNet was developed: Faster R-CNN. However, Faster R-CNN was unable to compress state-of-the-art identification organizations, which is when YOLO was put into action. The layers in the YOLO networks, the real-time object reorganization technique, and the CNN architecture are discussed in this piece of content.

2 Literature Review

The movement toward conceptually complicated computer vision methods that incorporate feature extraction plus predictive components into integrated pipelines is being driven by deep learning's current popularity, according to Bruckner et al. [4]. Although a few of these simulations have produced ground-breaking outcomes in domains including object identification, they are challenging to construct and modify, which hinders advancement. We believe that visual analysis might be a potent tool to help incremental deep model pipeline growth.

By using a CNN approach that can take use of object co-occurrences in real-world photos, Petrov et al.'s [5] study aims to enhance object detection results. Real-world sceneries usually display a coherent arrangement of objects in terms of the likelihood that they will occur together. For instance, dishes, bottles, and glasses are commonplace in restaurants. In the identification task, we hope to use this kind of information as a trigger for deciphering item labels.

Thakar and others [6] this study describes the implementation of a shorter delay intensity-based picture registration approach using the embedded platform DevKit-8500D. DevKit-8500D is employed because it supports higher-level programming languages like C/C++ and features an on-board DSP, which makes it suitable for processing images and videos. To shorten the overall process' computational time, template matching on scaled images using cross-correlation computing is employed for registration. In order to highlight the decreased amount of time-delay and to highlight faults caused by the resizing operation, a contrast among the regular and resized photos is also displayed.

We created a unique framework called DeepQoE for predicting video quality of experience (QoE), which was prompted by Zang et al.'s [7] research on the effectiveness of deep learning (DL) related techniques in anticipating, generalization, and representational learning. The end-to-end system first extracts generalized characteristics using a variety of DL methods (such as word embeddings). Then, for representational learning, these features are aggregated and supplied into a neural network. These representations act as the starting point for regression and classification tasks.

Manikandan et al. [8] with the help of the clustering process, feature selection is important to choose the most important features within the previously identified information set and to eliminate irrelevant and unused features for the best approach. Crucial required feature selection techniques are recovered through big data based on the criteria of time effectiveness provides complexity and effectiveness as per the data value. The feature selection reduces the computational complexity of learning and computing algorithms, while a non-selected characteristic saves measuring costs [9, 10].

3 Convolutional Neural Network

By utilizing Convolutional Neural Networks, Deep Neural Networks are often created for identifying objects and image processing. Despite having an infinite number of predetermined important variables for the source image, CNN methods are applied. CNN algorithms are made possible by the ease with which a model may be trained as well as the enormous quantity of knowledge that is accessible online. The methodology used by CNN algorithms to pinpoint and retrieve features from the computer document is wholly mathematical. This technique uses a weight-sharing approach to identify and distinguish information with comparable characteristics [11]. Through this approach, networks may look at large data dimensions and ultimately produce fantastic classification. The processing power of currently available hardware and the range of characteristics in datasets are two apparent barriers to employing CNN models to recover findings.

CNN was created in 1998 with LeNet and reached its peak in 2012 with AlexNet, both of which had error rates of 15.3%. ZF-net came in second. The mistake rate has decreased over time thanks to the development of Google LeNe and VGGNet. ResNet's achievement in 2015 of surpassing the error rate of 3.6%, which was lower than that generated by the human eye (5.1%), marked an exceptional turning point in this period and demonstrated that deep learning systems might outperform human talents [12].

3.1 Structure of CNN

An output layer follows an input layer, a layer of convolution, a working layer, a pooling layer, and a layer that is completely connected. Additional layers may be included in certain CNN model types for additional functions. Due to the complexity of it's the layers, error back propagation computations and forward pass during proficiency testing. A set of image data and associated labels are needed for the guided process of retraining the structure to develop into a model. The best weights to utilize during the testing phase will ultimately be determined at the conclusion of the training session. These layers, which are shown in Fig. 1, might be explained more in this order.

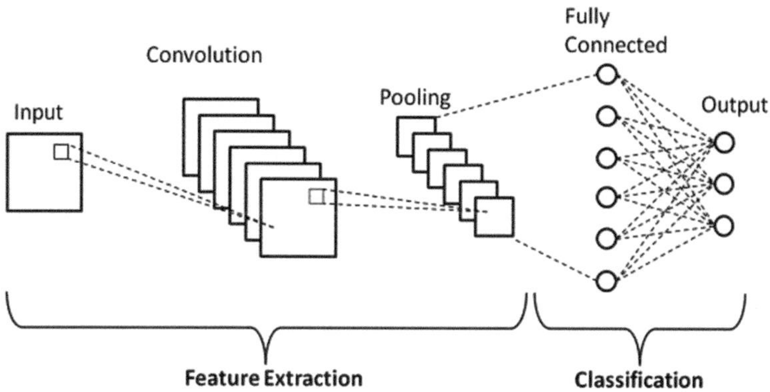

Fig. 1 CNN structured with multiple layers

3.2 *Input Layer*

This layer is used to initialize the source picture data and make each of the corresponding dimensions zero-centered. Employing this layer, the scales are normalized to fall from 0 to 1. This would aid in quickening the convergence rate. By brightening the data, this process of normalization also aids in eliminating redundancy. Principal component analysis, or PCA, is a technique used to minimize and embellish the various retrieved data dimensions while concentrating on important characteristics.

3.3 *Convolutional Layer*

The Convolutional layer, the most prevalent layer in a CNN structure, is how CNN got its name. Each neuron is made to untangle neighboring attributes of various spots in the previous layer and is covered by multiple element maps and different neurons within them. A filter known as the CONV kernel that slides on the initial image fed to it is used by several mutual attributes. While being applied to the Convolutional result, the CONV kernel multiplies and adds the numerical values of every pixel of the local data that is correlated under it in order to determine the element's portrayal of the image. The CONV kernel can be used for obtaining the image's features thanks to this alleged Convolutional rule. The fact that this speaks of shared weights is the justification for screening each of the parts of a picture employing the same CONV kernel. By using shared weights, neutral cells containing similar properties can be identified and categorized as belonging to the same item type. These might be inputted with variables like kernel size, depth, stride, zero-padding, and filter amount as indicated in the illustration 2 Fig. 2.

Due to under fitting, the layer that is currently active is the layer that usually resolves the disappearing gradient issue. The preceding Convolutional layer is to

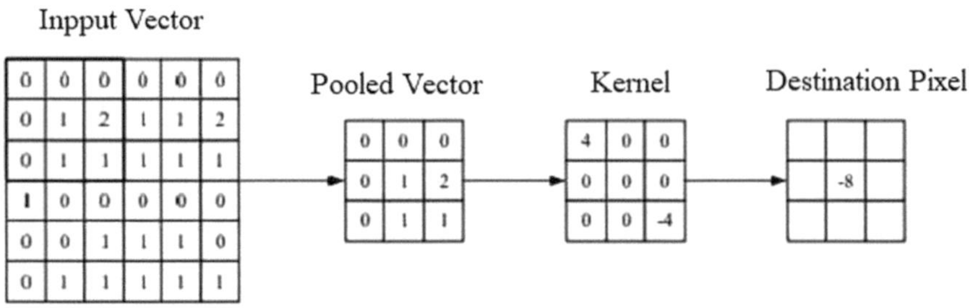

Fig. 2 Visualizing the CNN layers active layer

blame for this inadequately fitting, nonlinear issue. Under fitting is resolved by using any of the active layer functions, such as Sigmoid, Tanh, the rectified long measure (ReLu), the exponential linear measure (ELU), Leaky ELU, or Maxout. Due to their effectiveness and straightforwardness, sigmoid and Tanh function are frequently utilized. Given the speed of convergence, the ReLu functional became the most widely used.

3.4 Pooling Layer

To successfully decrease the size of the outcomes delivered through the Convolutional layer is the task of the pooling layer. This is frequently accomplished by combining the neurons' results from one layer into one neuron in the subsequent layer, which reduces the climate of the element maps and increases the strength of certain extractions. Pooling Layers Can Be Divided Under Three Different Types Based on Their Width: General Pooling, Overlapping Pooling, and Spatial Pyramid Pooling (Spp). Pooling Layers Are Typically Sandwiched Among Two Convolutional Layers. Since a pooling layer's width is mostly determined by its stride, it is referred to as a general pooling layer. Max Pooling and Normal Pooling Are Examples Of General Pooling Activities. Max Pooling is the process of using the most intense impulses from every neuron group in the previous layer. The conventional incentives are referred to be "normal pooling" if the situation is in hot water. Whenever the width is more than the stride, there is continuous pooling. As a result, by structuring certain Convolutional layers in addition to a final pooling layer, unusual attributes within the input layer can be obtained and purchased.

3.5 Fully Connected Layer

The fully connected layer, which is between the CNN layers and is frequently the final one reaching the output layer, transmits data to the output layer. It streamlines and accelerates the data calculation process by employing every neuron from the previous layer and connecting them to each neuron separately. It is an entirely allied layer that never stores any geographic data and is always followed by a yield layer.

Extra Layers Some CNN models require additional layers to be added to the various layers stated above that are required to structure a CNN model in order to provide the desired results. This includes layers like dropout layers and regression layers. Over fits is frequently addressed via applying the technique of dropout layers that avoid very subjective weights by updating neural cell knotted weights using a predetermined probability (determined by the stochastic strategy). Regression layer, on the other hand, is employed to categorize features using techniques like logistic regression (LR), Bayesian linear regression (BLR), and Gaussian Processes for Regression (GPR). The probabilities of all potential object categories are the result of a regression layer.

3.6 Algorithm You Only Look Once (YOLO)

YOLO is a cutting-edge method for quickly identifying several items in an image while creating bounding boxes around them. The CNN algorithm is only applied to the image once to get the result, hence the name. YOLO realistically operates a lot quicker than quicker R-CNN while being somewhat comparable to R-CNN and having a simpler design. YOLO, in contrast to Faster R-CNN, can classify and carry out bounding box regression simultaneously. With the class label YOLO, which contains objects, their location may be determined with just a quick glance. YOLO addresses object identification as an issue of regression by spatially isolating bounding boxes and their associated class probabilities that can be forecasted employing a single neural network. This completely departs from the conventional CNN pipeline. YOLO first presented a unified network architecture that performs bounding box forecasting and class probability computations simultaneously.

The speed and precision of YOLO's real-time object identification are most likely due to the extension of the GoogLeNet equations employed by the detection of YOLO method [13], which is used as their basic forwarding transport calculation. In contrast to R-CNN architectures, YOLO finds bounding boxes and class possibility score for bounding boxes simultaneously, as opposed to first applying a classifier to a prospective bounding box and then computing the probability scores. This slows down the YOLO algorithm, which is a key factor in its ability to recognize and categorize objects in real time while having very low mistake rates. The architecture of YOLO is more akin to the standard convolutional neural network used within the GoogleLeNet model to identify and categorize objects in images.

The design of YOLO is based on the GoogLeNet model for image categorization and is comparable to that of a normal convolutional neural network. The primary layer of the network first collects the features of the image, then the fully connected layers forecast the probabilities and coordinates of the output. The whole YOLO network model was developed using 24 convolutional layers, 2 fully connected layers, 1 × 1 reduction layers, and 3 × 3 convolutional layers.

As distinct elements combine into a single neural network due to the ultimate pipeline, YOLO is presented as a unitary algorithm. The characteristics of the whole image are reasoned globally in order to anticipate all of the boundaries concurrently. The optimal method to use YOLO is for it to perform its own end-to-end training in real time yet retain high levels of average precision. YOLO initially divides the input image into grids of the S × S size in order to understand unitary detection. If the center of an object is positioned inside a grid cell, the grid cell will attempt to identify objects on its own. In order to estimate a bounding box and its confidence ratings across every category trained to predict, each grid cell makes this attempt. The predicted confidence ratings will show how confident we are in our ability to assign everything a label and bounding box. The formal definition of the boldness scores is Pr (Object) × IOU truth red. The point of intersection over union (IOU) among the bottom truth and the predicted box indicates if an object was actually discovered inside the cell. Otherwise, the boldness rating could be as low as 0. Every confidence score is produced by the unified recognition and has five variables: w, y, w, h, and assurance. The box's centroid with respect to the borders of the grid cell is represented by the (x, y) coordinates. As was previously stated, if the middle of the box does not fall within the grid cell, the cell is not in control of its forecast. The predicted object's height and breadth are determined in relation to the entire image, as illustrated in Figs. 3 and 4, with every coordinate being adjusted to be included within an interval of 0 and 1.

Input Image

Output Image

Fig. 3 Input image

Fig. 4 Output image

As shown the Figs. 3 and 4 are the input and out images.

As was previously discussed, the output layer in a CNN is a fully connected layer where the input from the other layers is flattened and transferred to turn the output into the necessary number of classes.

a. Precision,
b. sensitivity, and
c. specificity.
d. F1 ranking.

Figure 5 has a y label with values, a title for performance, and legends for a proposed and an existing model.

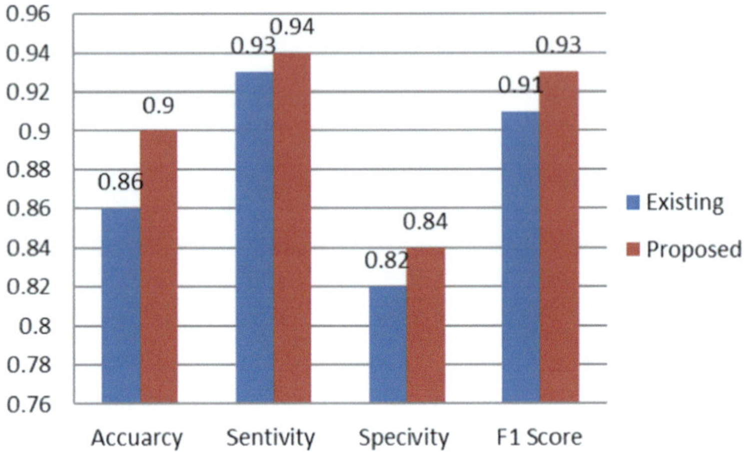

Fig. 5 Comparison of existing proposed values

4 Conclusion

This study summarizes the real-time object identification technique from YOLO as well as the fundamental CNN algorithm structure. CNN architecture models are able to eliminate highlights and identify objects in any image. When implemented appropriately, CNN models may tackle issues like deformity recognition, creating educational or instructive applications, etc. In practice, YOLO offers numerous benefits over other CNN algorithms. By using a combined object detection model that is easy to build and train because to its straightforward loss function, YOLO may train the entire model similarly. The optimum trade-off among rapidity and precision for object identification is offered by the state-of-the-art YOLOv2 version, which is the second major release. As a result of its cutting-edge object identification technique, YOLO is also more effective compared to other object detection models at expanding Object representations. Since of this, it may be suggested for real-time object detection. Considering these accomplishments, it is clear that the field of detecting objects offers a bright future.

References

1. Kong T et al (2017) Ron: Reverse connection with objectness prior networks for object detection. In: 2017 IEEE conference on Computer Vision and Pattern Recognition (CVPR). IEEE. Liu W et al Ssd: Single shot multibox detector
2. Howard AG et al (2017) Mobilenets: Efficient convolutional neural networks for mobile vision applications. arXiv preprint arXiv:1704.04861
3. Tang S, Yuan Y Object detection based on conventional neural network

4. Bruckner D (2014) Ml-o-scope: a diagnostic visualization system for deep machine learning pipelines. No. UCB/EECS-2014-99. California Univ Berkeley Department of Electrical Engineering and Computer Sciences
5. John Wiley & Sons, 2018. Petrov Y (2017) Improving object detection by exploiting semantic relations between objects. M S thesis. Universitat P olitècnica de Catalunya
6. Thakar K et al (2017) Implementation and analysis of template matching for image registration on DevKit- 8500D. Opt-Int J Light Electron Opt 130:935–944
7. Zhang H, Hu H, Gao G, Wen Y, Guan K (2018) Deepqoe: A unified framework for learning to predict video QoE. In: Multimedia and Expo (ICME) 2018 IEEE international conference on, pp 1–6
8. Manikandan RPS, Kalpana AM (2017) A study on feature selection in big data. In: Computer Communication and Informatics (ICCCI) 2017 international conference on, pp 1–5
9. Babu PA, Rao BS, Reddy YV, Kumar GR, Rao JN, Koduru SK, Kumar GS (2023) Optimized CNN-based brain tumor segmentation and classification using artificial bee colony and thresholding. Int J Comput Commun & Control 18. https://doi.org/10.15837/ijccc.2023.1.4577
10. Aelgani V, Gupta SK, Narayana VA (2023) Local agnostic interpretable model for diabetes prediction with explanations using XAI. https://doi.org/10.1007/978-981-19-8563-8_40
11. Nikouei, SY et al (2018) Intelligent surveillance as an edge network service: from HarrCascade. SVM to a Lightweight CNN. arXiv preprint arXiv:1805.00331
12. Jagadeeshwar K, Sreenivasarao T, Pulicherla P, Satyanarayana KN, Lakshmi KM, Kumar PM (2022) ASERNet: Automatic speech emotion recognition system using MFCC-based LPC approach with deep learning CNN. Int J Model, Simul, Sci Comput:2341029
13. Veiga, FJL (2016) Image process detect veh motion. In: European conference on computer vision. Springer, Cham

Random Forest Classifier Assessment on Liver Disease Estimating Through Smote-ENN Balancing for Precision and Complexity Matrix

Pinamala Sruthi, R. Suhasini, V. Narasimha, S. Rao Chintalapudi, D. T. V. Dharmajee Rao, and P. Mounika

Abstract The liver, the body's greatest inner tissue, is one of the significant diseases that, for several causes, impair an individual's usual, healthful proportions. Fatty liver, cirrhosis, hepatitis, chronic liver disease, liver cancer, and liver tumors are just a few of the many disorders that may harm the liver. Excessive triglyceride accumulation of fat results in fatty liver. Machine learning algorithms are designed in this study by investigating with various data balancing methods, including SMOTE + Encoded Nominal and Continuous (ENC), SMOTE + TOMEK, Synthetic Minority Oversampling Technique (SMOTE), KMEANS SMOTE, SMOTE + Edited Nearest Neighborhood (ENN), and SVM SMOTE, with Random Forest providing the best precision with SEMOT-ENN balancing technique. The collection of data is also cleaned, and the number of characteristics is decreased using tilt and correlate techniques. Subsequently to standardize characteristics, the PCA (principal component analysis) algorithm is applied. Lastly, the Random Forest classifier using the SEMOTE-ENN balance algorithm performs the prediction function. The Liver Disease information set from the KAGGLE or UCI websites has been used. The effectiveness of every approach is assessed based on its exactness, precision, and complexity matrix.

P. Sruthi · R. Suhasini · V. Narasimha
Department of CSE, CMR College of Engineering and Technology, Hyderabad, Telangana, India
e-mail: dr.p.sruthi@cmrcet.ac.in

R. Suhasini
e-mail: suhasini@cmrcet.org

V. Narasimha
e-mail: narasimhacse@cmrcet.org

S. R. Chintalapudi (✉)
Department of CSE (AI&ML), CMR Technical Campus, Hyderabad, Telangana, India
e-mail: srao.chintalapudi@gmail.com

D. T. V. D. Rao
Department of CSE, CMR Technical Campus, Hyderabad, Telangana, India

P. Mounika
Department of CSE (Data Science), CMR Technical Campus, Hyderabad, Telangana, India

V. K. Gunjan et al. (eds.), *Cybernetics, Human Cognition, and Machine Learning in Communicative Applications*, Cognitive Science and Technology,
https://doi.org/10.1007/978-981-97-8533-9_18

Keywords Liver diseases · Edited nearest neighborhood · Encoded nominal and continuous · Synthetic minority oversampling technique · SVM-SMOTE · KMEANS-SMOTE · SMOTE-TOMEK · Plant diseases · Deep Learning (DL) · Transfer Learning (TL) · Food security

1 Introduction

The liver, a vital organ, is located in the right upper abdomen, immediately beneath the rib cage. It helps the human body get rid of impurities and maintain appropriate blood sugar readings. Even while the body's organs are capable of healing on themselves, heavy drinking, contact with contaminated air, and pollution of the water injure the liver and raise the possibility of liver malfunction. The solution is a transplant of the liver, despite the fact that it is more costly and less successful. The possibility of liver failure can be reduced by early identification of liver illness. The machine-learning method can forecast diseases based on a data set that incorporates significant indicators of health for both individuals with and without ailments. An efficient data set with accurate illness categorization presentation is required for model development. As the liver continues to work correctly even when it is largely wiped out, issues are hard to identify earlier on. If a person's liver ailment is detected early on, their prognosis is improved. Indians are more likely to experience a failing liver. By 2025, India is predicted to be the world's epicenter for liver disorders. In India, a lack of physical activity, greater use of alcohol, and tobacco are all risk factors for liver illness. Liver diseases come in more than a hundred different kinds. As an outcome, the medical business would greatly benefit from the invention of a device that would help with disease detection. By using automatic liver diagnosis methods (likely mobile or web enabled), these innovations will help doctors make informed choices for patients and shorten patient wait times at liver specialists like endocrinologists. Healthcare providers must get samples from patients in order to diagnose disease, this can be expensive in terms of both money and time. To get all the data required for a more accurate diagnosis, it is frequently essential to perform multiple types of tests or collect numerous examples by the patient. The most popular ones includes the complete blood count (CBC), comprehensive metabolic panel (CMP), and urinalysis. Evaluations still provide a wealth of valuable information, but they are usually less expensive. Numerous metabolic functions are carried out by the liver, such as the creation and retention of glucose, the process of detoxification the production of digestive enzymes, erythrocyte regulation, the manufacture of proteins, and other metabolic functions. A few chronic liver diseases are cirrhosis, inflammation, and chronic hepatitis. An autoimmune response or an infection with a viral agent (including variables such the hepatitis C virus) can result in hepatitis. Hepatitis infection-related inflammatory can result in liver tissue loss and scars. Fibrosis, an ordinary kind of scarring, is distinguished from cirrhosis, a severe form of liver damage and scarring. Both alcoholic and also non-alcoholic fatty liver disorder have the potential to result in the growth of fibrosis and cirrhosis. The prevention of liver damage can be achieved through timely identification of liver disorders, namely during the period among infections and inflammation, but prior to the onset of cirrhosis. Around the types of liver illness might be diagnosed with tests similar to a CMP and a biopsy. Alkaline phosphatase (ALP), aspartate amino-transferase (AST), Albumin (ALB), gamma glutamyl-transferase (GGT), creatine (CREA), total protein (PROT), alanine amino-transferase (ALT), and bilirubin (BIL) can all be found during a CMP by a liver activity area. By analyzing the patterns of distribution and proportions of circulation liver-associated compounds detected via the CMP test as well as evaluate by to figures normalized among a patient's age, sex, and BMI, a specific liver disease can be identified, as well as the cause of the condition. Aminotransferases, AST, as well as ALT are enzymes that help the process of converting alpha-amino molecules to ketoglutaric acid groups, which is a step in the gluconeogenesis process. Although AST isn't as particular to the liver as ALT, it can indicate secondary non-hepatic sources of liver dysfunction. Categorization methods are commonly employed by several automated healthcare diagnosis systems. This study predicts liver illness using five boosting algorithms: gradient boosting, XGBoost, CatBoost, LightGBM, and AdaBoost. A number of preprocessing techniques are applied, along with suitable hyperparameter tweaking and feature selection, to improve the accuracy of predictions [1]. Pre-processing, feature extraction, and categorization are the steps that are involved in liver disease prediction at different levels. This study suggests a hybrid categorization approach for the prediction of liver disease. Additionally, datasets pertaining to Indian liver patient records are gathered from the Kaggle database. The accuracy of the suggested model was 77.58%. The Spyder tool is used to apply the suggested method in Python, and the reliability, precision, and recall of the outcomes are examined [2]. Throughout 2009, cirrhosis-related mortality is rising in the United States. According to a recent learning, those in the 25–34 age range have had the most relative increases in mortality caused alcoholic cirrhosis. Data on death certificates provided by the Vital Statistics Collaboration and populace statistics about the US Census Bureau, gathered by the Centers for the Control of Disease and prevention Wide-ranging Online Data over Epidemiologic Research (1999–2016), have been incorporated in the prospective cohort study. The main outcome measure were the number of deaths from hepatocellular carcinoma (HCC) and cirrhosis, using joinpoint regression used to assess trends [3]. In order to classify people with liver diseases and help medical professionals diagnose them more accurately, the study compared distinct classifier algorithms using machine learning, such as artificial neural network, random forest (RF), as well as support vector machine. These algorithms be adapted to a publicly available liver disease data set. In order to control excessive fitting issues, the minority class was oversampled using a synthetic minority oversampling technique [4]. These systems take meaningful information out of massive medical datasets and assist doctors in diagnosing and predicting illnesses with precision and speed by utilizing the ideas of deep learning, machine learning, and big data. Regarding this, the study offers a thorough overview of the developments made in the submission of artificial intelligence to the prediction and diagnosis of liver illnesses. It then lists the studies' relevant limitations and recommends more research in this area [5].

2 Literature Survey

An essential component of automated disease detection is the mining of useful data from large medical datasets. Liver illness has become one of the most prevalent diseases worldwide in recent times. These studies propose a model for liver disease diagnosis that utilizes Convolutional Neural Networks (CNNs). Additionally, the effectiveness of CNN was contrasted with that of conventional machine learning techniques, such as Logistic Regression (LR), Naive Bayes (NB), Support Vector Machine (SVM), and K-nearest Neighbors (KNN). BUPA and ILPD were the two datasets selected for assessment [6]. By exploring with different data balancing methods, including Synthetic Minority Oversampling Technique (SMOTE), SMOTE + Encoded Nominal as well as Continuous (ENC), SMOTE + Edited Nearest Neighbourhood (ENN), SMOTE + TOMEK, KMEANS SMOTE, and SVM SMOTE, and in computations Random Forest providing the greatest precision using SEMOT-ENN balancing method, this work acquires algorithms that utilize machine learning to improve Liver Disease prediction. Furthermore, the dataset is cleaned and its features are reduced using correlation and skewness algorithms. The next step is to normalize the characteristics using the PCA (principal component analysis) method. Lastly, the Random Forest classifier using the SEMOTE-ENN balance algorithm performs the prediction operation. Used is the Liver Disease dataset, which is available for download from the KAGGLE website or UCI. The accuracy, precision, and confusion matrix are used to assess the performance of each technique [7]. A1DE, NB, MLP, SVM, KNN, CHIRP, CDT, Forest-PA, J48, and RF are among the 10 classifiers that are compared in this study in order to determine the best approach for the precise and early diagnosis of liver disease. The GitHub source and the UCI ML repository provided the datasets used in this investigation. RMSE, RRSE, recall, particularity, precision, G- and F-measures, MCC, accuracy, and recall are used to evaluate the results. The results of the exploratory analysis demonstrate a better result of RF using the UCI dataset. When RF is evaluated using RMSE as well as RRSE, the results are 0.4328 as well as 87.6766, respectively. Its accuracy is 72.1739%, which is also higher than the average of other classifications in use [8]. In this regard, this reading employ the Indian Liver Patient Dataset (ILPD) to propose an innovative design based on better preprocessing and ensemble learning to predict liver illness. The ILPD is subjected to six ensemble learning methods, and the outcomes are contrasted with those of previous research. The recommended model build around the number of data preparation techniques, as well as feature selection, data balance, and feature scaling, to increase accuracy with suitable imputations. The process of multivariate complaint is used to block in the blanks a variety of normalization, min–max scaling, minimum absolute scaling, and resilient scaling approaches; log1p transformation was applied to skewed columns [9]. Employing a hybrid eXtreme Gradient Boosting representation that includes hyperparameter adjustment, we provide a more effective approach for timely liver disease predictions in order to help improve early on detection, diagnosis, and lowering of risks and mortality related to the condition. The study's dataset comprised 416 individuals by liver problems and 167 with no at times past of liver problems. In accordance with the patient's disease state, the population was split into two sets. The attribute "is_patient" had this binary data [10]. Machine learning is the act of verdict patterns in massive amounts of data so as to make decisions and enabling machines to learn (i.e. supervised, unsupervised, semi-supervised or reinforced). The UCI Repository's Liver Patient data set—which includes supervised learning—was utilized in this study. Numerous patient medical examination data sets are available in hospitals. Information about liver patients has been retrieved from these data sets, and this information may be utilized to further enhance the patients' situations as time goes on [11]. Additionally, the leave-one-out cross valuation is worn to diminish the impact of data unpredictability by dividing the data set between training and testing sets. Additionally examined are the correlations between various features and the significance of each feature. The outcomes of the experiment demonstrated the effectiveness of machine learning models in forecast consumption of alcohol. In regards to accuracy, the random forest performs the best between them (80.35%) [12].

3 Proposed System

The biggest internal organ in the human body is the liver, and liver illness is one of the serious conditions that, for a variety of reasons, impair a person's ability to maintain a normal, healthy stature. The phrase "liver disease" encompasses several medical disorders such as cirrhosis, hepatitis, fatty liver, persistent liver failure, liver cancer, including liver tumors. A fatty liver results from an excessive buildup of triglycerides. Alcohol abuse, poor eating habits, and other risk factors can all lead to hepatitis virus infection in the liver.

Acute and chronic infections are possible outcomes of hepatitis. Cirrhosis is a condition in which fibroid tissue takes the place of the liver's dead cells. Cancerous tumors by liver is to have spread from cancer that has impacted by alternate organs are known as metastatic tumors. With a death rate of roughly two million per year, chronic liver disease (CLD) is projected to affect up to 844 million individuals globally. As of to the World Health Organization (WHO), over a 20-year period, the mortality toll by cirrhosis and liver cancer enhanced to 50 million per year globally. According to international research conducted in the United Kingdom in 2015, there were 14.2 alcohol-related liver disease deaths for every 100,000 people worldwide. The schematic representation of the suggested methodology is shown in Fig. 1. The liver disease collection is represented here as a CSV file. Preprocessing functions are then performed to fill in blank characters and remove omitted spaces from the dataset. This preprocessing executing also ensures that the dataset is uniformly sized. PCA as well as skewness properties are subsequently utilized to obtain attributes by the dataset. Here, PCA holds the significant feature properties while reducing the dimensions, and subsequently the dataset is archived using the SMOTE-ENN balance technique. Both classes of dataset are evenly distributed via the SMOTE-

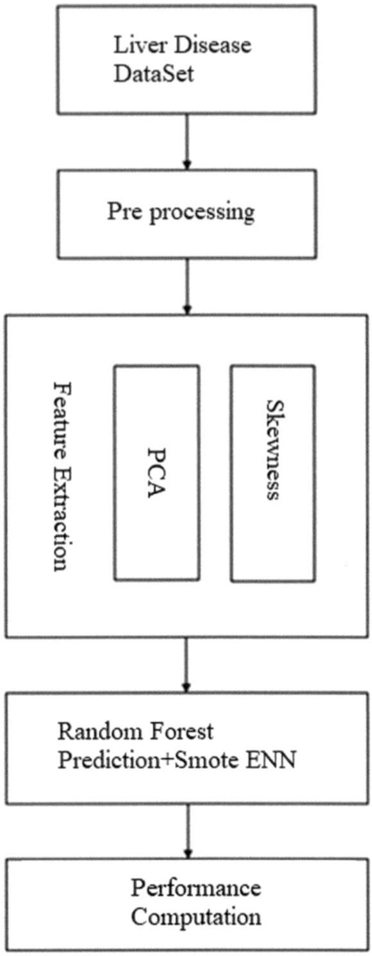

Fig. 1 Shows the recommended system's block diagram

ENN approach. To accomplish the various kinds of prediction activities, the random forest classification mechanism is used as shown the Fig. 1.

3.1 Dataset

The present study encompasses a dataset of 167 records of individuals without liver-related ailments and 416 records of individuals diagnosed with liver-related conditions. These records were collected throughout the northeastern part of Andhra Pradesh, India. The classification of groupings into liver patients (with liver illness) or non-patients (without illness) is determined by utilizing the "Dataset" field as the category label. This information collected comprises 142 clinical files pertaining to female patients and 441 medical records pertaining to male patients. Patients who have reached the age of 90 or over are commonly referred to as being in their ninetieth year.

Columns:

(a) Patient's Age
(b) Gender
(c) Total Bilirubin
(d) Direct Bilirubin
(e) Alkaline Phosphotase
(f) Alamine Aminotransferase
(g) Aspartate Aminotransferase
(h) Protien levels
(i) Albumin
(j) Albumin and Globulin Ratio.

Dataset: The category is employed to partition information into two distinct categories, namely patients afflicted with liver disorders and those without any manifestation of the condition.

3.2 *Preprocessing*

The process of transforming unprocessed information into a format suitable for utilization in a model for machine learning is sometimes referred to as gathering information. The initial and pivotal stage in constructing a machine learning algorithm is the primary focus. As of the context of machine learning development for projects, it is not invariably the case that one encounters pristine and well-organized details. Furthermore, it is imperative to engage in data cleaning and formatting whenever one is working with information. Consequently, an information preparation procedure is employed in this context.

3.3 *Need of Data Preprocessing*

In everyday situations, information often exhibits inherent noise, contains inaccurate information, and can be presented in formats that are not conducive to direct utilization in machine learning models. Data preparation is an essential step in the information analysis pipeline, as it involves the cleaning and transformation of the initial information to ensure its suitability for utilization in machine learning models. This procedure plays a crucial role in enhancing the precision and efficacy of the resulting algorithms.

Fig. 2 Feature scaling

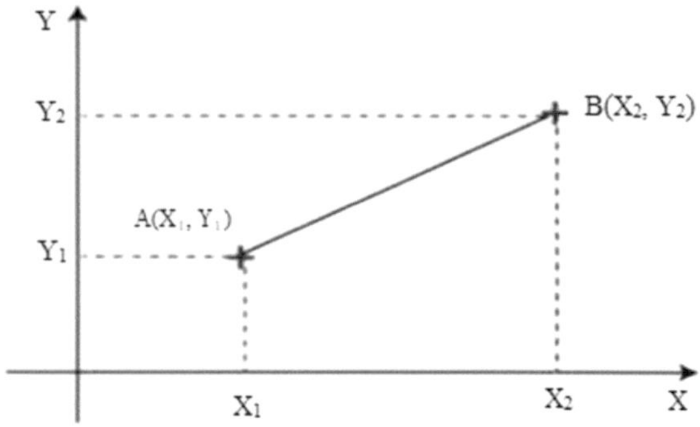

3.4 Feature Scaling

A feature scaling procedure is the concluding phase of information preparation in the field of machine learning. The technique employs the normalization of individual variables in an information set to a specified range in order to achieve standardization. As depicted in Fig. 2 of the characteristic scaling graph, the process involves scaling and standardizing every factor to a common range, so ensuring that no single factor dominates each other. The machine learning method relies on the utilization of Euclidean distance. Therefore, if we fail to appropriately scale the variables, our machine learning method may encounter certain challenges. Euclidean distance is given as:

$$\text{Euclidean Distance between A and B} = \sqrt{(x2 - x1)^2 + (y2 - y1)^2}$$

If we combine age and pay to compute any two values, the result will be inaccurate since the salary values will outweigh the age values. Therefore, scaling features for machine learning is necessary to solve this problem.

3.5 Splitting the Dataset

Throughout the machine learning data preprocessing stage, we divide our information set between a training set and a test set. One of the most crucial phases of data preprocessing since it enables us to enhance our machine learning model's performance. Assume for a moment that we utilize the same dataset to train and test our machine learning model. Thus, it might grow difficult for our model to understand the relationships between the models. If the algorithm is taught extensively and achieves a high level of correctness during the learning phase, its efficiency may deteriorate when presented with a new dataset. Hence, our ongoing endeavor is to design a

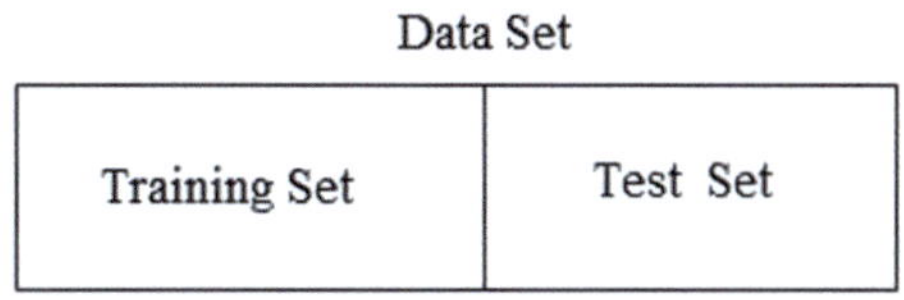

Fig. 3 Splitting the dataset

machine learning model that demonstrates exceptional performance on both the set used for training and the dataset being tested. Here, we can define these datasets as shown the Fig. 3.

Training Set: A portion of the dataset was used for training the machine learning model, and the results have already been determined.

Test set: A subset of an inventory is allocated for the purpose of evaluate the presentation of a machine learning system. Subsequently, the algorithm is employed to make predictions based on the test set.

3.6 PCA Feature Reduction

Principal component analysis (PCA) is a widely favored technique in the field of learning without supervision for the purpose of dimensionality reduction in information. By limiting the loss of data, it concurrently enhances comprehension. As depicted in Fig. 4, the utilization of this graphical representation facilitates the visualization of information in both two-dimensional and three-dimensional formats, hence assisting in the identification of salient features within a given dataset. Finding a series of linear mixtures of variables is made easier by PCA.

In the Fig. 4, multiple measurements are displayed on a two-dimensional coordinate grid. There exist two primary elements. The first principal component (PC1) is the key factor that accounts for the highest amount of variation in the dataset. Principle Component 2 (PC2) is an additional principle factor that exhibits orthogonality with respect to Principal Component 1 (PC1).

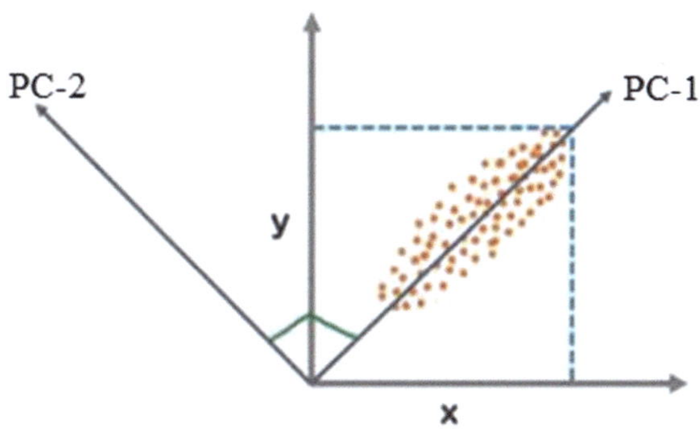

Fig. 4 PCA analysis

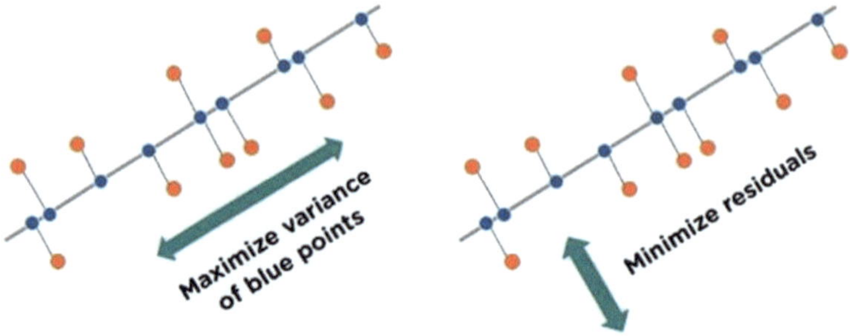

Fig. 5 PCA working

A popular method in the analysis and presentation of data is principal component analysis (PCA), which is used to represent high-dimensional data within a lower-dimensional environment.

The purpose of this technique is to decrease the dimensionality of medical information.

Principal Component Analysis (PCA) has the potential to facilitate the resizing of a picture.

The utilization of this tool in the field of finance enables the examination of stock data and the projection of future results.

Principal Component Analysis (PCA) is a widely used method in data analysis that facilitates the identification of structures within datasets characterized by a high number of dimensions.

As seen the Fig. 5 has following steps.

Step 1: **Normalize the data**: It is recommended to apply standardizing information prior to doing Principal Component Analysis (PCA). By using this procedure, it can be guaranteed that each attribute will possess an average value of 0 as well as an error value of 1.

$$Z = \frac{x - \mu}{\sigma}$$

Step 2: **Build the covariance matrix**: To convey the relationship among more than one feature in a multidimensional dataset, generate a square matrix.

Step 3: **Find the Eigenvectors and Eigen values**: Determine the Eigen values, unit vectors, and eigenvectors. The eigenvector of the covariance matrix is multiplied by eigen values that are scalars as shown the Fig. 6.

Step 4: Prefer the quantity of principal components after order the eigenvectors by highest to lowest.

Fig. 6 Covariance matrix formulation

$$\begin{bmatrix} Var(x_1) & \cdots\cdots & Cov(x_n, x_1) \\ \vdots & \ddots & \vdots \\ Cov(x_n, x_1) & \cdots\cdots & Var(x_n) \end{bmatrix}$$

3.7 Smote-ENN

We use the edited nearest neighbor (ENN) technique and the synthetic minority oversampling technique (SMOTE) to artificially enhance the LD samples because there's a discrepancy in sample size among non-LD and LD samples. SMOTE's main goal is to generate multiple new instances of the minority class by arbitrarily selecting one of its close neighbors and interpolating as explained. First, the k closest neighbors from other minority class samples are selected for each sample of class *Xp*. The instance *Xq* across the minority group is selected randomly between the k neighbors. In the final phase, the artificial sample *XNew* is produced through the process of interfering with *Xp* and *Xq*.

$$XNew = Xp + rand(0, 1)$$

where rand (0, 1) detects a generated random number among 0 and 1. The application of SMOTE can be viewed geometrically as an approximation among two LD samples. Thus, the LD samples' choice space is expanded. The quantity of each category can be balanced using the SMOTE approach. However, it may result in an overlap between the created minority class examples and the original general population samples, making them difficult to distinguish.

We employ the SMOTE + ENN approach to resolve this issue. The minority lass cases (X New) produced by SMOTE are located in the training set using the ENN technique. The instances of the minority class will be removed if these t cases fall under the majority class. In this situation, the SMOTE + ENN approach can help to clarify the boundaries of each class, enabling the suggested machine learning classifiers to perform better on predictions for unknown LD data. In this work, the twofold cross-validation process makes use of the SMOTE + ENN approach, as shown in Fig. 7.

Adaptive Synthetic Sampling: The concept of adaptive synthetic refers to the ability of a system or material to adjust or modify its properties or behavior in response to changing Dynamic samples is a technique that creates varying quantities of samples based on the spread of the data. The method's fundamental flow is outlined as follows:

Step 1: Determine the quantity of materials to be produced with the following calculation: The formula $G = (ml–ms) \times \beta$ represents a mathematical expression, where ml represents the quantity of major examples, ms represents the quantity of minority examples, and β is a constant. When β equals 1, the amount of samples with positive

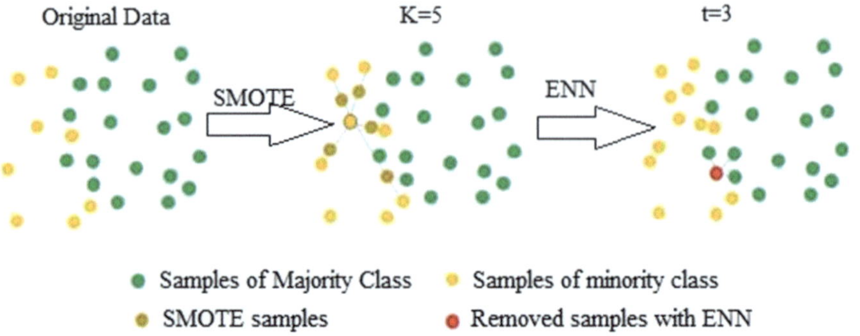

Fig. 7 It represents the SMOTE + ENN

and negative values becomes equal following the sampling process, suggesting that the information has equilibrium at this particular moment.

Step 2: To get the amount of K closest-neighbor values in each minority data, we estimate the quantity of majority examples inside the K neighbors. The equation for this calculation is as listed below: The equation ri = 1i/K represents the ratio of the quantity of majority examples (1i) to the total number of specimens in the K closest neighbors, $i = 1{,}2, 3., ms$

Step 3: To normalize ri, the formula is $\hat{r} = ri/Pms\ i = 1ri$

Step 4: Based on the sample's supplied objects, it is necessary to determine the quantity of fresh examples that must be produced for each minority example. The equation can be expressed as $g = \hat{r} \times G$. A single sample will be chosen from the K closest neighbors close to every information point with the label "1" for synthesis. The quantity of samples to be produced will be calculated using the above methods.

$$i = xi + (xzi - xi) \times \lambda$$

Let si represent the artificial sample, xi denote the ith minority study, and xzi represent an unknown integer among the minority sampling $\lambda \in [0, 1]$ chosen among the K closest neighbors of xi.

Two resampling techniques can be coupled, as opposed to just employing one undersampling or oversampling technique. For instance, the under-sampling technique SMOTE-ENN, ENN, which is included to the workflow after SMOTE to provide cleaner combined data, focuses on removing noise samples. The closest neighbors for each merged sample are determined using the Euclidean distance. Such samples will be eliminated if their majority of KNN samples differ from those of other classes.

4 Random Forest Algorithm

The Random Forest algorithm is widely recognized in the field of machine learning as a prominent member of the supervised training methodology. Machine learning (ML) can utilize this technique for tasks such as regression and classification.

The approach is founded upon the concept of batch instruction, a technique that combines multiple classifiers to tackle challenging problems and improve the efficiency of the framework. The algorithm known as "Random Forest" is so named because it consists of numerous decision trees which are constructed using distinct portions of the data provided. By averaging the predictions made by these decision trees, the Random Forest classifier aims to enhance the overall precision of its predictions for the supplied information. Rather than relying on a singular decision tree, the random forest algorithm leverages projections from multiple trees and makes its own estimate based on the majority of the vote among these projections. The presence of a greater number of trees inside a forest environment contributes to increased precision and mitigates the problem of excessive fitting.

As shown the Fig. 8 Random Forest algorithm has following steps are involved.

Step 1: In the Random Forest algorithm, a subset of n odd entries is sampled from an information set containing k entries.

Step 2: A separate decision tree is generated for every single instance.

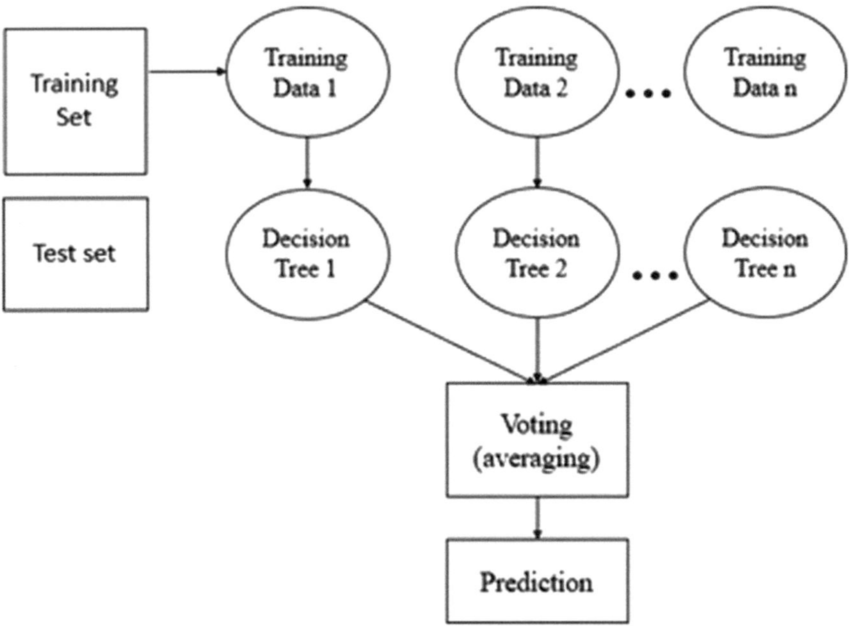

Fig. 8 Random forest algorithm

Step 3: Every decision tree is going to yield a result.

Step 4: The end result is determined through the application of Most Voting for categories and Average for regression analysis.

Important Features of Random Forest

Diversity—Not every traits, factors, or characteristics are taken into consideration during the construction of a single tree, resulting in variations between each tree.

Immune to the curse of dimensionality—As a result of individual trees not taking into account all characteristics, the dimension of features experiences a reduction.

Parallelization—Every tree is constructed autonomously using distinct information and properties. This implies that the central processing unit (CPU) can be effectively utilized for constructing random forests.

Train-Test split—In the circumstance of a random forest, it is not necessary to partition the information into separate training and testing sets, as a portion of the input (namely 30%) is consistently withheld from the decision tree.

Stability—Stability is achieved through the utilization of most voting or averaged as the underlying mechanism for determining the outcome. The user's text is too short to be rewritten academically.

5 Results and Discussion

This section provides a comprehensive examination of the simulation outcomes that were executed utilizing the "python environment". Additionally, the efficacy of the suggested approach is evaluated by comparing it with current techniques employing an identical information.

Figure 9's x-axis shows 1 (no disease) and 2 (disease present), while the y-axis shows the amount of records. As can be seen in the screen above, 1 class has an imbalance since it has more data than the other class, which has fewer records. Figure 10 shows the skewness of each property, with positive values denoting significant features and negative values denoting unimportant ones.

As shown the above Fig. 11 is to compare the various models performance with their existing and proposed models, it represents the Accuracy, precision, recall and F1 score values.

6 Conclusion

The machine learning techniques described in this work can supplement medical knowledge but are not a replacement when selecting from classification algorithms for diagnostic routes. These strategies can help to reduce a number of the problems

Fig. 9 Number of diseases

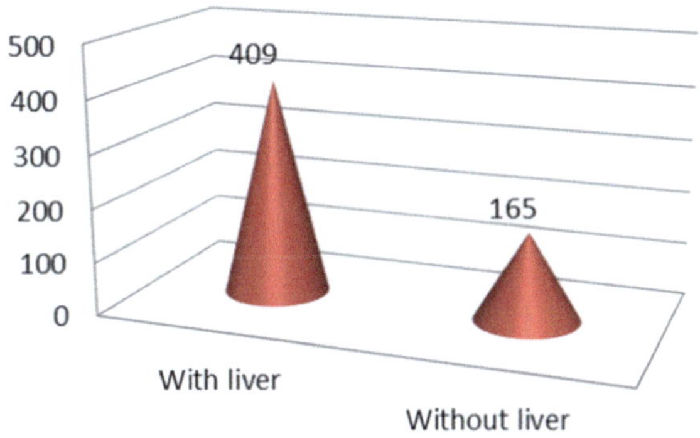

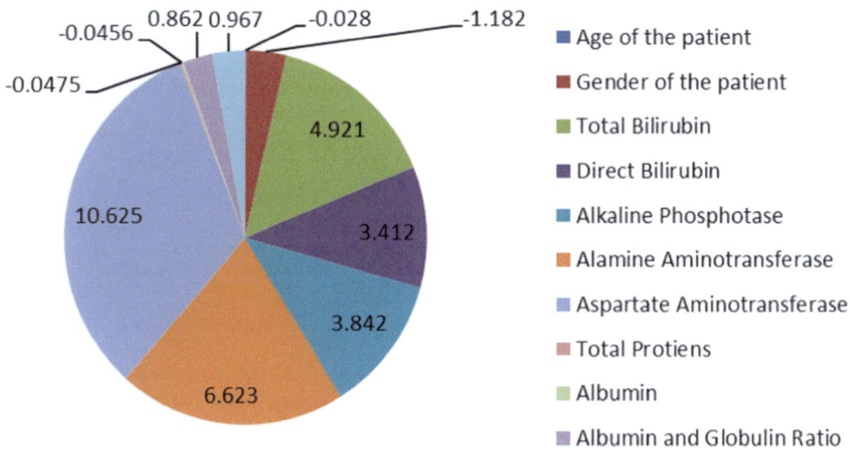

Fig. 10 Skewness feature of dataset

associated with healthcare related to incorrect diagnosis, lost data, expense, and time. The application of the RF-SMOTE-ENN approaches can contribute to reducing the overall impact of liver disorders on public health internationally by raising knowledge of hazards and diagnostic variables. More precisely, by detecting liver disorder at an early stage or in covert situations, RF-SMOTE-ENN may decrease liver-related mortality, transplants, and/or hospitalizations in patients with chronic liver disease. Early identification enhances prognosis since therapy can be started before the disease progresses to more advanced stages. Additionally, less invasive diagnostics, like biopsies, would be performed in this scenario. Although the focus of this study was on hepatitis and chronic liver disease variables for ML training, it is conceivable that the techniques may be applied to discriminate between people with other types of liver disease and healthy people. All of the aforementioned techniques could be used in additional medical specialties to enable AI/ML-assisted diagnosis.

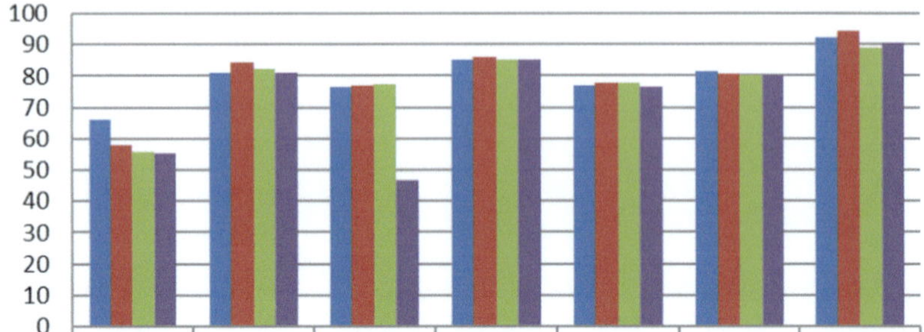

	Existing RF without data balancing	Existing RF with SMOTE balancing	Existing RF with SMOTE ENC balancing	Existing RF with SMOTE Tomek balancing	Existing K means SMOTE balancing	Existing SVM with SMOTE balancing	Proposed RF with SMOTE ENN balancing
Accuracy	66.22	80.92	76.65	85.2	76.84	81.45	92.3
Precision	58	84.3	76.92	85.95	77.7	80.45	94.2
Recall	55.62	82.4	77.45	85.3	77.8	80.24	88.96
F1 Score	55.42	80.97	46.77	84.96	76.55	80.32	90.7

Fig. 11 Performance comparison of various models

References

1. Ganie SM, Pramanik PKD (2024) Predicting chronic liver disease using boosting technique. In 1st International conference on artificial intelligence for innovations in healthcare industries (ICAIIHI-2023). Raipur, India
2. Shaheamlung G, Kaur H (2021) The diagnosis of chronic liver disease using machine learning techniques. Inform Technol Ind 9(2):554–565
3. Tapper EB, Parikh ND (2018) Mortality due to cirrhosis and liver cancer in the United States, 1999–2016: observational study. BMJ 362:k2817. https://doi.org/10.1136/bmj.k2817
4. Mostafa F, Hasan E, Williamson M, Khan H (2021) Statistical machine learning approaches to liver disease prediction. Livers 1(4):294–312
5. Tanwar N, Rahman FK (2021) Machine learning in liver disease diagnosis: Current progress and future opportunities. In: IOP Conf Series: Mater Sci Eng (ICCRDA 2020) 1022:012029. https://doi.org/10.1088/1757-899X/1022/1/012029
6. Mutlu EN, Devim A, Hameed AA, Jamil A (2022) Deep learning for liver disease prediction. In: Djeddi C, Siddiqi I, Jamil A, Ali Hameed A, Kucuk İ (eds) Pattern Recognition and Artificial Intelligence (MedPRAI 2021). Communications in computer and information science. Springer, Cham, pp 95–107
7. Lin RH (2009) An intelligent model for liver disease diagnosis. Artif Intell Med 47(1):53–62. https://doi.org/10.1016/j.artmed.2009.05.005. Epub 2009 Jun 21 PMID: 19540738
8. Naseem R, Khan B, Shah MA, Wakil K, Khan A, Alosaimi W, Uddin MI, Alouffi B (2020) Performance assessment of classification algorithms on early detection of liver syndrome, J Health Eng 2020 (Article ID 6680002)
9. Quadir MDA, Kulkarni S, Joshua CJ, Vaichole T, Mohan S, Iwendi C (2023) Enhanced preprocessing approach using ensemble machine learning algorithms for detecting liver disease. Biomedicines 11(2):581

10. Dalal S, Onyema EM, Malik A (2022) Hybrid XGBoost model with hyperparameter tuning for prediction of liver disease with better accuracy. World J Gastroenterol 28(46):6551–6563. https://doi.org/10.3748/wjg.v28.i46.6551
11. Gupta K, Jiwani N, Afreen N, Divyarani D (2022) Liver disease prediction using machine learning classification techniques. In 11th international conference on communication systems and network technologies (CSNT). Indore, India
12. Zhao J, Wang P, Pan Y (2022) Predicting liver disorder based on machine learning models. J Eng 2022(10):978–984
13. Shilaskar S, Ghatol A (2019) Diagnosis system for imbalanced multi-minority medical dataset. Soft Comput 23:4789–4799

Breast Cancer Evaluation and Prevention Assessment by Employing ML and SVM Algorithms

C. H. Rajakishore Babu, Golla Saidulu, K. Venkateswara Rao, Mahesh Kotha, Ratnababu Jyothi, and Mudimela Madhusudhan

Abstract According to the most recent research, cancer is a worldwide problem affecting individuals of all ages and levels of income. Most women will get breast cancer at one point in their lives. So, every change in how cancer is found and how it will be treated makes it more likely that you won't live a long life. The use of machine learning methods could have a big effect on predicting cancer and finding it early. Present work, the Wisconsin Cancer dataset is put into groups using two of the most popular machine learning methods. Using exactness, accuracy, recall, and ROC Area scores, the way these methods put things into groups was compared. The best results came from the Support Vector Machine method, proving to be accurate. With a 99.1% success rate, SVM is the best way to figure out what a prophet is saying. We think that this study leads us to the conclusion that SVM is the best method for prediction, and that DT works well with SVM.

Keywords Breast cancer detection · Benign · Malignant · NB · KNN · SVM · RF · DT

C. H. R. Babu · G. Saidulu · K. V. Rao
Department of CSE, CMR College of Engineering and Technology, Hyderabad, Telangana, India
e-mail: chrajakishorebabu@cmrcet.org

G. Saidulu
e-mail: gsaidulu@cmrcet.org

K. V. Rao
e-mail: kvenkateswarrao@cmrcet.org

M. Kotha (✉)
Department of CSE (AI&ML), CMR Technical Campus, Hyderabad, Telangana, India
e-mail: kotha.mahesh528@Gmail.com

R. Jyothi
Department of EEE, CMR Technical Campus, Hyderabad, Telangana, India

M. Madhusudhan
Department of CSE, CMR Technical Campus, Hyderabad, Telangana, India

V. K. Gunjan et al. (eds.), *Cybernetics, Human Cognition, and Machine Learning in Communicative Applications*, Cognitive Science and Technology,
https://doi.org/10.1007/978-981-97-8533-9_19

1 Introduction

Cancer was the next-greatest reason of death around the world in 2018, killing more than 9.6 million people. Cancer is the reason of one death every six around the world. Cancer kills more than 70% of people in nations with low or middle incomes. Colon, lung, and breast cancer are the three utmost ordinary types of cancer in women. Together, they report for fifty percent of all cases of cancer. Also, cancer is linked to 30% of them new cancer cases in women. ML algorithms are used to look at a dataset and find information and important correlations. A computer graphic is also made so that the data can be understood better. Examiners say that it is especially helpful for ML methods used to treat cancer [1]. SVM, kNN, NB, and DT (C4.5) were used on the Wisconsin Breast Cancer (Original) dataset. While the measure results are evaluated, the SVM classification technique is the most accurate (97.13%) and had the least number of mistakes. Breast Cancer was the data set for this study, and Weka was the machine learning tool. The key presentation characteristics of machine learning categories are judged based on their exactness, recall, precision, and ROC area. They say that BN has the finest recall and exact numbers and that the RF approach is the best ROC district. Ahmad et al. used ML methods to figure out how often breast cancer comes back every 2 years. The ICBC program provided the data, which covers the years from 1997 to 2008. The dataset has information about the community and 22 input factors, as well as 1189 cases of breast cancer in women [2]. ANN, SVM, and DT are all used by SVM, which gives the greatest outcomes with regard to precision and error rate. Summary of the participants of the NCCN Breast Cancer Panel contributed to the creation of these NCCN Clinical Guidelines for Practice in Oncology (NCCN Guidelines) for the treatment of breast cancer. The calculations and text make note of the supporting data and consensus classifications that were evaluated. The ideal course of treatment for patients at all levels of breast cancer is to enroll them in randomized clinical trials, even though this isn't mentioned clearly in the NCCN Guidelines at every single point [3].

2 Related Work

The purpose of the current study was to evaluate the Gail model's effectiveness for Turkish women. The length of breastfeeding was also looked at as a potential risk factor. 650 people suffering from invasive breast cancer (group 1) and 640 women who visited a mammary care unit as well as had mammograms performed for screening but had negative results were included in our research. The length of breastfeeding as well as the particular risk factors contained in the Gail model was compared between the two groups. Each woman's 5-year risk was predicted using the Gail model [4]. The principle of deep learning is to observe the key elements authority in the identification and management of severe illnesses. For instance, histopathological imaging or genetics can be used to identify breast cancer. Since genetic testing can be quite costly, the most popular method for detecting breast cancer is histopathological imaging. In this investigation, we conducted an exhaustive assessment of previous research on the identification and management of breast cancer through genome sequencing or histopathological imaging using machine learning as well as deep learning. We also offer suggestions to scientists who will do research in this area [5]. Through an emphasis on feature selection and dataset size, this study examines the variables affecting the effectiveness of machine learning methods in the field of BC prevention. The purpose of the learning is to investigate how feature, model, and dataset cardinality affect analytical performance as measured by exactness and the area under the curve (AUC) [6]. This review addresses the possibility of recognizing particular molecules against tumor cells or tumor microenvironments to prevent the growth of metastatic disease and enhance the prognosis of breast cancer patients. It also summarizes recent molecular mechanisms and novel ideas on the metastatic heterogeneity of breast cancer [7]. This learning worn a machine learning that comes up to find distinct gene causal networks linked to lung, brain, and bone metastases of breast cancer. Genes with variations in expression were found in primary breast cancer tissues, bone, brain, and lung breast cancer metastasis tissues, as well as in patient clinicopathological characteristics derived through the Gene Expression Omnibus microarray datasets using Bayesian network analysis. We investigated the causal Bayesian networks of female breast metastasis to distant locations (bone, brain, or lung) by (i) calculating the degree of how well the structures of each type of breast cancer metastasis fit the data, (ii) contrasting the constructions about acknowledged clinical data, and (iii) expressing the capacity for prediction of the structures [8]. In order to forecast breast cancer metastases to the sending lymph node (SLN), it is important to look into the potential utility of a machine learning model based on radioman parameters from CE-MRI. Methods at the initially affiliated hospital about Soochow University, a review was performed on the clinical, pathological, as well as MRI data of 177 patients with pathologically showed breast cancer (96 about SLN negative as well as 81 about SLN positive), who had conventional DCE-MRI before surgery among January 2015 and May 2021. Employing a 7:3 ratio, the samples were randomly divided into either a training set ($n = 123$) as well as a validation set ($n = 54$). The radiomic features, which consist of 1,316 original eigenvectors standardized by the highest and lowest normalization, were obtained from DCE-MRI phase 2 pictures. The optimal features were obtained by applying the optimal feature filter and selection operator (LASSO) method. Depending on the chosen features, five machine learning models—Random Forest, Gradient Boosting Decision Tree, Support Vector Machine, Logistic Regression, and Decision Tree—were built [9]. The suggested breast cancer detection model has a greater precision when using the BCP-SVM. The suggested BCP-T1F-SVM expert system provides a greater precision rate when it comes to breast cancer. Initial breast cancer diagnosis is being performed using the proposed BCP-T1F expert system. The BCP-T1F expert system handles breast cancer by taking into account the various stages of the disease. Based on the evaluation and computations performed in this study, BCP-SVM outperforms BCP-T1F. The accuracy percentage for the BCP-T1F is 96.56%, while the accuracy % for the BCP-SVM is 97.06% [10]. Our deep learning algorithm is valuable and can identify breast cancer in digital mammograms with different densities. Three different feature selection modules were used in our research: recursive feature elimination, univariate feature selection, and low-variance feature removal. Included are the medial–lateral and craniocaudally seen mammograms [11]. This learning suggests a PC facilitate termination (computer aided designing) technique that divides patients into three groups: non-dangerous, no harm, and frightening. It does this by using AI calculations. The review focuses on Irregular Backwoods (RF), Backing Vector Machines (SVM), as well as Convolutional Brain Organizations (CNN) as efficient classifiers. Interestingly, the study goes so much so far as to improve classification accuracy by preprocessing mammography images [12].

3 Proposed System

This method compares SVM, DT, RF, NB, and KNN search as ML techniques. The datasets from Wisconsin are used to put together the data gathering. In order to use the machine learning methods, the information is split up into testing sets. All five methods will be put side by side and looked at. PyCharm, a tool for working with Python, would get an example of the method that gives the best results. This data set can be found in the UCI Machine Learning storage area. It is composed of 32 real-world qualities that have more than one value. There are 569 cases in this set of data, and none of them is missing values.

Design Methodology

The system architecture for identifying the user as genuine is shown in Fig. 1.

In this, the information from the input register is pre-processed to get rid of any errors or null numbers. The data is then split into testing and training parts so that the analyzer can use it. The SVM Algorithm is taught with trainee data and then tested with analyzed data to figure out if it is right. Because of this, the person is considered to be real.

This study combines five machine learning (ML) methods: Random Forest (RF), K-Nearest Neighbor (KNN), Support Vector Machine (SVM), Naive Bayes (NB), and Decision Tree (DT). The Wisconsin Diagnostic Breast Cancer (WDBC) dataset is made up of digital images from an MRI. For the Machine Learning methods to work, the information set is making it to a few parts: the testing part and the training part. Methods that have the best results are used to decide whether a cancer is good or bad. How effective each algorithm is evaluated and evaluated as how accurate the consequences are the code for these methods is written in Python, and pyCharm, an experimental tool for Python, is used to run them.

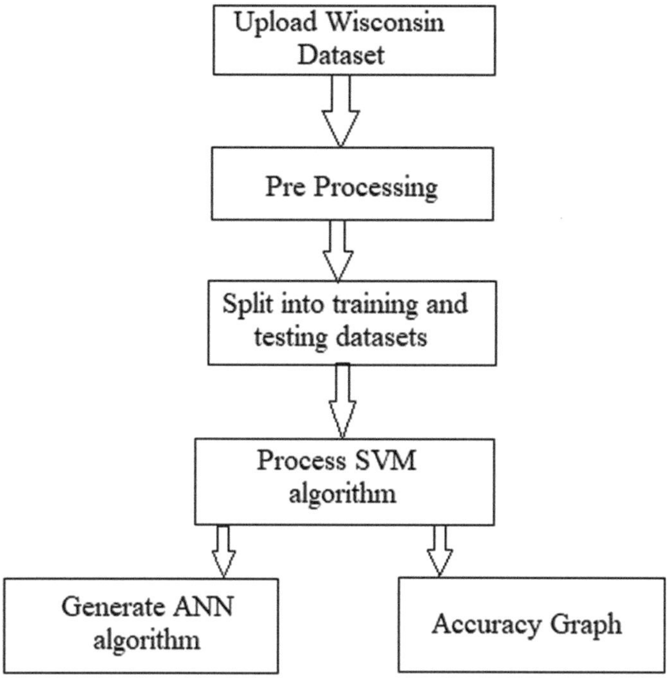

Fig. 1 System model

3.1 Upload Dataset

Upload Dataset is a way to share raw data sets that are important to your investigation plan. It can be found in databases, remote data, programming languages, NoSQL storage, and other places. To post a dataset, you have to find data sets, get data from them, and ask questions about the data. The work data set came from the Wisconsin Data set. We used other tools, like computer country with who is, to find out other things. The final dataset is made up of about 1,780 values that are used to teach Machine Learning models.

The Wisconsin Breast Cancer Dataset (WBCD) is employed in this research. This list was put together by Dr. William H. Wolberg of the University of Wisconsin. It came from UC Irvine's Machine Learning Repository. To look at the features of the information set, a digital picture of a fine needle aspirate (FNA) of a breast crowd is worn. They talk about what the image's cell nuclei look like. This dataset has 699 cases, two groups (good and bad), and nine integer-valued attributes, which are given in Table 1. The number 10 means that the case is mostly unusual. We tidy up the information by getting rid of the 16 places where there is no value. We have been given a new set of 683 cases because of this. The distribution of classes in the new sample looks like this: benevolent = 444 (65.01%) and malevolent = 239 (34.99%). Look at Fig. 3 to see how the classes and traits are spread out.

Table 1 Wisconsin breast cancer data sets

Attribute	Value
Sample code number	
Clump thickness	1–10
Uniformity of cell size	1–10
Marginal adhesion	1–10
Uniformity of cell shape	1–10
Single epithelial cell size	1–10
Bland chromatin	1–10
Bare nuclei	1–10
Normal nucleoli	1–10
Class	
Mitoses	1–102 = benign, 4 = malignant

3.2 Support Vector Machine (SVM)

Support Vector Machines (SVMs) are a way of education representation which is often utilized by additional learning systems to classify and weaken investigations. In the multidimensional setting, SVM works like a linear split among two information points to identify two different modules, as shown in Fig. 2. This approach makes a line between the two sections by figuring out a linear classifier. This division is known as the optimum hyper plane partition. That is picked by the set of hyper planes to group shapes that maximize the hyper plane boundary, or the space that exists between the hyper plane and the patterns nearest end. SVM makes 2 parallel hyper planes on one and other surface of the ideal hyper plane divider. It is thought that the greater the margin or distance among the parallel hyper planes, the less mistake the categorizer will make.

Fig. 2 SVM representation

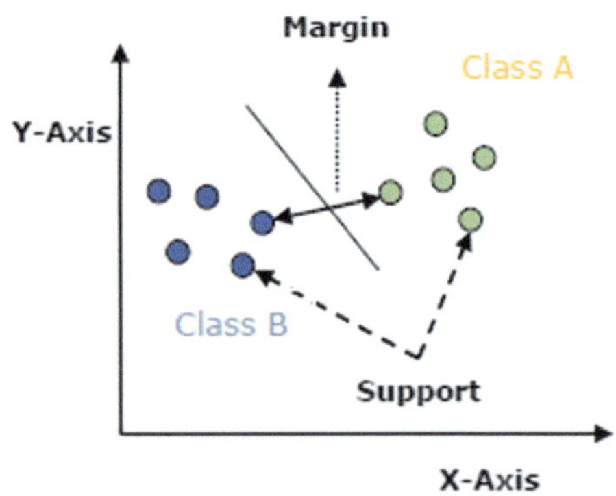

Fig. 3 Flowchart

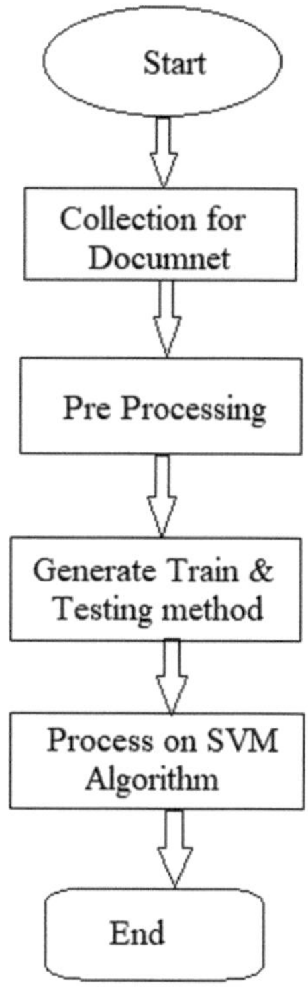

3.3 Flowchart

Figure 3 shows a diagram, which shows how an approach, structure, or computing method works. Here, we use a lot of different types of field to write, evaluate, sketch, get better at, and match up with common composite methods in clear, simple figures. Flowcharts, also called flow charts, are diagrams that use rectangular shapes, ovals, diamonds, and maybe even new shapes to show the type of step and connect them with lines to show how things move along.

3.4 Expected Outcomes and Conversation

Present work assessment and evaluation of every technique is executed depending on the amount of accuracy, exactness, and evoke as shown in Fig. 4.

The quantity of information points that was forecasted properly is divided by the entire amount of objects in the information collection. Then, this amount is broken down by the quantity of tested designs that made accurate forecasts by the quantity of testing procedures in the information set.

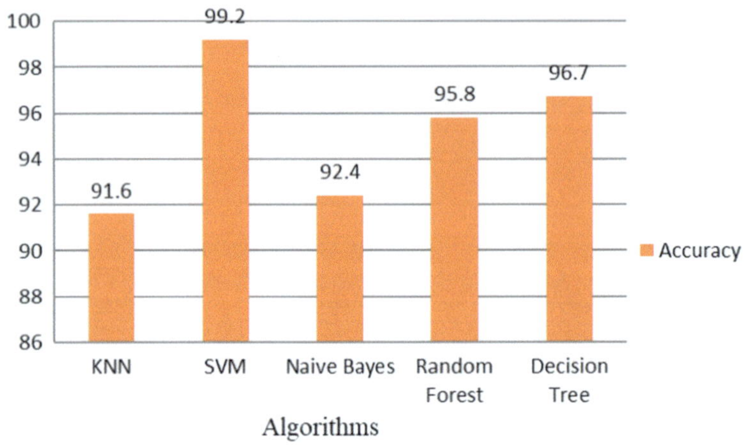

Fig. 4 Accuracy values for various algorithms

4 Conclusion

This research looks at two popular ways to categorize Wisconsin breast cancer using machine learning. In the health business of today, using analytical data to predict and find breast cancer becomes more and more important. The death rate from breast cancer might be lowered if it is found early. Recent study shows that machine learning methods are important for finding breast cancer early. In this study, three recognized types of machine learning are worn to locate breast cancer. Three examples are the MLP, the SVM, and the Random Forest. The Wisconsin Breast Cancer Diagnostic information set is utilized to compare the suggested strategies. When using the k-fold cross reasoning method, the outcomes of this study show that the SVM algorithm is the most accurate.

5 Future Scope

The WBC (Original) dataset is put into groups using ML techniques like ANN and SVM in the WEKA tool. The usefulness of different ML methods is linked based on important presentation factors like precision, exactness, evokes, and ROC region. Relying on the presentation measures of the ML methods used, the WBC dataset shows that SVM (the Sequential Minimal Optimization method) is the best for analysis and prediction with a 96.99% accuracy rate.

References

1. Siegel RL, Miller KD, Jemal A (2018) Cancer statistics. Ca-a Cancer J Clin 68(1):7–30
2. Maity NG, Das S (2017) Machine learning for improved diagnosis and prognosis in healthcare. In: 2017 IEEE aerospace conference, pp 1–9
3. Asri H, Mousannif H, Al Moatassime H, Noel T (2016) Using machine learning algorithms for breast cancer risk forecasting and diagnosis. Procedia Computer Science 83:1064–1069
4. Bazazeh D, Shubair R (2016) Comparative study of machine learning algorithms for breast cancer detection and diagnosis. In: 2016 5th international conference on electronic devices, systems and applications, pp 1–4
5. Huang MW, Chen CW, Lin WC, Ke SW, Tsai CF (2017) SVM and SVM ensembles in breast cancer forecasting. PloS One, 12 (1)
6. Kourou K, Exarchos TP, Exarchos KP, Karamouzis MV, Fotiadis DI (2015) Machine learning applications in cancer prognosis and forecasting. Comput Struct Biotechnol J 13:8–17
7. Umadevi S, Marseline KJ (2017) A survey on data mining classification algorithms. In: 2017 International conference on signal processing and communication, pp 264–268
8. Padmapriya S, Devika M, Meena V, Dheebikaa SB, Vinodhini R (2016) Survey on breast cancer detection using weka tool. Imp J Interdiscip Res (IJIR) 2(4)
9. Turgut S, Dagtekin M, Ensari T (2018) Microarray breast cancer data classification using machine learning methods. In: Int conf on electric electronics, computer science, biomedical engineerings meeting. https://doi.org/10.1109/EBBT.2018.8391468
10. Amrane M, Oukid S, Gagaoua I, Ensari T (2018) Breast cancer classification using machine learning. In: Int Conf on Electric Electronics, Computer Science, Biomedical Engineerings Meeting. https://doi.org/10.1109/EBBT.2018.8391453
11. Ak MF (2020) A comparative analysis of breast cancer detection and diagnosis using data visualization and machine learning applications. Healthc 8(2):111. https://doi.org/10.3390/healthcare8020111
12. Livingston SJ, Selvi BST, Thabeetha M, Grena CP, Jenifer CS (2019) A neural network based approach for sentimental analysis on amazon product reviews. Int J Innov Technol Explor Eng 8(6S):469–473
13. Tholkapiyan M, Aruna Devi B, Bhatt D et al (2023) Performance analysis of rice plant diseases identification and classification methodology. Wireless Pers Commun 130:1317–1341. https://doi.org/10.1007/s11277-023-10333-3
14. Lakshmi KL, Muthulakshmi P, Nithya AA et al (2023) Recognition of emotions in speech using deep CNN and RESNET. Soft Comput. https://doi.org/10.1007/s00500-023-07969-5
15. Thaseen A, Unnisa R, Sultana N, Madhavi KR, NagaJyothi G, Kirubakaran S (2023)Breast cancer detection using deep learning model. In: Proceedings of third international conference on advances in computer engineering and communication systems: ICACECS 2022. Springer Nature Singapore, Singapore, pp 669–677

Analysis on Classical Image Forgery Identification Approach with Higher Accuracy Using Fusion-Based Testing Methodology

K. Archana, Meka Kamala, V. A. Narayana, G. Vinoda Reddy, Raj Kumar Patra, and G. Menaka

Abstract This research proposes that image forgery detection is the main significant obstacle encountered in systems that function in actual time, is social networking sites and internet-based data portals. The scope of traditional methods of identification depending on the remnants of picture alterations is limited by preconceived assumptions, such as size, contrast, and hand-crafted features. In this paper, we offer a fusion-based judgment technique for identifying picture fraud. Three lightweight deep learning models—Squeeze Net, Mobile NetV2, and Shuffle Net—serve as the framework for the decision fusion. The fusion decision system's implementation consists of two steps. The retained weights of the lightweight deep learning models are used to determine whether the photos are fabricated in the outset. The outcomes of the forged photos are then contrasted with those of the previously trained models using the fine-tuned weights. The experimental findings imply that, in comparison to cutting-edge methods, the fusion-based decision technique achieves higher accuracy.

Keywords CNN · Deep learning · Image forgery detection · Image forensics

---

K. Archana · M. Kamala · V. A. Narayana
Department of CSE, CMR College of Engineering & Technology, Hyderabad, Telangana, India
e-mail: karchana@cmrcet.org

M. Kamala
e-mail: mkamala@cmrcet.org

G. V. Reddy
Department of CSE (AI&ML), CMR Technical Campus, Hyderabad, Telangana, India

R. K. Patra (✉)
Department of CSE, CMR Technical Campus, Hyderabad, Telangana, India
e-mail: rajkumarpatra.cse@cmrtc.ac.in

G. Menaka
Department of IT, CMR Technical Campus, Hyderabad, Telangana, India

V. K. Gunjan et al. (eds.), *Cybernetics, Human Cognition, and Machine Learning in Communicative Applications*, Cognitive Science and Technology,
https://doi.org/10.1007/978-981-97-8533-9_20

1 Introduction

In the current digital era, pictures and videos are used as convincing sources of witness in a variety of contexts, such as trial evidence, fraud in insurance, and social media [1]. The convenience with which editing tools can be used on digital pictures casts suspicion on their reliability even in the absence of evident indicators of alteration [1]. The development of technology innovations that would be able to distinguish false images is the responsibility of experts in picture forensics. The majority of manipulations or forgery detectors up until this point have been classified into one of three groups: those that accepted attributes descriptors, those that enabled inconsistent shadows, and afterwards those that permitted double compression of JPEG [1]. It is trivial to use sophisticated software to alter an image's contents in order to change how other people perceive it. The two main categories of image fabrication techniques are copy-move and cutting [2]. The zone containing the information of a picture is delineated and blurred within an identical picture in a case of copy-move forgery, whereas in a splicing forgery, the picture content area is smudged from other photographs. Over the last few years, a number of image fraud detection approaches have been presented in an effort to reconstruct the confidence in images. Numerous other investigations have attempted to extract entirely distinct image characteristics, such as illumination, shades, sensors element noise, and lens reflections, in order to detect the copying or sewing of forged sections [3].

Researchers [4–9] assessed the image's believability wherever it is thought to be either original or faked. There are numerous methods currently available to detect faked regions [7–15] that make use of the artifacts left behind by repeated compression of JPEG images and other methods of image alteration. There have also been studies of camera-based methods [16], where the detection is based on demos icing uniformity or sensing element pattern noise, where the abnormalities of the sensing element pattern are extracted and examined for anomalies [17]. Pictures that have been altered or falsified might mislead viewers and even pose a threat to life. Instead of using feature engineering or human feature extraction, this work [2] tries to identify the modified images by automating feature extraction. Deep learning uses nearby highly linked pixels and takes aggregated native connections into account [18].

The goal of using lightweight models is to avoid over-fitting convolutional neural network (CNN) architectures. These models are simple to implement on hardware with limited resources and have the ability to learn enriched representations [19–23]. Shuffle Net Assisting in the encoding of extra data and is particularly crucial to the effectiveness of tiny networks because it creates additional feature map pathways for a given computational budget [24]. Deep-separable convolutions are used by Mobile Net [21] to get cutting-edge outcomes, and [21] showed that Mobile Net is effective when used for a variety of tasks. Squeeze Net [25] uses 50 percent less parameters than Alex Net while maintaining the same level of accuracy. It does this by optimizing the design for fast processing performance. On hardware with limited resources, lightweight models can be used successfully and can develop enhanced representation.

For the detection of image counterfeiting, the article proposes the utilization of fusion techniques for compact deep learning-based systems. The suggested methodology is divided into two parts that focus on the relevant and improved lightweight deep learning models Squeeze Net [25], MobileNetV2 [22], and Shuffle Net [24]. In the initial stage, features from the photos are extracted without regularization using lite deep learning models. The detection of image forgery done in the subsequent phase involves the utilization of optimized, compact deep learning techniques that incorporate fusing and normalization techniques. The primary outcomes of this study are as follows.

The lightweight for the detection of image forgery is proposed as a decision fusion-based system technique. The Squeeze Net, Mobile Net V2, and Shuffle Net lightweight models were used to make the fusion decision.

Two phases are used to implement the decision system fusion. First, the forgery detection of the photos is assessed using the pertained weights for the lightweight models. Second, the outcomes of the counterfeit identification of the photos are compared with the pre-trained structures using the fine-tuned weights.

The use of lightweight models results in a decrease in erroneous matches, a decrease in false positive rates, and eventually an improvement in approach accuracy. The remainder of the paper is structured as follows. The related research on techniques and deep learning models for image forgery detection is covered in Sect. 2 of this article. The fusion model is suggested in Sect. 3, and then regularization is performed on the fusion model. The experiments and their findings are detailed in Sect. 4, and the conclusion is in Sect. 5.

2 Literature Review

Amerini et al. [1] proposed that convolutional neural networks (CNNs) have been employed to disclose and locate a once or twice JPEG compression, which has been presented as a step in this regard. A multitude of tests were undertaken with the aim of identifying prospective issues that warrant further examination, taking into account various CNN input types.

Xiao et al. [2] proposed a splicing for gery identification approach that consists of two steps: diluted adaptive clustering and a coarse-to-refined convolutional neural network (C2RNet). The proposed C2RNet extracts the variations in the image attributes among unaltered and tampered parts from image patches of various scales by cascading a coarse convolutional neural network (C-CNN) and a refined CNN (R-CNN). Additionally, an image-level CNN is implemented to replace patch-level CNN in C2RNet in order to reduce computational complexity. To ensure a steady detection performance, the suggested detection approach learns the distinctions between various image attributes, and the photo-level CNN significantly reduces its computing time.

Zhang et al. [3] in this work used a Stacked autoencoder approach to learn the complicated feature for each unique patch after having previously studied the first stage. To enhance the precision of the subsequent phase identification, this paper incorporates contextual data for every repair.

Goh et al. [4] conducted a quantitative study to assess all aspects of picture altering for the best feature set, and a hybrid evolutionary framework was presented. The categorization mechanism must be adjusted for good performance after feature evaluation and selection. To optimize classification efficiency for image tampered identification, the hybrid approach can discover the most effective several classification bands and the optimal set of characteristics for classifiers. This approach aims to minimize complication while maximizing reliability.

Sutthiwan et al. [5] presented that Markovian-Rake transform is used to create the statistical aspects of the suggested photo which are derived from the brightness portion of the picture. The Markovian rake transformation utilizes the principles of Markov processes to analyze variance arrays. These arrays are derived from two-dimensional arrays that have undergone quantized block discrete cosine transformations with different block sizes. The efficacy of these characteristics has been validated by the utilization of a recently developed extensive picture collection specifically designed for the purpose of detecting manipulation, with which several pertinent problems have been addressed and appropriate corrective actions have been made. The first tests utilizing these created classifiers on a few real-world forged photographs found online demonstrate the potential of the suggested attributes as well as the difficulty of image tampering detection for the research community.

He et al. [6] suggested using a Markov-based method to find this specific artifact. In the beginning, Shi et al.'s original Markov features, which were created from the transition probability matrices in the DCT domain, are expanded to include both the intra-block and inter-block correlation between block DCT coefficients. The three types of dependence between wavelet coefficients across roles, scales, and orientations are then characterized by additional features built in the DWT domain. The work of feature reduction is then carried out using the feature selection method SVM-RFE, which reduces the number of features and lowers the computational cost. The legitimate and spliced photos are finally classified using the final dimensionality-reduced feature vector using support vector machine (SVM).

Change et al. [7] proposed one of the most successful methods for image alteration, and a novel forgery detection algorithm has been proposed to identify tampering in painting images. The proposed technique consists of two main processes: forged region identification and suspicious region detection. In order to locate suspicious regions in a picture, suspicious region identification uses an analogous vector field to filter out false positives brought on by uniform areas. To distinguish the forged sections from the suspicious areas, forged region identification uses a novel technique called multi-region relation (MRR). Even for photographs with a homogeneous background, the suggested method can successfully determine if a picture is forged and point out the forged portions. In order to speed up calculation, this research also suggested a two-stage searching technique based on weight transformation.

Rhee et al. [8] created a succinct characteristic vector with three different feature set classifications. The 3-D length in the gradient variation of the intensity values of the adjacent column and row line pairs in the image, respectively, is what distinguishes the starting set from other sets. The second set is defined by the variation of the Fourier transform coefficient difference, which stands for the 3-D length in each of the line pairings. The gradient created by solving Poisson's equation describes the image that is created when an image is combined with its reconstructed image, which is also the 3-D length. The final set is retrieved from the image's residual portion, whereas the first two sets are taken from its spatial and spectral domains, respectively. The completely generated 9-D feature vector is then used to train the support vector machine classifier for MFD.

Lamba et al. [9] proposed that for the purpose of identifying duplicated portions in the image, a discrete fractional wavelet transform-based method has been presented. The test visual is divided into rectangular, overlapping chunks of different sizes. The features of each image block are then extracted using a discrete fractional wavelet transform. Following the process of block match and block filtering procedures to identify replicated blocks, if any, all vectors of features are systematized lexicographically. The suggested method is effective at finding both single and numerous duplicated regions.

Lin et al. [10] suggested identifying manipulated photos by analyzing the double quantization effect buried in the discrete cosine transform (DCT) coefficients. This research represents the sole existing method capable of proactively identifying altered regions. Additionally, it offers several advantages such as accurate recognition at the level of Discrete Cosine Transform (DCT) barriers, resilience against diverse forgery methods (such as alpha matting and in painting), simple detection of photo copying and pasting operations, the capacity to perform without completely reducing JPEG images, and efficient processing rate. JPEG picture experiments have shown positive outcomes.

3 Proposed Fusion Model

As depicted in Fig. 1, the structure of the suggested selection fusion relies on the portable deep learning models. Squeeze Net [25], Mobile NetV2 [22], and ShuffleNet [24] are the lightweight deep learning methods selected. Deep learning methods that have been pre-trained and fine-tuned are used to implement the proposed system in two stages. Regularization is not employed in the pre-trained model's implementation; instead, the pre-trained weights are used; in the fine-tuned execution, regularization is used to identify image counterfeiting. Three processes, namely, data pre-processing, categorization, and fusion, make up each phase. The picture provided in the request is pre-processed according to the dimensions needed by the deep learning models during the data pre-processing stage. The categorization of a picture as fabricated or not is done using SVM. We first talk about lightweight deep learning models, and in the following parts, we talk about the regularization technique.

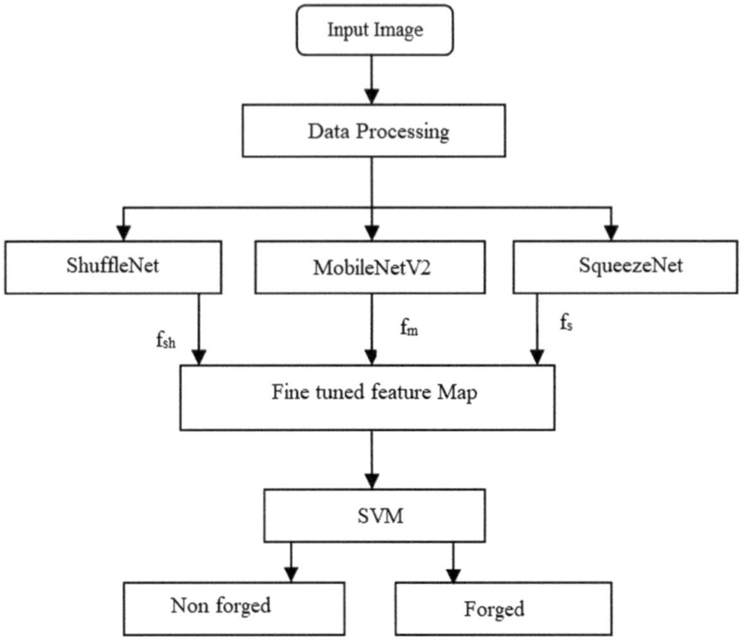

Fig.1 A fusion-based decision framework for the identification of forgeries

4 Data Preprocessing

The picture included in a query that needs to be determined as to if it is forged or not is preprocessed at this level. The image's minimum needed dimensions for Squeeze Net are 227 × 227. MobileNetV2 requires an image with dimensions of 224 × 224. Shuffle Net requires an image with dimensions of 224 × 224. According to the proportions needed for every one of the models, the input picture is first pre-processed. The input image is then used by each model to create feature vectors in subsequent phases [27].

4.1 Lightweight Deep Learning Models

Squeeze Net, Mobile NetV2, and Shuffle Net are three distinct lightweight deep learning models that are being taken into consideration for fusion. Numerous picture categorization issues are solved using these models. These models are briefly discussed in this section. Figure 4 summarizes the depth, characteristics, and image input size requirements for the three lightweight models1 that were taken into account: Squeeze Net, MobileNetV2, and Shuffle Net.

4.2 Classifier

As a classifier, SVM is utilized. SVM is a well-liked and effective binary classification method. By evaluating performance parameters including precision, recall, also known as true positive rate (TPR) [28], false positive rate (FPR) [29, 30], F-score, and accuracy, the suggested approach's effectiveness is assessed at the picture level.

4.3 Fusion Model and Regularization

The proposed method is initially developed using basic deep learning algorithms that utilize previously trained weights for the purpose of detecting picture counterfeiting. Subsequently, the utilization of compact designs, as previously discussed, is integrated with the envisaged system deployment. The image that is input is first transmitted to the portable models for the purpose of generating their corresponding map features. The list of features obtained from the SqueezeNet architecture is represented as f_s, while the feature map obtained from the MobileNet V2 architecture is marked as f_m. Similarly, the feature map obtained from the ShuffleNet architecture is indicated as f_{sh}. The fusion system utilizes the resultant attribute mapping f_p of a pre-trained lightweight deep learning algorithm. The feature map f_p is formed by aggregating the feature maps derived from the lightweight models, as depicted in Calculation (1).

$$f_p = f_s + f_m + f_s \tag{1}$$

The fusion method employs the characteristic map f_p as a regional identifier for obtaining the characteristics of an photo from an input region. The fusion model's picture is denoted as a function $Y_{fusion} = f(x)$, where x represents the patch inside the supplied picture. To determine the regional descriptor for a picture of size $m \times n$, a window that slides of width p × p is utilized. The computation of Y_{fusion} is demonstrated in formula (2), where Y_1, Y_2, …, Y_3 indicate the identifiers of the picture regions acquired from the models using deep learning. The fresh image depiction, denoted as f_{fusion}, is created by concatenating all the input patches xi. This model is derived using calculation (3), in which the stride size, denoted as s, is used to modify the input application. The resulting *ffusion* is then utilized as the feature graph for the analysis performed by the Support Vector Machine (SVM), distinguishing between forged and non-forged images.

$$Y_{fusion} = [Y_1 + Y_2 + \ldots + Y_T] \tag{2}$$

$$X_{fusion(Xi)=}m - \mathrm{w} + 1 * n - \mathrm{w} + 1 \tag{3}$$

The setup of the weighted kernels is employed, as demonstrated in Formula (4), for the purpose of fine-tuning the settings of the fusion models. The formula includes variables W_f, W_s, W_m and W_{sh}, which respectively reflect the weights of the fusion method, Squeeze Net method, MobileNetV2 method, and Shuffle Net design. The mass of the fusion method, denoted as W_f, is initialized according to the equation presented as Formula (5). The weight introduction serves as a form of normalization and aids the fusion model in acquiring strong characteristics for the detection of fraud, instead of focusing on intricate picture descriptions.

$$W_f = \left[W_{sj} W_{mj} W_{shj}\right] j = 1, 2, 3 \quad (4)$$

$$W_f = \left[W_f^{4k} -^2 W_m^{4k} -^2 W_{sh}^{4k}\right] wherek = \left[[j+1] mod 11\right] + 1 \quad (5)$$

5 Results and Analysis

5.1 Dataset

The benchmark MICC-F220 [26] data set, which consists of 110 authentic photos and 110 fake images with three channels, or color images with a size between 722 480 and 800 600 pixels, was used for the experiment. Figure 2 illustrates this. Pictures 2a–j depict altered pictures that have been subjected to 10 distinct topological and translational alterations, whereas Fig. 2k represents an unaltered picture. A total of 154 pictures were chosen in an arbitrary way from the collection for the purpose of instruction, while the other pictures were allocated for assessment.

As seen in the confusion matrix graph in Fig. 3 above, the x-axis corresponds to predicted labels, while the y-axis corresponds to true labels. As can be seen, both the X and Y boxes contain a greater number of correctly predicted classes.

5.2 Baseline Models

The following is a summary of the baseline models that were used to compare the fusion model.

a. Submit MICC-F220 Dataset: We will upload the dataset to the program using this module.
b. Pre-process Dataset: With this module, all photos will be read, their pixel values will be normalized, and they will all be resized to be the same size.
c. Generate and Load Fusion Model: In this module, we train the Squeeze Net, Mobile NetV2, and Shuffle Net algorithms before using the characteristics we

Fig. 2 The information set has ten distinct variations on geometrical and transformation attacks; **a–j**, forged; **k**, non-forged images

gleaned to train the fusion model. On test data, the prediction accuracy of each method will be calculated, as illustrated in Fig. 4.

d. By utilizing this section, we will integrate the characteristics derived from each of the methods in order to construct a fusion form. The fusion information is subsequently subjected to training using Support Vector Machines (SVM), and the algorithm's predictive accuracy is assessed.

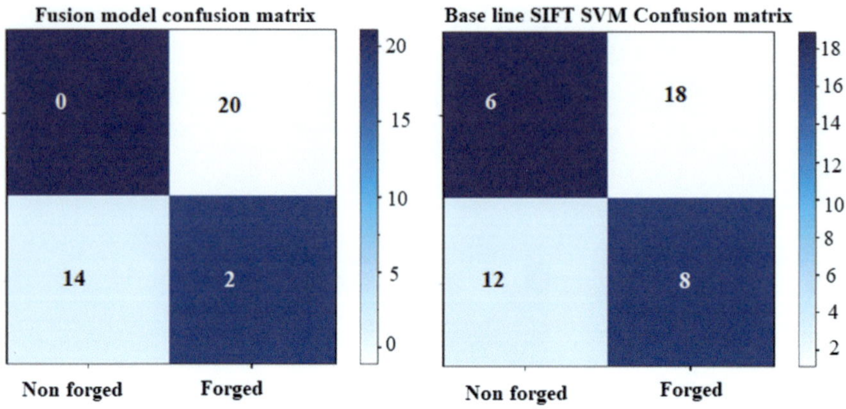

Fig. 3 Confusion matrixes of fusion method and base line SIFTSVM

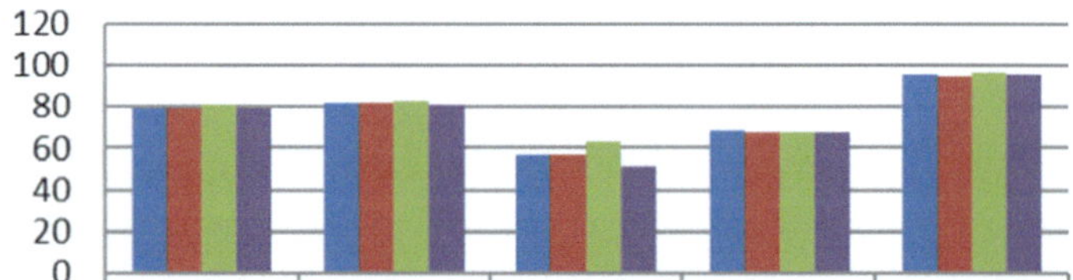

	Squeeze Net	Mobile Net V2	Shuffle Net	SIFT SVM	Fusion model SVM
Accuracy	79.8	81.9	56.9	68.4	95.6
Recall	79.5	81.8	56.8	67.6	95
Precision	81.2	82.9	62.7	68	96.2
F Score	79.3	81.6	51.2	67.6	95.5

Fig. 4 Performance comparisons

e. Run the Baseline SIFT method: In this section, we will utilize the SIFT-based strategy for collecting characteristics from photographs. These characteristics will then be utilized to train a Support Vector Machine (SVM) model. Subsequently, we will assess the precision of the predictions made by the model.
f. Accuracy Comparison Graph: We will use this method to plot a precision graph for each algorithm.
g. Performance Table: In this section, we will show the performance table for each algorithm.

6 Conclusion

Identifying picture forgery helps to identify between real images and ones that have been manipulated or are fake. In this study, a decision fusion framework is constructed utilizing minimal deep learning algorithms for the purpose of detecting fraudulent images. In order to assess whether a photo was falsified, the three lightweight deep learning models—Squeeze Net, MobileNetV2, and Shuffle Net—were to be combined. Regularity of the weights of the pertinent models is used to identify the forgeries. The trials' findings demonstrate that the fusion-based approach offers more precise than the cutting-edge methods. Other weight initialization techniques can be used in the future to enhance the fusion decision for the purpose of detecting image counterfeiting.

References

1. Amerini, Uricchio T, Ballan L, Caldelli R (2017) Localization of JPEG double compression through multi-domain convolutional neural networks. In: 2017 IEEE conference on Computer Vision and Pattern Recognition Workshops (CVPRW), pp 1865–1871. https://doi.org/10.1109/CVPRW.2017.233
2. Xiao B et al (2020) Image splicing forgery detection combining coarse to refined convolutional neural network and adaptive clustering. Information Sci 511:172–191
3. Zhang Y, Goh J, Win LL, Thing VL (2016) Image region forgery detection: a deep learning approach. SG-CRC 2016:1–11
4. Goh J, Thing VL (2015) A hybrid evolutionary algorithm for feature and ensemble selection inimage tampering detection. Int J Electron Secur Digit Forensics 7(1):76–104
5. Sutthiwan P, Shi YQ, Zhao H, Ng TT, Su W (2011) Markovian rake transform for digital image tampering detection. In: Shi YQ, Emmanuel S, Kankanhalli MS, Chang S-F, Radhakrishnan R (eds) Transactions on data hiding and multimedia security VI. Lecture notes in computer science, vol 6730. Springer, Berlin, Germany, pp 1–17
6. He Z, Lu W, Sun W, Huang J (2012) Digital image splicing detection based on Markov features in DCT and DWT domain. Pattern Recogn 45(12):4292–4299
7. Chang IC, Yu JC, Chang CC (2013) A forgery detection algorithm for exemplar-based in painting images using multi-region relation. Image Vis Comput 31(1):57–71
8. Rhee KH (2017) Median filtering detection based on variations and residuals in image forensics. Turk J Electr Eng & Comput Sci 25(5):3811–3826
9. Lamba AK, Jindal N, Sharma S (2018) Digital image copy-move forgery detection based on discrete fractional wavelet transform. Turk J Electr Eng & Comput Sci 26(3):1261–1277
10. Lin Z, He J, Tang X, Tang CK. Fast, automatic, and fine-grained tampered JPEG image detection via DCT coefficient analysis. PatternRecognition2009; 42(11):2492–2501.
11. Chen YL, Hsu CT (2011) Detecting recompression of JPEG images via periodicity analysisof compression artifacts for tampering detection. IEEE Trans Inf Forensics Secur 6(2):396–406
12. Bianchi T, Piva A (2012) Image forgery localization via block-grained analysis of JPE Gartifacts. IEEE Trans Inf Forensics Secur 7(3):1003–1017
13. Zach F, Riess C, Angelopoulou E (2012) Automated image forgery detection through classification of JPEG ghosts. In: Springer 2012 Joint DAGM (German Association for Pattern Recognition) and OAGM Symposium, Berlin, Heidelberg, pp 185–194
14. Thing VL, Chen Y, Cheh C (2012) An improved double compression detection method forJPEG image forensics. In: IEEE International Symposium on Multimedia. Irvine, CA, USA, pp 290–297

15. Wang W, Dong J, Tan T (2014) Exploring DCT coefficient quantization effects for localtampering detection. IEEE Trans Inf Forensics Secur 9(10):1653–1666
16. Amerini I, Caldelli R, Cappellini V, Picchioni F, Piva A (2010) Estimate of PRNU noise based on different noise models for source camera identification. Int J Digit Crime Forensics 2(2):21–33
17. Popescu AC, Farid H (2005) Exposing digital forgeries in color filter array interpolated images. IEEE Trans Signal Process 53(10):3948–3959.https://doi.org/10.1109/TSP.2005.855406
18. Hadji I, Wildes RP What do we understand about convolutional networks? arXiv2018; preprint arXiv:1803.08834
19. Khan A, Sohail A, Zahoora U, Qureshi AS A survey of the recent architectures of deep convolutional neural networks. arXiv2019; preprint arXiv:1901.06032
20. Rao Y, Ni J, Zhao H (2020) Deep learning local descriptor for image splicing detection andlocalization. IEEE Access 8:25611–25625
21. Howard AG et al (2017) Mobilenets: Efficient convolutional neural networks for mobile vision applications. arXiv preprint arXiv:1704.04861
22. Sandler M et al (2018) Mobilenetv2: Inverted residuals and linear bottlenecks. In: Proceedings of the IEEE conference on computer vision and pattern recognition
23. Phan-Xuan H, Le-Tien T, Nguyen-Chinh T, Do-Tieu T, Nguyen-Van Q et al (2019) Preserving spatial information to enhance performance of image forgery classification. In: IEEE ATC international conference on advanced technologies for communications. Hanoi, Vietnam, pp 50–55
24. Zhang X, Zhou X, Lin M, Sun J (2018) ShuffleNet: an extremely efficient convolutional neural network for mobile devices. In: IEEE 2018 conference on Computer Vision and Pattern Recognition (CVPR). SaltLakeCity, UT, USA, pp 6848–6856
25. Iandola FN, Han S, Moskewicz MW, Ashraf K, Dally WJ SqueezeNet: AlexNet-level accuracy with 50x fewer parameters and< 0.5 MB model size. arXiv 2016; preprint arXiv:1602.07360
26. Amerini I et al (2011) A sift-based forensic method for copy–move attack detection and transformation recovery. Electr Power Compon Syst 6(3):1099–1110
27. Kumar S, Baruah P, Kirubakaran S, Kumar AS, Singh K, Reddy MVJ (2023) Remora optimization with machine learning driven churn prediction for business improvement. In: 2023 International conference on intelligent data communication technologies and internet of things (IDCIoT), Bengaluru, India, pp 416–420. https://doi.org/10.1109/IDCIoT56793.2023.10053554
28. Skandha S, Saba L, Gupta SK, Kumar VK, Johri AM, Khanna NN, Mavrogeni S, Laird JR, Pareek G, Sfikakis PP, Protogerou A (2022) Magnetic resonance based Wilson's disease tissue characterization in an artificial intelligence framework using transfer learning. https://doi.org/10.1088/978-0-7503-2244-7ch5
29. Pradeepthi KV, Saxena A (2022) Botnet attack classification with deep learning models. In: Smart intelligent computing and applications, vol 2: Proceedings of fifth international conference on Smart Computing and Informatics (SCI 2021). Springer Nature Singapore, Singapore, pp 567–572
30. Dimlo UF, Umanesan R, Narasimharao J, Senthamilarasi N, Ranjit PS, Balaji B, Thamarai I, Dwivedi VK (2023) Optimal configuration planning of multi-energy systems using optimization-based deep learning technique. Electr Power Compon Syst:1–16

Determination of Covid-19 Disease Using Hybrid CNN Based Classification Using Deep Learning Applications

Vijaya Kumar Koppula, Siva Skandha Sanagala, K. L. S. Soujanya, Avala Raji Reddy, Goda Gangaram, and G. Vinesh Shankar

Abstract One of the most deadly and destructive illnesses affecting individuals today is the corona virus attack, often known as COVID-19. Due to communal transmission, the Corona virus infection has spread across the entire world. The mortality rate of the patient might be decreased by early sickness identification, yet in asymptomatic settings, via correct diagnosis. As a result, it's essential to create an independent recognition system that, by its swift and accurate results, halts the corona virus's spread. Medical imaging techniques including computed tomography (CT) and chest X-ray images are frequently used to diagnose COVID-19 patients and provide an early prognosis. To identify hidden patterns, deep learning techniques from the medical field are applied. In order to make predictions, Convolutional neural networks (CNNs) are used to extract the chest's X-ray properties. Predictions are made in the patient data utilizing pattern generation to enhance the health plan. To improve categorization, the chest X-ray picture attributes are combined with CNN model training. Extended data are considered throughout the model efficiency evaluation's testing phase. The proposed CNN-based algorithms outperform other state-of-the-art classification methods in terms of categorization and predicting diseases.

V. K. Koppula · S. S. Sanagala · K. L. S. Soujanya
Department of CSE, CMR College of Engineering and Technology, Hyderabad, Telangana, India
e-mail: sivaskandha@cmrcet.org

K. L. S. Soujanya
e-mail: klssoujanya@cmrcet.org

A. R. Reddy (✉)
Department of Mechanical Engineering, CMR Technical Campus, Hyderabad, Telangana, India
e-mail: avala.rajireddy@gmail.com

G. Gangaram
Department of CSE(AI&ML), CMR Technical Campus, Hyderabad, Telangana, India
e-mail: gangaram.aiml@cmrtc.ac.in

G. V. Shankar
Department of CSE, CMR Technical Campus, Hyderabad, Telangana, India

V. K. Gunjan et al. (eds.), *Cybernetics, Human Cognition, and Machine Learning in Communicative Applications*, Cognitive Science and Technology,
https://doi.org/10.1007/978-981-97-8533-9_21

Keywords Feature extraction · Prediction · Covid-19 · CNN

1 Introduction

The COVID-19 outbreak that started in December 2019 has executed added by 800,000 people worldwide and contaminated by 20 million more [1]. On January 30, 2020, the World Health Organization declares COVID-19 a worldwide health emergency. (WHO). CT imaging of the chest can help in the clinical diagnosis of COVID-19. However, the number of patients suffering from the corona virus is rapidly increasing due to a shortage of radiologists and medical professionals. The classification of COVID-19-assisted CT images of the chest requires the creation of automated algorithms [2]. The COVID-19 classification with chest classification makes use of a number of beneficial machine learning techniques [3]. In certain instances, an effective risk forecasting system can offer precision medicine suggestions for tailoring clinical care to every patient's requirements, delaying the procedure to full recovery. Lower wait times are achieved by enabling the emergency department to maximize patient flow [4].

Predicting patient outcomes requires extensive research employing a variety of data sources, including radiographic, laboratory, and clinical information. In accordance with a new study on COVID-19 prognosis/risk prediction methodologies, a great deal of writers are prejudiced for one of two reasons, even if some have provided some hopeful findings [5].

However, a lack of clinical follow-up data for numerous studies about the incorrect class employed by machine learning algorithms in the primary instance since major complications typically arise in patients and early clinical contact used for ML [6]. The subsequent problem is that numerous studies are limited in their ability to employ recent clinical data for their research; rather, they trust in procedures of the majority of fresh predictors that are currently accessible through electronic health records. Artificial intelligence (AI) is employed in its many forms to address a wide range of complex problems [7].

The thoughts, knowledge, organizing, in rank participation, and penetrating are among the AI subfields. The Deep Learning (DL) and Machine Learning (ML) algorithms, that are branches of AI, offer intelligent models for the aim of recognizing a particular assignment. The main branch of AI that requires the least amount of expertise to resolve problems is ML, which is characterized by statistical models and an innovative way of algorithmic modeling. The DL subset of ML is the next important model following this ML. Several individuals have employed DL algorithms in the past years.

Yet a lot of information is required to solve it. The fact that these ML and DL methods are being applied in a number of studies in the technical, medical, and military areas [8] is encouraging. The areas involved in the battle also launched ML and DL techniques utilizing cutting-edge AI not long before COVID-19 was destroyed. The goal of this work is to develop a computational technique depending on machine learning which emergency divisions can use to rapidly and precisely categorize CT images and calculate COVID-19 risk for patient's levels.

2 Literature Review

Using a unique Joint Classification and Segmentation (JCS) approach, Shang et al. [1] provide a real-time, comprehensible COVID-19 chest CT diagnosis. For the purpose of training our JCS system, we generate an extensive COVID-19 Identification and Segmentation (COVID-CS) dataset of 144,167 chest CT images comprising 400 COVID-19 patients and 350 untreated cases. 3,855 chest CT scans of 200 individuals have had fine-grained pixel-level labeling of pacifications—greater dampening of the lung parenchyma—added.

Yan et al.'s study [3] used machine learning polynomial regression under limited data conditions to reasonably calculate and investigate the trend of the number of coVID-19 established in the recent outbreak in Xinjiang, with a focus on the coVID-19 outbreak in Xinjiang in July.

In the last several days, the amount of COVID-19 cases has soared [7], putting enormous pressure on government and health professionals. Therefore, in these dire circumstances, the need for prophecy models to estimate the amount of novel infections is critical. In the current study, an artificial neural network (ANN), a machine learning method, is presented to predict the COVID-19 epidemic in India for the primary time. Additionally, in our study, we also made an effort to apply a mathematical curve fitting model to determine how well the projected ANN-based machine learning model performed.

Sentiment and emotion recognition in social media interactions is still difficult, and in this historic COVID-19 era, assessing people's emotions has become a crucial duty. Lockdowns, social isolation, travel, working as of home, exhausting a mask, and analysis of social media posts all have an impact on people's feelings and moods. The majority of them are melancholy, angry, or sad; nevertheless, some of them are neutral or even pleased. The results of the most current Sentiment Analysis (SA) studies, which use the Twitter dataset's short-text and rule-based (sentiment lexicon) methodology, do not consistently forecast how people will feel about COVID-19.

3 Implementation

Methods of Machine Learning with Covid-19

The past ten years have seen incredible advancement in the field of machine learning. Computational power is significantly increased when ML and deep learning approaches are combined [9]. As a consequence, machine learning is being adopted by many applications. The main use of machine learning in the battle against it might be the COVID-19 pandemic. Through the help of machine learning, numerous researchers are investigating numerous tactics to tackle COVID-19 [10]. The numerous machine learning programs address an extensive range of virus-related difficulties. As an illustration, COVID-19 diagnostic uses ML and DL to evaluate medical images, safeguard medical staff against those who have been affected, and assess the extent of the patient's illness to decide the best course of therapy. Disease transmission algorithms are created to precisely predict disease outbreaks. The models in question study the transmission list and the effect of the diffusion path by means of social networking data and COVID-19 case data [11]. Similar to airport security checks, the ML and DL are heavily used in public surveillance and epidemic prevention when patients become apparent and epidemics become apparent. [12].

In the COVID-19, four different types of machine learning algorithms are employed.

a. Drug development
b. Forecasting
c. Contact tracking
d. Medical diagnostics.

The amount of fresh infections was appropriately predicted by the deep learning systems. Recurrent neural networks outperform conventional methods including ARIMA models while forecasting time series. Recurrent neural networks were used by researchers to predict the frequency of novel infections and the dissemination of diseases as well as to take into consideration changing long-short-term memory networks. As a consequence, it is the machine learning model's main significant submission [13].

By means of computer vision, medical diagnosis can be improved. Human skills are nearly on par with machine learning models for numerous picture recognition tasks. Consequently, utilizing the patient's chest X-ray images from COVID-19, the image recognition computer certainly recognizes the signs of a virus. The merely way to find or diagnose the virus is through a pricey technique for chest X-ray scans, but many countries have a lack of sufficient healthcare resources for testing.

Based on earlier research, using deep learning approaches, COVID-19 may be identified based on a chest X-ray scan with approximately 99% accuracy [14]. To discover those who have been exposed to a virus, police deploy artificial intelligence smart contact tracing. Utilizing a variety of artificial intelligence-based applications to track the spread of infections. Researchers have employed algorithms for machine learning as a road map for upcoming discoveries in a variety of fields. In order to evaluate chemical compositions and create novel medicines, variation auto encoders are used. By combining the auto encoders with the presently accessible flu shots, the COVID-19 vaccine might be produced.

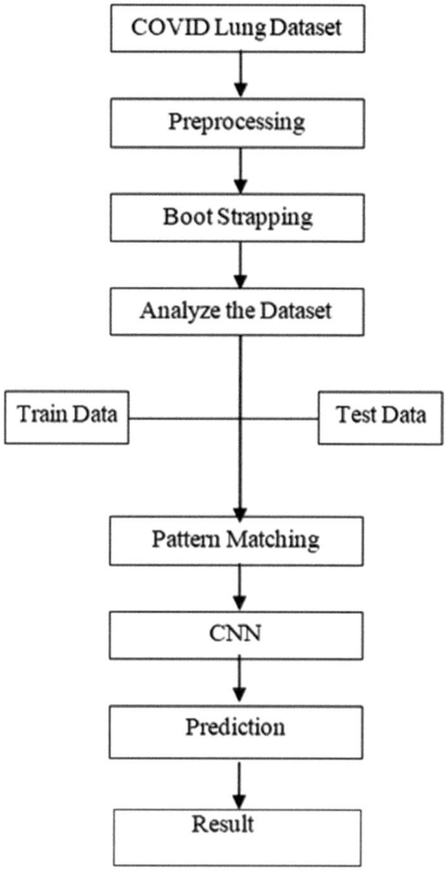

Fig. 1 Hybrid CNN based categorization

3.1 Hybrid CNN Based Classification

The suggested CNN that is retrieved from the COVID Lung dataset makes use of X-ray images. The hybrid CNN-based COVID-19 illness classification and prediction framework is depicted in Fig. 1 less than. The recommended system contains two modules: CNN and preprocessing.

3.2 COVID Lung Dataset

In this method, the X-ray datasets are obtained utilizing two datasets. In collection 1, the 79 images are split among the viral pneumonia and the bacterial pneumonia. 28 photographs of typical people and 78 photographs of COVID-19 patients make up the second data set on Kaggle. The dimensions of the training dataset have a crash on how healthy deep learning models execute. Calculating the generalization and robustness of these algorithms cannot be done because the COVID-19 collection of data is quite small in comparison to the vast dataset employed by deep learning models. The Keras Image Data Producer class [15] is worn to determine this problem by blending the CNN network using enormous X-ray pictures. The settings for image enhancement are defined by the image generating class made available by Keras. Shift, random rotation, dimension reordering, whitening, and fips are some of Keras' features.

3.3 Data Pre-processing

Both missing and noisy data are there in large amounts in the real-life data. As a result, in order to produce accurate forecasts, it is important to prepare the noisy data. The structure of the recommended model is shown in Fig. 1. Noise and data shortages were also present in the amount of information that was collected. Results are fast and exact when the missing data are filled in and the noise is removed. Aggregation, normalization, and smoothing are transformational processes that change a particular kind of data into a different kind of data.

3.4 Bootstrapping

The block exposes the programmable parameters that can be changed among 0.1 and 1.0.

3.5 Analyze the Dataset

Daily sliding window examination of confirmed cases is required for the first and second peaks as well as starting point detection. The ending of the prospect distribution parameters is based on the work done at the simulator's core. Through supervised learning, the intervention efficacy score is calculated, and it is also utilized to decide the populace of revision data as shown in Fig. 2.

3.6 Pattern Matching

The three sections that are separated from the photos are validation, testing, and training, in addition to this may be completed in the prototype identical block. This prototype matching may detect the patterns of picture data by pattern generating on the data that was trained and then contrasting the designs of the test information

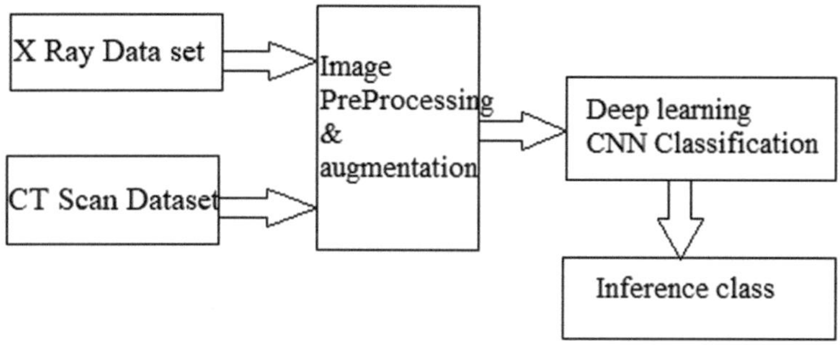

Fig. 2 Detection and analyzing Covid-19 medical image

against the trained history data. The validation set, which is divided into 5 folds after being divided into 5 little portions, is used to test the effectiveness of the trained model. Pick one of these five folds to serve as test data, and the other four to serve as training data.

In the proposed CNN model, there are a total of 38 layers: 6 dropout layers, 6 max pooling levels, 6 convolutional (Conv2D) layers, 8 batch normalization layers, 8 activation function layers, 3 completely connected layers, and 1 flatten layer. For the CNN model's input, a picture with the shape (150, 150, 3)—i.e., an RGB image 150 by 150—is employed. All Con2D levels use a 3*3 size kernel after increasing it in each two Con2D layers. The loss function of binary cross entropy (BCE) [16] is employed in the binary categorization of CNN. Because it is a binary classification from two classes that are provided, only one output is needed for data categorization. In the event of a BCE loss function, provide the production value to the sigmoid activation function. The production of the sigmoid activation function falls among 0 and 1. The sigmoid activation function finds the actual class mistakes and projected class errors. The "Adam" optimizer is used to adjust the learning rate and attribute weight so as to minimize learning model losses [17].

3.7 Predictions

The proposed approach is simulated using the intervention effectiveness scores so as to determine the intervention's future comparative crash. Subsequent to appropriate familiarity by means of the simulation parameters, this can be done for an exacting nation or population. Scores for efficiency are based on disease progression patterns throughout time, for instance the second peak in June.

4 Results and Analysis

Performance indicators like exactness, precision, F-score, recall, particularity, sensibility, and AUC were employed to gauge the study's effectiveness. The words True Negative (TN), True Positive (TP), False Negative (FN), and False Positive (FP) are used to describe the above evaluations of the presentation. Here context, positive represents the individual's test outcome is positive, and negative represents that the patient's testing consequence is negative. It follows that the terms "FP" and "TP" refer to patients who are healthy yet whose test outcomes are positive while "FP" and "TP" refer to patients who are unhealthy but their tests are negative. In the same way that FP denotes individuals who have adverse outcomes who lack any diseases and FN identifies patients having negative findings who exhibit this.

The projected technique of CNN presentation study for classifies and forecasting COVID-19 disorders is shown in Fig. 3.

It can be demonstrated that by 3 scenarios—i.e., for the 2 classes and 3 classes—more than 90% accuracy is attained. Figure 4 illustrates how the improved accuracy and F-score values were also attained.

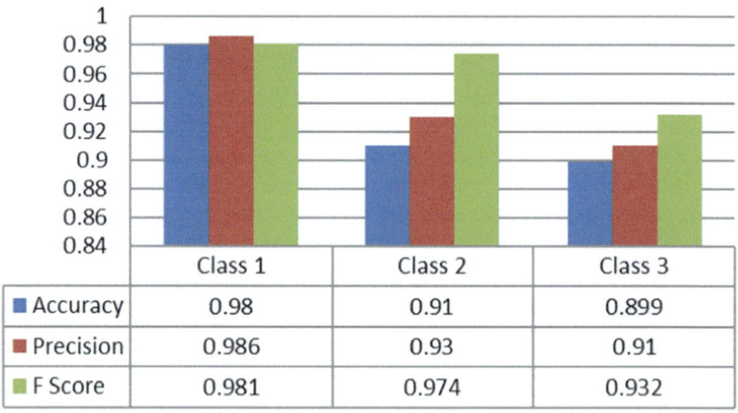

Fig. 3 Performance analysis for CNN approach

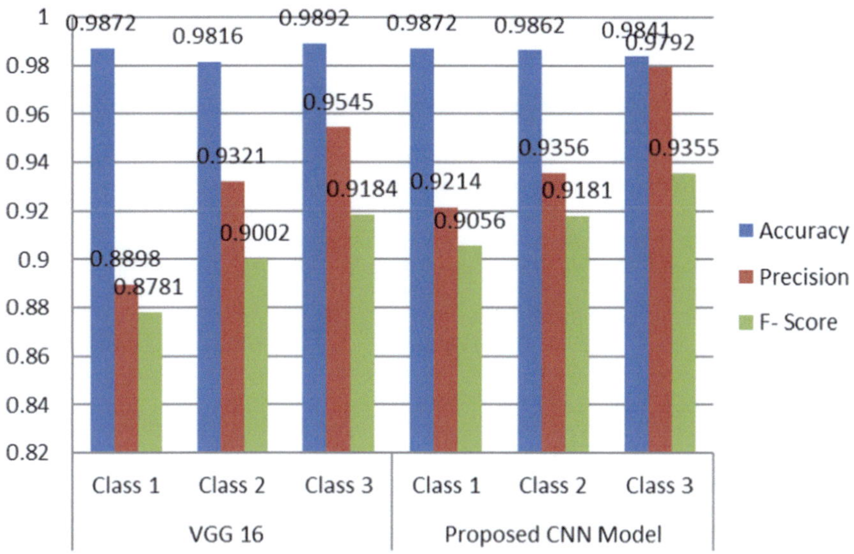

Fig. 4 Comparison study of presentation metrics for the classifiers

5 Conclusion

The present research employs deep learning and classification employing convolutional neural networks to distinguish COVID-19 patients among healthy individuals employing chest X-ray pictures. Since the COVID-19 data set is so little in evaluation of the huge dataset employed by deep learning models, it is hard to calculate the generalization and robustness of these models. The Keras Image Data Generator group could be used to combine CNN's programming using large amounts of X-ray pictures to tackle this problem. Numerous classes (as 2, 3, and 4) are employed to assess CNN's efficiency, comprising healthy persons, COVID-19 patients, people suffering from bacterial pneumonia, and people suffering from viral pneumonia. A number of the metrics utilized in models like AlexNet, VGG16, and the proposed CNN model for performance assessment include F-score values, precision, and accuracy. The outcomes clearly show that the CNN model outperforms other models.

References

1. Gao SH, Wu YH, Fan DP, Mei J, Xu J, Cheng MM, Zhang RG (2021) JCS: An explainable COVID-19 diagnosis system by joint classification and segmentation. IEEE Trans Image Process 30
2. Gómez-Ríos A, Tabik S, Sevillano- García I, Martín-Rodríguez JL, Charte D, Guirado E, Rey-Area M, Luengo J, Suarez JL, García-Villanova P, Valero- González MA, Herrera F,

Olmedo-Sánchez E (2020) COVIDGR Dataset and COVID-SDNet methodology for predicting COVID-19 based on chest X-ray images. IEEE J Biomed Health Inform 24(12)
3. Liu Y, Xiao Y (2020) Analysis andprediction of COVID-19 in Xinjiang based on machine learning. In: 2020 5th International Conf on Information Sci., Computer Technology and Transportation (ISCTT)
4. Mondal MR, Podder P (2020) Machinelearning to predict COVID-19 and ICU requirement. In: 2020 11th International Conf on Electrical and Computer Engg. (ICECE)
5. Choi GS, Reshi AA, Rustam F, Ullah S, Mehmood A, Aslam W, On BW (2020) COVID-19 future forecasting using supervised ML models. IEEE Access 8
6. Toshniwal D, Kumari P (2020) Real-timeestimation of COVID-19 cases using machine learning and mathematical models—The case of India. In: 2020 IEEE 15th International Conf on Industrial and Information Sys (ICIIS)
7. Shimpi R, Parashar V, Mishra M (2020) Development and evaluationof an AI System for early detection of Covid-19 pneumonia using X-ray (Student Consortium). In: 2020 IEEE sixth international conf on multimedia big data (BigMM)
8. İlhan AYDIN, Sevi M (2020) COVID-detection using deep learning methods. In: 2020 International conf on data analytics for business and industry: Way towards a sustainable economy (ICDABI)
9. Nasseri MA, Eslami A, Sarhan MH, Maier M, Zapp D, Lohmann CP, Navab N (2020) Machine learning techniques for ophthalmic data processing: A review. IEEE J Biomed Health Infor 24(12)
10. Alaei M, Ghorvei M, Rezaeijo SM (2020) A machine learning method based on lesionsegmentation for quantitative analysis of CT radiomics to detect COVID-19. In: 2020 6th Iranian Conf on Signal Proce and Inte Sys (ICSPIS)
11. Chinnalagu A, Kumar DA (2020) Sentiment and emotion in social media COVID-19 conversations: SAB-LSTM approach. In: 020 9th International conf sys modeling and advancement in research trends (SMART)
12. Delcea C, Milne RJ, Ioanăş C, Cotfas LA (2020) Evaluation of boarding methods adapted for social distancing when using apron buses. IEEE Access 8
13. Saba L, Skandha SS, Gupta SK, Puvvula A, Koppula VK, Johri AM, Khanna NN, Mavrogeni S, Laird JR, Pareek G, Miner M, Sfikakis P, Protogerou A, Misra D, Agarwal V, Sharma A, Viswanathan V, Rathore V, Turk M, Suri J (2022) Artificial intelligence based carotid plaque tissue characterisation and classification from ultrasound images using a deep learning paradigm. https://doi.org/10.1088/978-0-7503-2244-7ch6
14. Raviprasad B, Mohan CR, Devi GN, Pugalenthi R, Manikandan LC, Ponnusamy S (2022) Accuracy determination using deep learning technique in cloud-based IoT sensor environment. Measurement: Sensors 24:100459. https://doi.org/10.1016/j.measen.2022.100459
15. Khan TM, Baig MAA, Reddy AR, Raju KS (2023) Significance of critical thinking learning enhancement in fluid mechanics in Covid-19 era. In: AIP conference proceedings, vol 2477, no 1. AIP Publishing
16. Subramani P, Srinivas K, Parameshachari BD (2021) Prediction of muscular paralysis disease based on hybrid feature extraction with machine learning technique for COVID-19 and post-COVID-19 patients. Pers Ubiquitous Comput:1–14
17. Velliangiri S, Pandiaraj S, Muthubalaji S (2022) Multiclass recognition of AD neurological diseases using a bag of deep reduced features coupled with gradient descent optimized twin support vector machine classifier for early diagnosis. Concurr Comput: Pract Exp 34(21):e7099

Enhancing Precision Agriculture with Machine Learning and Image Processing: A Comparative Evaluation of YOLO and RCNN for Weed Identification and Detection

Satya Kiranmai Tadepalli, Shobarani Salvadi, Akhil Kambhatla, and Kadiyala Ramana

Abstract Agriculture is one of the origins of mortal food in this world. Currently due to the growing population, we need the lesser productive capability of the husbandry to meet the demands. In olden days, humans employed natural styles as a kind of industrial poison, just like they did with cow manure in fields. This resulted in a rise in output sufficient to suit the needs of the populace. However, subsequently, people were permitted to generate more profits through further development. A revolt known as the "Green Revolution" therefore began. After this period operation of deadly venoms as dressings and fungicides have increased to a drastic position. By doing so we were successful in adding productivity, but we've forgotten the damage done to the terrain, which will raise a mistrustfulness in our food on this beautiful earth. So, in this design, we're going to apply a model using image processing and deep literacy that will be suitable to separate whether a crop is weed or not and lets us to spot the dressings on only weed crops and not prompt the surroundings where good crops may be present, by this we can enhance crop yield as we're scattering the dressings on only weed crops and not on other areas of the field.

Keywords Deep learning · Image processing · YOLO3 · RCNN

1 Introduction

One of the most severe obstacles to crop productivity is weeds, In order to increase crop output and supply more food for a growing global population, weed control is essential. Nonetheless, weed control could have a harmful impact on the ecosystem.

S. K. Tadepalli · S. Salvadi · A. Kambhatla · K. Ramana (✉)
Department of Artificial Intelligence and Data Science, Chaitanya Bharathi Institute of Technology, Hyderabad, India
e-mail: ramana.it01@gmail.com

V. K. Gunjan et al. (eds.), *Cybernetics, Human Cognition, and Machine Learning in Communicative Applications*, Cognitive Science and Technology,
https://doi.org/10.1007/978-981-97-8533-9_22

The use of herbicides has the potential to pollute the environment since, in the majority of situations, only a very small amount of the sprayed chemicals really reach their intended targets, while the majority of herbicides fall to the ground, and some may even drift away. Mechanical weed removal may cause erosion and harm beneficial creatures including soil-dwelling earthworms and soil-surface spiders. Some weed management techniques have additional drawbacks and frequently have a severe impact on the environment. Sustainable weed management techniques must be created to solely effect weed plants and cause the least amount of disruption to the environment. Before using any control methods, weeds should be discovered and recognized in real-time, which would improve and increase the sustainability of weed management. In order to quickly identify weeds in crop fields, convolutional neural networks are used to develop a deep learning model to detect weeds with high accuracy.

2 Literature Review

Many authors have claimed development credit for the creation of weed detectors. The list of recent innovations that have been published in papers is as follows:

In paper [1] the author describes "Deep Neural Networks to detect weeds from crops in Agricultural Environments in Real-time". The author began by going over many methods. The neural network was trained using backpropagation, the hyper-parameters were adjusted using the validation set, and the final model was evaluated using the testing set. After pre-processing the photographs, they used a pre-trained convolutional neural network named VGG16 to extract features from the pictures. A new fully connected layer was then trained to execute the classification job on top of the VGG16 network. Finally, they implemented the trained neural network on a Raspberry Pi computer equipped with a camera module to detect weeds in the field in real-time. This technology's primary flaws are its expensive price and limited dataset.

In paper [2] "Algorithm of Weed Detection in Crops by Computational Vision" is described by the author. The authors provided a computer vision-based approach for weed detection in crops. To capture pictures of the crops and weeds, they employed a digital camera mounted on a moving platform, such as a cart or a tractor. After that, filters were used to pre-process the images. Then, in 2020, they removed. These articles use features from the segmented regions of interest, such as colour, texture, and shape, to demonstrate how computer vision can precisely locate objects inside the home. Using a machine learning algorithm that was trained using a set of annotated photographs, scientists identified the separated areas of interest as either crops or weeds. Post-processing procedures, such as morphological operations to exclude small items and smoothing techniques to remove noise, were used to further the categorization findings. The system's main flaws include limited scalability and environmental concerns.

"Weed Identification Using Deep Learning and Image Processing in Vegetable Plantation" is the topic of paper [3] by the author. The authors discussed several approaches, such as employing a camera to take pictures of the weeds and plants, pre-processing the images to enhance their quality, and identifying the weeds using a convolutional neural network (CNN). The CNN, trained on a big dataset of image data, is more accurate thanks to transfer learning. After the categorization results are shown on a user interface, farmers can either manually or automatically identify and remove the weeds. Overall, the application of this technology may result in a reduction in the amount of time and labour required for weed management in vegetable plantations. The system's flaws are its expensive cost, small dataset, and ineffectiveness in all vegetable crops.

In paper [4] "Weed Detection in Farm Crops Using Parallel Image Processing" is discussed by the author. In order to begin the procedure, photos of the crops are taken using a high-resolution camera placed on a tractor. The photographs are then pre-processed to raise their quality and reduce noise. The pre-processed pictures are then divided into different sections using a clustering technique. The segmented areas are separated into crop and weed categories using a support vector machine (SVM) classifier. The SVM classifies the regions using characteristics such as colour, texture, and form after being trained on a dataset of tagged pictures. Farmers can use the resulting weed map to find and remove weeds. The system is appropriate for large-scale farming operations since it is created to be effective, precise, and scalable. The system's two main drawbacks are that it uses a lot of computer resources and that not all types of farm crops can be detected for weeds using it.

In paper [5] the study "Deep Learning-Based Approach for Weed Detection in Potato Crops" aims to develop an automated weed detection system using deep learning techniques for potato crops. The procedure comprises gathering photos of weeds and potato plants from several fields, pre-processing the photos, and using a Convolutional Neural Network (CNN) to train a deep learning model. A sizable collection of photos with either "potato" or "weed" labels is used to train the model. The trained model is then put to the test on fresh photos to gauge how well it can identify weeds. The findings demonstrated that the suggested method obtained a high level of weed detection accuracy, making it a potentially useful tool for managing weeds in potato fields.

In the manuscript [6] the goal of the project "Deep Learning-Based Approach for Weed Detection in Potato Crops" is to create an automated system for weed identification in potato crops by utilizing deep learning methods. The technique entails collecting images of weeds and potato plants from several fields, preparing the images, and then training a deep learning model with CNN. The model is trained using a large set of images labelled as "potato" or "weed". Next, the trained model is evaluated on new images to see how effectively it recognizes weeds. The results showed that the recommended strategy achieved a high degree of accuracy in weed detection, indicating that it might be a helpful tool for controlling weeds in potato fields.

3 Proposed Work

Weed detection is highly valued in agriculture since it is crucial to maintaining the health and yield of crops. Large-scale weed identification is challenging since conventional techniques require a lot of physical labour and take a long time. Deep learning and image processing are currently used in promising weed detection techniques. The suggested method utilizes deep learning and image processing to identify and classify weeds successfully and automatically. To determine whether the input image contains weeds, convolutional neural networks (CNN) are used to extract features from the image.

Figure 1 depicts the proposed system The dataset, which is in CSV format, needs to be converted into a text file format that YOLO can read after we manually identify each object of interest with a bounding box and extract the photographs from the dataset. The YOLO model is then trained using the dataset, which entails changing the model's weights in response to errors it makes. The model can be used to extract features from fresh photos after it has been trained, and these features can then be used to build a scan window or box around each object that is spotted. Last but not least, the YOLO model delivers a precise output that includes the object's label and the coordinates for its enclosing box. Throughout this process, a convolutional neural network (CNN) is used to learn to extract relevant features from the input images, which is crucial for accurate object detection.

Fig. 1 Proposed system architecture

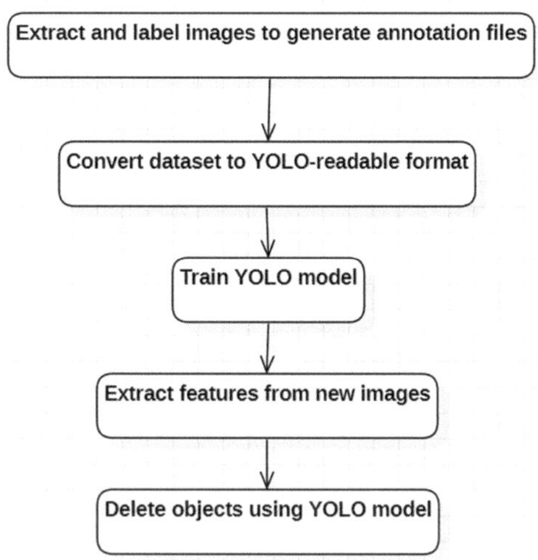

4 Convolutional Neural Network

Convolutional neural networks (CNNs), a type of deep learning architecture, were developed especially for analyzing images.

Tasks include segmentation, object detection, and image classification. It has a number of layers, including convolutional layers, pooling layers, and fully connected layers, and uses the convolution approach to extract attributes from the input image. The core idea behind CNNs is to train a set of filters—sometimes called kernels—that can identify important patterns or features in a picture. These filters are applied using convolution to the input picture, resulting in a collection of output feature maps that depict different facets of the image. In addition, CNNs use fully connected layers to complete the final classification task and pooling layers to down sample the output feature maps and lower the spatial dimensionality of the data (Fig. 2).

Different Layers of CNN are:

1. Convolutional Layer: A convolutional layer is often the first layer in a CNN. This layer applies a variety of filters on the input image to extract features. The input image is convolved with the weight matrix of each filter to produce a feature map.
2. Pooling Layer: A pooling layer is frequently added following each convolutional layer. By picking the highest or average value within a frame of pixels, this layer decreases the feature map's spatial dimensions. This makes the features more resilient to little changes in the input and lowers the computational cost of the network.
3. Activation Function: The output of each layer is subjected to an activation function to introduce nonlinearity to the network. The most common activation function, Rectified Linear Unit (ReLU), zeros out all negative values.
4. Fully Connected Layer: Similar to the layers in a conventional neural network, the final layers of a CNN are frequently completely linked layers. These layers

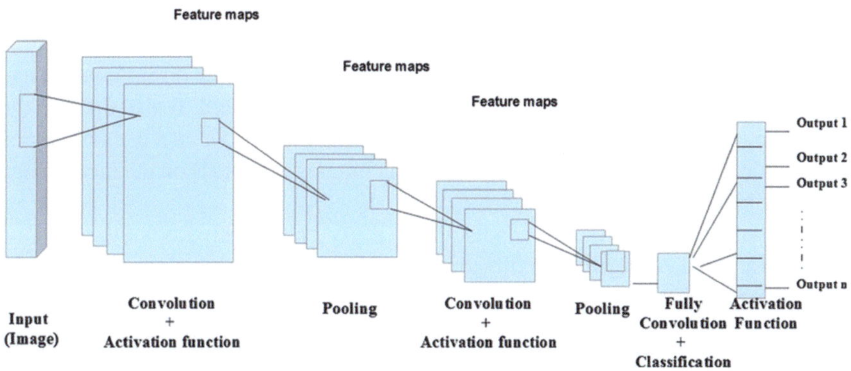

Fig. 2 CNN architecture

apply a set of weights to the previous layers' flattened output to create the final output.

5. Dropout Layer: After the completely connected layer, a dropout layer may be added to avoid overfitting. In order to lessen the co-adaptation of neurons, this layer randomly removes part of the neurons during training.
6. Softmax Layer: Similar to the layers in a conventional neural network, the final layers of a CNN are frequently completely linked layers. These layers apply a set of weights to the previous layers' flattened output to create the final output.

In general, a CNN's architecture is created to draw out useful elements from the input image and use them to generate predictions. The network may learn progressively complicated characteristics that are more and more invariant to minute changes in the input by stacking numerous convolutional and pooling layers.

5 Proposed Algorithm for Weed Detection

Using the YOLO Algorithm, we attempt to distinguish the weed images from the other crops for this project. Convolutional networks are used by YOLO, which was chosen because of its outstanding performance in item and sample recognition. This is why, in addition to monitoring moving objects, it now has real utility in areas like transportation and animal identification.

YOLOV3 Algorithm

First, we process the input image and create a blob from it by swapping the red and blue colour channels, shrinking the image to a fixed size of (512, 512), and scaling pixel intensities to the range [0, 1]. The blob is then designated as the YOLO network's input. The YOLO network is then operated to provide a list of outputs from the network's output layers called layer Outputs. Each output layer is in charge of identifying items of various sizes and corresponds to a particular scale in the image. The class ID and confidence ratings of the discovered objects are extracted when we cycle through each layer's output and each detection contained inside it. The method scales the bounding box coordinates back in relation to the size of the input image if the confidence score is higher than a predetermined threshold. It also updates the lists of bounding boxes, confidences, and class IDs. Finally, we use the class name and confidence score to create bounding box rectangles and labels on the input image for each object that was detected.

A one-stage object detection model, the YOLOV3 (You Only Look Once version 3) algorithm predicts object bounding boxes and class probabilities from input photos in a single pass. This is how it goes:

- Input image: At first it takes an input image and divides it into a grid of cells.
- Feature extraction: To extract information from the input image, a deep convolutional neural network (CNN) architecture processes the image. A feature pyramid

network (FPN) is used by the YOLOV3 architecture to detect objects at various scales and resolutions.

- Information about the object's width, height, centre coordinates, and a confidence level that expresses how likely it is that the box contains an object. The possibility that an object falls into one of several predetermined categories is represented by the class probabilities.
- Non-maximum suppression: The predicted bounding boxes are then subjected to a non-maximum suppression (NMS) algorithm by the YOLOV3 algorithm in order to remove redundant detections and keep only the most precise bounding boxes.
- Output: The final output of the YOLOV3 algorithm is a set of predicted bounding boxes and class probabilities for each object detected in the input image.

RCNN Algorithm

The second approach we've employed is known as RCNN (Regions with Convolutional Neural Networks), and it identifies objects in an image by combining region suggestions and convolutional neural networks (CNNs). It entails picking areas of an image that could contain an item, using CNNs to extract features, and then classifying the object with an SVM.

First, using a selective search method, we pre-process the images and annotations by scaling them, changing the format of the comments, and creating region suggestions. To make the data compatible with the RCNN architecture, which necessitates inputs of a specific size and format, this data pre-processing step is required. After creating the region suggestions, we used intersection over union (iou) to compare the generated regions with the labels on the ground truth. Any region with iou > 0.5 is saved as a positive example (it may object), and any region with iou 0.2 is preserved as a negative example.

There are three parts in RCNN:

1. CNN fine-tuning: First, using my created region suggestions, we adjusted the VGG16 model, which has three output classes (Crop, Weed, and Background) and an input size of $224 \times 224x3$. By updating the weights of the fully connected layers of the CNN while maintaining a weight-free state for the convolutional layers, this procedure entails training the CNN on the detection task. The CNN is taught to categorize object suggestions as either foreground (object) or background (background) and to fine-tune their placements during fine-tuning. The input image's features must be extracted in this stage in order to be utilized later on by the RCNN.
2. CNN + SVM training: The retrieved features from the improved CNN are used to train a support vector machine (SVM) classifier in the second stage. From the refined model, we eliminated the final two fully linked layers and used the CNN model as a feature extractor. Each object suggestion is represented as a fixed-length feature vector in the CNN output, which is a feature map that represents the image. In order to categorize each proposed object as a certain object class, the SVM classifier is trained on these feature vectors. A multi-class SVM or multiple

binary SVMs can be used to train the SVM to categorize different object classes. The SVM can be used to classify object suggestions and eliminate false positives once it has been trained.

3. Bounding box regression: It is a technique for enhancing the bounding boxes generated by early object proposals. Each sample includes an image, item suggestions, and the bounding box for the associated ground truth. The model is trained using a set of labelled data. By minimizing a loss function that penalizes the difference between the anticipated offsets and the actual offsets, the model learns to improve the bounding boxes. By better localizing items in an image, bounding box regression can considerably increase the accuracy of object detection. Even though we didn't apply this tactic during the training method, we still got good accuracy.

6 Results

We have used Sesame crops and weed dataset which was obtained from Kaggle and also we captured some images from the field and each image is of 512×512 coloured image (Fig. 3).

Fig. 3 Dataset images

YOLOV3 Weed Detection Performance

The graph in Fig. 4 shows the relation between the number of iterations and the average loss value. As the number of iterations increases the average loss value decreases the less the average loss value the more is the accuracy. We put the no of batches as 4000 which represents 4000 iterations and we run our training model up to 3200 iterations, we got the current average loss as 0.3496 which is low and as we increase the number of batches; the average loss value will decrease.

Once the model is trained, it can be used to extract features from new images, which are then used to create a scan window or box around each detected object. Finally, the YOLO model provides an accurate result, including the label of the object and the coordinates of its bounding box.

By considering several factors and based on the output obtained in order to assess the performance of the suggested YOLOV3 algorithm, the following counts were reached after considering 100 images (Table 1).

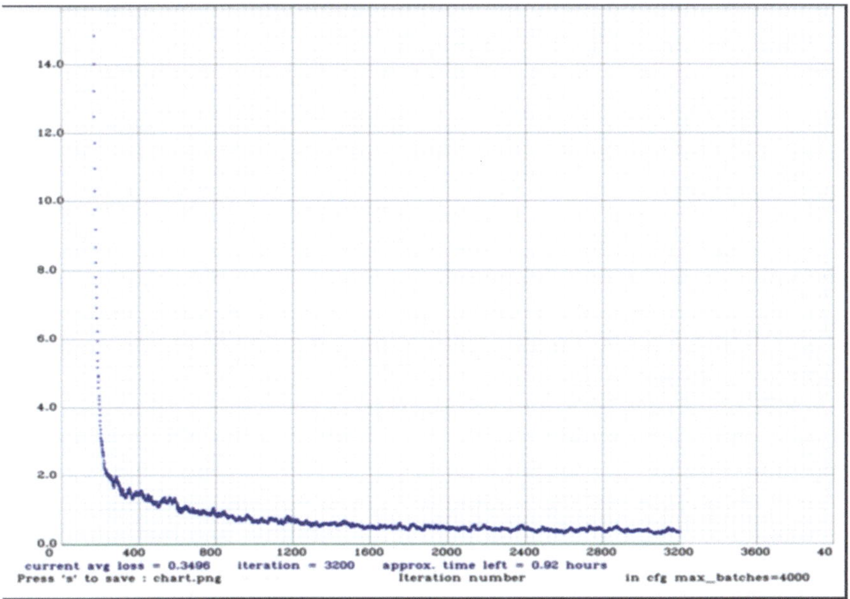

Fig. 4 Analysis of results

Table 1 Truth Table for YOLOv3

		Actual (True) values	
		Positive	Negative
Predicted values	Positive	49	2
	Negative	2	47

Table 2 Truth table for RCNN

		Actual (True) values	
		Positive	Negative
Predicted values	Positive	48	3
	Negative	2	47

With the use of the aforementioned variables, we discovered that the average accuracy was 96%, which is good and gives the YOLOV3 algorithm a good indicator of how well it is performing in comparison to other algorithms.

RCNN Weed Detection Performance

With the use of the aforementioned variables, we discovered that the average accuracy was 95%, which is good and gives the RCNN algorithm a solid indicator of how well it is functioning in comparison to other algorithms (Table 2).

7 Comparison of YOLOV3 and RCNN

Popular object detection techniques used in computer vision to find items in images include YOLOv3 and RCNN. They take different approaches to object detection, though.

A single-stage object detection method called You Only Look Once v3 (YOLOv3) divides the input image into a grid of cells before starting to look for things. Each cell forecasts a particular number of bounding boxes based on their class probability and confidence scores. Since the system only employs a single convolutional neural network (CNN) to generate predictions, it is speedier and more efficient than earlier object detection systems.

RCNN, on the other hand, is a two-stage object identification method that creates region proposals (i.e. potential item locations) before using a CNN to classify each proposal and enhance its bounding box. As a result, RCNN is able to handle a wider range of object sizes and forms. However, it is slower and uses more computer resources.

Both systems had excellent results for weed and crop detection. In our case, YOLOV3 provided an accuracy score of 96% and RCNN provided a score of 95%, both of which are excellent. The Detailed Comparison can be seen as shown in Table 3 which shows the different parameters.

Both YOLOv3 and RCNN can be efficient for weed and crop detection, although how well they perform may vary depending on the application and dataset. While RCNN may be preferable for smaller-scale applications where accuracy is more crucial, YOLOv3 may be more appropriate for real-time detection in large-scale areas where speed is crucial. To decide which model performs better for the current application, it is advised to compare the two models on the relevant task and dataset.

Table 3 Comparison of YOLOv3 and RCNN

Performance metrics	YOLOv3	RCNN
Accuracy	96%	95%
Precision	96%	94.11
Recall	96%	96
Specificity	95.91%	94
F1 score	0.96	0.95

8 Conclusion and Future Enhancements

In precision agriculture, weed detection is a crucial activity that can assist farmers in minimizing the effect of weeds on crop yields and reducing the usage of herbicides. Farmers can target weed infestations more efficiently and apply targeted weed control techniques for accurate weed detection. Weed detection is significant because it has the ability to increase agricultural yields, decrease the usage of herbicides, and lessen the environmental effects of weed management techniques.

We compared the effectiveness of YOLOv3 and RCNN for weed and crop detection in agricultural environments in this work. Using a collection of aerial photos of crops and weeds, we assessed both algorithms' performance in terms of accuracy and speed. Our findings demonstrated that both algorithms had superior sensitivity to small item detection. Overall, both models showed excellent performance in identifying crops and weeds, demonstrating the potential of both for use in precision agriculture.

Our study emphasizes the significance of choosing a suitable model for the particular application and dataset. While RCNN may be preferable for real-time applications where accuracy is more crucial, especially when recognizing tiny objects, YOLOv3 may be more appropriate for real-time applications where speed is crucial.

Future advancements can be done by combining deep learning models with other technologies like GPS, robotics, and sensors to provide a complete weed management system. Future research can focus on developing integrated systems that can provide real-time weed detection, mapping, and treatment.

References

1. Rakhmatulin I, Kamilaris A, Andreasen C (2021) Deep neural networks to detect weeds from crops in agricultural environments in real-time: a review. Remote Sensing 13(21):4486
2. Tejeda AI, Castro RC (2019) Algorithm of weed detection in crops by computational vision. In: 2019 International Conference on Electronics, Communications and Computers (CONIELECOMP). IEEE, pp 124–128
3. Jin X, Che J, Chen Y (2021) Weed identification using deep learning and image processing in vegetable plantation. IEEE Access 9:10940–10950

4. Umamaheswari S, Arjun R, Meganathan D (2018) Weed detection in farm crops using parallel image processing. In: 2018 Conference on Information and Communication Technology (CICT). IEEE, pp 1–4
5. Khan F, Zafar N, Tahir MN, Aqib M, Saleem S, Haroon Z (2022) Deep learning-based approach for weed detection in potato crops. Environ Sci Proc 23(1):6
6. Phung, Rhee Figure 1. Schematic diagram of a basic convolutional neural network. ResearchGate, 03-Jun2021. [Online]. Available https://www.researchgate.net/figure/Schematic-diagram-of-a-basicconvolutional-neural-network-CNN-architecture-26_fig1_336805909. [Accessed: 23-Dec-2021]
7. Zhao L, Li S (2020) Object detection algorithm based on improved yolov3. MDPI. [Online]. Available: https://www.mdpi.com/2079-9292/9/3/537. [Accessed: 24-Dec-2021]
8. Adarsh P, Rathi P, Kumar M (2020) YOLO v3-Tiny: Object detection and recognition using one stage improved model
9. Pham T, Dao S (2020) Plant leaf disease classification based on feature selection and deep neural network
10. Elstone L, How KY, Brodie S, Ghazali MZ, Heath WP, Grieve B (2020) High speed crop and weed identification in lettuce fields for precision weeding. Sensors 20(2):455

Gesture Based User Interface Access Using Convolution Neural Network

M. Mohamed Iqbal, Chintam Sravan Kumar, Mukkoti Maruthi Venkata Chalapathi, A. Vijaya Krishna, and P. Purushotham

Abstract Gesture-based UI Interaction is a process that facilitates user-system interaction without the need for physical touching. This method allows users to input characters or numbers and execute corresponding actions through gestures. The system displays symbols representing specific characters or numbers, and by presenting a particular action symbol, the associated action is performed. Particularly during viral crises, there is a considerable demand for touchless systems that prioritize minimal physical contact, thereby reducing transmission risks, promoting hygiene practices, and upholding public confidence. The implementation of touchless systems during viral crises involves the utilization of a Convolutional Neural Network (CNN) algorithm that enables touchless detection and interpretation of hand gestures performed by the user. The CNN accurately identifies hand gestures and translates them into characters or actions, with training phase accuracies of 0.912 for alphabet, 0.922 for numeric, and 0.933 for actions datasets, and testing phase accuracies of 0.924 for alphabet, 0.941 for numeric, and 0.953 for actions datasets. The touchless system achieved a combined accuracy of 0.922 during training and 0.939 during testing.

Keywords Convolution neural network · Hand gesture · UI interaction · Contactless interaction · Image recognition

M. M. Iqbal
Department of Information Technology, Sri Sivasubramaniya Nadar College of Engineering, Tamil Nadu, Chennai, India

C. S. Kumar · M. M. V. Chalapathi (✉)
School of Computer Science & Engineering (SCOPE), VIT-AP University, Amaravati, Andhra Pradesh, India
e-mail: mmv.chalapathi@gmail.com

A. V. Krishna
Department of Information Technology, G. Narayanamma Institute of Technology and Science, Hyderabad, Telangana, India

P. Purushotham
Department of Computer Science and Technology, G. Narayanamma Institute of Technology and Science, Hyderabad, Telangana, India

V. K. Gunjan et al. (eds.), *Cybernetics, Human Cognition, and Machine Learning in Communicative Applications*, Cognitive Science and Technology,
https://doi.org/10.1007/978-981-97-8533-9_23

1 Introduction

There is a huge need for a service that makes contactless interaction with a device because of some viruses which do not allow us to interact with devices by touching them. There is an increase in diseases that come through contacting other human beings or devices which are used by them. There is a high need for a device that can do all its processes or transactions by taking input without any contact with others. So, GUI-based interaction can help in doing contactless transactions with the hand gestures of human beings. This proposed work allows users to input commands using hand gestures, and it performs further actions based on the user's needs. This touchless interaction is particularly useful during situations when physical contact should be avoided, such as during a pandemic like COVID-19. Gesture-based applications have made significant impacts across various fields. One such application is Sign Language Translation, where innovative hand gesture-based UI designs swiftly interpret sign language into understandable spoken or written language [1]. This advancement holds great promise in improving communication and inclusivity for the deaf and hearing-impaired communities. Additionally, gesture-based interactions have gained popularity, allowing users to engage with virtual worlds using motion-sensing controllers. These controllers, like Microsoft Kinect or Leap Motion Controller [2], provide users with immersive and intuitive experiences in gaming and other industries. These examples highlight the transformative power of gesture-based technology in enhancing user interactions and experiences in diverse settings.

Furthermore, gesture-based applications find significant utility in Human–Robot Interaction, as they facilitate seamless communication and interaction between humans and robots. Moreover, the integration of hand gesture-based UI design can be observed in wearable devices, such as smartwatches or fitness trackers. In these devices, advanced deep learning-based algorithms are employed to detect specific hand gestures, enabling users to execute various actions on the respective UI design effectively. One of the deep learning algorithms CNN (Convolution Neural Network) is frequently utilized in so many researches based on image detection. The proposed study employs the Convolutional Neural Network (CNN) algorithm for the purpose of detecting and recognizing the displayed hand gestures. A CNN may be a kind of arranged engineering for profound learning calculations and is particularly utilized for picture acknowledgment and assignments that include the preparation of pixel data. This makes them profoundly appropriate for computer vision (CV) assignments and for applications where question acknowledgment is crucial, such as self-driving vehicles and facial acknowledgment. There are many applications of hand gestures but the GUI-based Interaction is not available. It is very much necessary for an Interaction to be accomplished without any touching. Also, it increases the comfort of handling the interaction between a system and a human.

2 Literature Survey

In recent years, numerous investigators have conducted investigations on gesture-based analysis. Gupta et al. [3] proposed framework is utilized for keeping money secure and can be utilized for companies or at personally secured places for security purposes. They have proposed a security-based motion acknowledgment system using a web camera. They have utilized skin discovery and canny location calculation to distinguish a motion from a picture.

The authors of [4] will utilize the camera for motion detection, while a non-vision-based technique will involve the use of sensors. In this context, an innovative vision-based approach was implemented to identify and track hand movements, creating a continuous channel of communication with others. By employing convolutional neural networks and custom-made neural frameworks, this ingenious system effectively translates gestures into signals. This investigation explores the advantages and limitations of hand movement recognition, shedding light on its potential and challenges as a reliable communication medium.

In [5] they created a dataset and after that utilized a lightweight CNN that appears to recognize and classify hand improvements capably. Furthermore, they evaluated the classification accuracy by limiting the number of frames in a video sequence. A comparison was made between the video signal recognition performance of the primary camera and the RGB camera. The performance of the suggested model on edge computing devices was assessed and compared against benchmark models in terms of accuracy and inference time.

The authors [6] aim to aid individuals with disabilities by enabling them to communicate using both American Sign Language and spoken language. Their research focuses on proposing gestures or signs for searching learned words using the YOLOv3 algorithm, which recognizes gestures and languages with equal letter representation.

In another study [7], the authors efficiently extract information from the skin field by introducing a model built on the centroid basin algorithm (ICWA) and dual-channel convolutional neural network (DCCNN). The efficiency of this approach lies in the accurate segmentation of the raw image achieved through ICWA.

The author [8] presented a real-time navigation system based on the TSL (color, saturation, lightness) adjustment zone of a single camera for detection and distance signatures. Here, first, skin correction is used in the TSL color space, which can process information from the skin to the hand area. Second, distant features are derived from hand contours and finger points are obtained, which reduces cognitive problems in finding the peaks of one-dimensional features.

In the [9] article, the focus is on exploring the application of edge detection and histogram equalization in pre-processing to enhance the overall performance of motion recognition. The system utilizes adjacent neighbors as classifiers for recognition purposes. Designed to categorize incoming images into six distinct classes, each representing a different command for the machine, the system effectively processes gesture images captured under controlled conditions using a webcam and a white face mask.

Yu et al. [10] proposes a point feature extraction method based on a multilayer perceptron. Hand tan features in the YCbCr color space are used to detect feet. By binarizing and improving the contrast of the hand image, the contour and features of the hand can be accurately extracted.

Yewale and Bharne [11] provides an overview of different methods for motion recognition using MATLAB. It also provides a working summary of the recognition process using edge detection and skin detection algorithms. The above pieces of literature described well contributed to Hand gesture detection but there is no literature that gives information about a service that makes interaction with the device using hand gestures only. This service enables interaction with devices through hand detection by presenting appropriate symbols to the camera. It proves particularly valuable in situations where individuals need to maintain distance and avoid physical contact due to health concerns. Notably, there are currently no existing citations or references related to contactless interaction with a device, indicating the novelty and uniqueness of this concept.

3 Proposed Methodology

3.1 Contribution

This proposed work is mainly used in the period of a critical era where the viruses spread widely due to touching the services. This proposed work will help users be touched free where the person needs to touch anything in order to perform any action. This proposed work will read the gesture of the hand and convert it into an appropriate symbol of character or action and perform it. At first, the model is trained with all the symbols of characters and actions then with appropriate coding the model is used in an application where it can be used to enter characters and perform actions on the system.

3.2 Convolution Neural Network

The core of a CNN is the convolutional layer, crucial for the algorithm's effectiveness. In this layer, learnable filters, also called kernels or feature detectors, are applied to the input image. These filters convolve over the image, calculating dot products among filter weights and localized patches, which is fundamental in the CNN's operation. Through this process, a collection of feature maps is generated, adeptly capturing distinct visual patterns inherent in the input.

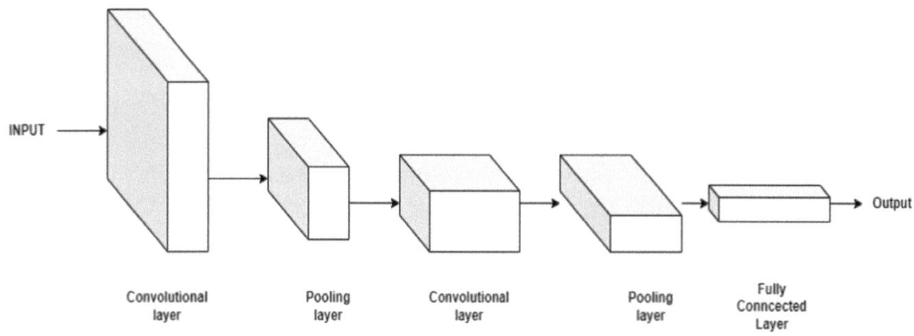

Fig. 1 CNN architecture

Referring to Fig. 1, it becomes apparent that the primary objective is to capture increasingly interactive features. CNNs commonly employ multiple convolutional layers. As information traverses deeper into the network, the neurons' receptive fields expand, enabling them to encapsulate more extensive and abstract features. These intermediate feature maps are then subjected to further processing by other layer types, such as pooling layers and fully connected layers.

Pooling layers play a critical role in decreasing the spatial dimensions of the feature maps while maintaining crucial details. An extensively used technique in pooling layers is max pooling, which selectively preserves the highest value within a specific pooling window, effectively downsizing the feature maps. This strategy ensures that vital features are retained while simultaneously reducing the computational load of the network.

At the conclusion of the network, fully connected layers assume responsibility for making predictions based on the extracted features. These layers establish dense connections between neurons from the previous layer and those in the subsequent layer, forming a comprehensive network. Activation functions, such as soft-max or sigmoid, are utilized within these layers to generate class probabilities or regression values, contingent upon the specific task at hand.

3.3 Proposed Architecture

Let us now explore the architectural diagram referenced in Fig. 2. At first, the hand gesture is captured. Following pre-processing, which involves converting raw data into a suitable format for analysis, hand gesture detection is performed using the CNN. CNNs, with their interconnected layers, automatically learn and extract crucial features from the input data. The Key components of a CNN include convolutional layers, pooling layers, activation functions, fully connected layers, a loss function, and an optimizer. These elements collectively contribute to the CNN's ability to recognize and interpret hand gestures effectively. After detecting hand gestures using CNN now convert that hand gesture to a particular character or action and perform

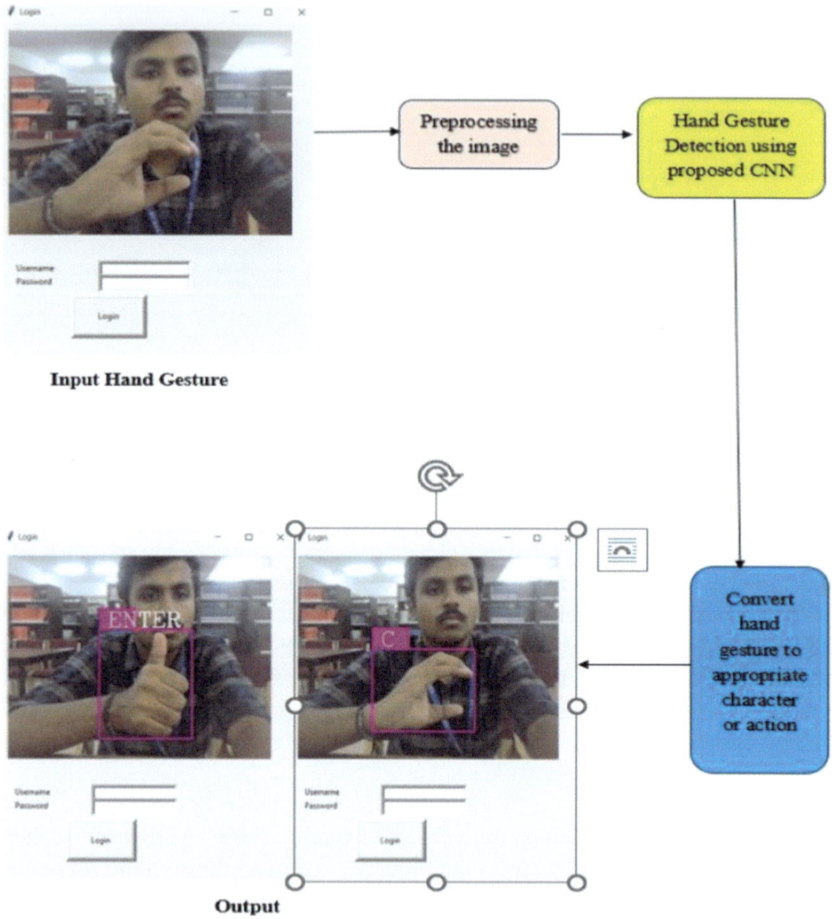

Fig. 2 Proposed architecture

it. After performing the appropriate action or showing the appropriate symbol the user can see the output on the screen.

3.4 *Training of Proposed CNN Model*

Let us now proceed to the training phase of the model, as illustrated in Fig. 3. There is a set of image datasets of hand gestures defined with proper character or action. Then the device has to read the images from the given dataset. Then the images should be pre-processed in order to get the finest images. After pre-processing the data is sent to the combination of the convolution and pooling layers. Here, four combinations of convolution and polling operation are applied. In convolution layer, the convolution operation between input image and kernel filters will take place. Finally, it will generate the feature map in high dimensional space.

The feature map with high dimensional space is given to the max pooling layer. In max pooling the dimensions of the feature map are reduced. In pooling the most conspicuous features are taken from the previous feature map and converted to the current feature map. Now the flattening of features is done where the 2-dimensional feature vector map is converted to a linear vector. Now divide the image into training and testing parts. The fully connected network allows for efficient pattern recognition and classification tasks. By connecting all neurons in a layer to all neurons in the

Fig. 3 Training flow on proposed model

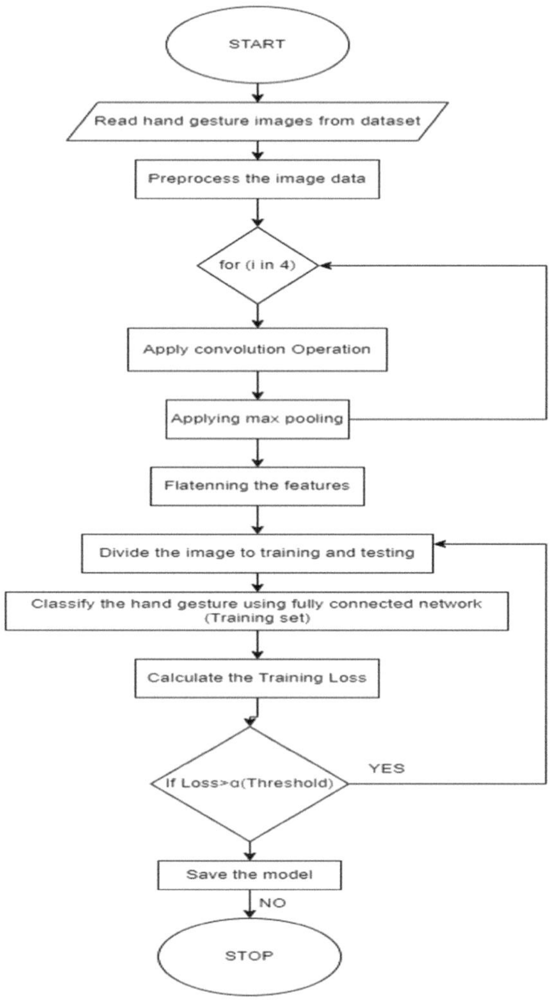

subsequent layer, the network can learn complex relationships and extract relevant features from the input data, ultimately leading to accurate classification results.

Now calculation of training loss takes place where the difference between the generated outcome and the anticipated outcome is taken. Conversely, if the loss is below or equal to the specified threshold (alpha), indicating that the model has achieved satisfactory performance, the system saves the trained model to retain its current state. After saving the model, the flow stops, and the system is ready to be deployed for practical use.

3.5 Testing of Trained Model on Real-Time Interface

We shall now transition to the testing phase, which involves capturing hand gestures and mapping them to corresponding actions, as indicated in Fig. 4. Let's start the process by taking the hand gesture input through the device camera. After capturing the hand gesture image then pre-process the image where the image data is cleaned properly and gets best out of worst. After pre-processing is done then the refined image of the hand gesture is classified into appropriate symbols using the trained model. If the symbol is an action then perform the appropriate action like enter, backspace, next, and submit, where enter option enters the shown character into the text box, the backspace option backspaces a character once, the next option tabs from the current text box to the upcoming text box, and submit option submits the user input. After performing action or showing character if the user wants to continue the operations then the loop will go to image hand gesture input otherwise it stops the execution. So this is the testing part of the proposed work.

Table 1 presents a comprehensive symbol table, which includes 26 distinct symbols allocated for representing each of the 26 alphabets in the English language. Moreover, the symbol table extends its utility to encompass numeric data as well, with 10 unique symbols assigned to represent the numbers 0 through 9. It also defines four hand gestures for specific actions, enabling touchless interaction with devices and applications described in Table 2.

4 Results and Discussions

4.1 Dataset Details

The dataset utilized in this proposed study has three types of image datasets. The alphabet dataset contains the images of different symbols representing English alphabets from A to Z. The numeric data set contains the images of different symbols representing numeric digits 0 to 9. The action data set contains the images of 4 different actions described in Tabel 2. The images from these data sets are collected from

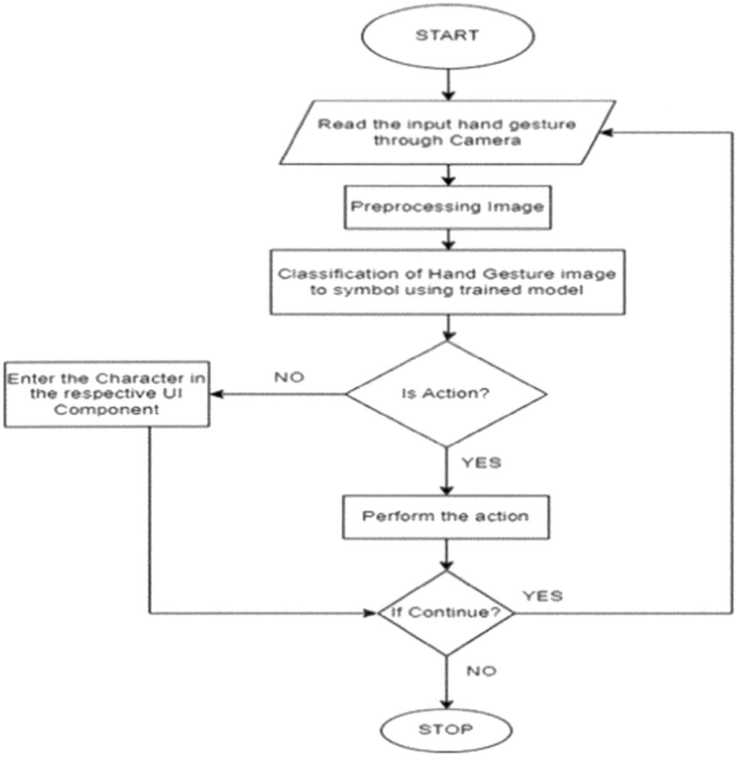

Fig. 4 Process flow of proposed model in real time

Kaggle open source. Since images collected from Kaggle dataset are not sufficient, the proposed work utilizes image augmentation techniques to generate sufficient number of images. For training, 12,988 images are used from alphabet dataset, 5177 images from numeric dataset, and 2000 images from actions dataset making a total count of 20,165 images for training. For testing 1000 images each are used from alphabet dataset, numeric dataset, and actions dataset making a total count of 3000 images for testing (Table 3).

4.2 Results Analysis

The parameters and layer details of the proposed CNN model are described in Table 4. The model utilizes four convolutional layers and four max pooling layers to extract the high level features for classification. These extracted features are flattened through flattening layer and generate feature vector of 25,088 features. These features vector is passed through two dense layers to predict the final outcome. In Table 4, the

Table 1 Symbols for each character

Symbol	Char	Symbol	Char	Symbol	Char
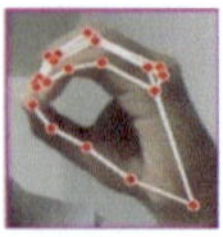	0	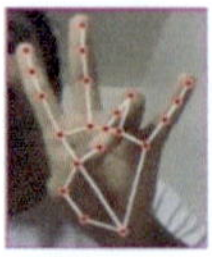	9	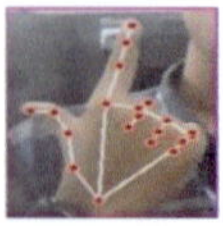	I
	1	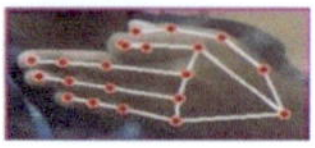	A	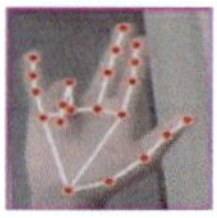	J
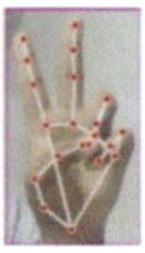	2	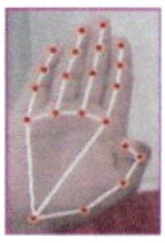	B		K
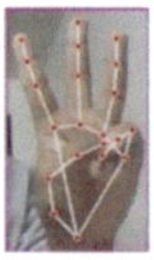	3	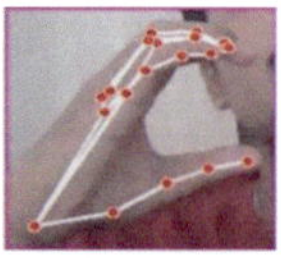	C	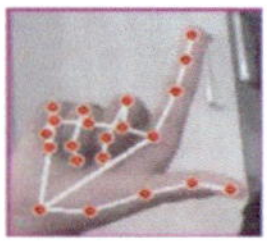	L
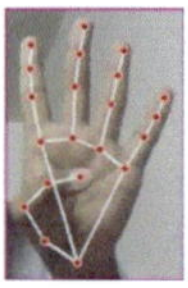	4	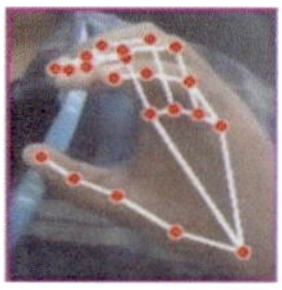	D	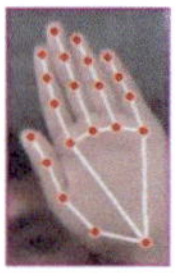	M
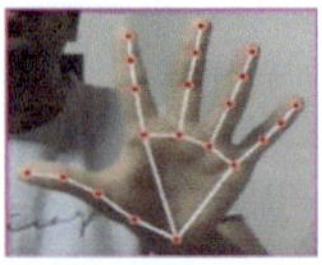	5	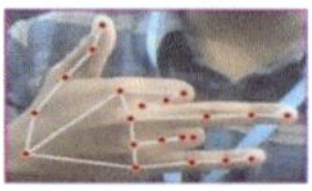	E	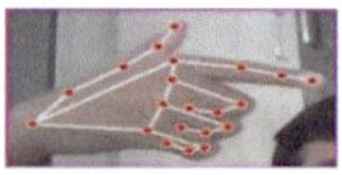	N
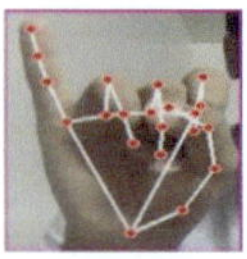	6	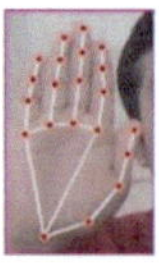	F		O
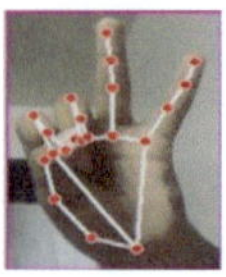	7	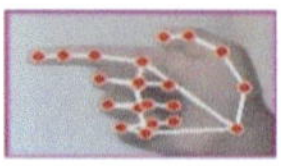	G		P

(continued)

Table 1 (continued)

Symbol	Char	Symbol	Char	Symbol	Char
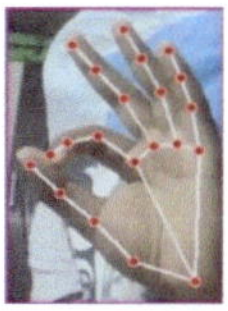	8	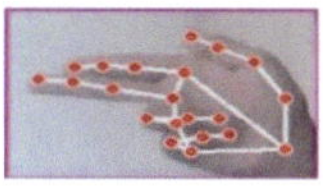	H		Q
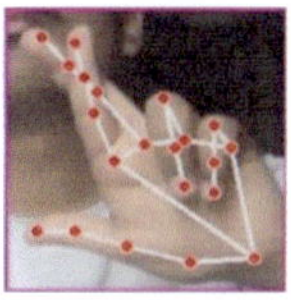	R	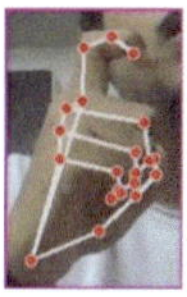	X	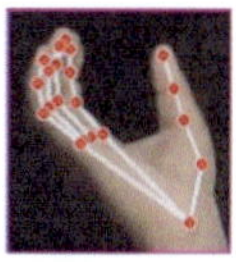	U
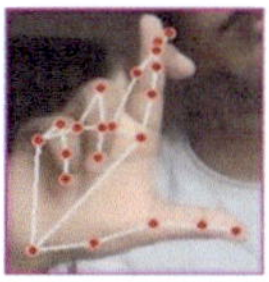	S	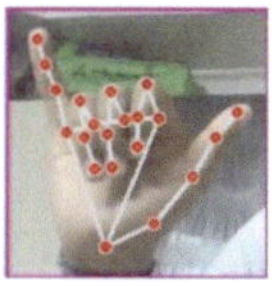	Y	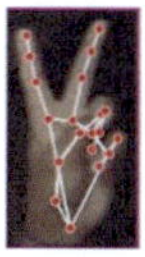	V
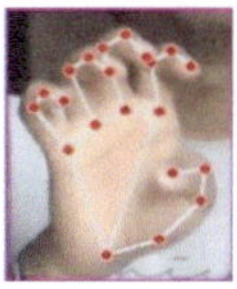	T	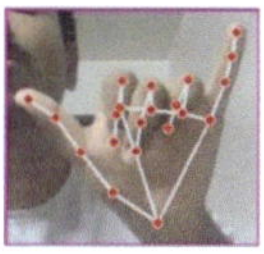	Z	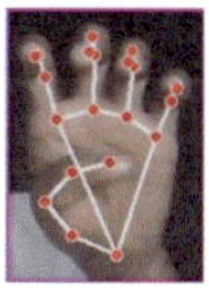	W

output dimensions of each layer is displayed in the second column, for instance, convolution2d_2 has an output shape of width and height of 60 each, and output channels are 64. The number of parameters to be trained is described in the third column in that table.

Table 5 presents the confusion matrix for the alphabets dataset. Table 6, on the other hand, displays the confusion matrix for the numeric dataset. Lastly, Table 7 represents the confusion matrix for the actions dataset, demonstrating the model's performance in recognizing actions within that dataset. Each of these tables provides valuable insights into the model's ability to classify and differentiate among the respective datasets.

Figure 5 provides the precision, recall, and F1 measures for each of the alphabet, numeric, and action datasets. For the alphabet dataset, the precision is 0.953, the recall value is 0.943, and the F1 measure is 0.948. Regarding the numeric dataset, the precision is 0.967, the recall value is 0.953, and the F1 measure is 0.960. Lastly, for the actions dataset, the precision is 0.975, the recall value is 0.961, and the F1 measure is 0.968.

Table 8 and Fig. 6 displays the accuracy results for three different datasets. The datasets are labeled as "Alphabet Dataset," "Numeric Dataset," and "Actions

Table 2 Symbols for each action

Symbol	Action
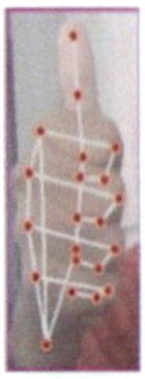	Enter
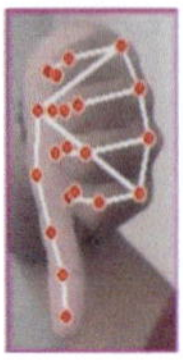	Backspace
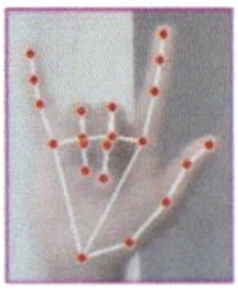	Next
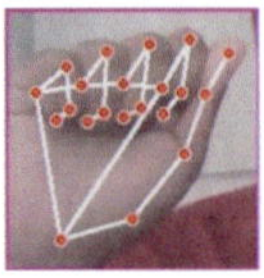	Submit

Table 3 Dataset details

Dataset name	Training	Testing
Alphabet dataset	12,988	1000
Numeric dataset	5177	1000
Actions dataset	2000	1000

Dataset." For each dataset, two types of accuracy metrics have been recorded: "Training Accuracy" and "Testing Accuracy". The Alphabet Dataset attained a training accuracy of 0.912 and a validation accuracy of 0.924. The Numeric Dataset displayed a training accuracy of 0.922 and a validation accuracy of 0.941. Finally, the Actions Dataset showcased a training accuracy of 0.933 and an outstanding validation accuracy of 0.953. These remarkable accuracy scores highlight the effectiveness of the models in precisely recognizing and classifying hand gestures for touchless interactions. By examining the tabulated data, it becomes evident that the "Actions Dataset" attains the most elevated training and testing accuracy figures, with the "Numeric Dataset" ranking second and the "Alphabet Dataset" coming in

Table 4 Parameter details of the proposed CNN model

Layer (Type)	Output shape	Param#
Input	(224, 224, 3)	0
Convolution2d_1 (convolution)	(None, 128, 128, 32)	
Max_Pooling2d_1	(None, 64, 64, 32)	0
Convolution2d_2	(None, 60, 60, 64)	
Max_Pooling2d_2 (max pooling)	(None, 30, 30, 128)	0
Convolution2d_3	(None, 28, 28, 128)	
Max_Pooling2d_3	(None, 15, 15, 256)	0
Convolution2d_4	(None, 13, 13, 256)	
Max_Pooling2d_4	(None, 7, 7, 512)	0
Flattening_1 (Flatten)	(None, 250, 88)	0
Dense_1 (Dense)	(None, 128)	32,11,392
Dense_2 (Dense)	(None, 40)	5160

Table 5 Confusion matrix for alphabet dataset

		Predicted	
	Classes	Positive	Negative
Actual	Positive	701	42
	Negative	34	223

Table 6 Confusion matrix for numeric dataset

		Predicted	
	Classes	Positive	Negative
Actual	Positive	706	35
	Negative	24	235

Table 7 Confusion matrix for action dataset

		Predicted	
	Classes	Positive	Negative
Actual	Positive	715	29
	Negative	18	238

third. This implies that the model exhibits superior predictive capabilities for the "Actions Dataset" when compared to the other two datasets.

In Fig. 7, the chart illustrates the model's training and validation precision over different epochs. The horizontal axis represents epoch numbers, ranging from 0 to 14, while the vertical axis displays the corresponding accuracy values. Blue line represents the model's training precision, starting from 0 and steadily increasing with each epoch. This positive trend indicates the model's capacity to effectively

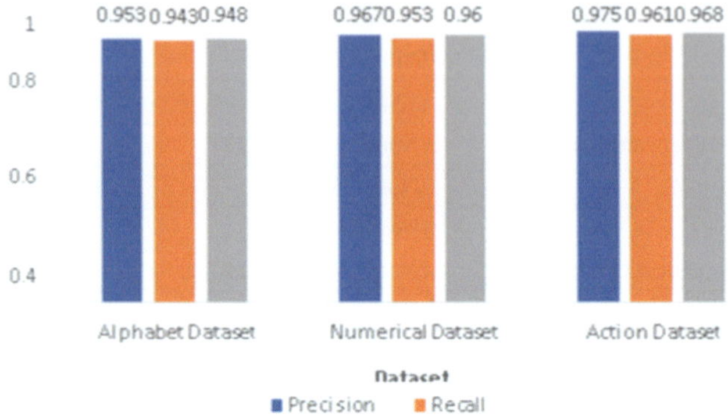

Fig. 5 Precision recall and F-measure on 3 datasets

Table 8 Accuracy details

Dataset	Training accuracy	Testing accuracy
Alphabet dataset	0.912	0.924
Numeric dataset	0.922	0.941
Actions dataset	0.933	0.953

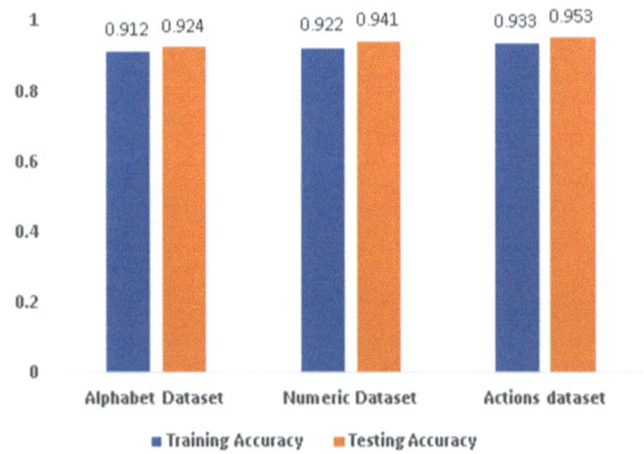

Fig. 6 Accuracy of proposed method

learn from the training data, leading to improved accuracy as the training progresses. Conversely, the orange line represents the model's validation precision, which also starts at 0 and shows a similar upward trend with increasing epochs. This observation suggests that the model not only does well on the training data but also demonstrates a strong ability to generalize effectively on unseen data. Throughout the graph, both accuracy lines show consistent and steady progress, indicating a positive learning curve. This implies that the model is progressively fine-tuning its predictions, leading to enhanced accuracy on both the training and testing datasets.

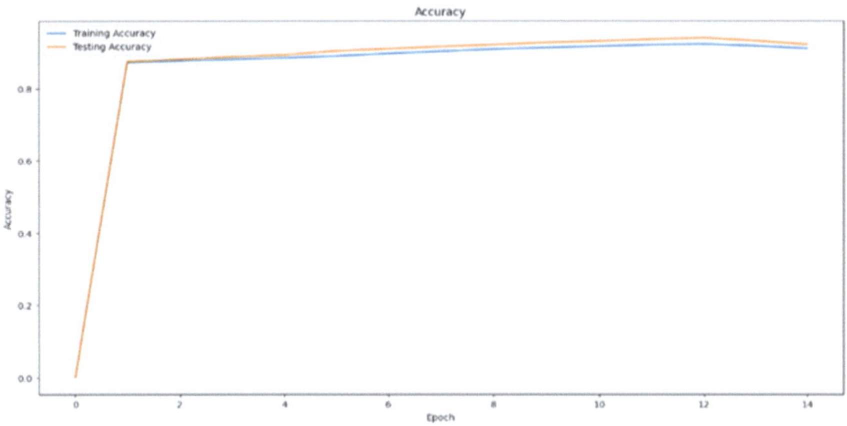

Fig. 7 Accuracy against epoch of proposed method

In Fig. 8, the graph represents the training and testing loss values of a model over various epochs. The x-axis denotes the epochs, ranging from 0 to 14, and the y-axis displays the corresponding loss values. The blue line on the graph represents the model's training loss at each epoch. As the epochs advance, the training loss consistently decreases, indicating that the model is effectively learning from the training data. This downward trend in the training loss suggests that the model is continuously improving its predictions and minimizing errors as it learns from the training data during the training process. The downward trajectory of the line suggests that the model is continuously refining its predictions and minimizing loss during the training process. The orange line signifies the model's testing loss at each epoch. Like the training loss, the testing loss also demonstrates a downward trend. This suggests that the model generalizes well to new, unseen data, as indicated by the reduction in loss values during the testing phase. The ability to reduce testing loss reflects the model's capability to make accurate predictions on data it has not encountered during training. In conclusion, the line graph provides valuable insights into the model's learning progress, showcasing the reduction in loss values throughout the training and testing phases. The graph's trends reveal the model's ability to learn from the data, adapt its predictions, and achieve higher accuracy on both training and testing datasets, making it an essential visualization for assessing the model's overall performance.

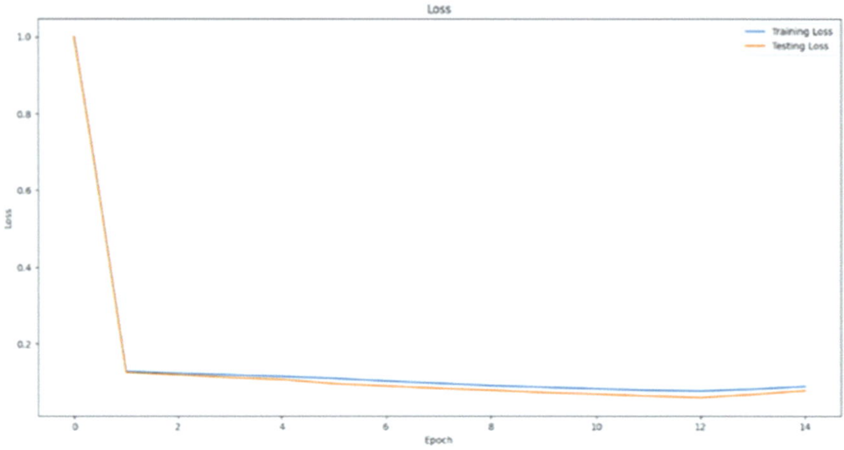

Fig. 8 Loss values of a model over various epochs

5 Conclusion

There are huge demands for touchless gesture-based user interface access especially during COVID-19-like diseases. The proposed method allows users to input characters or numbers and execute corresponding actions through their hand gestures. The proposed work designed the convolution neural network with the required layers and was well trained and tested on three real-time datasets. The proposed work yields a good accuracy with 93% as average on all three data sets. The proposed study trained with 30 different symbols including English alphabets (A-Z), digits (0–9), and some actions. In future, this method can be extended to support a greater number of symbols including lower- and upper-case alphabets (A-Z and a-z) and special characters, and also accuracy will be improved further.

References

1. Kunjumon J, Megalingam RK (2019) Hand gesture recognition system for translating indian sign language into text and speech. In: 2019 International Conference on Smart Systems and Inventive Technology (ICSSIT), Tirunelveli, India, pp 14–18. https://doi.org/10.1109/ICSSIT46314.2019.8987762
2. Ali A, Elmisery F, Mostafa R, Hussein M (2014) Motion control of robot by using kinect sensor. Res J Appl Sci Eng Technol 8:1384–1388. https://doi.org/10.19026/rjaset.8.1111
3. Gupta A, Kumar Y, Malhotra S (2015) Banking security system using hand gesture recognition. In: 2015 international conference on Recent Developments in Control, Automation and Power Engineering (RDCAPE), Noida, India, pp 243–246. https://doi.org/10.1109/RDCAPE.2015.7281403
4. Surekha P, Vitta N, Duggirala P, Ambadipudi VSS (2022) Hand gesture recognition and voice, text conversion using. In: 2022 Second International Conference on Artificial Intelligence and

Smart Energy (ICAIS), Coimbatore, India, pp 167–171. https://doi.org/10.1109/ICAIS53314.2022.9743064

5. León DG et al (2022) Video hand gestures recognition using depth camera and lightweight CNN. In: IEEE Sensors Journal, vol. 22, no. 14, pp. 14610–14619. https://doi.org/10.1109/JSEN.2022.3181518
6. Alon HD, Ligayo MAD, Melegrito MP, Franco Cunanan C, Uy II EE (2021) Deep-hand: A deep inference vision approach of recognizing a hand sign language using American alphabet. In: 2021 International Conference on Computational Intelligence and Knowledge Economy (ICCIKE), Dubai, United Arab Emirates, pp 373–377. https://doi.org/10.1109/ICCIKE51210.2021.9410803
7. Dong X et al (2018) A static hand gesture recognition model based on the improved centroid watershed algorithm and a dual-channel CNN. In: 2018 24th International Conference on Automation and Computing (ICAC), Newcastle Upon Tyne, UK, pp 1–6. https://doi.org/10.23919/IConAC.2018.8749063
8. Gu L, Yuan X, Ikenaga T (2012) Hand gesture interface based on improved adaptive hand area detection and contour signature. In: 2012 International symposium on intelligent signal processing and communications systems, Tamsui, Taiwan, pp 463–468. https://doi.org/10.1109/ISPACS.2012.6473534
9. Lionnie R, Timotius IK, Setyawan I (2011) An analysis of edge detection as a feature extractor in a hand gesture recognition system based on nearest neighbor. In: Proceedings of the 2011 international conference on electrical engineering and informatics, Bandung, Indonesia, pp 1–4. https://doi.org/10.1109/ICEEI.2011.6021611
10. Yu C, Wang X, Huang H, Shen J, Wu K (2010) Vision-based hand gesture recognition using combinational features. In: 2010 Sixth international conference on intelligent information hiding and multimedia signal processing, Darmstadt, Germany, pp 543–546. https://doi.org/10.1109/IIHMSP.2010.138
11. Yewale SK, Bharne PK (2011) Hand gesture recognition using different algorithms based on artificial neural network. In: 2011 international conference on Emerging Trends in Networks and Computer Communications (ETNCC), Udaipur, India, pp 287–292. https://doi.org/10.1109/ETNCC.2011.6255906

Privacy Preserving One Time Decryption Method for Distributed Environment

Kadiyala Ramana, R. Madana Mohana, A. Vijaya Krishna, S. Shobarani, and T. Satya Kiranmai

Abstract Privacy preservation during information exchange between distributed parties is a tedious task where extensive research has been going on for the past decade. Though anonymization, randomization, and generalization techniques protect the privacy of the data, they equally reduce the utilization of the data while applying data mining functions. So, protecting the privacy as well as not degrading the utility of the data has been a major challenge in Privacy Preservation Data Mining (PPDM). Cryptography in PPDM is one of the many available methods to protect privacy efficiently during information exchange. This paper describes a novel method called One Time Decryption (OTD) method which uses Advanced Encryption Standard (AES) for protecting the privacy of the owner's sensitive data as well as preserving the utility of the data in a distributed environment. The protocol only uses encryption and decryption algorithms over sensitive attributes and not all the attributes, the time consumed by the protocol is reduced to a very great extent.

Keywords AES · Cryptography · Data mining · Distributed environment · k-anonymity · Diversity · Perturbation · PPDM · Privacy · Randomization

1 Introduction

Every data mining function must satisfy an important property called Privacy Preserving Data Mining in order to protect the secrecy of the data. Let's assume a distributed environment where there are two parties. The parties are two separate medical institutions where each institution contains their own patient databases which

---

K. Ramana · R. M. Mohana (✉) · S. Shobarani · T. S. Kiranmai
Department of Artificial Intelligence and Data Science, Chaitanya Bharathi Institute of Technology, Hyderabad, Telangana, India
e-mail: rmmnaidu@gmail.com

A. V. Krishna
Department of Information Technology, G. Narayanamma Institute of Technology and Science, Hyderabad, Telangana, India

V. K. Gunjan et al. (eds.), *Cybernetics, Human Cognition, and Machine Learning in Communicative Applications*, Cognitive Science and Technology,
https://doi.org/10.1007/978-981-97-8533-9_24

have to be combined (Union) for certain data mining operations to get efficient results from the data. To get the results they choose a trusted third party (data miner) where he/she will apply data mining function to the combined data and retrieve the results. In this scenario, the privacy of the individuals in the database has to be protected from any parties involved in the system. The condition will fail if the combined databases do not hold the PPDM property. In some cases, after the data mining process has been completed if a party knows that about 50% of the overall patients in the combined database have some disease and if he also knows that more than 40% of the patients from his own database have the same disease then he can conclude that above 50% of the patients from the other party also have the same disease [1]. This does not leak any personal information of individuals and one can say that these kinds of leaks are unavoidable. But spilling of any personal or sensitive information from the data can be avoided if we use proper privacy preserving data mining property.

Generally, in privacy preservation, there will be a clear tradeoff between privacy and knowledge discovery. If the accuracy of the mining result is increased then there will be an obvious decline in the privacy of the data and vice versa. Three methods are used to protect privacy in data mining. The first method is called heuristic-based method and it can be applied normally in a centralized database system. The second method, a cryptographic approach, is used in a distributed environment. And the third approach called reconstruction method can be applied both in centralized as well as in distributed environments [2].

Privacy preserving data mining algorithms are designed to protect privacy even if there are unwanted adversaries present in the system and Taxonomy of PPDM Algorithms is shown in Fig. 1. Based on how adversaries learn secret information, we divide the adversaries into three types, semi-honest, malicious, and covert adversaries. Semihonest adversary follows the protocol honestly but tries to learn more information other than what he/she is intended to know. Malicious adversary completely deviates from the protocol and always learns new information from input data. In case of covert adversary, one cannot determine if the adversary is a semi-honest or malicious one. The following example describes semihonest and malicious adversaries. Let us assume that, in a distributed environment one of the parties needs to choose a random number. Now if it is a semihonest adversary then one can tell that the number was randomly chosen, but in the case of malicious adversary then he may choose the number to his advantage that will give him some extra information [1]. Normally it is simple to find a result for a semihonest adversary when compared to a malicious adversary. In OTD model, we consider the adversaries can either be semihonest or malicious, but protecting the privacy of the individual in any case is our final goal. PPDM for Data Mining Tasks summary is shown in Table 1.

Cryptography is a method used to encrypt the data into an unreadable format so even if the data got compromised there is no way for the intruder to understand the data without knowing the key for decryption. This method can also be used for protecting the privacy leakage while sharing the information in a distributed environment. Symmetric key cryptography and public key cryptography are the two types of cryptographic approaches available. Same key is used in both encryption and decryption in symmetric cryptosystems (Example: DES and AES). Two different

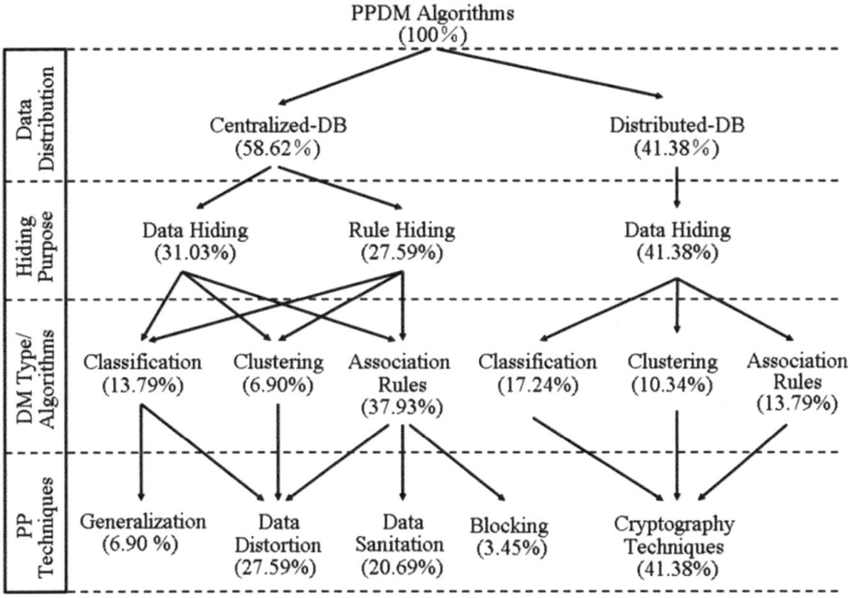

Fig. 1 PPDM algorithms taxonomy

Table 1 Summary of PPDM for data mining tasks

Data mining techniques	References	
	Centralized DB	Distributed DB
Classification	[3–6]	[7–11]
Association rules	[12–21]	[22, 8, 10, 23]
Clustering	[24, 25]	[26–28]

keys are used for encryption and decryption in asymmetric cryptosystems (Example: RSA and ECC). This method is also known as public key cryptography because from the two keys used one of the keys (Encryption key) is made public to everyone for encryption. This is not harmful for the data owners because only the private key is used to decrypt the ciphered data.

In our proposed OTD method, Advanced Encryption Standard (AES) cryptographic method plays an important role in protecting the privacy. AES is a block cipher cryptographical method, so it encrypts data by blocks. One other method available other than block cipher is called stream cipher. Both methods have advantages but when security is concerned block cipher provides better security than stream cipher. In block cipher cryptography there are other methods like DES, Triple DES, Blowfish, etc. DES is a traditional cryptographical method where it uses a small key of size 56 bit for encrypting the data. As it uses a small key, it can be broken easily by brute force attack itself [29]. Triple DES is a method which uses DES three times each time with a different key. This method may provide better security as it uses three different keys in each instance of DES but will consume a lot of time to compute the entire process in software implementation because DES was created to work with hardware [29]. Now while comparing AES with another symmetric key cryptographic techniques like Blowfish, Twofish, or Serpent, AES is the widely used method from top-secret government documents to online banking transactions. Although techniques have been implemented to break AES cryptographic methods like Biclique cryptanalysis, these techniques cannot be practically applied in the real world because a technique like this still takes trillion years to recover the AES key [30]. When compared to symmetric key cryptography, public key cryptographic system uses two different keys for encryption and decryption. In public key cryptography method, the public key can be distributed to everyone, but the private key needs to be with the data owner.

AES is a classic and powerful encrypting technique designed in 1997 by National Institute of Standard and Technology (NIST). US government's National Security Agency (NSA) uses AES to protect classified information. AES is called as symmetric key cryptographic system because same key is used in both encryption and decryption process. There are three types of AES, AES-128, AES-192 and AES-256. These types are based on the key size used in AES algorithm. Several rounds of process are included in AES algorithm, 10 rounds in AES-128-bit algorithm, 12 rounds in AES-192-bit algorithm and 14 rounds in AES-256-bit algorithm. Each round contains three functions namely substitution, transposition, and mixing of input plaintext. There is no separate algorithm for decryption of the cipher text. The same encryption algorithm is used in the reverse order for decryption.

In our work, AES is an important component for encrypting the sensitive information. The key size we use here is a 256-bit key. The larger the key size the higher the security. AES encryption is carried out in the first phase of the work itself. The parties who possess the actual data encrypt their data using the AES algorithm and only the cipher data is sent to the data miner for computing the function.

Cryptography makes privacy possible by limiting the information to be intelligible for only the authorized person. Hence cryptography can be used extensively for protecting the privacy in a distributed environment. This paper explains many available techniques for achieving PPDM and the implementation of a simple and effective cryptographic PPDM model for protecting the privacy leakage as well as maintaining the utility of the data to a greater extent.

Motivation

- In the distributed environment for protecting the secrecy of the data and privacy leakage, the Privacy Preserving Data Mining algorithm role is important for every data mining function.
- Usually while preserving the privacy there will be a clear tradeoff between privacy and knowledge discovery. If the accuracy of the mining result is increased then there will be an obvious decline in the privacy of the data and vice versa.
- Even in the presence of a lot of adversaries Privacy Preserving Data Mining algorithms are intended to protect the privacy.

- In the distributed environment to protect the privacy cryptographic approach is used in conjunction with Privacy Preserving Data Mining algorithms.

Contribution

- One Time Decryption Method is proposed.
- A 256-bit key-based AES cryptographic method is incorporated into the proposed privacy preserving technic in a distributed environment.
- Privacy is preserved and information is securely transmitted in the distributed environment.
- Many available Privacy Preserving Data Mining techniques are discussed in this paper.
- Cryptographic Privacy Preserving Data Mining Technique is proposed to deal with leakage in the privacy and protection of data and to a greater extent data utility is maintained.

The rest of the paper is organized as follows. Section 2 describes the related work in the area of privacy preserving data mining. Section 3 explains the definition of the problem. Section 4 describes the contribution of our work to PPDM. Section 5 explains the experimentation and results and finally, we conclude in Section 8.

2 Related Work

In this digital era, data collection has increased enormously. Extraction of information from the collected data is essential for many useful purposes. Publishing microdata such as medical data for information extraction is risky because it may leak sensitive information about the individual. Microdata is a dataset that contains attributes about an individual such as identity attributes, sensitive attributes, and Quasi Identifiers. Quasi Identifiers are a combination of attributes (Age, Gender, Postal Code) which may reveal individual identity. Prior to publishing the individual identity attribute (Name or Social Security Number) must be removed or masked from the microdata to avoid revealing sensitive information about any individual. Even though the identity information is masked or removed the microdata may disclose sensitive information if it undergoes a linking attack. Table 2 provides an example for masked microdata.

Masked microdata will not reveal the identity of any individual directly even though it does not hide the sensitive attributes. But combining the microdata with publically available databases like voter's database may disclose sensitive data of an individual. Public datasets such as voter's dataset does not have any sensitive attributes, and so there are no privacy constraints to protect the dataset. Tables 3 and 4 explain about linking attack.

The medical microdata in Table 3 contains sensitive information Disease, but as it does not have identity attribute, it is not possible for the intruder to discover which individual has which disease. Table 4 mentions voter's database which is available online publically. As it does not contain any sensitive information this dataset is

Table 2 Microdata

ID	SSN	Name	Zip code	Race	Age	Gen	Bal	Sal	Int
1	****	****	629,166	Indian	30	F	9 K	80 K	1 K
2	****	****	629,166	Indian	40	M	14 K	65 K	16 K
3	****	****	603,202	Black	40	F	49 K	110 K	32 K
4	****	****	603,202	White	36	F	44 K	21 K	11 K
5	****	****	629,166	Indian	40	F	79 K	100 K	12 K
…	…	…	…	…	…	…	…	…	…

Table 3 Medical microdata

ID	Age	Gender	Zip code	Disease
1	20	Male	11,000	Pneumonia
2	40	Male	13,000	Flu
3	50	Male	12,000	Flu
4	60	Female	54,000	Pneumonia

Table 4 Voter's data

ID	Name	Age	Gender	Zip code
1	Ramamoorthy	20	Male	11,000
2	Sathish	30	Male	11,000
3	Jose	50	Male	12,000
4	Cathy	60	Female	25,000

not masked or protected in any case. Now if the intruder combines this publically available database with the medical microdata he/she can find which individual has which disease because the quasi-identifier attribute (Age, Gender, Zip) of both the datasets are the same. For example, the first record of medical microdata says that a male of age 20 living in postal code 11000 has pneumonia, but it does not say which particular person has the disease. The first record of voter data says that a male of age 20 living in postal code 11000 is Ram. Comparing both the record the intruder can conclude that Ram is suffering from the disease pneumonia. To prevent this kind of information leakage scientists found out a new notion called Privacy Preserving Data Mining.

Agarwal and Srikant [31] and Lindell and Pinkas [32] were the scientists who first introduced Privacy Preserving notion in data mining. We can classify the existing PPDM techniques into two broad categories which are data perturbation and data distribution. Agarwal and Srikant described the technique of data perturbation and reconstruction of data at an aggregate level to do the mining operations. From the trained data a decision tree classifier model will be created, and every individual value of the record will be applied with the perturbation technique.

Sweeney in 2002 [33] addressed the linking attack problem explained using the Tables 2 and 3 by developing a method called k-anonymity. K-anonymity is achieved if every record in the table is indistinguishable from k-1 other records with respect to the quasi-identifier attribute. K-anonymity is obtained by applying generalization and suppression methods over the data. Generalization is the method where the data is transformed from being more person specific to less person specific. Generalization is guaranteed based on hierarchy. Generalization for k-anonymity is explained using Tables 5 and 6.

Table 5 describes an education microdata where it says which grade students need to buy a computer and who needs not. This is more person specific because any intruder can easily find out who buys computer and who does not. Figure 2 provides us with the hierarchy of education which is used to generalize Table 6 to make it less person specific data [34]. The education microdata is generalized based on the education attribute and is provided in Table 6. Generalization sometimes makes the data less usable if the generalization is done inside the data to the maximum extent. For example, in Table 6 the first two records can be further generalized to secondary level according to Fig. 2 (hierarchy of education). But doing that makes the data analyzer confused whether the students belong to junior level or senior level. So minimal generalization is preferred in k-anonymization.

K-Anonymized data for the medical microdata provided in Table 3 is given in Table 7. The age and zip attributes are generalized to achieve a three anonymous table. Table 3 is not anonymous because an intruder can conclude that a male of age 20 living in 11000 is suffering from pneumonia. But in the anonymized table, there are two distinct diseases that come under for a male of age 20 living in 11000. So, the intruder can't finalize which disease he/she has.

Table 5 Education microdata

Education	Gender	Working hours	Buys computer
9	Male	30	No
9	Male	30	No
10	Female	32	No
10	Female	32	No
Masters	Male	40	Yes
Masters	Male	40	Yes
Doctorate	Female	44	Yes

Table 6 Generalized education microdata

Education	Gender	Working hours	Buys computer
Junior	Male	30	0Y2N
Junior	Female	32	0Y2N
Grad school	Male	40	2Y0N
Grad school	Female	44	1Y0N

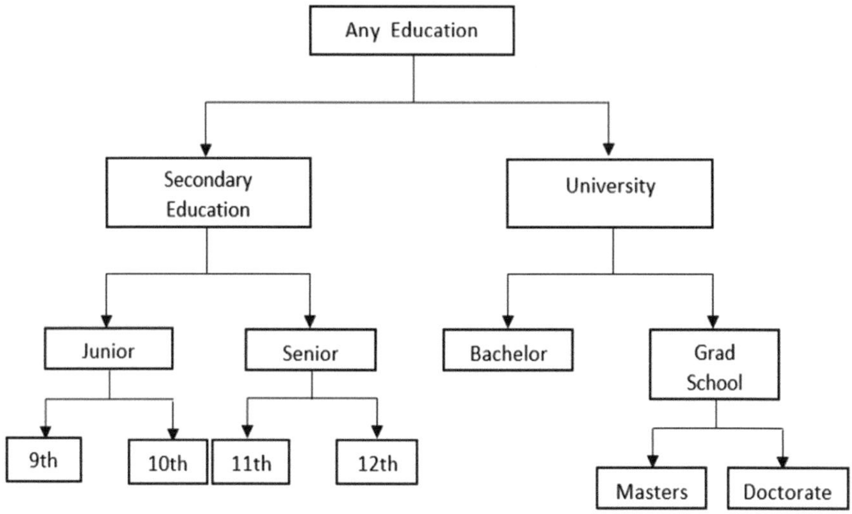

Fig. 2 Hierarchy of education

Table 7 Anonymized medical microdata

ID	Age	Gender	Zip code	Disease
1	[20, 50]	Male	[11000, 55000]	Pneumonia
2	[20, 50]	Male	[11000, 55000]	Flu
3	[20, 50]	Male	[11000, 55000]	Flu
4	[60, 80]	Female	[11000, 55000]	Pneumonia
5	[60, 80]	Female	[11000, 55000]	Pneumonia
6	[60, 80]	Female	[11000, 55000]	Flu

Sometimes the k-anonymized table often falls for two kinds of attack named as homogeneity attack and background knowledge attack. If the sensitive values in an equivalence class lack diversity then it is called homogeneity attack. If the attacker cracks the sensitive value of an individual by knowing some background information, then it is called background knowledge attack. Tables 8 and 9 describe homogeneity and background knowledge attack respectively.

In a microdata, if records have same quasi-identifier values then all the records which come under the same quasi-identifier values are considered to be in an equivalence class. In Table 8 the first three records come under an equivalence class (say, equivalence class 1) and the last three records come under another equivalence class (say, equivalence class 2). In this table, all the individuals from equivalence class 1 are suffering from Flu. So, the intruder can easily come to the conclusion that any male between age 20 to 50 living under the zip code 11000 or 55000 is suffering from the flu.

Table 8 Homogeneity attack

ID	Age	Gender	Zipcode	Disease
1	[20, 50]	Male	[11000, 55000]	Flu
2	[20, 50]	Male	[11000, 55000]	Flu
3	[20, 50]	Male	[11000, 55000]	Flu
4	[60, 80]	Female	[11000, 55000]	Pneumonia
5	[60, 80]	Female	[11000, 55000]	Pneumonia
6	[60, 80]	Female	[11000, 55000]	Flu

Table 9 Background knowledge attack

Id	Zipcode	Age	Nation	Disease
1	13,000, 14,000	< 30	****	Heart problem
2	13,000, 14,000	< 30	****	Heart problem
3	13,000, 14,000	< 30	****	Viral fever
4	13,000, 14,000	< 30	****	Viral fever
5	14,001, 15,000	40	****	Cancer
6	14,001, 15,000	> 40	****	Heart problem
7	14,001, 15,000	> 40	****	Viral fever
8	14,001, 15,000	> 40	****	Viral fever

Even after applying k-anonymity solution to the microdata sometimes the attacker can find the sensitive data if the attacker has some background knowledge about the individual. For example, in Table 9 the last three records come under an equivalence class. It says that a person either from zip 14001 or zip 15000 with age greater than 40 might be suffering from heart problem or viral fever.

Now let us assume that the attacker needs to find the disease of a particular individual called Umeo from that equivalence class and the attacker knows that he is from Japan. Normally people from Japan are less prone to Heart problems. With this kind of statistics, the attacker can conclude that Umeo is suffering from viral fever.

Gehrke et al. [35] in 2006 derived a method called l-diversity which brings a solution for the homogeneity attack and background knowledge attack. An equivalence class is said to be l-diverse if it has l well-represented values for the sensitive attributes. A table is said to be l-diverse if all the equivalence class in that table is l-diverse. Tables 10, 11 and 12 explain the principle of l-diversity.

Table 10 provides a medical data, and gives details about the disease of individuals. The anonymized data for health care microdata is provided in Table 10. Even after anonymization the data in Table 10 leaks sensitive information in some way. For example, consider two neighbors Alice and Bob. Bob knows that Alice is suffering from a serious disease and he needs to find out what disease Alice is suffering from. Bob finds out the recent anonymized inpatient data (Table 11) released by the hospital where Alice is getting the treatment from. As Alice and Bob are neighbors Bob knows

Table 10 Health care microdata

Id	Zipcode	Age	Nation	Disease
1	12,051	47	Indian	Heart problem
2	12,067	48	African	Heart problem
3	12,067	42	Chinese	Viral fever
4	12,051	44	African	Viral fever
5	15,852	69	Russian	Tumor
6	15,852	74	Indian	Heart problem
7	15,851	66	African	Viral fever
8	15,851	68	African	Viral Fever
9	15,051	52	African	Tumor
10	15,051	56	Russian	Tumor
11	12,067	55	Chinese	Tumor
12	12,067	54	African	Tumor

Table 11 4-anonymous health care microdata

ID	Zip code	Age	Nation	Disease
1	120**	< 50	*****	Heart problem
2	120**	< 50	*****	Heart problem
3	120**	< 50	*****	Viral fever
4	120**	< 50	*****	Viral fever
5	1585*	> = 60	*****	Tumor
6	1585*	> = 60	*****	Heart problem
7	1585*	> = 60	*****	Viral fever
8	1585*	> = 60	*****	Viral fever
9	120**	5*	*****	Tumor
10	120**	5*	*****	Tumor
11	120**	5*	*****	Tumor
12	120**	5*	*****	Tumor

that Alice is a 52-year-old female living in the postal code 12051. Knowing these details, Bob knows that the information he wants is in the records 9th to 12th of Table 11. Even though the table is anonymized the sensitive values are same in all the four records. So, Bob can conclude that Alice is suffering from cancer. To overcome these disadvantages L-diversity principle is applied on the same dataset to protect the sensitive data and is given in Table 11. L-diversity says that, a dataset is l-diverse if all the equivalence class in the dataset must have l distinct values. With Table 12 in hand, Bob knows that the information he needs is in the first four records. But now there are three distinct values in the sensitive value column and so the attacker can't identify which disease Alice is suffering from. In Table 12 all the

Table 12 3-diverse health care data

ID	Zip code	Age	Nation	Disease
1	1205*	< = 60	*****	Heart problem
2	1205*	< = 60	*****	Viral fever
3	1205*	< = 60	*****	Tumor
4	1205*	< = 60	*****	Tumor
5	1585*	> 60	*****	Tumor
6	1585*	> 60	*****	Heart Problem
7	1585*	> 60	*****	Viral fever
8	1585*	> 60	*****	Viral fever
9	1206*	< = 60	*****	Heart Problem
10	1206*	< = 60	*****	Viral fever
11	1206*	< = 60	*****	Tumor
12	1206*	< = 60	*****	Tumor

equivalence classes have atleast three distinct values. Hence it is called as 3-Diverse Health Care Data.

Li et al. [36] in 2007 proposed an attack called similarity attack on the l-diversity applied datasets. Similarity attack is that if the sensitive values in an equivalence class of a dataset are distinct but similar in meaning then there are possibilities for the attacker to learn secret information. Table 13 explains about similarity attack.

Table 13 shows an Income/Disease dataset which satisfies l-diversity principle by having three distinct values in both the sensitive attributes of Income and Disease. In this dataset both the sensitive attributes get leaked to the attacker. Let's consider the same scenario Alice and Bob which was mentioned earlier. Bob knows that Alice's data comes under the last three records according to the background knowledge he has. The income varies from 2K to 4K in these records. From this Bob can conclude that Alice's salary is relatively less. Disease attribute also have three distinct values

Table 13 Diverse income/disease data

ID	Zip code	Age	Income	Disease
1	586**	3*	2000	Ulcer
2	586**	3*	3000	Diarrhea
3	586**	3*	4000	Stomach Cancer
4	5890*	> = 50	6000	Gastritis
5	5890*	> = 50	12,000	Fever
6	5890*	> = 50	9000	Lung disease
7	586**	4*	8000	Lung disease
8	586**	4*	10,000	Pneumonia
9	586**	4*	11,000	Stomach Cancer

Ulcer, Diarrhea, and Stomach cancer. But even though there are three distinct values all these diseases are related to stomach disease. So, Bob can come to the conclusion that Alice is suffering from stomach related disease.

Authors of [36] developed a technique called t-closeness to overcome similarity attack. t-closeness says that the distance between probability distribution of sensitive attributes in an equivalence class and the distance between the probability distribution of sensitive attribute values in the whole table does not differ more than a threshold t. Table 14 gives the t-closeness dataset of Table 12 which protects the data against similarity attack. As we see the first equivalence class (First three records) in Table 14 has three distinct values and not all the values are semantically similar. So Bob can't predict the disease of Alice.

Agarwal and Aggarwal [37] in 2001 introduced a new method based on perturbation for preserving the privacy of the mined data. Perturbation and discretization are their important methods. The authors use an algorithm called Expectation Maximization algorithm for distribution reconstruction which reduces the loss of information during reconstruction to a greater extent. [38, 39] explain how categorical and binary data can be efficiently reconstructed. Association rule mining is the algorithm used in [38, 39]. These papers use randomization techniques to preserve the privacy of the data and to make the utility of the data effective. Evfimieski, Srikant, Agrawal, and Gehrke in [38] deal with a special randomization method called select-a-size method for protecting the privacy in association rule mining. Paper [3] authored by Zhan and Du builds up decision trees using randomization technique for protecting privacy while mining data from randomized data.

Randomization is a technique where some of the original pieces of information are removed from the data and some unwanted information gets introduced into the data. Additive noise-based randomization is a technique which is used to perturb numerical data and is given in (1),

$$Y = X + E \tag{1}$$

Table 14 t—closeness income/disease dataset

Id	Zip code	Age	Income	Disease
1	5867*	< = 50	2000	Ulcer
2	5867*	< = 50	4000	Stomach cancer
3	5867*	< = 50	10,000	Pneumonia
4	5890*	> = 50	6000	Gastritis
5	5890*	> = 50	12,000	Fever
6	5890*	> = 50	9000	Lung disease
7	5860*	> = 50	3000	Diarrhea
8	5860*	> = 50	8000	Lung disease
9	5860*	> = 50	10,000	Stomach cancer

where Y is the randomized data, X is the original data and E is the noise added. The noise added is independent of the original value X. An example for additive noise-based randomization is provided in Tables 15 and 16 respectively.

The data distribution approach is mainly based on cryptographic approach [41, 42]. Cryptography is also another major area where privacy of data can be preserved during data mining. Lots of cryptographic methods are available for Privacy Preserving Data Mining to find out solution for secure multiparty computation (SMC) [43–45]. SMC is that at the end of the computation no person involved knows anything except their own input. There are more than two parties involved in secure multiparty computation and each party will have their own secret input. They need to combine their inputs to get some important result but each party is not willing to compromise their own secret input to other party. The main challenge here is to proceed with the computation by protecting the privacy of the secret input. Benny Pinkas [1] used oblivious transfer protocol for privacy preserving data mining. Let's see a simple 1 out of 2 oblivious transfer protocol in the following example.

1 out of 2 oblivious transfer protocol is that, only one of the inputs is received by the receiver and the sender does not know which input the receiver received. Consider two entities Alice and Bob. Alice has two inputs w_0 and w_1 and he needs to transfer one of these inputs to Bob. For this, initially, Alice creates two random inputs X_0 and $X_{1,}$ and sends it to Bob. This protocol uses RSA cryptosystem where Alice generates a public key (e) and a private key (d) using a value n, where n is the product of two prime numbers. Assume Bob knows the public key generated by Alice before the protocol starts. Bob picks a value b where b Є {0,1}. Now, with 'b' in hand, Bob picks either x_0 or x_1 and is denoted as x_b. Bob generates another random value r and encrypts x_b with Alice's public key e using (2).

Table 15 Loan dataset

Id	Balance	Income	Loan (L)	Interest
1	10,000	87,000	10	8000
2	15,000	75,000	5	4000
3	50,000	120,000	20	16,000
4	30,000	95,000	12	9000
5	40,000	100,000	15	12,000

Table 16 Randomized loan dataset

ID	Balance	Income	Loan	Interest
1	17.334	88.759	10.099	9.150
2	19.199	82.537	12.030	4.030
3	59.199	128.447	23.678	18.341
4	39.048	100.692	18.318	10.178
5	48.920	103.681	20.132	12.999

$$v = (x_b + r^e) \bmod n \tag{2}$$

Bob sends the value v to Alice, who then decrypts v to generate two values, K_0 (3) and K_1 (4) with the help of his private key d and the random value x_0 and x_1 respectively.

$$K_0 = (v - x_0)^d \bmod n \tag{3}$$

$$K_1 = (v - x_1)^d \bmod n \tag{4}$$

One of the values, K_0 and K_1, one of the values is equal to k which is generated by Bob but Alice does not know which value is r. Now, Alice adds his secret messages w_0 and w_1 with K_0 and K_1 and sends the value to Bob. These equations are represented by (5) and (6).

$$W'_0 = K_0 + w_0 \tag{5}$$

$$W'_1 = K_1 + w_1 \tag{6}$$

Bob knows which of these two values contains the actual secret message. He then finds $w_b = w_b$'—k, which is the value Alice needs to transfer to Bob.

In [46], Craig Gentry proposed a new solution called fully homomorphic encryption scheme which can apply function over encrypted data and can produce the result which is the same as the function applied over original data without being decrypted. Later, by using this method scientists in [47, 48] developed homomorphic based k-nearest neighbor classification algorithm and Naïve-Bayesian classification algorithm which can perform secure classification operation over distributed environment without decrypting the encrypted data for the protection of sensitive information.

Azimi et al. have shown a k-implied common security saving for grouping calculation. This technique includes two calculations for protection. Firstly, every participant is used to discover the close cluster while the bunch focuses remain discreet to their members; secondly, the cluster focuses on the cluster without releasing any bunch focus data to the members, while keeping each member from making different individuals' sense in a similar cluster [49]. In addition, Nayahi and Kavitha proposed a privacy-conserving high-order c-meaning algorithm with cloud computing for big data clustering. This produced better privacy results compared to other methods with the Possibilistic c-Means algorithm [50]. Similarly, Yang et al. introduced Future Internet Route Decision Modeling Data-driven Model. Here, they analyze the decision-making process for inter-domain routing and then present a prefix level route prediction model [51].

Tian et al. clarified real-time lateral motion detection based on the edge computing environment evidence reasoning network. Firstly, the concept of the association of vulnerability is discussed. Taking into account the vulnerability and ecological data of the system architecture, the evidence thinking system is developed and the proof-thinking system's sidelong creation thinking capability is then employed [52]. Likewise, Tian et al. using the block chain to solve the big data and privacy problems with a loose coupling architecture, explained a stable digital proofs system. Proof and evidence knowledge were stored separately in this process. Moreover, information is stored in the blockchain evidence and trusted storage platform evidence is stored [53]. In addition, Tian et al. have clarified the modeling of potential Internet route decision-making methods, using a deep learning system, which learns, understands, and models the route decision process without prior knowledge [54].

In our work, we have provided a cryptographic approach for privacy preserving data mining using AES algorithm. In this method, there are four parties. Two parties who have the actual data where they encrypt and send the data to another party (data miner) who is going to apply the function over the encrypted data and provide the mined results to the data owners. The final party is named as the administrator who generates the key for AES algorithm and keeps the computation algorithm. Whether the adversary is a semi-honest or malicious party, protecting the privacy from any of them is our strong goal.

3 Problem Definition

Consider two entities, Alice and Bob, who have their own private datasets (say D_1 and D_2). They need to perform a data mining function by combining the two datasets so that they will get some useful results from the data. Let's say that the datasets D_1 and D_2 consist of n number of records $t_1, t_2, \ldots. t_n$ and m number of attributes. We assume that both datasets D_1 and D_2 have same number of attributes and also same set of attributes.

The privacy constraint here is that the two entities, Alice and Bob, are not willing to disclose their private inputs, D_1 and D_2, to each other. They only share the result c_q that they get after applying the function, where c_q is the result the data miner gets after applying the function. For this kind of problem, we have created a novel system which will take two encrypted datasets, D_1'and D_2' as inputs and produce results c_q without compromising the actual data or any sensitive data of any individual.

In case of classification data mining algorithm there is one more input provided called query data set, D_q. Query dataset can have any number of records $t_{q1}, \ldots. t_{qn}$, but it should have the same set of attributes as D_1 and D_2 respectively. For classification there will be (m+1) attributes in D_1, D_2, where the (m+1)th attribute is called as the class label. Initially, a model will be created using the datasets D_1 and D_2 and then this model will be used to find the class value (c_q) of the query dataset D_q.

4 Proposed Method

In this research, we propose a novel system, a secure protocol to find the result of a data mining function without compromising the input of the users. In our protocol, the encrypted data is outsourced to a third party (Data Miner). The Data Miner performs the function over encrypted data and gets the final result, c_q. Finally, the result c_q is provided to the desired users. In this protocol, four parties are involved during the entire process. Two parties (party1 and party2) are considered as the users of the system. The other two parties (party3 and party4) are involved in securely computing the inputs of the first two parties and producing the output. Even though party3 and party4 participate in secure computation, both play different roles and they too are not eligible to read the original input of the users. Party3 is the administrator of the whole system where he/she holds the algorithm of the secure computation. Another job of party3 is that he/she produces a unique key for the users of the system which is then used by party1 and party2 to encrypt their sensitive attributes of input data. The secure computation algorithm kept by the administrator is given to the Data Miner (party4), when he needs. The Data Miner, after getting the computation algorithm from the administrator, processes the encrypted inputs from the user to produce the result. The protocol mentioned in this paper meets the following privacy requirements:

1. Party1 does not read the input of party2 and vice versa.
2. The administrator will not get hold of any input information because he/she is involved only in providing the encryption key and the computation algorithm.
3. The Data Miner can't read the input data from the users because they are in the encrypted form.

The users of the system encrypt their input data individually and send the encrypted datasets, D_1'and D_2' directly to the Data Miner. So, party1 does not receive the input of party2 and also party2 does not receive the input of party1 in any form. Administrator creates the encryption key each time the user needs to apply data mining function and provides the key to party1 and party2 respectively. Also, administrator passes on the computation algorithm to the Data Miner for securely computing the output. During this process, the input datasets are not given to the administrator in any form, so that the input data is secure from party3. The Data Miner receives only the encrypted input from the users and the computation algorithm he/she uses for getting the result c_q, will not disclose the original dataset D_1 and D_2 to the Data Miner.

4.1 OTD Protocol

The One Time Decryption protocol we are using in this research employs cryptographic algorithm as a main component to protect the privacy of the inputs provided by the users. The traditional cryptographic system that we use in our model is called Advanced Encryption Standard. The existing models that we discussed in Section (2), do not use any kind of traditional cryptographic technique because the encryption and decryption algorithms are considered as time-consuming. In our model, we consider only a few attributes from a set of m attributes as sensitive. Hence we only encrypt the sensitive attributes, thus reducing the time consumed by the encryption and decryption algorithm. The model we propose protects the privacy of the inputs for any data mining functions that we apply over the data sets. For experimentation, we tried implementing classification algorithm. Classification algorithm includes one more input which is called a query dataset (D_q), for which the class labels have to be predicted using the model created from the remaining inputs D_1 and D_2. Before providing the query data set to the data miner, the data set should be encrypted using the key provided by the administrator, so that the sensitive attribute from the query data set is not seen by the data miner.

Figure 3 explains the complete architecture of our model. This system includes four parties (party1, party2, party3, party4). Party1 and party2 hold the input data (D_1 and D_2). Party3 (administrator) generates the key used for AES cryptosystem and provides it to party1 and party2, and also provides the computation algorithm to party4 (data miner). Party4 is responsible for applying the function over the data and gives the results to party1 and party2 respectively. The users of the system (party1 and party2) need to combine their inputs to get some useful results but they are not willing to share the secret information from their data with each other. To use OTD protocol initially, users request for a key to encrypt their data using Advanced Encryption Standard from the administrator. The key the administrator generates is a 256-bit key. After getting the key from the administrator, party1 and party2 encrypt the sensitive attributes from their data (D_1 and D_2) using AES cryptographic system. After encryption, the data D_1 and D_2 are denoted as D_1'and D_2' respectively. Now, D_1'and D_2' will be given to the data miner (party4).

4.2 *Threat Model*

We have considered the definitions of security from the literature we have surveyed about secure multiparty computation [45, 55] to show the various possible attacks on our protocols. The threat models available in SMC are semi-honest, covert, and malicious adversaries. In this protocol, we assume that adversary is either semi-honest or malicious. To demonstrate that our system is secure from any adversary, we ought to prove that our protocol does not spill any sensitive private information of an individual from the input data provided by the parties.

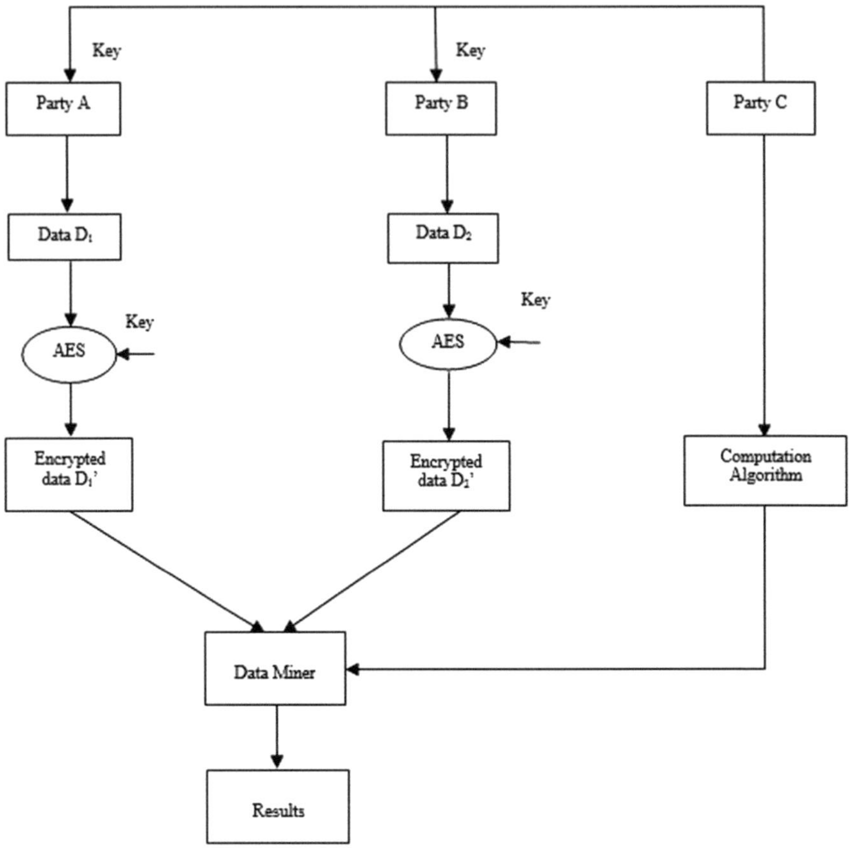

Fig. 3 Architecture

4.3 AES Encryption

Figure 4 shows various stages in AES algorithm and each round of the AES encryption process is composed of the following four transformation functions:

- SubBytes: This is a nonlinear byte substitution that operates independently on each byte of the State. Here, we denote this function as SB, and we denote Inverse SubBytes as SB-1.
- ShiftRows: Each row of the State is cyclically shifted with different offsets. We denote ShiftRows and its inverse as SR and SR-1, respectively.
- MixColumns: This is a linear transformation of each column of the State. We denote this function as MC and its inverse as MC-1.
- AddRoundKey: This is a bitwise XOR operation with each 128-bit round key.

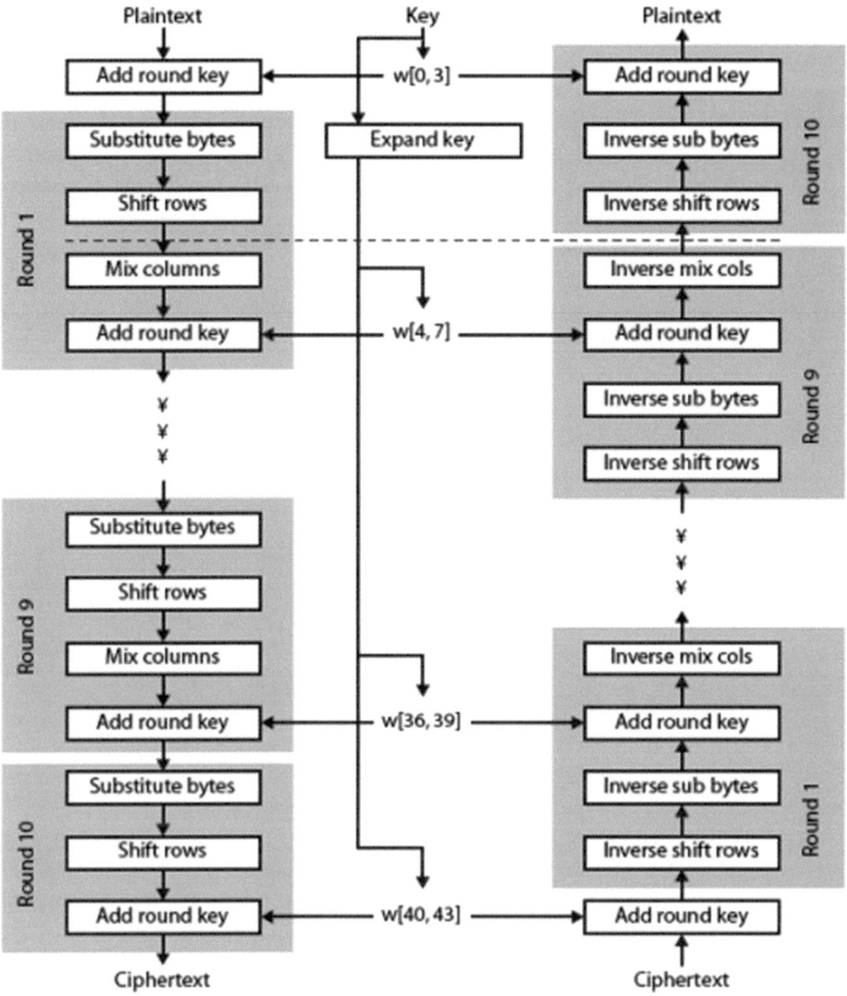

Fig. 4 AES algorithm

The 128-bit AES algorithm takes the 128-bit master secret key, denoted as SK, and performs a key expansion process to generate 11 round keys. Each 128-bit round key is composed of 4-byte words, denoted W[i]. The AES key expansion process consists of three transformation functions as follows:

- RotWord: This function takes a word $[a_0, a_1, a_2, a_3]$ as an input, performs a cyclic permutation, and returns the word $[a_1, a_2, a_3, a_0]$.
- SubWord: This is a function that takes a word composed of 4 bytes and applies the SB to each byte.
- Rcon[i]: This is a round constant word given by $[x^{i-1}, \{00\}, \{00\}, \{00\}]$. Here, is representing powers of x (x is denoted as {02}) in the field $GF(2^8)$.

The data miner has all the inputs now and because all the sensitive information from the data is encrypted, he/she won't get any useful information from the data. The input data given to the data miner should be equal both in the number of attributes as well as in their labels. After getting the masked input data miner combines the data sets D_1'and D_2' horizontally for applying the data mining function. To apply the function, party4 requests the computation algorithm from the administrator. Administrator encloses the key that has been generated for encrypting the original inputs D_1 and D_2 with the computation algorithm and sends it to the data miner. Computation algorithm is a package of key-enclosed AES decryption algorithm and the intended data mining function. Now, data miner gives the encrypted inputs D_1'and D_2' to the computation algorithm to get the intended result. Computation algorithm first decrypts the encrypted sensitive data before applying the data mining function. As the computation algorithm is a closed one, data miner cannot acquire the key embedded in the computation algorithm as well as he cannot see the process occurring inside the computation algorithm thereby protecting the input data given by the user. Data miner can only see the mining results coming out of the computation algorithm. Finally, the results will be provided to the intended users (party1 and party2). The OTD model we have proposed is fully capable of satisfying the privacy requirements mentioned in Sect. 4.

$$f(x) = b_{n-1}x^{n-1} + b_{n-2}x^{n-2} + \cdots + b_1 x + b_0 = \sum_{i=0}^{n-1} b_i x^i \tag{7}$$

where each b_i is having the value of 0 or 1.

$$b_{i'} = b_i \oplus b_{(i+4)} \bmod \cdot 8 \oplus b_{(i+5) \bmod 8} \oplus b_{(i+6) \bmod 8} \cdot \oplus b_{(i+7) \mathrm{mod} 8} \oplus c_i, \text{ for } 0 \le i < 8 \tag{8}$$

where b_i is the ith bit of the byte and c_i is the ith bit of a byte c with the value 63 or 01,100,011. Thus the input byte x is related to the output y of the S-box by the relation, $y = A \cdot x^{-1} + B$, where A and B are constant matrices.

$$s'_{r,c} = s_{(r,(c+shift(r+Nb)) \bmod Nb)} \tag{9}$$

for $0 < r < 4$ and $0 < c < Nb$

Where Nb is the number of words in the state matrix (each column in the state matrix is considered as word). In AES, Nb = 4 always as the input size is 128 bits and arranged in state matrix of size 4 × 4. Each cell in the state matrix is denoted as s with the index of row r and column c.

$$a(x) = \{03\}x^3 + \{01\}x^2 + \{01\}x^1 + \{02\} \tag{10}$$

$$s'(x) = a(x) \otimes s(x) \tag{11}$$

$$\left[s'_{0,c}, s'_{1,c}, s'_{2,c}, s'_{3,c}\right] = \left[s_{0,c}, s_{1,c}, s_{2,c}, s_{3,c}\right] \oplus \left[w_{round \times Nb+c}\right] \tag{12}$$

where $\oplus$ is the bitwise XOR and round is the round number at which round key is added and $0 \leq$ round $< Nr$.

5 Experiments and Analysis

Before starting our actual work, we tried to find the running time of the Naïve Bayesian and J48 classification algorithms for different cases such as classification with original data (without encryption), encryption of numerical attributes, categorical attributes, and all the attributes in a dataset. The data set we used to find the running time is the Adult dataset from UCI machine repository which contains totally 32,560 records. We have used two test cases, one containing 2560 records and another having 5560 records. We have trained the above-mentioned classification algorithms using the first 30000 records of adult dataset for the test case 2560 and first 27000 records for the test case 5560. Each test case is applied five times assuming the class labels as Age, Education, Occupation, Marital Status, and Income separately for Naïve Bayesian classification algorithm. The Age attribute is alone not used for J48 algorithm since numerical data cannot be classified by this algorithm. The running time of Naïve Bayesian and J48 classification algorithm for the test case with 2560 records is shown in Figs. 4, 5, 6 and 7 respectively. Similarly, the running time of both the algorithms for the test case having 5560 records are given in Tables 16 and 17 and (Fig. 8).

Figure 4 displays the running time of both the classification algorithms for adult dataset with income as class label for encrypting various types of attributes such as numerical only, categorical only, both numerical and categorical, and no encryption. Similarly, Figs. 5, 6, and 7 display the running time of the classification algorithms for adult dataset with different class labels such as education, marital, and occupation for encrypting various types of attributes as in Figure 4. In Table 16 the column with heading Original contains the time taken (in seconds) by Naïve Bayesian classification algorithm for the five class labels without encrypting any data in the dataset. The third column, Only Numerical, represents the same as the second column but by encrypting all the numerical attributes in the dataset. The fourth column in the table is obtained by encrypting only the categorical attributes in the dataset. The last column of the table is got by encrypting all the attributes. Table 17 follows the same structure as Table 16.

We can infer from Figures 4, 5, 6, and 7 and Tables 16 and 17 that the time taken by the two algorithms for classification with all the attributes encrypted is significantly greater than the time taken for classification without encrypting any data or encrypting partial data. But classification without encryption leads to leaking of private data to unwanted parties. So, we choose to encrypt only the sensitive

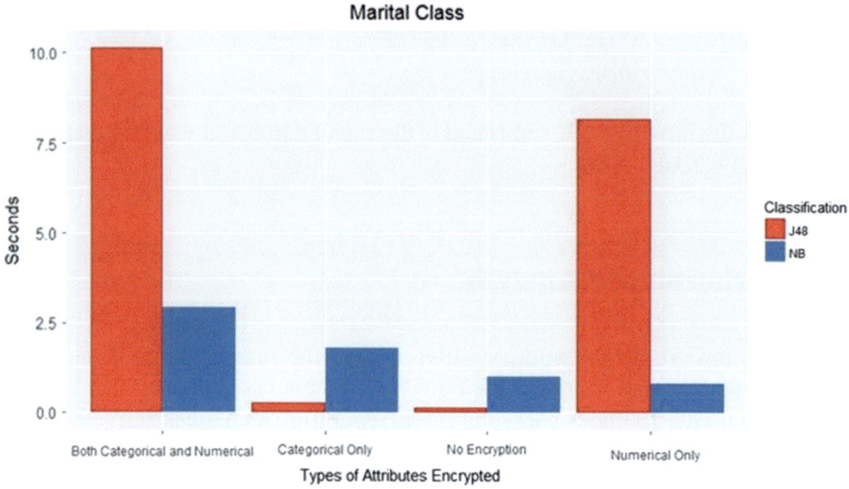

Fig. 5 Running time comparison for NB and J48 for income class with 2560 records

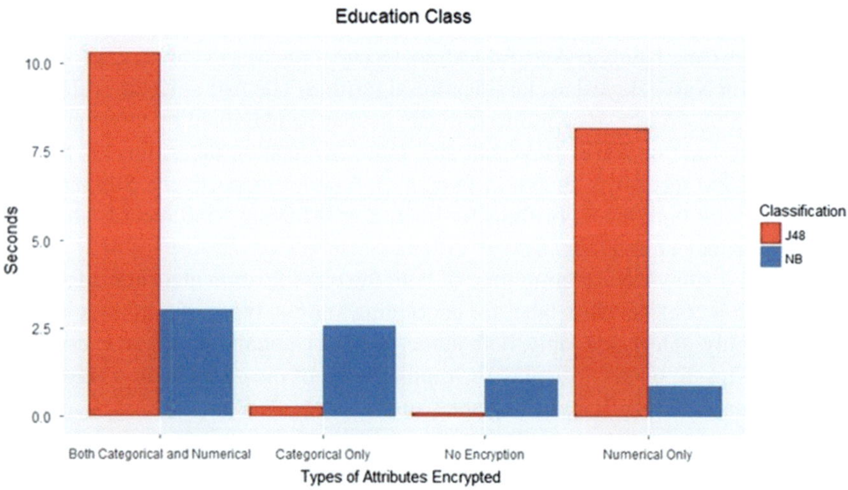

Fig. 6 Running time comparison for NB and J48 for marital class with 2560 records

data so that the privacy of the individual is protected and the time consumed by the algorithm is also reduced significantly when compared to encrypting the complete data. Tables 17 and 18 represent Naïve Bayesian and J48 running time with test case 5560 records. From Figures 4, 5, 6, and 7 and Tables 17 and 18, it is evident that an increase in test record size also increases the running time of the algorithm.

Table 19 shows the privacy preservation techniques summary, Table 20 presents the various PPDM components performance and Table 21 presents the comparison of current work with state-of-art literature.

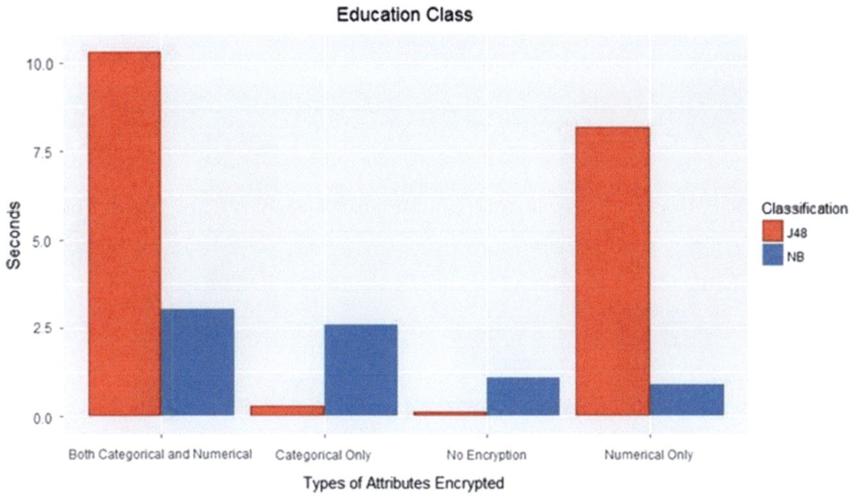

Fig. 7 Running time comparison for NB and J48 for education class with 2560 records

Table 17 Naïve Bayesian running time for 5560 records

Class label	Original (without encryption)	Numerical only	Categorical only	Both numerical and categorical
Income	1.82	1.73	2.38	2.72
Education	1.32	2.14	5.54	5.98
Age	3.98	3.92	4.13	5.32
Marital	2.06	1.87	3.73	4.54
Occupation	2.25	2.11	5.3	5.58

Assessment Criteria

In the state-of-art research, the most widely used measures are

Effectiveness

Effectiveness measures the data mining results accuracy and the sensitive information hiding capability [3–5, 12–15, 17–19, 15, 17, 18], and [6, 11, 20, 21]. For instance, hiding failures is used as a metric to measure data sanitation techniques' accuracy. Which involves accidental missing any hidden patterns then the situation is called misses cost and not hiding the sensitive information is considered in accuracy calculation.

Scalability

Within a stipulated period of time, the capability of the algorithm to solve the problem and the problem size are of more concern here and computational cost and scalability are related to one another. When scalability is low and computational cost is high

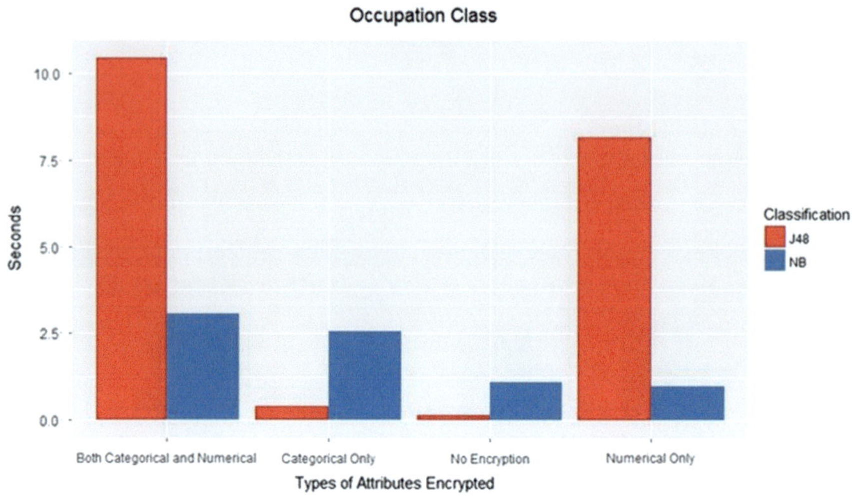

Fig. 8 Running time comparison for NB and J48 for occupation class with 2560 records

Table 18 J48 running time for 5560 records

Class label	Original (without encryption)	Numerical only	Categorical only	Both numerical and categorical
Income	0.20	8.25	0.36	10.33
Education	0.22	8.33	0.31	10.46
Marital	0.25	9.58	0.38	10.22
Occupation	0.24	8.64	0.45	10.59

Table 19 Outline of methods for preserving privacy

	Methods for preserving privacy	References
Centralized DB	Sanitation of data	[14–16, 56, 19, 20]
	Blocking	[18]
	Data distort	[3, 12, 13, 4, 24, 25, 17, 21]
	Generalization	[5, 6]
Distributed DB	Cryptography techniques	[7–9, 11, 22, 23, 26–28, 57]

then such algorithms are not used in the instance of solving problems with bigger size.

Efficiency

Efficiency of an algorithm depends on the capability of an algorithm to offer better performance when the requested resources provided to the algorithm. Computational cost is used to measure the performance [14, 15, 17, 18, 25, 56], and [20] and along

Table 20 Relative performance of PPDM components

Techniques	PPDM elements	Computational cost	Privacy preserving	Mining accuracy	Scalability
Hiding	Data hiding	Low	Contingent	Contingent	High
	Rule hiding	High	Contingent	Contingent	Low
Data mining	Classification	Low	N/A	Contingent	High
	Clustering	High	N/A	Contingent	Low
	association rule	Low	N/A	Contingent	High
Techniques for preserving privacy	Cryptography	High	High	High	Low
	Generalization	Low	High	Medium	High
	Blocking	Medium	Low	Medium	Low
	Distortion	Low	High	Low	High
	Sanitation	Medium	Medium	Medium	Low

Table 21 Privacy level and data quality metrics

Privacy level metrics		Data quality metrics	
Data metrics	Results metrics	Data metrics	Results metrics
Privacy model specific Entropy variance Average conditional Confidence level	Failures hidden	Information loss Discernibility metric Minimal distortion	Misses cost Artifactual patterns Misclassification error Classical data mining metrics

with computational cost, cost for communication is also involved in the distributed environment [7–9, 22, 26, 27], and [6, 10, 11, 20, 21, 23, 28, 57].

Assessing Performance

It is very difficult to measure the Privacy Preserving Data Mining Techniques' performance because in all possible situations different tasks will be handled differently by different algorithms.

In Table 20 a rating is provided representing the performance of different techniques and the rating falls into 3 different levels, low, medium, and high.

As said in Table 21 the proposed system is going to deliver optimal values for data metrics (Table 22).

Table 22 Comparison of various state-of-art works

Reference	Description	Remarks
[9]	The technological possibility of implementing privacy preserving data mining was investigated in this paper	For categorical characteristics, the efficacy of randomization with reconstruction is still being investigated
[57]	The authors of this paper created a privacy-preserving association rule mining method	Non-trivial extensions, non-categorical characteristics, and quantitative association rule mining are all much more difficult issues for it to solve
[28]	A standard data mining algorithm was created that has proven privacy and information disclosure characteristics	This hasn't been tried on a variety of database models
[23]	A new method for extracting common itemsets from remote databases We opted to transport only a tiny piece of each local model, the local MFIs, rather than sending data or significant chunks of the model across the locations. Our method becomes very communication efficient as a result of this approach, requiring just one cycle of message transmission and one reduction operation to aggregate final results	The distributed data mining work has technical issues, such as skewed distributions, in addition to performance and scalability
[11]	In this paper, a unique security paradigm for safe multiparty computing investigations is suggested. Rather than attaining perfect security, as most studies of SMC issues do, the suggested article's objective is to create acceptable (and customizable) security	They need to figure out a means to quantify the amount of security obtained by each protocol so that they may compare them in terms of security

(continued)

Table 22 (continued)

Reference	Description	Remarks
	A secure collaborative association rule mining protocol based on homomorphic encryption scheme	This work did not address privacy measures to quantitatively measure the privacy level achieved by the proposed secure protocols
Proposed work	In this research, we have developed a novel model called One Time Decryption protocol which helps in protecting the privacy of individuals while applying data mining functions over the data	• The loss of data in our model is very minimal as we do not add any noise or disturb the actual data in any way • The time consumed for performing the protocol depends on the size of the input • As we use encryption and decryption algorithms for protection, the time consumed by the protocol increases as the size of the input increases • In spite of that, as the protocol only uses encryption and decryption algorithms over sensitive attributes and not all the attributes, the time consumed by the protocol is reduced to a very great extent • The privacy preserving OTD protocol is an efficient technique for protecting the privacy leakage while sharing information in a distributed environment

6 Conclusion

The goal of privacy preserving data mining is to promote algorithms to conceal data or offer privacy to some sensitive information such that they do not get revealed to unapproved parties or invaders. In this research, we have developed a novel model called One Time Decryption protocol which helps in protecting the privacy of individuals while applying data mining functions over the data. The loss of data in our model is very minimal as we do not add any noise or disturb the actual data in any way. The time consumed for performing the protocol depends on the size of the input. As we use encryption and decryption algorithms for protection, the time consumed by the protocol increases as the size of the input increases. In spite of that, as the protocol only uses encryption and decryption algorithms over sensitive attributes and not all the attributes, the time consumed by the protocol is reduced to a very great extent. The privacy preserving OTD protocol is an efficient technique for protecting the privacy leakage while sharing information in a distributed environment.

References

1. Pinkas B (2002) Cryptographic technique for privacy preserving datamining. ACM SIGKDD Explorations 4(2):12–19
2. Bertino E, Fovino IN, Verykios VS, Provenza LP, Saygin Y, Theodoridis Y (2004) 'State-of-the-art in privacy preserving data mining'. SIGMOD Record 33(1):50–57
3. Du W, Zhan Z (2003) Using randomized response techniques for privacy—preserving data mining. In: Proc of the Ninth ACM SIGKDD int conf on knowledge discovery and data mining, Washington, D.C
4. Islam MZ, Brankovic L (2004) A framework for privacy preserving classification in data mining. In: Proc.of the 2nd workshop on Australasian information security, data mining and web intelligence, and software internationalization (AISW'04, AWDM&WI'04, AWSI'04), Dunedin, New Zealand, pp 163–168
5. Natwichai J, Li X, Orlowska M (2005) Hiding classification rules for data sharing with privacy preservation. In: Proc of 7th international conference on Data Warehousing and Knowledge Discovery (DaWaK'05), Copenhagen, Denmark, pp 468–477
6. Wang K, Yu S, Chakraborty S (2004) Bottom-up generalization: A data mining solution to privacy protection. In: Proc the 4th IEEE Int Conf on Data Mining (ICDM'04), Brighton, United Kingdom, pp 249–256
7. Du W, Zhan Z (2002) Building decision tree classifier on private data. In: Proc of the IEEE ICDM workshop on Privacy, Security and Data Mining (PSDM'02), Maebashi City, Japan, pp 1–8
8. Kantarcıoglu M, Clifton C (2003) Assuring privacy when big brother is watching. In: Proc of the 8th ACM SIGMOD workshop on research issues in data mining and knowledge discovery, privacy & security, pp 88–93
9. Lindell Y, Pinkas B (2000) Privacy preserving data mining. In: Advances in cryptology—CRYPTO 2000, Springer-Verlag, pp 36–54
10. Vaidya J, Clifton C (2005) Privacy-preserving decision trees over vertically partitioned data. In: Proc of the 19th Annual IFIP WG 11.3 working conf on Data and Applications Security (DAS'05), Storrs, CT, USA, pp 139–152

11. Wenliang D, Zhijun Z (2003) A study of several practical approach to solve secure multiparty computation problems. In: Pro of the Int Conf on Computer Networks and Mobile Computing (ICCNMC'03)
12. Evfimievski A, Gehrke J, Srikant R (2003) Limiting privacy breaches in privacy preserving data mining. In: Proc of the Twenty-second ACM SIGMODSIGACTSIGART symposium on principles of database systems, San Diego, California, pp 211–222
13. Evfimievski A, Srikant R, Agarwal R, Gehrke J (2004) Privacy preserving mining of association rules. In: Proc of the 8th ACM SIGKDD Int conf on Knowledge Discovery in Databases and Data Mining (KDD'02), Edmonton, Alberta, Canada, pp 217–228
14. Oliveira SRM, Zaïane OR (2002) Privacy preserving frequent itemset mining. In: Proc of the IEEE international conference on Privacy, Security and Data Mining (PSDM'02), Maebashi City, Japan, pp 43–54
15. Oliveira SRM, Zaïane OR (2003) Algorithms for balancing privacy and knowledge discovery in association rule mining. In: Pro of the 7th Int Database Engineering and Applications Symposium (IDEAS'03), Hong Kong, China, pp 54–65
16. Oliveira SRM, Zaïane OR (2003b) Protecting sensitive knowledge by data sanitization. In: Proc of the 3rd IEEE Int Conf. on Data Mining (ICDM'03), Melbourne, Florida, USA, pp 613–616
17. Rizvi J, Haritsa R (2002) Maintaining data privacy in association rule mining. In: Pro of the 28th Very Large Data Base Conf (VLDB'02), Hong Kong, China, pp 682–693
18. Saygin Y, Verykios V, Clifton C (2001) Using un-knowns to prevent discovery of association rules. ACM SIGMOD Rec 30:4
19. Saygin Y, Verykios V, Elmagarmid A (2002) Privacy preserving association rule mining. In: Proc of 12th int workshop on Research Issues in Data Engineering (RIDE)
20. Verykios S, Elmagarmid K, Elisa B, Saygin Y, Elena D (2003) Association rule hiding. IEEE Trans Knowl Data Eng
21. Xia Y, Yang Y, Chi Y, Muntz RR (2004) Mining association rules with nonuniform privacy concerns. Technical Report CSD-TR No. 040015, University of California
22. Kantarcioglu M, Clifton C (2002) Privacy-preserving distributed mining of association rules on horizontally partitioned data. In: Proc of ACM SIGMOD workshop on research issues in Data Mining and Knowledge Discovery (DMKD)
23. Veloso A, Meira Jr W, Parthasarathy S, de Carvalho M (2003) Efficient, accurate and privacy-preserving data mining for frequent itemsets in distributed databases. In: Proc of the 18th Brazilian symposium on databases, Manaus, Amazonas, Brazil, pp 281–292
24. Oliveira SRM, Zaïane OR (2003c) Privacy preserving clustering by data transformation. In: Pro of the 18th Brazilian symposium on databases, Manaus, Amazonas, Brazil, pp 304–318
25. Oliveira SRM, Zaïane OR (2004a) Achieving privacy preservation when sharing data for clustering. In: Pro of the int workshop on Secure Data Management in a connected world (SDM'04), in conjunction with the 30th Very Large Data Base Conference (VLDB'04), Toronto, Canada, pp 76–82
26. Klusch M, Lodi S, Moro G (2003) Distributed clustering based on sampling local density estimates. In: Proc of the 18th Int Joint Conf on Artificial Intelligence (IJCAI'03), Acapulco, Mexico, pp 485–490
27. Merugu S, Ghosh J (2003) Privacy-preserving distributed clustering using generative models. In: Proc of the 3rd IEEE Int Conf on Data Mining (ICDM'03), Melbourne, FL, USA, pp 211–219
28. Vaidya J, Clifton C (2003) Privacy-preserving k-means clustering over vertically partitioned data. In: Proc of the 9th ACM SIGKDD Int Conf on Knowledge Discovery in Data (KDD'03), Washington D.C., USA, pp 206–215
29. Erat J (2015) Security comparison of 3Des and AES. https://www.security.stackexchange.com/questions/26179/security-comparison-of-3des-and-aes
30. Goodin D (2011) AES crypto broken by 'groundbreaking' attack. http://www.theregister.co.uk/2011/08/19/aes_crypto_attack
31. Agarwal R, Srikant R (2000) Privacy preserving datamining. In: Proc SIGMOD conference on management of data, vol 29, no 2, pp 439–450

32. Lindell Y, Pinkas B (2000) Privacy-preserving data mining. In: Proc advances in Cryptology-CRYPTO, pp 36–54
33. Sweeney L (2002) Achieving k-anonymity privacy protection using generalization and suppression. Int. J. Uncertain., Fuzziness Knowl. Based Syst., 10(5):571–588
34. Fung BCM, Wang K, Yu PS (2007) Anonymizing classification data for privacy preservation. IEEE Trans. Knowl. Data Eng 19(5):711–725
35. Gehrke J, Kifer D, Machanavajjhala A, Venkitasubramaniam M (2007) L-diversity: Privacy beyond K-anonymity. In: Proc 22nd international conference on data engineering, vol 1, no 1, pp 1–47
36. Li N, Li T, Venkatasubramanian S (2007) t-closeness: Privacy beyond k-anonymity and l-diversity. In: Proc IEEE 23th international conference on data engineering
37. Aggarwal CC, Agarwal D (2001) On the design and quantification of privacy preserving data mining algorithm. In: Proc ACM SIGMOD, pp 247–255
38. Agarwal R, Evfimieski A, Gehrke J, Srikant R (2002) Privacy preserving mining of association rules. In: Proc 8th ACM SIGKDD, pp 217–228
39. Haritsa JR, Rizi S (2002) Maintaining data privacy in association rule mining. In: Proc 28th VLDB conference, pp 682–693
40. Clifton C, Kantarcioglu M, Lin X, Vaidya J, Zhu MY (2002) Tools for privacy preserving distributed data mining. ACM SIGKDD Explor Newsl 4(2):28–34
41. Clifton C, Kantarcioglu M (2004) Privacy preserving distributed mining of association rules on horizontally partitioned data. IEEE Trans Knowl Data Eng 16(9):1026–1037
42. Goldreich O, Micali S, Wigderson A (1987) How to play any mental game—A completeness theorem for protocols with honest majority. In: Proc 19th Annual Symposium on the Theory of Computing (STOC), ACM, pp 218–229
43. Ben-Or M, Goldwasser S, Wigderson A (1988) Completeness theorems for non cryptographic fault tolerant distributed computation. In: Proc 20th annual Symposium on the Theory of Computing (STOC), ACM, pp 1–9
44. Chaum D, Crepeau C, Damgard I (1988) Multiparty unconditionally secure protocols. In: Proc 20th annual symposium on the theory of computing (STOC), ACM, pp 11–19
45. Gentry C (2009) Fully homomorphic encryption using ideal lattices. In: Proc 41st annual ACM symposium on theory of computing, pp 169–178
46. Elmehdw Y, Jiang W, Samunthala BK (2015) K-nearest neighbour classification over semantically secure encrypted relational data. IEEE Trans Knowl Data Eng 27(5):1261–1273
47. Chen L, Liu X, Lu R, Ma J, Qin B (2016) Privacy preserving patient- centric clinical decision support system on naïve bayesian classification. IEEE J Biomed Health Inform 20(2):655–668
48. Azimi R, Sajedi H, Ghayekhloo M (2017) A distributed data clustering algorithm in p2p networks. Appl Soft Comput 51:147–167
49. Nayahi JJV, Kavitha V (2017) Privacy and utility preserving data clustering for data anonymization and distribution on Hadoop. Futur Gener Comput Syst 74:393–408
50. Yang JJ, Li JQ, Niu Y (2015) A hybrid solution for privacy preserving medical data sharing in the cloud environment. Futur Gener Comput Syst 43:74–86
51. Tian Z, Shi W, Wang Y, Zhu C, Du X, Su S, Guizani N (2019) Real-time lateral movement detection based on evidence reasoning network for edge computing environment. IEEE Trans Industr Inf 15(7):4285–4294
52. Tian Z, Li M, Qiu M, Sun Y, Su S (2019) Block-DEF: A secure digital evidence framework using blockchain. Inf Sci 491:151–165
53. Tian Z, Su S, Shi W, Du X, Guizani M, Yu X (2019) A data-driven method for future Internet route decision modeling. Futur Gener Comput Syst 95:212–220
54. Yao AC (1982) Protocol of secure computation. In: Proc 23rd annual symposium on foundations of computer science, pp 160–164

55. Oliveira SRM, Zaïane OR, Saygin Y (2004b) Secure association rule sharing. In: PAKDD, pp 74–85
56. Vaidya J, Clifton C (2002) Privacy preserving association rule mining in vertically partitioned data. In: Proc of the 8th ACM SIGKDD int conf on knowledge discovery and data mining, pp 639–644
57. Zhan JZ, Matwin S, Chang L (2005) Privacy-preserving collaborative association rule mining. DBSec:153–165

Smart Healthcare Systems and Predictive Medicine Using Machine Learning

Aditya Guntupalli, Aravind Kumar Muddana, Phani Kumar Turlapati, Siddhartha Paturi, and Mukkoti Maruthi Venkata Chalapathi

Abstract Smart healthcare systems and predictive medicine are in the modern-day definition of transformed healthcare systems that utilize artificial intelligence, machine learning, and advanced data analytics as the part. Understanding the basis of these applications forms the focus of this chapter with special reference to their contribution to improving diagnostic capability, prognosis of disease, and the tailoring of therapeutic plans across all branches of medicine. These technologies beginning from medical image analysis and machine learning algorithms used in electronic health records, to wearable devices and expanded application of telemedicine have measured its advantages that include better patient outcomes, ease of operations, and cost optimization. But issues, such as information quality, compatibility, and the use of big data analytics while protecting the patient's privacy remain key concerns that must be dealt with fully. From the current trends, future trends like AI drug discovery, Precision medicine, and innovations in blockchain, and quantum computing will further transform the healthcare industry as more collaborative research and innovative efforts target healthcare optimization worldwide.

1 Introduction

Overview of Smart Healthcare Systems and Predictive Medicine

IoT integrated smart healthcare systems incorporate various technologies and involve IoT, AI, and big data in the delivery of healthcare services. These systems can be used to meet the goal of delivering personalized care and services through the use of technology and timely information. Smart healthcare is a part of the smart city system and one of its important subtopics is predictive medicine which is based on the use of machine learning algorithms and statistical models aiming at the prediction of

A. Guntupalli · A. K. Muddana · P. K. Turlapati · S. Paturi · M. M. V. Chalapathi (✉)
School of Computer Science and Engineering (SCOPE), VIT-AP University, Amaravati, Andhra Pradesh, India
e-mail: mmv.chalapathi@gmail.com

V. K. Gunjan et al. (eds.), *Cybernetics, Human Cognition, and Machine Learning in Communicative Applications*, Cognitive Science and Technology,
https://doi.org/10.1007/978-981-97-8533-9_25

the disease outcomes and the individual treatment plans and prognosis. Altogether, a combination of these innovations will drastically change traditional healthcare practices by increasing their efficiency, reliability, and responsiveness.

Significance and Consequence of Machine Learning in the Healthcare Sector

Artificial intelligence is now an essential aspect of contemporary medicine due to its ability to drive improvements in diagnostic systems, solutions for patient care, and more. It can process large amounts of data within the shortest time possible and hence in diagnosing diseases, treatment regimens, and prognosis, is efficient. For example, it can be used to diagnose illnesses from medical images where the human brain is unable to see such images in better detail than the machine. Further, predictive analytics could predict the incidence of diseases and patients' worsening conditions and take proactive measures that help enhance patient survival while lowering expenditures. Machine learning applications in healthcare not only have a positive effect on the patient outcomes and their quality of life but also lead to increasing effectiveness of healthcare processes and decreasing the load on the healthcare professionals.

The following chapter aims at a comparison of several existing models of theories of planned behavior with the view of arriving at a more effective explanation of the intended behavior.

This chapter is designed to present the reader with an understanding of smart healthcare systems and predictive medicine, their importance, and the revolutionizing role of machine learning in the context of these concepts. To achieve this, the content is divided into four sections that outline the basics of these innovations, the technologies upon which they are based, examples of how they can be applied, the advantages of adopting them, and the limitations of their implementation. Specifically, the chapter will explore: Specifically, the chapter will explore [1].

The principles and aspects of smart healthcare systems

An analysis of the position and methods associated with machine learning to support the field of predictive medicine.

The real-life implementation of the proposed methods in the context of medical imaging, EHRs, wearable devices and technologies, and telemedicine.

The advantages of these technologies as well as the issues and obstacles that may be encountered in the process of their adoption.

Policies, ethical and privacy issues in data handling and analytical models.

Possible directions of its development in the future and possible innovations or improvements.

Through the material of this chapter, readers will obtain clear concepts of smart healthcare systems and predictive medicine as well as look at the future of the field and potential advancements and future studies in that line [2].

2 Foundations of Smart Healthcare Systems

Definition and Characteristics of Smart Healthcare Systems

Smart healthcare is one of the modern trends in health management and care, which implies using various technologies, including IoT, AI, Big Data, and Cloud Computing. It is noteworthy that the mentioned systems are developed to advance the existing and conventional healthcare systems in the aspects of patient care, healthcare processes, and, therefore, population health. Some of the characteristics of smart healthcare systems include integrated systems, patient-oriented, data-based, self-learning, cost-effective, and timely.

Interconnected systems help in acfacthieving coordination and the of information and communication between different healthcare equipment, equipment, and professionals, in a real-time setting. The that patient-centered systems are used guarantees that the care is going to be individualized and tailored to meet the particular needs and wants of the patient. Evidence-based practice utilizes large quantities of health information to make knowledgeable decisions and modify treatment approaches accordingly. Adaptive systems are dynamic in that within a short span of time they are capable of learning from new data and changing trends in healthcare. Last, effectiveness is attained by ensuring that services are delivered with the least costs possible, minimizing mistakes that might be made and lastly using resources in the right manner.

3 Components of Smart Healthcare Systems

The foundational components of smart healthcare systems include sensors, data processing, and decision-making mechanisms.

Sensors

Another aspect of smart healthcare is real-time data acquisition through the use of sensors that monitor different parameters. Smartwatches and fitness trackers, for example, monitor signals, including heartbeat rates, physical movement, and sleep cycles. Such implanted devices include insulin pumps, pacemakers, and neurological monitors that observe different internal body processes including blood sugar levels, heart rates, and neural firing patterns, respectively. Also, environmental sensors measure conditions outside the human body that may affect the mentioned health state, such as the quality of the air and the temperature in the environment. All of these sensors, collectively, give a holistic, and real-time view of the client's health condition to foster early intercessions as well as individual consideration [3].

Data Processing

Smart healthcare data processing has several steps starting from data acquisition from different sources such as EHRs, wearable devices, diagnosing apparatus, etc.

This data is then stored more securely through the use of scalable solutions like cloud based storage to enhance data availability and yet at the same time protect sensitive data information. In this step, data mining and a set of data analysis techniques such as machine learning algorithms and statistical analysis models are used to analyze the collected data and gain insights on the data. Last is data presentation where health information is arranged in an understandable manner in tools like a dashboard and graphical presentation that can assist the healthcare givers in their decision-making process.

Decision Making

Decision-making is one of the most important aspects that are incorporated into efficient healthcare management. CDSS implements AI and ML to provide the best clinical knowledge and help healthcare professionals to be efficient in their work. It includes immediate notification of patients and providers of important health events, such as changes in vital signs or medication timing. Moreover, individual care plans are designed based on big data analysis, making treatment more effective for each patient and improving healthcare services in general.

All these components: sensors, data processing, or decision-making mechanisms, collectively and cohesively form a smart healthcare system that fosters an effective and responsive healthcare system. This integration is not only beneficial to patients in enhancing outcomes but also to the delivery of healthcare because it becomes more agile and responsive to the dynamic needs of stakeholders and consumers of healthcare services.

4 Machine Learning in Predictive Medicine

Machine learning acts as a basis within the sphere of predictive medicine; powerful tools help to analyze a great amount of information and make a plethora of determinations to enhance the level of care. This transformation makes healthcare smarter and brings powerful diagnostic capabilities, disease prognosis, and customized therapeutic strategies according to patient parameters.

4.1 Role of Machine Learning in Predictive Medicine

In fact, machine learning algorithms are very efficient in handling different modalities of healthcare data such as EHRs, genomic data, medical imaging, and wearable biometrics. Machine learning models, therefore, use algorithms to analyze and learn from multiple datasets in order to create new patterns that determine the chances of a specific patient outcome by a certain time faster than statistical analysis. For instance, in cancer care, machine learning techniques consider the patients' genetic makeup and their tumor markers to forecast the nature of the tumor and determine the most appropriate and personalized treatment strategies. The capability also works to enhance treatment outcomes while reducing complications through providing the right forms of treatment. [4]

4.2 Key Techniques and Models Used

Regression Models: These models involve predicting values like blood glucose level or size of the tumor from the past records for the disease's management and control.

Classification Models: Sometimes utilized in the assessment of risk factors and disease severity, these models divide patients into various risk categories (e.g., low, moderate, high risk) for certain diseases. For instance, cardiovascular risk predictors involve determining a patient's cholesterol levels, blood pressure, and other factors to evaluate their potential risk of heart disease.

Neural Networks: Common deep learning methodologies include convolutional neural networks (CNN) and recurrent neural networks (RNN): these are used in analyzing medical images and processing natural language, respectively. CNNs, for instance, outperform image recognition tasks and come up with patterns when interpreting medical images such as X-rays and MRI, thus helping radiologists in identifying diseases [5].

Natural Language Processing (NLP): NLP tools help in retrieving useful data and insights from less-structured sources of clinical records and documentation along with the patients' reports found in the research literature. It is further applied to identify factors, hands, treatment responses, and clinical outcomes related to the diseases aiding in decision-making and risk prognosis [6].

These methods facilitate this power by providing healthcare providers with contextually-rich data for the purposes of precision medicine, enhanced patient care, and the future of medicine.

4.3 Applications of Predictive Medicine

Predictive medicine has transformative applications across various healthcare domains: Predictive medicine has transformative applications across various healthcare domains:

Risk Prediction: Diagnosis of the susceptibility of an individual patient for accomplishing certain diseases by the usage of factors including genetics, lifestyle, and history of diseases like diabetes, cancer, and Alzheimer's. When care is given early, this development seeks and monitors those in the higher risk categories to slow the progress of it and enhance their status.

Personalized Treatment Plans: An indication of adopting customized welfare approaches based on genetics and molecular profile as well as clinical phenotypes.

It also guarantees highly effective treatments, reduced side effects, and increased compliance among patients.

Early Disease Detection: Monitoring client transitional shifts that indicate that the patient is ill but not yet symptomatic, so as to control precursors and give prompt treatment. For example, predictive models may point to vital biomarkers of diseases which may be diagnosed early and treatment provided before irreparable complications set in.

Population Health Management: They use population data to diagnose disease patterns, to determine the healthcare need, and to foster prevention healthcare goals. Their uses include the identification of the population most likely to require intervention based on their condition and the subsequent directing of resources towards addressing the issues of those patients.

Therefore, the use of machine learning in the field of predictive medicine is a paradigm shift that leads to the approach of patient directed, prominent Data-Driven healthcare. Through the use of Big Data, sophisticated calculations, and large datasets, healthcare providers are able to diagnose conditions more accurately, estimate disease progression, and administer specialized care that would yield better outcomes and quality of life for the patients. Further improvement of studies in this direction is expected to enhance the development of the concept of predictive medicine for improved healthcare supply at a global level.

5 Applications of Smart Healthcare

Medical Imaging

The advent of smart healthcare systems has applied outcomes driven by preprogrammed procedures and superior algorithms in medical image interpretation to boost accuracy, reliability, and availability of diagnostic methods. Computer aided diagnosis models are used to analyze medical images which include; X-ray images, CT scans, and MRI scans, in order to identify any anomalies and to assist radiologists in arriving at an accurate interpretation of the images. Features and boundaries of organs and tissues in the human body can be well defined and this has application in planning for treatment and operations. Furthermore, advanced healthcare systems utilize AI in analyzing the images and offering instant information that aids in clinicians' decision-making and hence enhances patients' status [7].

Electronic Health Records (EHR)

EHRs are at the heart of smart health systems, as they help to effectively manage and process patients' data and ensure the effective organization of their treatment. Embedded data analytics tools in EHR systems analyze patient data as it is generated in real-time given the capacity to make clinical decisions on the same data. It simplifies activities like scheduling appointments, prescriptions, or billing hence improving efficiency in the execution of healthcare services. Moreover, the implementation of interoperable EHR systems would allow health information transfers between various practitioners, hence improving patient care coordination and decreasing adverse impacts due to medical errors [8].

Wearable Health Devices

Mobile health devices are used as an important component of smart health systems for accurately monitoring physiological signs, activity level, and health status in real-time. Wearables like smartwatches, fitness trackers, and portable sensors which monitor heart rate, blood pressure, glucose level, and sleep of the patient or the patient's behavior offer valuable information regarding the patient's health. Self-monitoring functions enable people to control their health status on their own and take prompt actions or notify healthcare providers of the development of certain conditions. Also, wearable devices enhance self-management and enable timely intervention in case of deterioration of an individual's health condition as they monitor patients and provide recommendations to encourage adherence to healthy lifestyles, hence enhancing patient satisfaction.

Telemedicine and Remote Care

Telemedicine and remote care services use the application of artificial intelligence to provide care services and consultation without in-person physical contact and this erases the geographical divide that hampers the patient from accessing medical personnel. AI-based consultations allow caregivers to agree on appointments and remote patient exams, condition identification, and suggested therapies. Patient sorting algorithms are applied as the cases with extreme need for the doctors' attention are identified and treated first while less severe situations do not lead to emergency room visits. Telemedicine platforms applied to follow-up care allow for better care compliance and patient support, which serves to optimize the quality of care and patients' satisfaction in the context of smart healthcare systems [9].

6 Benefits and Challenges

The Advantages of Intelligent & Prognostic Health Care and Medicine.

There are countless advantages of implementing smart health care systems and predictive medicine technology that transform the system of providing health care services. The advantages include better patient health, financial gains, increase in productivity, and patient-centered care.

Improved Outcomes

First, there is an opportunity for enhanced health since some of the activities are determined by health concerns. Real-time and analytics, which is the core of big data, can help healthcare providers in making better diagnosis, early intervention, and even track the progress of their patients. They assist in the early diagnosis of ailments, which can decrease the outbreak of severe health complications and increase the life quality of people diagnosed with diseases. Individualized treatment programs based on risk analysis will mean that the treatment programs given to patients are suited to the individual's needs, hence encouraging compliance and the effectiveness of the treatments.

Cost Savings

While the increased use of smart solutions in healthcare will lead to additional costs in the beginning, it is a small price to pay for improved efficiency and the elimination of unnecessary spending on resources. For instance, maintenance of health facilities equipment can be done before they develop faults, so that they are not a source of added costs due to breakdowns, and decisions on orders of tests and other procedures can be made from data analysis as opposed to the physician's bias. Since many diseases can be predicted and diagnosed at an early stage, the cost of the treatments and hospitalisations is reduced meaning that the providers of these services and patients benefit from the innovations.

Enhanced Efficiency

Higher levels of technology make healthcare delivery smooth and facilitate organizational operations. Digital tools minimize paperwork by providing options for automatic data acquisition and processing, which is beneficial for medical practitioners. By constantly observing and controlling the conditions of the patients through telemonitoring, healthcare providers are able to handle more patients within a given time span without adversely affecting their health. In addition to sharing information across different healthcare settings, communication between diverse healthcare facilities and devices makes it easier for multiple healthcare providers to work together.

Personalized Care

It is possible to note that the modern approach to the development of smart healthcare systems is to focus on the individual patient. Employing patient data, health risk assessment, genetic factors, lifestyle factors, and other relevant information, the models are capable of generating precise treatment plans. Not only does it increase the chances of positive outcomes for patients, but also engages the individuals and makes them happy to take their respective treatments. People become compliant with their prescribed treatment plans if the physician develops those plans according to the patient's conditions and needs.

7 Challenges in Implementation

However, there are several drawbacks in smart healthcare systems and predictive medicine that need to be mentioned: These challenges include those that have to do with data quality, compatibility, and some regulatory measures.

Data Quality

The information quality that has been used in developing smart healthcare systems plays a critical role in determining the level of accuracy and reliability of the systems. If the data available is either incomplete or incorrect or contains inconsistencies, the model will learn the wrong things and so will the decision maker. As can be seen, high-quality data must be collected using sound techniques, updated periodically, and have to go through extensive validation. Moreover, it is crucial to balance the data, avoid population bias, and make sure that the models are inclusive of the population that will be targeted.

Interoperability

One of the major problems is to ensure effective integration of the different systems and devices in health care facilities. Each system may well adopt a different format or standard of data, and so there can be problems in exchanging information. Lack of interoperability may cause such essential patient data to be locked in one EHR system, creating gaps in care and poor coordination. To increase the chances of successful exchange and reduce the identified barriers, healthcare organizations need to embrace standards and protocols that the entire ecosystem can adopt.

Regulatory Hurdles

Another challenge is the legal environment which is quite dynamic and challenging to deal with especially due to the ever changing laws and regulations. There are certain specific and rigid rules that healthcare technologies must adhere to protect the health of the patients, and their right to privacy and protection of their data. The U. S. Food and Drug Administration (FDA) and the European Medicines Agency (EMA) have specific guidelines against which any medical device and software must be tested and validated. Also, the legal requirements regarding data protection that have been put in place in different countries include the Health Insurance Portability and Accountability Act (HIPAA) in the United States and the General Data Protection Regulation (GDPR) in Europe must be adhered to so as to ensure that patients' information is protected. Due to the dynamic and innovative approaches to the current and future development of these regulations, it is crucial for both healthcare players and technology creators to stay vigilant and update themselves as and when required.

In conclusion, smart healthcare systems and the concepts of predictive medicine add value to modern healthcare, but the barriers relating to data quality, interfaces, and compliance issues have to be solved. However, for these innovations to be effectively and fully implemented across the healthcare system to enhance patient care and healthcare delivery, the following obstacles need to be overcome.

8 Ethical and Privacy Considerations

8.1 *The Ethical Challenges for Smart Healthcare*

In this context, the following ethical issues arise, which the stakeholders need to consider in order to ensure that the application of smart healthcare systems and predictive medicine is done responsibly: Some of the critical ethical questions involve patient consent, issues arising from the use of bias in developing the algorithms, and the question of equality in terms of access to sophisticated solutions in the management of health conditions.

Patient Consent

It is crucial to preserve patient autonomy as a noble ethical principle in the healthcare sector. As innovation in smart healthcare systems advances, it is important to ensure that patients properly comprehend how their data will be collected, processed, and shared. As for the patients, they have to be provided with the information regarding the aims of data collection, the kind of data that is to be collected, as well as the possible advantages and disadvantages of data collection. Customer care and ensuring that patient's consent are explicit and crucial in ensuring that patients have trust in doctors.

Bias in Algorithms

In particular, machine learning algorithms are only as accurate as the data which is fed into them with the purpose of being analyzed. In particular, the authors of the preprint noted that if the training data contains bias, then algorithms can deepen this bias or even make it even worse, which means making decisions based on prejudice and discrimination. For example, if the data used to build a predictive model is skewed towards a particular demographic group, then the modeling will be less robust to that population type. Bias can be tackled through the proper representation of the datasets, proper validation of the algorithm so that it does not show any prejudices, and lastly, the application of techniques that correct any prejudices that are observed.

Unequal Access

Smart healthcare systems, and other technologies such as predictive medicine, yield advantages and disadvantages unequally. It can be a concern that these innovative technologies increase the existing gaps in health equity if these technologies can only be afforded by a particular class or nations that have better facilities for healthcare. Promoting equal use of these technologies means investment to make them easily accessible and within the reach of the underprivileged groups. Hence, the policies and actions related to the disparities and the effort to close them and support inclusive healthcare innovation.

8.2 *Privacy and Data Security Concerns*

These are perhaps two of the most significant pain points in the modern organization.

The greater use of digital tools in healthcare raises concerns regarding the privacy and protection of sensitive data. Preventing access to the rich information concerning a patient's health and ensuring that privacy rights of the patients have been observed in the handling of the data is usually very important.

HIPAA Compliance

HIPAA: Health Insurance Portability and Accountability Act is the best defined practice in the United States of America for the confidentiality of patient's information. Medical care professionals and developers of technology must then make sure that their systems and procedures are up to HIPAA standards, which contain rules on the security and privacy of data including issues such as encryption, access, and reporting of breaches. Due to the nature of patient information, it is crucial for a healthcare setting to adhere to the rules and regulations under HIPAA to prevent the misuse of such information and to promote secure and appropriate use of such data.

Data Breaches

Due to the high value of medical data, the healthcare sector has emerged as one of the primary areas under consideration for cyber attacks. It is crucial to address data breaches as they can cause huge loss and damage such as identity theft, financial loss, and at times compromised patient safety. The use of strong measures of protecting or securing the health information is essential when putting into practice the intended objectives of meaningfully using health information. This encompasses security audits, staff awareness, and embracing of newer trending structures like blockchain and multi-factor authentication to safeguard the data. However, the establishment of an overall incident response process is critical to compensate for any breaches and act on them immediately.

Data Ownership and Control

Issues regarding the ownership and control over collected data are some of the most pertinent issues addressed in the privacy discourse of smart healthcare. A dearth of patients' rights for their health data means patients should be given full control to access their data, modify the information, and delete it if necessary. Handing over authority to the patient regarding ownership and control of data results in improved trust and accountability. New and existing protocols designed by healthcare providers and technology firms require from the patient the consent to use and share data collected throughout the course of treatment.

Therefore, mitigating ethical and privacy concerns is critical to enhancing smart health systems and, therefore, the concept of predictive medicine. A responsible, ethical healthcare should require, among others, the following key measures: Obtaining patient's informed consent, avoiding biased AI, providing equal access, and preserving patients' data confidentiality and integrity. If the above considerations are given priority, then the healthcare industry can ensure that they gain the trust of the patients and therefore fully benefit from the realization of technological advancement.

9 Future Trends and Innovations

Emerging Trends in Smart Healthcare

Smart healthcare trends that are already on the horizon can quickly and drastically change the current state of affairs by utilizing such elements as AI and precision medicine for drug discovery. Conventional drug discovery involves a lengthy process and coordinating the pharmaceutical composition, and it demands a vast amount of data for machines to process, hence the use of AI in the drug discovery process can save time and money. This technology enables the faster identification of new potential drugs or the reuse of existing drugs for new treatments, this ranks as a faster and efficient medical discovery. Precision medicine, conversely, directly targets an individual patient utilizing his or her genetic-environmental interactome, and life choices and opportunities. Technological advancement in the areas of genomics and data analysis ensures that healthcare providers formulate extremely relevant treatment regimes, and enhance the effectiveness of therapies, especially among the discipline specialties including oncology.

The Potential Impact of Future Technologies

Smart healthcare systems have been greatly enhanced by IT and other technologies, with future technologies like blockchain and quantum computing holding promise. Blockchain technology brings the security and integrated database to a new level increasing data availability and transparency of patient records. This cuts the number of hacking incidents, it makes work adhere to the set laws on privacy and also; in other aspects such as billing and claims. Quantum computing is the logical next step in the line of quantum information processing with significantly increased computational capabilities, effectively allowing for the processing of huge healthcare data sets that are impossible for classical computers. This technology can also speed up the process of the discovery of new drugs and contribute to the development of more efficient treatment plans by essentially mimicking molecular interactions and other complex biological phenomena, expanding the possibilities for advancement in the pharmaceutical field [10].

Future Outlook on the Role of AI and Machine Learning in Healthcare

As it has been seen throughout this paper, the future of healthcare will inevitably be framed by the use of AI and Machine Learning in different areas of healthcare. Advanced diagnostics will be another area that is enhanced by the use of AI, so healthcare professionals will be able to better diagnose patients at an earlier stage and with a higher degree of accuracy. ML models will predict the expected outcome in advance of a patient's condition with good accuracy so that required preventive measures can be taken. Additionally, the prevention and diagnosis of diseases along with the treatment methods, such as the targeted therapy based on the efficiency and the side effects, especially in chronic diseases and cancer, will depend on AI technology. Advancements in telemedicine with AI integration will improve the efficiency of care delivery through electronic personal medical companions and chatters to increase the patient's level of engagement. To optimize these benefits, the present research remains continuous, and stronger cooperative efforts must be integrated among interdisciplinary intensification healthcare frameworks, guaranteeing that all healthcare workers possess adequate knowledge and abilities to utilize these sophisticated technologies suitably.

10 Conclusion

Recap of Key Points Discussed

Here in this chapter, we have been able to comprehensively discuss the advancement and nature of both smart healthcare systems and predictive medicine. Their implementation was presented as the application of AI, machine learning, IoT, and big data concepts and technologies in transforming healthcare provision. Some of the parts which are the sensor, the processing of data, and the decision-making were discussed in detail since they contribute to the better delivery of care, the efficiency of procedures, and the patronization of individualized treatments. The use of smart healthcare systems in applications from diagnosis and treatment information technology healthcare records among others clearly illustrates the advantages of its use.

Summary of Potential for Smart Healthcare and Predictive Medicine

Artificial intelligence and big data analytics, along with its ability to predict and promote optimal healthcare are some of the concepts which have the capacity to reshape the industry in the future. The application of real-time data analysis and predictive modeling allows the early identification of disease, individually tailored solutions, and prompt healthcare actions. There is a potential for optimization of performance in managing patients' care and reducing the overall costs of healthcare for patients and practitioners through the integration of AI-driven innovations. However, the consequent evolution of technologies comprising blockchain and quantum computation is expected to compound the advantages, promising secure storage of data and unparalleled computation.

Suggestions for Future Research and Collaboration

It means that further investments in research and cooperation in the fields of smart healthcare and predictable medicine are condensed. The collaborative relationships between healthcare vendors, researchers, and policymakers will spur technological advancement, provide solutions to issues such as data accuracy and integrity, compatibility, and accept legal standards. Education and training to support healthcare workers will be made to make them prepared to apply and use advanced technologies in the delivery of healthcare. In this case, I am emphasizing that such measures should be a priority as they help to address the most pressing ethical issues affecting healthcare while maintaining patients' privacy and ensuring that more people can have access to innovative treatment solutions.

Altogether, smart healthcare systems and predictive medicine can be described as points of the new stage in the development of effective healthcare based on individual approaches, improved efficiency, and the use of big data. If such technologies are incorporated responsibly and in collaboration with stakeholders, it will be possible to provide a higher quality of care to the clients, as well as manage the available resources to improve the patients' quality of life and prepare for the development of technologies in the future.

References

1. Bhardwaj R, Nambiar AR, Dutta D (2017) A study of machine learning in healthcare. In: 2017 IEEE 41st annual computer software and applications conference (COMPSAC), vol 2, pp 236–241. IEEE. A Study of Machine Learning in Healthcare|IEEE Conference Publication|IEEE Xplore
2. Catarinucci L, De Donno D, Mainetti L, Palano L, Patrono L, Stefanizzi ML, Tarricone L (2015) An IoT-aware architecture for smart healthcare systems. IEEE Internet Things J, 2(6):515–526. An IoT-Aware Architecture for Smart Healthcare Systems | IEEE Journals & Magazine | IEEE Xplore
3. Formica D, Schena E (2021) Smart sensors for healthcare and medical applications. Sensors, 21(2):543. Sensors|Free Full-Text|Smart Sensors for Healthcare and Medical Applications (mdpi.com)
4. Alanazi HO, Abdullah AH, Qureshi KN (2017) A critical review for developing accurate and dynamic predictive models using machine learning methods in medicine and health care. J Med Syst, 41:1–10. ACritical Review for Developing Accurate and Dynamic Predictive Models Using Machine Learning Methods in Medicine and Health Care—PubMed(nih.gov)
5. Amato F, López A, Peña-Méndez EM, Vaňhara P, Hampl A, Havel J (2013) Artificial neural networks in medical diagnosis. J Appl Biomed, 11(2):47–58. Artificial neural networks in medical diagnosis—ScienceDirect
6. Nadkarni PM, Ohno-Machado L, Chapman WW (2011) Natural language processing: an introduction. J Am Med Inform Assoc, 18(5):544–551. Naturallanguageprocessing:anintroduction-PubMed(nih.gov)
7. Suetens P (2017) Fundamentals of medical imaging. Cambridge university press. FundamentalsofMedicalImaging(cambridge.org)
8. Evans RS (2016) Electronic health records: then, now, and in the future. Yearb Med Inform, 25(S 01):S48–S61. ElectronicHealthRecords: Then,Now,andintheFuture-PubMed(nih.gov)
9. Field MJ, Grigsby J (2002) Telemedicine and remote patient monitoring. Jama 288(4):423–425. Telemedicineandremotepatientmonitoring-PubMed(nih.gov)
10. Hölbl M, Kompara M, Kamišalić A, Nemec Zlatolas L (2018) A systematic review of the use of blockchain in healthcare. Symmetry 10(10):470. Symmetry|FreeFull-Text|ASystematicReviewoftheUseofBlockchaininHealthcare(mdpi.com)

Smart Object Integration in Neonatal Health: Leveraging RFID and Explainable AI for Mortality Risk Prediction

A. Anny Leema, P. Balakrishnan, Vijaya Krishna Akula, S. Ramacharan, and N. Jothiaruna

Abstract The healthcare sector has seen remarkable progress with the integration of Radio Frequency Identification (RFID) technology and sensors to monitor infants and mitigate mortality rates. By combining RFID technology with sensors, healthcare facilities gain access to a range of applications that streamline workflows, enhance patient care, and improve overall efficiency. Furthermore, when augmented with Machine Learning (ML), this synergy opens avenues for data-driven insights, predictive analytics, and finely-tailored healthcare management. In India, neonatal deaths remain a significant concern, particularly within the first week of life and alarmingly within the initial 24 h, urgent measures are needed to address this issue. To tackle neonatal mortality effectively, our proposed solution involves deploying an interpretable machine learning framework integrated with an RFID-based hospital monitoring system. This integrated system allows for real-time analysis, enabling early detection and intervention, ultimately leading to a reduction in neonatal mortality rates. Through the collection of real-time data on pregnant women's vital signs utilizing RFID technology and sensors, coupled with an interpretable ML ensemble classifier, healthcare providers can swiftly identify at-risk newborns, facilitating timely intervention and potentially saving numerous lives. We conducted an analysis on 40,000 infant records from the National Center for Health Statistics (NCHS) using logistic regression, aiming to predict mortality risk based on 30 attributes.

A. A. Leema · P. Balakrishnan (✉)
Analytics Department, School of Computer Science and Engineering, VIT Vellore, Vellore, India
e-mail: balakrishnan.p@vit.ac.in

V. K. Akula · S. Ramacharan
Department of Information Technology, G. Narayanamma Institute of Technology and Science, Hyderabad, Telangana, India

N. Jothiaruna
School of Computer Science and Engineering, VIT Bhopal, Bhopal, India

V. K. Gunjan et al. (eds.), *Cybernetics, Human Cognition, and Machine Learning in Communicative Applications*, Cognitive Science and Technology,
https://doi.org/10.1007/978-981-97-8533-9_26

While this approach yielded good precision (0.83) for survival prediction, it struggled to accurately identify non-survivors, prompting the need for further investigation. While ensemble models like BaggingClassifier and RandomForest classifier achieved high accuracy rates, the Bagging Classifier with Logistic Regression demonstrated outstanding performance in determining whether a newborn is alive at the time of report. Its remarkable performance in terms of accuracy (99.68%), precision (99.68%), recall (100%) and F1 score (99.84%) highlights how effective it was for this classification job. Finally, the SHapley Additive exPlanations (SHAP) analysis is employed to interpret the correlation between the risk features and the outcome.

1 Introduction

Child mortality is a daily tragedy and one of the more serious socioeconomic issues, particularly in India. Around 1 million babies pass away during the first 24 h, and 75% of all neonatal deaths take place during the first week of life. The first 28 days are considered to be a child's most vulnerable phase, and in India, 6, 00,000 babies pass away during this time each year from disorders linked to inadequate prenatal care or poor postpartum care. The infant mortality rate (IMR), also known as the neonatal mortality rate (NMR), is a vital public health indicator that counts the number of infants who pass away before turning one year old. The global IMR in 2021 was 30 fatalities for every 1,000 live births. IMR varies significantly between nations, with high-income nations having an average IMR of 5 deaths per 1,000 live births as opposed to low-income nations' 74 deaths per 1,000 live births. Prematurity, birth asphyxia, congenital deformities, poor maternal health, lack of access to excellent healthcare, and infections are the main causes of newborn mortality. The infant mortality rate (IMR), also known as the neonatal mortality rate, is a critical public health indicator. It measures the number of infants who die before reaching one year of age per 1,000 live births in a given population, typically within a year.

Numerous causes of neonatal mortality can be avoided with the help of competent medical care, immunizations, better nutrition, and availability of clean water. Therefore, lowering IMR is frequently considered an important and attainable public health objective. IMR reduction initiatives frequently include, among other things, treating malnutrition, boosting prenatal and postnatal care, guaranteeing access to safe delivery techniques, and introducing immunizations. Improving access to quality antenatal and postnatal care, Promoting breastfeeding and other healthy infant feeding practices. Immunizing infants against preventable diseases, and treating infections and other neonatal illnesses promptly and effectively.

Technology plays a pivotal role in revolutionizing healthcare, continually evolving to enhance patient outcomes, streamline processes, and lower costs. Among the myriad advancements, RFID-based technology and machine learning stand out as crucial tools in reducing infant mortality rates (IMR). These innovations not only improve healthcare quality but also enhance its efficiency and accessibility. In the realm of diagnostics, cutting-edge technologies like medical imaging, genetic testing, and laboratory automation enable early detection of health issues, including those pertaining to pregnancy and neonatal health. This early detection is paramount in addressing potential complications and improving outcomes. Moreover, continuous patient monitoring facilitated by wearable devices and remote sensors provides real-time data on vital signs, fetal health, and neonatal well-being. These insights empower healthcare professionals with timely information for intervention, ensuring prompt care delivery.

In the neonatal ward, the integration of various types of sensors further enhances monitoring capabilities. Sensors for tracking vital signs such as heart rate, respiratory rate, temperature, and oxygen saturation provide comprehensive data on the health status of infants. By continuously monitoring these parameters, healthcare providers can detect abnormalities or signs of distress early on, enabling swift intervention to mitigate risks and prevent adverse outcomes. In essence, the synergy between RFID technology, machine learning, and a plethora of sensors in healthcare underscores a transformative approach towards reducing infant mortality rates. By leveraging these technologies, healthcare systems can not only improve the quality of care but also save countless young lives, ushering in a new era of healthcare excellence.

2 Literature Review

Infant mortality is categorized into three types: perinatal, post-neonatal, and neonatal. Prenatal death, also known as fetal death, refers to the loss of a fetus before birth. This sadly occurs more frequently than most people realize. Chromosomal abnormalities are the most common cause of miscarriage, which happens in the first trimester (first 12 weeks) of pregnancy. This can be due to errors in egg or sperm formation, or during cell division after fertilization. These errors can lead to developmental problems that are incompatible with life. Placental abruption, Placental insufficiency and Placenta previa (placenta covering the cervix) deprive the baby of oxygen and nutrients, leading to fetal death. Pre-existing health problems in the mother, like uncontrolled diabetes, High blood pressure (chronic hypertension or pregnancy-induced hypertension), Autoimmune diseases, infections (e.g., cytomegalovirus, rubella), and severe malnutrition can create an unhealthy environment for the fetus or increase the risk of complications during pregnancy. Post-neonatal deaths are caused by factors such as malnutrition, infectious diseases, and the home environment. Babies born too early (premature) are more at risk due to underdeveloped organs, especially their lungs. Complications like breathing problems and infections can be life-threatening. Neonatal mortality refers to the death of a newborn within 28 days of delivery. Inadequate access to basic medical care during pregnancy and after delivery is a major contributor to neonatal deaths. Premature births and low birth weight (35.9%), birth asphyxia (lack of oxygen leading to brain damage), birth trauma, and malnutrition are the leading causes of neonatal mortality. This type of mortality accounts for nearly 40–60% of all infant deaths in developing countries. While recent technological advancements have helped to reduce the prenatal death rate, the probability of neonatal death remains high.

2.1 Overview of Neonatal Mortality Rates and Challenges

IndiaSpend, India's premier data-driven, public-interest journalism organization, analyzed health data from 13 Indian states to investigate the alarming infant and child mortality rates across the country. According to Census and UNICEF records, states such as Madhya Pradesh, Assam, Rajasthan, Odisha, Uttar Pradesh, and Assam have consistently reported high infant mortality rates. Conversely, Goa, Kerala, and Tamil Nadu have the lowest infant mortality rates in the country. While there are significant efforts by governments to reduce infant mortality rates in developing countries, the results have not been as desired, and India is no exception. In fact, Madhya Pradesh has seen an increase in its infant mortality rate, which remains the highest in the country at 48 deaths per 1,000 live births.

Numerous studies have shown that malnutrition considerably increases the risk of disease and mortality in the short time span of life [1], and that India is particularly active in the battle against malnutrition. Child malnutrition is impacted by the Socio Democratic Index connection. Economic disparities are getting worse in India on a national basis, within and between states, as well as in rural and urban areas [2]. Low birth weight, child stunting, child wasting, child underweight, anemia in children, and anemia in women between the ages of 15 and 49 are the top five indications of malnutrition [3]. Additionally, the National Nutrition Mission (NNM), an effort by the Government of India that was started in 2018 [4], makes monitoring child undernutrition at the district level extremely crucial. Despite accounting for 36% of all under-five deaths globally, 3.9 million neonatal deaths occur annually. In India alone, nearly 1.2 million newborns die within the first four weeks of life each year [5]. Therefore, the key to achieving child survival in India today lies in saving a significant number of newborn lives in homes, healthcare facilities, villages, towns, and cities. India possesses the necessary health system foundation and technical expertise to embark on an extraordinary effort to achieve low levels of neonatal mortality. Female infant mortality in India has been found to be higher than male infant mortality in studies of the subject [6]. Boys were more likely than girls to pass away during the neonatal period (0–1 month) in India, according to an analysis of nationally representative 2005–06 data and 2015–16 data [7, 8]. Malnutrition in children and pregnant women was one of the risk factors that contributed to 68.2% (65.8–70.7) of under-5 deaths and 83.0% (80.6–25.0) of neonatal deaths in India in 2017 [9]. The Indian government has started programs aimed at enhancing the nutritional status of kids, expectant women, and nursing mothers, access to clean water and sanitation, and public health facilities' hygiene. Child mortality will decrease with local involvement and planning [10, 11]. The most frequent causes of mortality in preterm very low birth weight babies were sepsis, birth asphyxia, and respiratory distress syndrome (RDS), according to a study from Haryana, India [12, 13]. Infant mortality rates have been found to rise when diabetes occurs during pregnancy. According to this study, patients with poor glycaemic control have a higher risk of infant mortality, necessitating monitoring and prenatal care [14]. Children born to mothers who had more births and fewer gaps between them had a higher likelihood of being underweight, as well as a higher risk of stunting, anemia, and mortality [15]. Even the healthiest women can find pregnancy to be a difficult time. Many women still pass away during and after pregnancy despite notable medical advancements due to a variety of causes, including excessive bleeding, infection, high blood pressure, anemia, and heart disease. Women with diabetes are more likely to develop complications during pregnancy, including miscarriage, renal failure, and premature delivery. Due to the final brain and immune system development occurring in the final few months of pregnancy, premature babies may be born with a variety of health issues.

2.2 *Analyzing the Global Dataset*

Nearly half of all deaths in children under the age of five are caused by undernutrition, which increases their risk of death. Undernutrition is the most common risk factor for death in children under the age of five. Researchers have also used machine learning algorithms to predict the risk of infant mortality based on the following attributes and also analyzed the relationship between infant mortality and malnutrition indicators. The malnutrition indicators that are associated with an increased risk of infant mortality are shown in Table 1. Hence the indicators like age of the mother, the number of gestations, the number of weeks of pregnancy, the gender and weight of the child, and the method of delivery. Characteristics, such as gender, marital status, pregnancy, age, number of live births, number of children born dead, gestational week, education, doctor visits, delivery, Apgar 1, Apgar 5, localization, anomaly, weight, and pregnancy type are monitored.

Table 1 Malnutrition indicators that cause infant mortality

S. no	Indicators	2015 (%)	2016 (%)	2017 (%)	2018 (%)	2019 (%)
1	Low birth weight	20.4	17	21.4	25	20.4
2	Child stunting (low height for their age)	38.4	35	39.3	35	37.8
3	Child wasting (significant of food shortage or disease)	20.8	15	15.7	17.3	20.8
4	Child underweight	35.7	36	32.7	23.8	47.8
5a	Anemia in children	57.6	57.3	58.8	42	42
5b	Ananemia in women 15–49 years	55.1	55.4	59.3	53.8	40

2.3 *Existing Strategies for Reducing Neonatal Mortality*

Saving newborns is a top priority in global health. Many strategies work together to achieve this goal. Early and regular checkups during pregnancy (prenatal care) are key to identifying and managing potential problems that could harm the mother or baby. Skilled doctors, midwives, or other qualified healthcare providers attending births are essential for handling emergencies and ensuring a safe delivery. Educating mothers and their families about healthy habits, hygiene, breastfeeding, and warning signs empowers them to make informed choices and seek medical attention when needed.

Vaccinations are another crucial weapon against neonatal mortality, protecting newborns from preventable illnesses. Kangaroo Mother Care, a method where mothers hold preemies skin-to-skin for warmth and breastfeeding, has been proven to improve their survival rates. Additionally, healthcare professionals are trained in newborn resuscitation to revive babies who struggle to breathe at birth. Infection prevention plays a critical role, with proper hygiene, sanitation, and clean birthing practices being essential. Management of preterm births is another focus area, with neonatal intensive care units providing specialized care for these fragile babies. Easy access to emergency obstetric and neonatal care, alongside community health workers who educate families and identify health problems early, is vital. Finally, proper nutrition for both mothers and newborns, along with family planning to reduce high-risk pregnancies, contributes significantly to better health outcomes for both mothers and babies.

3 Potential of RFID Technology with Sensors in Maternal Health Monitoring

This paper delves into the potential of RFID (Radio Frequency Identification) technology to enhance infant mortality rates in healthcare. Passive tags, favored for their affordability and durability, find widespread use in tracking patients, equipment, and medications within healthcare settings. Maternal complications, such as pre-existing conditions like diabetes and high blood pressure, infections, and placental issues, significantly contribute to both prenatal (fetal death before birth) and postnatal (infant death within the first 28 days) mortality rates. These conditions often impede oxygen and nutrient flow to the fetus, heightening the risks of miscarriage, premature birth, and delivery complications. Although current practices rely on regular checkups, blood tests, fetal monitoring, and ultrasounds, RFID technology harbors the potential to revolutionize maternal health monitoring.

Imagine expectant mothers wearing RFID-tagged wristbands that monitor their body temperature and blood pressure. These tags allow for real-time monitoring and quick identification of possible issues by wirelessly sending data to a central system. Combining this data with machine learning analysis makes it possible to identify babies who are more likely to be in danger, which can result in timely treatments and possibly save many lives. But before widespread adoption, issues including price, technological constraints, and privacy concerns need to be resolved.

RFID technology, when coupled with machine learning, presents a wealth of opportunities in healthcare. By leveraging data collected through RFID, machine learning algorithms can identify patterns and trends, aiding in predictive maintenance. The integration of sensors further enhances these capabilities, particularly in monitoring maternal health during pregnancy. Pregnant individuals equipped with RFID-tagged bracelets or stickers embedded with sensors can have vital signs like blood pressure, temperature, and even blood sugar levels (in the case of diabetes) collected during their hospital stay. Strategically placed RFID readers in the hospital enable the collection of this data, facilitating the identification of potential complications such as preeclampsia, infections, or other conditions. Additionally, sensors allow for the assessment of the baby's heartbeat, well-being during pregnancy and delivery, and visualization of the baby's development, enabling the early detection of potential problems. Wireless transmission of collected data to a central system triggers alerts for medical staff in cases as shown in Fig. 1, where readings deviate from healthy ranges, facilitating prompt intervention and management of potential complications.

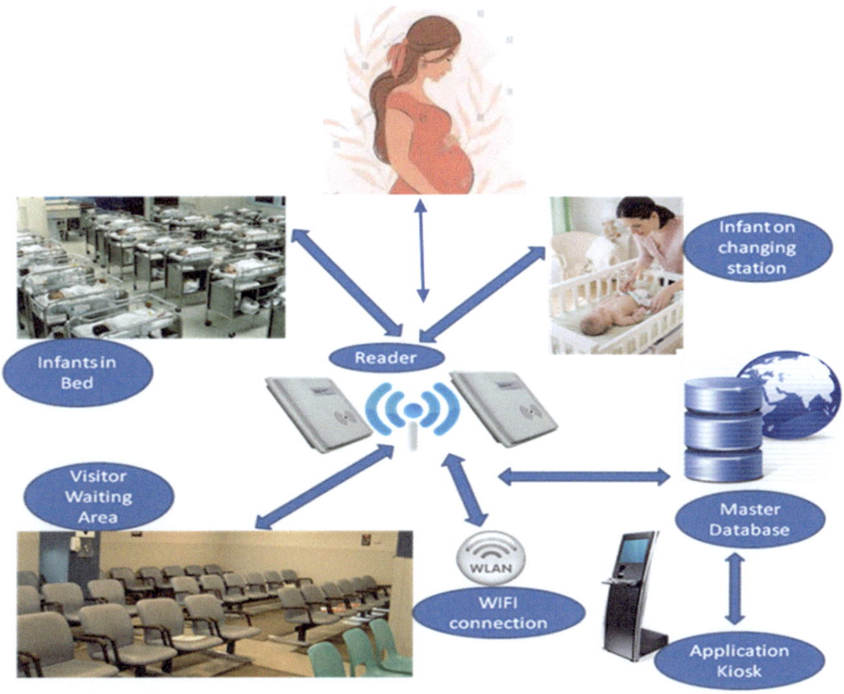

Fig. 1 Maternal and infant health monitoring

3.1 Tracking Infant Movements and VITAL Signals with RFID

Our suggested system monitors mother and baby vital signs using sensor-enabled battery-powered RFID tags (RFID Sensors). RFID technology with additional sensors (e.g., temperature, humidity, motion, light) to collect environmental data that provide both identification and the ability to monitor and transmit sensor data. RFID sensors have an internal power source (usually a battery) to operate not only the RFID communication but also the sensor components. It has more extensive data storage capacity to store sensor data over time, which can be later retrieved and analyzed. RFID sensors can continuously collect and transmit real-time sensor data, and in some cases, this data can be updated as conditions change. It plays a crucial role in tracking infant movements and monitoring vital signs in a neonatal or infant care environment. Continuous monitoring of these parameters allows for the early detection of any abnormalities, enabling healthcare providers to respond promptly to any signs of distress or health issues. It can be programmed to trigger alerts and notifications when vital signs fall outside the normal range or when infants move into restricted or prohibited areas. These alerts can be sent to healthcare providers' devices, enabling immediate intervention. It collects and stores historical data related to infant movements and vital signs. This data can be valuable for medical staff and physicians to assess trends and make informed decisions about an infant's health and care. The overall safety and security of the neonatal care unit are enhanced by preventing unauthorized movement and quickly identifying any discrepancies or abnormalities in infant behavior. It also helps in matching infants with their mothers accurately, ensuring that each infant receives the appropriate care and preventing any mix-ups. By automating the tracking of infant movements and vital signs, RFID sensors can optimize workflow efficiency in the care unit. Healthcare providers can focus more on patient care and less on manual record-keeping. It provides reassurance to parents and family members by offering real-time updates on their infant's location and well-being. This can reduce anxiety and stress, especially in situations involving premature infants or those with special medical needs.

3.2 Data Integration with EHR

Our suggested system uses sensor-enabled battery-powered RFID tags (RFID Sensors) to monitor the mother and baby's vital signs. RFID sensors combine RFID technology with additional sensors (e.g., temperature, humidity, motion, light) to collect environmental data. Continuous monitoring of these parameters allows for the early detection of any abnormalities, enabling healthcare providers to respond promptly to any signs of distress or health issues. The hospital's electronic health records (EHR) system can easily be integrated with data gathered by RFID sensors. This guarantees that important patient data is available to healthcare professionals right away. Then it is subjected to analysis, where data analytics come into play. After preprocessing the data, a machine learning algorithm is selected to carry out our task. The type of data and the desired result will determine which machine learning algorithm is best for a given task. For the machine learning model's input features, feature engineering entails choosing and modifying pertinent attributes from the RFID data. The equipment ID, timestamp, location, and sensor readings are a few examples of these features. The prepared data or historical data must be used to train the machine learning algorithm.

4 Methodology of the Proposed System

4.1 Analyzing RFID Data with ML Algorithms

Machine learning is essential for raising infant care during emergencies by identifying patterns in collected data. It can be used to spot early indications of infant health issues, assisting medical professionals in taking quick action to avert serious consequences. Leveraging IoT, EHRs, and Environmental Monitoring for Mother and Baby Wellbeing is depicted in Fig. 2.

4.2 Logistic Regression

Our research focuses on using machine learning to predict infant mortality risk. We're analyzing data from the National Center for Health Statistics (NCHS) containing over 40,000 records with 30 attributes for each case. Some of the important attributes are Month Prenatal Care Began Recode, Number of Prenatal Visits Recode, Cigarettes Before Pregnancy, Cigarettes 1st trimester, Cigarettes 2nd trimester, Cigarettes 3rd trimester, Prepregnancy Diabetes, Gestational Diabetes, Prepregnancy Hypertension, Gestational Hypertension, Eclampsia, Previous Preterm Birth, Infertility Treatment, no of previous Cesarean Deliveries, Chronic Hypertension, Pregnancy Associated Hypertension, Gonorrhea, Hepatitis B, and Hepatitis C. Initially we implemented logistic regression, a linear model used for binary classification tasks we obtained the class 0 represents infants who did not survive (negative class), and 1 represents infants who survived (positive class). For class 0 (not survive), precision is 0.00, which means none of the models' predictions for non-survival were correct. For class 1 (survive), precision is 0.83, indicating that out of all the infants the model predicted to survive, 83% actually did survive. For class 0 (not survive), recall is 0.00, which again means the model did not correctly identify any infants who didn't survive. Logistic regression doesn't handle complex interactions between features and may suffer from underfitting if the relationships are non-linear. So we decided

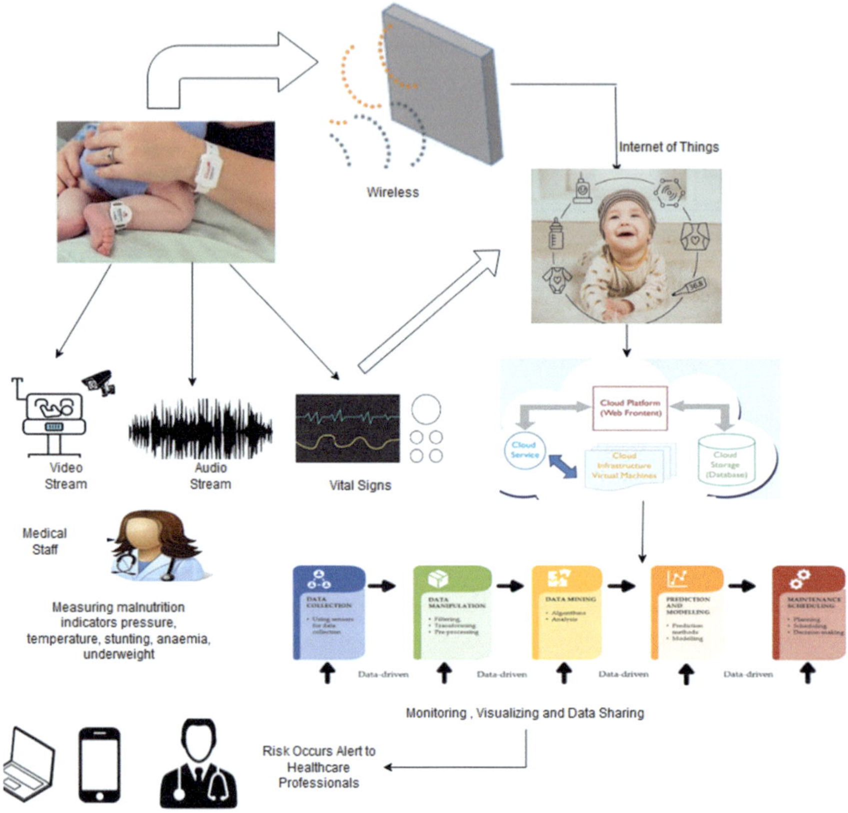

Fig. 2 Leveraging IoT, EHRs, and environmental monitoring for mother and baby

to proceed with Ensemble models that combine multiple base models to improve predictive performance.

4.3 RandomForest Classifier

The RandomForest classifier is trained on the training data and then evaluated using the validation set to assess its performance. The evaluation metrics include accuracy, precision, recall, and F1 score. The obtained validation metrics indicate high performance, with an accuracy of 99.67% precision of 99.68%, recall of 99.99%, and an F1 score of 99.83%. These metrics suggest that the classifier performs exceptionally well in predicting whether infants were living at the time of a report indicating a robust balance between precision and recall.

4.4 Ensemble Classifier

The objective is to construct models utilizing three ensemble machine learning methodologies: Decision Tree, Logistic Regression, and K-Nearest Neighbors (KNN). Employing an ensemble method known as Majority Voting, the algorithm aggregates predictions from three distinct classifiers (Decision Tree, K-Nearest Neighbors, and Logistic Regression) to generate a final prediction.

Algorithm for Ensemble Classifer

1. Splitting the Dataset: The dataset is loaded from a CSV file (train_infant.csv), and the features (x) and target variable (y) are separated. The dataset is split into training and testing sets using the train_test_split function from sklearn.model_selection.
2. Models Creation: Three different classifiers are instantiated: Decision Tree Classifier (model1), K-Nearest Neighbors Classifier (model2), and Logistic Regression Classifier (model3).
3. Model Training: Each classifier is trained on the training data (x_train, y_train) using the fit method.
4. Prediction: Predictions are made on the test data (x_test) using each trained classifier (model1, model2, model3), resulting in pred1, pred2, and pred3.
5. Final Prediction (Majority Voting): For each instance in the test data, the mode of the predictions from the three classifiers is calculated using the mode function from the statistics module. The final predictions are stored in the final_pred array.

This ensemble approach leverages the predictions from multiple classifiers to make a final decision, aiming to improve predictive performance compared to using individual classifiers alone. Since the majority voting ensemble method is used, the final prediction for each instance is determined by taking the mode (most common) prediction among the predictions made by the three individual classifiers. The obtained majority voting result is 1, means that at least two out of the three classifiers predicted class label 1 for that instance, leading to the final prediction being class label 1. Then import shapash library, and initialize a SmartExplainer object (xpl) with the list of models ([model1, model2, model3]), test features (x_test), and test target (y_test). Generate a summary of the model's interpretation. By using Shapash, we enhanced the explainability of the models and gained insights into the factors influencing the predictions, which can be crucial for decision-making and model debugging. The output includes the predictions made by each classifier (Decision Tree, K Neighbors, and Logistic Regression) and their respective accuracy scores. The accuracy score represents the proportion of correctly predicted samples out of the total samples in the test set. Model evaluation metrics are shown in Table 2, indicating effective performance in predicting whether an infant is living or not based on the given features. The obtained accuracy result is depicted in Fig. 3.

Table 2 Model evaluation metrics of the ensemble classifiers

S. no	Algorithm technique	Accuracy (%)	Precision (%)	Recall (%)	F1 score (%)
1	Decision tree classifier	99.17	99.70	99.45	99.58
2	K neighbors classifier	99.66	99.67	1	99.83
3	Logistic regression	99.66	99.67	1	99.83

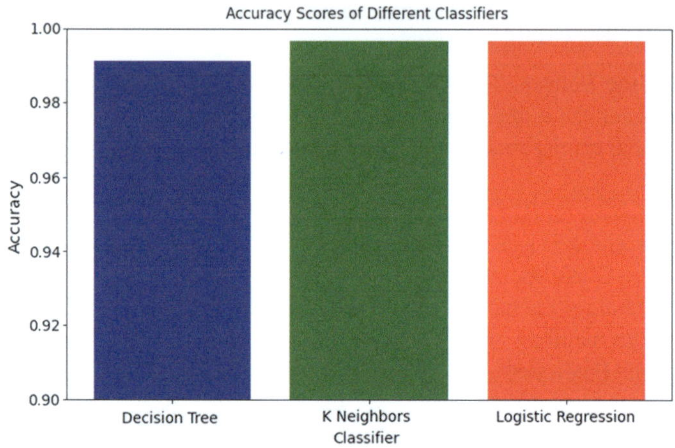

Fig. 3 Accuracy scores of the ensemble classifiers

4.5 Constructing an Ensemble with Hard Voting Classifier

In this stage, we instantiated three different classifiers logistic regression, decision tree, and k-nearest neighbours. Each classifier is then trained on the training data. Then, a voting classifier ensemble is created, incorporating the three individual classifiers with a 'hard' voting strategy. The ensemble model is trained on the training data as well. Predictions are made using the ensemble model on the test data, and the accuracy of the ensemble model is calculated and printed. It emphasizes the potential benefits of using voting classifier ensembles to improve model performance by combining the strengths of different models and mitigating individual weaknesses. An output accuracy of 99.67% indicates that the ensemble model achieved a high level of accuracy on the test data, suggesting that the combination of the three classifiers effectively contributed to making accurate predictions.

4.6 Ensemble of Weighted Soft Voting Classifiers

Here we construct an ensemble of classifiers using a weighted soft voting strategy. It begins by loading a dataset containing information about infants' status at the time of a report and splits it into features and target variables for both training and testing sets. Three distinct classifiers, namely logistic regression, decision tree, and k-nearest neighbours, are instantiated and trained on the training data independently. Following this, a Voting Classifier ensemble is formed, employing soft voting where predictions from each classifier are weighted according to their performance. The weights assigned to each classifier are manually set, with logistic regression being given greater weight compared to the other models. The ensemble model is then trained on the training data, and predictions are made on the test data. The resulting accuracy, standing at 99.67%, indicates a high level of predictive performance, suggesting that this ensemble model effectively combines the strengths of the individual classifiers to make accurate predictions regarding infants' living status at the time of the report.

4.7 Bagging Logistic Regression Classifier and Feature Importance Assessment

The SHAP (SHapley Additive exPlanations) model plays a crucial role in understanding feature importance in predictive models like the Bagging classifier. Finally, we trained a model using a technique called Bagging, which combines multiple logistic regression models for better performance. Created a Bagging classifier that combines 10 Logistic Regression models and trained the model on the training data. Using the trained model we made predictions on the validation set. The model accuracy is then evaluated and calculated using the metrics of precision and recall. Then we used an explainer tool (Smart Explainer object) to understand the impact of each feature on individual predictions as shown in Fig. 4. It clearly depicts the contribution of each feature in the prediction of mortality risk as shown in Fig. 5.

SHAP values help understand the contribution of each feature to the model's prediction for each data point. It considers the interaction between features and their impact on predictions. For example, SHAP values reveal that "Number of Prenatal Visits Recode" contributes 25% to the model's prediction of mortality risk. This means that, on average, this feature has a significant influence on the model's decision-making process. "Cigarettes 2nd trimester" consistently contributes to predictions across many data points, it suggests that this feature is generally important for the model's performance. It also reveals the different features contribute varying percentages to the model's predictions like "Month Prenatal Care Began Recode" at 11.43%. Additionally, "Cigarettes 1st trimester" accounts for 8% of the prediction, while "Mothers Single Year of Age" contributes 7%. "Number of previous Cesarean Deliveries" and "BMI" contribute 4 and 3%, respectively. Lastly, "chronic hypertension" and "prepregnancy diabetes" each contribute 1%. This breakdown helps understand the relative importance of each feature in predicting mortality risk.

We obtained an accuracy of 99.68% refers to the overall correctness of the model's predictions across all classes (survival and non-survival of infants). It measures the proportion of correct predictions (both true positives and true negatives) out of the

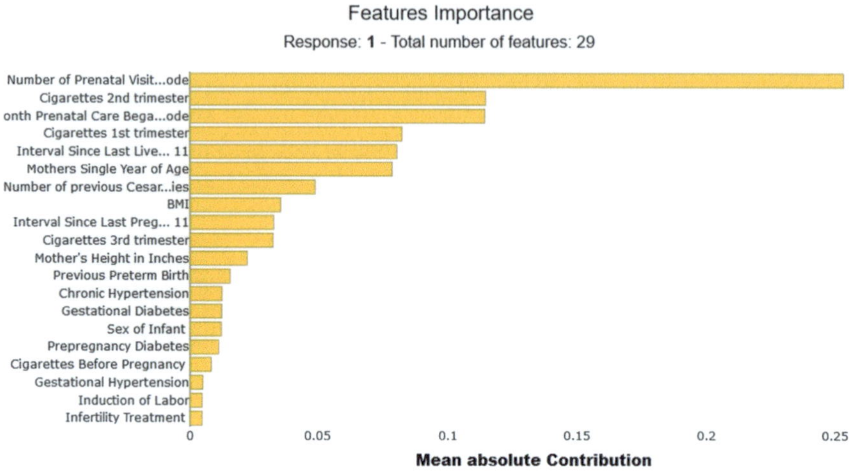

Fig. 4 Feature importance analysis with smart explainer

total number of predictions made by the model, with precision as 99.68% deals with false positives. It measures the proportion of true positive predictions among all positive predictions made by the model. A perfect score of 1.0 (recall) indicates the model identified all the infants who did not survive in the validation set (no false negatives). F1 score (0.9984) this metric combines precision and recall, providing a balanced view.

5 Results and Discussion

5.1 Comparative Study of Machine Learning Models for Predicting Infant Mortality Risk

Our research aimed to predict infant mortality risk using machine learning techniques, analyzing data from the National Center for Health Statistics (NCHS) with over 40,000 records and 30 attributes per case. Initially, logistic regression was employed, yielding a precision of 0.83 for surviving infants but lacking in identifying non-surviving infants. To address this, ensemble models were explored. The RandomForest classifier exhibited exceptional performance with an accuracy of 99.66%, precision of 99.67%, recall of 99.98%, and an F1 score of 99.83%, indicating a strong balance between precision and recall. The ensemble model demonstrated robust performance and attained high accuracy with hard and weighted soft voting leveraging the strengths of individual classifiers to achieve 99.67% accuracy. Finally, we tried Bagging Classifier with Logistic Regression which achieves exceptional accuracy of 99.68% in predicting infant mortality, corroborated by feature

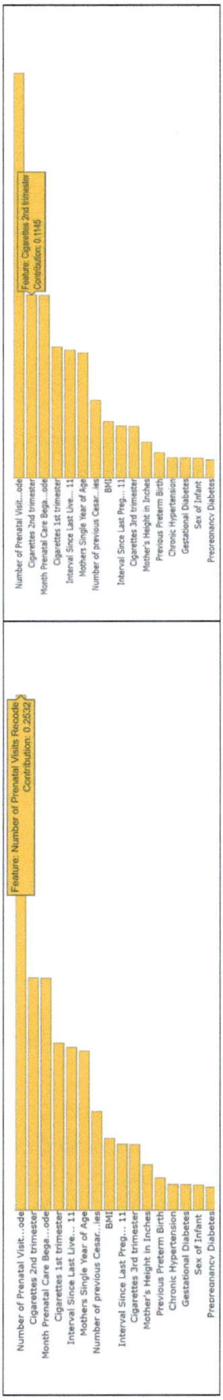

Fig. 5 Feature contributions in predicting infant mortality risk

Table 3 Performance of the various methods on the dataset

Classification model	Performance metric			
	Accuracy%	Precision%	Sensitivity (%)	F1 score (%)
Logistic regression (for survival)	82.75	83	1	91
Random forest classifier	99.66	99.67	99.98	99.83
Ensemble classifier (hard voting)	99.67	99.67	1	99.83
Ensemble weighted soft voting classifier	99.67	99.67	1	99.83
Bagging classifier with logistic regression	99.68	99.68	1	99.84

importance insights provided by SHAPASH. Moreover, a Bagging Logistic Regression Classifier combining 10 models achieved the best accuracy of 99.68%, with a precision and recall of 99.68% and 1.0 respectively, indicating no false negatives and providing a balanced view of the model's performance with an F1 score of 99.84%. The performance of all these models on the infant mortality dataset is shown in Table 3. These models not only accurately predict infant mortality but also provide insights into the importance of features, enhancing their applicability in medical decision-making and improving healthcare outcomes for infants.

6 Conclusion

The integration of Radio Frequency Identification (RFID) technology and sensors in healthcare has led to significant advancements, particularly in monitoring infants and reducing mortality rates. By combining these technologies with Machine Learning (ML), we have developed an interpretable framework capable of early detection and intervention, thus addressing the pressing issue of neonatal mortality, especially within the critical first week of life. Our analysis of infant records using logistic regression highlighted the need for improved accuracy in identifying non-survivors, which was effectively addressed by ensemble models such as the BaggingClassifier and RandomForest classifier. These models exhibited exceptional performance, with accuracy rates exceeding 99%, precision and recall scores indicating minimal false positives and false negatives, and high F1 scores reflecting a balanced evaluation of the models. Utilizing the SHAPASH Smart Explainer object enabled us to discern the significance of features influencing infant mortality prediction, thereby enhancing the interpretability and practicality of our framework. The insights gleaned from these models not only advance our comprehension of infant mortality risk factors but also furnish invaluable guidance for medical decision-making, ultimately fostering improved healthcare outcomes for infants.

References:

1. Victora CG, Adair L, Fall C, Hallal PC, Martorell R, Richter L, Maternal and Child Undernutrition Study Group (2008) Maternal and child undernutrition: consequences for adult health and human capital. The Lancet 371(9609):340–357
2. Subramanyam MA, Kawachi I, Berkman LF, Subramanian SV (2011) Is economic growth associated with reduction in child undernutrition in India? PLoS Med 8(3):e1000424
3. Chalasani S, Rutstein S (2014) Household wealth and child health in India. Popul Stud 68(1):15–41
4. Liou L, Kim R, Subramanian SV (2020) Identifying geospatial patterns in wealth disparity in child malnutrition across 640 districts in India. SSM-Popul Health 10:100524
5. Sankar MJ, Neogi SB, Sharma J, Chauhan M, Srivastava R, Prabhakar PK, Khera A, Kumar R, Zodpey S, Paul VK (2016) State of new born health in India. J Perinatol 36(Suppl 3):S3–S8. https://doi.org/10.1038/jp.2016.183
6. Kumar P, Patel R, Chauhan S, Srivastava S, Khare A, Patel KK (2020) Does socio-economic inequality in infant mortality still exists in India? An analysis based on National Family Health Survey 2005–06 and 2015–16. Clin Epidemiol Glob Health
7. Raj A, Johns NE, McDougal L, Trivedi A, Bharadwaj P, Silverman JG, Singh A (2019) Associations between sex composition of older siblings and infant mortality in India from 1992 to 2016. EClinicalMedicine 14:14–22
8. Mishra SK, Ram B, Singh A, Yadav A (2018) Birth order, stage of infancy and infant mortality in India. J Biosoc Sci 50(5):604–625
9. Pongou R (2013) Why is infant mortality higher in boys than in girls? A new hypothesis based on preconception environment and evidence from a large sample of twins. Demography 50(2):421–444
10. Dandona R, Kumar GA, Henry NJ, Joshua V, Ramji S, Gupta SS, Kassebaum NJ (2020) Subnational mapping of under-5 and neonatal mortality trends in India: The Global Burden of disease study 2000–17
11. Kumar P, Singhal N (2020) Mapping neonatal and under-5 mortality in India. The Lancet 395(10237):1591–1593
12. Bhakoo ON, Kumar P (2013) Current challenges and future prospects of neonatal care in India. Indian J Pediatr 80(1):39–49
13. Raikwar P (2018) A study of neonatal admission pattern and outcome from rural Haryana. Mortality 19:12–16
14. Arora S, Yadav P, Bajaj H, Thakur AS, Mittal M, Gupta MR, Arora R (2019) Improving clinical outcomes of very low birth weight infants: Implementation of standardized management guidelines in tertiary care hospital in Haryana. Int J Pediatr Adolesc Medicine
15. Sunjaya AP, Sunjaya AF (2018) Diabetes in pregnancy and infant mortality: link with glycemic control. Diabetes Metab Syndr 12(6):1031–1037

MIX
Papier aus verantwortungsvollen Quellen
Paper from responsible sources
FSC® C105338

If you have any concerns about our products,
you can contact us on
ProductSafety@springernature.com

In case Publisher is established outside the EU,
the EU authorized representative is:
Springer Nature Customer Service Center GmbH
Europaplatz 3, 69115 Heidelberg, Germany

Printed by Libri Plureos GmbH
in Hamburg, Germany

Milestones in Drug Therapy
MDT

COX-2 Inhibitors

Edited by M. Pairet and J. van Ryn

Springer Basel AG

Editors

Michel Pairet
Boehringer Ingelheim Pharma GmbH & Co. KG
Department of Pulmonary Research
Birkendorfer Str. 65
D-88397 Biberach a. d. Riss
Germany

Joanne van Ryn
Boehringer Ingelheim Pharma GmbH & Co. KG
Department Drug Discovery Support
Birkendorfer Str. 65
D-88397 Biberach a. d. Riss
Germany

Library of Congress Cataloging-in-Publication Data
COX-2 inhibitors / edited by M. Pairet and J. van Ryn.
p. ; cm. -- (Milestones in drug therapy)
Includes bibliographical references and index.
ISBN 3-7643-6901-9 (alk. paper)
1. Cyclooxygenase 2--Inhibitors--Therapeutic use. I. Pairet, M. (Michel), 1959-
II. Van Ryn, J. (Joanne), 1958- III. Series.
[DNLM: 1. Anti-Inflammatory Agents, Non-Steroidal--therapeutic use. 2. Cyclooxygenase Inhibitors--pharmacokinetics. QV 95 C877 2004]
RM666.C977C69 2004
616'.0473--dc22 2003063560

Bibliographic information published by Die Deutsche Bibliothek
Die Deutsche Bibliothek lists this publication in the Deutsche Nationalbibliografie; detailed bibliographic data is available in the Internet at <http://dnb.ddb.de>.

ISBN 978-3-0348-9607-8 **ISBN 978-3-0348-7879-1 (eBook)**
DOI 10.1007/978-3-0348-7879-1

Part of Springer Science+Business Media
Printed on acid-free paper produced from chlorine-free pulp. TCF ∞
Cover illustration: This view of the COX-2 channel is from the bottom looking up from the membrane binding domain. COX-2 is shown as a ribbon diagram and the amino acid residues of COX-2 that define the binding site are represented as sticks. The COX-2 inhibitor, colored red, is bound in the channel and represented as sticks. The editors wish to thank Günter Trummlitz for the picture and Gerald Birk for the stylized rendition.

ISBN 3-7643-6901-9
9 8 7 6 5 4 3 2 1 www.birkhäuser-science.com

Contents

List of contributors

Phillip R. Bennett, Imperial College Parturition Research Group, Institute of Reproductive and Developmental Biology, Hammersmith Hospital Campus, Du Cane Road, London W12 ONN, UK; e-mail: pbennett@ic.ac.uk

Jack H. Botting, The William Harvey Research Institute, Charterhouse Square, London EC1M 6BQ, UK; e-mail: r.m.botting@qmul.ac.uk

Regina M. Botting, The William Harvey Research Institute, Charterhouse Square, London EC1M 6BQ, UK; e-mail: r.m.botting@qmul.ac.uk

Christianne J. Buskens, Department of Surgery, Academic Medical Center, University of Amsterdam, 1105 AZ Amsterdam, The Netherlands; e-mail: c.j.buskens@amc.uva.nl

Frank Degner, CDept. Medical Affairs, Boehringer Ingelheim Pharma GmbH, Binger Str. 173, 55216 Ingelheim, Germany; e-mail: degner@ ing.boehringer-ingelheim.com

Tiina-Liisa Erkinheimo, Department of Obstetrics and Gynecology and Helsinki University Central Hospital and Molecular and Cancer Biology Research Program, Biomedicum Helsinki, University of Helsinki, P.O. Box 63, FIN-00014 Helsinki, Finland; e-mail: erkinhei@mappi.helsinki.fi

Caj Haglund, Department of Surgery, Biomedicum Helsinki, University of Helsinki, P.O. Box 63, FIN-00014 Helsinki, Finland; e-mail: caj.haglund@hus.fi

Koichi Hashimoto, Department of Microbiology, School of Medicine, Fukushima Medical University, 1st Hikariga-oka Fukushima, Fukushima, Japan 960-1295; e-mail: don@fmu.ac.jp

Amit S. Kalgutkar, Pharmacokinetics, Dynamics and Metabolism Department, Pfizer Global Research and Development, Groton, CT 06340, USA

Lawrence J. Marnett, Department of Biochemistry, Vanderbilt Institute of Chemical Biology, Vanderbilt-Ingram Cancer Center, and Center in Molecular Toxicology, Vanderbilt University School of Medicine, Nashville, Tennessee 37232-0146, USA; e-mail: larry.marnett@vanderbilt.edu

Brendan F. McAdam, Division of Cardiovascular Medicine, Vanderbilt University Medical Center, Nashville, TN 37232-8802, USA; e-mail: brendan.mcadam@vanderbilt.edu

Aarthi R. Mohan, Imperial College Parturition Research Group, Institute of Reproductive and Developmental Biology, Hammersmith Hospital Campus, Du Cane Road, London W12 ONN, UK; e-mail: a.mohan@imperial.ac.uk

Rahul Nayak, Division of Cardiovascular Medicine, Vanderbilt University Medical Center, Nashville, TN 37232-8802, USA

G. Johan A. Offerhaus, Department of Pathology, Academic Medical Center, University of Amsterdam, 1105 AZ Amsterdam, The Netherlands; e-mail: g.j.offerhaus@amc.uva.nl

Michel Pairet, Department of Pulmonary Research, Boehringer Ingelheim Pharma GmbH & Co. KG, Birkendorfer Str 65, D - 88397 Biberach, Germany; e-mail: Michel.Pairet@bc.boehringer-ingelheim.com

R. Stokes Peebles, Jr., Center for Lung Research, T-1217 MCN, Vanderbilt University Medical Center, Nashville, TN 37232-2650, USA; e-mail: stokes.peebles@vanderbilt.edu

Leena Pohjola, Department of Pathology and Helsinki University Central Hospital and Molecular and Cancer Biology Research Program, Biomedicum Helsinki, University of Helsinki, P.O. Box 63, FIN-00014 Helsinki, Finland; e-mail: lkpohjol@mappi.helsinki.fi

Kim D. Rainsford, Biomedical Research Centre, Sheffield Hallam University, Howard Street, Sheffield S1 1WB, UK; e-mail: k_d.rainsford@shu.ac.uk

Johanna Rintahaka, Department of Pathology and Helsinki University Central Hospital and Molecular and Cancer Biology Research Program, Biomedicum Helsinki, University of Helsinki, P.O. Box 63, FIN-00014 Helsinki, Finland; e-mail: rintahak@mappi.helsinki.fi

Ari Ristimäki, Department of Pathology and Helsinki University Central Hospital and Molecular and Cancer Biology Research Program, Biomedicum Helsinki, University of Helsinki, P.O. Box 63, FIN-00014 Helsinki, Finland; e-mail: ari.ristimaki@hus.fi

Kirsi Saukkonen, Department of Pathology and Helsinki University Central Hospital and Molecular and Cancer Biology Research Program, Biomedicum Helsinki, University of Helsinki, P.O. Box 63, FIN-00014 Helsinki, Finland; e-mail: kisaukko@mappi.helsinki.fi

Anna Sivula, Department of Pathology and Helsinki University Central Hospital and Molecular and Cancer Biology Research Program, Biomedicum Helsinki, University of Helsinki, P.O. Box 63, FIN-00014 Helsinki, Finland; e-mail: anna.sivula@helsinki.fi

Dirk O. Stichtenoth, Institute of Clinical Pharmacology, Medizinische Hochschule Hannover, 30623 Hannover, Germany; e-mail: Stichtenoth.Dirk@mh-hannover.de

Guenter Trummlitz, Boehringer Ingelheim GmbH, Birkendorferstr. 65, 88397 Biberach, Germany; e-mail: trummlitz@t-online.de

J. Jan B. van Lanschot, Department of Surgery, Academic Medical Center, University of Amsterdam, 1105 AZ Amsterdam, The Netherlands; e-mail: j.j.vanlanschot@amc.uva.nl

Bastiaan P. van Rees, Department of Pathology, Academic Medical Center, University of Amsterdam, 1105 AZ Amsterdam, The Netherlands; e-mail: b.van.rees@slingeland.nl

Joanne van Ryn, Department of Drug Discovery Support, Boehringer Ingelheim Pharma GmbH & Co. KG, Birkendorferstr. 65, 88397 Biberach, Germany; e-mail: Joanne.vanRyn@bc.boehringer-ingelheim.com

Timothy D. Warner, TheWilliam Harvey Research Institute, Bart´s and the London Queen Mary's School of Medicine and Dentistry, London EC1M 6BQ, UK; e-mail: t.d.warner@qmul.ac.uk

Preface

The pioneer work by Sir John Vane and collaborators demonstrating that aspirin-like drugs work by blocking cyclooxygenase (COX) had a huge impact on our understanding of the mechanism of action of the so-called non steroidal anti-inflammatory drugs (NSAIDs) and on their rational use. The subsequent discovery that two forms of COX exist, COX-1 and COX-2, had further massive implications. It helped us understand the mechanism of the wanted *versus* unwanted effects of NSAIDs. It was also the basis for a generation of new drugs, the COX-2 inhibitors. In addition, the discovery reactivated scientific work on the physiological and pathological roles of prostanoids and on the potential use of COX-2 inhibitors in indications other than pain and arthritis, such as cancer, pre-term delivery and Alzheimer's disease.

Much has been written in this field over the past 10 years; the danger of writing again on this topic is to come out with "just another book on COX-2". We are convinced, however, that more than 10 years after the discovery of COX-2 and with the first selective COX-2 inhibitors now on the market for several years, we may just now have the needed objectivity to answer a few key questions, based on both scientific and clinical data and far from the loud marketing communications which prevailed during the launch phases of the first selective COX-2 inhibitors.

We know for example that the theory of a bad COX-2 and a good COX-1 needs to be qualified, that COX-2 is not solely an inducible enzyme during inflammation, but can also be constitutively expressed in various tissues and is essential for physiological functions such as reproduction. In addition, a large clinical experience on the gastrointestinal tolerability of COX-2 inhibitors and new experimental data on the respective roles of COX-1 and COX-2 in mucosal protection and ulcer healing, allow us to better assess the level of improvement achieved by COX-2 inhibitors.

We also know that COX-2 inhibitors lack efficacy in patients with Alzehimer's disease but have a clear effect in Familial Adenomatous Polyposis and have therapeutic potential in various forms of cancer. All this could not have been predicted a few years ago based only on scientific data on the expression and function of COX-2 in tissues of interest. We also now realise that COX-2 inhibitors may have side-effects standard NSAIDs do not have, such as an increase in thrombotic risk. This is still a matter of intensive debate, but if confirmed may significantly change our risk/benefit analysis between COX-2 inhibitors and NSAIDs. Finally, the recent discovery of a COX-3 isoenzyme will further widen and possibly complicate our understanding of the physiopathology of COX.

We realise that this book will not answer all the questions the reader may have on the true therapeutic benefit of COX-2 inhibitors, on their potential in cancer prevention and/or therapy, or on the respective roles of the various COXs in pain. We simply hope that this book will help the reader to better understand the facts, limitations and hypotheses on COX and its inhibitors.

Michel Pairet and Joanne van Ryn

Biberach a.d. Riss,
October 2003

COX-2 Inhibitors
Edited by M. Pairet and J. van Ryn

The discovery of COX-2

Regina M. Botting and Jack H. Botting

The William Harvey Research Institute, Charterhouse Square, London EC1M 6BQ, UK

History of cyclooxygenase

For over 3000 years various herbal medicines, such as extracts of willow bark, or dried leaves of myrtle or meadowsweet have been used with apparent anecdotal success for the treatment of the pain and swelling of inflammatory disease. By the middle of the nineteenth century the active constituent of these various herbal remedies was found to be salicylate, and the better tolerated acetylated derivative, acetylsalicylate was introduced to medicine as aspirin over a hundred years ago. The manifest efficacy of aspirin resulted in its wide use as an antipyretic and as an analgesic for the pain associated with inflammatory conditions [1]. However, it soon became apparent that aspirin could cause gastrotoxicity, a fact demonstrated unequivocally in 1938 by the endoscopic studies of Douthwaite who observed erosions and ulcers in the gastric mucosa of patients on long-term aspirin therapy [2].

The search for an aspirin-like drug without gastrotoxicity resulted in the synthesis of many compounds of diverse chemical nature with anti-inflammatory actions. About 30 have entered the market over the last 30–40 years, but none of these so-designated nonsteroidal anti-inflammatory drugs (NSAIDs) was totally free of a gastrotoxic effect. Gastrotoxicity due to NSAID therapy is a significant problem. It has been estimated that in the USA alone between 10,000 and 20,000 patients die each year from NSAID-associated toxicity such as perforations and bleeding from the stomach [3].

The interesting phenomenon that chemicals of diverse structure should share the common therapeutic effects of antipyresis, analgesia and anti-inflammatory actions (these themselves of a diverse nature), and also share a common toxic action, suggested that a single biochemical effect could explain all the pharmacological actions of NSAIDs. Over the years, many theories for this common mechanism were proposed and subsequently rejected. Eventually, the enigma was resolved in 1971 by the elegantly simple experiments of Vane, using guinea pig lung homogenates who showed that aspirin and other NSAIDs inhibited the synthesis of the lipid mediators, prostaglandins (PGs) [4]. PGs were known to be pyrogenic, and were subsequently shown to be hyperalgesic and present at inflammatory foci, thus explaining the therapeutic effects of NSAIDs. That the common side-effects of NSAIDs were also due to inhibition of PG synthesis was apparent when PGs were shown to be cytoprotective on the gastric mucosa, and could maintain renal blood flow when renal circulation was compromised [1].

PGs are formed from arachidonic acid mobilised from membrane phospholipids by a phospholipase enzyme. Arachidonic acid is acted upon by the microsomal enzyme cyclo-oxygenase (COX) to form the cyclic endoperoxides PGG_2 and PGH_2. These unstable endoperoxides are then isomerised, enzymatically or non-enzymatically, into various prostanoids such as thromboxane A_2 (TxA_2), prostacyclin (PGI_2) and PGD_2, E_2 and $F_{2\alpha}$ (Fig. 1).

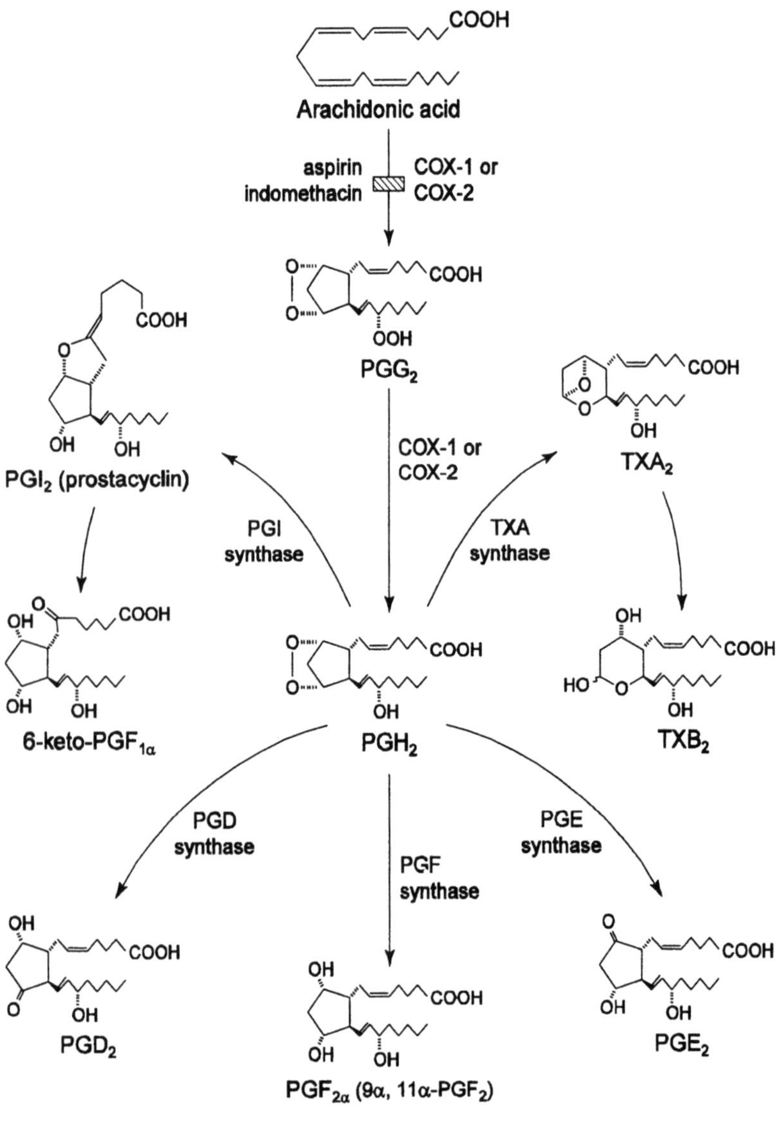

Figure 1. The arachidonic acid cascade.

Implicit in the establishment of the mechanism of action of NSAIDs was the assumption that since all the pharmacological actions depended on the inhibition of COX, it would be impossible to divorce the therapeutic effects from the toxic actions of these agents. However, there were some minor inconsistencies in this theory. Epidemiological studies showed that the severity of gastric toxic effects of different NSAIDs varied when the drugs were used at comparable anti-inflammatory doses. Ibuprofen, for example, causes less damage to the stomach than ketoprofen [5]. Similarly, the NSAID nimesulide was effective in models of inflammatory disease such as the carrageenan injected rat paw, yet was poorly active in inhibiting PG synthesis in conventional systems [6].

The existence of COX isozymes was a likely explanation for these inconsistencies, and over 30 years ago Flower and Vane suggested that paracetamol, which lacks anti-inflammatory actions, exerted its antipyretic effect by inhibition of a distinct isoform of COX in brain tissue [7]. More recently a number of workers showed that PG synthesis could be upregulated in inflammatory conditions by induction of synthesis of more COX enzyme. Needleman and colleagues, for example, showed that in a model of the inflammatory condition of hydronephrosis, produced in perfused rabbit kidney by prior ureteric obstruction, there was a marked increase in PGE_2 release following injection of bradykinin. This increased release of PG could be prevented by prior treatment with the protein synthesis inhibitors cycloheximide or actinomycin D. Of significance, aspirin inhibited the release of PG during the early stages of perfusion but became less effective as the experiment progressed. Needleman concluded that the increase in PG production in this model of inflammation was secondary to increased synthesis of COX enzyme [8].

Similar results were obtained from experiments using isolated cells stimulated with various inflammatory cytokines or growth factors. Human umbilical vein endothelial cells treated with interleukin-1 (IL-1) or tumour necrosis factor resulted in increased production of PGI_2 together with an increase in COX protein and mRNA [9], effects reduced by treatment with actinomycin D. Induction of COX protein accompanied by an increase in PG production was also demonstrated in murine smooth muscle cells and osteoblasts treated with epidermal growth factor [10, 11], 3T3 cells treated with PDGF [12], and in human dermal fibroblasts treated with IL-1 [13].

Needleman and his group [14–16] extended their earlier observations by demonstrating increased PG synthesis in human monocytes *in vitro* and in mouse macrophages *in vivo* due to activation with lipopolysaccharide (LPS). The increase in PG release was inhibited by prior treatment with dexamethasone, yet the basal level of COX activity was not affected. Needleman concluded that there may be various "pools" of COX within cells, one constitutive, another inducible, being upregulated by inflammatory stimuli [17]. The existence of such an inducible COX enzyme was firmly established by the discovery by Simmons and his colleagues at Brigham Young University of a novel PG synthase, encoded by an entirely different gene, induced in chick embryo fibroblasts by v-src, serum or phorbol esters [18, 19]. Subsequently, Herschman and his colleagues [20] found a similar gene in the mouse, as did Simmons et al. who, by cross hybridisation, cloned and characterised the murine homologue of the v-src-inducible COX [21] and showed the mRNA to be induced in an immediate early fashion by various mutagens. The deduced protein structure of the inducible enzyme was found to be 60% homologous to constitutive COX, and the two enzymes were designated COX-1 (constitutive) and COX-2 (inducible).

Although COX-1 is constitutive in most cells, very little COX-2 is present in resting cells, however, it is powerfully induced in fibroblasts, endothelial cells and vascular smooth muscle by mitogens or some cytokines, and in monocytic cells by LPS. These observations have led to the general hypothesis that the constitutive COX-1 is involved in normal physiological processes. This includes synthesising thromboxane A_2 involved in platelet aggregation, PGE_2 and PGI_2 to protect the gastric mucosa and PGI_2 in the vascular endothelium to maintain dilatation of blood vessels and to inhibit formation of inappropriate platelet thrombi. Although recent work suggests that formation of PGI_2 from endothelial cells may be due to COX-2 constantly induced in the cells by shear stress [22].

The inducible COX-2, on the other hand, is believed to be primarily responsible for the production of PGs in inflamed tissues, which cause the swelling and hyperalgesia associated with inflammatory disease (see Fig. 2). Certainly, increased COX-2 expression, which parallels increased PGE_2 synthesis, is demonstrable in models of inflammatory disease such as the murine air pouch, whilst COX-1 expression is unchanged [23]. Increased expression of COX-2 has also been demonstrated in synovial tissue taken from patients with rheumatoid arthritis, compared to tissues from patients with osteoarthritis or no arthritic pathology [24].

The acceptance that the COX-2 isoform was responsible for the synthesis of PGs at pathological foci and COX-1 for the synthesis of the beneficial, "housekeeping" PGs, implied that selective inhibitors of COX-2 would exert the therapeutic actions of NSAIDs without the previously regarded inevitable side-effect of gastrotoxicity. This led many workers to undertake studies to establish the relative activity of established NSAIDs on COX-1 and COX-2 by comparison of the concentration of NSAID necessary to inhibit the activity of COX by 50% (IC_{50}). Although simple in concept these measurements of selectivity were problematic in practice. Differences in source of enzyme, time of incubation, method of induction of COX-2 and protein concentration resulted in variation in ratios and even differences in IC_{50} values. However, the human whole blood assay described by Patrignani et al. [25] and subsequently modified by Warner [5], provided a reliable method with similar incubation times, appropriate human target cells and performed in the presence of plasma proteins. This method was reproducible and the selectivity of established NSAIDs such as nimesulide, meloxicam and etodolac towards inhibition of COX-2 rather than COX-1 was shown and this correlated well with data from large scale clinical trials and post-marketing surveillance studies, showing the substances to be accompanied by relatively low gastrotoxicity.

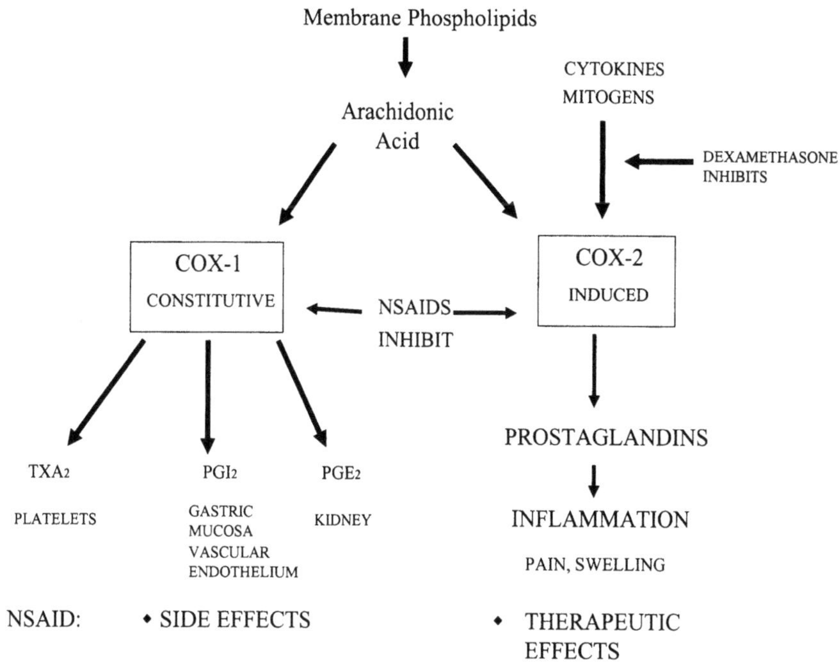

Figure 2. The side-effects of NSAIDs are caused by inhibition of the constitutive enzyme COX-1, which synthesises prostaglandins that serve essential physiological functions such as causing appropriate platelet aggregation, protection of the gastric mucosa, inhibition of thrombogenesis and maintenance of renal function. The therapeutic effects of NSAIDs are due to inhibition of COX-2, an enzyme induced by various factors released by bacteria, the vascular endothelium or other cells involved in the inflammatory response.

Structural basis for COX-2 selectivity

The three-dimensional structures of both COX-1 [26] and COX-2 [27] have been determined and are remarkably similar. Each enzyme is composed of three distinct folding units: an epidermal growth factor-like domain, a membrane binding section and a C-terminal enzymatic domain containing both the COX and peroxidase active sites. Through the membrane-binding domain the enzymes integrate into a single leaflet of the membrane lipid bilayer, thus positioning the COX site to allow access of the arachidonic acid substrate that is mobilised from membrane phospholipids.

Whereas NSAIDs generally act competitively to prevent access of arachidonic acid to the active site, aspirin acts in a unique manner to acetylate a serine residue at position 530 in COX-1, or the serine at an analogous position (516) in COX-2. While acetylation with aspirin will completely prevent the metabolism of arachidonic acid by COX-1, with COX-2 the substrate can reach the peroxidase site and be metabolised to 15-hydroperoxyeicosatetraneoic acid (15-HPETE) [28].

Although the active sites of COX-1 and COX-2 are similar, there are differences which have been utilised by medicinal chemists to synthesise molecules which have a selective action on COX-2. Of crucial significance is position 523, which in COX-2 is valine and in COX-1 isoleucine. This difference of a single methyl group is sufficient to allow access of a potential inhibitor to a side pocket in COX-2. In addition, proximity to an arginine residue at position 513 (histidine in COX-1), which provides hydrogen bonding for an inhibitor of a structure that enables it to extend into the side pocket [29].

A second significant difference between COX-2 and COX-1 is at position 503, which is the aromatic amino acid phenylalanine in COX-1 but the relatively small, nonaromatic leucine in COX-2; this allows leucine at position 384 to reorient its methyl side chain away from the enzymatic site and thus extend the space available for a larger inhibitor molecule at the active site of the COX-2 enzyme [30]. The differences in the structure of the COX channels of the COX isozymes are illustrated diagrammatically in Figure 3, in which the same numbering of the amino acids in the two enzymes is adopted for simplicity.

Drugs specifically designed to inhibit COX-2 have entered the market and two such have undergone comprehensive clinical investigation. Celecoxib is the prototype of a number of highly selective COX-2 inhibitors produced by Pharmacia (Pfizer). Celecoxib was shown to be an effective analgesic as assessed in patients after tooth extraction, and endoscopic studies after 3

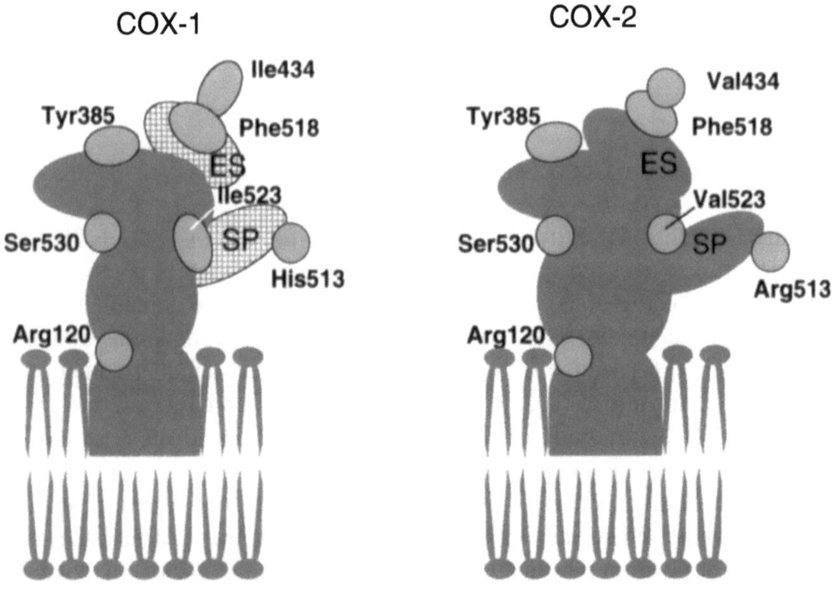

Figure 3. Amino acid substitutions between COX-1 and COX-2. The hatched areas in COX-1 are those that are more accessible in COX-2 due to the amino acid substitutions. SP: side pocket, ES: extra space.

months chronic usage showed no more gastric damage than with placebo. In a large scale clinical trial (the Celecoxib Long-term Arthritis Safety Study or CLASS, in 8000 patients) it was shown that, at 6 months, celecoxib produced less gastrotoxicity than the comparator drugs ibuprofen and diclofenac (though this failed to reach statistical significance at 12 months) [31]. Celecoxib is approved both for osteo- and rheumatoid arthritis in many countries.

Rofecoxib, produced by Merck, Sharp and Dohme, has been approved by many regulatory authorities for arthritis and acute pain. Rofecoxib in doses of 25–50 mg was an effective analgesic for dental pain, dysmenorrhoea and osteoarthritis. In a 6-month endoscopic study, with a dose of rofecoxib 10-fold greater than that required for an effective anti-inflammatory effect, gastric damage was no greater than that seen with placebo. In a large-scale, 9-month trial involving 8000 patients (the Vioxx Gastrointestinal Outcome Research, VIGOR) rofecoxib was compared with naproxen [32]. Rofecoxib was shown to produce substantially less gastrotoxicity than naproxen. Of possible significance, in the VIGOR study the incidence of myocardial infarction in the rofecoxib group was 0.4%, compared with 0.1% in the naproxen treated patients. This observation requires further study and is addressed in Chapter 7.

Newer even more selective COX-2 inhibitors such as valdecoxib [33], etoricoxib [34] and lumiracoxib [35] are currently either in Phase III development or have been introduced on the market. While there is less clinical experience with these agents to date, they all seem to be effective anti-inflammatory agents with reduced gastrointestinal toxicity.

Additional uses for selective COX-2 inhibitors

The availability of NSAIDs with clearly a vastly reduced propensity for causing gastrointestinal toxicity raised the possibility of the chronic use of these drugs for additional indications. Thus far, selective COX-2 inhibitors have been postulated to have potential in some cancers, premature labour and Alzheimer's disease.

COX-2 inhibitors and cancers

There is persuasive epidemiological evidence for an association of chronic use of NSAIDs with reduced incidence of gastrointestinal carcinomas [36–42]. Regular use of aspirin and other NSAIDs can halve the risk of development of colon cancer. Clinical and experimental studies have provided further evidence to justify this beneficial effect of NSAIDs; colonic tumour tissue has been shown to contain high levels of COX-2 protein but contiguous, normal tissue contains only COX-1 [43]. Patients with the inherited disorder of familial adenomatous polyposis (FAP), a condition characterised by a plethora of colonic polyps with eventual progression to tumour, manifest a marked reduction in polyp number on treatment with the NSAID sulindac [44, 45]. Similarly, COX inhibitors reduce tumours in azoxymethane-induced colorectal carcinogenesis in rat [46] and selective COX-2 inhibitors reduce tumour or polyp number in animal models of FAP such as minimal intestinal neoplasias (min) mice or APC (adenomatous polyposis coli gene mutant) knockout mice.

COX-2 is expressed in tumour tissue in many cancers other than colorectal carcinomas. Increased levels of COX-2 have been shown in human lung squamous cell and adenocarcinomas, squamous skin carcinomas and in tumours of the breast, liver, pancreas, prostate and cervical lymph nodes (reviewed in [47] and further discussed in Chapter 11).

Premature labour

The content of COX-2 mRNA in the uterus and the production of PGs increases just prior to and during labour, indicating an involvement of PGs (particularly $PGF_{2\alpha}$) in parturition. Premature labour, perhaps induced by inappropriate upregulation of COX-2 due to intrauterine infection, can be suppressed by COX inhibitors such as indomethacin, but inhibition of COX-1 will result in early closure of the ductus arteriosus which is maintained patent until birth by PGs synthesised by constitutive COX-1 [48]. Selective COX-2 inhibitors were thus thought to have a use in premature labour, producing a tocolytic effect without the problem of early closure of the ductus. The COX-2-selective inhibitor nimesulide has been used with some success in premature labour [49] but has been shown subsequently to produce oligohydramnios in 50% of exposed fetuses [50]. The side-effects of nimesulide are reversible, however, it is accepted that the use of this drug for premature labour should be reserved for patients at high risk of early preterm delivery and where adequate patient surveillance can be maintained. Since some of the renal side-effects of nimesulide may be due to actions other than COX inhibition, such as calcium channel blockade, thus other selective COX-2 inhibitors are at present under trial for preterm labour ([51], see Chapter 10).

Alzheimer's disease

Retrospective epidemiological studies suggest that long-term use of standard NSAIDs reduces the incidence of Alzheimer's disease (AD) [52], a benefit not seen with paracetamol or low-dose aspirin. Due to lack of suitable animal models there is little direct experimental confirmation, but it has been assumed that there is an inflammatory component in the pathogenesis of AD and *post mortem* studies have shown an increase of COX-2 expression in the cortex of AD patients [53]. Prolonged treatment with ibuprofen reduced the number of amyloid plaques in the brain of APP transgenic mice secondary to a lowered brain content of soluble Aβ peptide, which on aggregation forms the amyloid plaques [54]. Large placebo controlled trials of the highly selective COX-2 inhibitors celecoxib and rofecoxib have apparently shown no beneficial effect on the progress of AD [55, 56], which suggests that COX-2 may not be an appropriate target. It has been suggested that some NSAIDs may lower plaque formation through a mechanism independent of COX inhibition, possibly by inhibition of γ-secretase, which is the enzyme that cleaves the amyloid precursor protein into the plaque-forming Aβ42 peptide [57]. Such a prophylactic effect of NSAIDs would be in accord with the Cache County epidemiological study, which showed that general NSAID therapy does reduce incidence of AD if the drugs are used well before the onset of symptoms of dementia, whereas more recent exposure offered little protection [58].

Other cyclooxygenases

Although the discovery of COX-2 resolved the puzzle of why some NSAIDs manifested marked gastrotoxicity yet others, more selective for COX-2 were relatively less toxic, some enigmas remained. As mentioned above, acetaminophen, like conventional NSAIDs, has marked antipyretic and analgesic activity yet lacks the typical anti-inflammatory and antiplatelet activities of this group of drugs. Twenty years ago it was suggested that acetaminophen might inhibit a cyclooxygenase in the brain, and that various, perhaps tissue-specific, cyclooxygenases may exist that are selectively inhibited by different NSAIDs [7].

Some light has recently been shed upon this field by Simmons and his colleagues who have identified a new COX-1 variant in dog and human brain which is sensitive to acetaminophen and which has been designated COX-3 [59]. COX-3 has the same structure as COX-1 except that the mRNA has retained intron-1, which is translated into a 30 amino acid extension to the enzyme protein. COX-3 was ectopically expressed in insect cells and found to be more sensitive to inhibition by acetaminophen than either COX-1 or COX-2. At least two other partial COX-1 mRNAs were characterised but only one, PCOX-1a, had retained intron-1 in transcription (see Fig. 4). The partial COX-1 proteins did not form prostanoids and their functions remain unknown.

Human COX-3 was found in highest concentrations in the cerebral cortex followed by the heart and skeletal muscle. The possible function of COX-3 in organs other than the brain is as yet unknown. Simmons had previously found another COX isoform, sensitive to inhibition by acetaminophen, formed by a murine macrophage cell line (J774.2) treated with high concentrations of diclofenac [60]. However, unlike COX-3, this enzyme was insensitive to aspirin, and was considered to be a variant of COX-2, since a protein immunoreactive to COX-2 was co-induced together with the cyclooxygenase activity. Thus there may be yet other acetaminophen-sensitive cyclooxygenases derived from COX-2.

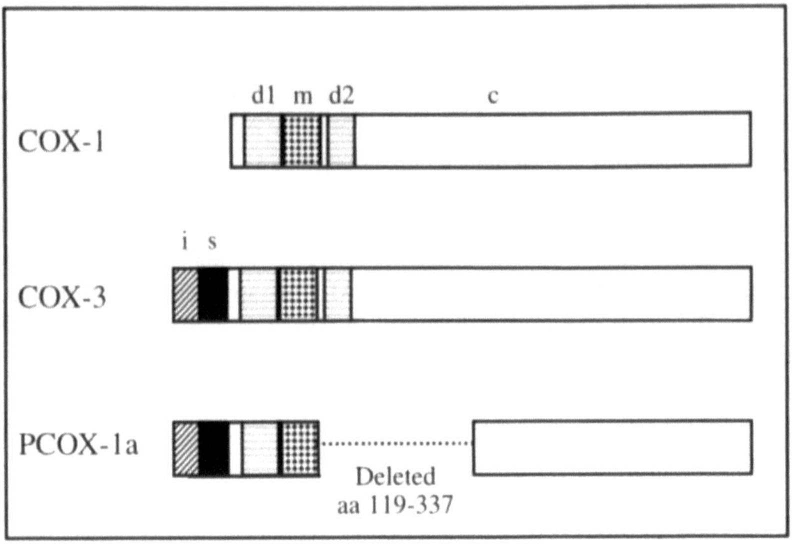

Figure 4. Structure of COX-3 and PCOX-1a. A schematic representation of the domains of COX-3 and PCOX-1a in comparison to COX-1. s: signal peptide; d1: dimerization domain/EGF-like domain; d2: dimerization domain 2; m: membrane binding domain; c: catalytic domain; i: 90-bp sequence encoded by intron 1. PCOX-1b is identical to PCOX-1a except that it lacks intron 1. Amino acid numbering is according to residues in sheep seminal vesicle COX-1. (Reproduced from [59] with permission.)

Conclusions

Clearly, the establishment of the existence of the isoform COX-2 has provided a means of reaping the long-recognised benefits of aspirin-like drugs without the sometimes lethal side-effects. Further research will no doubt provide refined treatments for inflammatory conditions and pain, possibly by establishing the optimal selectivity for COX-2, or by targeting novel isoforms of cyclooxygenase.

References

1 Vane JR, Botting RM (1996) The history of anti-inflammatory drugs and their mode of action. In: N Bazan, J Botting, J Vane (eds): *New Targets in Inflammation Inhibitors of COX-2 or Adhesion Molecules*. Kluwer Academic Publishers and William Harvey Press, London, 1–12
2 Douthwaite A, Lintott CA (1938) Gastroscopic observations of the effect of aspirin and certain other substances on the stomach. *Lancet* 2: 1222
3 Fries J (1996) Towards an understanding of NSAID-related adverse events: the contribution of longitudinal data. *Scand J Rheumatol* 25: 3–8
4 Vane JR (1971) Inhibition of prostaglandin synthesis as a mechanism of action for the aspirin-like drugs. *Nature* 231: 232–235
5 Warner TD, Guiliano F, Vojnovic I, Bukasa A, Mitchell JA, Vane JR (1999) Nonsteroid drug selectivities for cyclo-oxygenase-1 rather than cyclo-oxygenase-2 are associated with human gastrointestinal toxicity: a full *in vitro* analysis. *Proc Natl Acad Sci USA* 96: 7563–7568

6 Carr DP, Henn R, Green JR, Böttcher I (1986) Comparison of the systemic inhibition of thromboxane synthesis, anti-inflammatory activity and gastro-intestinal toxicity of non-steroidal anti-inflammatory drugs in the rat. *Agents and Actions* 19: 374–375
7 Flower RJ, Vane JR (1972) Inhibition of prostaglandin synthetase in brain explains the antipyretic activity of paracetamol (4-acetamidophenol). *Nature* 240: 410–411
8 Morrison AR, Moritz H, Needleman P (1978) Mechanism of enhanced renal biosynthesis of prostaglandins in ureter obstruction. Role of *de novo* protein biosynthesis. *J Biol Chem* 253: 8210–8212
9 Kawakami M, Ishibashi S, Ogawa H, Murase T, Takaku F, Shibata S (1986) Cachectin/TNF as well as interleukin-1 induces prostacyclin synthesis in cultured vascular cells. *Biochem Biophys Res Commun* 141: 482–487
10 Bailey JM, Muza B, Hla T, Salata K (1985) Restoration of prostacyclin synthase in vascular smooth muscle cells after aspirin treatment: regulation by epidermal growth factor. *J Lipid Res* 26: 54–61
11 Yokota K, Kusaka M, Ohshima T, Yamamoto S, Kurihara N, Yoshino T, Kumegawa M (1986) Stimulation of prostaglandin E_2 synthesis in cloned osteoblastic cells of mouse (MC3T3-E1) by epidermal growth factor. *J Biol Chem* 261: 15410–15415
12 Habenicht AJR, Goerig M, Grulich J, Rothe D, Gronwald R, Loth U, Schettler G, Kommerell B, Ross R (1985) Human platelet-derived growth factor stimulates prostaglandin synthesis by activation and by rapid *de novo* synthesis of cyclooxygenase. *J Clin Invest* 75: 1381–1387
13 Raz A, Wyche A, Siegel N, Needleman P (1988) Regulation of fibroblast cyclooxygenase synthesis by interleukin-1. *J Biol Chem* 262: 3022–3028
14 Honda A, Raz A, Needleman P (1990) Induction of cyclo-oxygenase synthesis in human promyelocytic leukemia (HL-60) cells during monocytic or granulocytic differentiation. *Biochem J* 272: 259–262
15 Fu J-Y, Masferrer JL, Seibert K, Needleman P (1990) The induction and suppression of prostaglandin H_2 synthase (cyclooxygenase) in human monocytes. *J Biol Chem* 265: 16737–16740
16 Masferrer JL, Zweifel BS, Seibert K, Needleman P (1990) Selective regulation of cellular cyclooxygenase by dexamethasone and endotoxin in mice. *J Clin Invest* 86: 1375–1379
17 Seibert K, Masferrer JL, Jiyi F, Hond A, Raz A, Needleman P (1990) The biochemical and pharmacological manipulation of cellular cyclooxygenase (COX) activity. *Adv Prostaglandin Thromboxane Leukotriene Res* 21: 45–51
18 Simmons DL, Levy DB, Yannoni Y, Erikson RL (1989) Identification of a phorbol ester-repressible v-src-inducible gene. *Proc Natl Acad Sci USA* 86: 1178–1182
19 Xie W, Chipman JG, Robertson DL, Erikson RL, Simmons DL (1991) Expression of a mitogen-responsive gene encoding prostaglandin synthase is regulated by mRNA splicing. *Proc Natl Acad Sci USA* 88: 2692–2696
20 Kujubu DA, Fletcher BS, Varnum BC, Lim RW, Herschman HR (1991) TIS10, a phorbol ester tumor promoter-inducible mRNA from Swiss 3 T3 cells, encodes a novel prostaglandin synthase/cyclooxygenase homologue. *J Biol Chem* 266: 12866–12872
21 Simmons DL, Xie W, Chipman J, Evett G (1991) Multiple cyclooxygenases: cloning of a mitogen-inducible form. In: M Bailey (ed.): *Prostaglandins, Leukotrienes, Lipoxins and PAF*. Plenum Press, London, 67–68
22 McAdam BF, Catella-Lawson F, Mardini IA, Kapoor S, Fitzgerald GA (1999) Systemic biosynthesis of prostacyclin by cyclooxygenase (COX-2): The human pharmacology of a selective inhibitor of COX-2. *Proc Natl Acad Sci USA* 96: 272–277
23 Willoughby DA, Tomlinson A, Gilroy D, Willis D (1996) Inducible enzymes with special reference to the inflammatory response. In: JR Vane, JH Botting, RM Botting (eds): *Improved Non-steroid Anti-inflammatory Drugs. COX-2 Enzyme Inhibitors*. Kluwer and William Harvey Press, London, 67–83
24 Sano H, Hla T, Maier JAM, Crofford LJ, Case JP, Maciag T, Wilder RL (1992) *In vivo* cylooxygenase expression in synovial tissues of patients with rheumatoid arthritis and osteoarthritis and rats with adjuvant and streptococcal cell wall arthritis. *J Clin Invest* 89: 97–108
25 Patrignani P, Panara NR, Greco A, Fusco O, Natoli C, Iacobelli S, Cipollone F, Ganci A, Creminon C, Maclouf J et al (1994) Biochemical and pharmacological characterisation of the cyclooxygenase activity of human blood prostaglandin endoperoxide synthases. *J Pharmacol Exp Ther* 272: 1705–1710

26 Picot D, Loll PJ, Garavito RM (1994) The x-ray crystal structure of the membrane protein prostaglandin H_2 synthase-1. *Nature* 367: 243–249
27 Luong C, Miller A, Barnett J, Chow J, Ramesha C, Browner MF (1996) Flexibility of the NSAID binding site in the structure of human cyclooxygenase-2. *Nat Struct Biol* 3: 927–933
28 Mancini JA, O'Neill GP, Bayly C, Vickers PJ (1994) Mutation of serine 516 in human prostaglandin G/H synthase-2 to methionine or aspirin or aspirin acetylation of this residue stimulates 15-R-HETE synthesis. *FEBS Lett* 342: 33–37
29 Wong E, Bayly C, Waterman HL, Reindeau D, Mancini JA (1997) Conversion of prostaglandin G/H synthase-1 into an enzyme sensitive to PGHS-2-selective inhibitors by a double His^{513} to Arg and Ile^{523} to Val mutation. *J Biol Chem* 272: 9280–9286
30 Browner MF (1998) The structure of human COX-2 and selective inhibitors. In: J Vane, J Botting (eds): *Selective COX-2 Inhibitors. Pharmacology, Clinical Effects and Therapeutic Potential.* Kluwer and William Harvey Press, London, 19–26
31 Silverstein FE, Faich G, Goldstein JL, Simon LS, Pincus T, Whelton A, Makuch R, Eisen G, Agrawal NM, Stenson WF et al (2000) Gastrointestinal toxicity with celecoxib vs nonsteroidal anti-inflammatory drugs for osteoarthritis and rheumatoid arthritis: the CLASS study: a randomised controlled trial. Celecoxib Long-term Arthritis Safety Study. *J Am Med Assoc* 284: 1247–1255
32 Bombardier C, Laine L, Reicin A, Shapiro D, Burgos-Vargas R, Davis B, Day R, Ferraz MB, Hawkey CJ, Hochberg MC et al, VIGOR Study Group (2000) Comparison upper gastrointestinal toxicity of rofecoxib and naproxen in patients with rheumatoid arthritis. *N Engl J Med* 343: 1520–1528
33 Ormrod D, Wellington K, Wagstaff AJ (2002) Valdecoxib. *Drugs* 62: 2059–2071
34 Gottesdiener K, Schnitzer T, Fisher C, Bockow B, Markenson J, Ko A, DeTora L, Curtis S, Geissler L, Gertz BJ (2002) Results of a randomized, dose-ranging trial of etoricoxib in patients with osteoarthritis. *Rheumatology (Oxford)* 41: 1052–1061
35 Sorbera LA, Castaner J, Bayes M, Silvestre JS (2002) Lumiracoxib. *Drugs of the Future* 27: 740–747
36 Giovannucci E, Egan KM, Hunter DJ (1995) Aspirin and the risk of colorectal cancer in women. *N Engl J Med* 333: 609–614
37 Greenberg ER, Baron JA, Freeman DH, Mandel JS, Haile R (1993) Reduced risk of large bowel adenomas among aspirin users. The Polyp Prevention Study Group. *J Natl Cancer Inst* 85: 912–916
38 Thun MJ, Namboodiri MM, Heath CWJ (1991) Aspirin use and reduced risk of colon cancer. *N Engl J Med* 325: 1593–1596
39 Thun MJ, Namboodiri MM, Calle EE, Flandes WD, Heath CW (1993) Aspirin use and risk of fatal cancer. *Cancer Res* 53: 1322–1327
40 Peleg II, Maibach HT, Brown SH, Wilcox CM (1994) Aspirin and nonsteroidal antiinflammatory drug use and the risk of subsequent colorectal cancer. *Arch Int Med* 154: 394–399
41 Giavannucci E, Rimm EB, Stampfer MJ, Colditz GA, Ascherio A, Willett WC (1994) Aspirin use and the risk of colorectal cancer and adenoma in male health professionals *Ann Intern Med* 121: 241–246
42 Eberhart CE, Coffey RJ, Radhika A (1994) Up-regulation of cyclooxygenase 2 gene expression in human colorectal adenomas and adenocarcinomas. *Gastroenterology* 107: 1183–1188
43 Giardello FM, Hamilton SR, Krush AJ (1993) Treatment of colonic and rectal carcinomas with sulindac in familial adenomatous polyposis. *N Engl J Med* 328: 1313–1316
44 Waddell WR, Loughry RW (1983) Sulindac for polyposis of the colon. *J Surg Oncol* 24: 83–87
45 Waddell WR, Gasner GF, Cerise EJ, Loughry RW (1989) Sulindac for polyposis of the colon. *Am J Surg* 157: 175–178
46 Reddy BS, Rao CV, Rivenson A, Kellof G (1993) Inhibitor effect of aspirin on azoxymethane-induced colon carcinogenesis in F344 rats. *Carcinogenesis* 14: 1493–1497
47 Ristimaki A, Narko K, Nieminen, Saukkonen K (2001) Role of cyclooxygenase-2 in carcinogenesis other than colorectal cancer. In: JR Vane, RM Botting (eds): *Therapeutic Roles of Selective COX-2 Inhibitors*. William Harvey Press, London, 418–441
48 Bennett PR, Elder MG (1992) The mechanisms of preterm labour: common genital tract pathogens do not metabolise arachidonic acid to prostaglandins or other eicosanoids. *Am J Obstet Gynecol* 166: 1541–1545
49 Sawdy R, Slater D, Fisk N, Edmonds DK, Bennett P (1997) Use of a cyclo-oxygenase type-2-

selective non-steroidal anti-inflammatory agent to prevent preterm delivery. *Lancet* 350: 265–266
50 Peruzzi L, Gianoglio B, Porcellini MG, Coppo R (1999) Neonatal end-stage renal failure associated with maternal ingestion of cyclo-oxygenase-type-1 selective inhibitor nimesulide as tocolytic. *Lancet* 354: 1615
51 Allport V, Bennett P (2001) Cyclooxygenase enzymes and human labour. In: JR Vane, RM Botting (eds): *Therapeutic Roles of Selective COX-2 Inhibitors*. William Harvey Press, London, 252–273
52 Stewart WF, Kawas C, Corrada M, Metter EJ (1997) Risk of Alzheimer's disease and duration of NSAID use. *Neurology* 48: 626–632
53 Pasinetti GM, Aisen PS (1998) Cyclooxygenase-2 expression is increased in frontal cortex of Alzheimer's disease brain. *Neuroscience* 87: 319–324
54 Lim GP, Yang F, Chu T, Chen P, Beech W, Teter B, Tran T, Ubeda O, Ashe KH, Frautschy SA et al (2000) Ibuprofen suppresses plaque pathology and inflammation in a mouse model for Alzheimer's disease. *J Neurosci* 20: 5709–5714
55 McGeer PL (2000) Cyclo-oxygenase-2 inhibitors: rationale and therapeutic potential for Alzheimer's disease. *Drugs Aging* 17: 1–11
56 Sainati SM, Ingram DM, Talwalker S, Geis GS (2000) Results of a double-blind, randomized, placebo-controlled study of celecoxib in the treatment of progression of Alzheimer's disease. Abstracts of the Sixth International Stockholm/Springfield Symposium on Advances in Alzheimer Therapy. April 5–8 (abstract)
57 Weggen S, Eriksen JL, Das P, Sagi SA, Wang R, Pietrzik CU, Findlay KA, Smith TE, Murphy MP, Bulter T et al (2001) A subset of NSAIDs lower amyloidogenic Aβ42 independently of cyclooxygenase activity. *Nature* 414: 212–216
58 Zandi PP, Anthony JC, Hayden KM, Mehta K, Mayer L, Breitner JC (2002) Reduced incidence of AD with NSAID but not H2 receptor antagonists. The Cache County study. *Neurology* 59: 880–886
59 Chandrasekharan NV, Dai H, Roos KL, Evanson NK, Tomsik J, Elton TS, Simmons DL (2002) COX-3, a cyclooxygenase-1 variant inhibited by acetaminophen and other analgesic/antipyretic drugs: cloning, structure, and expression. *Proc Natl Acad Sci USA* 99: 13926–13931
60 Simmons DL, Botting RM, Robertson PM, Madsen ML, Vane JR (1999) Induction of an acetaminophen-sensitive cyclooxygenase with reduced sensitivity to nonsteroid antiinflammatory drugs. *Proc Natl Acad Sci USA* 96: 3275–3280

COX-2 Inhibitors
Edited by M. Pairet and J. van Ryn

Structural diversity of selective COX-2 inhibitors

Lawrence J. Marnett[1] and Amit S. Kalgutkar[2]

[1] *Department of Biochemistry, Vanderbilt Institute of Chemical Biology, Vanderbilt-Ingram Cancer Center, and Center in Molecular Toxicology, Vanderbilt University School of Medicine, Nashville, Tennessee 37232-0146, USA*
[2] *Pharmacokinetics, Dynamics and Metabolism Department, Pfizer Global Research and Development, Groton, CT 06340, USA*

Introduction

Non-steroidal anti-inflammatory drugs (NSAIDs) are among the most widely used agents in the treatment of pain and inflammation, especially rheumatoid arthritis [1]. The first NSAID aspirin was introduced over a century ago, and continues to be a popular over-the-counter medication in the treatment of the symptoms of pain and inflammation. The global production of aspirin is approximately 50,000 tons annually [2]. The pharmacological effects of NSAIDs arise from their inhibition of the prostaglandin synthase or cyclooxygenase (COX) isozymes [3]. COX isozymes catalyze the rate-limiting step in the arachidonic acid cascade involving the conversion of arachidonic acid to prostaglandin H_2 (PGH_2), the common biosynthetic precursor to prostaglandins and thromboxane (Fig. 1). These bioactive lipids mediate numerous physiological and pathophysiological effects, including pain, fever, inflammation, hemostasis, and regulation of renal function and maintenance of mucosal integrity in the stomach.

In vitro evaluation of COX-2 selectivity

Because of the numerous and widely differing experimental protocols available for the *in vitro* evaluation of COX-1 and COX-2 activity, comparison of

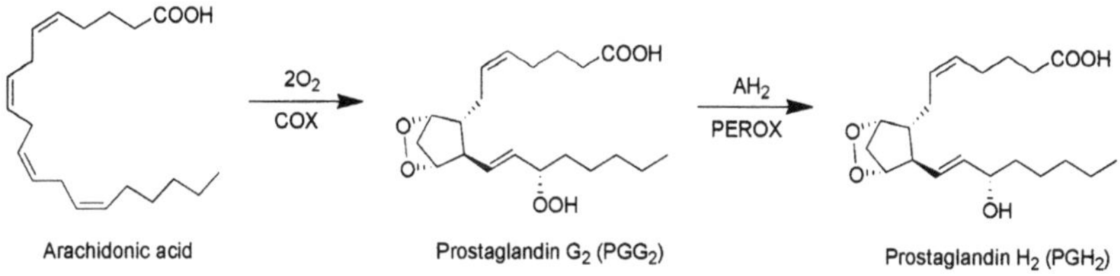

Figure 1. Biosynthesis of prostaglandin H_2.

potency and selectivity data for NSAIDs and selective COX-2 inhibitors from different laboratories has been somewhat of a challenge. Recombinant enzyme systems [4–6], microsomal membranes and cultured intact cells [7], insect cells [8], human whole blood [9–11], human monocytes [12], and various other mammalian cells and platelets [13–15] are routinely used in the screening of COX inhibitors. Examination of the COX-1/COX-2 selectivity ratio for indomethacin under various assay conditions reveals that this ratio varies from 2.0 to 14 [10, 14, 16, 17]. This highlights the influence different assay conditions can have and is discussed in more detail in Chapter 3.

Diarylheterocycles

Research within this class of compounds has culminated in the FDA approval and marketing of celecoxib, rofecoxib and valdecoxib [18–20]. Other agents from this class of compounds in various phases of clinical trials include etoricoxib and JTE-522 [21].

Historical perspective

The chemical structures of diarylheterocycles can be traced back to clinical candidates discovered over 30–40 years ago (Fig. 2). Examination of the structures of commercial selective COX-2 inhibitors from the diarylheterocycle series (e.g. celecoxib, rofecoxib, valdecoxib, and etoricoxib) reveals a striking resemblance to the older COX inhibitor phenylbutazone [22, 23]. The exceptional anti-inflammatory properties of phenylbutazone [24] undoubtedly provided much of the impetus that led chemists to exploit this type of structure. The discovery of the diphenyloxazole derivative oxaprozin as a clinical candidate for the treatment of inflammation is one example of the utility of the diarylheterocycle scaffold [25, 26]. Likewise, the 2,3-diarylindole indoxole was disclosed in the early 1960s as a potent anti-inflammatory agent; however, the drug was suspended from clinical development since it caused photosensitivity in humans [27–29]. Throughout the period from late 1960s to the early 1980s, a plethora of diarylheterocycles with variations on the central heterocyclic ring were evaluated as anti-inflammatory agents [30] and several were advanced into clinical trials based on their superior pharmacological and gastrointestinal profile in preclinical species compared with traditional NSAIDs [31, 32]. An example of such a compound is the 4,5-diarylthiazole derivative itazigrel that was shown to be less ulcerogenic than aspirin in subsequent clinical trials [33]. Relative to the unsubstituted congeners, structure-activity relationship (SAR) studies in the prototype diarylheterocycles indicated significant enhancement in anti-inflammatory properties when methoxy or halogen groups were incorporated in the *para* position on the two pendant phenyl rings (see Fig. 3) [34].

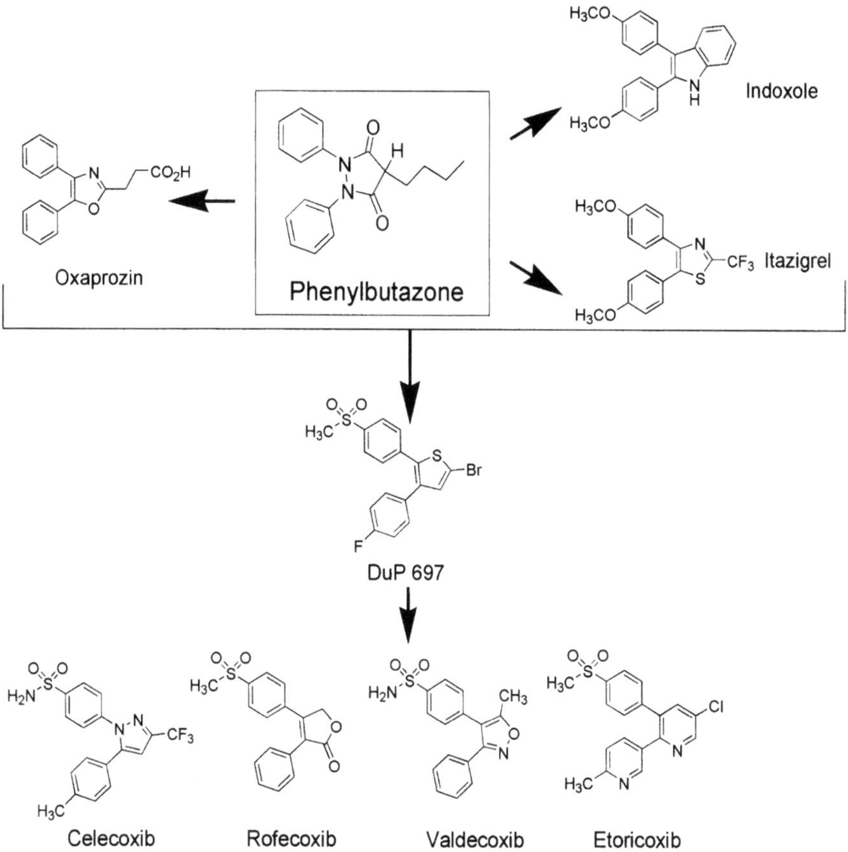

Figure 2. Evolution of selective COX-2 inhibitors.

General determinants of selective COX-2 inhibition by diarylheterocycles

The 4-methylsulfonyl (CH_3SO_2)-diarylheterocycle combination, commonly found in the most selective COX-2 inhibitors, was first observed in the structure of the 2,3-diarylthiophene, DuP 697, disclosed in the late 1980s (see Fig. 2) [35]. SAR studies on potential methylsulfone replacements in 2,3-diarylthiophenes indicated that only the 4-sulfonamido (SO_2NH_2) group maintained COX-2 inhibitory potency and selectivity observed with DuP697 [36, 37]. Incorporating the 4-sulfonamido group also results in increased COX-2 inhibitory potency, albeit with a loss of COX-2 selectivity. This SAR trend appears to be a common theme in all of the diarylheterocycles studied to date and is illustrated in Figure 3 with the 1,5-diarylpyrazole-based methylsulfone SC 58125 (COX-1/COX-2 selectivity > 1000) and the sulfonamide, celecoxib (COX-1/COX-2 selectivity ~325) [38]. Loss in COX-2 selectivity, however, is often compensated by an overall improvement in physiochemical and pharmacokinetic properties [38]. The oxidation state of the sulfur in the

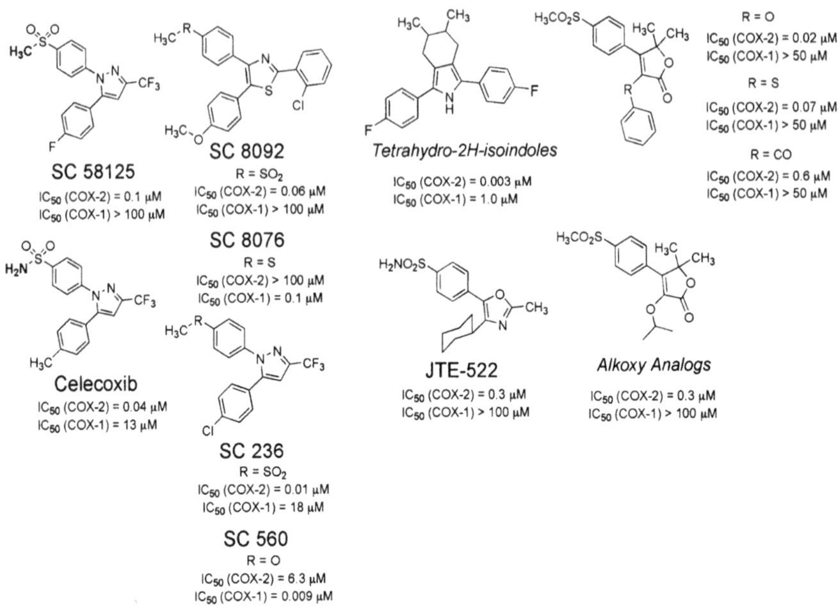

Figure 3. Structural requirements for COX-2 inhibitory potency and selectivity.

4-methylsulfone-containing compounds is extremely crucial for selective COX-2 inhibition; its reduction to sulfoxide or sulfide reverses isozyme selectivity. Thus, methylsulfone-containing 4,5-diarylthiazole (SC-8092) is a selective COX-2 inhibitor, whereas the corresponding methylthioether analog (SC-8076) exhibits COX-1-selective inhibitory properties (see Fig. 3) [17]. Likewise, replacement of the methylsulfonyl group in the selective COX-2 inhibitor SC-236 with a methoxy substituent also reverses isozyme specificity as highlighted with SC-560 [39].

In contrast with these observations, a series of novel 1,3-diaryltetrahydro-2H-isoindoles that lack the essential sulfonyl group were recently disclosed as potent and selective COX-2 inhibitors (IC_{50} = 0.6–100 nM for COX-2, 100–1000 nM for COX-1). SAR studies established that a bicyclic ring appended to the pyrrole nucleus and 4,4'-difluoro substitution on the phenyl rings was optimal for COX-2 potency (see Fig. 3) [40]. Interestingly, incorporation of the methylsulfonyl group in these compounds proved detrimental for COX-2 potency. The structural basis for COX-2 selectivity in this series remains unclear at the present time.

SAR studies also indicate that the phenyl substituent that does not contain the methylsulfonyl or the sulfonamide group can be replaced with a cyclohexyl substituent as illustrated with oxazole derivative JTE-522 (see Fig. 3). JTE-522 is currently under clinical evaluation as an anti-inflammatory agent [41, 42]. Introduction of a heteroatom (O and S) or a carbonyl link between the central ring system and the phenyl ring (lacking the methylsulfone or sulfonamide) also results in compounds that retain COX-2 potency and selectivity [43]. This is illustrated in Figure 3 with the 5,5-dimethylfuranone derivatives. The aryloxy moiety can also be replaced with alkoxy groups without any deleterious effect on potency and selectivity [43–46].

Overall, the common structural features essential for inhibition by most diarylheterocycles can be summarized as follows: (1) the presence of two lipophilic substituents (one of which has to be a phenyl ring) on adjacent positions within the central unsaturated ring system and (2) the presence of a methylsulfone/sulfonamide group on the essential phenyl ring. That the two lipophilic substituents need to be adjacent to one another is highlighted with 2,3- and the 3,4-diarylthiophenes, which are potent and selective COX-2 inhibitors [47], whereas the corresponding 2,4- or 2,5-diarylthiophenes suffer dramatic losses in COX-2 potency (Fig. 4).

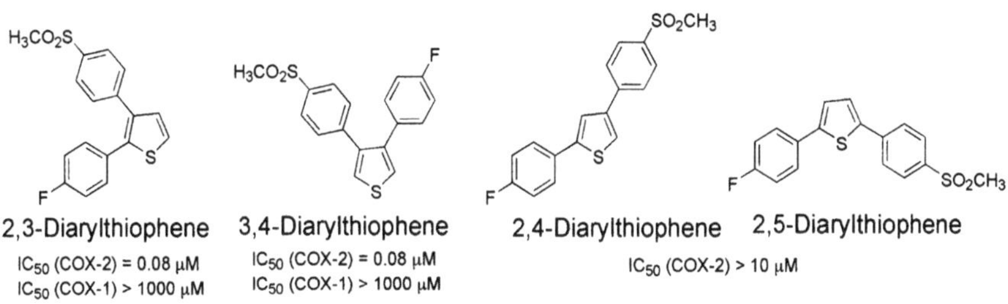

Figure 4. Effect of the phenyl group positioning on COX-2 inhibition by diarylthiophene isomers.

Variations on the central ring system

Numerous SAR studies have been focused on the alteration of the central heterocyclic ring and to date, virtually every imaginable carbocyclic (Fig. 5), heterocyclic (Fig. 6), and bicycloheterocyclic (Fig. 7) ring system has been incorporated.

Carbocycles

Examples of diarylcarbocycles include four-membered (e.g., cyclobutene and cyclobutenone) [48], five-membered (e.g., cyclopentene and cyclopentenone) [49–51], spirocyclic [52, 53] and substituted phenyl ring (terphenyl) systems [54, 55]. Although preliminary SAR studies suggested that the central ring was essential for favorable spatial orientation of the two pendant phenyl rings, a series of 1,2- and 1,1-diarylstilbenes were recently disclosed as potent and selective COX-2 inhibitors [56–58].

Heterocycles

Examples of central heterocycles include 5- or 6-membered rings that contain one (thiophenes [35–37], pyrroles [59–61], furans [62, 63] and pyridines

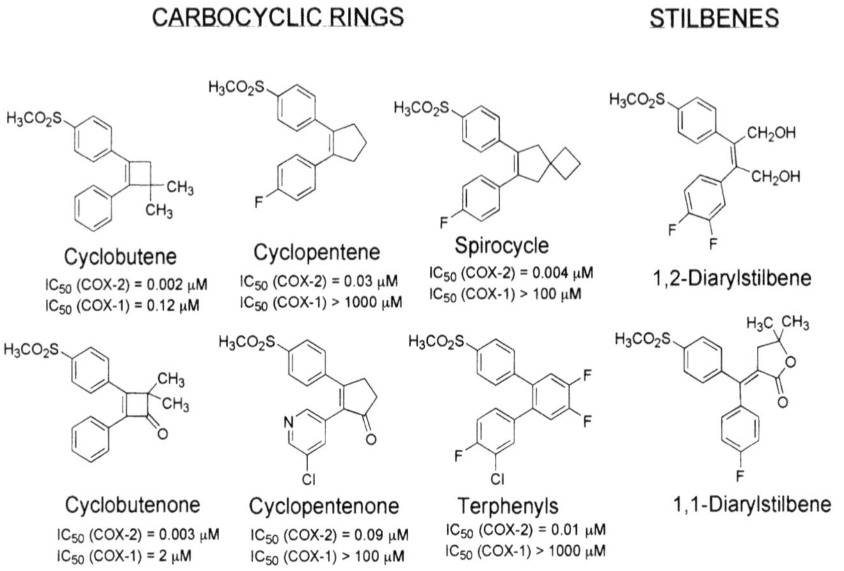

Figure 5. Diarylcarbocycles as selective COX-2 inhibitors.

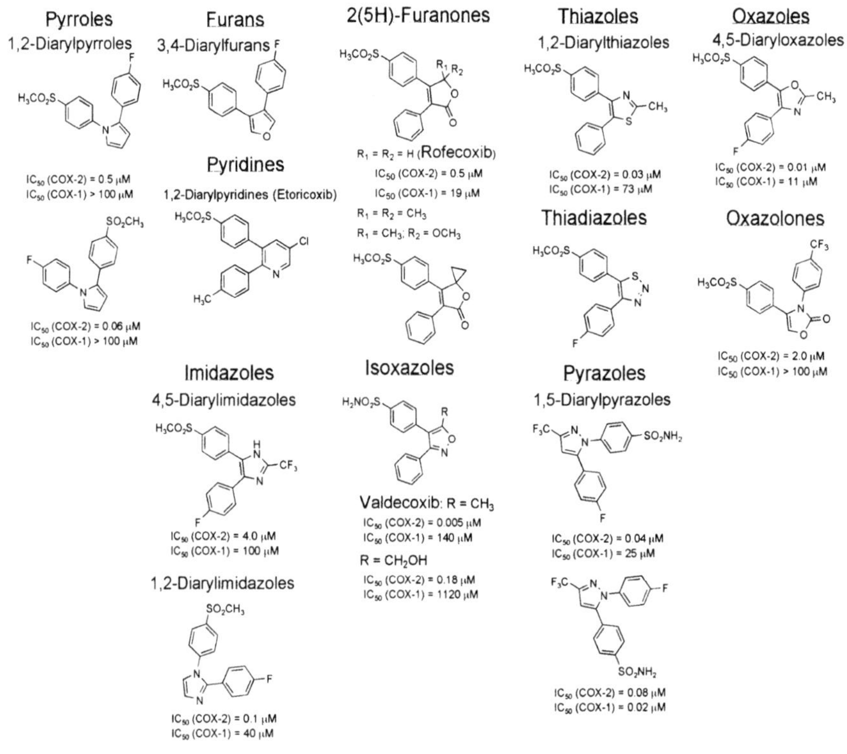

Figure 6. Diarylheterocycles as selective COX-2 inhibitors.

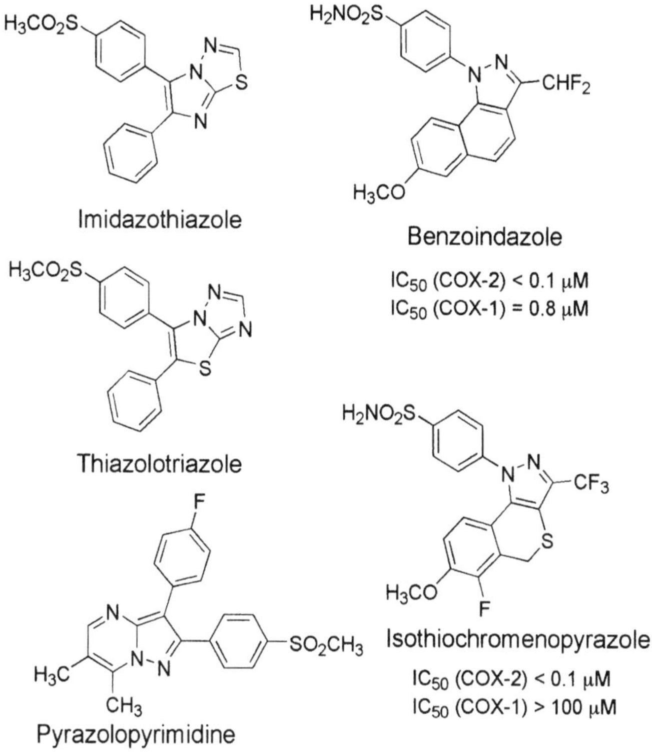

Figure 7. Diarylbicycloheterocycles as selective COX-2 inhibitors.

[64–66]), two (thiazoles [67, 68], oxazoles and the corresponding oxazolones [69, 70], imidazoles [71–73], isoxazoles [74–76], pyrazoles [38, 77–79]), or three (thiadiazole [80]) heteroatoms (Fig. 6).

As observed with thiophenes, both the 2,3- and 3,4-diarylfuran isomers demonstrate selective COX-2 inhibition [62, 63]. The inhibitory potency of diarylfurans, however, is inferior when compared with the corresponding thiophene derivatives. Structural alteration of the central furan ring to the 2(5H)-furanone scaffold however, results in substantial increases in COX-2 potency and selectivity as illustrated with the furanone rofecoxib (Fig. 6) [81, 82]. SAR analysis indicates that disubstitution at the 5-position in rofecoxib with alkyl, alkoxy or cycloalkyl groups leads to metabolically more stable compounds (see Fig. 6) with retention of COX-2 inhibitory potency and selectivity [45, 83–85].

Comparison of the SAR for selective COX-2 inhibition by terphenyls (see Fig. 5) and pyridines reveals that substitution other than in the 1- or 2-position in the central benzene ring in terphenyls with groups as small as fluorine affords inactive compounds, whereas such replacements are not problematic in the pyridine series as indicated with etoricoxib in Figure 6. SAR studies on diarylpyrroles indicate that COX-2 can be selectively inhibited by all possible isomers (1,2-diaryl, 2,3-diaryl and 3,4-diaryl) [59–61]. Furthermore, regardless of the 4-methylsulfonylphenyl-substitution pattern (N1 *versus* C-2), the 1,2-diarylpyrrole regioisomers maintained selective COX-2 inhibitory properties (see Fig. 6). In contrast, selective COX-2 inhibition by 1,2-diarylimidazoles and 1,5-diarylpyrazoles was only discernible with compounds in which the 4-methylsulfonylphenyl- or 4-sulfonamidophenyl groups were attached to the nitrogen atom. Imidazole or pyrazole analogs, in which the 4-methylsulfonylphenyl- or 4-sulfonamidophenyl groups were attached at the C-2 or the C-5 positions respectively, were nonselective or selective COX-1 inhibitors [38, 71–73]. This is depicted with the 1,5-diarylpyrazole derivatives in Figure 6.

Bicyclic or fused heterocycles

Examples of bicyclic or fused-ring systems as the central template have also been disclosed. Although many compounds demonstrate COX-2 potency and selectivity, they suffer from poor physiochemical and pharmacokinetic properties, which ultimately result in poor oral absorption. Examples of orally active analogs are few and include the imidazothiazole-, thiazolotriazole- and the pyrazolopyrimidine-containing diarylbicycloheterocycle derivatives (Fig. 7) [86–88]. Other variations in this class are highlighted with the benzindazole and the isothiochromenopyrazole ring system in which one of the pendant aryl rings was fused to the bicyclic central core (see Fig. 7) [89–91].

Alkylsulfonanilide derivatives

Investigations on the anti-inflammatory properties of alkylsulfonanilides such as nimesulide (Fig. 8) began in the 1960s [92]. Details on the pharmacologi-

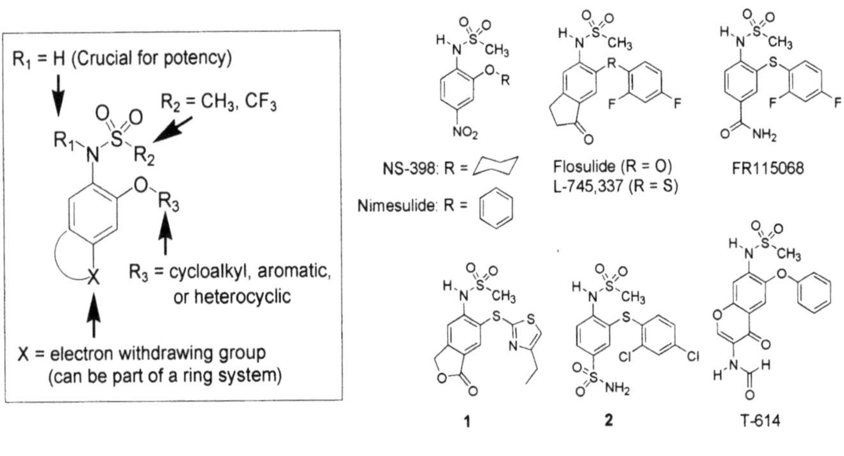

Figure 8. SAR studies on akylsulfonanilides as selective COX-2 inhibitors.

cal [93] and clinical [94] evaluation of nimesulide clearly demonstrated the excellent anti-inflammatory properties of this compound. The recent finding on the selective COX-2 inhibition by the structurally related sulfonanilide NS-398 (see Fig. 8) led to a renewed interest in this general class of anti-inflammatory agents [16, 95, 96]. Likewise, the nonulcerogenic anti-inflammatory activity of nimesulide [94] and the close structural analog flosulide [97] had been previously established, but re-evaluation of both compounds as COX inhibitors revealed their selective COX-2 inhibitory properties [98]. The common structural features of the alkylsulfonanilides are depicted in Figure 8. The alkyl substituent is typically methyl, but halogenated methyl substituents such as trifluoromethyl have also been reported [99]. N-Methylation of the sulfonamide nitrogen generates inactive compounds [99]. The *ortho*-substituent typically includes aryl (e.g., nimesulide), heterocyclic (thiazole analog **1**) [100, 101], or cycloalkyl ethers (e.g., NS-398) and thioethers (L-745,337 [99] and FR115068 [102], respectively). The thiazolyl-thioether analog **1** is reported to possess greater COX-2 potency and selectivity, superior *in vivo* pharmacokinetic and pharmacological profile, and a greater gastrointestinal safety profile than flosulide [100]. The *para*-substituent invariably bears an electron-withdrawing group that may be incorporated as part of a ring. A variety of methanesulfonanilides with different *para*-electron-withdrawing groups have been evaluated as selective COX-2 inhibitors and as orally active anti-inflammatory agents. Substituents include 4-acetyl, 4-cyano, 4-carboxamido (FR115068), nitro (NS-398 and nimesulide), trifluoromethyl, sulfonamido (compound **2**), sulfonyl and several others [102]. Another structural variant is the incorporation of the *para*-electron-withdrawing group as part of a ring system. For instance, the principal difference between nimesulide/NS-398 and flosulide/L-745,337 is the incorporation of the electron-withdrawing group as part of a five-membered cyclopentanone ring system in the latter compounds. The electron withdrawing substituent also may be incorporated as part of a heterocyclic ring as seen in T-614 [103–105]. Although the purpose of the electron-withdrawing group remains unclear, its presence is crucial for COX-2 inhibitory potency. Molecular modeling studies suggest that the electron-withdrawing group may function as an H-bond acceptor from the active site Tyr355 residue [106].

Structural modifications of NSAIDs into selective COX-2 inhibitors

Modifying nonselective NSAIDs into selective COX-2 inhibitors represents an interesting strategy and has met with variable degrees of success. Thus far, aryl acetic acid- (e.g., indomethacin, zomepirac, sulindac, diclofenac, and etodolac), fenamic acid- (e.g., meclofenamic acid), propionic acid- (e.g., flurbiprofen), enol-carboxamide- (e.g., piroxicam), and benzoic acid- (e.g., aspirin) containing NSAIDs have been successfully elaborated into potent and selective COX-2 inhibitors.

Aryl acetic acids

Indomethacin

Modifications on the N-substituted indole-3-acetic acid framework found in indomethacin have yielded tremendous success in the search for novel structural types as selective COX-2 inhibitors (Fig. 9). Some of the changes in indomethacin structure that imparts COX-2 potency and selectivity include: (1) replacement of the 4-chlorobenzoyl group with a 2,4,6-trichlorobenzoyl- or a 4-bromobenzyl group (examples **3** and **4**) [107, 108], (2) increasing the chain length of the 3-alkanoic acid group (example **5**) [108], (3) neutralization of the carboxylic acid group via conversion to the corresponding esters (compd. **6**) or amides (compd. **7**) [109–111] as well as the corresponding reverse esters/amides (compds. **8** and **9**) [112], and (4) replacing the carboxylic acid group with appropriate bioisosteres as exemplified with the iminoxy (**10**) or substituted thiazole (**11**) analogs, respectively [113, 114]. It is noteworthy to point out that the 2-methyl group in indomethacin and the neutral amide and ester derivatives is crucial for inhibitory potency, since its replacement with hydrogen results in compounds that are devoid of inhibitory potency against the COX isozymes [109, 110]. Exchanging the methoxy substituent in indomethacin with the methylsulfone group common in diarylheterocycles and

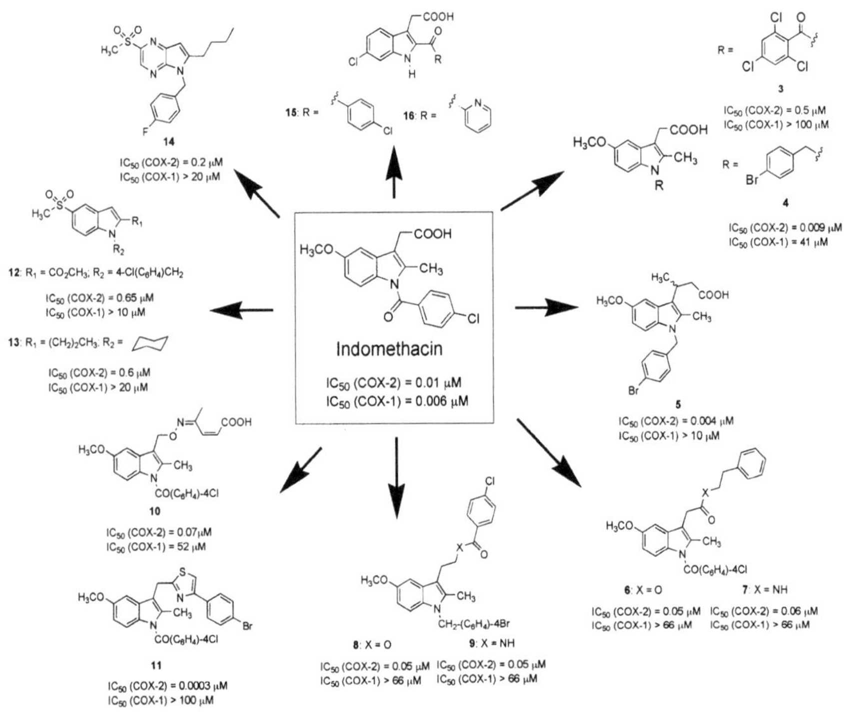

Figure 9. Strategies implemented in the conversion of indomethacin to a COX-2 selective inhibitor.

moving the alkanoic ester group from the 3- to the 2-position on the indole ring also results in reasonably selective and potent COX-2 inhibitors as highlighted with compound **12** [115]. Additional improvements in the COX-2 potency and selectivity of **12** have also been described as exemplified with the mono- and diazaindolyl derivatives **13** and **14** [116–118]. These latter compounds lack the carboxylic acid group that is a key determinant for COX inhibition by indomethacin and instead possess a methylsulfone group, an essential criterion for selective COX-2 inhibition by diarylheterocycles. Whether these analogs bind in the same region of the COX-2 enzyme, as diarylheterocycles remains unclear. An interesting variation on the utilization of the indomethacin template in the design of COX-2 inhibitors is shown with 2,3-disubstituted indolyl derivatives (compds. **15** and **16**) [119]. These analogs contain an aroyl group on the 2-position of the indole ring, unlike indomethacin, in which the aroyl group is attached to the nitrogen on the indole ring.

Zomepirac and sulindac

As observed with indomethacin, replacement of the carboxylic acid group in zomepirac and sulindac with certain bioisosteres also confers COX-2 selectivity in the resulting compounds (Fig. 10). For instance, exchanging the carboxylic group in sulindac with a tetronic acid moiety (compd. **17**) or in

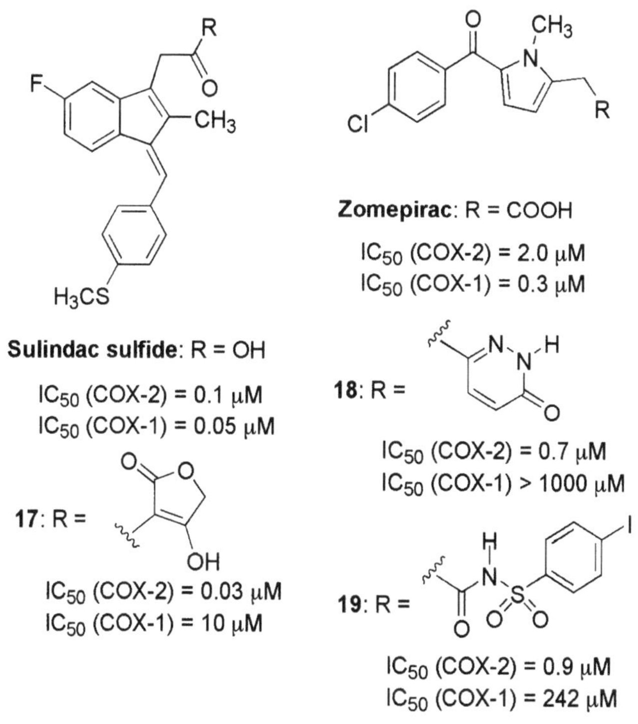

Figure 10. Carboxylic acid bioisosteres of sulindac and zomepirac as selective COX-2 inhibitors.

zomepirac with pyridazinone (compd. **18**) or an acylsulfonamide group (compd. **19**) affords potent and selective COX-2 inhibitors [120–122].

Diclofenac

Carboxylic acid bioisosteres of diclofenac also demonstrate potent and selective COX-2 inhibition. This is highlighted with the iminoxy analog **20** (Fig. 11) [123]. The observation that the active site in COX-2 is larger than in COX-1 has led to the hypothesis that introducing steric constraints in the NSAID scaffold should potentially generate compounds with COX-2 selectivity. This proposal has been validated in the case of diclofenac. Incorporation of *meta*-alkyl substituents in the arylacetic acid ring in diclofenac reverses isozyme specificity as shown with analog **21** [124]. Replacement of a chloro substituent from the dichlorophenyl ring in **21** with a fluoro group resulted in the discovery of Lumiracoxib (see Fig. 11) [125].

Etodolac

Evaluation of the COX isozyme selectivity for the NSAID etodolac (Fig. 12) led to the finding on the moderately selective COX-2 inhibitory properties of

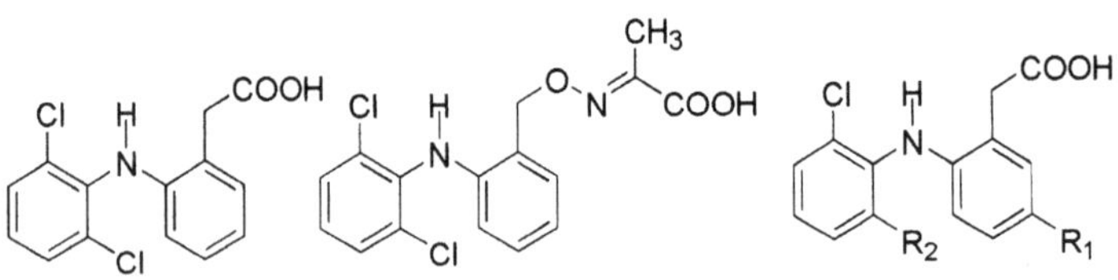

Diclofenac

20

IC_{50} (COX-2) = 0.1 μM

IC_{50} (COX-1) > 1 μM

21: R_1 = CH_3; R_2 = Cl

IC_{50} (COX-2) = 0.01 μM

IC_{50} (COX-1) > 10 μM

Lumiracoxib (R_1 = CH_3; R_2 = F)

IC_{50} (COX-2) = 0.007 μM

IC_{50} (COX-1) > 10 μM

Figure 11. Structural modifications on diclofenac that result in selective COX-2 inhibition.

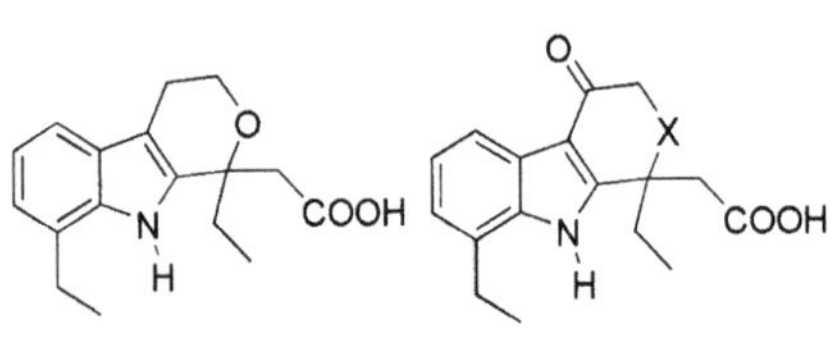

Etodolac

IC_{50} (COX-2) = 1.4 μM

IC_{50} (COX-1) = 15 μM

22: X = O

IC_{50} (COX-2) = 1.9 μM

IC_{50} (COX-1) = 270 μM

23: X = CH_2

IC_{50} (COX-2) = 2.0 μM

IC_{50} (COX-1) > 90 μM

Figure 12. Etodolac and related analogs as selective COX-2 inhibitors.

this NSAID [126]. These findings were consistent with the better gastrointestinal safety of etodolac compared with other NSAIDs. Adjustments in the oxidation state from a methylene at C-4 in the pyran ring of etodolac to the corresponding keto derivative **22** and removal of the oxygen atom in the pyran ring of **22** to afford **23** led to dramatic improvement in the COX-2 selectivity of the resulting compounds (see Fig. 12) [127, 128].

Fenamates

As observed with indomethacin, secondary amide derivatives of the fenamic acid NSAID, meclofenamic acid, also demonstrated potent and selective COX-2 inhibition [109, 129]. The 2-phenoxyethylamide derivative **24** was the most selective inhibitor in the series, with a COX-2 selectivity ratio of ~440 (Fig. 13). Unlike indomethacin SAR, only the amide derivatives of meclofenamic acid demonstrated COX-2 selectivity. The esters were either inactive or nonselective COX inhibitors. The reason(s) for this discrepancy remains unclear.

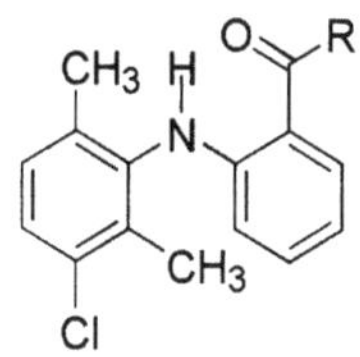

Meclofenamic acid: R = OH **24**: R = 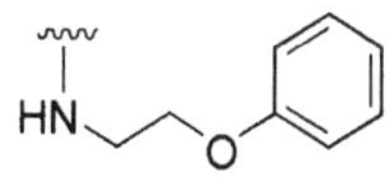

IC_{50} (COX-2) = 0.05 μM IC_{50} (COX-2) = 0.15 μM

IC_{50} (COX-1) = 0.04 μM IC_{50} (COX-1) = 66 μM

Figure 13. Meclofenamic acid amides as selective COX-2 inhibitors.

Propionates

Flurbiprofen is the only example in this structural class that has been successfully elaborated into a selective COX-2 inhibitor. Examination of the flurbiprofen-COX-1 and -COX-2 cocrystal structure revealed that introducing steric constraints in the 4-phenyl ring in flurbiprofen should in theory afford compounds devoid of COX-1 inhibitory properties without affecting COX-2 potency [94]. The finding that introduction of diethoxy substituents in the 4-phenyl ring of flurbiprofen led to selective COX-2 inhibition (compd. **25**, Figure 14) validated this hypothesis [130].

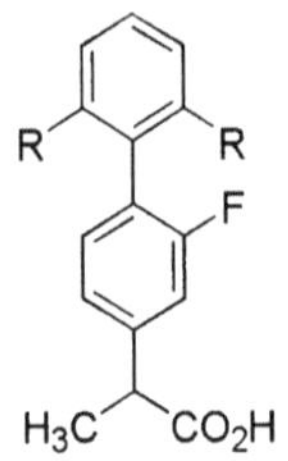

IC_{50} (COX-2) = 0.01 μM
IC_{50} (COX-1) = 0.01 μM

25: R = OC_2H_5

IC_{50} (COX-2) = 0.1 μM
IC_{50} (COX-1) = 7.7 μM

Flurbiprofen: R = H

Figure 14. Modification of the propionic acid NSAID flurbiprofen into a selective COX-2 inhibitor.

Enol-carboxamides

Replacement of the pyridine ring in the nonselective enolcarboxamide NSAID piroxicam with a 5-methylthiazoyl group resulted in the discovery of meloxicam, which demonstrated reasonable selectivity as a COX-2 inhibitor (Fig. 15) [131]. Further structural modifications on meloxicam resulted in isoquinolinediones such as **26** (see Fig. 15) that represent a new class of orally active COX-2-selective inhibitors [132, 133].

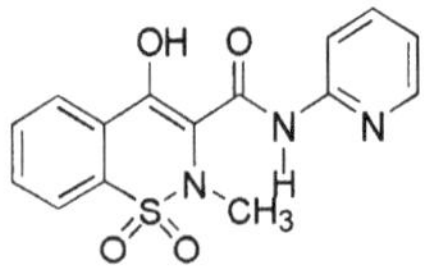

Piroxicam
IC_{50} (COX-2) = 0.5 μM
IC_{50} (COX-1) = 0.4μM

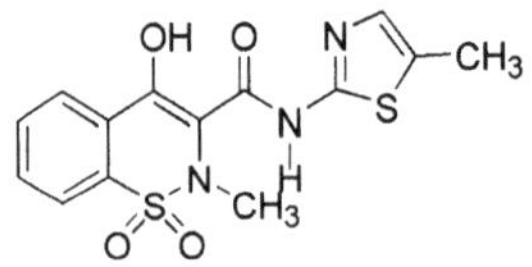

Meloxicam
IC_{50} (COX-2) = 0.49 μM
IC_{50} (COX-1) = 37 μM

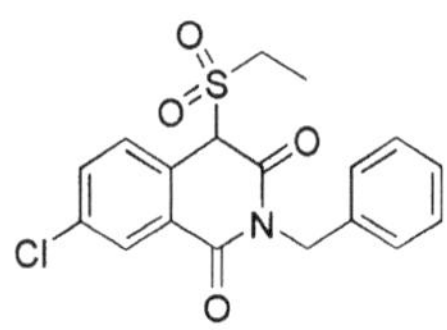

26
IC_{50} (COX-2) = 0.06 μM
IC_{50} (COX-1) = 1.4 μM

Figure 15. Structural modifications on the enol-carboxamide piroxicam that lead to selective COX-2 inhibition.

Benzoic acid

Recently, we demonstrated the structural alteration of aspirin to APHS, a selective COX-2 inhibitor (Fig. 16). Aspirin is the only NSAID that covalently modifies COX isozymes [134]. Covalent bond formation is achieved through a regioselective acetylation of the active site Ser530 residue in the two isozymes [135]. Replacement of the carboxylic acid group in aspirin with alkyl- or alkynylsulfide groups reverses isozyme specificity as shown with APHS [136, 137]. SAR studies indicated that the presence of the sulfur is crucial for COX inhibition; its replacement with oxygen, nitrogen, and carbon generates inactive compounds [137]. Furthermore, the acetoxy group in APHS

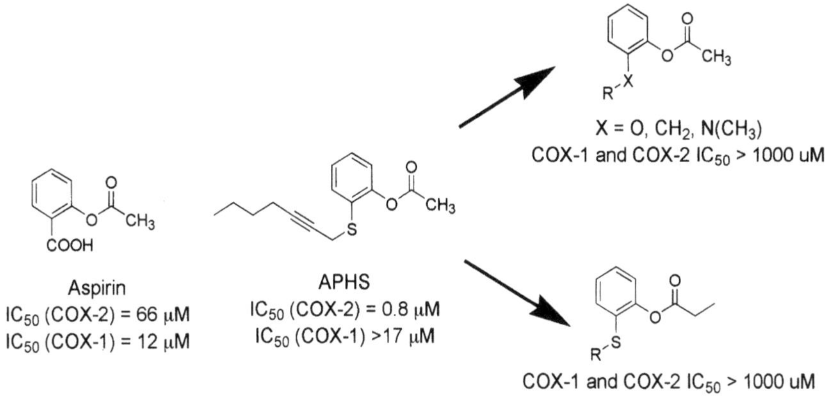

Figure 16. Transformation of aspirin to the selective COX-2 inhibitor APHS.

was optimal for inhibition since the corresponding propionoxy derivative was inactive as a COX inhibitor [137]. APHS acetylates the same serine residue on COX-2 (Ser530) that is acylated by aspirin [137]. Site-directed mutagenesis studies have indicated that the hept-2-ynyl chain in APHS probably binds in the same region of the protein as the ω-end of the fatty acid substrate arachidonic acid [138]. This binding mode of APHS contrasts with aspirin in which the carboxylate moiety is involved in electrostatic interactions with the positively charged Arg120 residue in the mouth of the active site [139]. Whether these differences in binding are also responsible for the COX-2 selectivity of APHS remain unclear.

Di-tert-butylphenols and dihydrobenzofurans

Certain di-*tert*-butylphenolic analogs containing five-membered heterocyclic rings *para* to the phenolic OH group were recently disclosed as a novel class of potent and selective COX-2 inhibitors [140–142]. In addition to inhibiting COX-2, these compounds also possess 5-lipoxygenase inhibitory properties [140]. Examples of five-membered cyclic heterocyclic substituent included the thiazolone, oxazolone, 1,3,4-thiadiazole, and 1,3,4-oxadiazole ring systems (Fig. 17). Thus, the thiazolone analog PD138387 indicated good *in vitro* COX-2 potency and selectivity (IC_{50} (COX-1) > 100 μM; IC_{50} (COX-2) = 0.98 μM). Upon oral administration, PD138387 demonstrated anti-inflammatory activity in the mouse footpad edema model (ED_{40} = 16 mg/kg) as well as analgesic properties in the acetic acid writhing test (ED_{40} = 0.1 mg/kg). The bioisosteric replacement of the sulfur atom in PD138387 with an oxygen atom resulted in the oxazolone **27**, which demonstrated selective COX-1 inhibition (IC_{50} (COX-1) = 7.3 μM; IC_{50} (COX-2) > 100 μM). Direct attachment of the di-*tert*-butylphenolic portion to heterocyclic rings resulted in improved COX-2 potency and selectivity as exemplified with the 1,3,4-thiadiazole PD164387 (IC_{50}

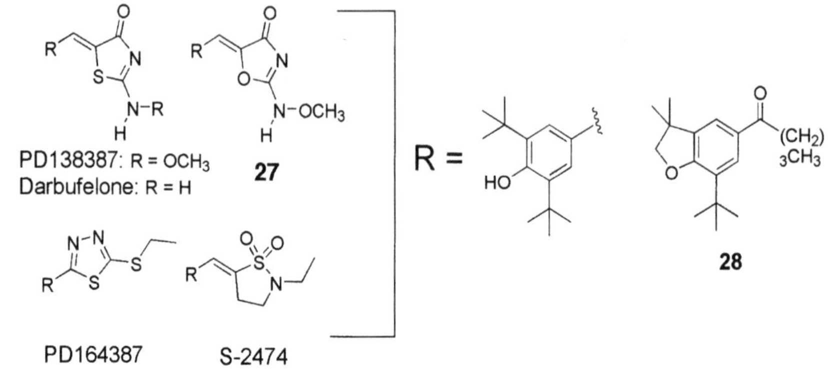

Figure 17. Di-*tert*-butylphenols and dihydrobenzofurans as selective COX-2 inhibitors.

(COX-1) = 0.14 μM; IC_{50} (COX-2) > 100 μM) [142]. The C-4 5-membered cyclic sulfonamide moiety in S-2474 serves as an additional example of structural diversity in the series. In addition to inhibiting COX-2 and 5-lipoxygenase, S-2474 also inhibited the production of IL-1 *in vitro*. S-2474 has also proven to be effective in several animal models of inflammation and is devoid of ulcerogenic liability [143]. Based on these attributes, S-2474 is under clinical investigation as an anti-inflammatory agent [143]. The recent disclosure on the selective COX-2 inhibition by the structurally related dihydrobenzofurans such as **28** (IC_{50} (COX-1) = 10 μM; IC_{50} (COX-2) = 0.09 μM) broadens the scope of the di-*tert*-butylphenol template [144–146]. As observed in the case of diarylheterocycles, di-*tert*-butylphenols such as darbufelone (see Fig. 17) also demonstrate slow, tight-binding inhibition of COX-2 [147]. Structural information is not yet available on the binding of this class of compounds to COX-2.

Chromenes

The discovery of novel structural types continues as evident from the potent and selective COX-2 inhibition by certain chromene derivatives (Fig. 18) [148]. *In vitro* inhibition data suggest that COX-2 potency and selectivity of

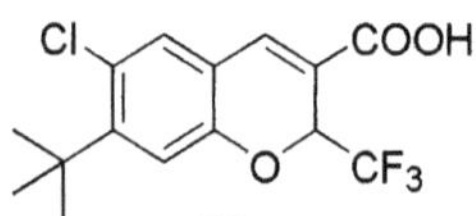

29

IC_{50} (COX-2) < 0.1 μM
IC_{50} (COX-1) = 61 μM

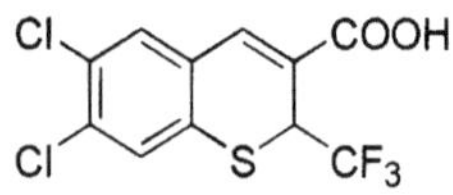

30

IC_{50} (COX-2) < 0.1 μM
IC_{50} (COX-1) = 0.8 μM

Figure 18. Chromene derivatives as selective COX-2 inhibitors.

chromenes is comparable to that of diarylheterocycles. The carboxylic acid and the trifluoromethyl substituents are major determinants of potency in the series as highlighted with analog **29**, whereas COX-2 selectivity is dependent on the presence of the oxygen atom in the chromene ring; its replacement with a sulfur atom results in dramatic loss of selectivity as shown with **30**.

Kinetic basis for selective COX-2 inhibition

COX-1 and COX-2 exist as homodimers with a molecular mass of ~70 kDa per monomer. COX-1 and COX-2 are bifunctional enzymes that carry out two sequential chemical reactions in spatially distinct but mechanistically coupled active sites. The cyclooxygenase activity *bis*oxygenates arachidonic acid to PGG_2 and the peroxidase activity reduces the hydroperoxyendoperoxide PGG_2 to PGH_2 (see Fig. 1) [149]. Both the cyclooxygenase and the peroxidase active sites are located in the catalytic domain, which accounts for 80% of the protein. All NSAIDs, as well as selective COX-2 inhibitors, bind in the COX active site but not the peroxidase site [3]. A general model for COX inhibition is emerging in which multiple equilibria are established between free enzyme, inhibitor and two or three enzyme-inhibitor complexes (Eq. 1) [150, 151]. The most potent COX inhibitors are slow, tight-binding inhibitors that form very stable complexes. The structural differences between COX-inhibitor complexes have been the subject of considerable debate and speculation. Crystallographic data on the slow, tight-binding NSAIDs (e.g., flurbiprofen and indomethacin) bound to COX-1 and COX-2 are assumed to reflect the most stable complexes [152–154]. However, flurbiprofen and the competitive inhibitor ibuprofen exhibit similar conformations in the COX active site and the structures of these cocrystals are virtually superimposible [155]. Thus, no structural differences are apparent in complexes of COX-1 bound to a slow, tight-binding inhibitor or a competitive inhibitor. Many selective COX-2 inhibitors from the diarylheterocycle or alkylsulfonanilide class appear to be competitive inhibitors of COX-1, but display time-dependent inhibition of COX-2 that is responsible for their selectivity [151, 156–158]. Fluorescence quenching analysis indicates that diarylheterocycles or neutral indomethacin amides binds to both enzymes and the kinetics of fluorescence decay reveal sequential bimolecular and unimolecular steps that occur at comparable rates [159, 160]. A third unimolecular step is only observed with COX-2. The first two steps in the binding are reversible, whereas the third step in the binding is pseudo irreversible. Recovery of fluorescence in the presence of a substantial excess of the nonfluorescent inhibitor flurbiprofen takes on the order of ~3 h with COX-2, whereas it is complete in ~40 s with COX-1. Of some interest in this context are the recent biochemical observations on the slow, tight-binding inhibition of COX-2 by the di-*tert*-butylphenol darbufelone [147]. Although, the biochemical mechanism for the selective inhibition of COX-2 is similar to that displayed by diarylheterocycles, darbufelone appears to be a selective noncompetitive inhibitor of COX-2, since arachidonic acid and various other COX inhibitors do not alter the binding affinity of this agent. It has been proposed that di-*tert*-butylphenols inhibit COX-2 by exploiting a previously unrecognized binding site on the enzyme [147].

$$E + I \underset{k_{-1}}{\overset{k_1}{\rightleftharpoons}} [EI] \underset{k_{-2}}{\overset{k_2}{\rightleftharpoons}} [EI^*] \underset{k_{-3}}{\overset{k_{intact}}{\rightleftharpoons}} [EX]$$

Equation 1. Kinetic scheme for COX inhibition. k_1, k_2, and k_{inact} represent forward rate constants for the three discrete steps of the overall reaction, whereas k_{-1}, k_{-2}, and k_{-3} represent the backward rate constants. E and I represent the free enzyme and inhibitor. Three distinct enzyme-inhibitor complexes, identified from enzyme kinetics, are shown by [EI], [EI*], and [EX]. The precise structure of each of these complexes remain unclear at the present time.

Concluding remarks

COX-2 is one of the most attractive and highly pursued targets in the history of the pharmaceutical industry. The depth and breadth of discovery efforts is attested to by the structural diversity of the inhibitor candidates summarized above. The clinical and commercial successes of celecoxib and rofecoxib validate the "COX-2 hypothesis" and guarantee that additional candidates will be advanced to the clinic. It is also likely that accumulating structural information on COX-inhibitor complexes will begin to have an impact on drug discovery and optimization. Despite the fact that COX enzymes have been the targets of enormous investigation over the past 30 years and that considerable structural information is available on COX-inhibitor complexes, remarkably little is known about the molecular details of the slow, tight binding that is responsible for the potency and selectivity of most selective or non-selective COX inhibitors. This suggests that functional studies must be performed in combination with structural studies to uncover the discrete interactions that lead to this these most fundamental of interactions responsible for COX inhibition.

Acknowledgements
Research in the Marnett laboratory is supported by research grants from the National Institutes of Health (CA89450 and GM15431).

References

1 Brune K (2002) Next generation of everyday analgesics. *Am J Ther* 9: 215–223
2 Kuhnert N (2000) A century of aspirin. The history of the most successful drug of the last century. *Pharmazie in unserer Zeit* 1: 32–39
3 Marnett LJ, Rowlinson SW, Goodwin DC, Kalgutkar AS, Lanzo CA (1999) Arachidonic acid oxygenation by COX-1 and COX-2. Mechanisms of catalysis and inhibition. *J Biol Chem* 274: 22903–22906
4 Barnett J, Chow J, Ives D, Chiou M, Mackenzie R, Osen E, Nguyen B, Tsing S, Bach C, Freire J

et al (1994) Purification, characterization and selective inhibition of human prostaglandin G/H synthase 1 and 2 expressed in the baculovirus system. *Biochim Biophys Acta* 1209: 130–139
5 Laneuville O, Breuer DK, Dewitt DL, Hla T, Funk CD, Smith WL (1994) Differential inhibition of human prostaglandin endoperoxide H synthases-1 and -2 by nonsteroidal anti-inflammatory drugs. *J Pharmacol Exp Ther* 271: 927–934
6 Meade EA, Smith WL, DeWitt DL (1993) Differential inhibition of prostaglandin endoperoxide synthase (cyclooxygenase) isozymes by aspirin and other non-steroidal anti-inflammatory drugs. *J Biol Chem* 268: 6610–6614
7 Mitchell JA, Akarasereenont P, Thiemermann C, Flower RJ, Vane JR (1993) Selectivity of nonsteroidal antiinflammatory drugs as inhibitors of constitutive and inducible cyclooxygenase. *Proc Natl Acad Sci USA* 90: 11693–11697
8 Cromlish WA, Kennedy BP (1996) Selective inhibition of cyclooxygenase-1 and -2 using intact insect cell assays. *Biochem Pharmacol* 52: 1777–1785
9 Warner TD, Giuliano F, Vojnovic I, Bukasa A, Mitchell JA, Vane JR (1999) Nonsteroid drug selectivities for cyclo-oxygenase-1 rather than cyclo-oxygenase-2 are associated with human gastrointestinal toxicity: a full *in vitro* analysis. *Proc Natl Acad Sci USA* 96: 7563–7568
10 Patrignani P, Panara MR, Greco A, Fusco O, Natoli C, Iacobelli S, Cipollone F, Ganci A, Creminon C, Maclouf J (1994) Biochemical and pharmacological characterization of the cyclooxygenase activity of human blood prostaglandin endoperoxide synthases. *J Pharmacol Exp Ther* 271: 1705–1712
11 Panara MR, Greco A, Santini G, Sciulli MG, Rotondo MT, Padovano R, di Giamberardino M, Cipollone F, Cuccurullo F, Patrono C et al (1995) Effects of the novel anti-inflammatory compounds, N-[2-(cyclohexyloxy)-4-nitrophenyl] methanesulphonamide (NS-398) and 5-methanesulphonamido-6-(2,4-difluorothio-phenyl)-1-inda none (L-745,337), on the cyclo-oxygenase activity of human blood prostaglandin endoperoxide synthases. *Br J Pharmacol* 116: 2429–2434
12 Kato M, Nishida S, Kitasato H, Sakata N, Kawai S (2001) Cyclooxygenase-1 and cyclooxygenase-2 selectivity of non-steroidal anti-inflammatory drugs: investigation using human peripheral monocytes. *J Pharm Pharmacol* 53: 1679–1685
13 Chulada PC, Langenbach R (1997) Differential inhibition of murine prostaglandin synthase-1 and -2 by nonsteroidal anti-inflammatory drugs using exogenous and endogenous sources of arachidonic acid. *J Pharmacol Exp Ther* 280: 606–613
14 Grossman CJ, Wiseman J, Lucas FS, Trevethick MA, Birch PJ (1995) Inhibition of constitutive and inducible cyclooxygenase activity in human platelets and mononuclear cells by NSAIDs and Cox 2 inhibitors. *Inflamm Res* 44: 253–257
15 Kargman S, Wong E, Greig GM, Falgueyret JP, Cromlish W, Ethier D, Yergey JA, Riendeau D, Evans JF, Kennedy B et al (1996) Mechanism of selective inhibition of human prostaglandin G/H synthase-1 and -2 in intact cells. *Biochem Pharmacol* 52: 1113–1125
16 Futaki N, Takahashi S, Yokoyama M, Arai I, Higuchi S, Otomo S (1994) NS-398, a new anti-inflammatory agent, selectively inhibits prostaglandin G/H synthase/cyclooxygenase (COX-2) activity *in vitro*. *Prostaglandins* 47: 55–59
17 Kalgutkar AS, Crews BC, Marnett LJ (1996) Kinetics of the interaction of nonsteroidal antiinflammatory drugs with prostaglandin endoperoxide synthase-1 studied by limited proteolysis. *Biochemistry* 35: 9076–9082
18 McKenna F, Arguelles L, Burke T, Lefkowith J, Geis GS (2002) Upper gastrointestinal tolerability of celecoxib compared with diclofenac in the treatment of osteoarthritis and rheumatoid arthritis. *Clin Exp Rheumatol* 20: 35–43
19 Weaver AL (2001) Rofecoxib: clinical pharmacology and clinical experience. *Clin Ther* 23: 1323–1338
20 Ormrod D, Wellington K, Wagstaff AJ (2002) Valdecoxib. *Drugs* 62: 2059–2071
21 Gottesdiener K, Schnitzer T, Fisher C, Bockow B, Markenson J, Ko A, DeTora L, Curtis S, Geissler L, Gertz BJ (2002) Results of a randomized, dose-ranging trial of etoricoxib in patients with osteoarthritis. *Rheumatology (Oxford)* 41: 1052–1061
22 Flower R, Gryglewski R, Herbaczynska-Cedro K, Vane JR (1972) Effects of anti-inflammatory drugs on prostaglandin biosynthesis. *Nat New Biol* 238: 104–106
23 Lombardino JG (1974) In: RA Sherrer, MW Whitehouse (eds): *Antiinflammatory agents, chemistry and pharmacology*. Academic Press, New York, 131–164
24 Frey HH, Nielsen CK (1966) Contributions to the evaluation of antiphlogistic test results. 2. Comparison of the effects of phenylbutazone with various models of inflammation.

Arzneimittelforschung 16: 87–89
25 Brown K, Cavalla JF, Green D, Wilson AB (1968) Diaryloxazole and diarylthiazolealkanoic acids: two novel series of non-steroidal anti-inflammatory agents. *Nature* 219: 164
26 Rosenthale ME, Begany AJ, Dervinis A, Malis JL, Shriver DA, Datko SL, Gluckman MI (1974) Anti-inflammatory properties of 4,5-diphenyl-2-oxazolepropionic acid (oxaprozin). *Agents Actions* 4: 151–159
27 Szmuszkovicz J, Glenn EM, Heinzelman RV, Hester JB Jr, Youngdale GA (1966) Synthesis and antiinflammatory activity of 2,3-bis(p-methoxyphenyl)indole and related compounds. *J Med Chem* 9: 527–536
28 Glenn EM, Bowman BJ, Kooyers W, Koslowske T, Myers ML (1967) The pharmacology of 2,3-bis-(p-methoxyphenyl)-indole (indoxole). *J Pharmacol Exp Ther* 155: 157–166
29 Kaiser DG, Glenn EM, Johnson RH, Johnston RL (1967) Correlation of serum indoxole [2,3-bis(p-methoxyphenyl)-indole] levels with biologic activity. *J Pharmacol Exp Ther* 155: 174–180
30 Talley JJ (1999) Selective inhibitors of cyclooxygenase-2 (COX-2). *Prog Med Chem* 36: 201–234
31 Tanaka A, Sakai H, Motoyama Y, Ishikawa T, Takasugi H (1994) Antiplatelet agents based on cyclooxygenase inhibition without ulcerogenesis. Evaluation and synthesis of 4,5-bis(4-methoxyphenyl)-2-substituted-thiazoles. *J Med Chem* 37: 1189–1199
32 Hsyu PH, Koets MD, Luderer JR (1994) Pharmacokinetics and cyclooxygenase inhibition of itazigrel in normal volunteers after single oral doses. *J Pharm Sci* 83: 1747–1750
33 Demke DM, Luderer JR, Wakefield LK, Euler AR, Brouwer RD, Metzler CM (1987) The effect of itazigrel and aspirin on the mucosa of the esophagus, stomach, and duodenum of normal subjects. *J Clin Pharmacol* 27: 916–920
34 Sharpe TR, Cherkofsky SC, Hewes WE, Smith DH, Gregory WA, Haber SB, Leadbetter MR, Whitney JG (1985) Preparation and antiarthritic and analgesic activity of 4,5-diaryl-2-(substituted thio)-1H-imidazoles and their sulfoxides and sulfones. *J Med Chem* 28: 1188–1194
35 Gans KR, Galbraith W, Roman RJ, Haber SB, Kerr JS, Schmidt WK, Smith C, Hewes WE, Ackerman NR (1990) Anti-inflammatory and safety profile of DuP 697, a novel orally effective prostaglandin synthesis inhibitor. *J Pharmacol Exp Ther* 254: 180–187
36 Leblanc Y, Gauthier JY, Ethier D, Guay J, Mancini J, Riendeau D, Tagari P, Vickers P, Wong E, Prasit P (1995) Synthesis and biological evaluation of 2,3-diarylthiophenes as selective Cox-2 and Cox-1 inhibitors. *Bioorg Med Chem Lett* 5: 2123–2128
37 Bertenshaw SR, Talley JJ, Rogier DJ, Graneto MJ, Rogers RS, Kramer SW, Penning TD, Koboldt CM, Veenhuizen AW, Zhang Y et al (1995) 3,4-diarylthiophenes are selective COX-2 inhibitors. *Bioorg Med Chem Lett* 5: 2919–2922
38 Penning TD, Talley JJ, Bertenshaw SR, Carter JS, Collins PW, Docter S, Graneto MJ, Lee LF, Malecha JW, Miyashiro JM et al (1997) Synthesis and biological evaluation of the 1,5-diarylpyrazole class of cyclooxygenase-2 inhibitors: identification of 4-[5-(4-methylphenyl)-3-(trifluoromethyl)-1H-pyrazol-1-yl]benzenesulfonamide (SC-58635, celecoxib). *J Med Chem* 40: 1347–1365
39 Carter JS (2000) Inhibitors of cyclooxygenase-2. *Exp Opin Ther Pat* 10: 1011–1020
40 Portevin B, Tordjman C, Pastoureau P, Bonnet J, De Nanteuil G (2000) 1,3-Diaryl-4,5,6,7-tetrahydro-2H-isoindole derivatives: a new series of potent and selective COX-2 inhibitors in which a sulfonyl group is not a structural requisite. *J Med Chem* 43: 4582–4593
41 Hashimoto H, Imamura K, Haruta J, Wakitani K (2002) 4-(4-cycloalkyl/aryl-oxazol-5-yl) benzenesulfonamides as selective cyclooxygenase-2 inhibitors: enhancement of the selectivity by introduction of a fluorine atom and identification of a potent, highly selective, and orally active COX-2 inhibitor JTE-522. *J Med Chem* 45: 1511–1517
42 Hashimoto H, Maeda K, Ozawa K, Haruta J, Wakitani K (2002) 4-Aryl/cycloalkyl-5-phenyloxazole derivatives as selective COX-2 inhibitors. *Bioorg Med Chem Lett* 12: 65–68
43 Lau CK, Brideau C, Chan CC, Charleson S, Cromlish WA, Ethier D, Gauthier JY, Gordon R, Guay J, Kargman S et al (1999) Synthesis and biological evaluation of 3-heteroaryloxy-4-phenyl-2(5H)-furanones as selective COX-2 inhibitors. *Bioorg Med Chem Lett* 9: 3187–3192
44 Leblanc Y, Roy P, Boyce S, Brideau C, Chan CC, Charleson S, Gordon R, Grimm E, Guay J, Leger S et al (1999) SAR in the alkoxy lactone series: the discovery of DFP, a potent and orally active COX-2 inhibitor. *Bioorg Med Chem Lett* 9: 2207–2212
45 Leblanc Y, Roy P, Wang Z, Li CS, Chauret N, Nicoll-Griffith DA, Silva JM, Aubin Y, Yergey JA, Chan CC et al (2002) Discovery of a potent and selective COX-2 inhibitor in the alkoxy lactone

series with optimized metabolic profile. *Bioorg Med Chem Lett* 12: 3317–3321
46 Li CS, Black WC, Brideau C, Chan CC, Charleson S, Cromlish WA, Claveau D, Gauthier JY, Gordon R, Greig G et al (1999) A new structural variation on the methanesulfonylphenyl class of selective cyclooxygenase-2 inhibitors. *Bioorg Med Chem Lett* 9: 3181–3186
47 Pinto DJP, Copeland RA, Covington MB, Pitts WJ, Batt DG, Orwat MJ, Lam GN, Joshi A, Chan Y-C, Wang S et al (1996) Chemistry and pharmacokinetics of diarylthiophenes and terphenyls as selective COX-2 inhibitors. *Bioorg Med Chem Lett* 6: 2907–2912
48 Friesen RW, Dubé D, Fortin R, Frenette R, Prescott S, Cromlish W, Greig GM, Kargman S, Wong E, Chan CC et al (1996) Novel 1,2-diarylcyclobutenes: selective and orally active cox-2 inhibitors. *Bioorg Med Chem Lett* 6: 2677–2682
49 Reitz DB, Li JJ, Norton MB, Reinhard EJ, Collins JT, Anderson GD, Gregory SA, Koboldt CM, Perkins WE, Seibert K et al (1994) Selective cyclooxygenase inhibitors: novel 1,2-diarylcyclopentenes are potent and orally active COX-2 inhibitors. *J Med Chem* 37: 3878–3881
50 Li JJ, Anderson GD, Burton EG, Cogburn JN, Collins JT, Garland DJ, Gregory SA, Huang HC, Isakson PC, Koboldt CM et al (1995) 1,2-Diarylcyclopentenes as selective cyclooxygenase-2 inhibitors and orally active anti-inflammatory agents. *J Med Chem* 38: 4570–4578
51 Black WC, Brideau C, Chan CC, Charleson S, Chauret N, Claveau D, Ethier D, Gordon R, Greig G, Guay J et al (1999) 2,3-Diarylcyclopentenones as orally active, highly selective cyclooxygenase-2 inhibitors. *J Med Chem* 42: 1274–1281
52 Reitz DB, Huang HC, Li JJ, Garland DJ, Manning RE, Anderson GD, Gregory SA, Koboldt CM, Perkins WE, Seibert K et al (1995) Selective cyclooxygenase inhibitors: novel 4-spiro 1,2-diarylcyclopentenes are potent and orally active COX-2 inhibitors. *Bioorg Med Chem Lett* 5: 867–872
53 Huang HC, Li JJ, Garland DJ, Chamberlain TS, Reinhard EJ, Manning RE, Seibert K, Koboldt CM, Gregory SA, Anderson GD et al (1996) Diarylspiro[2.4]heptenes as orally active, highly selective cyclooxygenase-2 inhibitors: synthesis and structure-activity relationships. *J Med Chem* 39: 253–266
54 Pinto DJ, Batt DG, Pitts WJ, Petraitis JJ, Orwat MJ, Wang S, Jetter JW, Sherk SR, Houghton GC, Copeland RA et al (1999) Terphenyl cyclooxygenase-2 (COX-2) inhibitors: optimization of the central ring and o-biphenyl analogs. *Bioorg Med Chem Lett* 9: 919–924
55 Li JJ, Norton MB, Reinhard EJ, Anderson GD, Gregory SA, Isakson PC, Koboldt CM, Masferrer JL, Perkins WE, Seibert K et al (1996) Novel terphenyls as selective cyclooxygenase-2 inhibitors and orally active anti-inflammatory agents. *J Med Chem* 39: 1846–1856
56 Atkinson J, Wang Z (1998) Stilbene derivatives useful as cyclooxygenase-2 inhibitors. Merck Frosst Canada Inc US5849943.
57 Nicolai E, Teulon J-M (2000) Preparation of diarylmethylidine furan derivatives as anti-inflammatory, analgesic and chemopreventive agents. Lab UPSA PCT WO 2000014082.
58 Nicolai E, Teulon J-M (1997) Preparation of 3-diarylmethylidene-2-furanones and analogs as cyclooxygenase-2 inhibitors. Lab UPSA PCT WO9737984.
59 Khanna IK, Weier RM, Yu Y, Collins PW, Miyashiro JM, Koboldt CM, Veenhuizen AW, Currie JL, Seibert K, Isakson PC (1997) 1,2-Diarylpyrroles as potent and selective inhibitors of cyclooxygenase-2. *J Med Chem* 40: 1619–1633
60 Wilkerson WW, Galbraith W, Gans-Brangs K, Grubb M, Hewes WE, Jaffee B, Kenney JP, Kerr J, Wong N (1994) Antiinflammatory 4,5-diarylpyrroles: synthesis and QSAR. *J Med Chem* 37: 988–998
61 Wilkerson WW, Copeland RA, Covington M, Trzaskos JM (1995) Antiinflammatory 4,5-diarylpyrroles. 2. Activity as a function of cyclooxygenase-2 inhibition. *J Med Chem* 38: 3895–3901
62 Bertenshaw SR, Collins PW, Penning TD, Reitz DB, Rogers RS (1994) Novel 3,4-diaryl thiophenes and analogs thereof having use as antiinflammatory agents. GD Searle & Co PCT WO9415932.
63 Ducharme Y, Gauthier JY, Prasit P, Leblanc Y, Wang Z, Leger S, Therien M (1995) Phenyl heterocycles as COX-2 inhibitors. Merck Frosst Canada Inc US5474995.
64 Dube D, Brideau C, Deschenes D, Fortin R, Friesen RW, Gordon R, Girard Y, Riendeau D, Savoie C, Chan CC (1999) 2-heterosubstituted-3-(4-methylsulfonyl) phenyl-5-trifluoromethyl pyridines as selective and orally active cyclooxygenase-2 inhibitors. *Bioorg Med Chem Lett* 9: 1715–1720
65 Friesen RW, Brideau C, Chan CC, Charleson S, Deschenes D, Dube D, Ethier D, Fortin R, Gauthier JY, Girard Y et al (1998) 2-Pyridinyl-3-(4-methylsulfonyl) phenylpyridines: selective and orally active cyclooxygenase-2 inhibitors. *Bioorg Med Chem Lett* 8: 2777–2782

66 Riendeau D, Percival MD, Brideau C, Charleson S, Dube D, Ethier D, Falgueyret JP, Friesen RW, Gordon R, Greig G et al (2001) Etoricoxib (MK-0663): preclinical profile and comparison with other agents that selectively inhibit cyclooxygenase-2. *J Pharmacol Exp Ther* 296: 558–566
67 Carter JS, Kramer S, Talley JJ, Penning T, Collins P, Graneto MJ, Seibert K, Koboldt CM, Masferrer J, Zweifel B (1999) Synthesis and activity of sulfonamide-substituted 4,5-diaryl thiazoles as selective cyclooxygenase-2 inhibitors. *Bioorg Med Chem Lett* 9: 1171–1174
68 Carter JS, Rogier DJ, Graneto MJ, Seibert K, Koboldt CM, Zhang Y, Talley JJ (1999) Design and synthesis of sulfonyl-substituted 4,5-diarylthiazoles as selective cyclooxygenase-2 inhibitors. *Bioorg Med Chem Lett* 9: 1167–1170
69 Talley JJ, Bertenshaw SR, Brown DL, Carter JS, Graneto MJ, Koboldt CM, Masferrer JL, Norman BH, Rogier DJ Jr, Zweifel BS et al (1999) 4,5-Diaryloxazole inhibitors of cyclooxygenase-2 (COX-2). *Med Res Rev* 19: 199–208
70 Puig C, Crespo MI, Godessart N, Feixas J, Ibarzo J, Jimenez JM, Soca L, Cardelus I, Heredia A, Miralpeix M et al (2000) Synthesis and biological evaluation of 3,4-diaryloxazolones: A new class of orally active cyclooxygenase-2 inhibitors. *J Med Chem* 43: 214–223
71 Barta TE, Stealey MA, Collins PW, Weier RM (1998) Antiinflammatory 4,5-diarylimidazoles as selective cyclooxygenase inhibitors. *Bioorg Med Chem Lett* 8: 3443–3448
72 Khanna IK, Yu Y, Huff RM, Weier RM, Xu X, Koszyk FJ, Collins PW, Cogburn JN, Isakson PC, Koboldt CM et al (2000) Selective cyclooxygenase-2 inhibitors: heteroaryl modified 1,2-diarylimidazoles are potent, orally active antiinflammatory agents. *J Med Chem* 43: 3168–3185
73 Khanna IK, Weier RM, Yu Y, Xu XD, Koszyk FJ, Collins PW, Koboldt CM, Veenhuizen AW, Perkins WE, Casler JJ et al (1997) 1,2-Diarylimidazoles as potent, cyclooxygenase-2 selective, and orally active antiinflammatory agents. *J Med Chem* 40: 1634–1647
74 Talley JJ, Brown DL, Carter JS, Graneto MJ, Koboldt CM, Masferrer JL, Perkins WE, Rogers RS, Shaffer AF, Zhang YY et al (2000) 4-[5-Methyl-3-phenylisoxazol-4-yl]-benzenesulfonamide, valdecoxib: a potent and selective inhibitor of COX-2. *J Med Chem* 43: 775–777
75 Talley JJ, Bertenshaw SR, Brown DL, Carter JS, Graneto MJ, Kellogg MS, Koboldt CM, Yuan J, Zhang YY, Seibert K (2000) N-[[(5-methyl-3-phenylisoxazol-4-yl)-phenyl]sulfonyl]propanamide, sodium salt, parecoxib sodium: A potent and selective inhibitor of COX-2 for parenteral administration. *J Med Chem* 43: 1661–1663
76 Chavatte P, Yous S, Marot C, Baurin N, Lesieur D (2001) Three-dimensional quantitative structure-activity relationships of cyclo-oxygenase-2 (COX-2) inhibitors: a comparative molecular field analysis. *J Med Chem* 44: 3223–3230
77 Tsuji K, Nakamura K, Konishi N, Tojo T, Ochi T, Senoh H, Matsuo M (1997) Studies on anti-inflammatory agents. IV. Synthesis and pharmacological properties of 1,5-diarylpyrazoles and related derivatives. *Chem Pharm Bull* 45: 987–995
78 Ochi T, Yamane-Sugiyama A, Ohkubo Y, Sakane K, Tanaka H (2001) The anti-inflammatory effect of FR188582, a highly selective inhibitor of cyclooxygenase-2, with an ulcerogenic sparing effect in rats. *Jpn J Pharmacol* 85: 175–182
79 Seibert K, Zhang Y, Leahy K, Hauser S, Masferrer J, Perkins W, Lee L, Isakson P (1994) Pharmacological and biochemical demonstration of the role of cyclooxygenase 2 in inflammation and pain. *Proc Natl Acad Sci USA* 91: 12013–12017
80 Lau CK (1997) Diphenyl-1,2,3-thiadiazoles as anti-inflammatory agents. Merck Frosst Canada Inc US5677318.
81 Prasit P, Wang Z, Brideau C, Chan CC, Charleson S, Cromlish W, Ethier D, Evans JF, Ford-Hutchinson AW, Gauthier JY et al (1999) The discovery of rofecoxib, [MK 966, Vioxx, 4-(4'-methylsulfonylphenyl)-3-phenyl-2(5H)-furanone], an orally active cyclooxygenase-2-inhibitor. *Bioorg Med Chem Lett* 9: 1773–1778
82 Chan CC, Boyce S, Brideau C, Charleson S, Cromlish W, Ethier D, Evans J, Ford-Hutchinson AW, Forrest MJ, Gauthier JY et al (1999) Rofecoxib [Vioxx, MK-0966; 4-(4'-methylsulfonylphenyl)-3-phenyl-2-(5H)-furanone]: a potent and orally active cyclooxygenase-2 inhibitor. Pharmacological and biochemical profiles. *J Pharmacol Exp Ther* 290: 551–560
83 Shin SS, Noh MS, Byun YJ, Choi JK, Kim JY, Lim KM, Ha JY, Kim JK, Lee CH, Chung S (2001) 2,2-Dimethyl-4,5-diaryl-3(2H)furanone derivatives as selective cyclo-oxygenase-2 inhibitors. *Bioorg Med Chem Lett* 11: 165–168
84 Riendeau D, Percival MD, Boyce S, Brideau C, Charleson S, Cromlish W, Ethier D, Evans J, Falgueyret JP, Ford-Hutchinson AW et al (1997) Biochemical and pharmacological profile of a tetrasubstituted furanone as a highly selective COX-2 inhibitor. *Br J Pharmacol* 121: 105–117

85 Yergey JA, Trimble LA, Silva J, Chauret N, Li C, Therien M, Grimm E, Nicoll-Griffith DA (2001) *In vitro* metabolism of the COX-2 inhibitor DFU, including a novel glutathione adduct rearomatization. *Drug Metab Dispos* 29: 638–644
86 Therien M, Brideau C, Chan CC, Cromlish WA, Gauthier JY, Gordon R, Greig G, Kargman S, Lau CK, Leblanc Y et al (1997) Synthesis and biological evaluation of 5,6-diarylimidazo[2.1-b]thiazole as selective COX-2 inhibitors. *Bioorg Med Chem Lett* 7: 47–52
87 Roy P, LeBlanc Y, Ball RG, Brideau C, Chan CC, Chauret N, Cromlish W, Ethier D, Gauthier JY, Gordon R et al (1997) A new series of selective COX-2 inhibitors: 5,6-diarylthiazolo[3,2-b][1,2,4]triazoles. *Bioorg Med Chem Lett* 7: 57–62
88 Almansa C, de Arriba AF, Cavalcanti FL, Gomez LA, Miralles A, Merlos M, Garcia-Rafanell J, Forn J (2001) Synthesis and SAR of a new series of COX-2-selective inhibitors: pyrazolo[1,5-a]pyrimidines. *J Med Chem* 44: 350–361
89 Farrerons Gallemi C, Miquel Bono I-J, Fernandez Serrat AM, Monserrat Vidal C, Lagunas Arnal C, Gimenez Guasch F, Fernandez Garcia A (2000) Substituted imidazo[1,2a]azines as selective inhibitors of COX-2. Lab SALVAT SA PCT WO200008024.
90 Talley JJ, Bartenshaw SR, Graneto MJ, Rogier DJ (1996) Preparation of cytokine-inhibiting benzopyranopyrazole antiinflammatories and analgesics. GD Searle & Co PCT WO9609304.
91 Talley JJ, Bertenshaw SR, Rogers RS (1997) Benz[G]indazolyl derivatives for the treatment of inflammation. GD Searle & Co US5696143.
92 Harrington J, Robertson JE, Kvam DC, Hamilton RR, McGurran KT, Trancik RJ, Swingle KF, Moore GG, Gerster JF (1970) Antiinflammatory agents. I. 3-Benzoylfluoroalkanesulfonamides. *J Med Chem* 13: 137
93 Swingle KF, Moore GG, Grant TJ (1976) 4-Nitro-2-phenoxymethanesulfonanilide (R-805): a chemically novel anti-inflammatory agent. *Arch Int Pharmacodyn Ther* 221: 132–139
94 Davis R, Brogden RN (1994) Nimesulide. An update of its pharmacodynamic and pharmacokinetic properties, and therapeutic efficacy. *Drugs* 48: 431–454
95 Masferrer JL, Zweifel BS, Manning PT, Hauser SD, Leahy KM, Smith WG, Isakson PC, Seibert K (1994) Selective inhibition of inducible cyclooxygenase 2 *in vivo* is antiinflammatory and nonulcerogenic. *Proc Natl Acad Sci USA* 91: 3228–3232
96 Futaki N, Takahashi S, Kitagawa T, Yamakawa Y, Tanaka M, Higuchi S (1997) Selective inhibition of cyclooxygenase-2 by NS-398 in endotoxin shock rats *in vivo*. *Inflamm Res* 46: 496–502
97 Wiesenberg-Bottcher I, Schweizer A, Green JR, Seltenmeyer Y, Muller K (1989) The pharmacological profile of CGP 28238, a highly potent anti-inflammatory compound. *Agents Actions* 26: 240–242
98 Vago T, Bevilacqua M, Norbiato G (1995) Effect of nimesulide action time dependence on selectivity towards prostaglandin G/H synthase/cyclooxygenase activity. *Arzneimittelforschung* 45: 1096–1098
99 Li CS, Black WC, Chan CC, Ford-Hutchinson AW, Gauthier JY, Gordon R, Guay D, Kargman S, Lau CK, Mancini J et al (1995) Cyclooxygenase-2 inhibitors. Synthesis and pharmacological activities of 5-methanesulfonamido-1-indanone derivatives. *J Med Chem* 38: 4897–4905
100 Ouimet N, Chan CC, Charleson S, Claveau D, Gordon R, Guay D, Li CS, Ouellet M, Percival DM, Riendeau D et al (1999) Substituted heterocyclic analogs as selective COX-2 inhibitors in the flosulide class. *Bioorg Med Chem Lett* 9: 151–156
101 Guay D, Li C-S, Quimet N (1996) Preparation of 5-methanesulfonamido-3H-isobenzofuran-1-one antiinflammatory inhibitors of cyclooxygenase-2. Merck Frosst Canada Inc PCT WO9623786.
102 Tsuji K, Nakamura K, Konishi N, Okumura H, Matsuo M (1992) Studies on antiinflammatory agents. I. Synthesis and pharmacological properties of 2'-phenoxymethanesulfonanilide derivatives. *Chem Pharm Bull* 40: 2399–2409
103 Tanaka K, Shimotori T, Makino S, Aikawa Y, Inaba T, Yoshida C, Takano S (1992) Pharmacological studies of the new antiinflammatory agent 3-formylamino-7-methylsulfonylamino-6-phenoxy-4H-1-benzopyran-4-one. 1st communication: antiinflammatory, analgesic and other related properties. *Arzneimittelforschung* 42: 935–944
104 Berg J, Fellier H, Christoph T, Kremminger P, Hartmann M, Blaschke H, Rovensky F, Towart R, Stimmeder D (2000) Pharmacology of a selective cyclooxygenase-2 inhibitor, HN-56249: a novel compound exhibiting a marked preference for the human enzyme in intact cells. *Naunyn-Schmiedeberg's Archives of Pharmacology* 361:363-372.
105 Kawasaki H, Tanaka K (1996) Cyclooxygenase 2 inhibitors containing phenoxybenzopyranone

derivative. Toyama Kagaku Kogyo KK JP8157361.
106 Park CH, Siomboing X, Yous S, Gressier B, Luyckx M, Chavatte P (2002) Investigations of new lead structures for the design of novel cyclooxygenase-2 inhibitors. *Eur J Med Chem* 37: 461–468
107 Black WC, Bayly C, Belley M, Chan CC, Charleson S, Denis D, Gauthier JY, Gordon R, Guay D, Kargman S et al (1996) From indomethacin to a selective COX-2 inhibitor: Development of indolalkanoic acids as potent and selective cyclooxygenase-2 inhibitors. *Bioorg Med Chem Lett* 6: 725–730
108 Lau CK, Black WC, Belley M, Chan CC, Charleson S, Denis D, Gauthier JY, Gordon R, Guay D, Hamel P et al (1997) From indomethacin to a selective COX-2 inhibitor. Development of indolalkanoic acids as potent and selective cyclooxygenase-2 inhibitors. *Adv Exp Med Biol* 407: 73–78
109 Kalgutkar AS, Crews BC, Rowlinson SW, Marnett AB, Kozak KR, Remmel RP, Marnett LJ (2000) Biochemically based design of cyclooxygenase-2 (COX-2) inhibitors: facile conversion of nonsteroidal antiinflammatory drugs to potent and highly selective COX-2 inhibitors. *Proc Natl Acad Sci USA* 97: 925–930
110 Kalgutkar AS, Marnett AB, Crews BC, Remmel RP, Marnett LJ (2000) Ester and amide derivatives of the nonsteroidal antiinflammatory drug, indomethacin, as selective cyclooxygenase-2 inhibitors. *J Med Chem* 43: 2860–2870
111 Kozak KR, Prusakiewicz JJ, Rowlinson SW, Marnett LJ (2002) Enantiospecific, selective cyclooxygenase-2 inhibitors. *Bioorg Med Chem Lett* 12: 1315–1318
112 Kalgutkar AS, Marnett LJ (2001) Esters derived from indolealkanols and novel amides derived from indolealkylamides that are selective COX-2 inhibitors. Vanderbilt University US6306890.
113 Woods KW, McCroskey RW, Michaelides MR (1998) Heterocyclic compounds as COX-2 inhibitors. Abbott Laboratories US5811425.
114 Woods KW, McCroskey RW, Michaelides MR, Wada CK, Hulkower KI, Bell RL (2001) Thiazole analogues of the NSAID indomethacin as selective COX-2 inhibitors. *Bioorg Med Chem Lett* 11: 1325–1328
115 Palomer A, Cabre F, Pascual J, Campos J, Trujillo MA, Entrena A, Gallo MA, Garcia L, Mauleon D, Espinosa A (2002) Identification of novel cyclooxygenase-2 selective inhibitors using pharmacophore models. *J Med Chem* 45: 1402–1411
116 Matsuoka H, Kato N, Takahashi T, Maruyama N, Ishizawa T, Suzuki Y (1999) Preparation of heterocyclic indole derivatives and mono- or diazaindole derivatives as cyclooxygenase-2 (COX-2) inhibitors. Chugai Seiyaku Kabushiki Kaisha PCT WO9961436.
117 Matsuoka H, Maruyama N, Kashiwagi H (1998) Preparation of indole, monoazaindole, and diazaindole derivatives as cyclooxygenase inhibitors. Chugai Seiyaku Kabushiki Kaisha (1998) PCT WO9851667.
118 Matsuoka H, Maruyama N, Kato Y (1998) Preparation of indene derivatives as COX 2 inhibitors. Chugai Seiyaku Kabushiki Kaisha PCT WO9857924.
119 Nakao K, Stevens RW, Kawamura K, Uchida C, Koike H, Caron S (1999) Preparation of 2,3-substituted indoles as COX-2 inhibitors. Pfizer Pharmaceuticals, Japan PCT WO9935130.
120 Failli AA (1999) Indene inhibitors of COX-2. American Home Products Corp US5869524.
121 Barnett JW, Dunn JP, Kertesz DJ, Miller AB, Morgans DJ Jr, Ramesha CS, Sigal CE, Sjogren EB, Smith DB, Talamas FX (1996) Pyrrole derivatives. F Hoffmann-La Roche AG EP-714895.
122 Luong C, Miller A, Barnett J, Chow J, Ramesha C, Browner MF (1996) Flexibility of the NSAID binding site in the structure of human cyclooxygenase-2. *Nature Struct Biol* 3: 927–933
123 Brooks CDW, Craig RA, Kolasa T, Stewart AO (1999) Iminoxyderivatives of fenamates as inhibitors of prostaglandin biosynthesis. Abbott Laboratories US5863946.
124 Fujimoto RA, Mcquire LW, Mugrage BB, Van Duzer JH, Xu D (1999) Preparation of 5-alkyl-2-arylaminophenylacetic acids as COX-2 cyclooxygenase inhibitors. Novartis AG PCT WO9911605.
125 Sorbera LA, Castaner J, Bayes M, Silvestre JS (2002) Lumiracoxib. *Drugs of the Future* 27: 740–747
126 Glaser K, Sung ML, O'Neill K, Belfast M, Hartman D, Carlson R, Kreft A, Kubrak D, Hsiao CL, Weichman B (1995) Etodolac selectively inhibits human prostaglandin G/H synthase 2 (PGHS-2) *versus* human PGHS-1. *Eur J Pharmacol* 281: 107–111
127 Kreft AF, Caufield CE, Failli AA, Caggiano TJ, Greenfield AA, Kubrak DM (1998) Pyranoindole inhibitors of COX-2. American Home Products Corp US5776967.

128 Kreft AF, Caufield CE, Failli AA, Caggiano TJ, Greenfield AA, Kubrak DM (1998) Carbazole inhibitors of COX-2. American Home Products Corp US5824699.
129 Kalgutkar AS, Rowlinson SW, Crews BC, Marnett LJ (2002) Amide derivatives of meclofenamic acid as selective cyclooxygenase-2 inhibitors. *Bioorg Med Chem Lett* 12: 521–524
130 Bayly CI, Black WC, Leger S, Quimet N, Quellet M, Percival MD (1999) Structure-based design of COX-2 selectivity into flurbiprofen. *Bioorg Med Chem Lett* 9: 307–312
131 Ogino K, Hatanaka K, Kawamura M, Katori M, Harada Y (1997) Evaluation of pharmacological profile of meloxicam as an anti-inflammatory agent, with particular reference to its relative selectivity for cyclooxygenase-2 over cyclooxygenase-1. *Pharmacology* 55: 44–53
132 Lazer ES, Sorcek R, Cywin CL, Thome D, Possanza GJ, Graham AG, Churchill L (1998) Antiinflammatory 2-benzyl-4-sulfonyl-4H-isoquinoline-1,3-diones: novel inhibitors of COX-2. *Bioorg Med Chem Lett* 8: 1181–1186
133 Lazer ES, Miao CK, Cywin CL, Sorcek R, Wong HC, Meng Z, Potocki I, Hoermann M, Snow RJ, Tschantz MA et al (1997) Effect of structural modification of enol-carboxamide-type nonsteroidal antiinflammatory drugs on COX-2/COX-1 selectivity. *J Med Chem* 40: 980–989
134 Lecomte M, Laneuville O, Ji C, DeWitt DL, Smith WL (1994) Acetylation of human prostaglandin endoperoxide synthase-2 (cyclooxygenase-2) by aspirin. *J Biol Chem* 269: 13207–13215
135 Roth GJ, Stanford N, Majerus PW (1975) Acetylation of prostaglandin synthase by aspirin. *Proc Natl Acad Sci USA* 72: 3073–3076
136 Kalgutkar AS, Crews BC, Rowlinson SW, Garner C, Seibert K, Marnett LJ (1998) Aspirin-like molecules that covalently inactivate cyclooxygenase-2. *Science* 280: 1268–1270
137 Kalgutkar AS, Kozak KR, Crews BC, Hochgesang GP, Marnett LJ (1998) Covalent modification of cyclooxygenase-2 (COX-2) by 2-acetoxyphenyl alkyl sulfides, a new class of selective COX-2 inactivators. *J Med Chem* 41: 4800–4818
138 Hochgesang GP, Nemeth-Cawley JF, Rowlinson SW, Caprioli RM, Marnett LJ (2003) Functional analysis of the molecular determinants of cyclooxygenase-2 acetylation by 2-acetoxyphenylhept-2-ynyl sulfide. *Arch Biochem Biophys* 409: 127–133
139 Hochgesang GP, Rowlinson SW, Marnett LJ (2000) Tyrosine 385 is critical for acetylation of cyclooxygenase-2 by aspirin. *J Am Chem Soc* 122: 6514–6515
140 Mullican MD, Wilson MW, Connor DT, Kostlan CR, Schrier DJ, Dyer RD (1993) Design of 5-(3,5-di-tert-butyl-4-hydroxyphenyl)-1,3,4-thiadiazoles, -1,3,4-oxadiazoles, and -1,2,4-triazoles as orally-active, nonulcerogenic antiinflammatory agents. *J Med Chem* 36: 1090–1099
141 Song Y, Connor DT, Doubleday R, Sorenson RJ, Sercel AD, Unangst PC, Roth BD, Gilbertsen RB, Chan K, Schrier DJ et al (1999) Synthesis, structure-activity relationships, and *in vivo* evaluations of substituted di-tert-butylphenols as a novel class of potent, selective, and orally active cyclooxygenase-2 inhibitors. 1. Thiazolone and oxazolone series. *J Med Chem* 42: 1151–1160
142 Song Y, Connor DT, Sercel AD, Sorenson RJ, Doubleday R, Unangst PC, Roth BD, Beylin VG, Gilbertsen RB, Chan K et al (1999) Synthesis, structure-activity relationships, and *in vivo* evaluations of substituted di-tert-butylphenols as a novel class of potent, selective, and orally active cyclooxygenase-2 inhibitors. 2. 1,3,4- and 1,2,4-thiadiazole series. *J Med Chem* 42: 1161–1169
143 Inagaki M, Tsuri T, Jyoyama H, Ono T, Yamada K, Kobayashi M, Hori Y, Arimura A, Yasui K, Ohno K et al (2000) Novel antiarthritic agents with 1,2-isothiazolidine-1,1-dioxide (gamma-sultam) skeleton: cytokine suppressive dual inhibitors of cyclooxygenase-2 and 5-lipoxygenase. *J Med Chem* 43: 2040–2048
144 Janusz JM, Young PA, Ridgeway JM, Scherz MW, Enzweiler K, Wu LI, Gan L, Darolia R, Matthews RS, Hennes D et al (1998) New cyclooxygenase-2/5-lipoxygenase inhibitors. 1. 7-tert-butyl-2,3-dihydro-3,3-dimethylbenzofuran derivatives as gastrointestinal safe antiinflammatory and analgesic agents: discovery and variation of the 5-keto substituent. *J Med Chem* 41: 1112–1123
145 Janusz JM, Young PA, Scherz MW, Enzweiler K, Wu LI, Gan L, Pikul S, McDow-Dunham KL, Johnson CR, Senanayake CB et al (1998) New cyclooxygenase-2/5-lipoxygenase inhibitors. 2. 7-tert-butyl-2,3-dihydro-3,3-dimethylbenzofuran derivatives as gastrointestinal safe antiinflammatory and analgesic agents: variations of the dihydrobenzofuran ring. *J Med Chem* 41: 1124–1137
146 Janusz JM, Young PA, Ridgeway JM, Scherz MW, Enzweiler K, Wu LI, Gan L, Chen J, Kellstein DE, Green SA et al (1998) New cyclooxygenase-2/5-lipoxygenase inhibitors. 3. 7-tert-butyl-2, 3-dihydro-3,3-dimethylbenzofuran derivatives as gastrointestinal safe antiinflammatory and anal-

gesic agents: variations at the 5 position. *J Med Chem* 41: 3515–3529
147 Johnson AR, Marletta MA, Dyer RD (2001) Slow-binding inhibition of human prostaglandin endoperoxide synthase-2 with darbufelone, an isoform-selective antiinflammatory di-tert-butyl phenol. *Biochemistry* 40: 7736–7745
148 Carter JS, Obukowicz MG, Devadas B, Talley JJ, Brown DL, Graneto MJ, Bertenshaw SR, Rogier DJ Jr, Nagarajan SR, Hanau CE, Hartmann SJ, Ludwig CL, Metz S (1998) Substituted benzopyran derivatives for the treatment of inflammation. G D Searle & Co PCT WO9847890.
149 Kurumbail RG, Kiefer JR, Marnett LJ (2001) Cyclooxygenase enzymes: catalysis and inhibition. *Curr Op Struct Biol* 11: 752–760
150 Rome LH, Lands WE (1975) Structural requirements for time-dependent inhibition of prostaglandin biosynthesis by anti-inflammatory drugs. *Proc Natl Acad Sci USA* 72: 4863–4865
151 Walker MC, Kurumbail RG, Kiefer JR, Moreland KT, Koboldt CM, Isakson PC, Seibert K, Gierse JK (2001) A three-step kinetic mechanism for selective inhibition of cyclo-oxygenase-2 by diarylheterocyclic inhibitors. *Biochem J* 357: 709–718
152 Picot D, Loll PJ, Garavito RM (1994) The X-ray crystal structure of the membrane protein prostaglandin H2 synthase-1. *Nature* 367: 243–249
153 Kurumbail RG, Stevens AM, Gierse JK, McDonald JJ, Stegeman RA, Pak JY, Gildehaus D, Miyashiro JM, Penning TD, Seibert K et al (1996) Structural basis for selective inhibition of cyclooxygenase-2 by anti-inflammatory agents. *Nature* 384: 644–648
154 Loll PJ, Picot D, Ekabo O, Garavito RM (1996) Synthesis and use of iodinated nonsteroidal anti-inflammatory drug analogs as crystallographic probes of the prostaglandin H2 synthase cyclooxygenase active site. *Biochemistry* 35: 7330–7340
155 Selinsky BS, Gupta K, Sharkey CT, Loll PJ (2001) Structural analysis of NSAID binding by prostaglandin H2 synthase: time-dependent and time-independent inhibitors elicit identical enzyme conformations. *Biochemistry* 40: 5172–5180
156 Copeland RA, Williams JM, Giannaras J, Nurnberg S, Covington M, Pinto D, Pick S, Trzaskos JM (1996) Mechanism of selective inhibition of the inducible isoform of prostaglandin G/H synthase. *Proc Natl Acad Sci USA* 91: 11202–11206
157 Gierse JK, Koboldt CM, Walker MC, Seibert K, Isakson PC (1999) Kinetic basis for selective inhibition of cyclo-oxygenases. *Biochem J* 339: 607–614
158 Lanzo CA, Sutin J, Rowlinson S, Talley J, Marnett LJ (2000) Fluorescence quenching analysis of the association and dissociation of a diarylheterocycle to cyclooxygenase-1 and cyclooxygenase-2: dynamic basis of cyclooxygenase-2 selectivity. *Biochemistry* 39: 6228–6234
159 Timofeevski SL, Prusakiewicz JJ, Rouzer CA, Marnett LJ (2002) Isoform-selective interaction of cyclooxygenase-2 with indomethacin amides studied by real-time fluorescence, inhibition kinetics, and site-directed mutagenesis. *Biochemistry* 41: 9654–9662
160 Kiefer JR, Pawlitz JL, Moreland KT, Stegeman RA, Hood WF, Gierse JK, Stevens AM, Goodwin DC, Rowlinson SW, Marnett LJ et al (2000) Structural insights into the stereochemistry of the cyclooxygenase reaction. *Nature* 405: 97–101

COX-2 Inhibitors
Edited by M. Pairet and J. van Ryn

The molecular and biological basis for COX-2 selectivity

Guenter Trummlitz[1], Joanne van Ryn[1] and Timothy D. Warner[2]

[1] *Boehringer Ingelheim Pharma GmbH & Co. KG, Birkendorfer Str. 65, D-88397 Biberach, Germany*
[2] *TheWilliam Harvey Research Institute, Bart's and the London Queen Mary's School of Medicine and Dentistry, London EC1M 6BQ, United Kingdom*

Introduction

Before the development of the first three-dimensional structure of cyclooxygenase (COX) [1], it was difficult to understand why very chemically diverse inhibitors acted via similar mechanisms at this same enzyme. The information from X-ray analyses and from mutagenesis experiments now allow us to use molecular modelling approaches to gain insight into the molecular mechanism of selective COX-2 inhibition by at least eight different structural classes of inhibitors. The particular advantage of molecular modelling approaches is that they allow us to visualize and rationalize the recognition process of substrates and inhibitors by COX isoenzymes [2].

In addition to the differential binding of NSAIDs the realisation by three independent groups that two COX isozymes exist (see Chapter 1 for details) further progressed understanding of the different efficacy/side-effect profiles of different NSAIDs. COX-1/-2 selectivity measurements have since become standard methods to determine the *in vitro* COX-2 selectivity of already known compounds, and to assist in the discovery of new compounds. Additionally, it is possible to correlate the *in vitro* COX-2 selectivity data with information from *in vivo* studies and so to build a continuum from molecular biology through to the clinic.

COX genes

Prostaglandin biosynthesis and the existence of COX isoenzymes have been well characterised in vertebrates. In mammals at least one COX isoenzyme is found in most cell types and organs. Selective synthesis of prostaglandins by one COX isoenzyme over the other has been observed in cells that express both isoforms. This indicates that the COX-1 and COX-2 pathways are segregated within cells permitting products derived from the two pathways to be channelled to different downstream signalling targets. This appears to be explained both by particular co-localisation of COX isoenzymes and prostanoid synthetic enzymes and by the preferential utilisation of low concentrations of free arachidonic acid by COX-2 [3]. Prostaglandins formed following the actions of COX-1 and COX-2 largely signal through G-protein linked receptors present on the cell surface. In addition, prostanoids, particularly COX-2 derived products, may signal through nuclear receptors. For example, prostacyclin can activate nuclear PPAR-δ receptors and so promote apoptosis [4]. Further refinement of these differences in production is supplied by the tissue differential expression of the COX-1 and COX-2 genes.

COX-1 is constitutively expressed in most tissues and produces prostaglandins to regulate physiological processes in response to hormonal and other stimuli. In contrast, COX-2 expression is highly regulated and increases in response to pathological stresses, such as inflammation. It must be noted, however, that this distinction is now appreciated to be somewhat blurred as it is clear that in certain tissues, e.g., the kidney and brain, COX-2 is constitutively expressed. Consistent with the general idea of COX-1 being constitutive and COX-2 inducible are the different natures of the COX-1 and COX-2 genes. The COX-1 gene, which is 22 kB in size and located on chromosome 9 in humans, has few binding sites for inducible transcription factors. The COX-2 gene in comparison is a so-called immediate-early gene – 8.3 kB in size and located on chromosome 1 – the expression of which is tightly controlled. The COX-2 gene is responsive to many regulators from various signalling pathways, such as growth factors and mediators of inflammation such as IL-1, TNF-α and LPS [3, 5].

COX enzyme structures

COX-1 and COX-2 are monotropic enzmyes bound on the membranes of the inner surface of the endoplasmic reticulum and on the inner and outer surface of the nuclear envelope. Both isoenzymes are homodimer proteins with similar structure and catalytic activity. The amino acid sequences of COX-1 and COX-2 from numerous mammalian, bird and fish species are known. COX-1 and COX-2 amino acid sequences are 60 to 65% identical and each isoform is conserved by 85% and 65% within mammalian and vertebrate species, respectively.

Both isoforms contain signal peptides of varying lengths, 22–26 amino acids in COX-1 and 16–20 in COX-2. After cleavage of the signal peptides the mature COX-1 and COX-2 proteins contain 576 amino acids and 587 amino acids, respectively in mammalians. Major structural differences between these two isoforms occur: (a) in the EGF-like domain, which is shorter in COX-2; (b) in the membrane binding domain of COX-2, in which a proline is inserted after Tyr106 and (c) in the catalytic domain of COX-2, in which 18 amino acids are inserted close to the C-terminus (Fig. 1). The high degree of sequence identity between the mature isoforms and between species allows one-to-one

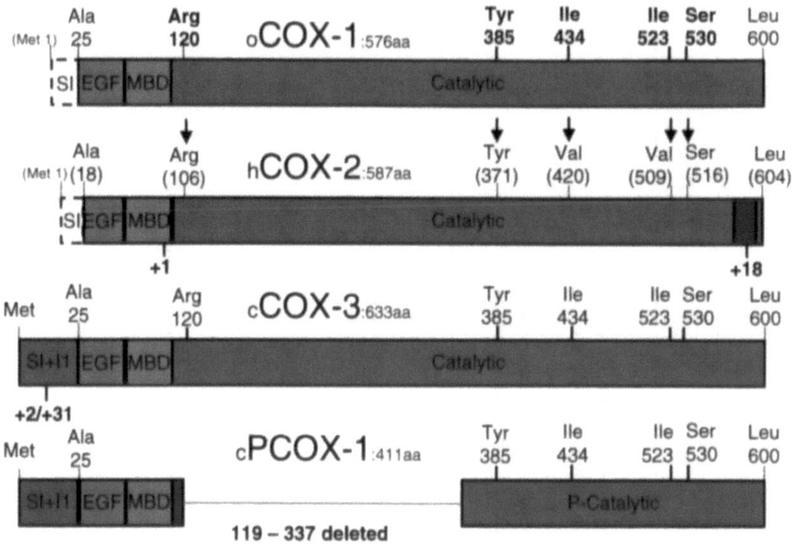

Figure 1. Amino acid sequences of ovine(o) COX-1, human(h) COX-2, canine(c) COX-3 and cPCOX-1a. SI = signal peptides, I1 = Intron 1, EGF = Epidermal growth factor-like domain, MBD = membrane binding domain. For cross-species and cross-isoform comparisons the numbering of oCOX-1 is used.

comparisons of the known sequences to be made. To aid in these cross-isoform and cross-species comparisons, the common consensus is to use the ovine COX-1 sequence numbering as the reference [5].

Recently it has been reported that there could be additional variant protein products from the COX-1 gene [6]. COX-3 and PCOX-1a (partial COX-1a), are derived from the COX-1 gene, but their mRNAs retain intron 1 yielding an insertion of 30–34 amino acids, depending on the species, into the signal peptide, which is not cleaved off. PCOX-1 proteins additionally contain a deletion of exons 5–8 of the COX-1 mRNA resulting in deletion of amino acids 119–337 in the protein (Fig. 1). The initial report also showed that COX-3, but not PCOX-1a, possesses COX activity. Most recently, however, it has been shown that the human COX-1 mRNA is such that COX-3 could not be transcribed [7].

The three-dimensional structure of ovine(o) COX-1 has been resolved by X-ray analyses in two crystal modifications in the presence of four different substrates and nine inhibitors. In addition, three other groups have used recombinant human(h) and murine(m) COX-2 complexed with eight different ligands to make structural determinations of the binding modes of COX-2 inhibitors (Tab. 1, [1, 8–20]).

Given the high degree of amino acid conservation between the two isoenzymes, it is not surprising that the structures of COX-1 and COX-2 are very similar. COX-1 and COX-2 crystallise as dimers, with each monomer consisting of three independent folding units: (1) an EGF-like region that forms a portion of

Table 1. X-ray structures of COX isoenzyme/ligand complexes and critical interaction sites

Ligand	COX	Reference	Critical interaction sites
Unliganded:	mCOX-2	Kiefer [8]	
Substrates:			Arachidonic acid:
Arachidonic acid	oCOX-1	Malkowski [9]	Tyr385: 13-pro-S-hydrogen abstraction
	mCOX-2	Kiefer [8]	Tyr348, Gly533: important positioning
	oCOX-1	Thuresson [10]	Arg120: ionic binding COX-1
Dihomo-γ-linolenic acid	oCOX-1	Malkowski [11]	Val349, Trp387, Leu534: steric position
Eicosapentaenoic acid	oCOX-1	Malkowski [11]	11 amino acids: catalysis optimization
Linoleic acid			
Salicylic acids:			Bromoaspirin:
Bromo-aspirin	oCOX-1	Loll [12]	Ser530: acetylation site
O-Acetylsalicylhydroxamic acid	oCOX-1	Loll [13]	Arg120: ionic binding to carboxylate
Arylacetic acids:			Indomethacin:
Indomethacin	mCOX-2	Kurumbail [14]	Arg120: ionic binding
Iodo-indomethacin	oCOX-1	Loll [15]	Ser530: hydrogen bonding
L-758048 (Indomethacin-analog)	hCOX-2	Merck & Co [16]	Leu384: steric interaction
Alclofenac	oCOX-1	Selinsky [17]	RS-104897:
RS-57067 (Zomepirac-analog)	hCOX-2	Luong [18]	Ser530: hydrogen bond
RS-104897 (Zomepirac-analog)	hCOX-2	Luong [18]	Arg120, Tyr355, (Gln524): hydrogen bonding

(Continued on next page)

Table 1. (Continued)

Ligand	COX	Reference	Critical interaction sites
Arylpropionic acids:			Flurbiprofen:
Flurbiprofen	oCOX-1	Picot [1]; Garavito [19]	Arg120: ionic + hydrogen interaction
	mCOX-2	Kurumbail [14]	Tyr355: hydrogen bond
Methylflurbiprofen	oCOX-1	Selinsky [17]	Arg120,Glu524,Tyr355,His90,H_2O:
Ibuprofen	oCOX-1	Selinsky [17]	ionic + hydrogen network
Iodosuprofen	oCOX-1	Loll [18]	Ser530: contact to distal phenyl
Enolcarboxamides:			
Meloxicam	oCOX-1	Trummlitz [20]	Phe518: steric interaction with 5'-methyl Arg120, Tyr355,Ser530:hydrogen bonds
Diarylheterocycles:			
SC-558 (Celecoxib-analog)	mCOX-2	Kurumbail [14]	Val523, Phe518: access to side pocket, Arg513, His90: hydrogen bond to sulfonamide)

the dimer interface and has a role in protein folding; (2) a membrane binding domain that is incorporated into the inner side of the membrane bilayer and (3) the catalytic domain (Fig. 2). The hydrophobic amino acid residues at the membrane binding domain are associated with the membrane monolayers, thereby allowing direct access from the interior of the bilayer to the active site or "channel" of COX. In this way the enzyme substrate, arachidonic acid, enters the COX channel and this is also the route of entry for competitive enzyme inhibitors. The catalytic domain contains both the COX and the peroxidase active sites. In order to pass into the upper part of the COX channel, substrate and inhibitors need to pass the constriction formed by Arg120 and Tyr355. The residue Tyr385 is located at the top of the channel and represents the active centre.

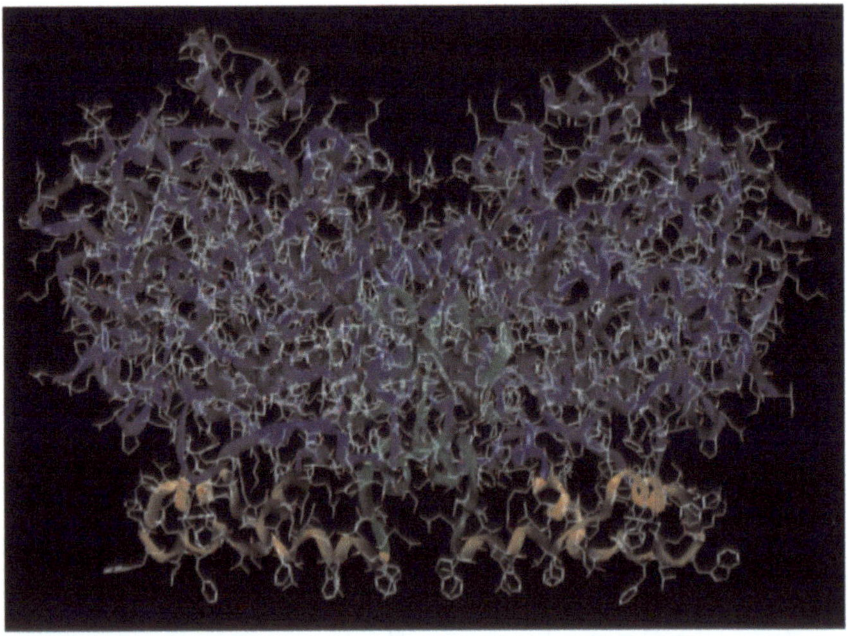

Figure 2. Three dimensional structure of the oCOX-1 dimer. Ribbon representation of the amino acid backbone is used. EGF-like domain: green; MBD: orange; catalytic domain: blue. Amino acid residues are represented as stick models. The coordinates are taken from Malkowski et al. [9].

Substrate binding

Recently, Malkowski et al. prepared a native-like, but functionally inert, Co^{3+} protoporphyrin reconstituted oCOX-1 [21] and determined the structure of arachidonic acid bound within the active site of COX [9]. Arachidonic acid is bound in an extended L-shaped conformation and makes two hydrophilic and a total of 48 hydrophobic contacts with the enzyme, involving 19 different amino acids. The interaction of arachidonic acid with the COX channel has been visualized (Fig. 3a) using an algorithm that represents channel and cavity surfaces by colour coding local hydrophobicity [2]. The carboxylate part of arachidonic acid binds with Arg 120 with the carbon 13 being oriented near the Tyr385, where the pro-S-hydrogen can be abstracted in order to initiate the COX reaction. Additionally, the binding to COX of other substrates has been studied (Tab. 1) and it has been shown, for instance, that the selective oxidation of 2-arachidonylglycerol by COX-2 has as a critical determinant interaction with Arg513 [22–24].

Selective COX-2 inhibitors

NSAIDs display a high degree of structural diversity (Tab. 2) and numerous structural classes with a wide range of physicochemical properties were already available when COX-2 was discovered [25]. To date, all newly discovered selective COX-2 inhibitors belong to one of these classes. Interestingly very slight modifications to a compound can drastically change its selectivity for one of the COX isoenzymes (see Chapter 2 for details). X-ray structure analyses (Tab. 1), assays of binding to COX mutants (Tab. 3, [26–42]) and molecular modelling of COX-2 inhibitors have supplied us with a tremendous amount of knowledge that allow us to rationalise this selectivity based upon the molecular features of the selective binding in each structural class. (See reviews: [2, 3, 5, 43]).

Aspirin is the only marketed COX inhibitor that covalently modifies the enzyme. Detailed structural data taken from the X-ray structure of the bromoaspirin/COX-1 adduct [12] show that aspirin acetylates Ser530. When COX-2 is acetylated by aspirin arachidonic acid can still be oxidised but 15*R*-HPETE is produced [26, 44]. Conversely, when Ser530 is acetylated in COX-1 the enzyme is completely inactivated. Structural alterations arising from mutations of Ser 530 confirmed the critical role of this amino acid (Tab. 3).

Replacement of the carboxylate moiety of aspirin with alkyl- and alkynylsulfide moieties results in COX-2 selective, irreversible inhibitors [30]. Mutational experiments revealed that the lead compound of this group, 2-(acetoxyphenyl)hept-2-ynyl sulphide (APHS), also acetylates Ser530, but does not interact with the side pocket (Val523, Arg513). Mutations that introduce steric

Figure 3a–d. (see pp 48, 49) The inside surface of the active side (COX channel) has been identified using the channel surface calculation of MOLCAD (Tripos Inc), and the local hydrophobicity is mapped on this surface (from brown: lipophilic areas to green-blue for hydrophilic areas). The COX coordinates used for modelling are taken from the PDB-files 1DIH, 6COX, 1PRH. For clarity, only the most important amino acid residues are shown as stick models and labelled: catalytic side: Tyr365, ASA acetylation side Ser530, hydrogen network at the gate: Arg120 and Tyr355, side pocket: Val/Ile523 and Arg/His513, flexible extra space: Phe518 and Val/Ile 434, extra space: Leu384 and Leu/Phe503. Arachidonic acid and inhibitors are shown as stick models with white carbon atoms and colored heteroatoms. (a) Arachidonic acid bound in the COX-1 channel, H-abstraction at C13 in yellow. (b) Flurbiprofen bound to the centre of the COX-1 channel. (c) Meloxicam bound at COX-2; the 5'-methyl group of meloxicam occupies the flexible extra space at the top of the active site. (d) Celecoxib bound to COX-2, the phenylsulfamoyl group occupies the side pocket. Figures 3a–d are adapted from [2].

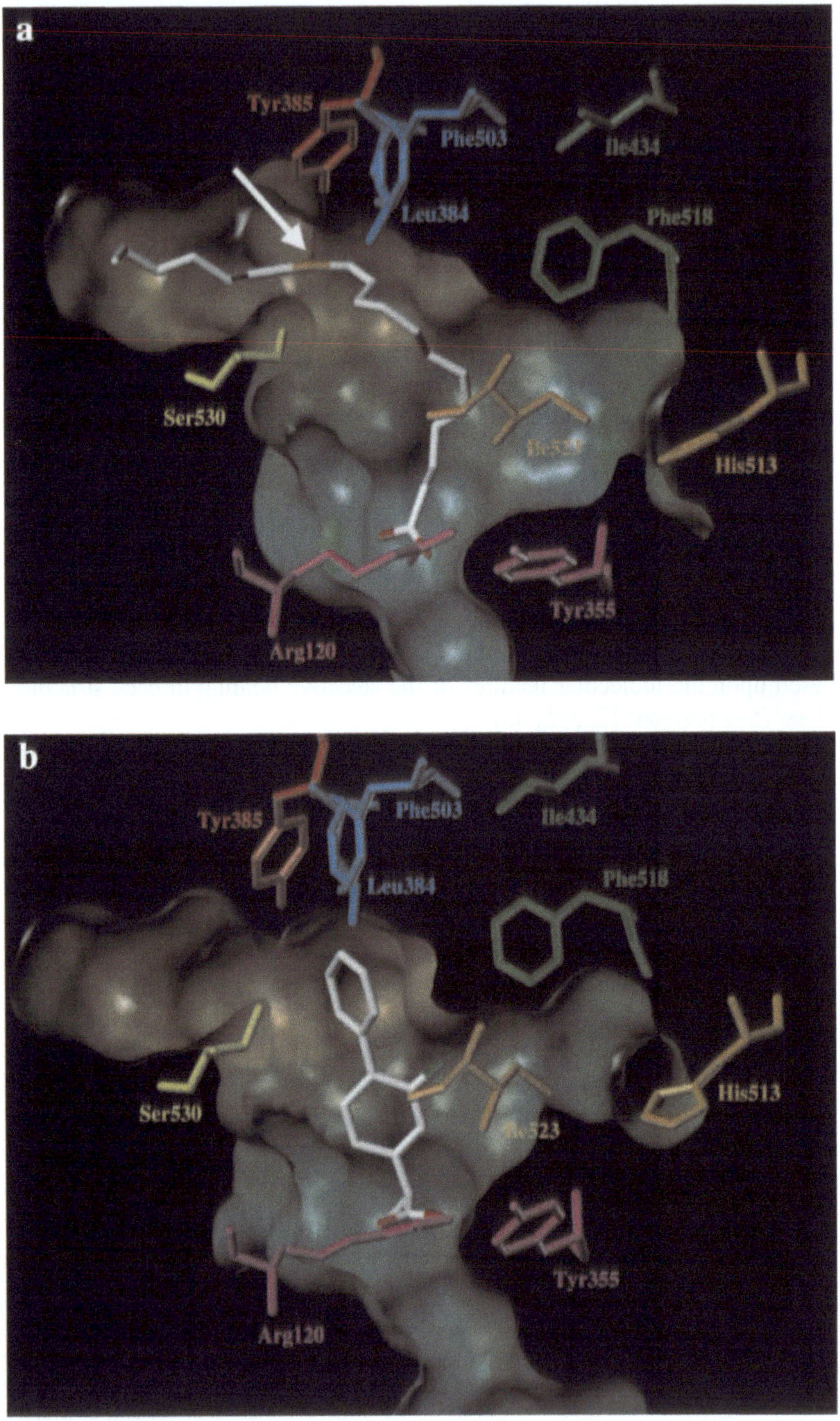

Figure 3 a, b (legend see p. 47)

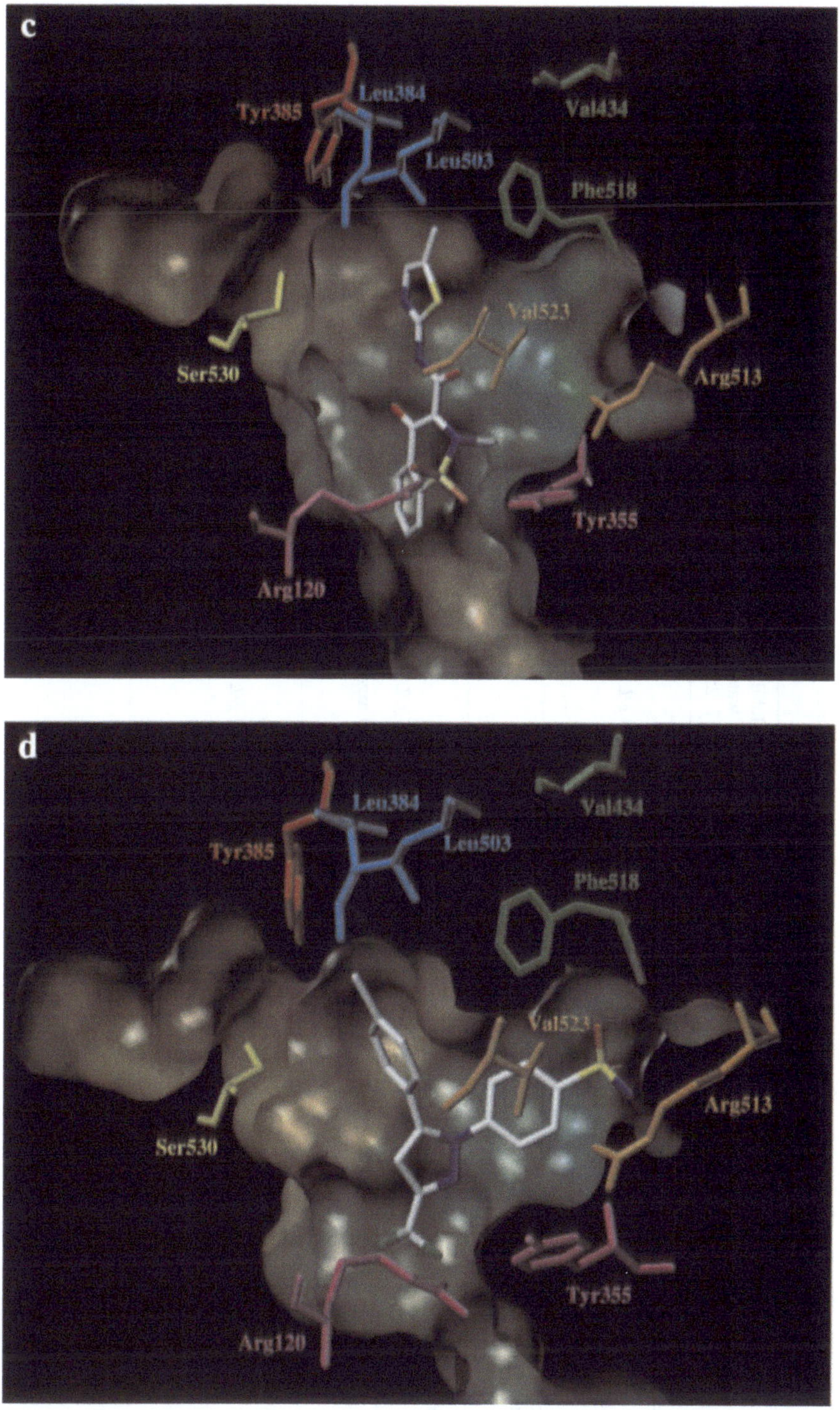

Figure 3 c, d (legend see p. 47)

Table 2. Structures and properties of selective COX-2 inhibitors and structurally related NSAIDs

Structural class (Non-selective inhibitors: *italics*; selective COX-2 inhibitors: **bold**)		Solubility in water (pH 7, 25°C)	Ionization (pKa value)	Lipophilicity (logP at pH 7.4)	Protein binding	Volume of distribution (Vd, ss)
1. Acetoxyphenyl-derivatives (aspirin):						
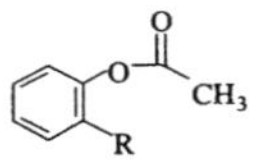	*aspirin* (R=COOH)	*~10 g/L*	*3.5 (acid)*	*−2.8*	*(~85%)*[1]	*(12 L)*[1]
	APHS (R=heptinylsulfide)	–	–	–	–	–
2. Anthranilic acids (fenamates):						
	ester of meclofenamic acid	–	–	–	–	–
3. Arylacetic acids (fenacs):						
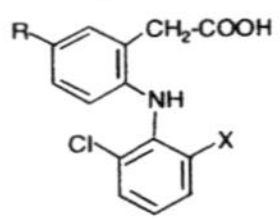	*diclofenac* (R=H, X=Cl)	*~5 g/L*	*4 (acid)*	*1.2*	*>99%*	*~10 L*
	COX-189 (R=CH_3, X=F)	**280 mg/L**	**4.3 (acid)**	**1.5**	**>99%**	**~10 L**
	others: etodolac, RS57067, L-761066					
4. Arylpropionic acids (profens):						
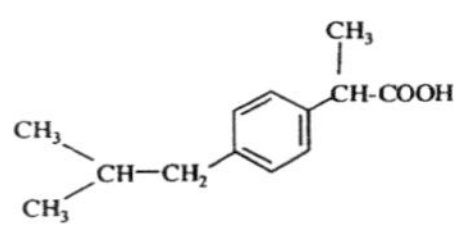	*ibuprofen*	*~5 g/L*	*4.5*	*1.5*	*>99%*	*~10 L*
	bisethoxyflurbiprofen	–	–	–	–	–

(Continued on next page)

Table 2. (Continued)

Structural class (Non-selective inhibitors: *italics*; selective COX-2 inhibitors: **bold**)		Solubility in water (pH 7, 25°C)	Ionization (pKa value)	Lipophilicity (logP at pH 7.4)	Protein binding	Volume of distribution (Vd, ss)
5. Enolcarboxamides (oxicams):						
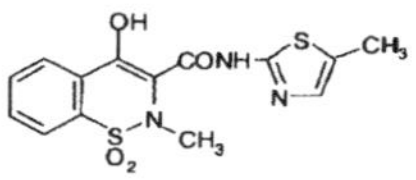	**meloxicam**	**250 mg/L**	**1.1 + 4.2 (zwitterion)**	**0.1**	**>99%**	**~10 L**
6. Sulfanilides (sulides):						
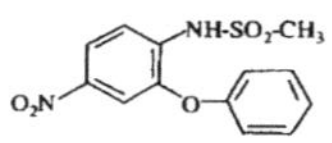	**nimesulide**	**20 mg/L**	**6.9 (weak acid)**	**2.6**	**99%**	**~20 L**
	others: **NS-398, flosulide, L-745337**					
7. Diarylheterocycles (coxibs):						
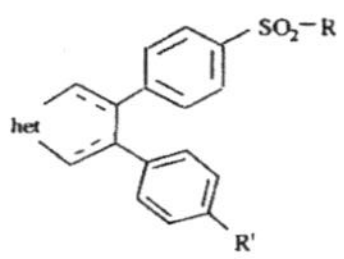	**celecoxib** (R=NH_2, R'=CH_3, het=CF_3-pyrazol)	**5 μg/L**	**11.1 (neutral)**	**>3**	**97%**	**~400 L (app VdSS)**
	rofecoxib (R=CH_3, R'=H, het=furanone)	**4 μg/L**	**12.5 (neutral)**	**>3**	**87%**	**100 L (app VdSS)**
others: **DUP697, SC58125, DFP, DFU, valdecoxib, etoricoxib**						
8. o,o-Dialkylphenols:						
	BF398, CI-1004, PD-138387, PD-136095	–	–	–	–	–

Physicochemical values were determined according to the methods described by Luger et al. [25]. Pharmacokinetic values were taken from the product information sheets. [1]Rapid transformation of aspirin to salicylic acid, values given are concentration dependent and are given for low doses. -: no data available

Table 3. Overview of the interactions of COX inhibitors with COX mutants

Ligand	COX-mutants [Reference]
Salicylic acids:	
Aspirin	*hCOX-2:* Ser530Met [26]; *hCOX-1:* Arg120Glu [27] *mCOX-2:* Tyr385Phe; Ser530Ala; Arg120Ala,Gln; Tyr348Phe; VRV-IHI [28] *mCOX-2:* Val434Ile; Arg 513His; Val523Ile; VRV-IHI; Val228Phe; Gly533Ala [29]
2-Acetoxyphenylhept-2-ynyl sulfide (APHS)	*mCOX-2:* Arg120Gln; Tyr355Ala; VRV-IHI [30] *mCOX-2:* Tyr385Phe; Ser530Ala; Arg120Gln; Gly533Ala,Val; Tyr348Phe; VRV-IHI [31]
O-Acetylsalicylhydroxamic acid (AcSHA)	*hCOX-1:* Ser530Ala, Arg120Gln [13]
Arylacetic acids:	
Indomethacin	*hCOX-1:* Arg120Glu [27]; *hCOX-2*: Arg120Gln [32] *hCOX-2:* Tyr355Phe [33]; *hCOX-2:* Ser530Met [34] *hCOX-1:* His513Arg;Ile523Val; HI-AV [36]; *hCOX-2:* Val523Ile [35]
Diclofenac	*hCOX-1:* Arg120Glu [27]; *hCOX-2:* Arg120Gln [32] *hCOX-2:* Ser530Met [34] *hCOX-1:* His513Arg;Ile523Val; HI-AV [36]; *hCOX-2:* Val523Ile [35]
Sulindac	*hCOX-2:* Ser530Met [34]; *hCOX-1:* His513Arg;Ile523Val; HI-AV [36]
RS-57067 (Zomepirac analog)	*hCOX-2:* Tyr355Phe [33]
Indomethacin analogs:	
L-588,983; L-761,066	*hCOX-2:* Arg120Gln [32]
LM-4143	*mCOX-2:* Arg120Gln,Ala; Tyr355Phe,Ala; Glu524Leu [37]
Arylpropionic acids:	
Ibuprofen	*hCOX-1:* Arg120Glu [27]; *oCOX-1:* Arg120Gln,Lys,Glu; Tyr355Phe D- and L-ibuprofen [38]
Flurbiprofen	*hCOX-1:* Arg120Glu [27]; *oCOX-1:* Arg120Gln,Lys,Glu [38]; *hCOX-2:* Arg120Gln [32, 39]; *hCOX-2:* Ser530Met [34]; *hCOX-2:* Val523Ile [35]; *hCOX-2:* Tyr355Phe [33]
Ketoprofen	*hCOX-1:* Arg120Glu [27]; *hCOX-2:* Ser530Met [34]
Naproxen	*hCOX-2:* Val523Ile [35]
Flufenamic acids:	
Meclofenamic acid	*hCOX-1:* Arg120Glu [27]; *hCOX-2:* Arg120Gln [32]; *hCOX-2:* Ser530Met [34]
N-Phenoxyethylamid of Meclofenamic acid	*mCOX-2:*Arg120Gln,Ala; Tyr355Phe, Ser530Ala; Glu524Leu; Leu503Phe; Val523Ile, VRV–IHI [40]
Mefenamic acid	*hCOX-2:* Val523Ile [35]

(Continued on next page)

Table 3. (Continued)

Ligand	COX-mutants [Reference]
Sulfanilides:	
NS 398	*hCOX-2:* Arg120Gln [32, 39]; *hCOX-1:* His513Arg; Ile523Val; HI-AV [36]; *hCOX-2:* Val523Ile [35]; Val523Ile,Ala,Glu,Lys [40]
Nimesulide	*hCOX-2:* Val523Ile,Ala,Glu,Lys [41]
Flosulide	*h-COX-2:* Arg120Gln [32]
L-745,337	*hCOX-2:* Ser530Met [34]
Diarylheterocycles:	
DUP-697	*hCOX-1:* Arg120Glu [27]; *hCOX-2:* Arg120Gln [32, 39]; *hCOX-2*: Ser530Met [34]); *hCOX-2:* Val523Ile [35]; Val523Ile,Ala,Glu,Lys [41]; *hCOX-1:* His513Arg; Ile523Val; HI-AV [36]
SC 58125	*hCOX-2:* Arg120Gln [39]; *hCOX-2:* Val523Ile; TSP-LLL [35]; *hCOX-2:* Val523Ile,Ala,Glu,Lys [41]; *hCOX-1:* His513Arg; Ile523Val; HI-AV [36]
Celecoxib, Valdecoxib	*mCOX-2:* Arg120Ala,Gln,Asn; Tyr355Ala; Val523Ile; VRV-IHI [42]
L-746,483 (DUP-697 analog)	*hCOX-1:* Arg120Glu [27]; *hCOX-2:* Arg120Gln [34]
L-745,296 (DUP-697 analog)	*hCOX-2:* Arg120Gln [34]; *hCOX-1:* His513Arg; Ile523Val; HI-AV [36]
DFU	*hCOX-1:* His513Arg; Ile523Val; HI-AV [36]

HI-AV: double mutant *COX-1* His513Arg + Ile523Val; VRV-IHI: triple mutant *COX-2* Val434Ile + Arg513His + Val523Ile; TSP-LLL: triple mutant *COX-2* Tyr101Leu + Ser105Leu + Phe343Leu.

bulk into the COX channel at the place where the ω-end of arachidonic acid binds (Gly533) lead to reduced COX activity [31].

Before the introduction of selective COX-2 inhibitors arylacetic acids, arylpropionic acids and fenamic acids represented the largest group of marketed NSAIDs. Detailed X-ray analyses (Tab. 1), site-directed mutagenesis experiments (Tab. 3) and comparative molecular modelling studies [2] provide us with a large amount of information as to the critical amino acid residues in COX-1 and COX-2 responsible for the selective interaction with compounds of these classes. In Figure 3b the interaction of the nonselective flurbiprofen within the COX channel has been visualized using an algorithm that represents channel and cavity surfaces by color coding local hydrophobicity.

Meloxicam was characterised during development as being COX-2 selective, and it was therefore launched as the first selective COX-2 inhibitor. By applying molecular modelling methods [45], it has subsequently been established that the enolate anion of meloxicam does not interact with Arg120 via an ionic bond but that the thiazine-sulfone can form hydrogen bridges to Arg120 and Tyr355. More importantly, the 5-methyl group of the thiazole ring exploits the "flexible extra space" at the top of the COX-2 channel. The substitution of the second-shell amino acid residue Ile434 in COX-1 for Val in COX-2 allows the side chain of Phe518 (a residue at the active side) to open an "extra space", which favours the binding of meloxicam to COX-2 (Fig. 3c). The binding conformation of meloxicam is less optimal in COX-1; for example, a steric interaction with Phe518 is most notable. X-ray analyses and newer molecular modelling studies have confirmed these interactions and have allowed even more precise evaluations to be made [20].

Nimesulide and the experimental compounds L-745,337 (Merck & Co), flosulide (Novartis AG) and NS-398 (Taisho Pharmaceutical Co) are members of the sulfanilide group of NSAIDs. Mutational analyses [35, 39] suggest that the methylsulfone group of these compounds interact particularly with Arg120 at the constriction of the COX channel.

Our knowledge about the molecular basis of the selective action of compounds such as celecoxib and rofecoxib is based on the X-ray analysis of SC 558 [14], a structural analogue of celecoxib, together with mutagenesis [42], kinetic [46, 47] and modelling experiments [48]. The phenylsulfonamide of celecoxib and the phenylmethylsulfone moiety of rofecoxib occupy the side pocket in COX-2, made accessible by the substitution of Val523 for Ile523, and are within binding distance of Arg513. One other amino acid substitution, Leu503 in COX-2 replacing Phe503 in COX-1, also extends the volume of the active site in COX-2 and this is exploited by the coxibs. In Figure 3d the binding of celecoxib into the COX-2 channel based on the X-ray data [14] is demonstrated.

Because of their structural distinctions, the COX-2 inhibitors have markedly different molecular properties, as demonstrated by their solubility, protonation behaviour and lipophilicities (Tab. 2). These different properties lead to differences in the pharmacokinetic behaviours of the COX-2 inhibitors, such as their protein binding, tissue distribution and metabolic pathways. For this reason adverse event profiles of selective COX-2 inhibitors should take into account not only the advantages of COX-2 selectivity (i.e., COX-1 sparing) but also differences in drug distribution and metabolism as compared to standard NSAIDs.

Enzyme kinetics and inhibitor complexes

Enzyme kinetics play an important role in the activation and inhibition of COX. Selective COX-2 inhibition is correlated with a time-dependent COX-2 inhibition and a time-independent reversible COX-1 inhibition [49]. Kinetic aspects of binding interactions are discussed in this chapter based on work from crystallographic studies that have led to insight into the enzyme-inhibitor complexes. Further kinetic studies are also described in Chapter 2.

Prior to the work of Selinsky et al. [17], it was believed that competitive inhibitors formed only a freely reversible complex (EI) and that time-dependent inhibitors could additionally slowly form tightly bound irreversible complexes (EI*):

$$E + I_{rev} \rightleftharpoons EI$$

$$E + I_{irr} \rightleftharpoons EI \longrightarrow EI^*$$

It had been postulated, therefore, that selectivity of COX-2 inhibitors may occur by the formation of EI* with COX-2 and of reversible EI with COX-1 [48, 46]. Selinsky et al. [17] proposed that both reversible time-dependent inhibitors and irreversible time-independent inhibitors bind by the same mechanism and differ by the activation energy (E^a) of the EI $\longrightarrow$ EI* transition:

$$E + I_{rev} \rightleftharpoons EI \overset{\text{low } E^a}{\rightleftharpoons} EI^*$$

$$E + I_{irr} \rightleftharpoons EI \overset{\text{high } E^a}{\rightleftharpoons} EI^*$$

With a low E^a separating EI and EI*, the transition to EI* will be rapid. One should observe only the presence of EI* in kinetic measurements and so crystal analyses will be of COX in the EI* state. The same conformational status (EI*) will be observed in X-ray experiments for slow, tightly-bound inhibitors, and due to the high E^a, the first step to EI could also be measured kinetically. COX-2 selective inhibitors, like meloxicam might gain their selectivity by having a low E^a for COX-1 and a high E^a for COX-2.

More recently Walker et al. [47] proposed that some selective COX-2 inhibitors of the diarylheterocyclic type (i.e., valdecoxib) could generate a three-step time-dependent irreversible COX-2 inhibition:

$$E + I_{irr} \rightleftharpoons EI \rightleftharpoons EI^* \overset{\text{low } E^a}{\longrightarrow} EI^{**}$$

COX assays

In vitro *assays*

Before the availability of enzyme modelling many *in vitro* assay systems were developed to investigate the COX-2 selectivity of NSAIDs. These have been reviewed in detail elsewhere [50] but in broad terms these tests systems can be classified into three groups: assays using animal enzymes, cells, or cell lines, which were the first to be developed; assays using human recombinant enzymes, cell lines, or blood cells (mainly platelets and monocytes), which are the current standards; and models using those human cells that are targets for the anti-inflammatory and adverse effects of NSAIDs, e.g., human gastric mucosa cells, chondrocytes, and synoviocytes.

Apart from variations in the sources of the COX enzymes there are wide variations in the experimental conditions in these various assays. For example, arachidonic acid can come from both endogenous (often in cell based assays) and exogenous sources (in isolated enzyme or broken cell assays). It is important to remember that the potencies of some NSAIDs, particularly those that act as competitive substrate inhibitors [46], are influenced by the supply of arachidonic acid. Also, expression systems and cell transfection targets vary considerably, including even non-mammalian cells such as those from insects. Other experimenters have used cells that express the COX enzymes constitutively, following stable transfection with the recombinant enzymes, to avoid the problems associated with variability in the types and concentrations of inducing agents used. As well as substrate, the time of incubation of the test drug with the COX enzyme systems can have a strong influence on the apparent potencies of drugs. In particular, a number of NSAIDs and COX-2 selective agents inhibit COX-2 in a time-dependent manner [46, 51]. Another important factor is the protein concentrations present in the medium. This is a critical issue for NSAIDs, which bind avidly to plasma proteins such that within the circulation, for instance, the concentration of free drug is usually only a few percent of the total drug concentration. In general terms it appears that an ideal assay of use in trying to predict the clinical effects of these agents should use native human enzymes that are present in whole cells; that the cells used should be target cells for the anti inflammatory and adverse effects of NSAIDs; that COX-2 should be induced, thereby simulating an inflammatory process, rather than being constitutively expressed; that prostaglandin synthesis should be measured from arachidonic acid released from endogenous stores rather than from exogenously added arachidonic acid; and, that the protein concentration in the medium should also closely mimic plasma protein concentrations.

Exploiting data from in vitro *assays*

The IC_{50} values of COX-1 and COX-2 inhibition obtained in *in vitro* assays are often used to calculate selectivity ratios. However, this is a very simplistic approach. It relies upon concentration curves being parallel, and takes little account of the plasma concentrations of drugs achieved following therapeutic dosing. Furthermore, if two drugs have similar selectivity ratios but very different slopes to their inhibitor curves the conclusions we can draw must be very different. In simple terms, steep curves allow a better dissociation between COX-1 and COX-2 inhibition than flat curves [50]. So, even though COX-2 selectivity can be expressed by selectivity ratios derived from *in vitro* assays we must be very cautious before extrapolating to clinical conditions.

Human whole-blood assay (hWBA)

For many of the reasons discussed above the human whole blood assay (hWBA) appears to offer many advantages over other *in vitro* tests. In this assay, COX-1 activity is assessed by the release of prostanoids from platelets following stimulation of human whole blood clotting, while COX-2 activity is assessed by the release over about 18 h of prostanoids from white blood cells (principally monocytes) following incubation of the whole blood with bacterial lipopolysaccharide (LPS). The hWBA has many advantages: intact human cells are used, which are target cells for the anti inflammatory effects (monocytes) and side-effects (platelets) of NSAIDs; plasma proteins are present; the whole blood used for both assays is taken from the same volunteer (or patient) at the same time, allowing a direct comparison of the results from each assay; the assay can be performed using blood from volunteers (or patients) who have been treated previously with NSAIDs (*ex vivo* assay). Results obtained by different laboratories, using the hWBA have demonstrated similar rank orders for COX-2 selectivity [50]. Standard NSAIDs range from being COX-1 selective to being approximately equally effective on both isoenzymes with diclofenac having the most favourable profile. Compounds such as etodolac, nimesulide, and meloxicam followed by celecoxib, flosulide, rofecoxib, etoricoxib, valdecoxib and COX-189 show increasing selectivity for COX-2. Another way to express this data, if COX-2 is inhibited by >80% (i.e., the therapeutic target) how much COX-1 inhibition can one expect *in vivo*. This is illustrated in Figure 4 for both traditional NSAIDs and COX-2 selective compounds. It indicates that traditional NSAIDs will inhibit COX-1 to the same extent as COX-2, while more selective inhibitors have less effect on COX-1. The new generation of COX-2 inhibitors do not inhibit COX-1 at concentrations that maximally inhibit COX-2.

A further development of the hWBA has been to mix with aliquots of human whole blood human A549 cells already induced to express COX-2 (human modified whole blood assay, hmWBA) [52]. This permits the COX-1 and COX-2 assays to be conducted over identical time courses. As for the standard hWBA the hmWBA has been used to generate ratios for the COX-1/COX-2 selectivities of a wide range of NSAIDs and newer COX-2 selective agents.

Other assays

Alternative and newer *in vitro* assays are always being developed. On the face of it test systems that appear the most interesting are those using human cells such as gastric mucosa cells, chondrocytes or synoviocytes which are target cells for the anti inflammatory or adverse effects of NSAIDs. These generate similar data to those seen with human recombinant enzymes and the hWBA [53–56] and it can be shown that the data from such assays correlates with the

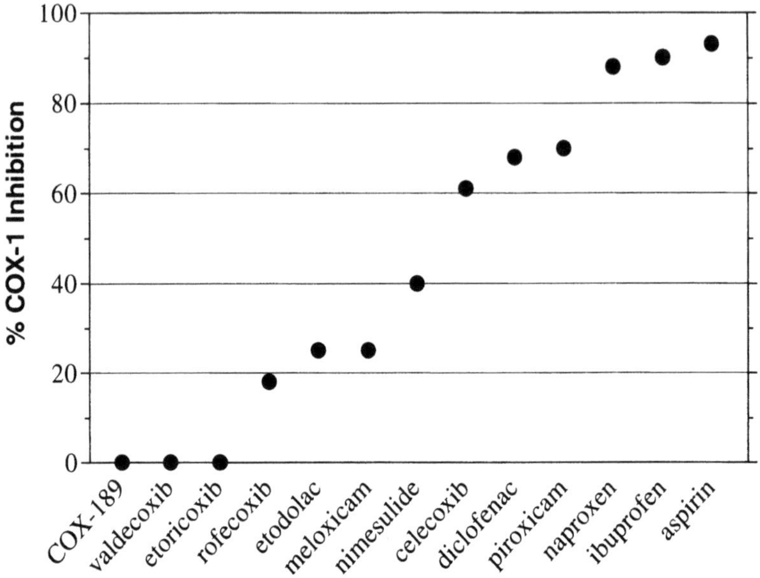

Figure 4. The degree of inhibition of COX-1 when COX-2 is inhibited by 80% (i.e., IC_{80}). The data were obtained from [52] using IC_{80} determination *in vitro* and unpublished results (valdecoxib, etoricoxib and COX-189).

idea that increased COX-2 selectivity is associated with reduced gastrointestinal toxicity [57] but still these assays have the problem that they do not mimic well the drug binding to proteins seen *in vivo*. Thus, one must be careful when trying to draw conclusions about the effects on tissue prostaglandin formation by comparing potencies against tissue and whole blood prostanoid formation [58].

In vivo relevance of *in vitro* results

Clearly the purpose of *in vitro* tests is to predict the usefulness of NSAIDs *in vivo*. In particular, following the idea that inhibition of COX-2 explains efficacy and inhibition of COX-1 explains side-effects, will a drug be efficacious without causing side-effects? Using the whole blood assays we can map blood therapeutic concentrations derived by pharmacokinetic studies (taking into account the differences between plasma and whole blood concentrations) and compare them to the concentrations of drug required to inhibit COX-1 and COX-2 in *in vitro* assays. Such composite figures have been derived from the hmWBA, examples being those for diclofenac (Fig. 5a), meloxicam (Fig. 5b) and rofecoxib (Fig. 5c). This modelling shows us that these three drugs (diclofenac, meloxicam and rofecoxib), which are established to be efficacious at standard doses, should produce similar inhibition of COX-2 when applied at

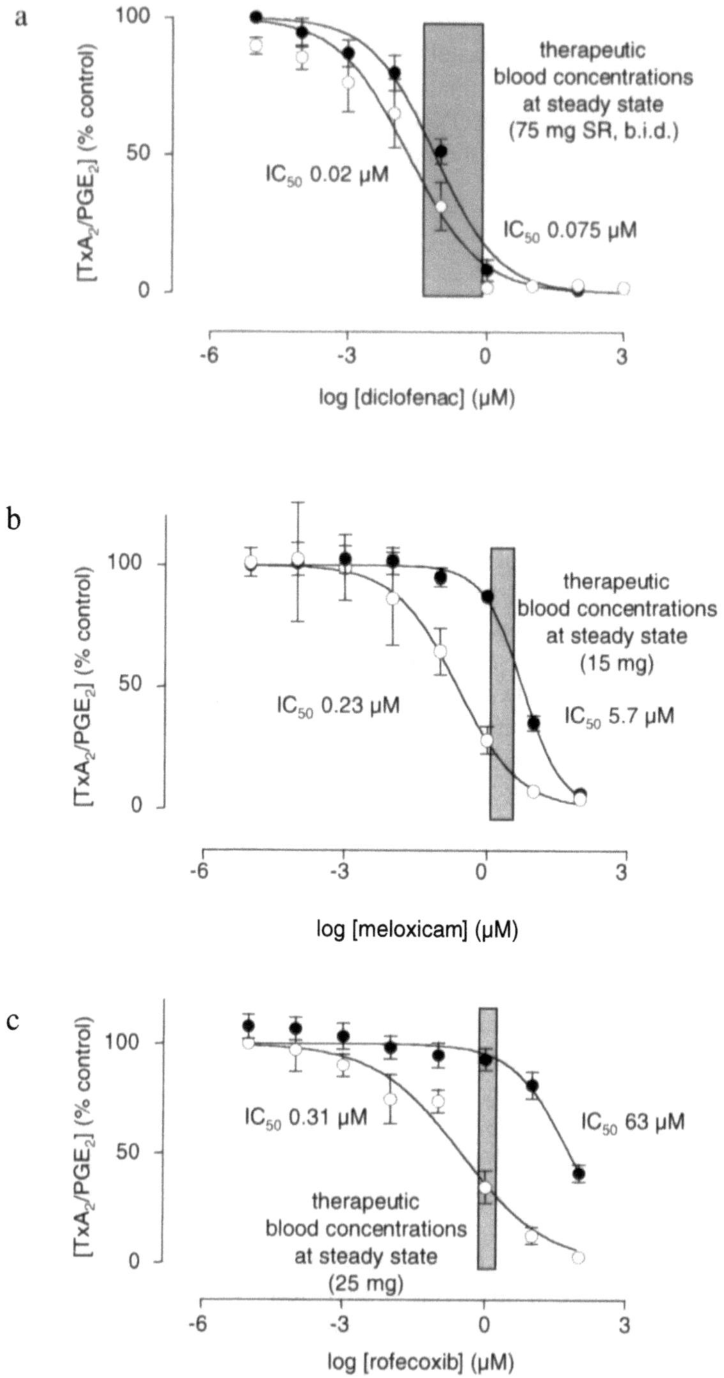

Figure 5a–c. Inhibition of COX-1 (●) and COX-2 (○) by diclofenac (a), meloxicam (b), and rofecoxib (c) in hmWBA [52]. Greyed bar indicates concentrations achieved *in vivo* following ingestion of therapeutic doses.

standard doses. However, the level of COX-1 inhibition varies greatly (from approximately 70% to approximately 10%). Once again, this supports the idea that it is inhibition of COX-2 that underlies the efficacious effects of COX inhibitors; i.e., if inhibition of COX-1 was required for drugs to be efficacious diclofenac would be considerably more efficacious than either meloxicam or rofecoxib.

Human pharmacology studies

The most relevant markers of COX-1 and COX-2 activity *in vivo* would be PGE_2 or PGI_2 production by the gastric mucosa, and PGE_2 production by sites of inflammation, most notably the inflamed synovium. Indeed, similar models have been developed for use in experimental animals. However, for practical reasons, such studies are much more difficult to perform in humans, since target tissues are not easily obtainable. Furthermore, prostaglandin production can be stimulated when tissues are harvested. Despite these reservations there are a number of studies that appear promising [59, 60]. Generally the differential inhibition of COX-l and COX-2 by NSAIDs is studied in humans by *in vivo* pharmacological studies. The effects of repeated administration of anti-inflammatory doses on COX-2 activity measured in a hWBA *ex vivo*, or on markers of COX-1 activity, such as platelet aggregation, plasma thromboxane B_2 (TxB_2) concentrations, or urinary excretion of PGE_2, are usually investigated. Of these serum TxB_2 appears to be the most sensitive measure of COX-1 inhibition by a drug, as platelet aggregation does not appear affected until plasma TxB_2 is reduced by more than 95% [61]. This is illustrated with meloxicam, which was given in doses of 7.5, 15 and 30 mg to healthy volunteers [62]. *Ex vivo* serum TxB_2 was inhibited dose-dependently by ~40, 60 and 75%, respectively, however, there was no effect on platelet aggregation or bleeding time at these therapeutic and supratherapeutic doses. Nimesulide, is also without effect upon bleeding time or *ex vivo* aggregation [63]. Studies performed with several non-selective "standard" NSAIDs showed a close relationship between plasma drug concentrations and inhibition of serum TxB_2 [64–68] (most likely to be COX-1 inhibition) and inhibition of synovial PGE_2 (most likely to be COX-2 inhibition) [59, 60]. Naproxen and ibuprofen both inhibit serum TxB_2 by more than 95% at therapeutic doses and inhibit platelet aggregation *ex vivo* [69, 70].

Celecoxib did not inhibit collagen, ADP, or arachidonate-induced platelet aggregation, and also had no effect on TxB_2 levels, 2 and 4 h after the last administration of 600 mg twice daily for 6 days [71]. Similarly, more studies using celecoxib at doses of up to 1200 mg in comparison to 800 mg ibuprofen or 1000 mg naproxen have shown that the traditional NSAIDs, but not celecoxib, inhibited *ex vivo* platelet aggregation induced by agonists such as collagen. Also, while naproxen but not celecoxib depressed circulating thromboxane A_2 levels, both ibuprofen and celecoxib inhibited LPS-induced COX-2 expression *ex vivo* [72]. Interestingly, however, the plasma TxB_2 concentration, a more sensitive measure of COX-1 inhibition than platelet aggregation (see above), was significantly reduced by approximately 30% after a single administration of the supratherapeutic dose of 800 mg celecoxib [73]. In the same studies, both ibuprofen and celecoxib were found to inhibit urinary excretion of the PGI_2 metabolite 2,3-dinor-6-keto-$PGF_{1\alpha}$ [73]. This may indicate that COX-2 is responsible for the formation of PGI_2 in the circulation. Indeed, further studies with rofecoxib have shown that like its comparator indomethacin this COX-2-selective inhibitor also reduces the urinary excretion of the PGI_2 metabolite 2,3-dinor-6-keto-prostaglandin $F_{1\alpha}$ [74]. At the same time even supra-therapeutic doses of rofecoxib of up to 375 mg for 14 days produces no inhibition of COX-1 activity *ex vivo*, as assessed by TxB_2 generation in clotting blood, whereas there is profound inhibition of *ex vivo* COX-2 activity [75]. Consistent with these results, the new generation of COX-2 selective agents such as etoricoxib [76, 77], valdecoxib [78], and COX-189 [79] all do not inhibit serum TxB_2 at therapeutic doses and do not affect platelet aggregation or bleeding time at therapeutic or supratherapeutic doses.

Acknowledgements
The authors wish to thank U. Werthmann for determining the physiochemical values of various NSAIDs. We also wish to thank H. Wittneben and C. Klein for performing the 3D-molecular modelling and to E. Haaksma and H. Nar for their helpful suggestions.

References

1 Picot D, Loll PJ, Garavito RM (1994) The X-ray crystal structure of the membrane protein prostaglandin H2 synthase-1. *Nature* 367: 243–249
2 Trummlitz G, van Ryn J (2002) Designing selective COX-2 inhibitors: Molecular modeling approaches. *Curr Opin Drug Discov Devel* 5: 550–561
3 Smith, WL, DeWitt, DL, Garavito, RM (2000) Cyclooxygenases: structural, cellular and molecular biology. *Annu Rev Biochem* 69: 145–182
4 Hatae T, Wada M, Yokoyama C, Shimonishi M, Tanabe T (2001) Prostacyclin-dependent apoptosis mediated by PPAR delta. *J Biol Chem* 276: 46260–46267
5 Garavito RM (2001) Prostaglandin endoperoxide H2 synthases-1 and -2. In: A Messerschmidt, R Huber, T Poulos (eds): *Handbook of Metalloproteins*. John Wiley and Sons Ltd, Chichester, UK, 245–263
6 Chandrasekharan NV, Dai H, Roos LT, Evanson NK, Tomsik J, Elton TS, Simmons DL (2002) COX-3, a cyclooxygenase-1 variant inhibited by acetaminophen and other analgesic/antipyretic drugs: Cloning, structure, and expression. *Proc Natl Acad Sci USA* 99: 13926–13931
7 Dinchuk JE, Liu RQ, Trzaskos JM (2003) COX-3: in the wrong frame in mind. *Immunol Lett* 86: 121
8 Kiefer JR, Pawlitz JL, Moreland KT, Stegeman RA, Hood WF, Gierse JK, Stevens AM, Goodwin DC, Rowlinson SW, Marnett LJ et al (2000) Structural insights into the stereochemistry of the cyclooxygenase reaction. *Nature* 405: 97–101
9 Malkowski MG, Ginell SL, Smith WL, Garavito RM (2000) The productive conformation of arachidonic acid bound to prostaglandin synthase. *Science* 289: 1933–1937
10 Thuresson ED, Malkowski MG, Lakkides KM, Rieke CJ, Mulichak AM, Ginell SL, Garavito RM, Smith WL (2001) Mutational and X-ray crystallographic analysis of the interaction of dihomo-gamma-linolenic acid with prostaglandin endoperoxide H syntheses. *J Biol Chem* 276: 10358–10365

11 Malkowski MG, Thuresson ED, Lakkides KM, Rieke CJ, Micielli R, Smith WL, Garavito RM (2001) Structure of eicosapentaenoic and linoleic acids in the cyclooxygenase site of prostaglandin endoperoxide H synthase-1. *J Biol Chem* 276: 37547–37555

12 Loll PJ, Picot D, Garavito RM (1995) The structural basis of aspirin activity inferred from the crystal structure of inactivated prostaglandin H2 synthase. *Nature Struct Biol* 2: 637–643

13 Loll PJ, Sharkey CT, O'Connor SJ, Dooley CM, O'Brien E, Devocelle M, Nolan KB, Selinsky BS, Fitzgerald DJ (2001) O-Acetylsalicylhydroxamic acid, a novel acetylating inhibitor of prostaglandin H2 Synthase: Structural and functional characterization of enzyme-inhibitor interactions. *Mol Pharmacol* 60: 1407–1413

14 Kurumbail RG, Stevens AM, Gierse JK, McDonald JJ, Stegeman RA, Pak JY, Gildehaus D, Miyashiro JM, Penning TD, Seibert K et al (1996) Structural basis for selective inhibition of cyclooxygenase-2 by anti-inflammatory agents. *Nature* 384: 644–648

15 Loll PJ, Picot D, Ekabo O, Garavito RM (1996) Synthesis and use of iodinated nonsteroidal anti-inflammatory drug analogs as crystallographic probes of the prostaglandin H2 synthase cyclooxygenase active site. *Biochemistry* 35: 7330–7340

16 Merck and Co. Inc. (Bayly CI, McKeever BM, Ouellet M, Pandrya SR, Percival MD, Roy P) (1998) Crystals of human cyclooxygenase-2 (hCOX-2). GB2319032A

17 Selinsky BS, Gupta K, Sharkey CT, Loll PJ (2001) Structural analysis of NSAID binding by prostaglandin H2 synthase: Time-dependent and time-independent inhibitors elicit identical enzyme conformations. *Biochemistry* 40: 5172–5180

18 Luong C, Miller A, Barnett J, Chow J, Ramesha C, Browner MF (1996) Flexibility of the NSAID binding site in the structure of human cyclooxygenase-2. *Nature Struct Biol* 3: 927–933

19 Garavito RM, Picot D, Loll PJ (1995) The 3.1 A X-ray crystal structure of the integral membrane enzyme prostaglandin H2 synthase-1. *Adv Prostaglandin Thromboxane Leukot Res* 23: 99–103

20 Trummlitz G, Wittneben H, van Ryn J, Harman C, Clayton G, Garavito RM (2003) X-Ray crystallographic analysis and computational docking studies of cyclooxygenase isoform inhibitors. *Ann Rheum Dis 62 (Suppl)*: Abs FRI0230

21 Malkowski MG, Theissen MJ, Scharmen A, Garavito RM (2000) The formation of stable fatty acid substrate complexes in prostaglandin H_2 synthetase-1. *Arch Biochem Biophys* 380: 39–45

22 Ringbom T, Huss U, Stenholm A, Flock S, Skattebol L, Perera P, Bohlin L (2001) COX-2 inhibitory effects of naturally occurring and modified fatty acids. *J Nat Prod* 64: 745–749

23 Kozak KR, Crews BC, Ray JL, Tai HH, Morrow JD, Marnett LJ (2001) Metabolism of prostaglandin glycerol esters and prostaglandin ethanolamides *in vitro* and *in vivo*. *J Biol Chem* 276: 36993–36998

24 Kozak KR, Prusakiewicz JJ, Rowlinson SW, Schneider C, Marnett LJ (2001) Amino acid determinants in cyclooxygenase-2 oxygenation of the endocannabinoid 2-aracidonylglycerol. *J Biol Chem* 276: 30072–30077

25 Luger P, Daneck K, Engel W, Trummlitz G, Wagner K (1996) Structure and physicochemical properties of meloxicam, a new NSAID. *Eur J Pharmac Sci* 4: 75–187

26 Mancini JA, O'Neill GP, Bayly C, Vickers PJ (1994) Mutation of serine-516 in human prostaglandin G/H synthase-2 to methionine or aspirin acetylation of this residue stimulates 15-*R*-HETE synthesis. *FEBS Letters* 342: 33–37

27 Mancini JA, Riendeau D, Falgueyret JP, Vickers PJ, O'Neill GP (1995) Arginine 120 of prostaglandin G/H synthase-1 is required for the inhibition by nonsteroidal anti-inflammatory drugs containing carboxylic acid moiety. *J Biol Chem* 270: 29372–29377

28 Hochgesang GP, Rowlinson SW, Marnett LJ (2000) Tyrosine-85 is critical for acetylation of cyclooxgenase-2 by aspirin. *J Amer Chem Soc* 122: 6514–6515

29 Rowlinson SW, Crews BC, Goodwin DC, Schneider C, Gierse JK, Marnett LJ (2000) Spatial requirements for 15-(*R*)-Hydroxy-5Z,8Z,11*Z*,13*E*-eicosatertraenoic acid synthesis within the cyclooxygenase active site of murine COX-2. *J Biol Chem* 275: 6586–6591

30 Kalgutkar AS, Crews BC, Rowlinson SW, Garner C, Seibert K, Marnett LJ (1998) Aspirin-like molecules that covalently inactivate cyclooxygenase-2. *Science* 280: 1268–1270

31 Hochgesang GP, Nemeth-Cawley JF, Rowlinson SW, Caprioli RM, Marnett LJ (2003) Functional analysis of the molecular determinants of cyclooxygenase-2 acetylation by 2-acetoxyphenyl-hept-2-ynyl sulfide. *Arch Biochem Biophys* 409: 127–133

32 Greig GM, Francis DA, Falgueyret JP, Ouellet M, Percival MD, Roy P, Bayly C, Mancini JA, O'Neill GP (1997) The interaction of arginine 106 of human prostaglandin G/H synthase-2 with inhibitors is not a universal component of inhibition mediated by nonsteroidal anti-inflammatory

drugs. *Mol Pharmacol* 52: 829–838
33 So OY, Scarafia LE, Mak AY, Callan OH, Swinney DC (1998) The dynamics of prostaglandin H synthase 2 Y355F unmask mechanisms of time-dependent inhibition and allosteric activation. *J Biol Chem* 273: 5801–5807
34 Mancini JA, Vickers PJ, O'Neill GP, Boily C, Falgueyret JP, Riendeau D (1997) Altered sensitivity of aspirin-acetylated prostaglandin G/H Synthase to inhibition by nonsteroidal anti-inflammatory drugs. *Mol Pharmacol* 51: 52–60
35 Gierse JK, McDonald JJ, Hauser SD, Rangwala SH, Koboldt CM, Seibert K (1996) A single amino acid difference between cyclooxygenase-1 (COX-1) and -2 (COX-2) reverses the selectivity of COX-2 specific inhibitors. *J Biol Chem* 271: 15810–15814
36 Wong E, Bayly C, Waterman HL, Riendeau D, Mancini JA (1997) Conversion of prostaglandin G/H synthase-1 into an enzyme sensitive to PGHS-2-selective inhibitors by a double His513 → Arg and Ile523 → Val mutation. *J Biol Chem* 272: 9280–9286
37 Timofeevski SL, Prusakiewicz JJ, Rouzer CA, Marnett LJ (2002) Isoform-selective interaction of cyclooxygenase-2 with indomethacin amides studied by real-time fluorescence, inhibition kinetics, and site-directed mutagenesis. *Biochemistry* 41: 9654–9662
38 Bhattacharyya DK, Lecomte M, Rieke CJ, Garavito RM, Smith WL (1996) Involvement of arginine 120, Glutamate 524, and tyrosine 355 in the binding of arachidonate and 2-phenylpropionic acid inhibitors to the cyclooxygenase active site of ovine prostaglandin endoperoxide H synthase-1. *J Biol Chem* 271: 2179–2184
39 Rieke CJ, Mulichak AM, Garavito RM, Smith WL (1999) The role of arginine 120 of human prostaglandin endoperoxide H synthase in the interaction with fatty acids and inhibitors. *J Biol Chem* 274: 17109–17114
40 Kalgutkar AS, Crews BC, Rowlinson SW, Marnett AB, Kozak KR, Remmel RP, Marnett LJ (2000) Biochemically based design of cyclooxygenase-2 (COX-2) inhibitors: facile conversion of nonsteroidal antiinflammatory drugs to potent and highly selective COX-2 inhibitors. *Proc Natl Acad Sci USA* 97: 925–930
41 Guo QP, Wang LH, Ruan KH, Kulmacz RJ (1996) Role of Val509 in time-dependent inhibition of human prostaglandin H synthase-2 cyclooxygenase activity by isoform-selective agents. *J Biol Chem* 271: 9134–19139
42 Hood WF, Gierse JK, Isakson PC, Kiefer JR, Kurumbail RG, Seibert K, Monahan JB (2003) Characterization of celecoxib and valdecoxib binding to cyclooxygenase. *Mol Pharmacol* 63: 870–877
43 Kurumbail RG, Kiefer JR, Marnett LJ (2001) Cyclooxygenase enzymes: catalysis and inhibition. *Curr Opin Struct Biol* 11: 752–760
44 Lecomte M, Laneuville O, Ji C, DeWitt DL, Smith WL (1994) Acetylation of human prostaglandin endoperoxide synthase-2 (cyclooxygnase-2) by aspirin. *J Biol Chem* 269: 13207–13215
45 Trummlitz G, Wittneben H (2001) Insight into the structural basis of selective cyclooxygenase-2 inhibition. *Ann Rheum Dis* 60 (Suppl I): 234 Abstr SAT0080
46 Gierse JK, Koboldt CM, Walker MC, Seibert K, Isakson PC (1999) Kinetic basis for selective inhibition of cyclo-oxygenase. *Biochem J* 339: 607–614
47 Walker MC, Kurumbail RG, Kiefer JR, Moreland KT, Koboldt CM, Isakson PC, Seibert K, Gierse JK (2001) A three-step kinetic mechanism for selective inhibition of cyclo-oxygenase-2 by diarylheterocyclic inhibitors. *Biochem J* 357: 709–718
48 Price MLP, Jorgensen WL (2000) Analysis of binding affinities for celecoxib analogues with COX-1 and COX-2 from combined docking and Monte Carlo simulations and insight into the COX-2/COX-1 selectivity. *J Am Chem Soc* 122: 9455–9466
49 Copeland RA, Williams JM, Giannaras J, Nurnberg S, Covington M, Pinto D, Pick S, Trzaskos JM (1994) Mechanism of selective inhibition of the inducible isoform of prostaglandin G/H synthase. *Proc Natl Acad Sci USA* 91: 11202–11206
50 Pairet M, van Ryn J (1998) Experimental models used to investigate the differential inhibition of cyclooxygenase-1 and cyclooxygenase-2 by non-steroidal anti-inflammatory drugs. *Inflamm Res* 47 (Suppl 2): S93–S101
51 Lanzo CA, Sutin J, Rowlinson S, Talley J, Marnett LJ (2000) Fluorescence quenching analysis of the association and dissociation of a diarylheterocycle to cyclooxygenase-1 and cyclooxygenase-2: dynamic basis of cyclooxygenase-2 selectivity. *Biochemistry* 39: 6228–6234
52 Warner TD, Giuliano F, Vojnovic I, Bukasa A, Mitchell JA, Vane JR (1999) Nonsteroid drug selec-

tivities for cyclo-oxygenase-1 rather than cyclo-oxygenase-2 are associated with human gastrointestinal toxicity: a full *in vitro* analysis. *Proc Natl Acad Sci USA* 96: 7563–7568
53 Blanco F, Guitian R, Moreno 1, Hernandez A, Freire M, Atanes A, Grana J, Toro FJ de, Galdo F (1997) NSAIDs effects on COX 1 and COX 2 activity in human articular chondrocytes. *Arthritis Rheum* 40 (Suppl 9): 347
54 Kawai S, Nishida S, Kato M, Furumaya Y, Okamoto R, Koshino T, Mizushima Y (1998) Comparison of cyclooxygenase-1 and -2 inhibitory activities of various nonsteroidal anti-inflammatory drugs using human platelets and synovial cells. *Eur J Pharmacol* 347: 87–94
55 Tavares IA (2000) The effects of meloxicam, indomethacin or NS-398 on eicosanoid synthesis by fresh human gastric mucosa. *Aliment Pharmacol Ther* 14: 795–799
56 Blanco FJ, Guitian R, Moreno J, de Toro FJ, Galdo F (1999) Effect of antiinflammatory drugs on COX-1 and COX-2 activity in human articular chondrocytes. *J Rheumatol* 26: 1366–1373
57 Kawai S (1998) Cyclooxygenase selectivity and the risk of gastro-intestinal complications of various non-steroidal anti-inflammatory drugs: a clinical consideration. *Inflamm Res* 47 (Suppl 2): S102–106
58 Cryer B, Feldman M (1998) Cyclooxygenase-1 and cyclooxygenase-2 selectivity of widely used nonsteroidal anti-inflammatory drugs. *Am J Med* 104: 413–421
59 Day RO, Francis H, Vial J, Geisslinger G, Williams KM (1995) Naproxen concentrations in plasma and synovial fluid and effects on prostanoid concentrations. *J Rheumatol* 22: 2295–2303
60 Bertin P, Lapicque F, Payan E, Rigaud M, Bailleul F, Jaeger S, Treves R, Netter P (1994) Sodium naproxen: Concentration and effect on inflammatory response mediators in human rheumatoid synovial fluid. *Clin Pharmacol* 46: 3–7
61 Reilly IA, FitzGerald GA (1987) Inhibition of thromboxane formation *in vivo* and *ex vivo*: implications for therapy with platelet inhibitory drugs. *Blood* 69: 180–186
62 Rinder HM, Tracey JB, Souhrada M, Wang C, Gagnier RP, Wood CC (2002) Effects of meloxicam on platelet function in healthy adults: a randomized, double-blind, placebo-controlled trial. *J Clin Pharmacol* 42: 881–886
63 Marbet GA, Yasikoff Strub ML, Macciocchi A, Tsakiris DA (1998) The effect of nimesulide *versus* placebo on hemostasis in healthy volunteers. *Eur J Clin Pharmacol* 54: 383–387
64 Rane A, Oelz O, Frölich JC, Seyberth HW, Sweetman BJ, Watson J, Wilkinson GR, Oates JA (1978) Relationship between plasma concentrations of indomethacin and its effect on prostaglandin synthesis and platelet aggregation in man. *Clin Pharmacol Ther* 23: 658–668
65 Cronberg S, Wallmark E, Sodeberg I (1984) Effect on platelet aggregation of oral administration of 10 non steroidal analgesics to humans. *Scand J Hacmatol* 33: 155–159
66 Vinge E (1985) Arachidonic acid induced platelet aggregation and prostanoid formation in whole blood in relation to plasma concentration of indomethacin. *Eur J Clin Pharmacol* 28: 163–169
67 Cox SR, Vanderlugt JT, Gumbleton TJ, Smith RB (1987) Relationships between thromboxane production, platelet aggregability, and serum concentrations of ibuprofen and flurbiprofen. *Clin Pharmacol Ther* 41: 510–521
68 Schafer AI (1995) Effects of nonsteroidal antiinflammatory drugs on platelet function and systemic hemostasis. *J Clin Pharmacol* 35: 209–219
69 Van Hecken A, Schwartz JI, Depre M, De Lepeleire I, Dallob A, Tanaka W, Wynants K, Buntinx A, Arnout J, Wong PH et al (2000) Comparative inhibitory activity of rofecoxib, meloxicam, diclofenac, ibuprofen, and naproxen on COX-2 *versus* COX-1 in healthy volunteers. *J Clin Pharmacol* 40: 1109–1120
70 Catella-Lawson F, Reilly MP, Kapoor SC, Cucchiara AJ, DeMarco S, Tournier B, Vyas SN, FitzGerald GA (2001) Cyclooxygenase inhibitors and the antiplatelet effects of aspirin. *N Engl J Med* 345: 1809–1817
71 Mengle Gaw L, Hubbard R, Karim A, Yu S, Talwalker S, Isakson P (1997) A study of the platelet effects of SC 58635, a novel COX 2 selective inhibitor [abstract]. *Arthritis Rheum* 40 (Suppl 9): 374
72 Leese PT, Hubbard RC, Karim A, Isakson PC, Yu SS, Geiss GS (2000) Effects of celecoxib, a novel cyclooxygenase-2 inhibitor, on platelet function in healthy adults: a randomised, controlled trial. *J Clin Pharmacol* 40: 124–132
73 McAdam BF, Catella-Lawson F, Mardini IA, Kapoor S, Lawson JA, FitzGerald GA (1999) Systemic biosynthesis of prostacyclin by cyclooxygenase (COX)-2: the human pharmacology of a selective inhibitor of COX-2. *Proc Natl Acad Sci USA* 96: 272–277
74 Catella-Lawson F, McAdam B, Morrison BW, Kapoor S, Kujubu D, Antes L, Lasseter KC, Quan

H, Gertz BJ, FitzGerald GA (1999) Effects of specific inhibition of cyclooxygenase-2 on sodium balance, hemodynamics, and vasoactive eicosanoids. *J Pharmacol Exp Ther* 289: 735–741

75 Depre M, Ehrich E, Van Hecken A, De Lepeleire, I, Dallob A, Wong P, Porras A, Gertz BJ, De Schepper PJ (2000) Pharmacokinetics, COX-2 specificity, and tolerability of supratherapeutic doses of rofecoxib in humans. *Eur J Clin Pharmacol* 56: 167–174

76 Dallob A, Depré M, De Lepeleire I, Agrawal N, Musser B, Kher U, Van Hecken A, Wildonger L, Matthews C, De Schepper P et al (2000) MK-0663 a highly selective inhibitor of COX-2 in humans. *Ann Rheum Dis* 59 (Suppl 1): 331

77 Riendeau D, Percival MD, Brideau S, Charleson S, Dubé D, Ethier D, Falgueyret J-P, Friesen RW, Gordon R, Greig G et al (2001) Etoricoxib (MK-0663): Preclinical profile and comparison with other agents that selectively inhibit cyclooxygenase-2. *J Pharm Exp Ther* 296: 558–566

78 Leese PT, Talwalker S, Kent JD, Recker DP (2002) Valdecoxib does not impair platelet function. *Am J Emerg Med* 20: 275–281

79 Sorbera LA, Canstañer J, Bayés M, Silvestre JS (2002) Lumiracoxib. *Drugs of the Future* 27: 740–747

COX-2 Inhibitors
Edited by M. Pairet and J. van Ryn

Pharmacology and toxicology of COX-2 inhibitors

K.D. Rainsford

Biomedical Research Centre, Sheffield Hallam University, Howard Street, Sheffield S1 1WB, UK

Introduction

The pharmacological premise upon which the prostaglandin G/H synthase II (PGHS-II) or cyclo-oxygenase-2 (COX-2) inhibitors was developed was essentially that founded in the identification of two different genes coding for the cyclo-oxygenases-1 and –2 [1–3]. COX-1 is considered to be a constitutive enzyme that is responsible for "housekeeping" or physiological functions (e.g., some gastrointestinal (GI) mucosal, renal and haemostatic functions and protective effects) while COX-2, which is induced by inflammatory stimuli, is considered to be important in inflammation, pain and fever [1–5]. Comprehensive reviews on the regulation and roles of these two isoenzymes in inflammation are found in [1–9] and Figure 1 summarizes their essential features.

This review examines the published studies which support the claims for COX-2 selective inhibitors having comparable therapeutic actions to conventional NSAIDs, and in standard laboratory and clinical models under controlled conditions has improved GI safety compared with some of the conventional NSAIDs.

Determination of COX selectivity

The determination of COX-2 selectivity of NSAIDs has relied upon a variety of different assay methods that evolved after the discovery of the COX-isoforms about a decade ago. However, the clinical relevance of many of these *in vitro* assays has been questioned. The pharmacokinetic relationships, especially the plasma concentrations of drugs, have been related to their *in vitro* effects in an attempt to more accurately define the clinical significance of relative COX-1 and COX-2 inhibitory effects to clinical actions [10–12]. An attempt to arrive at consensus criteria for defining the clinical significance of COX-2/COX-1 selectivity was made at an international meeting sponsored by one of the major coxib drug companies (Merck) [11]. This has proven useful in bringing together the essential pharmacokinetic and clinical criteria.

Understanding of the role of intracellular kinetics and physico-chemical parameters underlying drug actions *in vitro* has been elegantly explored [13,

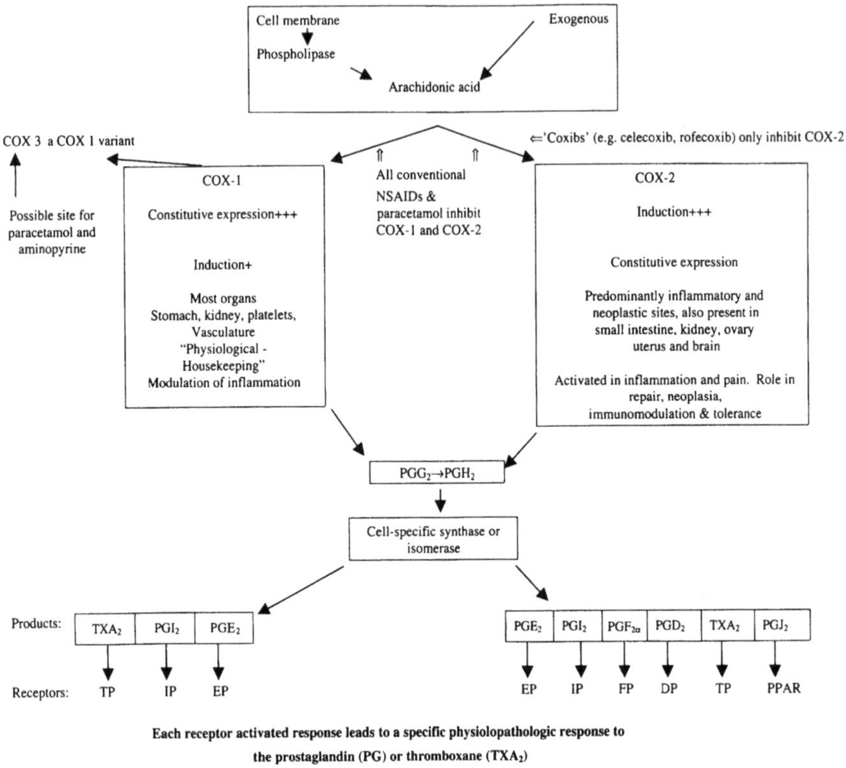

Figure 1. Roles of COX-1 and COX-2-derived prostanoids in regulating physiological (COX-1) or inflammatory (COX-2) reactions. COX-3 is a splice product of COX-1 found in the brain and some other organs and is presumed to have a role in pain. The inhibition of COX-3 by paracetamol (acetaminophen), dipyrone as well as some NSAIDs has been suggested to underlie their analgesic actions in the CNS.

14] and the importance of the nature of stimuli and the effects of the exogenous and endogenous supply of the substrate arachidonate on enzyme activity determined [15, 16]. Table 1 shows data on the *in vitro* effects of a range of NSAIDs and coxibs determined using the whole blood assay [17]. The relationship between the COX-1/COX-2 ratios for enzyme inhibition and occurrence of severe GI events is inferred from data shown in Figure 2 [18].

COX-2 selective NSAIDs

Etodolac

Etodolac ([±]-1,8-diethyl-1,3,4,9-tetrahydropyrano(3,4-b)indole-1-acetic acid Lodine®; Fig. 3) is chemically distinct from currently used NSAIDs owing to

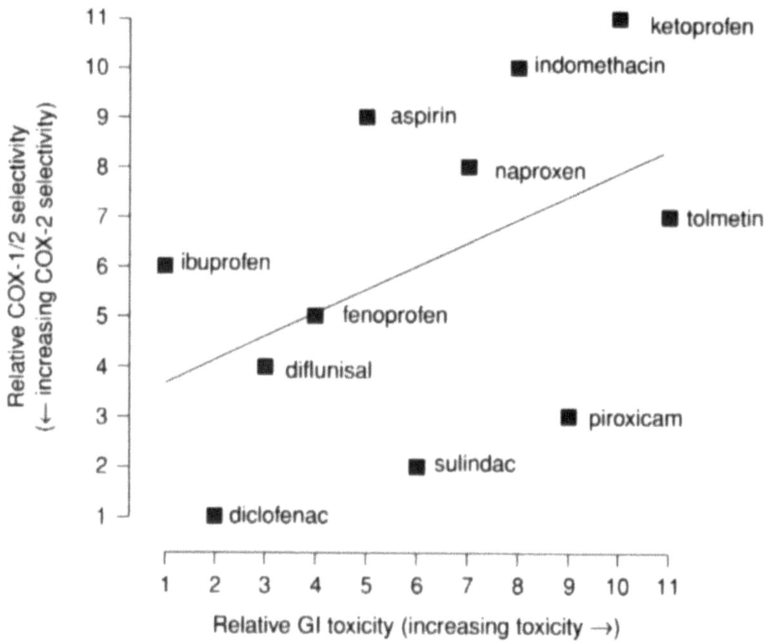

Figure 2. Relationship between COX-1/COX-2 inhibition ratios and the development of serious gastrointestinal events by unselective COX inhibitors. From Warner et al. [18] with permission of the publishers.

its tetrahydropyrano-indole nucleus. It was originally discovered in 1976 and developed by Ayerst Laboratories in Montreal (Canada) [19, 20]. Etodolac is thus among the NSAIDs whose COX-2 selectivity was established long after it was established in clinical use with a relatively favourable GI profile [20, 21]. Etodolac was originally selected in preclinical development on the basis of having the best therapeutic ratio compared with established NSAIDs and related pyranolone analogues in response to therapeutic effects in the adjuvant-induced chronic polyarthritis in rats in comparison with its gastro-ulcerogenic effects [19]. This was associated with a relative lack of acute effects of the drug on the rat gastric mucosa [19].

COX-2 selectivity

Earlier studies suggested that etodolac may spare the production of gastroprotective prostaglandins (PGs) in the gastric mucosa [22] so inferring that this may account for the low GI ulcerogenicity and bleeding of this drug in animals [19] and in human studies including those in arthritic subjects and those at high risk of serious NSAID-induced GI events [21]. In more recent studies etodolac has been shown to have COX-2 selectivity with values ranging from 10-fold in human recombinant enzymes systems, 24-fold in the modified whole blood

Table 1. Potencies of compounds as inhibitors of prostanoid formation determined in the COX-1 assay, WBA-COX-2

	COX-1 IC_{50}-µM	WBA-COX-2 IC_{50}-µM	WHMA-COX-2 IC_{50}-µM	IC_{50} ratios		Ranking at IC_{50} ratios	
				WBA COX-1	WHMA COX-1	WBA COX-1	WHMA COX-1
NSAIDs							
5-aminosalicylic acid	410	61	n.d.	0.15	n.d.	n.d.	n.d.
Aspirin	1.7	>100	7.5	>100	4.4	34	23
Diclofenac	0.075	0.038	0.020	0.5	0.3	10	9
Diflunisal	113	8.2	134	0.1	1.2	9	14
Flubiprofen	0.075	5.5	0.77	73	10	31	27
Flufenamate	3.0	9.3	n.d.	3.1	n.d.	13	n.d.
Ibuprofen	7.6	7.2	20	0.9	2.6	14	20
Indomethacin	0.013	1.0	0.13	80	10	29	24
Ketoprofen	0.047	2.9	0.24	61	5.1	31	25
Meclofenamate	0.22	0.7	0.2	3.2	0.91	22	11
Mefenamic acid	25	2.9	1.3	0.11	0.049	-	-
Naproxen	9.3	28	35	3.0	3.8	18	22
Piroxicam	2.4	7.9	0.17	3.3	0.1	17	13
Salicin	>100	>100	n.d.	-	n.d.	-	n.d.
Salicylaldehyde	>100	>100	n.d.	-	n.d.	-	n.d.
Sodium salicylate	4956	34440	482	6.9	0.10	16	15
Sulfasalazine	3242	2507	n.d.	0.8	n.d.	15	n.d.
Sulindac	>100	>100	58	-	-	-	-
Sulindac sulphide	1.9	55	1.21	29	0.64	20	10
Valeryl salicylate	42	2.3	n.d.	0.053	n.d.	-	n.d.

(Continued on next page)

Table 1. (Continued)

	COX-1 IC_{50}-μM	WBA-COX-2 IC_{50}-μM	WHMA-COX-2 IC_{50}-μM	IC_{50} ratios		Ranking at IC_{50} ratios	
				WBA COX-1	WHMA COX-1	WBA COX-1	WHMA COX-1
COX-2 Inhibitors							
Celecoxib	1.2	0.83	0.34	0.7	0.3	8	7
Etodolac	12	2.2	0.94	0.2	0.1	6	5
Meloxicam	5.7	2.1	0.23	0.37	0.040	11	6
Nimesulide	10	1.9	0.39	0.19	0.038	7	8
L745,337	>100	8.6	1.3	<0.01	<0.01	1=	1=
NS398	6.9	0.35	0.042	0.051	0.0061	5	4
Rofecoxib	63	0.84	0.31	0.013	0.0049	4	3
SC58125	>100	2.0	n.d.	>0.01	n.d.	1=	n.d.
Analgesics (non-narcotic)							
Aminopyrone	55	203	85	3.7	1.5	24	19
Paracetamol	>100	49	64	-	-	-	-

From Warner *et al.* [17].

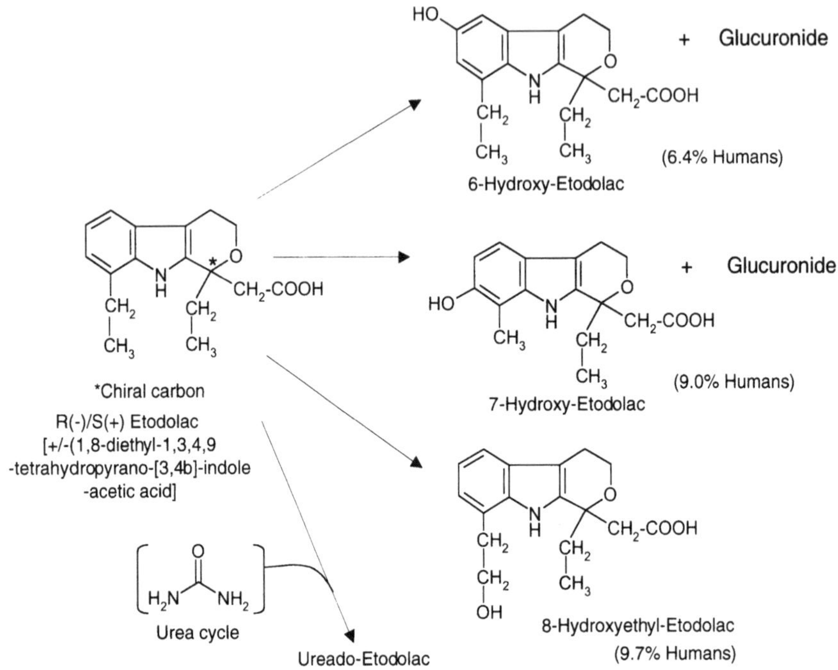

Figure 3. Pathways for the metabolism of R(–)/S(+) etodolac showing the prominent metabolites formed in humans. Based on data in [20, 26, 27].

assay up to 1000-fold in CHO cells transfected with human recombinant enzymes [21]. These data have, however, been derived from studies using the diastereoisomeric form of the drug that is commercially available. The component enantiomers of (±) etodolac (AY-24236) exhibit differing PG synthetase inhibitory activities, with the S(+) enantiomer (AY-28689) having greater potency than the R(–) (AY-28688) which is almost inactive in most assays) [23]. No attempts appear to have been made to establish if the S(+) enantiomer has the same degree of COX-2 selectivity as observed with the racemic mixture. With ibuprofen there are differing influences of the inactive R(–) enantiomers on the inhibitory effects of S(+) ibuprofen on COX-1 and COX-2 activity [24] and this may exist with the enantiomers of etodolac as well.

Despite the differences in PG synthesis inhibitory effects of the two enantiomers there is little difference in their acute gastro-irritant effects in the cholinomimetic-sensitized mouse assay [25]. However, the gastric irritancy of the racemate is lower than that of the S(+) or R(–) enantiomers [25] suggesting that there may be physical interactions between the two that in some way mask the potential irritancy of the individual enantiomers. There is no bioconversion of R(–)- to the S(+)-enantiomer of etodolac [25] in contrast to that of some propionic acids e.g., ibuprofen [24]. Thus, it is unlikely that there are any influences of bioconversion of R(–) to S(+) etodolac on gastric irritancy.

Pharmacokinetics

The pharmacokinetics of etodolac in humans, rats and dogs involves presumed cytochrome P_{450} mediated hydroxylation to the 5-, 6-, 7- and 8-hydroxymetabolites [20, 26, 27] and subsequent phenolic glucuronidation of most if not all of these hydroxyl groups (see Figure 3 for details of metabolism; Table 2 and [20, 26, 27] for details of pharmacokinetic parameters in humans and rats). There is little information available on the possible formation of the acyl glucuronides.

The main pathways of the metabolism of etodolac in humans appear similar to those in rats and dogs [19, 26–28] but there are quantitative differences in the rates and routes of elimination. Pronounced biliary excretion and subsequent enterohepatic recirculation of both unchanged drug and the glucuronides occurs in rats [26] and dogs [19] with consequent predominant excretion in the faeces and to a lesser extent the urine [19]. Thus, the expected intestinal injury in rats and dogs in long-term toxicity studies from etodolac is probably related to the entero-hepatic re-circulation of the drug [19]. However, in short-term studies this drug does not exhibit intestinal injury in these species [29–31]. There is less faecal excretion and greater urinary excretion of the drug in humans but, nonetheless, the rare occurrence of cases of colitis [31] and benign colonic strictures [32–34], which may be related to breakdown in the intestinal barrier and permeability changes that are induced by NSAIDs [35, 36]. Etodolac is rapidly absorbed after oral intake. Its pharmacokinetic profile in humans favours twice daily dosing having a plasma half-life for elimination ($t^{1/2}$) of 6–7 h after single or multiple dosages for 7 days [19]. None of the pharmacokinetic parameters are appreciably affected by age, moderate renal impairment, or in patients with stable liver cirrhosis or those with rheumatic conditions [19, 20, 37, 38]. Thus conditions in the high-risk groups of patients where drug accumulation occurs with some other NSAIDs poses less of a risk for systemic toxicity with etodolac, although obviously there would be need to have caution in these "at risk" subjects with the drug.

GI and other adverse effects

The drug also appears to have low risk for GI ulcers and haemorrhage in the elderly who are at risk for developing GI ulcers [19, 39]. Long-term clinical and epidemiological data show there is a lower risk for GI ulcers or bleeding with etodolac than other NSAIDs including ibuprofen [19, 40, 41]. It should be noted, however, that the rigor of data from such studies as compared to prospective, randomised, double blind trials with coxibs has been questioned [42]. The relative inhibitory effects of etodolac on platelet aggregation and COX-1 activity are low [43, 44] and low rates of microbleeding (from ^{51}Cr-red cell loss) suggest that even where there is evidence of mucosal injury from whatever cause there is likely to be a low risk of bleeding. Adaptation of the GI mucosa to repeated intake of etodolac 300 mg b.d. for up to 28 days has been observed and was similar to that of naproxen 500 mg b.d. but etodolac caused increased antral blood flow, which was reduced in naproxen-treated

Table 2. Principal pharmacokinetic properties of COX-2 selective drugs

Drug	C_{max}	T_{max}, hr	$T_{1/2}$, hr	V_D, litres	Renal $^{l/hr l}$	Dosage mg	PKa	Comments	Refs
Etodolac	9.5–18.6 μg/ml	1–2	6.0–7.4	~10		200–600 o.d.	4.8	Metabolism probably via CYP 2C9	[17, 158]
Meloxicam	0.9–2.5 μg/ml	4.9–5.6	20–22	10–15	7.19–11.7	7.5–15 o.d.	1.09, 4.18	Metabolism via CYP 2C9	[75, 79, 82, 87, 158]
Nimesulide	2.86–9.5 μg/ml	1.22–3.17	1.22–4.95	12.6–27.3	2.11–5.18	100–200 b.i.d.	6.4–6.8	Predominant metabolism via CYP 2C9	[156, 162]
Celecoxib	705 ng/ml	2.8	11.2	429	27.7	100–200 b.i.d.	11.1	Metabolism via CYP 2C9	
Rofecoxib	207 ng/ml	2–3	17	86–91		12.5–50 o.d.	11.0	Predominant cytosolic NADH-dependent metabolism with CYP3A minor route	
Lumiracoxib	5.96–6.64 μg/ml	1.22–3.17	1.22–4.95	12.6–27.3	2.11–5.18	100–200 b.i.d.	6.4–6.8	Predominant metabolism via CYP 2C9	[358, 360]

Most values of plasma C_{max}, T_{max} and $t_{1/2}$ after multiple dosage. Values of V_D approximated to 70 kg adult.

subjects [45]. This adaptive feature appears to be reduced in elderly subjects or those who are smokers when they take etodolac or other NSAIDs [46].

Etodolac inhibits mucosal cell proliferation [47], adhesion molecule expression [48] and induction of apoptosis [49, 50], as well as inhibition of cytokine [51, 52], *Helicobacter pylori-* [53, 54] and growth factor [53]-induced production of PGE_2 or other prostanoids, similar to other COX-2 inhibitors. Prevention of colon cancer and other cancers [55–57] may also represent potential inhibitors of regeneration or repair of the GI mucosa following injury by ulcerogens. Inhibition of bone fracture healing by etodolac via inhibition of COX-2 [58] may also be considered a potential negative effect of this and other NSAIDs though the clinical significance or conditions of limitation of the use of this drug and other COX-2 inhibitors in post-orthopaedic surgical conditions have not been evaluated.

Renal impairment and other renal syndromes are rare conditions with etodolac [19] even though etodolac has been found to affect mechano-sensory neuronal function in the renal pelvic wall via a presumed COX-2 mediated mechanism [59]. There may also be other consequences of inhibiting COX-2 in the macula densa and other cells of the kidney that may promote renal impairment in at risk subjects. Etodolac has been reported to cause fulminant hepatic failure [60] though this and other hepatic events are not frequent occurrences.

Meloxicam

The discovery that this enol-carboxamide NSAID had COX-2 selectivity arose at the time when this drug had been selected for, and was well in progress, in phase III clinical trials. Thus, selection of this drug as a clinical candidate was not originally based on its COX-2 selectivity, but on the basis of conventional pharmacokinetics, pharmacodynamic, pre-clinical and phase 1 studies.

Meloxicam was first granted marketing authorization in 1995 and is now available in more than 100 countries [61]. The drug has been studied extensively in clinical trials involving more than 30,000 patients [61, 62]. It is recommended for, and has demonstrated efficacy, in osteoarthritis (OA), rheumatoid arthritis (RA) and ankylosing spondilitis, and has been used in some pain models [61, 62]. Additionally, the drug has been employed for veterinary use principally for treating of pain and inflammation in osteoarthritis in dogs where it has been found effective in at least two studies in joint inflammation [63, 64] and it has a wide range of other veterinary uses. In addition to the conventional oral formulation of the drug (Mobic®), the drug has been shown to have acceptable tissue tolerability when administered by intravenous/intramuscular or subcutaneous injection as well as in eyedrops, gels and suppositories [65], and additionally has been shown to be efficacious when given by intramuscular injection in patients with RA or OA [66].

Discovery and development of meloxicam

As indicated above the discovery of meloxicam took place in the pre-COX-2 era and thus the pharmacological profiling of the drug was based on determination of inhibitory effects of PG synthesis using bull seminal vesicles (which are now known to be comprised of COX-1) in addition to kaolin oedema, acute paw oedema in rats and screening for anti-arthritic activity using the adjuvant arthritis model [67, 68]. The initial aims were to base the drug development on the earlier oxicam class of drugs, sudoxicam and piroxicam and create a third generation of oxicam NSAID in which the pharmacokinetic properties were optimised for once daily administration (the very long half-lives of other oxicams appearing to be related to the development of GI and other serious side-effects), rapid and high distribution to the synovial fluid and improved risk to efficacy relationships [68]. In addition later studies showed that meloxicam did not have the potential to develop the so-called metabolite "C" (Fig. 4) which was associated with the development of phototoxic reactions with piroxicam [69] and as well autoradiographic studies showed that meloxicam had a very low skin penetration compared with that of other oxicams, thus reducing the potential for the development of phototoxicity reaction [69]. These earlier studies together with insight into the structure and physicochemical properties of meloxicam [70, 71] indicated that the increased lipophilicity might result in a shorter plasma half-life of the drug and that the relatively low solubility in the gastric pH range of meloxicam [70] may give a relatively low GI irritancy of this drug compared with that of established NSAIDs [61, 67, 71]. Similar arguments have been made to explain the low acute and subchronic GI irritancy of another keto-enolic NSAID, azapropazone [72]. The keto-enolic function of meloxicam produces zwitterions at high solvent polarity. Since the pKa values of meloxicam are 1.09 and 4.18 respectively the keto-enolic form is predominant in the gastric milieu whereas the enolic form at the same pH range is more likely to exist in the highly lipophilic environment [70] such as is present in mucosal membranes; the presumption here is that the enolic form of the drug is unlikely to be present following absorption of the drug and therefore

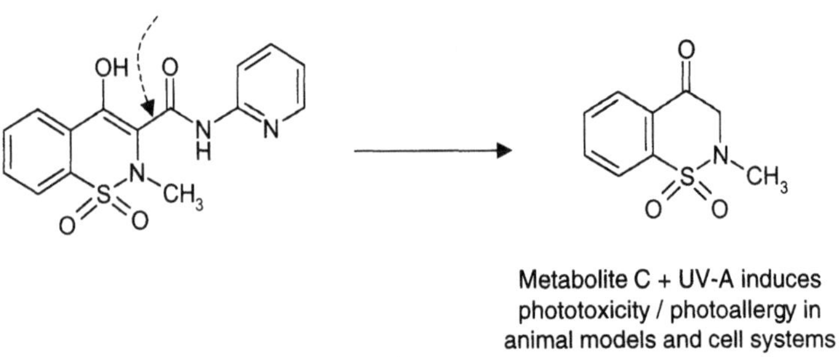

Figure 4. Formation of metabolite "C" from photo-activation of piroxicam which is presumed to underlie the development of photo-allergic reactions to this drug [69]. Inspection of the pathways of meloxicam metabolism (Fig. 4) shows that the formation of this metabolite is not possible with the latter drug.

would not, like acetylsalicylic acid, be capable of releasing a proton or protons to cause intracellular acidification and mucosal injury. Similar physicochemical explanations have been advanced to explain the differences in the gastric ulcerogenicity of low *versus* high gastrotoxic drugs [73].

Comparisons of the GI ulcerogenicity of meloxicam with standard NSAIDs as well as the therapeutic index relating ulcerogenicity to antiinflammatory activity in the adjuvant arthritis test is shown in Table 3 [74–76]. The ulcerogenicity studies were performed in normal rats and these showed a favourable therapeutic ratio of the order of about 20 compared with that of other NSAIDs which were all appreciably lower. In comparison with these data from 3 day dosage of rats the acute ulcerogenic effects of meloxicam showed high indexes (ED_{50} = 15.8 mg/kg) and in the intestine (ED_{20} = 19.7 mg/kg) and these were appreciably lower than the ulcerogenicity from indomethacin in the stomach, ED_{50} = 2.6 mg/kg and in the intestine 3.4 mg/kg [77].

In the Shay rat (in which acid secretion is measured in pyloric ligated animals), meloxicam produced a smaller increase in gastric acid production compared with diclofenac or piroxicam [77]. Furthermore, production of PGE_2 into the gastric juice of rats given meloxicam as well as gastric emptying was much lower than that of comparator NSAIDs.

Long-term toxicity studies undertaken with meloxicam compared with other similar, but not identical studies, with rofecoxib and celecoxib are shown in Table 4. These studies indicate that over a 50-week period meloxicam had a non-effect dose level of 0.8 mg/kg/day compared with that of rofecoxib 1.0 mg/kg/day while celecoxib did not achieve a non-effect dose level in a 2-year study. In comparison with the therapeutic effects in adjuvant arthritis the non-effect dose level was about 6–7 times higher than the ED_{50} dose, whereas that with rofecoxib was much closer to the therapeutic dose of 0.74 mg/kg/d for the adjuvant arthritis.

In the long-term toxicity studies the kidney was also a major target organ for the action of meloxicam, in addition to the GI mucosa, and the drug was found to delay parturition in pregnant rats. However, the dose required for

Table 3. Comparison of the gastro-intestinal effects of meloxicam with anti-arthritic effects in rats

	GI-erosions overall UD_{50} (mg/kg)	Inhibition of adjuvant arthritis ID_{50} mg/kg/d	Therapeutic ratio GI injury/anti-arthritic effects (UD_{50}/ID_{50})
Meloxicam	2.47	0.12	20
Diclofenac	2.71	1.24	2.2
Naproxen	11.1	11.8	0.9
Piroxicam`	1.07	0.76	1.4
Flubiprofen	1.4	0.96	0.3

Data from [77]. Similar ID_{50} values for inhibition of adjuvant-induced arthritis can be found in [74].

Table 4. Comparison of the GI Ulcerogenic potential of some COX-2 inhibitors in subacute and long-term toxicity studies in rats

	Meloxicam	Celecoxib	Rofecoxib
Sub-acute GI-ulcers	UD_{50}: 2.42 mg/kg[1]		No lesions at 300 mg/kg[3]
GI-bleeding ^{51}Cr-RBC	n.d.	not done	No effect at 300kg/kg[3]
Chronic GI-ulcers 52 weeks	<1 mg/kg[2] NOAEL: 0.8 mg/kg[2]	NOAEL not achieved in 2 y study[4]	<2 mg/kg[5] NOAEL: 1.0 mg/kg[5]
Therapeutic effects ID_{50} in adjuvant arthritic rats	0.12 mg/kg[1]	0.3 mg/kg[4]	0.74 mg/kg[1]

[1]Engelhardt et al. [74]
[2]Lehmann et al. [69]
[3]Chan et al. [319]
[4]Celecoxib summary basis of approval 20998, FDA [242]
[5]Rofecoxib summary basis of approval 021042/52 [318]

these toxic effects was well above those of the pharmacological effects [74]. In a 2-year carcinogenicity study in rats and mice there were no haematological/biochemical abnormalities, nor any evidence of major histopathological changes or evidence of mutagenicity.

Pharmacokinetics

The pharmacokinetic properties of meloxicam in humans (Tab. 2) are similar to that in rats [75, 78–80]. Thus, the values for plasma maximal concentration (Cp_{max}), time of maximal concentration (T_{max}), clearance (Cl), area under the plasma concentration curves (AUC's) and volumes of distribution (V_D) after single and multiple dose are similar in humans and rats ([75–80], Tab. 2). In addition the metabolic pathways (Fig. 5) are similar in those two species but for some additional minor metabolites whose identities have been established in radiolabel studies in rats [81] and where there is greater possibility experimentally to determine such minor metabolites (Fig. 5). These observations are of significance since the timing, dosing conditions and correlated pharmacological studies can be considered a good basis for relating studies in rats to those in humans – an unusual situation for NSAIDs, who often have differing patterns of metabolism and faster rates of clearances in rats compared with humans.

Oxidative metabolism of meloxicam in the human liver is by cytochromes P450 (CYP) 2C9 and 3A4 [76, 82]. There are indications from recent pharmacokinetic studies in Chinese subjects that differences in metabolism of meloxicam in "slow" *versus* "fast" metabolisers may be due to genetic polymorphisms of CYP 2C9 [83].

The synovial uptake of ^{14}C-meloxicam into the joints of arthritic rats has shown that the drug has a high rate of joint penetration [81] and this is confirmed in clinical studies [84].

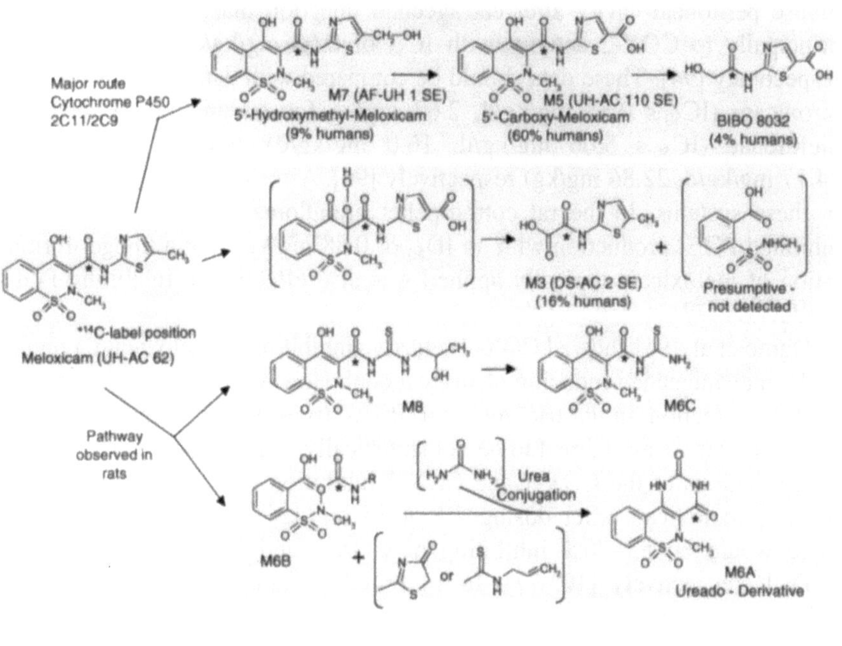

Figure 5. Pathways for the metabolism of meloxicam based on radio-label studies in rats [81] and conventional mass spectral identification of isolated in humans [75, 78–80]. The two upper pathways show the identified metabolites in humans with the compounds shown in brackets [ ] being presumed metabolites, while those in the low pathways have been additionally found in rats along with those compounds shown in the two upper pathways.

Metabolism and kinetic studies have been performed in a number of small and large animals as a basis for veterinary applications of the drug as well as being of interest in relation to comparative animal pharmacology [78, 80, 81, 85–91]. In addition to being available for use in dogs in several European countries, the oral suspension of meloxicam (Metacam®) was recently approved in the USA for use in the treatment of pain and inflammation associated with osteoarthritis [91].

Extensive clinical studies have been undertaken examining interactions with food, commonly co-administered drugs and differing routes of administration [92]. Pharmacokinetics following intramuscular injection showed that there was rapid uptake and onset of analgesia with low severity of reactions at the sites of injection [92] making this a particularly useful drug for rapid analgesia.

Pharmacological actions – COX inhibition

Meloxicam is a potent inhibitor of COX-2 *in vitro* (Tab. 1) while its metabolites do not appear to affect COX-pathways [77, 85, 93]. Confirmation of the selective inhibition of COX-2 by meloxicam has also been shown in animal models *in vivo*. Engelhardt et al. showed that orally administered meloxicam produced potent inhibition of PGE_2 production in the rat pleural cavity and mouse peritoneal cavity after carrageenan injection that is presumably due principally to COX-2 activity with IC_{50} of 0.65 mg/kg/d and 1.36 mg/kg respectively [94]. These data should be compared with data from dosing with piroxicam (IC_{50}'s 0.85 mg/kg/d, 2.62 mg/kg for rat and mouse systems) diclofenac (IC_{50}'s 5.06 mg/kg/d, 16.0 mg/kg/d) and naproxen (IC_{50}'s 14.17 mg/kg/d, 22.86 mg/kg) respectively [94]. Aspirin was virtually inactive in these systems. In the rat cotton-pellet granuloma assay meloxicam p.o. inhibited PGE_2 production with an ID_{50} of 0.88 mg/kg [77]; a lipogel formulation of meloxicam topically applied was also effective in inhibiting PGE_2 production [77].

Ogino et al. [95] showed COX-2 selective inhibition of meloxicam 3 mg/kg in the rat carageenan-inducing pleurisy model. However, they found there was a 50% reduction in gastric mucosal PGE_2 (presumably principally from COX-1) which was claimed to be not statistically significant from control.

Estimations of the COX-1 and COX-2 inhibition from plasma concentrations of meloxicam after dosing 7.5 or 15 mg/day to humans suggest that there would be 55–70% inhibition of COX-2 and 10–25% inhibition of COX-1, respectively [10, 77]. In 14 female volunteers given meloxicam 7.5 mg/day or indomethacin 25 mg t.i.d., platelet aggregation and thromboxane B_2 production were unaffected by meloxicam but almost completely inhibited by indomethacin [77]. Urinary production of PGE_2 was also not affected by meloxicam, whereas it was inhibited by 43% by indomethacin [77]. *Ex vivo* effects of meloxicam 7.5 or 15 mg given to human volunteers in the whole blood assay showed that thromboxane B_2 production was inhibited by 21 and 36% at the respective doses [77]. PGE_2 production (presumably COX-2 derived) was inhibited by 48 and 71% at 7.5 or 15.0 mg meloxicam [77].

Overall, these data from *in vivo*/*ex vivo* studies in animals and human volunteers show there is predominantly COX-2 inhibition by meloxicam, with a lesser, dose-dependent inhibition of COX-1. This could account for some inhibitory effects on gastric mucosal PGE_2 and may even be beneficial in control of COX-1 derived PGs in the analgesic pathways in the central nervous system (CNS). Enhanced uptake of meloxicam into the rat brain has been suggested to be related to binding to α_1-acid glycoprotein aiding penetration though endothelial cells of the blood-brain barrier [86]. The clinical significance of these studies is not clear but may be related to the central mechanisms of analgesia. The lack of cardiovascular events of meloxicam in clinical trials (see Chapter 7) which contrasts with the occurrence of these events in rofecoxib-treated patients, may be due some mild COX-1 inhibition and modest cardioprotection in susceptible patients and as well may contribute to the COX-1 component of central analgesia which appears of significance as well as that involving inhibition of COX-2.

A structural isomer of meloxicam, 4'-meloxicam, has effects on COX-1 as well as COX-2 in contrast to meloxicam itself [85].

Effects on oxyradical and nitric oxide (NO) production and actions

Human polymorphonuclear neutrophil (PMN) leucocytes stimulated with tumour necrosis factor-α (TNF_{α}) and formyl-methionyl-leucyl peptide (fMLP) and analysed by fluorescence activated cell sorting (FACS) for intracellular oxyradical production were found to be inhibited by 3–15 μM meloxicam in the presence of azide as an inhibitor of myeloperoxidase [96]. Inhibition of chemiluminesce from neutrophils stimulated with histone-opsonized *Streptococci* was also noted but at higher concentrations [96].

In the mouse carrageenan-induced pleurisy assay meloxicam, like that of indomethacin and nabumetone, inhibited exudate formation, PMN cell migration and myeloperoxidase at the first phase (4 h) but nitrate production was unaffected by meloxicam or indomethacin at the second peak at 48 h [97].

An *in vivo* study on adjuvant arthritic rats showed that meloxicam dose-dependently reduced the elevated levels of lipid peroxides in the plasma, glutathione peroxidase in the blood and restored to near-normal total plasma thiols and albumin, but did not alter erythrocyte superoxide dismutase [98]. It is not possible to establish from this study if there were direct effects of meloxicam on components of oxyradical production, actions as a consequence of the PG-related effects, inhibition of leucocyte recruitment or activation by the drug. No gastric lesions were observed in the stomachs of the arthritic rats with meloxicam although there was significant reduction in paw swelling [98].

Gastric lesions that develop following celiac artery clamp-induced ischaemia reperfusion in rats reach a maximum at 12 h and progress to deep ulcers at 3 days [99]. In this model, Brzozowski et al. found that there was initially reduced gastric mucosal blood flow and mucosal PGE_2 and elevated COX-2 mRNA [99]. Meloxicam, indomethacin, the COX-1 inhibitor resveratrol, rofecoxib and NS-398 increased the acute gastric lesions that develop at the early stage and delayed progression of these lesions to form ulcers [99]. The results suggested that both COX-1 and COX-2 derived PGs are important in ulcer healing [99] but there were other interpretations of these results, for example, effects on leucocyte accumulation/activation as well as effects on oxyradical production from accumulated leucocytes.

In a related study by the same group the protective effects of mild stress in rats from water immersion against subsequent severe stress were abolished by meloxicam 2 mg/kg p.o. and indomethacin 5 mg/kg i.p. as well as by NS-398 10 mg/kg p.o. coincident with reduction in protective hyperaemia, and in the case of meloxicam and indomethacin reduction in mucosal PGE_2 [100]. These results in what is a complex model of mucosal protection/ulcer formation imply blood flow/oxyradical effects as well as influence of inhibition of mucosal PG production that is elevated during the protective effects of mild stress. It is hard to discriminate the effects of meloxicam on a particular pharmacological target in these studies. Whether they have any clinical significance is obscure.

Acute gastric mucosal injury by meloxicam (3.75–30 mg/kg) in rats was found associated with increased xanthine oxidase activity but there was no increase in mucosal myeloperoxidase activity (an index of leucocyte infiltration) or the lipid peroxidase scavenging enzyme glutathione peroxidase [101].

These results suggest that mucosal injury from meloxicam may be independent of leucocytes but involve xanthine oxidase derived superoxide generation and, depending on the nature and type of stress factors, could involve inhibitory effects of the drug on mucosal PG production. Reduced mucosal glutathione concentrations may also play a role in meloxicam-induced injury [102].

Leucocyte infiltration/activation is an important target for NSAIDs including meloxicam during acute inflammation induced by agents, e.g., carrageenan, and these dugs combined with NFκB-inhibitors have been found to act synergistically on leucocytes [102, 103].

Analgesic, anti-inflammatory and antipyretic activities

Among the sites of action of meloxicam are the cytokines and leucocyte adhesion molecules. Inhibition of TNF_{α} production has been shown in lung inflammation bronchiolavage samples [104] and interleukin (IL)-8 levels in synovial fluids in antigen-induced arthritis in rabbits [105]. However, synovial production of IL-8 as well as IL-1 and IL-6 from human OA synovium in organ culture was not affected by meloxicam 0.1–10.0 μM [106]. Increased production of IL-1 which is associated with inhibition of PG production [107, 108] and which is considered a factor in promotion of cartilage damage by indomethacin in OA [109] is not evident with meloxicam.

The systemic inflammatory responses to endotoxin given i.v. were shown to be markedly reduced in heifers by prior i.v. meloxicam 0.5 mg/kg [110]. The drug produced inhibition of plasma $PGF_{2\alpha}$ metabolite, cortisol and alterations in Ca^{2+}, Fe^{2+} and Zn^{2+} induced by endotoxin but did not prevent the febrile response due to this inflammagen. Fever in guinea pigs induced by lipopolysaccharide (LPS) was reduced by meloxicam 5 mg/kg i.a. along with reduced plasma cortisol and IL-6, but TNF_{α} levels were increased [111].

Meloxicam has been found to have potent analgesic activity in conventional acute and chronic pain models in rodents [67, 75–77, 112–115]. Meloxicam has potent analgesic activity in OA or joint inflammation induced in dogs [63, 64], in chronic arthritic disease clinical trials [89, 116, 117], or in perioperative pain in dogs [118, 119] demonstrating its importance in veterinary applications in relief of pain [116–124]. Here meloxicam was comparable in analgesia with standard NSAIDs. Analgesic activity has been shown for meloxicam following surgery in rats, cats, dogs and in piglets [116–124].

It is generally considered that meloxicam has "chondroneutral" effects on cartilage in OA [125]. In organ culture of porcine, canine or human cartilage explants, meloxicam did not show any inhibitory effects on the synthesis of radiosulphate labelled proteoglycans with or without IL-1 as seen with indomethacin [64, 106], a drug which is known to accelerate cartilage destruction in OA [107]. The proportion of high molecular weight proteoglycans that were synthesized in human cartilage explants was shown to be increased by meloxicam 4 μM and substantially reduced by indomethacim 4 μM compared with controls [107]. Thus, meloxicam may be like nimesulide and aceclofenac in increasing matrix component synthesis and protecting chondrocytes against apoptosis [125]. Bovine chondrocyte production of mRNA for metalloproteinase (MMP)-1 was reduced by meloxicam, naproxen and dexamethasone but not by other NSAIDs [126]. Proteoglycanase and collagenase activities in chondrocytes were reduced by these two drugs [126]. In a rat molar model of lesion associated bone loss where IL-6, COX-2 derived PGs and MMP-1 are primarily implicated in the resorptive process, meloxicam reduced periapical bone resorption and the number of cells synthesizing IL-6 and MMP-1 [127]. These results may have relevance to the bone destruction that occurs in arthritic diseases although specific studies in animals and human models need to be undertaken to establish if these important actions of meloxicam can be confirmed along with the effects of meloxicam on oxyradicals, NO, apoptosis, COX-2, cytokines [128, 129] and uncoupling of oxidative phosphorylation [130] correlated with joint destructive changes in these models.

The mechanisms of pain control by meloxicam compared with other analgesics and NSAIDs have been studied in a number of animal models, principally in rodents. In the formalin model of nociception in rat tails which enables discrimination of central hyperalgesia from peripheral inflammation, meloxicam, lornoxicam and piroxicam produced effects on the hyperalgesia, although lornoxicam was the more effective drug of the three [131]. Single unit recordings of evoked potentials in the rat spinal dorsal horn neurons following noxious mechanical stimulation were reduced by meloxicam [131] suggesting there are direct spinal effects. In the second phase of the rat formalin test administration of the NO synthase inhibitor, N(G)-L-nitro-arginine methyl ester, (L-NAME) or the guanylate cyclase inhibitor, I*H*-(1,2,4)-oxadiazole-(4,2-a)-quinoxaline-1-one (ODQ) both blocked meloxicam induced analgesia suggesting that NO-cyclic GMP coupled reactions in the periphery contribute to the anti-nociceptic effects of meloxicam [112], this being of course in addition to its other COX-2/COX-1 effects in peripheral inflammation.

Other pharmacological effects

Like many NSAIDs, meloxicam has been found to have inhibitory effects on the growth and proliferation of rat carcinogen-induced colorectal cancer [132] and this may be related to the promotion of apoptosis [132]. Meloxicam was also found to inhibit the growth of the MAC13 tumour in mice and reduced cachexia [133] which is an important observation since not all NSAIDs exhibit this and certainly cytotoxic agents can further exacerbate weight and muscle loss in tumour bearing animals.

The relationship between inhibition of colon cancer and GI toxicity from orally-administered NSAIDs is a major issue which limits the interpretation of long-term use of NSAIDs for prevention of colon cancer. In a study in rats in which colon cancer aberrant crypt formation (ACF) was induced by 1,2-dimiethylhydrazine and a range of NSAIDs (including sulindac and its sulphone metabolite) given orally for 3 weeks it was found that meloxicam and sulindac were the most potent inhibitors of aberrant crypt formation (46%) while of the NSAIDs including sulindac sulphone and celecoxib were less potent [134]. The COX-2 inhibitors, meloxicam and celecoxib were the only drugs that were without significant gastrotoxicity [134]. Since the potency of meloxicam against ACF was twice that of celecoxib and equivalent to sulindac it would seem that meloxicam would be worth studying further in other colon cancer animal models and in humans as a potential prophylactic. Sulindac is considered by many to be among the most interesting NSAID for prevention of colon cancer because it is a pro-drug and the non-PG active sulphone metabolite is an active inhibitor of ACF. It has been suggested that sulindac itself could be a useful preventative in colon cancer. However, it is appreciably gastrotoxic and thus only meloxicam could be employed with low risk of GI toxicity. There is much debate about the mechanism of inhibition of colon cancer by NSAIDs and while COX-2 inhibition forms part of the action of some of these drugs effects on signalling pathways seem an important mechanism. This is discussed further in Chapter 11.

Neuroprotective effects of meloxicam have been shown in a few animal models. In the MPTP (1-methyl-4-phenyl-1,2,3,6-tetrahydropyridine)-induced model of Parkinson's disease in mice, a single dose of 7.5 and 50 mg/kg meloxicam i.p. significantly prevented the reduction in striatal dopamine by 36 and 40% respectively as well as almost completely preventing decline in tyrosine hydroxylase in the substantia nigra and loss of neurons in this region [135]. Aspirin 50 and 100 mg/kg i.p. had similar effects but this is the more gastrotoxic drug of the two. However, the dose of 50 mg/kg meloxicam will be well within the gastrotoxic range for this drug.

High doses of meloxicam, like that of aspirin and the TNF_{α} soluble inhibitor, etenercept, inhibited retinopathy in streptozotocin-rats by a mechanism that may involve reduction in TNF_{α}-stimulated leucocyte adhesion and COX-2 expression [136, 137]. Damage to the sciatic nerve in experimental autoimmune neuritis in rats was reduced by meloxicam, celecoxib and indomethacin but the latter was associated with significant GI and renal toxicity [138]. The authors of this study suggested that the two former drugs might be useful in the treatment of Guillain-Barre syndrome. However, the actions in this study suggest that these NSAIDs may have utility in general in those neurological conditions where there is involvement of TNF_{α} and leucocyte adhesion/activation. In conditions (including febrile reactions) where there is cytokine-mediated NO production [138, 139] as well as COX-2 derived PGs there may be neuro-protective effects, especially in conditions such as ischaemic stroke where meloxicam like other NSAIDs has been found to have protective effects in rats [140].

Cardiac ischaemia-reperfusion derived myocardial vascular injury in guinea pigs was found to reduce thromboxane B_2 production but have little other effects from meloxicam 0.3–3.0 μM whereas aspirin 100 μM had marked injurious effects [141]. Anti-fibrinogen activity has also been observed with meloxicam 0.065 mg/kg/d given for 10 days to rats exposed to various injuries following multiple laparotomies [142]. This may have some relevance to prevention of fibrin formation in cardiovascular disease.

Exercise-induced apoptosis in rats has been found to be reduced by meloxicam 11 mg/kg [143] adding further protective properties to the list of those where ischaemia may, with anti-apoptotic effects, be an important basis for protective effects of the drug.

GI and other side-effects

Meloxicam has relatively low acute GI irritancy in rats in relation to its anti-inflammatory actions when compared with that of other NSAIDs [77, 144, 145]. The low acute gastric irritancy of meloxicam in rats has been found to be related to the low potential for the drug to inhibit gastric PGE_2 production [95] although inhibition of COX-2 at the ulcer margin and increase in pro-inflammatory cytokines may contribute to promotion of the injury from meloxicam [146]. While oxyradical generation from ischaemia-reperfusion injury enhances gastric mucosal injury from meloxicam and other NSAIDs [98] with intestinal injury this may be related to drug induced effects on xanthine as well as gastric oxidase activity [101, 145]. Coadministration of an α_2-adrenergic receptor agonist (tizanidine) with meloxicam and some other NSAIDs reduced gastric lesion development in rats [115] suggesting that there may be effects of these drugs on acid secretory functions that are controlled by α_2-adrenergic receptors in the CNS.

Sub-chronic treatment with meloxicam 3.75 or 5 mg/kg/d to rats (14 or 28 days) was not associated with neutrophil infiltration into the GI mucosa or elevations in plasma transaminases indicative of liver injury or creatinine or urea reflecting renal effects as seen with piroxicam 5 mg/kg/d [147]. As with other NSAIDs, meloxicam uncouples oxidative phosphorylation [130] and the consequent reduction in mucosal or renal ATP may contribute to development of damage in these organs. Like other COX-2 selective or unselective inhibitors, meloxicam can impair the PG-regulated renal perfusion and this drug and indomethacin have been shown to reduce cortical blood flow but not mean arterial pressure in euvolemic rats [148]. However, meloxicam 1–2 mg/kg i.v had no effects on renal and cortical blood flow in furosemide Na^+-deficient hypovolemic rats whereas indomethacin 5 mg/kg i.v. compromised both renal components of blood flow [148]. These results are suggestive of some renal haemodynamic effects of meloxicam compared with other NSAIDs although their effects on renal cortical and medullary blood flow may contribute to renal effects of this drug in susceptible individuals.

In accordance with the properties of being a preferential COX-2 inhibitor, meloxicam shows a low incidence of severe adverse GI events in large-scale clinical trials and in a large-scale meta-analysis [61, 62, 149, 150].

A recent prescription-event monitoring study comparing the GI toxicity in general practice in the UK of meloxicam with rofecoxib showed that both drugs had the same incidence of gastric peptic ulcer bleeds (0.4%) and had

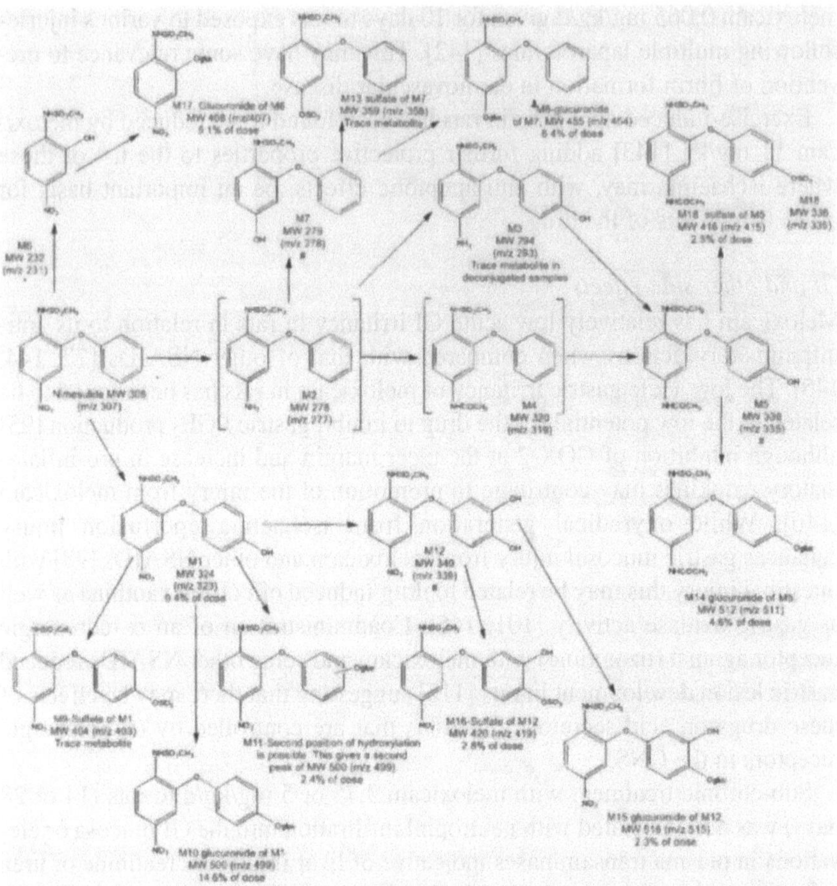

Figure 6. Pathways for the metabolism of nimesulide in humans based on studies with the radiolabelled drug [166, 176] (unpublished studies by MacPherson et al.). Taken from [166], with permission:

7.2% and 7.4% incidence of acid/peptic symptomatic events respectively [151]. These results are of interest in view of the substantial data on rofecoxib [8, 12, 152, 153], a low GI risk NSAID, since they show that under general practice conditions the preferential COX-2 inhibitor meloxicam has the same GI ulcer/bleeding/symptomatic profile as the highly selective COX-2 inhibitor, rofecoxib.

Nimesulide

This widely-used drug belongs to the class of sulphonamide COX-2 inhibitors [154–156] that were progenitors of some of the modern coxibs [155]. Its chemical structure (Fig. 6) is 4-nitro-2-phenoxy methane sulphonanilide and it has a pKa of 6.4–6.8 [154, 156, 157]. Nimesulide (Aulin™, Mesulid™, Nimed™ and several other trade mark names) was first developed by Riker Laboratories Inc., USA as R-805 based on their pioneering research on NSAIDs with antioxidant activities [157, 158]. In rodents and pigs this drug was found to have very low GI mucosal injury and this was a major factor in its selection for further development [158]. It was first introduced in Italy in 1985 (Helsinn Healthcare SA, Switzerland) and is now marketed successfully in over 50 countries (except in North America, Japan and UK), where in some it is a product leader being the fifth largest selling NSAID worldwide [159]. A large number of generic formulations of the drug have been developed for sale especially in Italy and India, the pharmacokinetic properties of some which are not of as high standard as the main marketed drug. A considerable number of reviews have been published on the pharmacology, pharmaceutics, pharmacokinetics, uses and safety/toxicology of nimesulide in rheumatology and other painful or inflammatory states [154, 157, 159–169]. Nimesulide is principally indicated for use in relief of pain and inflammation associated with OA and other musculo-skeletal conditions (except RA) as well as in acute pain (e.g., dental, dysmennorhea) [159, 161, 162, 166–174]. Being of relatively low cost this drug has particular advantages from a safety-pharmaco-economic viewpoint [175]. Nimesulide was shown to be the first drug to have COX-2 selectivity in humans *ex vivo* and low GI injury observed by endoscopy and in intestinal permeability studies [171].

Pharmacokinetics

The clinical pharmacokinetics in humans of nimesulide (see Tab. 2) have been comprehensively reviewed by Bernareggi [164]. The drug is rapidly absorbed with a mean C_{max} of 2.86–6.50 μg/ml being achieved 1.22–2.75 h after an oral dose of 100 mg tablets. This rapid absorption accounts for the rapid onset of analgesia of the drug [160]. The values for the volume of distribution, V_D, range from 0.1–0.2L/kg and the drug is bound to plasma proteins to 96–99% principally to albumin but also there is weak binding to α_1-acid glycoprotein and γ-globulin. The drug has a short mean elimination half-life, $t^1/_2$ = 1.8–4.73 h and low clearance with a hepatic extraction ratio Cl/F of 0.1 reflecting low impact of metabolic load in the liver despite diverse metabolic pathways (Fig. 6). The low $t^1/_2$ may be a factor accounting for the low GI adverse reactions from the drug along with lack of accumulation in the elderly [164, 173].

The principle metabolism of the drug is initially hydroxylation to 4'-hydroxynimesulide by the activity of the cytochromes P_{450} 2C9 and 2C19. This is the main metabolite detected in plasma and in a free or conjugated form in the urine [164, 166, 176]. Phenolic glucuronides and sulphates are formed [176]. Minor metabolic routes include reductive amination of the nitro-group and subsequent N-acetylation [166, 176] and it has been speculated that nitroso- and hydroxylamine metabolites may be formed during the reductive amination with subsequent covalent binding of these metabolites [166, 177, 178] but without any evidence. Further speculation that these covalent metabolites may account for liver injury from the drug [177, 178] is also without evidence. Nimesulide, like that of several other NSAIDs, has been shown to inhibit human liver phenol sulphotransferases [179].

Hepatic, but not moderate renal insufficiency or impairment, affects the kinetics of nimesulide [164]; the limited effects of mild-moderate renal impairment on the kinetics of the drug may reflect the capacity of about half of the drug to be eliminated in the faecal route [164]. Interestingly, there are no apparent or clinically significant interactions of nimesulide with commonly co-administered drugs (e.g., antacids, cimetidine, glibenclamide, furosemide, theophylline, warfarin or digoxin [164].

The kinetics and distribution of ^{14}C-nimesulide in rats is similar to that in humans [164, 180] indicating that as with meloxicam this enables comparison of effects of the drug in rats with that in humans even though the clearance rates and formation of minor hydroxylation metabolites may be greater in rats that in humans.

Pharmacology

The pharmacological effects of nimesulide have been comprehensively reviewed elsewhere [154–156, 158–160, 162, 164, 166, 173]. A comparison of acute anti-inflammatory effects in the carrageenan paw oedema model and acute gastric irritancy in rats of nimesulide with that of other established NSAIDs is shown in Table 5 [156]. It is apparent that there is a favourable therapeutic index with nimesulide compared with other NSAIDs in these acute models and this is confirmed in more recent studies [160, 170]. *In vitro* effects

Table 5. Acute anti-inflammatory and gastric irritancy effects of nimesulide on rats and calculated therapeutic indices

Drug	ED50 (mg kg^{-1})^{a}	LD50 (mg kg^{-1})^{b}	Therapeutic index
Nimesulide	1.25	324 (295–356)	260
Naproxen	2.1	395 (281–557)	190
Ibuprofen	13.5	923 (833–1020)	68
Difumidone	38	750 (694–811)	20
Flufenamic acid	14.7	249 (221–280)	17
Phenylbutazone	29.5	406 (375–440)	14
Acetylsalicylic acid	135	1520 (1360–1710)	11
Indomethacin	2.95	21 (19–23)	7

Data from [156, 157]. [a]Calculated in carrageenan-induced rat paw oedema from the 3-point regression of response on dose, and [b]Calculated by the method of Litchfield and Wilcoxon (95% confidence limits are given in parentheses).

of nimesulide and its metabolites have been compared in relation to the therapeutic concentrations of the drug achieved in plasma in patients given oral doses of the drug (Tab. 6) [154, 160].

Based on assessments of the actions of nimesulide at therapeutic concentrations or slightly higher it is apparent that this drug has multifactorial activities (Tab. 6) [154, 160, 164, 166, 167]. Aside from having mostly COX-2 activity with some COX-1 inhibitory effects which may be relevant for analgesia [154] nimesulide inhibits (a) the adherance and migration of neutrophils and their

Table 6. Pharmacological effects of nimesulide relevant to anti-inflammatory activity

I. Actions of nimesulide at normal doses or concentrations, or at therapeutically relevant plasma concentrations *in vitro* (up to 0.06 $\mu g\ ml^{-1}$; ~0.2 μM in the absence of albumin). All the effects listed are inhibitory.

Pathway	Man	Lab animals	Cells *in vitro*	Dose/conc
COX-2 activity	leucocytes *ex vivo*			100 mg b.d.
			human leucocytes	Therapeutic range
Superoxide formation	leucocytes *ex vivo*			
Histamine action	skin *in vivo*			200 mg
Histamine release		guinea-pigs		1.6 $\mu mol\ kg^{-1}$
Histamine release		guinea-pigs		0.1–1 $mg\ kg^{-1}$ i.v.
Cytokine action		Rats		7 $mg\ kg^{-1}$
COX-2 formation			human synovocytes	0.03 $\mu g\ ml^{-1}$
Metalloprotease formation			human synovocytes	0.03 $\mu g\ ml^{-1}$
Chondrocyte apoptosis		Rat	chondrocytes	1 pM–1 μM

II. Actions of nimesulide *in vitro* at a moderately high concentration[1]

Pathway	Cells/tissue	Nimesulide	Serum %	Effect %
Urokinase synthesis	Synovial fibroblasts	0.3 $\mu g\ ml^{-1}$	1	50
Interleukin 6 synthesis	Synovial fibroblasts	0.3 $\mu g\ ml^{-1}$	1	50
Plasminogen activator inhibitor	Synovial fibroblasts	0.3 $\mu g\ ml^{-1}$	1	+50
Histamine action	Bronchial muscle	0.3 $\mu g\ ml^{-1}$	0	–20
Cyclic AMP	PMN leukocytes	1 μM	0	+40

[1] Experiments are required with lower concentrations *in vitro* or in human subjects with therapeutic doses, to investigate whether the above effects can be put into the 'therapeutically relevant' class as defined in this chapter.
From Bennett and Villa [159, 160].
Data from references in this review.

MPO enzyme and oxyradical activities, (b) release of histamine from mast cells, (c) release of leukotriene C_4 and platelet activating factor (PAF) from eosinophils, (d) production by neutrophils of the "second messenger", cyclic AMP, (e) production of cytokines (IL-1, IL-6 and possible TNF_{α}) by synovial cells, (f) production by synovial cells of NO, (g) inhibition of caspase-3 mediated apoptosis and (h) displacement of tryptophan from binding sites of albumin which, being a precursor to 5-HT (serotonin), may affect the production of increased 5-HT in the efferent neural pathways that mediate analgesia in the CNS [154, 180].

High concentrations (100 μM) of nimesulide have been found to protect against apoptosis and inhibit cytokine-mediated cartilage resorption but there are no effects on the synthesis of proteoglycans by this drug in the presence or absence of pro-inflammatory cytokines (IL-1, TNF_{α}) [181–183]. Other cartilage protective effects of nimesulide include stimulation of the synthesis of plasminogen activator inhibitor, decreased production of collagenase and other metalloproteinases, urokinase and inactivation of α_1-antitrypsin [154]. Cartilage uptake of radiolabelled nimesulide in the presence and absence of pro-inflammatory cytokines [183] is appreciable and may account for any putative protective effects of the drug on cartilage destruction.

Thus, the pharmacological actions of nimesulide that are relevant to its anti-inflammatory, analgesic and antipyretic effects involve effects on the production of a wide range of inflammatory mediators and those cells involved in the focus of inflammatory changes that occur in the joints of arthritic patients. The anti-histamine and inhibitory effects on leukotrienes and PAF are suggestive of some anti-allergic activity. While this has not been directly demonstrated it is interesting that there is evidence of some tolerance to the drug in NSAID-sensitive asthmatics [184, 185]. It is possible this is related to the effects of the drug on leukotrienes and PAF, as well as resulting in a balanced inhibition of inflammatory reactions that are manifest in allergic states. Further studies would be required to establish if there is some tolerance or antiallergic effects of nimesulide.

Molecular models of the interactions of docking of nimesulide with the active site of COX-2 have established the key roles played by valine-509 in COX-2, which is analogous to an isoleucine-523 residue in COX-1, as well as some other key amino acids [186–188]. Another interesting feature about the interactions of nimesulide with COX-isoforms was highlighted in the earlier studies by Rufer et al. [189] and work from the Riker group [190] on the interactions of nimesulide with the peroxidative (PEROX) system. Based on the original studies with phenolic compounds, e.g., MK-447 with seminal vesicle PGHS preparations, it was proposed that these act to scavenge or capture the oxyradical that is released in the conversion of PGE_2 to PGH_2 while nimesulide and some other sulphonanilides additionally act to block the initial cyclooxygenase reaction according to their respective pKa values [188]. These dual actions of nimesulide on both COX and PEROX reactions may be of significance in relation to its actions in inflammation and its low GI irritancy.

The antioxidant character of nimesulide which has been demonstrated in a number of systems [190–194] may contribute to its effects on the PEROX system. The mechanism of anti-oxidant effects of nimesulide is possibly related to the formation of its 4'-hydroxy metabolite which has been proposed to have a chain-breaking action with peroxy-radicals, as well as $OH^{\cdot}$ and $O_2^{\cdot -}$ [191].

In addition to having inhibitory effects on LTC_4 production [154], nimesulide also inhibits LPS-stimulated LTB_4 production in human leucocytes in the range of 10–100 μM as well as in human gastric explants [195]. This balanced inhibition of both 5-LOX as well as COX pathways by nimesulide may as with other dual COX-LOX inhibitors be important for its benefits as an anti-inflammatory as well as being of significance in its low GI irritancy [196].

Variable concentration-dependent effects of nimesulide have been shown on platelet aggregation [197]. At low concentrations (0.01–0.1 μM) nimesulide potentiated aggregation induced by sub-threshold concentrations of adrenaline, the effects of which were blocked by calcium antagonists and the NO-donor SNAP. At higher concentrations (1.0–100 μM) nimesulide blocked platelet aggregation induced by adrenaline and inhibited thromboxane B_2 production but the drug did not affect platelet aggregation induced by collagen, ADP, calcium ionophore or arachidonic acid. Whether or not these effects of nimesulide on adrenaline-induced platelet aggregation have any consequence for potential cardioprotective effects of the drug or in potentiating gastric bleeding in susceptible individuals is not known but should be investigated further.

Particular interest has focussed in recent years on the potential for NSAIDs to have central as well as peripheral effects in analgesia. In the formalin test in rats the injection into the formalin treated paws of the NO synthase inhibitor, L-NAME, or the soluble guanulate cyclase inhibitor, ODQ (H-1-(1,2,4)-oxadiazolo (4,2-a) quinoxalin-1) blocked the antinociceptive response to nimesulide [198]. Furthermore, the analgesic effect was potentiated by the NO-donor, SIN-1 (3-morpholino-syndnonimine). The results suggest that the peripheral analgesic effects of nimesulide and possibly also those in the CNS mediated by NO may be affected by the drug at the level of NO activity. Nociceptive effects of nimesulide have been shown from both intrathecal as well as intraperitoneal administration to mice [199] indicating this drug, like other NSAIDs, has central as well as peripheral analgesic effects [200].

The renal haemodynamic effects of noradrenaline have been shown to be potentiated by COX-2 inhibition from nimesulide and inhibition by L-NAME suggesting there are interactions between COX-2 and NO [201]. Whether these vascular actions have significance in other systems, e.g., vascular headache has yet to be determined. In a model of post-operative ileus in rats nimesulide like other COX-2 inhibitors overcame the reduction in GI transit from laparotomy but L-NAME and an iNOS inhibitor blocked the effects of the COX-2 inhibitors [202]. These results, overall, suggest that interactions between COX-2 and NO vary according to the system under investigation.

Uncoupling of oxidative phosphorylation has been shown with nimesulide, as with other NSAIDs [203–205]. Mitochondria from the livers of adjuvant arthritic rats showed greater sensitivity to the uncoupling effects of nimesulide [203]. The uncoupling effects of nimesulide evident with an IC_{50} of 5 μM were paralleled by Ca^{2+} efflux [204]. The reduction of the nitro group to an amine in nimesulide abolished the uncoupling effects of nimesulide [204]. It is not known if these uncoupling effects of nimesulide with consequent reduction in cellular ATP have any relationship to its anti-inflammatory actions. However, it has been suggested that the hepatotoxicity of nimesulide could be related to reduction in ATP production by the drug [205]. Unfortunately, the cell viability studies performed in the latter study were demonstrated at concentrations of 0.1 to 1.0 mM which exceeds concentrations that have been regarded as within the pharmacological or realistic toxicologic ranges. Other cytotoxicity studies performed at concentrations up to 100 μM did not reveal any cytotoxic effects of the drug, its major metabolites or manufacturing impurities [206].

Production of aspirin-triggered 15-epi-lipoxin A_4, which has been implicated in the anti-inflammatory actions of aspirin, and which is produced by liver cytochrome P_{450} activity, was not found to be produced by nimesulide [207].

Nimesulide has been shown to initiate apoptosis in a number of tumour cell lines [208–212] which may be related to its inhibitory effects on the growth and proliferation of chemically-induced tumours in rodents [213–216], in cell lines [217], in choline deficient F-344 male rats in whom hepatocarcinomas develop [218] and xenotransplanted melanomas in mice [219]. Paradoxically, however, nimesulide as well as ibuprofen has been shown to protect isolated chondrocytes against staurosporine-induced apoptosis [181, 220]. Thus, the state of cellular controls may dictate whether or not apoptosis is induced by this drug, or indeed possibly any other NSAID.

Membrane calcium flux is affected by nimesulide [221] and this with its inhibitory effects on peroxisomal proliferator-activated (PPAR) induction of COX-2 [222] may account for effects of the drug on PG production aside from the direct inhibition of enzyme activity. The linkage of this effect of nimesulide on PPAR induction does not appear to translate to effects on apoptosis [49]. Nimesulide is also a moderately effective inhibitor of neutrophil L-selection [49].

Cartilage protective effects of nimesulide may involve direct inhibition of cytokine-induced COX-2 [223] and the range of previously mentioned inhibitory actions on IL-1, TNF_{α}, IL-6 (but not IL-8), oxyradical and NO production, MMP's and cyclic nucleotide production [154, 159, 160, 181, 183, 190–194]. Translation of these *in vitro* effects into changes observed by magnetic resonance imaging in cartilage of patients with osteoarthritis does not appear to have revealed any direct effects of 24 weeks of treatment with nimesulide 100 mg b.i.d. compared with piroxicam 20 mg or placebo [224]. The small scale of this trial and limited duration of treatment did not allow for clinically-significant changes in cartilage structure to be identified, but there were significant changes in joint function observed with both drugs, with greater efficacy evident with the patients treated with nimesulide.

Adverse events and toxicity

A considerable number of studies have examined the pattern of adverse reactions attributed to nimesulide including regional pharmaco-epidemiological or pharmacovigilance studies [225–227] and critical evaluation of case reports to the central database of the company producing and marketing the drug [165, 173]. The relatively low incidence or frequency of cases of severe GI adverse reactions compared with those from other NSAIDs is shown in this data. The occurrence of hepatic reactions which have been reported in a number of published case reports appears no more than that of other NSAIDs [165, 173, 225–227] and is probably a class effect of the drug. Most reports are unspecific increases in the plasma levels of liver transaminases and less frequently alterations in liver function tests. Often concurrent intake of other hepatotoxic drugs (e.g., clavulanic acid, cephalosporin antibiotics, paracetamol, diclofenac) or prior or concurrent liver disease including that related to alcohol abuse has been implicated in many of the cases [165, 173, 228–230]. Skin reactions occur infrequently and are mostly minor rashes, urticaria and eruptions; very rarely have there been reports of Stevens-Johnson or Lyell's syndromes [231–234]; which are typical of those from NSAIDs [232, 234, 235], including the development of atopy being a risk factor [232]. Cases of lichenoid photodermatitis have been reported [236]. Cutaneous reaction rates to challenge test with nimesulide in patients with NSAID-related urticaria and angioedema have been observed in 21.3% of the NSAID-sensitive patients, but were 17.3% with meloxicam, 33.3% with celecoxib and 3.0% with rofecoxib [237], showing that there are appreciable differences in cross-reactivity among the COX-2 selective drugs in NSAID-sensitive patients. Using the standard local lymph node assay nimesulide has not been found to be a sensitizer [238].

Asthma and renal injury have also been observed infrequently with nimesulide [226, 239] and can be regarded as class effects of these drugs with the latter being related to effects on PGE_2-related renal haemodynamics and renal excretion [240].

The low incidence of GI ulcers and bleeding with nimesulide has been related to the relative sparing of COX-1 inhibition by the drug in the stomach and intestinal mucosa [171, 226]. This may also be related to the relatively high pKa of the drug and differences in the pattern of uptake into gastric mucosal cells compared with the more acidic NSAIDs [73].

Aside from the prevention of carcinomas, there may be potential beneficial effects of nimesulide on Alzheimer's disease [241] and other neuroprotective effects but these have yet to be extensively explored in animal models and humans with the diseases.

Coxibs

The COX-2 selective inhibitors known as coxibs were developed with the objective of reducing the risks associated with COX-1 inhibition in the GI tract, kidneys and some other systems where COX-1 inhibition was suspected as being the basis of the development of adverse reactions. This section reviews the clinical and experimental pharmacology and pharmacokinetics of this group of drugs and highlights principal toxicological and adverse reactions that have been associated with the coxibs.

Celecoxib

Celecoxib (4-[5-(4-methylphenyl)-3-(trifluoromethyl)-1*H*-pyrazol-1-yl]benzenesulphonamide; Celebrex™ (SC-58635); Figure 7) is a diaryl substituted pyrazole with a pKa of approximately 11.1 and a high degree of lipophilicity. This drug was the first of the coxibs to be introduced in 1999 by Searle/Pharmacia (now part of Pfizer Inc, USA) and was approved by the Food and Drug Administration in the USA and other countries for the treatment of signs and symptoms of RA, OA and the management of acute and chronic pain [242, 243]. The selection of celecoxib was based on an extensive structure-activity analysis of sulphonamide derivatives that had been optimised to once or twice day dosage [244]. The basis for COX-2 selectivity has been described in a number of studies [17, 242], and as with a number of other COX-2 selective and nonselective NSAIDs the enhanced supply of arachidonic acid *in vivo* can overcome some of the inhibitory effects of these drugs [245].

In vivo pharmacology was undertaken by the company that developed celecoxib (Searle) and the anti-inflammatory, analgesic, anti-pyretic and chemopreventive effects were studied in standard models principally in rodents and can be found in the NDA submission to the FDA [242].

Celecoxib has acute anti-inflammatory effects in the carrageenan paw oedema model, with an ED_{50} 7 mg/kg p.o. Chronic anti-inflammatory activity was tested in the adjuvant induced arthritis model in male Lewis rats dosed with 0.03–3 mg/kg b.i.d. p.o. with celecoxib for 10 days and a dose-dependent reduction in paw swelling with an ED_{50} of 0.3 mg/kg/d was obtained. Reduction in cartilage destruction, bone lysis, bone proliferation, soft tissue, swelling and synovial inflammation was observed in animals dosed with 1 mg/kg p.o. for 10 days. The greatest effects of the drug were evident in soft tissue oedema, cartilage destruction and bone lysis, the other effects being of lower intensity.

In vivo reduction by celecoxib of PGE_2 and 6-keto $PGF_{1\alpha}$ was observed in the carrageenan-induced air pouch in male Lewis rats with an ED_{50} of 0.2 mg/kg. Another study in the same model in fasted Lewis rats, showed dose-dependent reduction in PGE_2 in the pouch with an ED_{50} of 0.97 mg/kg, which is nearly five times less than that in the fed animals in the previous study. No explanation has been given for these differences but they perhaps highlight the role of fasting in drug bioavailabity. In the same study there was a "non-dose dependent" reduction in gastric mucosal PGE_2 production by 20–40% over the entire dose range of 0.1–200 mg/kg. Such a substantial reduction in PGE_2 production in the stomach may represent a partial inhibition of COX-1 inhibition by celecoxib.

Thus, the results obtained in the *in vivo* assays that formed part of the NDA [242] in the carrageenan air pouch and gastric mucosal assay suggest that there is some marked COX-1 inhibition although it is not as great as COX-2 inhibition by celecoxib. The inhibition of PGE_2 production in adjuvant arthritic rats, paw exudate or synovial fluid indicated that a higher dosage of 3–10 mg/kg was required to achieve between 50% and 60% inhibition of PGE_2 production [242]. Thus, differences in potency may be related to the degree of inflammation and chronicity, for in the acute carrageenan air pouch model the IC_{50} was 0.2 mg/kg in contrast to the much higher doses that were required in the chronic inflammatory model adjuvant arthritis in rats, where even doses of 10 mg/kg did not completely inhibit PGE_2 production. In comparison with published studies of NSAIDs in the standard carrageenan and adjuvant arthritis models it could be estimated that the inhibitory potency of celecoxib ranks approximately to that of naproxen.

Analgesic activity of celecoxib was investigated in the Hargreave's hyperalgesia model in Sprague-Dawley rats, which showed a dose-dependent reduction with an ED_{50} of 35 mg/kg. In the formalin paw injection model in Swiss-Webster mice, celecoxib showed an inhibition of between 67% and 88% in the dose range of approximately 30–50 mg/kg. In the phenyl benzoquinone (PBQ)-induced doxoflexion model in Swiss-Webster mice there was a 56% inhibition with 5 mg/kg p.o. There was a dose-dependent reduction in doxoflexion by 50–500 mg/kg p.o. celecoxib, in the acetic acid-induced writhing model in ICR mice.

Overall, these results show that the doses required for inhibition of PGE_2 production *in vivo* following inflammatory stimuli are much lower than those that are required for exhibiting analgesic activity. In another study by Smith et al. [246] the dose for reduction of carrageenan-induced paw oedema by celecoxib appeared to be similar to that for hyperalgesia. It is clear, therefore, that there are differences in the relation between anti-inflammatory compared with analgesic responses that depend on the model and type of stimuli used.

Anti-pyretic effects were investigated using the LPS-induced *in vivo* model in rats wherein it was found that celecoxib 30 mg/kg reduced the LPS fever by 42% but did not affect normal body temperature [247].

As evidence of chemoprotective activity celecoxib was administered to F344 rats that had been treated with azoxymethane to induce colonic aberrant cryptic foci and tumours in comparison with the effects of sulindac (which is regarded as an effective drug for the treatment of this condition) [242]. Following 52 weeks treatment with celecoxib in the diet there was a marked reduction in the numbers of adenocarcinomas and their volume in the treated rats but the numbers of crypts formed after 12 weeks treatment with celecoxib appeared unchanged; there also appeared to be no changes affected by sulindac.

Of the safety pharmacology that was undertaken as part of the NDA [242] there were some minor behavioural changes exhibited at relatively high doses of the drug (500 mg/kg), effects on increasing the hexobarbital induced sleep, reduction in electroshock and chemical induced convulsions but no effects on body temperature. There are other minor effects on the autonomic nervous system and smooth muscle involving reduction in acetylcholine and histamine induced contractions in the isolated guinea pig ileum, no effects on the cardiac or respiratory systems in dogs but there was a slight reduction in urine volume with increased urinary osmolality in rats concomitant with a reduction in renal PGE_2 production.

Single and repeated doses of celecoxib were evaluated in standard toxicology models with the GI tract and kidneys as the primary end organs of toxicity (Tab. 4) [242]. No mutagenic activity was observed *in vivo* or in the Ames' tests and there were no indications of untoward effects in special toxicology (e.g., antigenic assays).

Pharmacokinetics

The pharmacokinetics of celecoxib in humans has been reviewed by Davies et al. [248] (see Tab. 2). The essential features of drug pharmacokinetics are that it is well and completely absorbed with peak plasma concentrations being reached within 2–4 h after oral administration. The drug has a long and variable beta-elimination half-life that can range between 4.5 and 16 h. Like all NSAIDs, celecoxib is extensively protein-bound (<97%) to plasma albumin principally. However, the volume of distribution (V/F) is 455 L (5.7–7.1 L/kg) which is some five to seven times greater than most NSAIDs. This extensive volume of distribution may be related to the lipophilic nature of celecoxib as well as to its high pKa (11.1). The rate of absorption of the drug in humans is similar to that in rats and dogs.

The metabolism of celecoxib in humans and animals is principally through cytochrome P_{450} 2C9 mediated metabolism to the hydroxymethyl and subsequent carboxyl derivatives (Fig. 7) [249–253]. The Phase II metabolites include four positional isomers of the acyl glucuronide conjugates of the carboxylic acid metabolites [253]. It appears these metabolites are formed in appreciable quantities in rats [251] and rabbits [253]. The principle excretion

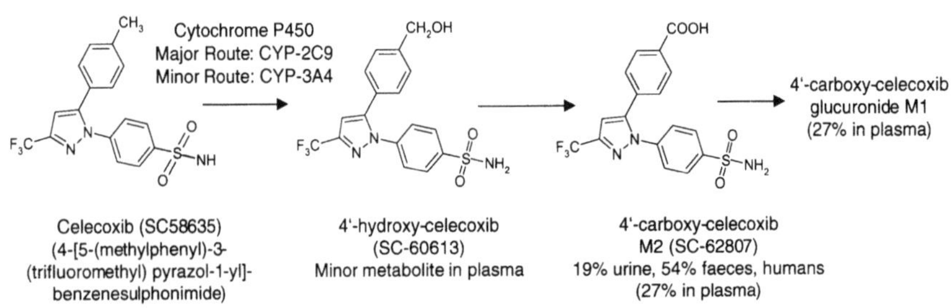

Figure 7. Pathways for the formation of the principal metabolites from celecoxib in humans [248, 249]. Oxidative metabolism via cytochromes P-450 2C9 and 3A4 is the major route of metabolism of this drug.

in the rat is in the faeces from both the intravenously administered and the orally administered radiolabelled compound [251]. In humans the carboxylic acid metabolite (SC-62807) constitutes 19% of the total dose excreted in the urine and 54% in the faeces, respectively. The hydroxymethyl and carboxylic acid metabolites constitute about one-quarter each of the total metabolites observable in plasma while unmetabolised celecoxib constitutes about one-half of the total drug present in plasma. In contrast to the situation in rats there is a much lower rate of excretion of the drug and its metabolites in the faeces in humans. However, the predominant faecal excretion of the drug, especially observed in rats, which occurs with a number of acidic NSAIDs, raises the possibility that intestinal injury may be a feature of celecoxib. To establish if this occurred Tibble et al. [254] compared celecoxib with the effects of orally administered indomethacin, which is well-known to cause intestinal mucosal changes in permeability, lesions and ulcers proceeding to peritonitis in the rat. Indomethacin also causes uncoupling of mitochondrial oxidative phosphorylation which is considered to part of its "topical" effect. Celecoxib did not cause any changes in intestinal permeability, uncoupling of oxidative phosphorylation or mucosal inflammation-injury, whereas indomethacin did [254]. These studies are important for highlighting the absence of the potential for developing intestinal injury even though some lesions were observed in long-term toxicity studies as cited in the NDA submission [242]. Since the systemic bioavailability of celecoxib in dogs when instilled into the duodenum, jejunum or colon was the same as that when given intragastrically, it appears that the absence of intestinal inflammation-injury cannot be due to poor absorption of the drug from the intestinal tract [255]. The absence of intestinal injury which is an important safety feature of both celecoxib as well as rofecoxib appeared to be related to the sparing of COX-1 in the intestine since there were no effects on PGE_2 content of the intestinal mucosa 24 h after oral administration of the drug or of rofecoxib [256]. It is interesting that these studies showed that the selective COX-1 inhibitor SC560 also failed to cause mucosal injury but when SC560 and rofecoxib were given simultaneously intestinal injury occurred [256]. It appeared from these studies that the administration of SC560 alone caused upregulation of COX-2 in the small intestine and this may have had a partial protective effect even though COX-1 induced injury may have been evident with this drug [256].

The predominance of celecoxib metabolism through cytochrome P_{450} 2C9 raises issues about the potential for drug interactions and the possibility of unequal metabolism through different genetic variants of this cytochrome [248, 250]. The occurrence of polymorphisms of cytochrome P_{450} 2C9 may account for the variable hepatotoxicity that has been reported with this drug. Interestingly slow and fast metabolisers of celecoxib in beagle dogs have been related to novel variants of the cytochromes P_{450} 2D15, 3A12 and variance of the 2D15 predominantly accounted for the polymorphism in canine celecoxib metabolism [250]. The relationship of this to hepatic injury which has been reported with this drug [257] has not been investigated, at least in dogs.

Pharmacology and toxicology

Some of the aspects concerning the pharmacology and toxicology of celecoxib have been discussed in the earlier part of this section but here it is useful to illustrate aspects relating to the mechanisms of celecoxib in inflammation, algesia and the toxicity of the drug.

An interesting study comparing the effects of celecoxib or rofecoxib with those of indomethacin, dexamethasone and the COX-1 inhibitor SC560 was performed in rats in which acute inflammation was induced by injection of carrageenan into the hind paws or pleural cavity [258]. There was differentiation of the effects of COX-1 and COX-2 inhibition inasmuch as the paw volumes, though reduced by all the drugs except the COX-1 inhibitor SC560, were accompanied by a reduction in PGE_2 content in the pleural exudates, and a reduction in NO by dexamethasone and indomethacin but not by the COX-2 inhibitors. Also, in the rat hind paw dexamethasone, indomethacin and SC560, but not the two COX-2 inhibitors inhibited myeloperoxidase activity, an index of leucocyte infiltration in the hind paw. Of the two pro-inflammatory cytokines investigated, indomethacin increased IL-1 production while dexamethasone decreased IL-1 and TNF_α whereas the two COX-2 inhibitors had no effect on the concentration of these cytokines in the carrageenan-induced pleural exudate. In contrast, in the paws indomethacin, dexamethasone and celecoxib, but not rofecoxib, decreased the concentration of IL-1 whereas indomethacin and rofecoxib increased TNF_α production, while dexamethasone decreased its production. These somewhat complex results reflect potentially differences in the actions of these drugs on different components of the acute inflammatory process. It must be said, however, that the lack of full dose-response studies makes it difficult to provide an overview of what relative effects these drugs have on the different components of acute inflammation. In addition, the selective nature of the experiments in rat paws as compared with those in the carrageenan-induced pleural inflammation has undoubtedly contributed to the complexity of the results that have been obtained. Furthermore, the lack of time-dependent studies also mitigates against clear interpretation of the effects of the drugs. Nonetheless there are indications from these studies that there may be weaker effects of celecoxib and rofecoxib on acute inflammation compared with that of potent established drugs such as indomethacin and dexamethasone but this obviously requires further investigation.

In an acute model of lung injury induced by carrageenan administration in rats, celecoxib administration reduced the infiltration of PMN leucocytes and subsequent lipid peroxidation, reduced PGE_2, TNF_α and IL-1β production and the upregulation of adhesion molecules ICAM and P-selectin and the production of nitro-tyrosine [259]. This is an interesting study for highlighting the multiple actions of celecoxib which may be related in part to its properties as a sulphonamide rather like that of nimesulide which is also a sulphonamide derivative.

An interesting differentiation between the anti-oedemic effects of COX-2 inhibitors and other NSAIDs and the effects on hyperalgesia or hypoalgesia were shown in the study by Francischi et al. [260] using a model in which carrageenan was injected into the hind paw and the hyperalgesic response was measured 24 h later concomitant with anti-oedemic effects being measured by plethysmography. It was shown that celecoxib abolished the inflammatory hyperalgesia and induced hypoalgesia without reducing the oedema at the doses given. Similar differentiation was established with the other COX-2 inhibitors, rofecoxib and SC236, but the results were less marked than for celecoxib. In the formalin footpad nociceptive model, peripheral administration of celecoxib did not produce anti-nociceptive effects whereas diclofenac and the COX-1 inhibitor, resveratrol, did produce dose-dependent anti-nociceptive effects during the second phase of the formalin-induced nociception [261]. Again, as with the studies by Francischi et al. [260] it appears that celecoxib has less effect on the inflammatory as compared with the direct neural effects of pain. This may be due to more significant effects of the drug in inhibiting COX-2, and possibly even to a small extent COX-1, in the CNS pathways of pain and may be linked to the high lipophilicity of the drug and its extensive volume of distribution. Unfortunately, studies linking the actions of celecoxib and other drugs with localised uptake into the CNS have not been established to verify this point. The possibility that there may be direct effects of celcoxib like that of other NSAIDs on ion currents mediated by glutamate or glycine sensitive neurones has been indicated from the studies of Shin et al. [262] using the niastatin perforated patch clamp method for measurement of ion currents in the periacqueductal gray (PAG) neurones. It was found that application of celecoxib resulted in decreased glycine induced ion currents and increased glutamate induced ion currents. The latter results are puzzling in view of the established effects of NSAIDs on glutamate sensitive neurons and clearly further investigation is required.

Other potentially important pharmacological effects of celecoxib have been extensively investigated including anti-cancer activity, demonstrated in models of colon and other cancer [263–276, 278, 279]. These anti-tumour effects are probably multifactorial but in a large number of these studies appear to be related to anti-angiogenic effects, the consequences of inhibition of COX-2 in general inflammatory responses and possibly (though not convincingly) due to NFκB and other intracellular transductional events [277, 278]. These extensive studies in a variety of tumour models supported by the studies showing the role of COX-2 inhibition in head and neck xenograph tumours [278] are impressive for showing consistently the relatively potent effects of celecoxib like that of many other NSAIDs in the inhibition of carcinogenesis and tumour growth and proliferation which with ongoing clinical studies especially in colorectal cancer may provide proof of principle of the clinical efficacy of celecoxib [279].

Neuroprotective effects have been indicated in a number of model systems including the age-related degenerative changes and accompanying production in the hippocampus of IL-1, TNF_{α} and PGE_2 [280]. Repeated oral treatment with celecoxib on these degenerative neural markers and the accompanying behavioural changes were effective, providing the treatment was started early on in the development of the aging process [280]. In the model of MMP-induced neurodegeneration with accompanying development of degenerative changes in dopamine sensitive neurones, celecoxib 2.5–50 mg/kg was ineffective [281]. This was shown to be related to the lack of effect on hydroxylamine formation; in contrast salicylate 50 and 100 mg/kg was very effective. Diclofenac 5–10 mg/kg, which also was ineffective in controlling oxyradical production, likewise was ineffective in this model [281]. Thus, celecoxib would be unlikely to be effective in the adjuvant therapy in Parkinson's disease as has been proposed for aspirin or salicylate.

In a transgenic mouse model of amyotrophic lateral sclerosis (ALS) which accompanies glutamate mediated excitatory activity, oxyradical damage and apoptosis with accompanying elevations of COX-2-derived PGE_2, it was found that celecoxib markedly inhibited the production of PGE_2 in the spinal cords of the ALS mice and significantly delayed the onset of weakness and weight loss and loss of survival which develops in these mice [282]. It appeared that the spinal cords of these mice showed significant preservation of spinal neuronal integrity and the prevention of astrogliosis and microglial activation by this COX-2 inhibitor. While these results are at variance with the lack of effects on oxyradical production [281] it is clear that in the spinal cord there are protective effects of celecoxib that may be related to its multiple actions as well as COX-2 inhibition [282].

Renal toxicity with selective COX-2 inhibitors has attracted considerable interest since these effects appear to be mediated by COX-2 inhibition in the same way as experienced with non selective COX inhibitors [283–287]. The consequent effects of actions of COX-2 inhibitors on alterations in glomerular filtration rate, sodium retention, the development of peripheral oedema, hypertension and long-term rare effects in causing nephrotic syndrome, acute interstitial nephritis and chronic interstitial nephritis, hypercalaemia, are of particular concern especially since some of these can be related to the development of hypertension [286]. Given that these are probably class effects common with all NSAIDs there do appear to be some differences between the relative effects of celecoxib compared with that of rofecoxib. In the study by Welton et al. (as part of the Success VI Study Group), which was a double-blind trial for 6 weeks in patients with OA greater than 65 years of age and taking antihypertensive agents, there were half as many patients that had received celecoxib who experienced oedema, systolic and diastolic blood pressure changes and mean arterial blood pressure alterations compared with those on rofecoxib during the same period [282]. Thus, there may be significant differences between the relative cardio-renal effects of different coxibs but this requires independent investigation and detailed experimental studies. While renal prostacyclin and thromboxane production is known to be altered by celecoxib and other NSAIDs, alterations in the expression of iNOS in the cortex do not seem to be implicated in the development of renal damage [288]. Alterations in the production of renin in the activity of the renin-angiotensin system have been known to be affected by aspirin and some other NSAIDs but apparently in WKY hypertensive and normotensive rats COX-2 does not appear to be related to the stimulation of renin production [289]. There are indications that COX-2 inhibition by celecoxib is not related to alterations in renin response in this model. In another study in which hypertension was induced by the chronic administration of *N*-omega-nitro-L-arginine methylester it was found that celecoxib significantly elevated blood pressure in both normal and hypertensive rats and this was related to elevation of plasma arginine-vasopressin levels, plasma urea and reduction in plasma renin [290]. Increased leucocyte adherence to the endothelium of both normal and hypertensive rats were also observed [290]. In ureteric obstructive rats significant reduction in creatinine levels occurred with celecoxib and accompanying urine volume and urine sodium values were increased together with certain morphological alterations [291]. In normotensive salt-depleted subjects where there is a compensatory change in sodium and potassium alterations, it was found that celecoxib, while having no effect on systemic blood pressure, caused a short-term transient decrease in renal blood flow and glomerular filtration rate with evident sodium retention within the first 3 days of treatment [292]. It appears, therefore, that there are functional alterations in sodium and water retention that can be achieved even in normotensive individuals [292]. Furthermore, in rats with experimental membranous glomerular nephritis there were alterations in albumin urea, mesangiolysis and refraction of glomerular endothelial cells whereas proliferation and activation of mesangial cells was unaffected [293]. Thus, selective COX-2 inhibitors including rofecoxib may impair glomerular capillary repair after mesangiolysis and thus these inhibitors should be used in caution with patients exhibiting endocapillary glomerular disorders [293].

The hepatic effects of celecoxib have already been discussed and related to the cytochrome P_{450} metabolism of the drug and the potential interactions highlighted with other drugs that affect the cytochromes or other hepatotoxic drugs. Of particular interest however, since NSAIDs are frequently taken with paracetamol (acetaminophen) is the potential for interactions between COX-2 inhibitors and paracetamol since both are potentially hepatotoxic. In COX-2 "knock-out" (–/–) and –/+ mice, paracetamol induced hepatotoxicity and lethality were markedly increased compared with COX-2 (+/+) mice and this appeared to be unrelated to alterations in drug protein adduct formation [294]. The COX-2 deficiency in COX knock-out mice appeared to be related to profoundly impaired induction of heat shock proteins and this suggested that PGs may be important endogenous stress signals regulating heat shock protein production. The increased liver toxicity in mice treated with celecoxib also appeared to be related to these drug effects [294] suggesting that there may be important molecular and regulatory alterations that take place during COX-2 inhibition linking COX-2 regulated PG production to that of liver injury.

One of the major issues which confounds the identification of NSAIDs with greater GI safety are the kinds of experimental models which are employed to discover new drugs and screen them for GI toxicity as well as the use of long-term toxicity studies [295, 296]. The unfortunate fact is that most of the drug screening takes place in essentially normal animals and since drugs are being used for arthritic disease it would be more appropriate if the drugs were initially screened in animal models that replicate the chronic inflammatory diseases that are experienced in patients [295]. Thus, we have advocated the use of a range of disease and physical stress models in order to quickly screen for the GI toxicity of NSAIDs and these models have been used extensively and are well-defined [295, 296]. It is not surprising, therefore, that in studies in arthritic rats the acute gastric toxicity of both celecoxib and rofecoxib has been shown to be markedly enhanced compared with that in normal animals [297, 298]. It is also not surprising that the expression of COX-2 mRNA and COX-2 immunoreactivity was enhanced in arthritic but not all normal animals [297]. The inference from the study of Kato et al. [297, 298] was that the inhibition of elevated expression of COX-2 by celecoxib and rofecoxib would somehow be responsible for the enhanced gastric injury.

This is a naïve interpretation since it is most logical that the arthritic process has undoubtedly compromised mucosal protective functions and integrity. The extensive effects that are known to occur with adjuvant arthritic rats on liver and other metabolism and detoxification [295] were not considered in this study. Delayed ulcer healing and impaired angiogenesis, both influenced by COX-2, also contribute to the ulcerogenic effects of COX-2 inhibitors including celecoxib [299, 300].

The beneficial GI effects of celecoxib (5 mg/kg/d) and other COX-2 inhibitors have been explored in the inhibition of the development of colitis induced by dinitrobenzene sulphonic acid in rats [301]. In addition, celecoxib reduced haemorrhagic diarrhoea and weight loss [301]. Celecoxib also reduced the myeloperoxidase activity in the mucosa suggesting that there was a reduction in infiltration of PMN leucocytes. The malondialdehyde activity, which is a measure of oxyradicals, was reduced as well as levels of nitrotyrosine which is formed from peroxynitrite reaction with tyrosine [301]. Despite these potentially important effects on GI inflammation models it is, however, evident that celecoxib like that of other NSAIDs can decrease epithelial proliferation, and this may have consequences in treating inflammatory bowel diseases.

Increasing concerns about the relative safety of NSAIDs for female reproduction have been of ongoing concern [302]. COX-1 as well as COX-2 inhibition appears to underlie deficient abnormalities and ovulation, fertilisation, implantation and decidualisation in mice [303]. While non-selective as well as selective inhibitors of COX-2 have been employed as potent tocolytic agents they unfortunately have potential risks for constriction of the foetal ductus arteriosus [304]. In studies in sheep the closure of the lamb ductus arteriosus and ductal blood flow was more profoundly affected by indomethacin than celecoxib [304], implying that the use of celecoxib as a tocolytic agent may have some safety advantages. However, studies in rabbits indicate that when the newborn animals are exposed to hyperoxia there was a lower thromboxane B_2 to 6 keto $PGF_{1\alpha}$ which was linked to the effect of celecoxib on reducing

8-epi-$PGF_{2\alpha}$ and this may have indirect influences on thromboxane production [305]. Clearly more detailed investigations are required and this is addressed further in Chapter 10.

Valdecoxib and parecoxib

In the search for other diarylisosoxisole COX-2 inhibitors having sulphonamide moieties, Searle identified valdecoxib (Bextra™) (4-[5methyl-3-phenyloxisol-4-yl]-benzene sulphonamide) as an orally effective COX-2 inhibitor [306]. Parecoxib (*N*-2[[(5-methyl-3-phenyloxisol-4yl)-phenyl]sulphonyl] propane amide) is the pro-drug precursor of valdecoxib for use as an injectable form in acute pain relief [307]. These two second generation coxibs are particularly interesting therapeutic developments. In the human *ex vivo* whole blood assay valdecoxib (SC65872) appeared to have approximately 26-fold selectivity for COX-2 and the hydroxymethyl metabolite was virtually devoid of any COX-1 activity [306]. The overall potency of COX-1/COX-2 inhibition of valdecoxib was lower than that of the progenitor celecoxib in the whole blood assay but it appeared to have significantly greater selectivity in the assay involving human recombinant isoenzymes. The potency of valdecoxib in rat adjuvant arthritis was some 10-fold greater than that of celecoxib; the IC_{50} being 0.032 mg/kg/d compared with that of celecoxib 0.37 mg/kg/d. Valdecoxib was slightly less potent in the carrageenan oedema assay.

Parecoxib was shown to form valdecoxib and the half-life of valdecoxib is approximately 7–8 h while that of the pro-drug precursor is approximately 30 min to 1 h [307]. The drug was shown to reverse hyperalgesia [307] and is the first injectable COX-2 selective inhibitor that is being employed clinically.

Valdecoxib has been recently introduced for the treatment of OA and RA, pain and treatment of primary dysmenorrhea and has all the conventional contraindications and special considerations as with any other NSAID or coxib [308]. It appears to undergo cytochrome P_{450} 2C9 and 3A4 mediated metabolism and thus not surprisingly any inducers or inhibitors of these isoforms will inevitably influence the plasma concentrations and relative safety of valdecoxib. In the treatment of RA and OA, valdecoxib 10, 20 or 40 mg once daily was as effective as naproxen 50 mg twice daily and superior to placebo [309, 310]. In a randomised 12-week double-blind study in patients with osteoarthritis of the hip, valdecoxib 5 and 10 mg once daily was as effective as 500 mg twice daily naproxen but had fewer GI adverse events [310]. In an upper GI endoscopy study performed in 1052 osteoarthritis patients, valdecoxib 10 mg once daily or 20 mg once daily was compared with ibuprofen 800 mg t.i.d. or diclofenac 75 mg t.i.d. over a 12-week period. The incidence of gastroduodenal ulcers over the 12-week period was 5% in the patients receiving 10 mg once daily valdecoxib, 4% in those receiving 20 mg once daily, 7% in those receiving placebo, 16% in those with ibuprofen and 17% in those with diclofenac [311]. With there being serious concerns about the potential for upper GI mucosal injury in the elderly, a study was recently performed in subjects between 65–76 years old, who received valdecoxib at the supratherapeutic dosage of 40 mg twice daily compared with naproxen 500 mg twice daily. [312]. Upper GI mucosal injury was evaluated by oesophago-gastroduodenal endoscopy at day 7. The patients who received valdecoxib had no evident gastroduodenal ulcers while those that received placebo had an incidence of 3% and those that received naproxen 18%. These studies are indicative of there being a favourable GI safety with valdecoxib in elderly healthy subjects [312].

Increasing concerns about the risk of cardiovascular thrombotic events from coxibs [313] has led to studies being undertaken with valdecoxib to compare the incidence of myocardial infarction, myocardial ischaemia and unstable angina, cerebrovascular disorders or peripheral arterial ischaemia in patients receiving valdecoxib 10–18 mg once daily compared with that of diclofenac 75 mg b.i.d. ibuprofen 800 mg t.i.d. or naproxen 500 mg b.i.d. [314]. These studies have shown that in 4531 patients that received valdecoxib with a total exposure of 1340 patient years there was no difference in the total number of serious thrombotic events or any of the other individual events as compared with traditional NSAIDs but these did, however, show a slightly increased risk of any event above placebo [314].

Thus, valdecoxib may represent a new second generation sulphonamide coxib that can be given once daily (this compares with the twice daily dosage of celecoxib) and this may be associated with a lower potential risk of cardiovascular thrombotic events than possibly experienced with other coxibs, but this aspect has yet to be clarified.

Rofecoxib

Rofecoxib (MK996; VIOXX™; Merck Inc; 4-(4'-methylsulphonylphenyl)-3-phenyl-2,5*H*)-furanone is probably among the most potent of the coxib inhibitors that has been employed clinically so far. This drug being a furanone is of a different chemical class than celecoxib, valdecoxib and parecoxib. The initial discovery and the overall clinical pharmacokinetics and pharmacodynamic properties of rofecoxib have been previously reviewed [315–317] and the details of the NDA and clinical studies can be found at the FDA website [318]. The initial development of rofecoxib at Merck Frosst Laboratories, Montreal [318–320], followed an extensive and fundamental programme of other eicosanoid inhibitors and clearly benefited from an enormous effort that was made in those laboratories to understand the molecular, biological and cellular aspects of inflammatory processes. The database from which rofecoxib was developed was probably amongst the largest in the pharmaceutical industry devoted to the discovery of new NSAIDs. In the standard benchmark whole blood COX-1 and COX-2 assays, rofecoxib has amongst the highest ranking of all COX-2 selective drugs for selectivity; with COX-1-COX-2 selectivity ratios of 0.013 (Tab. 1).

In the rat carrageenan paw oedema model, the drug exhibited an ED_{50} of 1.5 mg/kg p.o. in the carrageenan-induced paw hyperalgesic model this anti-oedemic effect overlaps that of the hyperalgesia where the ED_{50} is 1 mg/kg [319]. In the adjuvant arthritis model rofecoxib had an ED_{50} of 0.74 mg/kg/d and showed significant protective effects against cartilage and bone destruction measured radiographically 14 or 21 days after induction of arthritis in Lewis rats. The inhibition of secondary paw swelling paralleled that of the reduction in radiographical evidence of joint damage. The antipyretic effects of rofecoxib were investigated in the LPS-induced pyresis in rats with ED_{50} of 0.24 mg/kg being obtained [319].

Using the radiochromium labelled red cell loss technique for GI bleeding in rats and in squirrel monkeys rofecoxib was shown to have no effects up to 200 mg/kg per day for 5 days [319]. Regrettably in the radiochromium GI blood loss studies there was no information provided about the integrity of the gastrointestinal mucosa. It is well known that radiochromium blood loss is an ineffective measure of GI injury and only adequately reflects the actions of aspirin and other drugs that inhibit platelet aggregation and so may potentiate blood loss resulting from mucosal injury [321]. Since rofecoxib does not affect platelet function [321], it is not surprising that there is little evident blood loss from the GI tract.

A considerable number of GI safety studies were performed with rofecoxib [152, 153, 318, 322–324] including GI endoscopy in patients with RA, which was the initial primary indication for this drug [324–325]. All these studies showed that rofecoxib had somewhat lower gastric ulcer incidence and incidence of peptic ulcer bleeds than the lower ulcerogenic group of NSAIDs but, as highlighted earlier, the relative reductions in these severe ulcer events are probably relatively small when reviewed from long-term data [326]. Considerable concern has, however, been expressed about the potential for rofecoxib to be a serious cause of renal injury; this aspect having been reviewed in the previous section [285–287]. The conclusion overall is that renal injury is probably a COX-2 mediated event but is probably potentiated in long half-life coxibs that are very potent because of the high binding to COX-2 in the macula densa which is an underlying feature of all inhibitors of cyclooxygenases [327, 328].

Rofecoxib has had a reasonably favourable safety profile in general practice with predictable adverse reactions [329]. In a long-term randomised control trial in 784 patients with OA of the hip or knee randomised to receive rofecoxib 12.5 mg or 25 mg once daily or diclofenac 15 mg t.i.d., it was shown that rofecoxib was effective in inhibiting the symptoms of pain and WOMAC scores [330]. An interesting feature about this study was that the liver transaminases present in the plasma of patients that received diclofenac were all consistently elevated from 2 weeks to the end of the study whereas there were no changes in transaminases evident with rofecoxib or placebo over this period [330]. While the elevation of liver transaminases is considered to be an unspecific event associated with intake of all NSAIDs interestingly there were no indications of any liver changes evident with rofecoxib.

Pharmacokinetics

The pharmacokinetics of rofecoxib in humans (see Tab. 2) are essentially quite complex and may be variable depending on patients and other conditions. The drug has a variable GI absorption with the maximum concentration occurring between 2 and 9 h. The mean oral bioavailability is 93% following acute administration [317] and it has a terminal half-life of elimination of approximately 17 h at steady-state. Like celecoxib, the volume of distribution of rofecoxib is between 86–91 L and much greater than normally expected with NSAIDs (their average being 7–10 L). This may be related to the high lipophilicity of this drug and like celecoxib, it has a high pKa value. Rofecoxib has a lower percentage of binding to human plasma proteins, this being 87%. The drug also crosses the blood brain barrier in rats and has been shown to cross the placenta in rats and rabbits thus raising the possibility of foetotoxic events. The metabolic routes of rofecoxib are shown in Figure 8 [317]. An interesting feature about rofecoxib, which is not evident with celecoxib (which should be recalled has a predominant cytochrome P_{450}-based metabolism), is that the drug is primarily metabolised by cytosolic reductases in the liver with only a minor role for the cytochromes P_{450} [317]. There is apparently a reversible metabolism of rofecoxib to the 5-hydroxy derivative at least in rats, but in rats the 5-hydroxymetabolite is the major metabolite. The complex

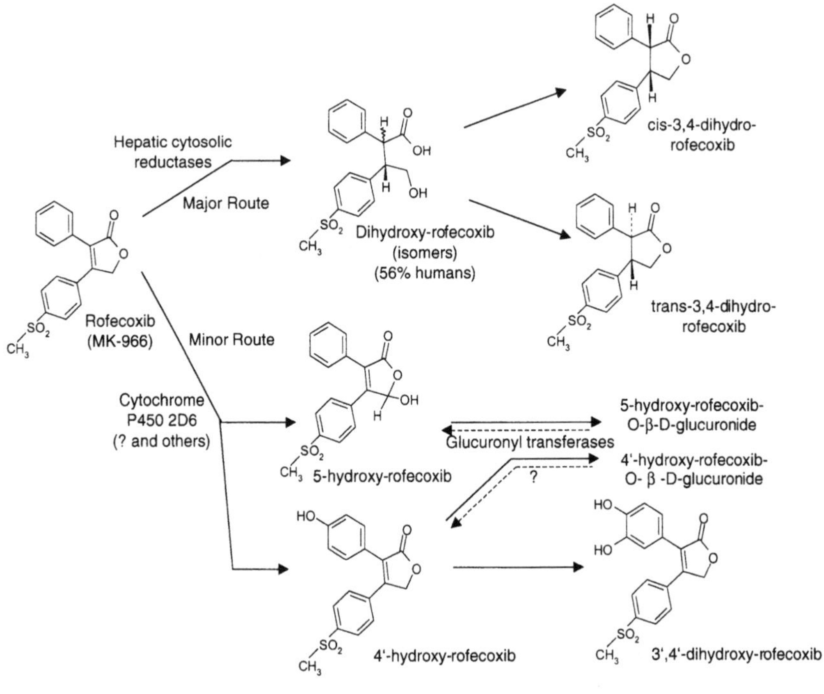

Figure 8. Pathways of the metabolism of rofecoxib in humans. The principal route is by oxidation via cytosolic NADPH oxido-reductases in the liver, with the involvement of the cytochromes P-450 being relatively minor [317, 318].

metabolism of rofecoxib progresses to the formation of the 3-4-dihydrodiol and 4-hydroxy derivatives that can then progress to form glucuronides. Relatively little unchanged drug occurs in the urine as it is primarily metabolised in the liver. The potential drug interactions at the level of the minor cytochrome P_{450} system probably involves intestinal cytochrome P_{450} 3A but not that of the liver; the intestinal enzymes being inhibited by concurrent administration of midazolam [317]. Cimetidine, which is an inhibitor of cytochrome P_{450} 2D6, slightly increases the maximum plasma concentration of rofecoxib but this does not lead to the necessity to reduce dosage. However, potent inducers of cytochrome P_{450} among them rifampin, carbamazepine, phenobarbitone and phenytoin may decrease levels of rofecoxib, thus implying cytochrome metabolism even by the minor route can have small but nonetheless clinically significant effects on the pharmacokinetics of the drug.

The ADME and general pharmacokinetics of rofecoxib have been investigated in rats and dogs [318]. The oral absorption of rofecoxib was relatively rapid in both species with peak levels achieved at 0.5 h in rats and 1.5 h in dogs which is appreciably shorter than that in humans [317, 318]. The bioavailability was poor in dogs being 26% and the drug was primarily excreted by the biliary route in both species [318]. The metabolism of rofecoxib is quite extensive in both species forming the major metabolites as identified in Figure 8. The distribution of radiolabelled drug following intravenous administration was high in most tissues at 5–30 mins with predominant amounts being in the liver, skin, fat, prostate and bladder [318]. Using synthetic metabolites of rofecoxib it was found using the whole blood assay *in vitro* that none of the metabolites had any effects on COX-2 or COX-1 activity [320].

The mechanisms underlying the biliary excretion of rofecoxib and its 5-hydroxymetabolite were investigated in intact and biliary-canulated rats following administration of ^{14}C- or ^{18}O-labelled compounds. The biliary excretion of both rofecoxib and its 5-hydroxy metabolite were dependent on uninterrupted bowel flow since this was blocked by ligation of the bowel duct. The secondary peak of drug which occurs in the plasma at approximately 10 h after administration of rofecoxib or its 5-hydroxymetabolite was abolished by bile duct canulation. The reversible reaction of 5-hydroxy rofecoxib to the parent drug was also demonstrated. These results suggested that a metabolic sequence involving 5-hydroxylation of rofecoxib followed by biliary elimination of the corresponding glucuronide and de-conjugation of the glucuronide in the lower intestinal tract presumably by β-glucuronidases from enteric bacteria yielded the corresponding hydroxy-acid. This spontaneously cyclised to regenerate parent rofecoxib and was reabsorbed into the systemic circulation [331]. The biliary excretion in rats following oral administration is about 58–76% and that following intravenous administration 37–94%, showing there is a considerable degree of overlap when the radiolabelled drug is administered by these two routes [318, 331]. The biliary excretion in dogs is appreciably lower than that from orally administered drug being 26.5% and following intravenous administration 28.2% [318, 331]. The urinary excretion correspondingly is increased and 16.8% following oral dosage and 32% following intravenous administration. This contrasts with the markedly lower urinary excretion of the drug in rats which is about 1–2% [331, 332].

In vitro studies on the cytochromes P_{450} show that there was no impairment of testosterone metabolism, thus indicating that there was no likely effect of rofecoxib on cytochrome P_{450} 3A4 even though it appeared that rofecoxib could be metabolised by this isoform and cytochrome P_{450} 1A2 to a weak extent. Overall, it appeared that rofecoxib only had a very weak ability to inhibit cytochromes P_{450}.

As indicated earlier, rofecoxib is primarily metabolised by cytosolic enzymes and investigations indicated that this was an NADPH-dependent system. It appears that the oxidative pathway of metabolism of rofecoxib predominates in the rat whereas the reductive pathway (involving NADPH) predominates in humans as well as in monkeys and is most marked in dogs [317, 332]. These important metabolic differences may have considerable significance for the interpretation of the effects of rofecoxib in rats as compared with dogs. It appears that the metabolism in dogs is closer to that of humans.

A full dose ranging pharmacokinetic study in which the drug was administered at oral doses in healthy men, covering the range of therapeutic and supratherapeutic doses (25 to 375 mg/d) was undertaken over a 14-day period and these data were compared with the effects on COX-1 and COX-2 activity using the whole blood assay [321]. The kinetics of rofecoxib were found to be complex and non-linear. The elimination half-life ranged from 9.9–17.5 h after multiple dosages with an accumulation ratio of approximately two overall doses. The C_{max} after 14 days with the standard 25 mg dose of the drug was 382 ng/ml and this ranged up to 3,169 ng/ml following the last dose of 375 mg per day. There appeared to be no differences in the area under the curve (AUC_{0-24}) at the 250 and 375 mg dosage levels suggesting that there is a limitation to absorption at the high doses. Furthermore, the geometric mean of the C_{max} values was not different for the doses of 100–150 and the 375 mg dose but they were all greater than the 25 mg dose following single oral administration of the drug. For each of the doses the C_{max} values were significantly greater after multiple doses than single dose administration of rofecoxib. While there were no effects of any of the dosages on COX-1 derived serum thromboxane production there were markedly different results obtained with the dosages on LPS stimulated PGE_2, which is presumed to be from COX-2. The 25 mg dose, given as a single dose or after multiple doses, did not cause complete inhibition of COX-2 activity. There was an ~40–50% inhibition over the time course of COX-2 derived PGE_2 following single dosage and 50–70% inhibition following multiple dosage of 25 mg rofecoxib. The higher doses, however, caused almost complete inhibition of COX-2 activity. These results are of interest in view of the wide-spread usage of the 25 mg dose for control of pain and inflammation, where it is clear that the effects on COX-2 activity in the circulation are not complete.

Pharmacology and toxicology

The pre-clinical anti-inflammatory, analgesic and antipyretic activity as described in the NDA and in the earlier section are indicative of a classical NSAID in terms of the properties in standard model systems. Toxicokinetic studies formed in rats, mice, rabbits and dogs and the long-term (53 weeks) studies at the maximum performed in dogs and rats have been described in the NDA [318] and all revealed a classical pattern of NSAID related toxicity with primary toxic effects occurring in the GI tract and renal systems. There was no evident genotoxicity [318]. Aspects of safety pharmacology were investigated and there were no influences on cardiovascular function, autonomic responses, respiratory function or gastric secretion in dogs following administration of 25 mg/kg of rofecoxib. GI motility, motor function, autonomic responses and spinal reflexes were not affected in mice when given 1–100 mg/kg p.o. of the drug.

In female dogs administration of 5 or 25 mg/kg rofecoxib caused a significant decrease in urine volume, sodium excretion and urinary PGE_2 levels. Glomerular filtration rate, renal plasma flow, plasma renin activity and antidiuretic hormone levels were unaltered. The results suggest there may be some effect of rofecoxib at the level of tubular functions but not at the glomerulus in dogs. The changes in urinary volume and sodium excretion appear to be secondary to the reduction in renal PGE_2.

In volume depleted normotensive male Wistar rats given 10 mg/kg of rofecoxib p.o., there was a reduction in glomerular filtration rate, renal blood flow, sodium excretion and urine volume. Overall, the study suggested that rofecoxib had similar effects to that of many NSAIDs on renal functions and that these were primarily at the level of tubular functions.

The renal effects of rofecoxib as well as those of other COX-2 inhibitors have received much attention since there have been a considerable number of reports of renal toxicity associated with these drugs in humans. In a study undertaken in healthy elderly adults on fixed sodium intake, rofecoxib 50 mg once daily, indomethacin 50 mg t.i.d. or placebo was compared over a 2-week period [333]. Glomerular filtration rate was unaffected by rofecoxib whereas it was inhibited by indomethacin thus the conclusions of a lack of a glomerular effect suggested from the preclinical toxicology studies were confirmed in humans. Rofecoxib caused a clinically insignificant and transient retention of sodium (a feature seen with most NSAIDs) and this appeared to be related to the selective inhibition of COX-2. Since glomerular filtration was impaired by indomethacin it was suggested that this was a COX-1 related effect, though it must be said that no studies with direct acting COX-1 inhibitors have been undertaken to support this view in humans.

There have been extensive investigations of the safety profile of rofecoxib in controlled clinical trials sponsored by the company marketing the drug (Merck Inc., Review and Evaluation of Pharmacology and Toxicology; [334]). These studies showed that in elderly, osteoarthritic patients having varying degrees of alterations in creatinine clearance, rofecoxib 50 mg but not 25 mg or the lower dose of 12.5 mg significantly altered creatinine clearance, but this was within the range of effects seen by ibuprofen 2.4 g daily and diclofenac 150 mg t.i.d. [327]. Some of the OA patients reported peripheral oedema and mean changes in systolic and diastolic blood pressure occurred with rofecoxib, with systolic blood pressure increasing ~2 mm over all the dosage groups whereas there was lower degree of increase in systolic blood pressure with ibuprofen and diclofenac. These results have suggested that under controlled clinical trial conditions there may be effects of rofecoxib, like that of other COX-2 inhibitors, on components of renal function and cardiovascular activity.

In the prescription event monitoring studies performed in 15,268 patients that received rofecoxib primarily for OA (24% total patients) under general practice conditions in the UK there were three reports of acute renal failure [329]. It is notable that despite claims that the drug rofecoxib was primarily developed to have appreciably lower GI toxicity in humans that there were 110 reports of serious GI events including 76 upper GI bleeds and peptic ulcers and one perforated colon. There were also 111 thromboembolic events reported, one case of Stephen Johnson's syndrome, severe anaphylaxis and angio-oedema [329]. Cardiovascular complications have been a well-known problem with the COX-2 inhibitors [335] and it has been suggested that this may relate in part to the inhibitory effects of the COX-2 inhibitors on the vascular production of COX-2 derived prostacyclin [336].

Investigating the mechanistic aspects of the actions of rofecoxib in relationship to anti-inflammatory analgesic and anticancer effects has largely implicated the actions of the drug on the COX-2 system. However, there are indications that there may be some actions of the drug on signal transduction pathways that may be of considerable significance. Thus, the binding of the transcription factor NFκB to DNA was inhibited at high concentrations of 10–100 μmol/l whereas the binding of AP1 was markedly increased at 100 μmol/l concentrations of the drug [337, 338]. These concentrations are relatively high and thus caution should be noted about the interpretation of these effects in relationship to known pharmacologically active concentrations of the drug.

Reduction in polyps in APCΔ716 mice were shown with both rofecoxib and sulindac so that in relation to colon cancer there may be inhibitory effects of rofecoxib like that of sulindac on vascular angiogenesis and transcription factor activation linked to the development of human adenomatous polyposis [339]. In another study 10 μmol/l rofecoxib attenuated NFκB activation and the expression of NFκB dependent genes in mouse peritoneal macrophages [338]. The drug was also found to protect against LPS-induced apoptosis by the impairment of the inflammatory response presumably mediated by the NFκB system as well as that involving mitogen activated protein kinase p38 and the extracellular regulated kinase p44 as well as the cJUN-JUN N-terminal kinase [338].

The relationship of the signal transduction effects of rofecoxib in causing inhibition of NFκB binding to DNA and the paradoxical increase in activating protein 1 (AP1) binding high concentrations of the drug [337] have been related to the effects of rofecoxib *in vivo* in the spinal cord of rats in whom oedema had been induced by injection of zymosan [340]. At doses of 1 and 10 mg/kg the spinal cord induction of iNOS was increased but COX-2 protein expression was decreased. This suggested that the opposite effects of rofecoxib in activating AP1 and inhibiting NFκB may explain the paradoxical changes in the spinal cord on the nitric oxide and COX systems.

In the uric acid induced pain model in the rat prior treatment with rofecoxib or the selective COX-1 inhibitor SC560 showed similar analgesic effects although the effects of the latter were somewhat lower than the effects of rofecoxib [340]. Paradoxically, however, when SC560 was given after the induction of the pain response by uric acid, it produced greater inhibition than that with rofecoxib [340]. Thus there may be differential effects on COX-1 and COX-2 responses that explain some of the analgesic effects of NSAIDs [338, 340].

In the hot plate test in rats the prior treatment with p-chlorophenylalanine inhibited the analgesic activity of rofecoxib [341]. It was found that rofecoxib significantly increased 5HT concentrations as well as $5HT_2$ receptors in the frontal cortex suggesting that, as with many other NSAIDs and analgesics, rofecoxib may influence the serotonergic pathway and that this may contribute to the central analgesic effects of this drug [341].

Interestingly the anti-nociceptive and anti-inflammatory effects of melatonin which have been shown to be potentiated by indomethacin are only slightly affected by rofecoxib [342]. This suggests that the major effect of the interaction may be due to COX-1 inhibition or perhaps the linked inhibition of COX-1 or COX-2 by indomethacin. Rofecoxib was also found to attenuate the neurogenic component of adjuvant induced temporomandibular chronic inflammation in Freund's adjuvant injected rats [343]. As with other NSAIDs and celecoxib, rofecoxib has been shown to inhibit a number of tumour models [344, 345] and potentiation of cytokine-related angiogenic responses in combination with radiation exposure [339].

As noted previously, coxibs have been shown to be more ulcerogenic in the stomach and intestine in adjuvant arthritic rats [298]. It was suggested by Kato and Takeuchi [298] that the increased ulcerogenic response in arthritic rats was due to upregulation of iNOS and COX-2 but it is also noted that these authors did not consider the other extensive effects of arthritic processes in this model.

Impairment of ulcer healing has been implicated with rofecoxib administration in rats, possibly by impairment of growth factors such as hepatocyte growth factor and gastrin [346]. Laudanno and co-workers observed that the ulcerogenic effects of rofecoxib were revealed to be markedly increased in rats that had been exposed to stress, duodenal ulceration by injection of cysteamine or gastric ulcers induced by acetic acid [347]. An interesting phenomenon observed by these authors was that in unstressed animals there was an increased neutrophilia but not leucocyte accumulation in the mucosa in rats treated with rofecoxib or celecoxib [347]. Thus there may be some systemic effects that are caused by these coxibs that predispose to the development of mucosal injury when the animals are exposed to stress or other ulcerogens. There may also be some indirect effects on "ischaemic preconditioning" relating to the actions of PGs, NO, adenosine and activation of sensory nerves during the development of gastric mucosal injury that is induced by alterations in blood flow [348].

The potential for COX-2 rebound to be an important factor relating to wound healing as well as the ulcerogenic effects following chronic administration of coxibs has been highlighted by studies in rats where prolonged administration of coxibs has been investigated [349]. Colville-Nash and Gilroy have speculated that COX-2 may help resolve pathology, e.g., GI ulceration, inflammation, wound healing, etc. possibly by generation of an alternative series of PGs such as the cyclopentenone PGs [349]. The exact influence of these and the relationship to resolve resolution of injury has yet to be established but this is an interesting concept.

In terms of untoward effects of COX-2 inhibition it has been suggested that COX-2 function is important for bone fraction healing [350] using implanted tibiae into the ear chambers then given 4 weeks oral administration of rofecoxib, 12 mg/kg/d or naproxen 110 mg/kg. It was found that both drugs decreased bone growth as determined by antibody staining for vitronectin receptor reflecting osteoclast-like cells [351]. Whether these studies in animal model systems implicating inhibition of bone growth by coxibs as well as NSAIDs has any clinical significance has yet to be determined.

Etoricoxib

Etoricoxib (MK-O663) (5-chloro-2-(6-methylpyridine-3-yl)-3-(4-methylsulphonylphenyl) pyridine is another methyl sulphonylphenyl derivative that was developed by Merck and recently introduced as a second generation coxib. The pharmacokinetic and pharmacological properties of etoricoxib have been reviewed [352] and the main claim for unique pharmacology with etoricoxib is that it has a greater selectivity for COX-2 in relationship to COX-1 than either that of its progenitor rofecoxib, celecoxib or valdecoxib [352]. Indeed the selectivity in the whole blood assay is indicated by the ratio of COX-2 to COX-1 being 106 for etoricoxib compared with 35 for rofecoxib, 30 for valdecoxib, 7.6 for celecoxib, 7.3 for nimesulide, 2 for meloxicam, 2.4 for etodolac and various other ranges of 0.2–3.0 for other NSAIDs [352]. It is claimed that etoricoxib is least likely to interfere with the antiplatelet effects of aspirin via the irreversible inactivation of COX-1, an issue which has been raised with some other NSAIDs. The GI blood loss from etoricoxib 120 mg once daily was compared with ibuprofen 800 mg t.i.d. and placebo for 28 days and it was found that etoricoxib did not cause any significant blood loss measured by radiochromium technique [352]. Like that of rofecoxib, etoricoxib is recommended for once a day administration due to its long elimination half-life of ~22 h [352]. Like that of the other coxibs, etoricoxib has a large volume of distribution, approximately 120 L but is more extensively bound to plasma proteins (92%) in comparison with rofecoxib [352]. In contrast to rofecoxib, etoricoxib undergoes hydroxylation by cytochromes P_{450} 3A4 and 2D6, 2C9, 1A2 and 2C19 may also play a role [352]. Thus any drugs interacting with these cytochromes will inevitably lead to some alterations in pharmacokinetic or pharmacodymaic properties and safety.

The effects of etoricoxib in various cyclooxygenase systems were extensively investigated and the selectivity confirmed for COX-2 [353]. In the carrageenan paw oedema in rats, etoricoxib was more potent having an ED_{50} of 0.6 mg/kg compared with that of rofecoxib of 1.5 mg/kg and indomethacin 2 mg/kg [353]. In the carrageenan induced paw hyperalgesia, etoricoxib showed a three-fold greater potency compared with rofecoxib whereas in the adjuvant arthritis model rofecoxib (ED_{50} 0.6 mg/kg/d) was almost identical to that of rofecoxib (0.7 mg/kg/d) [17]. Antipyretic activity administration of etoricoxib in rats was approximately three- to four-fold less than with rofecoxib but within the range of that observed with indomethacin [353].

The combined pharmacokinetics and influences on circulating COX-1 and COX-2 enzymatic activity were investigated in healthy volunteers who received doses of 5, 25, 125 and 500 mg of etoricoxib or 10, 50 and 250 mg once daily of etoricoxib [354]. The drug was found to be rapidly absorbed with a maximum concentration at 1–1.5 h and following multiple oral dosage the accumulation ratio averaged approximately 1.6 [354]. There was no evident inhibition of thromboxane production using the whole blood assay *ex vivo* but the percent inhibition of LPS induced PGE_2 in the whole blood assay correlated with the plasma concentrations of the drug. The lower doses of 25 and 50 mg etoricoxib only produced partial inhibition of LPS induced PGE_2 production from COX-2 whereas the higher doses of 100 mg or greater did achieve at least 80% inhibition after single or repeated dosage of the drug [354]. In dose ranging studies performed in patients with OA doses of 30 to 90 mg etoricoxib were approximately equally effective to diclofenac 50 mg t.i.d. over a 6 week treatment period [355] and the 19 mg dose of etoricoxib employed in two studies in RA [356, 357] appeared to be similar to that in efficacy to naproxen 500 mg b.i.d. It would appear, therefore, in relationship to COX-2 inhibition *ex vivo* [298] that although there is about 70–80% inhibition of COX-2 derived PGE_2 at the dose range for effects in OA and RA [355–357] that it may not be that there is complete inhibition.

Lumiracoxib

Lumiracoxib is a structurally distinct COX-2 selective inhibitor that does not have the sulphur containing moiety but instead possesses a carboxylic acid, giving it weakly acidic properties (pKa 4.7) [358, 359] and is structurally similar to that of diclofenac. The drug is currently under development by Novartis (Switzerland, USA) but relatively little basic information is available on this drug to date. The pharmacokinetics of the drug have been reported recently at a number of international meetings [359–366]. The drug is rapidly absorbed with peak drug values appearing in the plasma at 2 h and with a relatively short elimination half-life of 5.8 h [359]. In patients with RA there is a marked uptake of the drug and the 4-hydroxy or 5-carboxy metabolite into the synovial fluid following lumiracoxib 400 mg for 7 days. The values of this metabolite in synovial fluid were approximately that of the plasma concentrations, lumiracoxib peaked at about 6 h in synovial fluid at about half the concentration seen in plasma [359]. It is clear that there is a relatively high uptake of the drug into the synovial fluid and this may have therapeutic advantages for the control of pain and inflammation in joints. There does not appear to be any accumulation of the drug in elderly OA patients as the clearance rates are identical after 28 days treatment with lumiracoxib 200 or 400 mg once daily [360]. Antacids or the proton pump inhibitor omeprazole do not affect the pharmacokinetics of lumiracoxib nor does the administration of warfarin or an oral contraceptive [361]. Lumiracoxib also does not influence the pharmacokinetic parameters of methotrexate in patients with RA. The GI absorption of lumiracoxib was investigated in healthy male volunteers and the regions of uptake and bioavailability were investigated [363]. It was shown that lumiracoxib was rapidly and efficiently absorbed from all regions of the GI tract with no particular region being favoured [363]. The longer median maximum time of absorption in the stomach relative to that of the small bowel and colon probably reflects dissolution effects in the local pH environment. Daily doses of 100 to 400 mg lumiracoxib once daily were shown to be effective in controlling the symptoms of pain and inflammation in patients with OA of the hand, hip or knee [364–366]. Further studies including pre-clinical investigations on lumiracoxib are awaited with much interest. This acidic coxib may present with interesting properties that separate it from the established coxibs.

Conclusions

In this chapter a review of established and newly available COX-2 selective drugs has revealed that there are important pharmacokinetic and pharmacodynamic differences among these drugs. The eventual safety and clinical significance of many of these studies especially with the newer COX-2 selective inhibitors has yet to be fully evaluated as further extensive long-term pharmacoepidemiological and clinical studies are awaited with interest especially to establish if the preclinical investigations of these drugs have actually translated into favourable therapeutic profiles.

Acknowledgements
My thanks to Mrs Veronica Rainsford-Koechli and Mrs Marguerite Lyons for assistance with the typing of this manuscript and to Drs Guenther Trummlitz and Joanne van Ryn for the preparation of Figures 3–5, 7 and 8.

References

1 Smith WM, Meade EA, De Witt DL (1994) Interactions of PGH synthase isozymes-1 and –2 with NSAIDs. *Ann NY Acad Sci* 755: 50–57
2 Vane JR, Botting, RM (1995) A better understanding of anti-inflammatory drugs based on isoforms of cyclooxygenase (COX-1 and COX-2). In: B Samuelsson et al (eds): *Advances in Prostaglandin, Thromboxane and Leukotriene Research, Volume 23.* Raven Press Ltd., New York, 41–48
3 Vane JR, Botting RM (1995) New insights into the mode of action of anti-inflammatory drugs. *Inflamm Res* 44: 1–10
4 Dubois RN, Abramson SB, Crofford L, Gupta RA, Simon LS, Van de Putte LBA, Lipsky PE (1998) Cyclooxygenase in biology and disease. *FASEB J* 12: 1063–1073
5 Kam PCA, See AU-L (2000) Cyclo-oxygenase isoenzymes: physiological and pharmacological role. *Anaesthesia* 55: 442–449
6 Jouzeau J-Y, Terlain B, Abid A, Nédélec E, Netter P (1997) Cyclo-oxygenase isoenzymes. How recent findings affect thinking about nonsteroidal anti-inflammatory drugs. *Drugs* 53: 563–582
7 Kulkarni SK, Jain NK, Singh A (2000) Cyclooxygenase isoenzymes and newer therapeutic potential for selective COX-2 inhibitors. *Methods Findings Exp Clin Pharmacol* 22: 291–298
8 Vane J, Botting J, Botting R (eds) (1995*) Improved non-steroid anti-inflammatory drugs COX-2 enzyme inhibitors.* Kluwer Academic Publishers, Dordrecht and William Harvey Press, London
9 Vane J, Botting RM (eds): (2001) *Therapeutic roles of selective COX-2 inhibitors.* William Harvey Press, London
10 Fenner H (1997) Differentiating among nonsteroidal anti-inflammatory drugs by pharmacokinetic and pharmacodynamic profiles. *Semin Arthritis Rheum* 26: 28–33
11 Brooks P, Emery P, Evans JF, Fenner H, Hawkey CJ, Patrono C, Smolen J, Breedveld F, Day R, Dougados M et al (1999) Interpreting the clinical significance of the differential inhibition of cyclooxygenase-2 and cyclooxygenase-2. *Rheumatology* 38: 779–788
12 Fitzgerald GA, Patrono C (2001) The coxibs, selective inhibitors of cyclooxygenase-2. *N Engl J Med* 345: 433–442
13 Herbette LF, Vecchiarelli, Trummlitz G (1996) NSAID mechanism of action: the role of intracellular pharmacokinetics. In: J Vane, J Botting, R Botting (eds): *Improved Non-Steroid Anti-Inflammatory Drugs COX-2 Enzyme Inhibitors.* Kluwer Academic Publishers, Dordrecht, William Harvey Press London, 85–102
14 Trummlitz G, van Ryn J (2002) Designing selective COX-2 inhibitors: molecular modelling approaches. *Curr Opin Drug Discovery Development* 5: 550–561
15 Chulada PC, Langenbach R (1997) Differential inhibition of murine prostaglandin synthase-1 and –2 by nonsteroidal anti-inflammatory drugs suing exogenous and endogenous sources of arachidonic acid. *J Pharmacol Exp Ther* 280: 606–613
16 Hulkower KI, Otis ER, Wernimont AK, Bell RL (1997) Stimulus dependence of non-steroidal anti-inflammatory drug potency in a cellular assay of prostaglandin H synthase-2. *Eur J Pharmacol* 331: 79–85
17 Warner TD, Giuliano F, Vojnovic I, Bukasa A, Mitchell JA, Vane JR (1999) Nonsteroid drug selectivities for cyclo-oxygenase-1 rather than cyclo-oxygenase-2 are associated with human gastrointestinal toxicity: a full *in vitro* analysis. *Proc Natl Acad Sci USA* 96: 7563–7568
18 Warner TD, Pairet M, Van Ryn J (2001) Test systems for inhibitors of cyclooxygenase-1 and cyclooxygenase-2. In: JR Vane, RM Botting (eds): *Therapeutic roles of selective COX-2 inhibitors.* William Harvey Press, London, 76–94
19 Sanda M, Jacob GB, Fliedner Jt L, Kennedy J, Gotz M (1987) Etodolac. In: AJ Lewis, DE Furst DE (eds): *Nonsteroidal Anti-inflammatory Drugs, Mechanisms and Use.* Marcel Dekker Inc, New York, 349–370
20 Jones R (2001) Etodolac: Clinical profile of an established cyclooxygenase-2 inhibitor. In: JR Vane, RM Botting (eds): *Therapeutic Roles of Selective COX-2 inhibitors.* William Harvey Press, London, 482–497
21 Glaser KB (1995) Cyclooxygenase selectivity and NSAIDs: cyclooxygenase-2 selectivity of etodolac (Lodine). *Inflammopharmacology* 3: 335–345
22 Dvornik D, Lee DKH (1989) Theoretical mechanism for the gastrointestinal safety of etorolac: selective sparing of cytoprotective prostaglandins. *Clin Rheumatol* 8 (Suppl. 1): 16–24
23 Denurson CA, Humber LG, Abraham NA, Schilling G, Martel RR, Pace-Asciak C (1983)

Resolution of etodolac and anti-inflammatory and prostaglandin synthetase inhibiting properties of the enantiomers. *J Med Chem* 26: 1778–1780
24 Rainsford KD (1999) Pharmacology and toxicology of ibuprofen. In: KD Rainsford (ed.): *Ibuprofen: A Critical Bibliographic Review.* Taylor and Francis, London, 143–275
25 Rainsford KD (1995) Gastric ulcerogenicity in mice of enantiomers of non-steroidal anti-inflammatory drugs having differing potency as prostaglandin synthesis inhibitors. *Pharmaceutical Sci* 1: 169–171
26 Ogiso T, Kitagawa T, Iwaki M, Tanino T (1997) Pharmacokinetic analysis of enterohepatic circulation of etodolac and effect of hepatic and renal injury on the pharmacokinetics. *Biol Pharm Bull* 20: 405–410
27 Strickmann DB, Blaschke G (2001) Isolation of an unknown metabolite of the non-steroidal anti-inflammatory drug etodolac and its identification as 5-hydroxy etodolac. *J Pharmaceutical Biomedical Analysis* 25: 977–984
28 Brocks DR, Jamali F (1990) The pharmacokinetics of etodolac enantiomers in the rat. Lack of pharmacokinetic interaction between enantiomers. *Drug Metabolism Disp* 18: 471–475
29 Laudanno OM, Cesolari JA, Esnarriaga J, Flaherty P, Vada J, Guastalli G, San Miguel P, Bedini OA (1998) *In vivo* selectivity of nonsteroidal anti-inflammatory drugs on COX-1-COX-2 and gastrointestinal ulcers, in rats. *Acta Gastroenterol Latinoam* 28: 249–255
30 Reimer ME, Johnston SA, Leib MS, Duncan RB Jr, Reimer DC, Marini M, Gimber TK (1999) The gastroduodenal effects of buffered aspirin, carprofen, and etodolac in healthy dogs. *J Vet Intern Med* 13: 472–477
31 Laudanno OM, Cesolarij JA, Esnarriaga J, San Miguel P, Bedini OA (2000) *In vivo* selectivity of nonsteroidal anti-inflammatory drugs and gastrointestinal ulcers in rats. *Dig Dis Sci* 45: 1359–1365
32 Wilcox GM, Porensky RS (1997) Acute colitis associated with etodolac. *J Clin Gastroenterol* 25: 367–368
33 Eis MJ, Watkins BM, Philip A, Welling RE (1998) Nonsteroidal-induced benign strictures of the colon: a case report and review of the literature. *Am J Gastroenterol* 93: 120–121
34 Gopal DV, Katon RM (1999) Endoscopic balloon dilation of multiple NSAID-induced colonic strictures: case report and review of literature on NSAID-related colopathy. *Gastrointest Endosc* 50: 120–123
35 Bjarnason I, Zanelli G, Smith T, Prouse P, Williams P, Smethurst P, Delacey G, Gumpel MJ, Levi AJ (1987) Nonsteroidal anti-inflammatory drug-induced intestinal inflammation in humans. *Gastroenterology* 93: 480–489
36 Bjarnason I, Hayllar J, MacPherson AJ, Russell AS (1993) Side-effects of nonsteroidal anti-inflammatory drugs on the small and large intestine in humans. *Gastroenterology* 104: 1832–1847
37 Svendsen KB, Bech JN, Sorensen TB, Pedersen EB (2000) A comparison of the effects of etodolac and ibuprofen on renal haemodynamics, tubular function, renin, vasopressin and urinary excretion of albumin and alpha-glutathione-S-transferase in healthy subjects: a placebo-controlled cross-over study. *Eur J Clin Pharmacol* 56: 383–388
38 Boni JP, Korth-Bradley JM, Martin P, Simcoe DK, Richards LS, Rennebohm R, Walson PD (1999) Pharmacokinetics of etodolac in patients with stable juvenile rheumatoid arthritis. *Clin Ther* 21: 1715–1724
39 Rothstein R (1998) Safety profiles of leading nonsteroidal anti-inflammatory drugs. *Am J Med* 105 (5A): 39S–43S
40 Russell RI (1999) Defining patients at risk of non-steroidal anti-inflamamtory drug gastropathy. *Ital J Gastroenterol Hepatol* 31:S14–S18
41 Neustadt DH (1997) The influence of age, gender, Helicobacter pylori and smoking on gastric mucosal adaptation to non-steroidal anti-inflammatory drugs. *Aliment Pharmacol Ther* 11: 907–912
42 Micklewright R, Lane S, Linley W, McQuade C, Thompson F, Maskrey N (2003) NSAIDs, gastroprotection and cyclo-oxygenase-II-selective inhibitors. *Aliment Pharmacol Ther* 17: 321–332
43 Wilcox GM, Porensky RS (1997) Acute colitis associated with etodolac. *J Clin Gastroenterol* 25: 367–368
44 Kawai S, Nishida S, Kato M, Furumaya Y, Okamoto R, Koshino T, Mizushima Y (1998) Comparison of cyclooxygenase-1 and –2 inhibitory activities of various nonsteroidal anti-infalmamtory drugs using human platelets and synovial cells. *Eur J Pharmacol* 347: 87–94
45 Lipscomb GR, Wallis N, Armstrong G, Goodman MJ, Rees WD (1995) Gastric mucosal adapta-

tion to etodolac and naproxen. *Aliment Pharmacol Ther* 9: 379–385
46 Lipscomb GR, Campbell F, Rees WD (1997) The influence of age, gender, Helicobacter pylori and smoking on gastric mucosal adaptation to non-steroidal anti-inflammatory drugs. *Aliment Pharmacol Ther* 11: 907–912
47 Kusuhara H, Komatsu H, Sumichika H, Sugahara K (1999) Reactive oxygen species are involved in the apoptosis induced by nonsteroidal anti-inflammatory drugs in cultured gastric cells. *Eur J Pharmacol* 383: 331–337
48 Noda M, Tasumi Y, Tomizawa M, Takama T, Mitsufuji S, Sugihara H, Kashima K, Hattori T (2002) Effects of etodolac, a selective cyclooxygenase-2 inhibitor, on the expression of E-cadherin-catenin complexes in gastrointestinal cell lines. *J Gasdtroenterol* 37: 896–904
49 Kusunoki N, Yamazaki R, Kawai S (2002) Induction of apoptosis in rheumatoid synovial fibroblasts by celecoxib, but not by other selective cyclooxygenase 2 inhbitors. *Arthritis Rheum* 46: 3159–3167
50 Yamazaki R, Kusunoki N, Matsuzaki T, Hashimoto S, Kawai S (2002) Selective cycloxygenase-2 inhibitors show a differential ability to inhibit proliferation and induce apoptosis of colon adenocarcinoma cells. *FEBS Lett* 531: 278–284
51 Sato TA, Keelan JA, Blumenstein M, Mitchell MD (2002) Efficacy and specificity of non-steroidal anti-inflammatory drugs for the inhibition of cytokine-stimulated prostaglandin E(2) secretion by amnion-derived WISH cells. *Prostaglandins Leukotrienes Essential Fatty Acids* 66: 525–527
52 Sato TA, Keelan JA, Mitchell MD (2002) Regulation of prostaglandin production in an immortalized human myometrial cell line by cytokines and non-steroidal anti-inflammatory drugs. *Eur J Obstet Gynecol Reprod Biol* 100: 158–162
53 Takahashi M, Katayama Y, Takada H, Kuwayama H, Terano A (2000) The effect of NSAIDs and a COX-2 specific inhibitor on Helicobacter pylori-induced PGE2 and HGF in human gastric fibroblasts. *Aliment Pharmacol Ther* 14: 44–49
54 Laine L, Cominelli F, Sloane R, Casini-Raggi V, Marin-Sorensen M, Weinstein WM (1995) Interaction of NSAIDs and Helicobacter pylori on gastrointestinal injury and prostaglandin production: a controlled double-blind trial. *Aliment Pharmacol Ther* 9: 127–135
55 Kishimoto Y, Takata N, Jinnai T, Morisawa T, Shiota G, Kawasaki H, Hasegawa J (2000) Sulindac and a cyclooxygenase-2 inhibitor, etodolac, increase APC mRNA in the colon of rats treated with azoxymethane. *Gut* 47: 812–819
56 Chen WS, Wei SJ, Liu JM, Hsiao M, Kou-Lin J, Yang WK (2001) Tumor invasiveness and liver metastasis of colon cancer cells correlated with cyclooxygenase-2 (COX-2) expression and inhibited by a COX-2 selective inhibitor, etodolac. *In J Cancer* 91: 894–899
57 Kamijo T, Sato T, Nagatomi Y, Kitamura T (2001) Induction of apoptosis by cyclooxygenase-2 inhibitors in prostate cancer cell lines. *Int J Urol* 8: S35–S39
58 Endo K, Sairyo K, Komatsubara S, Sasa T, Egawa H, Yonekura D, Adachi K, Ogawa T, Murakami R, Yasui N (2002) Cyclooxygenase-2 inhibitor inhibits the fracture healing. *J Physiol Anthropol Appl Human Sci* 21: 235–238
59 Kopp UC, Cicha MZ, Smith LA, Haeggstrom JZ, Samuelsson B, Hokfelt T (2000) Cyclooxygenase-2 involved in stimulation of renal mechanosensitive neurons. *Hypertension* 35: 373–378
60 Mabee CL, Mabee SW, Baker PB, Kirkpatrick RB, Levine EJ (1995) Fulminant hepatic failure associated with etodolac use. *Am J Gastroenterol* 90: 659–661
61 Degner F, Richardson B (2001) Review of gastrointestinal tolerability and safety of meloxicam. *Inflammopharmacology* 9: 71–80
62 Degner F, Lanes S, van Ryn J, Sigmund R (2001) Pharmacological and clinical profile of meloxicam. In: JR Vane, RM Botting (eds): *Therapeutic roles of selective COX-2 inhibitors.* William Harvey Press, London, 498–523
63 Van Bree H, Justus C, Quirke JF (1994) Preliminary observations on the effects of meloxicam in a new model for acute intra-articular inflammation in dogs. *Vet Res Commun* 18: 217–224
64 Rainsford KD, Skerry TM, Chindemi P, Delaney K (1999) Effects of the NSAIDs meloxicam and indomethacin on cartilage proteoglycan synthesis and joint responses to calcium pyrophosphate crystals in dogs. *Vet Res Commun* 23: 101–113
65 Stei P, Kruss B, Wiegleb J, Trach V (1996) Local tissue tolerability of meloxicam, a new NSAID: indications for parenteral, dermal and mucosal administration. *Brit J Rheumatol* 35: 44–50
66 Ghozlan PR, Bernhardt M, Vélicitat P, Bluhmki E (1996) Tolerability of multiple administration of intramuscular meloxicam: a comparison with intramuscular piroxicam in patients with rheuma-

toid arthritis or osteoarthritis. *Brit J Rheumatol* 35: 51–55
67 Engelhardt G (1996) Pharmacology of meloxicam, a new non-steroidal anti-inflammatory drug with an improved safety profile through preferential inhibition of COX-2. *Brit J Rheumatol* 35: 4–12
68 Trummlitz G, Engelhardt G (1987) Meloxicam: Structural aspects of a new NSAID with improved therapeutic properties. Poster presented at the XIth European Congress of Rheumatology, Athens, 28 June to 4 July 1987
69 Lehmann HA, Baumeister M, Lützen L, Wiegleb J (1996) Meloxicam: a toxicology overview. *Inflammopharmacology* 4: 105–123
70 Luger P, Daneck K, Engel W, Trummlitz G, Wagner (1996) Structure and physicochemical properties of meloxicam, a new NSAID. *Eur J Pharm Sci* 4: 175–187
71 Schoenfeld P (1999) Gastrointestinal safety profile of meloxicam: a meta-analysis and systematic review of randomised controlled trials. *Am J Med* 107 (6A): 48S–54S
72 McCormack K (1982) Mathematical model for assessing risk of gastrointestinal reactions to NSAIDs. In: KD Rainsford (ed.): *Azapropazone – 20 years of clinical use.* Kluwer Academic Publishers, London, 81–93
73 Rainsford KD (1999) Profile and mechanisms of gastrointestinal and other side-effects of nonsteroidal anti-inflammatory drugs (NSAIDs). *Am J Med* 107 (6A): 27S–36S
74 Engelhardt G, Homma D, Schnitzler C (1995) Meloxicam: a potent inhibitor of adjuvant arthritis in the Lewis rat. *Inflamm Res* 44: 548–555
75 Schmid J, Busch U, Heinzel G, Bozler G, Kaschke S, Kummer M (1995) Pharmacokinetics and metabolic pattern after intravenous infusion and oral administration to healthy subjects. *Drug Metab Dispos* 23: 1206–1213
76 Chesne C, Guyomard C, Guillouzo A, Schmid J, Ludwig E, Sauter T (1998) Metabolism of meloxicam in human liver involves cytochromes P4502C9 and 3A4. *Xenobiotica* 28: 1–13
77 US Food and Drug Administration (2002) MELOXICAM, MOBIC™ Tablets, 7.5 mg. Boehringer-Ingelheim Pharmaceuticals Inc (Sponsor). New Drug Application – NDA 20–938
78 Schmid J, Busch U, Trummlitz G, Prox A, Kaschke S, Wachsmuth H (1995) Meloxicam: Metabolic profile and biotransformation products in the rat. *Xenobiotica* 25: 1219–1236
79 Turck D, Roth W, Busch U (1998) A review of the clinical pharmacokinetics of meloxicam. *Br J Rheumatol* 35 (Suppl 1): 13–16
80 Busch U, Schmid J, Heinzel G, Schmaus H, Baierl J, Huber C, Roth W (1998) Pharmacokinetics of meloxicam in animals and the relevance to humans. *Drug Metab Dispos* 26: 576–584
81 Busch U, Engelhardt G (1990) Distribution of (^{14}C) meloxicam in joints of rats with adjuvant arthritis. *Drugs Exp Clin Res* 16: 49–52
82 Ludwig E, Schmid J, Beschke K, Ebner T (1999) Activation of human cytochrome P-450 3A4-catalyzed meloxicam 5'-methylhydroxylation by quinidine and hydroquinidine *in vitro*. *J Pharmacol Exp Ther* 290: 1–8
83 Xu HY, Zhong DF, Zhao LM, Zhang YF, Zhang BJ (2001) Pharmacokinetics of meloxicam in healthy Chinese volunteers. *Yao Xue Xue Bao* 36: 71–73
84 Lapicque F, Vergne P, Jouzeau JY, Loeuille D, Gillet P, Vignon E, Thomas P, Velicitat P, Turck D, Guillaume C et al (2000) Articular diffusion of meloxicam after a single oral dose: relationship to cyclo-oxygenase inhibition in synovial cells. *Clin Pharmacokinet* 39: 369–382
85 Pairet M, van Ryn J, Schierok H, Mauz A, Trummlitz G, Engelhardt G (1998) Differential inhibition of cyclooxygenases-1 and -2 by meloxicam and its 4'-isomer. *Inflamm Res* 47: 270–276
86 Jolliet P, Simon N, Bree F, Urien S, Pagliara A, Carrupt PA, Testa B, Tillement JP (1997) Blood-to-brain transfer of various oxicams: effects of plasma binding on their brain delivery. *Pharm Res* 14: 650–656
87 Poulsen Nautrup B, Horstermann D (1999) Pharmacodynamic and pharmacokinetic aspects of the non-inflammatory non-steroidal agent meloxicam in dogs. *Dtsch Tierarztl Wochenschr* 106: 94–100
88 Lees P, Sedgwick AD, Higgins AJ, Pugh KE, Busch U (1991) Pharmacodynamics and pharmacokinetics of meloxicam in the horse. *Brit Vet J* 147: 97–108
89 Yabe T, Honma M, Katsuki S, Wiegleb J, Luetzen L, Pueschner H, Lehmann H (1997) Single and repeated dose toxicity studies of meloxicam by oral administration in minipigs. *Pharmacometrics* 53: 197–212
90 Baert K, De Backer P (2003) Comparative pharmacokinetics of three non-steroidal anti-inflammatory drugs in five bird species. *Comp Biochem Physiol C Toxicol Pharmacol* 134: 25–33

91 US Food and Drug Administration (2000) New Animal Drug Application NADA 141-213. Metacam® (meloxicam) 0.5 mg/mL and 1.5 mg/mL Oral Suspension "indicated for the control of pain and inflammation associated with osteoarthritis in dogs. Boehringer Ingelheim Vetmedica, Inc. (Sponsor). Approval date April 15 2003

92 Euller-Ziegler L, Velicitat P, Bluhmki E, Türck D, Scheuerer S, Combe B (2001) Meloxicam: a review of its pharmacokinetics, efficacy and tolerability following intramuscular administration. *Inflamm Res* 50 (Suppl 1): S5–S9

93 Engelhardt G, Boegel R, Schnitzer C, Utzmann R (1996) Meloxicam: influence on arachidonic acid metabolism, part I. *In vitro* findings. *Biochem Pharmacol* 51: 21–28

94 Engelhardt G, Boegel R, Schnitzer C, Utzmann R (1996) Meloxicam: influence on arachidonic acid metabolism, part II. *In vivo* findings. *Biochem Pharmacol* 51: 29–38

95 Ogino K, Hatanaka K, Kawamura M, Ohno T, Harada Y (2000) Meloxicam inhibits prostaglandin E(2) generation via cyclooxygenase 2 in the inflammatory site but not that via cyclooxygenase 1 in the stomach. *Pharmacology* 61: 244–250

96 Rainsford KD, Ginsburg I, Gadd SJ (1997) A comparison between the effects of meloxicam and other NSAIDs on the production of oxyradicals by human polymorphonuclear leucocytes. *Immunopharmacology* 5: 9–19

97 Saleh TS, Calixto JB, Medeiros YS (1999) Effects of anti-inflammatory drugs upon nitrate and myeloperoxidase levels in the mouse pleurisy induced by carrageenan. *Peptides* 20: 949–956

98 Agha AM, El-Khatib AS, Al-Zuhair H (1999) Modulation of oxidant status by meloxicam in experimentally induced arthritis. *Pharmacol Res* 40: 385–392

99 Brzozowski T, Konturek PC, Konturek SJ, Sliwowski Z, Drozdowicz D, Stachura J, Pajdo R, Hahn EG (1999) Role of prostaglandins generated by cyclooxygenase-1 and cyclooxygenase-2 in healing of ischemia-reperfusion-induced gastric lesions. *Eur J Pharmacol* 385: 47–61

100 Brzozowski T, Konturek PC, Konturek SJ, Drozdowicz D, Pajdo R, Pawlik M (2000) Expression of cyclooxygenase (COX-1 and COX-2) in adaptive cytoprotection induced by mild stress. *J Physiol Paris* 94: 83–91

101 Villegas I, Martin MJ, La Casa C, Motilva V, Alarcon de la Lastra C (2000) Effects of meloxicam on oxygen radical generation in rat gastric mucosa. *Inflamm Res* 49: 361–366

102 Villegas I, Martin MJ, La Casa C, Motilva V, De La Lastra CA (2002) Effects of oxicam inhibitors of cyclooxygenase on oxidative stress generation in rat gastric mucosa. A comparative study. *Free Radic Res* 36: 769–777

103 Frode-Saleh TS, Calixto JB (2000) Synergistic anti-inflammatory effect of NfkappaB inhibitors and steroidal or non-steroidal anti-inflammatory drugs in the pleural inflammation induced by carrageenan in mice. *Inflamm Res* 49: 330–337

104 Bednarek D, Szuster-Ciesielska A, Zdzisinska B, Kondracki M, Paduch R, Kandefer-Szerszen M (1999) The effect of steroidal and non-steroidal anti-inflammatory drugs on the cellular immunity of calves with experimentally-induced local lung inflammation. *Vet Immunol Immunopathol* 71: 1–15

105 Lopez-Armada MJ, Sanchez-Pernaute O, Largo R, Diez-Ortego I, Palacios I, Egido J, Herrero-Beaumont G (2002) Modulation of cell recruitment by anti-inflammatory agents in antigen-induced arthritis. *Ann Rheum Dis* 61: 1027–1030

106 Rainsford KD, Ying C, Smith FC (1997) Effects of meloxicam, compared with other NSAIDs, on cartilage proteoglycan metabolism, synovial prostaglandin E_2, and production of interleukins 1, 6 and 8, in human and porcine explants in organ culture. *J Pharm Pharmacol* 49: 991–998

107 Bonta IL, Elliott GR (1992) Non-steroidal anti-inflammtory drugs and the augmented lipoxygenase pathway: conceivable impact on joint conditions. In: KD Rainsford, GP Velo (eds): *Side-effects of Non-inflammatory Drugs 3*. Kluwer Academic Publishers, Dordrecht, 269–274

108 Rainsford KD (2003) Cytokines and eicosanoids in arthritis. In: P.B. Curtis-Prior (ed.): *Eicosanoids*. Wiley, Chichester, 347–358

109 Bayliss MT, Roughley PJ (1985) The properties of proteoglycan prepared from human articlular cartilage by using associative cesium chloride gradients of high and low starting densities. *Biochem J* 232: 111–117

110 Konigsson K, Odensvik K, Kindahl H (2002) Endocrine, metabolic and clinical effects of intravenous endotoxin injection after pre-treatment with meloxicam in heifers. *J Vet Med A Physiol Pathol Clin Med* 49: 408–414

111 Roth J, Hubschle T, Pehl U, Ross G, Gerstberger R (2002) Influence of systemic treatment with cyclooxygenase inhibitors on lipopolysaccharide-induced fever and circulating levels of

cytokines and cortisol in guinea-pigs. *Pflugers Arch* 443: 411–417
112 Aguirre-Banuelos P, Granados-Soto V (2000) Evidence for the participation of the nitric oxide-cyclic GMP pathway in the antinociceptive action of meloxicam in the formalin test. *Eur J Pharmacol* 395: 9–13
113 Laird JM, Carter AJ, Grauert M, Cervero F (2001) Analgesic activity of a novel use-dependent sodium channel blocker, crobenetine, in mono-arthritic rats. *Br J Pharmacol* 134: 1742–1748
114 Chidiac JJ, Rifai K, Hawwa NN, Massaad CA, Jurjus AR, Jabbur SJ, Saade NE (2002) Nociceptive behaviour induced by dental application of irritants to rat incisors: a new model for tooth inflammatory pain. *Eur J Pain* 6: 55–67
115 Jain NK, Kulkarni SK, Singh A (2002) Modulation of NSAID-induced antinociceptive and ant-inflammatory effects by alpha2-adrenoceptor agonists with gastroprotective effects. *Life Sci* 70: 2857–2869
116 Doig PA, Purbrick KA, Hare JE, McKeown DB (2000) Clinical efficacy and tolerance of meloxicam in dogs with chronic osteoarthritis. *Can Vet J* 41: 296–300
117 Moreau M, Dupuis J, Bonneau NH, Desnoyers M (2003) Clinical evaluation of a nutraceutical, carprofen and meloxicam for the treatment of dogs with osteoarthritis. *Vet Rec* 152: 323–329
118 Mathews KA, Pettifer G, Foster R, McDonell W (2001) Safety and efficacy of preoperative administration of meloxicam, compared with that of ketoprofen and butorphanol in dogs undergoing abdominal surgery. *Am J Vet Res* 62: 882–888
119 Lascelles BD, Henderson AJ, Hackett IJ (2001) Evaluation of the clinical efficacy of meloxicam in cats with painful locomotor disorders. *J Small Anim Pract* 42: 587–593
120 Nell T, Bergman J, Hoeijmakers M, Van Laar P, Horspool LJ (2002) Comparison of vedaprofen and meloxicam in dogs with musculoskeletal pain and inflammation. *J Small Anim Pract* 43: 208–212
121 Slingsby LS, Waterman-Earson AE (2002) Comparison between meloxicam and carprofen for postoperative analgesia after feline ovariohysterectomy. *J Small Anim Pract* 43: 286–289
122 Reyes L, Tinworth KD, Li KM, Yau DF, Waters KA (2002) Observer-blinded comparison of two nonopioid analgesics for postoperative pain in piglets. *Pharmacol Biochem Behav* 73: 521–528
123 Budsberg SC, Cross AR, Quandt JE, Pablo LS, Runk AR (2002) Evaluation of intravenous administration of meloxicam for perioperative pain management following stifle joint surgery in dogs. *Am J Vet Res* 63: 1557–1563
124 Singsby LS, Waterman-Pearson AE (2000) Postoperative analgesia in the cat after ovariohysterectomy by use of carprofen, ketoprofen, meloxicam or tolfenamic acid. *J Small Anim Pract* 41: 447–450
125 Din C (2002) Do NSAIDs affect the progression of osteoarthritis? *Inflammation* 26: 139–142
126 Sadowski T, Steinmeyer J (2001) Effects of non-steroidal anti-inflammatory drugs and dexamethasone on the activity and expression of matrix metalloproteinase-1, matrix metalloproteinase-3 and tissue inhibitor of metalloproteinases-1 by bovine articular chondrocytes. *Osteoarthritis Cartilage* 9: 407–415
127 Lin SK, Kok SH, Kuo MY, Wang TJ, Wang JT, Yeh FT, Hsiao M, Lan WH, Hong CY (2002) Sequential expressions of MMP-1, TIMP-1, IL-6, and COX-2 genes in induced periapical lesions in rats. *Eur J Oral Sci* 110: 246–253
128 Bezerra MM, de Lima V, Alencar VB (2000) *In vitro* effects of nonsteroidal anti-inflammatory drugs on cyclooxygenase activity in dogs. *Am J Vet Res* 61: 802–810
129 Gomez-Gaviro MV, Gonzalez-Alvaro I, Dominguez-Jimenez C, Peschon J, Black RA, Sanchez-Madrid F, Diaz-Gonzalez F (2002) Structure-function relationship and role of tumor necrosis factor-alpha-converting enzyme in the down-regulation of L-selectin by non-steroidal anti-inflammatory drugs. *J Biol Chem* 277: 38212–38221
130 Moreno-Sanchez R, Bravo C, Vasquez C, Ayala G, Silveira LH, Martinez-Lavin M (1999) Inhibition and uncoupling of oxidative phosphorylation by nonsteroidal anti-inflammatory drugs: study in mitochondria, submitochondrial particles, cells, and whole heart. *Biochem Pharmacol* 57: 743–752
131 Pitcher GM, Henry JL (2001) Meloxicam selectively depresses the after discharge of rat spinal dorsal horn neurones in response to noxious stimulation. *Neurosci Lett* 305: 45–48
132 Brown WA, Skinner SA, Malcontenti-Wilson C, Vogiagis D, O'Brien PE (2001) Non-steroidal anti-inflammatory drugs with activity against either cyclooxygenase 1 or cyclooxygenase 2 inhibit colorectal cancer in a DMH rodent. *Gut* 48: 660–666
133 Hussey HJ, Tisdale MJ (2000) Effect of the specific cyclooxygenase-2 inhibitor meloxicam on

tumour growth and cachexia in a murine model. *Int J Cancer* 87: 95–100
134 Brown WA, Skinner SA, Malcontenti-Wilson C, Misajon A, DeJong T, Vogiagis D, O'Brien PE (2000) Non-steroidal anti-inflammatory drugs with different cyclooxygenase inhibitory profiles that prevent aberrant crypt foci formatin but vary in acute gastrotoxicity in a rat model. *J Gastroenterol Hepatol* 15: 1386–1392
135 Teisman P, Ferger B (2001) Inhibition of the cyclooxygenase isoenzymes COX-1 and COX-2 provide neuroprotection in the MPTP-mouse model of Parkinson's disease. *Synapse* 39: 167–174
136 Joussen AM, Poulaki V, Mitsiades N, Kirchhof B, Koizumi K, Dohmen S, Adamis AP (2002) Nonsteroidal anti-inflammatory drugs prevent early diabetic retinopathy via TNF-alpha suppression. *FASEB J* 16: 438–440
137 Pop-Busui R, Marinescu V, Van Huysen C, Li F, Sullivan K, Greene DA, Larkin D, Stevens MJ (2002) Dissection of metabolic, vascular, and nerve conduction interrelationships in experimental diabetic neuropathy by cyclooxygenase inhibition and acetyl-L-carnitine administration. *Diabetes* 51: 2619–2628
138 Miyamoto K, Oka N, Kawasaki T, Miyake S, Yamamura T, Akiguchi I (2002) New cyclooxygenase-2 inhibitors for treatment of experimental autoimmune neuritis. *Muscle Nerve* 25: 280–282
139 Mohn C, Lomniczi A, Faletti A, Scorticati C, Elverdin JC, McCann SM, Rettori V (2001) Effects of aminoguanidine and meloxicam on nitric oxide and prostaglandin E production induced by lipopolysaccharide in the hypothalamus and anterior pituitary of the rat. *Neuroimmunomodulation* 9: 276–285
140 Gupta YK, Chaudhary G, Sinha K (2002) Enhanced protection by melatonin and meloxicam combination in a middle cerebral artery occlusion model of acute ischemic stroke in rat. *Can J Physiol Pharmacol* 80: 210–217
141 Heindl B, Becker BF (2001) Aspirin, but not the more selective cyclooxygenase (COX)-2 inhibitors meloxicam and SC 58125, aggravates postischaemic cardiac dysfunction, independent of COX function. *Naunyn Schmiedebergs Arch Pharmacol* 363: 233–240
142 Moya M, Campana V, Gavotto A, Spitale L, Simes J, Palma J (2002) Hyperfibrinogenemia in rats treated with meloxicam. *Jpn Heart J* 43: 559–566
143 Arslan S, Erdem S, Sivri A, Hascelik Z, Tan E (2002) Exercise-induced apoptosis of rat skeletal muscle and the effect of meloxicam. *Rheumatology International* 21: 133–136
144 Laudanno OM, Cesolari JA, Esnarriaga J, San Miguel P, Bedini OA (2000) *In vivo* selectivity of nonsteroidal antiinflamamtory drugs and gastrointestinal ulcers in rats. *Dig Dis Sci* 45: 1359–1365
145 Villegas I, Alarcon de la Lastra C, La Casa C, Motilva V, Martin MJ (2001) Effects of food intake and oxidative stress on intestinal lesions caused by meloxicam and prioxicam in rats. *Eur J Pharmacol* 414: 79–86
146 Brzozowski T, Konturek PC, Konturek SJ, Sliwowski Z, Pajdo R, Drozdowicz D, Ptak A, Hahn EG (2001) Classic NSAID and selective cyclooxygenase (COX)-1 and COX-2 inhibitors in healing of chronic gastric ulcers. *Microscopy Research and Technique* 53: 343–353
147 Villegas I, Alarcon de la Lastra C, Martin MJ, Motilva V, La Casa Garcia C (2002) Gastric damage induced by subchronic administration of preferential cyclooxgyenase-1 and cyclooxygenase-2 inhibitors in rats. *Pharmacology* 66: 68–75
148 Birck R, Krzossok S, Knoll T, Braun C, van Der Woude FJ, Rohmeiss P (2000) Preferential COX-2 inhibitor, meloxicam, compromises renal perfusion in euvolemic and hypovolemic rats. *Exp Nephrol* 8: 173–180
149 Hawkey C, Kahan A, Steinbrück K, Alegre C, Baumelou E, Bégaud B, Dequeker J, Isomäki H, Littlejohn G, Mau J et al on behalf of the Melissa Study Group (1998) Gastrointestinal tolerability of meloxicam compared to diclofenac in osteoarthritis patients. *Brit J Rheumatol* 37: 937–945
150 Dequeker J, Hawkey C, Kahan A, Steinbrück K, Alegre C, Baumelou E, Bégaud B, Isomäki H, Littlejohn G, Mau J et al on behalf of the Select Study Group (1998) Improvement in gastrointestinal tolerability of the selective cyclooxygenase (COX)-2 inhibitor, meloxicam, compared with piroxicam: results of the safety and efficacy large-scale evalution of COX-inhibiting therapies (select) trial in osteoarthritis. *Brit J Rheumatol* 37: 946–951
151 Layton D, Heeley E, Hughes K, Shakir SAW (2003) Comparison of the incidence rates of selected gastrointestinal events reported for patients prescribed rofecoxib and meloxicam in general practice in England using prescription-event monitoring data. *Rheumatology* 42: 622–631
152 Hawkey CJ (2001) NSAIDs and COX-2 inhibitors: what can we learn from large outcomes trials? The gastroenterologist's perspective. *Clin Exp Rheumatol* 19 (Suppl 25): S23–S30

153 Hawkey CJ (1999) COX-2 inhibitors. *Lancet* 353: 307–314
154 Ward A, Brogden RN (1988) Nimesulide. A preliminary review of its pharmacological properties and therapeutic efficacy in inflammation and pain states. *Drugs* 36: 732–753
155 Kalgutkar AS (1999) Selective cyclooxygenase-2 inhibitors as non-ulcerogenic anti-inflammatory agents. *Exp Opin Ther Patents* 9: 831–849
156 Singla AK, Chawla M, Singh A (2000) Nimesulide: some pharmaceutical and pharmacological aspects – an update. *J Pharm Pharmacol* 52: 467–486
157 Swingle KF, Moore GGI, Grant TJ (1976) 4-nitro-2-phenoxymethanesulfonanilide (R 805): a chemically novel anti-inflammatory agent. *Arch Int Pharmacodyn* 221: 132–139
158 Rainsford KD (1996) Mode of action, uses, and side-effects of anti-inflammatory drugs. In: KD Rainsford (ed.): *Advances in Anti-Rheumatic Therapy*. CRC Press, Boca Raton. 59–111
159 Bennett A (2001) Nimesulide: A well-established cyclooxygenase-2 inhibitor with many other pharmacological properties relevant to inflammatory diseases. In: JR Vane, RM Botting (eds): *Therapeutic roles of selective COX-2 inhibitors.* William Harvey Press, London, 524–540
160 Bennett A, Villa G (2000) Nimesulide: an NSAID that preferentially inhibits COX-2, and has various unique pharmacological activities. *Expert Opinion on Pharmacotherapy* 1: 277–286
161 Bennett A, Berti F, Ferreira SH (eds) (1993) Nimesulide: A Multifactorial Therapeutic Approach to the Inflammatory Process? A 7-year clinical experience. *Drugs* 46: 1–283
162 Davis R, Brogden RN (1994) Nimesulide. An update of its pharmacodynamic and pharmacokinetic properties, and therapeutic efficacy. *Drugs* 48: 431–454
163 Rabasseda X (1997) Safety profile of nimesulide: ten years of clinical experience. *Drugs of Today* 33: 1–10
164 Bernareggi A (1998) Clinical pharmacokinetics of nimesulide. *Clin Pharmacokinet* 35: 247–274
165 Rainsford KD (1998) An analysis from clinico-epidemiological data of the principal adverse events from the COX-2 selective NSAID, nimesulide, with particular reference to hepatic injury. *Inflammopharmacology* 6: 203–221
166 Rainsford KD (2001) Nimesulide: Overview of properties and applications. *Drugs of Today* 37 (Supp B): 3–7
167 Bennett A (2001) Clinical importance of the multifactorial actions of nimesulide. *Drugs of Today* 37 (Supp B): 9–14
168 Fitzgerald D, McCrory C (2001) Pharmacology of COX-2 inhibition in man: anti-inflammatory and analgesic effects of nimesulide. *Drugs of Today*, 37 (Supp B): 15–20
169 Huskisson EC (2001) Nimesulide, a balanced drug for the treatment of osteoarthritis. *Clin Exp Rheumatol* 19 (I): S21–S25
170 Tavares IA, Borrelli F, Welsh NJ (2001) Inhibition of gastric acid secretion by nimesulide: a possible factor in its gastric tolerability. *Clin Exp Rheumatol* 19: S13–S15
171 Shah AA, Thjodleifsson B, Murray FE, Kay E, Barry M, Sigthorsson G, Gudjonsson H, Oddsson E, Price AB, Fitzgerald DJ et al (2001) Selective inhibition of COX-2 in humans is associated with less gastrointestinal injury: a comparison of nimesulide and naproxen. *Gut* 48: 339–346
172 Huskisson EC (2001) Nimesulide, a balanced drug for the treatment of osteoarthritis. *Clin Exp Rheumatol* 19 (Suppl 22): S21–S25
173 Rainsford KD (1999) Relationship of nimesulide safety to its pharmacokinetics: assessment of adverse reactions. *Rheumatology* 38: 4–10
174 Wober W (1999) Comparative efficacy and safety of nimesulide and diclofenac in patients with acute shoulder, and a meta-analysis of controlled studies with nimesulide. *Rheumatology* 38: 33–38
175 Liaropoulos L (1999) Economic comparison of nimesulide and diclofenac, and the incidence of adverse events in the treatment of rheumatic disease in Greece. *Rheumatology* 38: 39–46
176 Carini M, Aldini G, Stefani R, Marinello C, Facino RM (1998) Mass spectrometric characterization and HPLC determination of the main urinary metabolites of nimesulide in man. *J Pharm Biomed Anal* 18: 201–211
177 Boelsterli UA (2002) Mechanisms of NSAID-induced hepatotoxicity: focus on nimesulide. *Drug Safety* 25: 633–648
178 Boelsterli UA (2002) Nimesulide and hepatic adverse effects: roles of reactive metabolites and host factors. *Int J Clin Pract Suppl* 128: 30–36
179 Vietri M, De Santi C, Pietrabissa A, Mosca F, Pacifici GM (2000) Inhibition of human liver phenol sulfotransferase by nonsteroidal anti-inflammatory drugs. *Eur J Clin Pharmacol* 56: 81–87
180 Carrasco Portugal MC, Granados Soto V, Camacho Vieyra GA, Parez Urizar J, Flores Murrieta FJ

(2000) A simple method for determination of nimesulide in rat blood samples by high performance liquid chromatography. *J Liquid Chromatography and Related Technologies* 23: 2237–2244
181 Mukherjee P, Rachita C, Aisen PS, Pasinetti GM (2001) Non-steroidal anti-inflammatory drugs protect against chondrocyte apoptotic death. *Clin Exp Rheumatol* 19: S7–S11
182 Sanchez C, Mateus MM, Defresne MP, Crielaard JM, Reginster JY, Henrotin YE (2002) Metabolism of human articular chondrocytes cultured in alginate beads. Longterm effects of interleukin 1beta and nonsteroidal antiinflamamtory drugs. *J Rheumatol* 29: 772–782
183 Rainsford KD, Omar H, Ashraf A, Hewson AT, Bunning RAD, Rishiraj R, Shepherd P, Seabrook RW (2002) Recent pharmacodynamic and pharmacokinetic findings on oxaprozin. *Inflammopharmacology* 10: 185–239
184 Bavbek S, Celik G, Ediger D, Mungan D, Demirel YS, Misirligil Z (1999) The use of nimesulide in patients with acetylsalicylic acid and nonsteroidal anti-inflammtory drug intolerance. *J Asthma* 36: 657–663
185 Sanchez-Borges M, Capriles-Hulett A (2000) Atopy is a risk factor for non-steroidal anti-inflammatory drug sensitivity. *Ann Allergy Asthma Immunol* 84: 101–106
186 Rowlinson SW, Kiefer JR, Prusakiewicz J, Pawlilz JL, Kozak KR, Kalgutkar AS, Stallings WC, Kurumbail RG, Marnett LJ (2003) A novel mechanism of cyclooxygenase-2 inhibition involving interactions with Ser-530 and Tyr-385. *J Biol Chem* Aug 18, 2003, in press
187 García-Nieto R, Pérez C, Checa A, Gago F (1999) Molecular model of the interaction between nimesulide and human cyclooxygenase-2. *Rheumatology* 38: 14–18
188 Julemont F, de Leval X, Michaux C, Damas J, Charlier C, Durant F, Pirotte B, Dogne JM (2002) Spectral and crystallographic study of pyridinic analogues of nimesulide: determination of the active form of methanesulfonamides as COX-2 selective inhibitors. *J Med Chem* 45: 5182–5185
189 Rufer C, Schillinger E, Böttcher I, Repenthin W, Herrmann Ch (1982) Non-steroidal anti-inflammatories – XII. *Biochem Pharmacol* 22: 3591–3596
190 Swingle KF, Bell RL, Moore GGI (1985) Anti-inflammatory activity of antioxidants. In: KD Rainsford (ed.): *Anti-Inflammatory and Anti-Rheumatic Drugs. Volume III: Anti-Rheumatic Drugs, Experimental Agents and Clinical Aspects of Drug Use.* CRC Press, Boca Raton, 105–126
191 Orhan H, Dogruer DS, Cakir B, Sahin G, Sahin MF (1999) The *in vitro* effects of new non-steroidal anti-inflammatory compounds on antioxidant system of human erythrocytes. *Exp Toxicol Pathol* 51: 397–402
192 Maffei Facino R, Carini M, Aldini G, Saibene L, Morelli R (1995) Differential inhibition of superoxide, hydroxyl and peroxyl radicals by nimesulide and its main metabolite. *Drug Res* 45: 10–17
193 Zheng SX, Mouithys-Mickalad A, Deby-Dupont GP, Deby CM, Maroulis AP, Labasse AH, Lamy ML, Crielaard JM, Reginster JY, Henrotin YE (2000) *In vitro* study of the antioxidant properties of nimesulide and 4-OH nimesulide: effects of HRP- and luminal-dependent chemiluminescence produced by human chondrocytes. *Osteoarthritis and Cartilage* 8: 419–425
194 Mouithys-Mickalad AM, Zheng SX, Deby-Dupont GP, Deby CM, Lamy MM, Reginster JY, Henrotin YE (2000) *In vitro* study of the antioxidant properties of non steroidal anti-inflammatory drugs by chemiluminescence and electron spin resonance (ESR). *Free Radic Res* 33: 607–621
195 Tavares IA, Bishai PM, Bennett A (1995) Activity of nimesulide on constitutive and inducible cyclooxygenases. *Drug Res* 45: 1–3
196 Shah AA, Murray FE, Fitzgerald DJ (1999) The *in vivo* assessment of nimesulide cyclooxygenase-2 selectivity. *Rheumatology* 38: 19–23
197 Saeed SA, Afzal MN, Shah BH (1998) Dual effects of nimesulide, a COX-2 inhibitor, in human platelets. *Life Sci* 63: 1835–1841
198 Islas Cadena M, Aguirre Banuelos P, Granados Soto V (1999) Evidence for the participation of the nitric oxide-cyclic GMP pathway in the antinociceptive effect of nimesulide. *J Pharmacol Toxicol Methods* 42: 87–92
199 Miranda HF, Lopez J, Sierralta F, Correa A, Pinardi G (2001) NSAID antinociception measured in a chemical and thermal assay in mice. *Pain Res Manag* 6: 190–196
200 Sandrini G, Tassorelli C, Cecchini AP, Alfonsi E, Nappi G (2002) Effects of nimesulide on nitric oxide-induced hyperalgesia in humans – a neurophysiological study. *Eur J Pharmacol* 450: 259–262
201 Lopez R, Llinas MT, Roig F, Salazar FJ (2003) Role of nitric oxide and cyclooxygenase-2 in regulating the renal hemodynamic response to norepinephrine. *Am J Physiol Regul Integr Comp Physiol* 284: R488–R493

202 Korolkiewicz RP, Ujda M, Dabkowski J, Ruczynski J, Rekowski P, Petrusewicz J (2003) Differential salutary effects of nonselective and selective COX-2 inhibitors in postoperative ileus in rats. *J Surg Res* 109: 161–169
203 Caparroz-Assef SM, Salgueiro-Pagadigorria CL, Bersani-Amado CA, Bracht A, Kelmer-Bracht AM, Ishii-Iwamoto EL (2001) The uncoupling effect of the nonsteroidal anti-inflammatory drug nimesulide in liver mitochondria from adjuvant-induced arthritic rats. *Cell Biochem Funct* 19: 117–124
204 Mingatto FE, dos Santos AC, Rodrigues T, Pigoso AA, Uyemura SA, Curti C (2000) Effects of nimesulide and its reduced metabolite on mitochondria. *Brit J Pharmacol* 131: 1154–1160
205 Mingatto FE, Ridrigues T, Pigoso AA, Uyemura SA, Curti C, Santos AC (2002) The critical role of mitochondrial energetic impairment in the toxicity of nimesulide to hepatocytes. *J Pharmacol Exp Ther* 303: 601–607
206 Rainsford KD, Seabrook RW, Spencer S, Hewson AT (2001) Effects of nimesulide and its metabolites or manufacturing intermediates on the viability and growth of the human hepatoma HepG2 cell line. *Life Sci* 69: 2965–2973
207 Titos E, Chiang N, Serhan CN, Romano M, Gaya J, Pueyo G, Claria J (1999) Hepatocytes are a rich source of novel aspirin-triggered 15-epi-lipoxin A(4). *Am J Physiol* 277: C870–C877
208 Tardieu D, Jaeg JP, Deloly A, Corpet DE, Cadet J, Petit CR (2000) The COX-2 inhibitor nimesulide suppresses superoxide and 8-hydroxy-deoxyguanosine formation, and stimulates apoptosis in mucosa during early colonic inflammation in rats. *Carcinogenesis* 21: 973–976
209 Hida T, Kozaki K, Muramatsu H, Masuda A, Shimizu S, Mitsudomi T, Sugiura T, Ogawa M, Takahashi T (2000) Cyclooxygenase-2 inhibitor induces apoptosis and enhances cytotoxicity of various anticancer agents in non-small cell lung cancer cell lines. *Clin Cancer Res* 6: 2006–2011
210 Zhang Z, DuBois RN (2000) Par-4, a proapoptotic gene, is regulated by NSAIDs in human colon carcinoma cells. *Gastroenterology* 118: 1012–1017
211 Tian G, Yu JP, Luo HS, Yu BP, Yue H, Li JY, Mei Q (2002) Effect of nimesulide on proliferation and apoptosis of human hepatoma SMMC-7721 cells. *World J Gastroenterol* 8: 483–487
212 Maria DA, Ribeiro OG, Pizzocaro KF, De Franco M, Cabrera WK, Starobinas N, Gallois V, Siqueira M, Seman M, Ibanez OM (2000) Resistance to melanoma metastases in mice selected for high acute inflammatory response. *Carcinogenesis* 22: 337–342
213 Nakatsugi S, Ohta T, Kawamori T, Mutoh M, Tanigawa T, Watanabe K, Sugie S, Sugimura T, Wakabayashi K (2000) Chemoprevention by nimesulide, a selective cyclooxygenase-2 inhibitor, of 2-amino-1-methyl-6-phenylimidazo[4,5-b]pyridine (PhIP)-induced mammary gland carcinogenesis in rats. *Jpn J Cancer Res* 91: 886–892
214 Kawamori T, Wakabayashi K (2002) COX-2 and prostanoid receptors: good targets for chemoprevention. *J Environ Pathol Toxicol Oncol* 21: 149–153
215 Kubatka P, Kalicka K, Chamilova M, Ahlersova E, Ahlers I, Bojkova B, Adamekova E (2002) Nimesulide and melatonin in mammary carcinogenesis prevention in female Sprague-Dawley rats. *Neoplasma* 49: 255–259
216 Furukawa F, Nishikawa A, Lee IS, Kanki K, Umemura T, Okazaki K, Kawamori T, Wakabayashi K, Hirose M (2003) A cyclooxygenase-2 inhibitor, nimesulide, inhibits post-initiation phase of N-nitrosobis (2-oxopropyl)amine-induced pancreatic carcinogenesis in hamsters. *Int J Cancer* 104: 269–273
217 Vardar S, Baylas H, Huseyinov A (2003) Effects of selective cyclooxygenase-2 inhibition on gingival tissue levels of prostaglandin E2 and prostaglandin F2alpha and clinical parameters of chronic periodontitis. *J Periodontol* 74: 57–63
218 Yamazaki R, Kusunoki N, Matsuzaki T, Hashimoto S, Kawai S (2002) Selective cyclooxygenase-2 inhibitors show a differential ability to inhibit proliferation and induce apoptosis of colon adenocarcinoma cells. *FEBS Lett* 531: 278–284
219 Denda A, Kitayama W, Murata A, Kishida H, Sasaki Y, Kusuoka O, Tsujiuchi T, Tsutsumi M, Nakae D, Takagi H et al (2002) Increased expression of cyclooxygenase-2 protein during rat hepatocarcinogenesis caused by a choline-deficient, L-amino acid-defined diet and chemopreventive efficacy of a specific inhibitor, nimesulide. *Carcinogenesis* 23: 245–256
220 Mukherjee P, Pasinetti GM (2002) Altered gene expression during nimesulide-mediated inhibition of apoptotic death in human chondrocytes. *Int J Clin Pract Suppl* 128: 20–23
221 Wang D, An SJ, Wang WH, McGiff JC, Ferrari NR (2001) CaR-mediated COX-2 expression in primary cultured in TAL cells. *Am J Physiol Renal Physiol* 281: F658–F654
222 Kalajdzic T, Faour WH, He QW, Fahmi H, Martel-Pelletier J, Pelletier JP, Di Battista JA (2002)

Nimesulide, a preferential cyclooxygenase 2 inhibitor, suppresses peroxisome proliferators-activated receptor induction of cyclooxygenase 2 gene expression in human synovial fibroblasts: evidence for receptor antagonism. *Arthritis Rheum* 46: 494–506

223 Fahmi H, He Y, Zhang M, Martel-Pelletier J, Pelletier JP, Di Battista JA (2001) Nimesulide reduces interleukin-1beta-induced cyclooxygenase-2 gene expression in human synovial fibroblasts. *Osteoarthritis Cartilage* 9: 332–340

224 Sharma S, Rastogi S, Gupta V, Rohtagi D, Gulati P (1999) Comparative efficacy and safety of nimesulide *versus* piroxicam in osteoarthritis with special reference to chondroprotection. *Am J Ther* 6: 191–197

225 Garcia Rodriguez LA, Cattaruzzi C, Tronçon MG, Agostinis L (1998) Risk of hospitalization for upper gastrointestinal tract bleeding associated with ketorolac, other nonsteroidal anti-inflammtory drugs, calcium antagonists and other antihypertensive drugs. *Arch Intern Med* 158: 33–39

226 Conforti A, Leone R, Moretti U, Mozzo F, Velo G (2001) Adverse drug reactions related to the use of NSAIDs with a focus on nimesulide: results of spontaneous reporting fgrom a Northern Italian area. *Drug Safety* 24: 1081–1090

227 Macia MA, Carvajal A, del Pozo JG, Vera E, del Pino A (2002) Hepatotoxicity associated with nimesulide: data from the Spanish Pharmacovigilance System. *Clin Pharmacol Ther* 72: 596–597

228 Villa G (1999) NSAIDs and hepatic reactions. *Lancet* 353: 846

229 Chitturi S, Farrell GC (2001) Drug-induced cholestasis. *Sem Gastrointest Dis* 12: 113–124

230 Elmalem E (2000) Nimesulide, clavulanic acid and hepatitis. *J Intern Med* 248: 168–169

231 Kanwar AJ, Kaur S, Thami GP (2000) Nimesulide-induced purpura. *Dermatology* 201: 376

232 Sanchez-Borges M, Capriles-Hulett A (2000) Atopy is a risk factor for non-steroidal anti-inflammatory drug sensitivity. *Ann Allergy Asthma Immunol* 84: 101–106

233 Nettis E, Marcandrea M, Ferrannini A, Tursi A (2001) Tolerability of nimesulide and paracetamol in patients with NSAID-induced urticaria/angioedema. *Immunopharmacol Immunotoxicol* 23: 343–354

234 Naldi L, Conforti A, Venegoni M, Troncon MG, Caputi A, Ghiotto E, Cocci A, Moretti U, Velo G, Leone R (1999) Cutaneous reactions to drugs. An analysis of spontaneous reports in four Italian regions. *Br J Clin Pharmacol* 48: 839–846

235 Anonymous (1993) Cutaneous reactions to analgesic-antipyretics and nonsteroidal anti-inflammatory drugs. Analysis of reports to the spontaneous reporting system of the Gruppo Italiano Studi Epidemiologici in Dermatologia. *Dermatology* 186: 164–169

236 Tursen U, Kaya TI, Kokturk A, Dusmez D (2001) Lichenoid photodermatitis associated with nimesulide. *Int J Dermatol* 40: 767–768

237 Sanchez Borges M, Capriles-Hulett A, Caballero-Foneseca F, Perez CR (2001) Tolerability to new COX-2 inhibitors in NSAID-sensitive patients with cutaneous reactions. *Ann Allergy Asthma Immunol* 87: 201–204

238 Kanikkannan N, Jackson T, Shaik MS, Singh M (2001) Evaluation of skin sensitization potential of melatonin and nimesulide by murine local lymph node assay. *Eur J Pharm Sci* 14: 217–220

239 Mangalvedhekar SS, Gogtay NJ, Phadke AV, Gore S, Shah JM, Shah SM, Kshirsagar NA (2000) Adverse drug reactions postal survey-bronchial asthma and angioedema with nimesulide. *J Assoc Physicians India* 48: 548

240 Llinas MT, Rodriguez F, Moreno C, Salazar FJ (2000) Role of cyclooxygenase-2-derived metabolites and nitric oxide in regulating renal function. *Am J Physiol Regul Integr Comp Physiol* 279: R1641-R1646

241 Aisen PS, Schmeidler J, Pasinetti GM (2002) Randomized pilot study of nimesulide treatment of Alzheimer's disease. *Neurology* 58: 1050–1054

242 Food and Drug Administration (USA) Celebrex® (celecoxib capsules). Centre for Drug Evaluation. (CDER, New and Generic Drug Approvals: 1998–2003), NDA Application # 20-998, approval date 12/31/98. Pharmacological Review, part 1-7, posted 6/9/99. *www.fda.gov/cder/approval/index.htm* (Accessed Dec 2003)

243 Clemett D, Goa KL (2000) Celecoxib: a review of its use in osteoarthritis, rheumatoid arthritis and acute pain. *Drugs* 59: 957–980

244 Penning TD, Talley JJ, Bertenshaw SR, Carter JS, Collins PW, Docter S, Graneto MJ, Lee LF, Malecha JW, Miyashiro JM et al (1997) Synthesis and biological evaluation of the 1,5-diarylpyrazole class of cyclooxygenase-2 inhibitors: identification of 4-[5-(4-methylphenyl)-3-(trifluoromethyl)-*1H*-pyrazol-1-yl]benzene sulfonamide (SC-58635, cele-

coxib). *J Med Chem* 40: 1347–1365
245 Hamilton LC, Mitchell JA, Tomlinson AM, Warner TD (1999) Synergy between cyclo-oxygenase-2 induction and arachidonic acid supply *in vivo*: consequences for nonsteroidal anti-inflammatory drug efficacy. *FASEB J* 13: 245–251
246 Smith CJ, Zhang Y, Koboldt CM, Muhammad J, Zweifel BS, Shaffer A, Talley JJ, Masferrer JL, Seibert K, Isakson (1998) Pharmacological analysis of cycloxygenase-1 in inflammation. *Proc Natl Acad Sci USA* 95: 13313–13318
247 Gierse JK, Koboldt CM, Walker MC, Seibert K, Isakson PC (1999) Kinetic basis for selective inhibition of cyclo-oxygenases. *Biochem J* 339: 607–614
248 Davies NM, McLachlan AJ, Day RO, Williams KM (2000) Clinical pharmacokinetics and pharmacodynamics of celecoxib. A selective cyclo-oxygenase-2 inhibitor. *Clin Pharmacokinet* 38: 225–242
249 Tang C, Shou M, Mei Q, Rushmore TH, Rodrigues AD (2000) Major role of human liver microsomal cytochrome P450 2C9 (CYP2C9) in the oxidative metabolism of celecoxib, a novel cyclooxgyenase-II inhibitor. *J Pharmacol Exp Ther* 293: 453–459
250 Paulson SK, Engel L, Reitz B, Bolten S, Burton EG, Maziasz TJ, Yan B, Schoenhard GL (1999) Evidence for polymorphism in the canine metabolism of the cyclooxygenase 2 inhibitor, celecoxib. *Drug Metabolism and Disposition* 27: 1133–1142
251 Paulson SK, Zhang JY, Breau AP, Hribar JD, Liu NW, Jessen SM, Lawal YM, Cogburn JN, Gresk CJ, Markos CS et al (2000) Pharmacokinetics, tissue distribution, metabolism, and excretion of celecoxib in rats. *Drug Metabolism and Disposition* 28: 514–521
252 Paulson SK, Zhang JY, Jessen SM, Lawal Y, Liu NW, Dudkowski CM, Wang YF, Change M, Yang D, Findlay JW et al (2000) Comparison of celecoxib metabolism and excretion in mouse, rabbit, dog, cynomolgus monkey and rhesus monkey. *Xenobiotica* 30: 731–744
253 Zhang JY, Wang Y, Dudkowski C, Yang D, Chang M, Yuan J, Paulson SK, Breau AP (2000) Characterization of metabolites of celecoxib in rabbits by liquid chromatography/tandem mass spectrometry. *J Mass Spectrom* 35: 1259–1270
254 Tibble JA, Sigthorsson G, Forster R, Bjarnason I (2000) Comparison of the intestinal toxicity of celecoxib, a selective COX-2 inhibitor, and indomethacin in the experimental rat. *Scand J Gastroenterol* 35: 802–807
255 Paulson SK, Vaughn MB, Jessen SM, Lawal Y, Gresk CJ, Yan B, Maziasz TJ, Cook CS, Karim A (2001) Pharmacokinetics of celecoxib after oral administration in dogs and humans: effect of food and site of absorption. *J Pharmacol Exp Ther* 297: 638–645
256 Tanaka A, Hase S, Miyazawa T, Takeuchi K (2002) Up-regulation of cyclooxygenase-2 by inhibition of cyclooxygenase-1: a key to nonsteroidal anti-inflammatory drug-induced intestinal damage. *J Pharmacol Exp Ther* 300: 754–761
257 O'Beirne JP, Cairns SR (2001) Drug points: Cholestatic hepatitis in association with celecoxib. *BMJ* 323: 23
258 Pinheiro RM, Calixto JB (2002) Effect of the selective COX-2 inhibitors, celecoxib and rofecoxib in rat acute models of inflammation. *Inflamm Res* 51: 603–610
259 Cuzzocrea S, Mazzon E, Sautebin L, Dugo L, Serraino I, De Sarro A, Caputi AP (2002) Protective effects of celecoxib on lung injury and red blood cells modification induced by carrageenan in the rat. *Biochem Pharmacol* 63: 785–795
260 Francischi JN, Chaves CT, Moura AC, Lima AS, Rocha OA, Ferreira-Alves DL, Bakhle YS (2002) Selective inhibitors of cyclo-oxygenase-2 (COX-2) induce hypoalgesia in a rat paw model of inflammation. *Br J Pharmacol* 137: 837–844
261 Torres-Lopez JE, Ortiz MI, Castaneda-Hernandez G, Alonso-Lopez R, Asomoza-Espinosa R, Granados-Soto V (2002) Comparison of the antinociceptive effect of celecoxib, diclofenac and resveratrol in the formalin test. *Life Sci* 70: 1669–1676
262 Shin MC, Jang MH, Chang HK, Kim YJ, Kim EH, Kim CJ (2003) Modulation of cyclooxygenase-2 on glycine- and glutamate-induced ion currents in rat periaqueductal gray neurons. *Brain Res Bull* 59: 251–256
263 Kawamori T, Rao CV, Seibert K, Reddy BS (1998) Chemoprotective activity of celecoxib, a specific cyclooxygenase-2 inhibitor, against colon carcinogenesis. *Cancer Res* 58: 409–412
264 Jacoby RF, Seibert K, Cole CE, Kelloff G, Lubet RA (2000) The cyclooxygenase-2 inhibitor celecoxib is a potent preventative and therapeutic agent in the min mouse model of adenomatous polyposis. *Cancer Res* 60: 5040–5044
265 Alshafie GA, Abou-Issa HM, Seibert K, Harris RE (2000) Chemotherapeutic evaluation of cele-

coxib, a cyclooxygenase-2 inhibitor, in a rat mammary tumor model. *Oncol Rep* 7: 1377–1381
266 Grubbs CJ, Lubet RA, Koki AT, Leahy KM, Masferrer JL, Steele VE, Kelloff GJ, Hill DL, Seibert K (2000) Celecoxib inhibits N-butyl-N-(4-hydroxybutyl)-nitrosamine-induced urinary bladder cancers in male B6D2F1 mice and female Fischer-344 rats. *Cancer Res* 60: 5599–5602
267 Grosch S, Tegeder I, Niederberger E, Brautigam L, Geisslinger G (2001) COX-2 independent induction of cell cycle arrest and apoptosis in colon cancer cells by the selective COX-2 inhibitor celecoxib. *FASEB J* 15: 2742–2744
268 Rao CV, Indranie C, Simi B, Manning PT, Connor JR, Reddy BS (2002) Chemopreventive properties of a selective inducible nitric oxide synthase inhibitor in colon carcinogenesis, administered alone or in combination with celecoxib, a selective cyclooxygenase-2 inhibitor. *Cancer Res* 62: 165–170
269 Leahy KM, Ornberg RL, Wang Y, Zweifel BS, Koki AT, Masferrer JL (2002) Cyclooxygenase-2 inhibition by celecoxib reduces proliferation and incudes apoptosis in anagiogenic endothelial cells *in vivo*. *Cancer Res* 62: 625–631
270 Medhurst SJ, Walker K, Bowes M, Kidd BL, Glatt M, Muller M, Hattenberger M, Vaxelaire J, O'Reilly T, Wotherspoon G, Winter J, Green J, Urban L (2002) A rat model of bone cancer pain. *Pain* 96: 129–140
271 Orengo IF, Gerguis J, Phillips R, Guevara A, Lewis AT, Black HS (2002) Celecoxib, a cyclooxygenase 2 inhibitor as a potential chemopreventive to UV-induced skin cancer: a study in the hairless mouse model. *Arch Dermatol* 138: 751–755
272 Lu S, Zhang X, Badawi AF, El-Sohemy A, Archer MC (2002) Cyclooxygenase-2 inhibitor celecoxib inhibits promotion of mammary tumorigenesis in rats fed a high fat diet rich in n-6 polyunsaturated fatty acids. *Cancer Letters* 184: 7–12
273 Wenger FA, Kilian M, Bisevac M, Khodadayan C, von Seebach M, Schimke I, Guski H, Muller JM (2002) Effects of celebrex and zyflo on liver metastasis and lipid peroxidation in pancreatic cancer in Syrian hamsters. *Clin Exp Metastasis* 19: 681–687
274 Howe LR, Subbaramaiah K, Patel J, Masferrer JL, Deora A, Hudis C, Thaler HT, Muller WJ, Du B, Brown AM, Dannenberg AJ (2002) Celecoxib, a selective cyclooxygenase 2 inhibitor, protects against human epidermal growth factor receptor 2 (HER-2)/neu-induced breast cancer. *Cancer Res* 62: 5405–5407
275 Trifan OC, Durham WF, Salazar VS, Horton J, Levine BD, Zweifel BS, Davis TW, Masferrer JL (2002) Cyclooxygenase-2 inhibition with celecoxib enhances antitumor efficacy and reduces diarrhea side-effect of CPT-11. *Cancer Res* 62: 5778–5784
276 Blomme EA, Chinn KS, Hardy MM, Casler JJ, Kim SH, Opsahl AC, Hall WA, Trajkovic D, Khan KN, Tripp CS (2003) Selective cycloxygenase-2 inhibition does not affect the healing of cutaneous full-thickness incisional wounds in SKH-1 mice. *Br J Dermatol* 148: 211–223
277 Neiderberger E, Tegeder I, Vetter G, Schmidtko A, Schmidt H, Euchenhofer C, Brautigam L, Grosch S, Geisslinger G (2001) Celecoxib loses its anti-inflammatory efficacy at high doses through activation of NFκB. *FASEB J* 15: 1622–1624
278 Zweifel BS, Davis TW, Ornberg RL, Masferrer JL (2002) Direct evidence for a role of cyclooxygenase 2-derived prostaglandin E2 in human head and neck xenograft tumors. *Cancer Res* 62: 6706–6711
279 Subbaramaiah K, Dannenberg AJ (2003) Cyclooxygenase 2: a molecular target for cancer prevention and treatment. *Trends Pharmacol Sci* 24: 96–102
280 Casolini P, Catalani A, Zuena AR, Angelucci L (2002) Inhibition of COX-2 reduces the age-dependent increase of hippocampal inflammatory markers, corticosterone secretion, and behavioural impairments in the rat. *J Neurosci Res* 68: 337–343
281 Sairam K, Saravanan KS, Banerjee R, Mohanakumar KP (2003) Non-steroidal anti-inflammatory drug sodium salicylate, but not diclofenac or celecoxib, protects against 1-methyl-4-phenyl pyridinium-induced dopaminergic neurotoxicity in rats. *Brain Res* 966: 245–252
282 Drachman DB, Frank K, Dykes-Hoberg M, Teismann P, Almer G, Przedborski S, Rothstein JD (2002) Cyclooxygenase 2 inhibition protects motor neurons and prolongs survival in a transgenic mouse model of ALS. *Ann Neurol* 52: 771–778
283 Appel GB (2001) COX-2 inhibitors and the kidney. *Clin Exp Rheumatol* 19 (Suppl 25): S37–S40
284 Boyd IW, Mathew TH, Thomas MC (2000) COX-2 inhibitors and renal failure: the triple whammy revisited. *Medical Journal of Australia* 173: 274
285 Brater DC, Harris C, Redfern JS, Bertz BJ (2001) Renal effects of COX-2 selective inhibitors. *Am J Nephrol* 21: 1–15

286 Swan SK, Rudy DW, Lasseter KC, Ryan CF, Buechel KL, Lambrecht LJ, Pinto MB, Dilzer SC, Obrda O, Sundblad KJ et al (2000) Effect of cyclooxygenase-2 inhibition on renal function in elderly persons receiving a low-salt diet. *Ann Int Med* 133: 1–9

287 Whelton A, Fort JG, Puma JA, Normandin D, Bello AE, Verburg KM (2001) Cyclooxygenase-2-specific inhibitors and cardiorenal function: a randomized, controlled trial of celecoxib and rofecoxib in older hypertensive osteoarthritis patients. *Amer J Ther* 8: 85–95

288 Kiyataka M, Rich KA, Ingram M, Yamamoto T, Bing RJ (2002) Nitric oxide, anti-inflammatory drugs on renal prostaglandins and cyclooxygenase-2. *Hypertension* 39: 785–789

289 Hocherl K, Wolf K, Castrop H, Ittner KP, Bucher M, Kees F, Grobecker HF, Kurtz A (2001) Renocortical expression of renin and of cyclooxygenase-2 in response to angiotensin II AT1 receptor blockade is closely coordinated but not causally linked. *Pflugers Arch* 442: 821–827

290 Muscara MN, Vergnolle N, Lobren F, Triggle CR, Elliott SN, Asfaha S, Wallace JL (2000) Selective cyclo-oxygenase-2 inhibition with celecoxib elevates blood pressure and promotes leukocyte adherence. *Br J Pharmacol* 129: 1423–1430

291 Ozturk H, Ozdemir E, Otcu S, Buyukbayram H, Ihsan Dokucu A (2002) Renal effects on a solitary kidney of specific inhibition of cyclooxygenase-2 after 24 h of complete ureteric obstruction in rats. *Urol Res* 30: 223–226

292 Rossat J, Maillard M, Nussberger J, Brunner HR, Burnier M (1999) Renal effects of selective cyclooxygenase-2 inhibition in normotensive salt-depleted subjects. *Clin Pharmacol Ther* 66: 76–84

293 Kitahara M, Eitner F, Ostendorf T, Kunter U, Janssen U, Westenfeld R, Matsui K, Kerjaschki D, Floege J (2002) Selective cyclooxygenase-2 inhibition impairs glomerular capillary healing in experimental glomerulonephritis. *J Am Soc Nephrol* 13: 1261–1270

294 Reilly TP, Brady JN, Marchick MR, Bourdi M, George JW, Radonovich MF, Pise-Masison CA, Pohl LR (2001) A protective role for cyclooxygenase-2 in drug-induced liver injury in mice. *Chem Res Toxicol* 14: 1620–1628

295 Rainsford KD (1991) Uncoupling the toxicological morass in the development of new antirheumatic drugs—is there any hope? *Br J Rheumatol* 30: 161–163

296 Rainsford KD (1989) Gastrointestinal side-effects. In: CY Chang, AJ Lewis (eds): *Pharmacological Methods in the Control of Inflammation.* Alan R Liss, New York, 343–362

297 Kato S, Ogawa Y, Kanatsu K, Okayama M, Watanabe T, Arakawa T, Takeuchi K (2002) Ulcerogenic influence of selective cyclooxygenase-2 inhibitors in the rat stomach with adjuvant-induced arthritis. *J Pharmacol Exp Ther* 303: 503–509

298 Kato S, Takeuchi K (2002) Alteration of gastric ulcerogenic and healing responses in rats with adjuvant-induced arthritis. *Jpn J Pharmacol* 89: 1–6

299 Ma L, del Soldato P, Wallace JL (2002) Divergent effects of new cyclooxygenase inhibitors on gastric ulcer healing: shifting the angiogenic balance. *Proc Natl Acad Sci USA* 99: 13243–13247

300 Tibble J, Sigthorsson G, Caldwell C, Palmer RH, Bjarnason I (2001) Effects of NSAIDs on cryoprobe-induced gastric ulcer healing in rats. *Aliment Pharmacol Ther* 15: 2001–2008

301 Cuzzocrea S, Mazzon E, Serraino I, Dugo L, Centorrino T, Ciccolo A, Sautebin L, Caputi AP (2001) Celecoxib, a selective cyclo-oxygenase-2 inhibitor reduces the severity of experimental colitis induced by dinitrobenzene sulfonic acid in rats. *Eur J Pharmacol* 431: 91–102

302 Ostensen M (2001) Drugs in pregnancy. Rheumatological disorders. *Best Pract Res Clin Obstet Gynecol* 15: 953–969

303 Reese J, Zhao X, Ma WG, Brown N, Maziasz TJ, Dey SK (2001) Comparative analysis of pharmacologic and/or genetic disruption of cyclooxygenase-1 and cyclooxygenase-2 function in female reproduction in mice. *Endocrinology*, 142: 3198–3206

304 Takahashi Y, Roman C, Chemtob S, Tse MM, Lin E, Heymann MA, Clyman RI (2000) Cyclooxygenase-2 inhibitors constrict the fetal lamb ductus arteriosus both *in vitro* and *in vivo. Am J Physiol Regul Integr Comp Physiol* 278: R1496-R1505

305 Ozaki N, Beharry K, Nishihara KC, Akmal Y, Ang JG, Modanlou HD (2002) Differential regulation of prostacyclin and thromboxane by dexamethasone and celecoxib during oxidative stress in newborn rabbits. *Prostaglandins Other Lipid Mediat* 70: 61–78

306 Talley JJ, Brown DL, Carter JS, Graneto MJ, Koboldt CM, Masferrer JL, Perkins WE, Rogers RS, Shaffer AF, Zhang YY et al (2000) 4-[5-methyl-3-phenylisoxazol-4-yl]-benzenesulfonamide, valdecoxib: a potent and selective inhibitor of COX-2. *J Med Chem* 43: 775–777

307 Talley JJ, Bertenshaw SR, Brown DL, Carter JS, Graneto MJ, Kellogg MS, Koboldt CM, Yuan J, Zhang YY, Seibert K (2000) N-[[(5-methyl-3-phenylisoxazol-4-yl)-phenyl]sulfonyl]propan-

amide, sodium salt, paracoxib sodium: a potent and selective inhibitor of COX-2 for parenteral administration. *J Med Chem* 43: 1661–1663
308 US Food and Drug Adminstration, Centre for Drug Evaluation and Research (2002) Bextra (valdecoxib) tablets. Company: G. D. Searle and Co. Application No.: 21-341. Approval Date: 11/16/01 *www.fda.gov/cder/foi/nda/2001/21-341_Bextra.htm* (Accessed 25/10/2002)
309 Benson W, Weaver A, Espinoza L, Zhao WW, Riley W, Paperiello B, Recker DP (2002) Efficacy and safety of valdecoxib in treating the signs and symptoms of rheumatoid arthritis: a randomized, controlled comparison with placebo and naproxen. *Rheumatology* 41: 1008–1016
310 Makarowski W, Zhao WW, Bevirt T, Recker DP (2002) Efficacy and safety of the COX-2 specific inhibitor valdecoxib in the management of osteoarthritis of the hip: a randomized, double-blind, placebo-controlled comparison with naproxen. *Osteoarthritis and Cartilage* 10: 290–296
311 Sikes DH, Agrawal NM, Zhao WW, Kent JD, Recker DP, Verburg KM (2002) Incidence of gastroduodenal ulcers associated with valdecoxib compared with that of ibuprofen and diclofenac in patients with osteoarthritis. *Eur J Gastroenterol Hepatol* 14: 1101–1111
312 Goldstein JL, Kivitz AJ, Verburg KM, Recker DP, Kent JD (2003) A comparison of the upper gastrointestinal mucosal effects of valdecoxib, naproxen and placebo in healthy elderly subjects. *Aliment Pharmacol Ther* 18: 125–132
313 Strand V, Hochberg MC (2002) The risk of cardiovascular thrombotic events with selective cyclooxygenase-2 inhibitors. *Arthritis and Rheumatism* 47: 349–355
314 White WB, Verburg KM, Whelton A (2003) The COX-2 specific inhibitor, valdecoxib, is not associated with an increased risk of cardiovascular thrombotic events in arthritis patients
315 Prasit P, Wang Z, Brideau C, Chan C-C, Charleson S, Cromlish W, Ethier D, Evans JF, Ford-Hutchinson AW, Gauthier JY et al (1999) The discovery of rofecoxib, (MK 966, VIOXX[∀], 4-(4'-methylsulfonylphenyl)-3-phenyl-2(5*H*)-furanone], an orally active cyclooxygenase-2 inhibitor. *Bioorganic and Medicinal Chemistry Letters* 9: 1773–1778
316 Scott LJ, Lamb HM (1999) Rofecoxib: new drug profile. *Drugs* 58: 499–506
317 Ahuja N, Singla AK, Singh A, Singh B (2003) Rofecoxib: an update on physicochemical, pharmaceutical, pharmacodynamic and pharmacokinetic aspects. *J Pharm Pharmacol* 55: 859–894
318 Food and Drug Adminstration (USA): Vioxx® (rofecoxib). Centre for Drug Evaluation (CDER, New and Generic Drug Approvals: 1998–2003), NDA Application # 21-042, approval date 5/20/99. Pharmacological Review posted 4/17/00. *www.fda.gov/cder/approval/index.htm* (Accessed Dec 2003)
319 Chan CC, Boyce S, Brideau C, Charleson S, Cromlish W, Ether D, Evans J, Ford-Hutchinson AW, Forrest MJ, Gauthier JY et al (1999) Rofecoxib (Vioxx, MK-0966); 4-(4'-methylsulfonylphenyl)-3-phenyl-2-(5*H*)-furanone]: a potent and orally active cyclooxygenase-2 inhibitor. Pharmacological and biochemical profiles. *J Pharmacol Exp Ther* 290: 551–560
320 Nicoll-Griffith DA, Yergey JA, Trimble LA, Silva JM, Li C, Chauret N, Gauthier JY, Grimm E, Leger S, Roy P et al (2000) Synthesis, characterization, and activity of metabolites derived from the cyclooxygenase-2 inhibitor reofecoxib (MK-0966, Vioxx). *Bioorg Med Chem Lett* 10: 2683–2686
321 Greenberg HE, Gottesdiener K, Huntington M, Wong P, Larson P, Wildonger L, Gillen L, Dorval E, Waldman SA (2000) A new cyclooxygenase-2 inhibitor, rofecoxib (VIOXX®), did not alter the antiplatelet effects of low-dose aspirin in healthy volunteers. *J Clin Pharmacol* 40: 1509–1515
322 Feldman M, McMahon AT (2000) Do cyclooxygenase-2 inhibitors provide benefits similar to those of traditional nonsteroidal anti-inflammatory drugs, with less gastrointestinal toxicity? *Ann Intern Med* 132: 134–143
323 Lanza FL, Rack MF, Simon TJ, Quan H, Bolognese JA, Hoover ME, Wilson FR, Harper SE (1999) Specific inhibition of cyclooxygenase-2 with MK-0966 is associated with less gastroduodenal damage than either aspirin or ibuprofen. *Aliment Pharmacol Ther* 13: 761–767
324 Bombardier C, Laine L, Reicin A, Shapiro D, Burgos-Vargas R, Davis B, Day R, Ferraz MB, Hawkey CJ, Hochberg MC et al (2000) Comparison of upper gastrointestinal toxicity of rofecoxib and naproxen in patients with rheumatoid arthritis. *N Engl J Med* 343: 1520–1528
325 Depré M, Ehrich E, Van Hecken A, De Lepeleire I, Dallob A, Wong P, Porras A, Gertz BJ, De Schepper PJ (2000) Pharmacokinetics, COX-2 specificity, and tolerability of supratherapeutic doses of rofecoxib in humans. *Eur J Clin Pharmacol* 56: 167–174
326 Schoenfeld P (2001) An evidence-based approach to the gastrointestinal safety profile of COX-2 selective ani-inflammatories. *Gastroenterology Clinics of North America* 30: 1027–1044
327 Daniels B, Gertz B, Morrison B, Seidenberg B (2003) Renal safety profile of rofecoxib, a spe-

cific inhibitor of COX-2, in controlled clinical trials. European League of Rheumatology Congress (EULAR), Lisbon, Portugal

328 Nantel F, Meadows E, Denis D, Connolly B, Metters KM, Giaid A (1999) Immunolocalization of cyclooxygenase-2 in the macula densa of human elderly. *FEBS Letters* 457: 475–477

329 Layton D, Riley J, Wilton LV, Shakir SAW (2003) Safety profile of Rofecoxib as used in general practice in England: results of a prescription-event monitoring study. *Br J Clin Pharmacol* 55: 166–174

330 Cannon GW, Caldwell JR, Holt P, McLean B, Seidenberg B, Bolognese J, Ehrich E, Mukhopadhyay S, Daniels B (2000) Rofecoxib, a specific inhibitor of cyclooxygenase 2, with clinical efficacy comparable with that of diclofenac sodium. *Arthritis and Rheumatism* 43: 978–987

331 Baillie TA, Halpin RA, Matuszewski BK, Geer LA, Chavez-Eng CM, Dean D, Braun M, Doss G, Jones A, Marks T et al (2001) Mechanistic studies on the reversible metabolism of rofecoxib to 5-hydroxyrofecoxib in the rat: evidence for transient ring opening of a substituted 2-furanone derivative using stble isotope-labeling techniques. *Drug Metab Dispos* 29: 1614–1628

332 Halpin RA, Geer LA, Zhang Kem Marks TM, Dean DC, Jones AN, Melillo D, Doss G, Vyas KP (2000) The absorption, distribution, metabolism and excretion of rofecoxib, a potent and selective cyclooxygenase-2 inhibitor, in rats and dogs. *Drug Metab Dispos* 28: 1244–1254

333 Catella-Lawson F, McAdam B, Morrison BW, Kapoor S, Kujubu D, Antes L, Lasseter KC, Quan H, Gertz BJ, Fitzgerald GA (1999) Effects of specific inhibition of cyclooxygenase-2 on sodium balance, hemodynamics, and vasoactive eicosanoids. *J Pharmacol Exp Ther* 289: 735–741

334 Langman MJ, Jensen DM, Watson DJ, Harper SE, Zhao P, Quan H, Bolognese JA, Simon TJ (1999) Adverse upper gastrointestinal effects of rofecoxib compared with NSAIDs. *J Am Med Assn* 282: 1929–1933

335 Strand V, Hochberg MCC (2002) The risk of cardiovascular thrombotic events with selective cyclooxygenase-2 inhibitors. *Arthritis and Rheumatism* 47: 349–355

336 McAdam BF, Catella-Lawson F, Maradini IA, Kapoor S, Lawson JA, FitzGerald GA (1999) Systemic biosynthesis of prostacyclin by cyclooxygenase (COX)-2. The human pharmacology of a selective inhibitor of COX-2. *Proc Natal Acad Sci USA* 96: 272–277

337 Niederberger E, Tegeder I, Schafer C, Seegel M, Grosch S, Geisslinger G (2003) Opposite effects of rofecoxib on nuclear factor kappaB and activating protein-1 activation. *J Pharmacol Exp Ther* 304: 1153–1160

338 Callejas NA, Fernandez-Martinez A, Castrillo A, Bosca L, Martin-Sanz P (2003) Selective inhibitors of cyclooxygenase-2 delay the activation of the nuclear factor kappa B and attenuate the expression of inflammatory genes in murine macrophages treated with lipopolysaccharide. *Mol Pharmacol* 63: 671–677

339 Oshima M, Murai N, Kargman S, Arguello M, Luk P, Kwong E, Taketo MM, Evans JF (2001) Chemoprevention of intestinal polyposis in the Apcdelta716 mouse by rofecoxib, a specific cyclooxygenase-2 inhibitor. *Cancer Res* 61: 1733–1740

340 Martinez RV, Reval M, Campos MD, Terron JA, Dominguez R, Lopez-Munoz FJ (2002) Involvement of peripheral cycloxygenase-1 and cyclooxygenase-2 in inflammatory pain. *J Pharm Pharmacol* 54: 405–412

341 Sandrini M, Vitale G, Pini LA (2002) Effect of rofecoxib on nociception and the serotonin system in the rat brain. *Inflamm Res* 51: 154–159

342 El-Shenawy SM, Abdel-Salam OM, Baiuomy AR, El-Batran S, Arbid MS (2002) Studies on the anti-inflammatory and anti-nociceptive effects of melatonin in the rat. *Pharmacol Res* 46: 235–243

343 Hutchins B, Patel H, Spears R (2002) Attenution of pro-inflammatory neuropeptide levels produced by a cyclooxygenase-2 inhibitor in an animal model of chronic temporomandibular joint inflammation. *J Orofac Pain* 16: 312–316

344 Kalgutkar AS, Zhao Z (2001) Discovery and design of selective cyuclooxygenase-2 inhibitors as non-ulcerogenic, anti-inflammatory drugs with potential utility as anti-cancer agents. *Curr Drug Targets* 2: 79–106

345 Dicker AP, Williams TL, Grant DS (2001) Targeting angiogenic processes by combination rofecoxib and ionizing radiation. *Am J Clin Oncol* 24: 438–442

346 Brzozowski T, Konturek PC, Konturek SJ, Pajdo R, Schuppan D, Drozdowicz D, Ptak A, Pawlik M, Nakamura T, Hahn EG (2000) Involvement of cyclooxygenase (COX)-2 products in acceleration of ulcer healing by gastrin and hepatocyte growth factor. *J Physiol Pharmacol* 51: 751–773

347 Laudanno OM, Cesolari JA, Esnarriaga J, Rista L, Piombo G, Maglione C, Aramberry L,

Sambrano J, Godoy A, Rocaspana A (2001) Gastrointestinal damage induced by celecoxib and rofecoxib in rats. *Dig Dis Sci* 46: 779–784
348 Pajdo R, Brzozowski T, Konturek PC, Kwiecien S, Konturek SJ, Sliwowski Z, Pawlik M, Ptak A, Drozdowicz D, Hahn EG (2001) Ischemic preconditioning, the most effective gastroprotective intervention: involvement of prostaglandins, nitric oxide, adenosine and sensory nerves. *Eur J Pharmacol* 427: 263–276
349 Colville-Nash PR, Gilroy DW (2001) Potential adverse effects of cyclooxygenase-2 inhibition: evidence from animal models of inflammation. *BioDrugs* 15: 1–9
350 Simon AM, Manigrasso MB, O'Connor JP (2002) Cyclo-oxygenase 2 function is essential for bone fracture healing. *J Bone Miner Res* 17: 977–978
351 Goodman S, Ma T, Trindade M, Ikenoue T, Matsuura I, Wong N, Fox N, Genovese M, Regula D, Smith RL (2002) COX-2 selective NSAID decreases bone in growth *in vivo*. *J Orthop Res* 20: 1164–1169
352 Cochrane DJ, Jarvis B, Keating GM (2002) Etoricoxib. *Drugs* 62: 2637–2651
353 Riendeau D, Percival MD, Brideau C, Charleson S, Dubé D, Ethier D, Falgueyret J-P, Friesen RW, Gordon R, Greig G et al (2001) Etoricoxib (MK-0663): preclinical profile and comparison with other agents that selectively inhibit cyclooxygenase-2. *J Pharmacol Exp Ther* 296: 558–566
354 Agrawal NG, Porras AG, Matthews CZ, Rose MJ, Woolf EJ, Musser BJ, Dynder AL, Mazina KE, Lasseter KC, Hunt et al (2003) Single- and multiple-dose pharmacokinetics of etoricoxib, a selective inhibitor of cyclooxygenase-2, in man. *J Clin Pharmacol* 43: 268–276
355 Gottesdiener K, Schnitzer T, Fisher C, Bockow B, Markenson J, Ko A, Detora L, Curtis S, Geissler L, Gertz BJ (2002) Results of a randomized, dose-ranging trial of etoricoxib in patients with osteoarthritis. *Rheumatology* 41: 1052–1061
356 Matsumoto AK, Melian A, Mandel DR, McIlwain HH, Borenstein D, Zhao PL, Lines CR, Gertz BJ, Curtis S (2002) A randomized, controlled, clinical trial of etoricoxib in the treatment of rheumatoid arthritis. *J Rheumatol* 29: 1623–1630
357 Collantes E, Curtis SP, Lee KW, Casas N, McCarthy T, Melian A, Zhao PL, Ridgers DB, McCormick CL, Lee M, Lines CR, Gertz BJ (2002) A multinational randomized, controlled, clinical trial of etoricoxib in the treatment of rheumatoid arthritis. *BMC Family Practice* 3: 2–10
358 Sorbera LA, Castaner J, Bayes M, Silvestre JS (2002) Lumiracoxib. *Drugs of the Future* 27: 740–747
359 Reynolds C, Scott G, Looby M, Milosavljev S, Huff JP, Ruff DA, Rordorf C (2003) Pharmacokinetics of lumiracoxib in synovial fluid and plasma of patients with rheumatoid arthritis. European Assoc. Clinical Pharmacology and Therapeutics (EACPT), Istanbul; P-195 (Abstr)
360 Scott G, Branson J, Miloszvljev S, Rordorf C, Haraoui B, Ouellet J-P, Schell E (2003) Lumiracoxib demonstrates dose-proportional and time-independent pharmacokinetics in patients with osteoarthritis of the knee. *Ann Rheum Dis* 62 (Suppl 1): FRI235 (Abstr)
361 Bonner J, Branson J, Milosavljev S, Rordorf C, Scott G (2003) Co-administration of lumiracoxib and warfarin does not alter the pharmacokinetic profile of *R*- or *S*-warfarin. *Ann Rheum Dis* 62 (Suppl 1): FRI0225 (Abstr)
362 Hartman S, Scott G, Rordorf C, Campesrini J, Branson J, Keller U (2003) Lumiracoxib demonstrates high absolute bioavailability in healthy subjects. European Assoc. Clinical Pharmacology and Therapeutics (EACPT), Istanbul; P-199 (Abstr)
363 Wilding IR, Connor AL, Carpenter P, Rordorf C, Branson J, Milosavljev S, Scott G (2003) Lumiracoxib shows similar bioavailability at different sites in the gastrointestinal tract. European Assoc. Clinical Pharmacology and Therapeutics (EACPT), Istanbul; P-196 (Abstr)
364 Grifka J, Zacher J, Brown J, Seriolo B, Lee A, Moore A, Gimona A (2003) Lumiracoxib is effective and well tolerated in patients with osteothritis of the hand: a randmomized placebo-controlled trial. *Ann Rheum Dis* 62 (Suppl 1): FRI0222 (Abstr)
365 Benevolenskaya L, Tüzün S, Hagin E, Moore A, Gimona A (2003) Lumiracoxib is effective in relieving symptoms of osteoarthritis of the hip or knee after 4 weeks of treatment: results from a randomized placebo-controlled trial. *Ann Rheum Dis* 62 (Suppl 1): FRI0246 (Abstr)
366 Fleischmann R, Sheldon E, Maldonado Cocco J, Yu S, Dutta D, Usiskin K (2003) A prospective randomized 13-week study revaluating the efficacy of lumiracoxib in patients with osteoarthritis of the knee. *Ann Rheum Dis* 62 (Suppl 1): FRI0233 (Abstr)

COX-2 Inhibitors
Edited by M. Pairet and J. van Ryn

Efficacy and gastrointestinal safety of selective COX-2 inhibitors

Frank Degner

CDept. Medical Affairs, Boehringer Ingelheim GmbH, Binger Str., D-55216 Ingelheim/Rhein, Germany

Introduction

Nonsteroidal anti-inflammatory drugs (NSAIDs) are among the most commonly prescribed drugs in the world and are used frequently as anti-inflammatory and analgesic agents in the treatment of musculoskeletal conditions. Their use is mainly limited by adverse effects, which are more frequently seen with NSAIDs than with any other prescription drug. These side-effects are of particular importance in the elderly who are more prone to NSAID complications [1].

The background rate of ulcer complications in patients not using NSAIDs varies with risk factors, such as age, but has been estimated to be between 0.2 and one event per 100 patient years. Epidemiological studies have fairly consistently shown a four-fold increase of ulcer complications in patients using NSAIDs compared with control individuals, with an absolute rate in the NSAID group between one and four events per 100 patient years. The fraction of disease attributable to NSAID use has been estimated to be between 25% and 35%, although increased use of low-dose aspirin may have enhanced this. In one study, NSAID use was associated with 1.25 extra hospitalizations per 100 patient years for ulcer complications. Based on figures from the Office of Population Census and Surveys, there are 4000 deaths per annum from peptic ulcer disease in the UK, implying that about 1200 of them are caused by non-aspirin NSAIDs. Recent estimates of 2000 deaths per annum in the UK and 16,500 deaths per annum in the USA are at the upper end of a range of estimates of the mortality attributable each year to these drugs.

The initial hypothesis concerning NSAID action was that they inhibited a single cyclooxygenase (COX) enzyme, and would, therefore, have inseparable efficacy and toxicity [2, 3]. Preliminary data, which suggested a variation in gastrotoxicity, were explained by NSAID use in different disease subsets and age groups, differences in the half-life of NSAIDs and their potency, and fixed-dose prescribing [4].

Subsequent large studies have, however, provided interesting information about relative NSAID toxicity. Singh et al. have developed a "toxicity index" to give an overall measure of the differential adverse effect profile of NSAIDs based on information obtained in patients with rheumatoid arthritis from the Arthritis, Rheumatism and Aging Medical Information System (ARAMIS) database [5]. Langman et al. and Garcia Rodriguez et al. have assessed NSAID use in cases of upper gastrointestinal bleeds and were able to calculate the odds ratios of commonly used NSAIDs for gastrointestinal bleeds [6, 7]. These studies demonstrate that azapropazone, piroxicam and indomethacin have the highest risk of gastrointestinal bleeding, with ibuprofen having a low risk. The risk factors for gastrointestinal adverse events were older age, previous history of gastrointestinal bleed and higher dose of NSAID. Langman et al. noted that peptic ulcer bleeding was strongly associated with the use of NSAIDs in the 3 months prior to admission, and that the risk for most NSAIDs tended to be greatest in those recently starting treatment. This group were also more likely to be on the maximum recommended dose. More recently, MacDonald et al. have demonstrated that gastrointestinal risk from NSAIDs seems to persist with continuous use, and there may even be carry-over toxicity after NSAID cessation [8]. The discovery of the two different forms of COX enzymes (COX-1 and COX-2) expanded on the original concepts of Sir John Vane [9–11], and led to the hypothesis that NSAID toxicity was related to COX-1 inhibition, and that selective COX-2 inhibitors would retain the efficacy of nonselective NSAIDs without their COX-1 mediated side-effects [12]. Many reports on the relative COX-1 and COX-2 inhibition of currently used NSAIDs have subsequently appeared [13], and while their clinical studies demonstrate a wide variation in the incidence of gastrointestinal side-effects with existing NSAIDs, COX-2 selectivity does appear to be associated with a decreased incidence of these side-effects [14–19].

The COX-1-selective drugs indomethacin and piroxicam are associated with high toxicity indices whereas ibuprofen, which is more selective for COX-2, is associated with a low index. Aspirin, however, is the most COX-1-selective drug in this series and has a low toxicity index, suggesting that other factors are also important in determining toxicity. In epidemiological studies, ibuprofen has consistently been associated with lower than average risks of ulcer complications, compared with other nonselective NSAIDs, while other agents, such as azapropazone and piroxicam, have generally been associated with high risks. The safety advantage of ibuprofen is, however, only seen with the lower doses of 1200 mg traditionally used in Europe, rather than the higher doses often used in the USA. Ibuprofen is also a drug with a short serum half-life that persists in joints for much longer, thereby achieving selectively enhanced joint exposure compared with gastric exposure. By contrast, in drugs with a long half-life (such as piroxicam or tenoxicam) no such selectivity is achieved.

Reducing the risks of current NSAIDs use

When prescribing NSAIDs, taking a careful drug history is important to assess the possibility of drug interactions, especially in the elderly. Over-the-counter

NSAIDs are increasingly available and their inadvertent co-use is something that all physicians need to be aware of. Concomitant aspirin, even in low dose, and corticosteroid use increases risk of peptic ulceration with NSAIDs. NSAIDs should be avoided wherever possible in those patients on anticoagulation treatment because of their anti-platelet and protein binding effects. NSAIDs should be used in the lowest possible doses, and it seems prudent initially to use those with lower toxicity indices [20]. There is no significant evidence that controlled release, intramuscular or rectal preparations reduce risks of gastrointestinal adverse effects.

In clinical practice, many patients with risk factors, and those who develop dyspepsia, are prescribed gastroprotective agents (even though the relationship between dyspepsia and peptic ulcer disease is unclear), making the drug combination less cost effective. In chronic NSAID use, standard dose histamine-type 2 (H2)-antagonists reduce the incidence of endoscopic duodenal ulcers, but not gastric ulcers, whereas double-dose H2-antagonists and standard dose proton-pump inhibitors reduce both duodenal and gastric ulcers [21, 22]. Misoprostol 400 mg/day and 800 mg/day reduce gastric and duodenal ulcers, with the higher dose being more effective for reducing gastric ulcers. Notably, only misoprostol at the higher dose has been demonstrated to reduce clinical adverse events, although diarrhoea is a common side-effect at this dose. Sucralfate heals duodenal ulcers as effectively as H2-antagonists, whether or not NSAIDs are continued [23]. Prophylactic gastroprotective agents should be considered in the elderly, those with a history of peptic ulcer disease, those on concomitant corticosteroids, those who develop dyspepsia and those with serious co-morbid conditions.

Efficacy of selective COX-2 inhibitors in rheumatology

Rheumatoid arthritis

Rheumatoid arthritis is a severe inflammatory polyarthritis, with a peak incidence in premenopausal women. It has a prevalence of 0.5–1.0%. Although the causes of rheumatoid arthritis are not fully understood, evidence suggests that pro-inflammatory cytokines play an important role in pathogenesis. It is clear that early on in disease, the major events are cytokine-driven prostaglandin (PG)-mediated inflammatory effects, and later on, cytokine-driven osteoclast activation with subsequent bone damage. In rheumatoid arthritis, COX-2 is markedly up-regulated in inflamed synovial tissue, with resultant increased PG production.

The common feature of disease-modifying anti-rheumatic drugs (DMARDs) is the blockade of cytokines. By doing so, the mediators of bone destruction and of stimulants to the hepatic acute phase response are blocked. With the exception of (nonlicensed) tenidap, the existing NSAIDs inhibit COX alone without any effect on systemic cytokine levels. NSAIDs, however, remain the cornerstone of symptomatic treatment in rheumatoid arthritis. Although clinical experience has shown that NSAIDs do not appear to influence radiological progression, evidence seems to suggest that they improve Health Assessment Questionnaire (HAQ) scores and reduce swollen and tender joint scores in rheumatoid arthritis. Imaging studies, using magnetic resonance imaging and ultrasonography, of early rheumatoid arthritis have also demonstrated the ability of NSAIDs to reduce synovial volume [24].

Selective COX-2 inhibition may well have extra benefits in the control of synovitis. Whether or not selective COX-2 inhibitors given at high doses in early disease could influence disease progression in rheumatoid arthritis remains to be tested in a formal programme. In rheumatoid arthritis, up-regulation of tissue necrosis factor (TNF) is associated with increased COX-2 production. One of the features of anti-TNF therapy is the quick reduction in synovitis. The speed of this response suggests that down-regulation of COX-2 may be involved. Another potential benefit is that COX-2 is expressed in endothelial cells, particularly in situations of shear stress and angiogenesis. If inhibitors have an effect on angiogenesis then clearly there would be a potential role in inhibiting synovial growth.

Osteoarthritis

Osteoarthritis is the most common joint disorder in the world and affects all joints, but most commonly the knees, hips, hands and spine. The prevalence of osteoarthritis increases with age and presents with use-related pain, loss of joint motion and disability. Patients with osteoarthritis have an inflammatory phase with synovitis that is important in the disease progression. Up-regulation of COX-2 has been detected in the synovium from patients with osteoarthritis, although to a lesser extent than in rheumatoid arthritis. There is also evidence of up-regulation of COX-2 and subsequent PG production in osteoarthritic cartilage.

Osteoarthritis presents practical difficulties with current NSAID use in that it is prevalent in that group of patients most at risk of gastrointestinal toxicity, namely the elderly. Patients often find NSAIDs to be the most effective pain relievers in osteoarthritis, creating a major conflict between efficacy and toxicity. It is worth noting that pain is the major reason for joint replacement to be undertaken, and that COX-2 is involved in pain transmission both peripherally and centrally.

Again, there are potential advantages to using selective rather than nonselective NSAIDs in osteoarthritis subgroups with synovitis. Although animal model studies suggest meloxicam, unlike indomethacin, does not inhibit cartilage proteoglycan synthase, the effects of COX-2 inhibition on PG-mediated bone turnover and chondrocyte biology are yet to be determined and further studies will help delineate the role of the COX isoenzymes in bone biology.

Other rheumatological conditions

The potential role of selective COX-2 inhibition in controlling synovitis may also be relevant in systemic lupus erythematosus (SLE), where high-dose inhibition may result in better symptom relief and importantly, a steroid-sparing effect. COX-2 is up-regulated in mesangial cells in lupus nephritis and COX-2 inhibition, in particular without any COX-1 related loss of glomerular filtration rate may well prove beneficial in this group [25].

Up-regulation of COX-2 has also been demonstrated in other inflammatory arthritides, such as ankylosing spondylitis and psoriatic arthritis. In ankylosing spondylitis, adequate analgesia coupled with exercise is the mainstay of treatment to minimise loss of spinal movement. NSAID use has been associated with flares of skin lesions in psoriasis, which is thought to be secondary to an increase in lipoxygenase pathway products as a result of COX inhibition. Selective COX-2 inhibition may change the balance of PGs and lipoxygenase products resulting in fewer exacerbations of skin lesions.

COX-2-selective agents registered for use in rheumatological conditions

Meloxicam is licensed for use in rheumatoid arthritis, osteoarthritis and ankylosing spondylitis at doses of 7.5–15 mg once daily and is superior to placebo and comparable with piroxicam and diclofenac in the treatment of osteoarthritis of the hip and knee [26–28]. In the symptomatic treatment of rheumatoid arthritis, meloxicam is superior to placebo and comparable with naproxen and diclofenac [29–31]. It is also superior to placebo and comparable with piroxicam for symptom relief in ankylosing spondylitis [32].

Celecoxib is licensed for use in osteoarthritis at 100–200 mg once daily and in rheumatoid arthritis at 100–200 mg twice daily, and is also licensed for use in the USA as an adjunct to other prophylactic therapy in familial adenomatous polyposis (FAP). Celecoxib is superior to placebo and as effective as naproxen and diclofenac for symptomatic relief in rheumatoid arthritis [33] and osteoarthritis of the knee [34–37].

Rofecoxib is licensed for use in osteoarthritis at doses of 12.5–25 mg once daily. It is superior to placebo and comparable with ibuprofen and diclofenac for the treatment of osteoarthritis of the hip and knee [38–44], although there are studies showing significantly greater dropouts due to lack of efficacy with rofecoxib than with diclofenac [45]. It is also licensed for use in rheumatoid arthritis and is superior to placebo [46], achieving American College Rheumatology (ACR) 20 Remission in 43.9% and 49.7% of patients taking 25 mg and 50 mg daily, respectively. The drug is not registered for long-term use in daily doses above 25 mg.

Gastrointestinal tolerabiliy and safety of selective COX-2 inhibitors

Although gastrointestinal side-effects of NSAIDs are very important, it should not be forgotten that there are other significant associated problems. Nephrotoxicity can result in acute and chronic renal failure, and hypertensive effects vary widely between patients. PGs in the kidney are important in maintaining glomerular filtration rate and renal blood flow, especially in salt-depleted states [47]. Since glomerular filtration rate is reduced with age, the elderly are more susceptible to these renal adverse effects. This is discussed in more detail in Chapter 8.

Hepatic transaminitis is not uncommon, but is rarely significant after drug cessation. Hypersensitivity reactions may occur in patients with asthma, resulting in bronchospasm and rhinitis. Bleeding complications can result from NSAIDs interfering with platelet function. Confusion and drowsiness have been reported in the elderly and there are rare reports of aseptic meningitis and blood dyscrasias.

Studies in healthy elderly patients suggest that selective COX-2 inhibitors may spare the persistent reduction in glomerular filtration rate seen with conventional NSAIDs, although in patients with salt-depleted states, such as congestive cardiac failure or diuretic use, this benefit seems to be lost. It is postulated that NSAID-associated hypersensitivity and exacerbations of asthma are COX-1 mediated, and this seems to be supported by reports of these patients tolerating selective COX-2 inhibitors.

As with conventional NSAIDs, prescribing physicians need to be aware of drug interactions. Even though selective COX-2 inhibitors have no antiplatelet effect, and their addition leads to little alteration in the international normalized ratio (INR) of patients taking a titrated dose of warfarin, careful monitoring of the INR is advisable especially in the elderly, if an NSAID is indicated. Concomitant aspirin use seems to remove the gastroprotective effect of these drugs, and prophylactic gastroprotective agents may be required as with conventional NSAIDs.

Meloxicam (dose range 7.5–15 mg/day)

Meloxicam was the first of the existing NSAIDs that was found to have COX-2 selectivity. Meloxicam at a dose of 7.5 mg causes little acute gastric mucosal injury. In two large studies involving over 8000 patients, meloxicam 7.5 mg was significantly better tolerated than slow release diclofenac 100 mg or piroxicam 20 mg with regard to symptomatic gastrointestinal adverse events. A combined analysis of studies suggests that the symptom benefits of meloxicam 7.5 mg also applies to meloxicam 15 mg [48].

There have been no prospective randomized trials specifically designed to investigate gastrointestinal complications. In the MELISSA [49] and SELECT [50] studies, however, no ulcer bleeding or perforation was observed with meloxicam, while four such cases were observed each with both diclofenac and piroxicam. Numerically fewer upper gastrointestinal ulcer events were reported with meloxicam compared with diclofenac (5 *versus* 7) or piroxicam (7 *versus* 16). Patients also appeared to spend less time in intensive care units. Analysis of the entire meloxicam database suggests a reported annual incidence of perforations, ulceration and bleeding (PUBs, including uncomplicated upper gastrointestinal ulcers) for meloxicam 7.5 mg and 15 mg of 1.2% *versus* 1.9% for diclofenac 100 mg, 4.8% for piroxicam 20 mg and 1.8% for meloxicam 22.5 mg/day [48]. Postmarketing analyses consistently show a favourable overall and gastrointestinal tolerability for both low risk as well as high risk patients (e.g., those with previous GI history) under actual conditions for meloxicam [51]. Although the effect of meloxicam on outcomes has not been systematically investigated in a prospective way, results of meta-analyses and observational studies are consistent and reassuring. Recently a comparative prescription event monitoring study on 19,087 patients treated with meloxicam and 15,268 patients treated with rofecoxib in England over a period of up to 9 months reported drop-outs for gastrointestinal symptoms in 15.4% with meloxicam and 20.2% patients with rofecoxib [52]. Perforations and gastrointestinal bleeds were observed in 0.4% in both groups. Adverse drug reactions attributed by the physician in the meloxicam group were 1.1% and 2.4% for rofecoxib. A total of 101 (0.66%) thromboembolic events were reported during treatment or within a month of stopping rofecoxib, no comparison data was published for meloxicam.

The differential and dose-dependent effects of meloxicam on platelet aggregation and thromboxane synthesis are remarkable in that studies show inhibition of platelet thromboxane in a dose-dependent fashion over a range of daily doses from 7.5 mg to 30 mg [53]. In these studies, however, it is demonstrated consistently that meloxicam does not inhibit platelet aggregation at doses up to 30 mg. It is under discussion whether small amounts of COX-1 inhibition are harmful or might even have some benefits on the endothelial, renal or cardiovascular system. These are issues that will require future resolution and are further discussed in Chapter 7.

Rofecoxib (dose range 12.5–25 mg/day)

In acute studies, rofecoxib 50 mg caused no depression of PG synthesis and at 250 mg/day caused no increase in acute injury compared with placebo. In endoscopic patient studies, ulcer rates associated with rofecoxib 12.5 mg and 25 mg were no different from those with placebo and were four-fold lower than those of ibuprofen 2.4 g/day [54, 55].

A supratherapeutic dose of rofecoxib 50 mg has been compared with naproxen 1 g/day over a median 9 months in 8076 patients with rheumatoid arthritis in the VIGOR study [56]. This study showed a 54% reduction in clinical ulcers and a 57% reduction in complicated upper gastrointestinal events with rofecoxib compared with naproxen. Rofecoxib 50 mg did not result in an overall reduction in adverse events and the rate of myocardial infarction on rofecoxib was higher than on naproxen. Whether this occurred by chance, because of a harmful effect of COX-2 inhibition or a beneficial effect of COX-1 inhibition with naproxen, or (because it was not seen with other studies involving normal clinical doses of rofecoxib) whether it was dose specific are all currently unclear.

The VIGOR study was large enough to allow identification of subgroups of patients in whom large absolute risk reductions with rofecoxib compared with naproxen were seen. For patients over 75 years of age or those with a past history, risk reductions were sufficiently great that only 10 patients per annum would be treated to avoid one gastrointestinal event on a COX-2 inhibitor compared with a nonselective NSAID.

Celecoxib (dose range 200–400 mg)

In acute studies, celecoxib 800 mg/day was at placebo levels in terms of acute gastric injury [57]. In chronic patient studies, in a dose range of 100–800 mg/day there was likewise no difference from placebo and/or a significant reduction compared with diclofenac or naproxen in five out of six studies [57–62]. In one study, the rate of ulceration on celecoxib did not differ from diclofenac [62].

As with rofecoxib, a supratherapeutic dose of celecoxib (800 mg/day) has been evaluated in the CLASS study [63, 64]. In this study, patients with osteoarthritis and rheumatoid arthritis were enrolled, aspirin less than 325 mg/day was allowed and used by 21% of the patients, and there were differences in trial conduct from rofecoxib. Over the full median 12 months of the study, celecoxib was not significantly different from comparators (ibuprofen or diclofenac) in the rate of ulcer complications, but after 6-month pooled data showed a ~50% reduction in patients receiving celecoxib though this failed to reach significance for each of the individual comparator NSAIDs. In part, this may have been because aspirin was used by 21% of patients, which may have independently increased rates of ulceration in both treatment groups.

Recently Chan et al. compared a strategy of a standard NSAID with a gastroprotective agent *versus* celecoxib alone in high risk arthritic patients who had suffered a recent gastrointestinal hemorrhage on NSAIDs [65]. Ulcer healing was confirmed by endoscopy in 290 subjects who were then randomized to receive celecoxib 400 mg/day alone or diclofenac 150 mg/day with omeprazole 20 mg/day. After 6 months of follow-up, 24 serious GI events occurred. Sixteen cases of recurrent ulcer bleeding occurred (seven with celecoxib and nine with diclofenac + omeprazole). All but one of these was due to recurrent bleeding from the previously identified gastric ulcer site. The 6-month probability of recurrent bleeding was 4.9% and 6.4% for the celecoxib and diclofenac/omeprazole groups respectively (not statistically different). Of the

260 patients who were not taking aspirin, the 6-month probability of recurrent bleeding was 4.5% and 5.6% for the celecoxib and diclofenac/omeprazole groups respectively (not statistically different). The authors conclude that the two strategies for recurrent ulcer prevention are equivalent. It should however, be pointed out that the annulized incidence of NSAID related ulcer bleeding in this study of nearly 10% per year is extremely high. In the MUCOSA study, high-risk patients taking standard NSAIDs without prophylaxis had a 6 month incidence of ulcer complications of 9% [66].

Comparison of randomised clinical trials of meloxicam, rofecoxib and celecoxib

No randomised clinical trials are available to compare adequately the gastrointestinal safety profiles of any of the selective COX-2 inhibitors. Therefore such comparisons can only be done indirectly by using meta-analytical techniques. Table 1 is based on the results of a comprehensive literature database search combined with a vigorous selection process performed by the Canadian Coordinating Office for Health Technology Assessment to summarize randomised clinical trials providing information on clinical ulcer complications of the selective COX-2 inhibitors compared with traditional NSAIDs [67–76]. Table 2 summarizes the results of a meta-analysis performed on these studies regarding clinical ulcer complications and symptomatic ulcers (PUBs). From a gastrointestinal toxicity perspective the data can be summarised that COX-2 selective NSAIDs are safer than naproxen and high-dose ibuprofen and are better tolerated then NSAIDs in general. Nevertheless, data to address the important clinical questions regarding comedication with low-dose aspirin and/or gastroprotective agents in cardiovascular and gastrointestinal risk patients are only beginning to become available and final conclusions are not available at this point in time.

While data from a slightly different set of studies were entered into the analysis performed by the National Institute for Clinical Excellence (NICE), similar results were obtained with regard to the conclusions [77]. NICE recommends restricting the use of selective COX-2 inhibitors to patients at high risk, namely patients with age of 65 years and over; previous clinical history of gastroduodenal ulcer, gastrointestinal bleeding or gastroduodenal perforation; concomitant use of medications that are known to increase the likelihood of upper gastrointestinal adverse events e.g., steroids and anticoagulants; presence of serious co-morbidity, such as cardiovascular disease, renal or hepatic impairment, diabetes and hypertension; requirement for the prolonged use of maximum recommended doses of standard NSAIDs. Existing data suggest that selective COX-2 inhibitors are associated with substantial reductions in risk compared with nonselective NSAIDs in older patients, those with a past history and those using corticosteroids. Although not specifically tested, the use of selective COX-2 inhibitors would seem wise in those with impaired

Table 1. Study characteristics of RCTs with COX-2 selective NSAIDs regarding clinical ulcer complications

Study and study drug	Dose of selective COX-2 inhibitor	NSAID	Number of patients	Mean age (yrs)	Indication	Follow up
Celecoxib						
Silverstein 2000 (CLASS) [63, 64]	400 mg bid	Diclofenac (75 mg bid) Ibuprofen (800 mg tid)	8059	60	OA & RA	4, 13, 26 wks (1 year-FDA)
Goldstein 2000 [69] (Combined analysis study)	25 mg bid to 400 mg bid 7 dose steps	Naproxen (500 mg bid) Diclofenac (75 mg bid) Ibuprofen (800 mg tid)	11008	59	OA & RA	2–24 wks
Success-1 Abstract [70]	100 mg bid 200 mg bid	Naproxen (500 mg bid)	13274	62	OA	6, 12 wks
Rofecoxib						
Bombardier 2000 (VIGOR) [56]	50 mg/day	Naproxen (500 mg bid)	8076	58	RA	6 wks, 4, 8, 12 mo
Langman 1999 [71] (Combined analysis study)	25 mg/day 50 mg/day	Ibuprofen (800 mg tid) Diclofenac (50 mg tid) Nabumetone (1500mg)	5435	63	OA	6wks, 4, 6, 12, 24 mo
Meloxicam						
Dequeker 1998 (SELECT) [50]	7.5 mg/day	Piroxicam	8656	61	OA	4 wks
Hawkey 1998 (MELISSA) [49]	7.5 mg/day	Diclofenac (100 mg/day)	9323	62	OA	4 wks
Yocum 2000 [27]	3.75–15 mg	Diclofenac (100 mg/day)	774	62	OA	12 wks
Goei 1997 [72]	15 mg	Diclofenac (100 mg/day)	258	71	OA	6 wks
Hosie 1997 [73]	15 mg	Piroxicam (20 mg)	455	67	OA	6 mo
Hosie 1996 [74]	7.5 mg	Diclofenac (100 mg/day)	335	64	OA	6 mo
Lindén 1996 [28]	15 mg	Piroxicam 20 mg	260	67	OA	6 wks
Wojtulewski 1996 [31]	7.5 mg	Naproxen 750 mg	379	57	RA	6 mo
Distel 1996 [75, 76] (combined analysis study)	7.5–15 mg	Piroxicam 20 mg/day Naproxen 1000 mg/day Diclofenac 100 mg/day	6129	59	OA & RA	Up to 1 yr

Table 2. Upper gastrointestinal perforations, ulcers and bleeds (PUBs) with COX-2 selective NSAIDs relative to traditional NSAIDs

COX – 2 inhibitor	Number of studies	Number of patients	RR	95% CI	Risk reduction
Celecoxib-6 month	3	30,306	0.41	0.21–0.78	59%
Celecoxib-12 month	3	30,306	0.43	0.21–0.88	57%
Rofecoxib	2	12,997	0.47	0.36–0.63	53%
Meloxicam	8	20,279	0.48	0.26–0.88	52%
Total	13	63,582	0.47	0.38–0.57	53%

coagulation, since they lack the anti-platelet effects of at least some nonselective NSAIDs.

NICE reviewed a total of 53 randomised controlled trials involving 61,731 patients with osteo-and rheumatoid arthritis were reviewed (14 rofecoxib, 15 celecoxib, 3 rofecoxib *versus* celecoxib, 9 meloxicam and 12 etodolac). These trials included both placebo and NSAID comparators (naproxen, diclofenac, ibuprofen, piroxicam, tenoxicam, nabumetone, nimesulide and aspirin). Evidence was found that the COX-2 selective inhibitors are more effective than placebo and of equivalent efficacy to other NSAIDs in their ability to reduce pain (resting and active) and to improve physical and global function in both osteoarthritis and rheumatoid arthritis patients. There was also evidence to suggest that, although the COX-2 selective inhibitors increase the incidence of gastrointestinal adverse events compared to placebo in both sets of patients, the magnitude of this effect is less than that of standard NSAID therapy. Rofecoxib and celecoxib have been shown to be associated with a reduced incidence of gastric erosions on endoscopy compared to standard NSAIDs in patients with arthritis. However, the importance of this finding as a surrogate for the development of peptic ulceration or other major gastrointestinal adverse effects in patients treated with these drugs in routine clinical practice is uncertain. Finally, indirect comparison of the effects of meloxicam, rofecoxib and celecoxib relative to other NSAIDs based on available evidence, does not demonstrate a difference in either efficacy or adverse events between these drugs.

COX-2 inhibition: potential risks

In view of their better safety profile, selective COX-2 inhibitors may well be used more frequently in high-risk patients and subsequent toxicity may not reflect the data from controlled studies. COX-2 is up-regulated in *Helicobacter* gastritis and inflammatory bowel disease, and whether COX-2 inhibition can be harmful in these conditions will depend on the relative contribution of COX-2 to PG synthesis. COX-2 is induced at the rim of ulcers in humans and

COX-2 inhibitors retard ulcer healing in animal studies [78]. It is important, therefore, that these drugs are demonstrated to be safe in this particular subset of patients. Fortunately, there is an increasingly large database of selective COX-2 inhibitor use in unselected patients in whom a favourable gastrointestinal profile is seen.

Selective COX-2 inhibitors will not have the beneficial anti-platelet effects of conventional NSAIDs and this, in combination with the reported role of COX-2 in sustaining vascular protacyclin production, may well have an impact on the incidence of vascular disease [79].

Conclusion

In summary, in the absence of evidence of significant differences in anti-inflammatory efficacy between the COX-2 selective drugs and standard NSAIDs, the avoidance of serious adverse effects, particularly on the gastrointestinal tract, becomes the most relevant factor when considering their use. For the most part the evidence suggests that COX-2 selective agents are less likely to produce symptomatic gastrointestinal tract adverse events. However, there remains some concern regarding the potential cardiovascular risks associated with COX-2 selective agents and therefore caution is needed, as it is for standard NSAID therapy, when prescribing in patients with pre-existing cardiovascular disease.

Overall, the arrival of selective COX-2 inhibitors heralds one of the biggest changes in rheumatological practice and with current knowledge it will be difficult to justify prescribing higher-risk drugs. Prescribers should be aware that gastrointestinal problems are not the only adverse consequences of NSAID use, they should be vigilant for fluid retention and its consequences and should remember that ulcer complications in patients taking a drug may be caused by background problems rather than by the drugs themselves. Our enthusiasm needs to be tempered by the reality that these drugs are still relatively new, and that possible advantages and disadvantages await further evaluation.

References

1 Girgis L, Brooks P (1994) Nonsteroidal anti-inflammatory drugs. Differential use in older patients. *Drugs Aging* 4: 101–112
2 Vane JR (1971) Inhibition of prostaglandin synthesis as a mechanism of action for aspirin-like drugs. *Nature* 231: 232–235
3 Roth GJ, Stanford N, Majerus PW (1975) Acetylation of prostaglandin synthetase by aspirin. *Proc Natl Acad Sci USA* 72: 3073–3076
4 Henry D, Lim LL, Garcia-Rodriguez LA, Perez Gutthann S, Carson JL, Griffin M, Savage R, Logan R, Moride Y, Hawkey C et al (1996) Variability in risk of gastrointestinal complications with individual non-steroidal antiinflammatory drugs: results of a collaborative meta-analysis. *BMJ* 312: 1563–1566
5 Singh G, Ramey DR, Morfeld D, Fries JF (1994) Comparative toxicity of non-steroidal anti-inflammatory agents. *Pharmacol Ther* 62: 175–191

6 Langman MJ (1988) Ulcer complications and nonsteroidal anti-inflammatory drugs. *Am J Med* 84: 15–19
7 Garcia Rodriguez LA, Jick H (1994) Risk of upper gastrointestinal bleeding and perforation associated with individual non-steroidal anti-inflammatory drugs. *Lancet* 343: 769–772
8 MacDonald TM, Morant SV, Robinson GC, Shield MJ, McGilchrist MM, Murray FE, McDevitt DG (1997) Association of upper gastrointestinal toxicity of non-steroidal anti-inflammatory drugs with continued exposure: cohort study. *BMJ* 315: 1333–1337
9 Kujubu DA, Fletcher BS, Varnum BC, Lim RW, Herschman HR (1991) TIS10, a phorbol ester tumor promoter-inducible mRNA from Swiss 3 T3 cells, encodes a novel prostaglandin synthase/cyclooxygenase homologue. *J Biol Chem* 266: 12866–12872
10 O'Banion MK, Sadowski HB, Winn V, Young DA (1991) A serum- and glucocorticoid regulated 4-kilobase mRNA encodes a cyclooxygenase-related protein. *J Biol Chem* 266: 23261–23267
11 Simmons DL, Levy DB, Yannoni Y, Erikson RL (1989) Identification of an ester repressible v-src-inducible gene. *Proc Natl Acad Sci USA* 86: 1178–1182
12 Vane JR, Bakhle YS, Botting RM (1998) Cyclooxygenases 1 and 2. *Annu Rev Pharmacol Toxicol* 38: 97–120
13 Warner TD, Giuliano F, Vojnovic I, Bukasa A, Mitchell JA, Vane JR (1999) Nonsteroidal drug selectivities for cyclo-oxygenase-1 rather than cyclo-oxygenase-2 are associated with human gastrointestinal toxicity: a full *in vitro* analysis. *Proc Natl Aca Sci USA* 96: 7563–7568
14 Brooks P, Emery P, Evans J, Fenner H, Hawkey CJ, Patrono C, Smolen J, Breedveld F, Day R, Dougados M et al (1998) Interpreting the clinical significance of the differential inhibition of cyclooxygenase-1 and cyclooxygenase-2. *Br J Rheumatol* 27 (Suppl. 1): 118
15 Donnelly MT, Hawkey CJ (1997) Review article: COX-2 inhibitors – a new generation of safer NSAIDs? *Aliment Pharmacol Ther* 11: 227–236
16 Emery P (1996) Clinical implications of selective cyclooxygenase-2 inhibition. *Scand J Rheumatol* 102 (Suppl 1): 23–28
17 Hawkey CJ (1999) COX-2 inhibitors. *Lancet* 353: 307–314
18 Hawkey CJ, Lanas AI (2001) Doubt and certainty about nonsteroidal antiinflammatory drugs in the year 2000: a multidisciplinary expert statement. *Am J Med* 110:S79–S100
19 Richardson C, Emery P (1996) The clinical implications of inhibition of the inducible form of cyclo-oxygenase. *Drug Safety* 15: 249–260
20 Rostom A, Wells G, Tugwell P, Welch V, Dube C, McGowan J (2000) The prevention of chronic NSAID induced upper gastrointestinal toxicity: a Cochrane collaboration metaanalysis of randomized controlled trials. *J Rheumatology* 27: 2203–2214
21 Yeomans ND, Tulassay Z, Juhasz L, Racz I, Howard JM, van Rensburg CJ, Swannell AJ, Hawkey CJ (1998) A comparison of omeprazole with ranitidine for ulcers associated with nonsteroidal antiinflammatory drugs. Acid Suppression Trial: Ranitidine *versus* Omeprazole for NSAID-associated Ulcer Treatment (ASTRONAUT) Study Group. *N Engl J Med* 338: 719–726
22 Hawkey CJ, Karrasch JA, Szczepanski L, Walker DG, Barkun A, Swannell AJ, Yeomans ND (1998) Omeprazole compared with misoprostol for ulcers associated with nonsteroidal antiinflammatory drugs. Omeprazole *versus* Misoprostol for NSAID-induced Ulcer Management (OMNIUM) Study Group. *N Engl J Med* 338: 727–734
23 McCarthy DM (1991) Sucralfate. *N Engl J Med* 325: 1017–1025
24 Emery P (1999) COX-1, COX-2: So what? *Scand J Rheumatol* 28: 6–9
25 Tomasoni S, Noris M, Zappella S, Gotti E, Casiraghi F, Bonazzola S, Benigni A, Remuzzi G (1998) Upregulation of renal and systemic cyclooxygenase-2 in patients with active lupus nephritis. *J Am Soc Nephrol* 9: 1202–1212
26 Lund B, Distel M, Bluhmki E (1998) A double-blind, randomized, placebo-controlled study of efficacy and tolerance of meloxicam treatment in patients with osteoarthritis of the knee. *Scand J Rheumatol* 27: 32–37
27 Yocum D, Fleischmann R, Dalgin P, Caldwell J, Hall D, Roszko P (2000) Safety and efficacy of meloxicam in the treatment of osteoarthritis: a 12-week, double-blind, multiple-dose, placebo-controlled trial. *Arch Intern Med* 160: 2947–2954
28 Lindén B, Distel M, Bluhmki E (1996) A double-blind study to compare the efficacy and safety of meloxicam 15 mg with piroxicam 20 mg in patients with osteoarthritis of the hip. *Br J Rheumatol* 35 (Suppl 1): 35–38
29 Furst DE, Kolba KS, Fleischmann R, Silverfield J, Greenwald M, Roth S, Hall DB, Roszko PJ (2002) Dose response and safety study of meloxicam up to 22.5 mg daily in rheumatoid arthritis:

a 12 week multicenter, double blind, dose response study *versus* placebo and diclofenac. *J Rheumatol* 29: 436–446

30 Lemmel EM, Bolten W, Burgos-Vargas R, Platt P, Nissilä M, Sahlberg D, Bjorneboe O, Baumgartner H, Valat JP, Franchimont P et al (1997) Efficacy and safety of meloxicam in patients with rheumatoid arthritis. *J Rheumatol* 24: 282–290

31 Wojtulewski JA, Schattenkirchner M, Barceló P, Le Loët X, Bevis PJR, Bluhmki E, Distel M (1996) A six-month double-blind trial to compare the efficacy and safety of meloxicam 7.5 mg daily and naproxen 750 mg daily in patients with rheumatoid arthritis. *Br J Rheumatol* 35 (Suppl 1): 22–28

32 Dougados M, Gueguen A, Nakache JP, Velicitat P, Veys EM, Zeidler H, Calin A (1999) Ankylosing spondylitis: what is the optimum duration of a clinical study? A one year *versus* a six weeks non-steroidal anti-inflammatory drug trial. *Rheumatology* 38: 235–244

33 Simon LS, Weaver AL, Graham DY, Kivitz AJ, Lipsky PE, Hubbard RC, Isakson PC, Verburg KM, Yu SS, Zhao WW et al (1999) Anti-inflammatory and upper gastrointestinal effects of celecoxib in rheumatoid arthritis: a randomized controlled trial. *JAMA* 282: 1921–1928

34 Bensen WG, Fiechtner JJ, McMillen JI, Zhao WW, Yu SS, Woods EM, Hubbard RC, Isakson PC, Verburg KM, Geis GS (1999) Treatment of osteoarthritis with celecoxib, a cyclooxygenase-2 inhibitor: a randomized controlled trial. *Mayo Clin Proc* 74: 1095–1105

35 McKenna F, Borenstein D, Wendt H, Wallemark C, Lefkowith JB, Geis GS (2001) Celecoxib *versus* diclofenac in the management of osteoarthritis of the knee. *Scand J Rheumatol* 30: 11–18

36 Williams GW, Hubbard RC, Yu SS, Zhao W, Geis GS (2001) Comparison of once-daily and twice-daily administration of celecoxib for the treatment of osteoarthritis of the knee. *Clin Ther* 23: 213–227

37 Zhao S, McMillen JI, Markenson JA, Dedhiya SD, Zhao WW, Osterhaus JT, Yu SS (1999) Evaluation of the functional status aspects of health-related quality of life of patients with osteoarthritis treated with celecoxib *Pharmacotherapy* 19: 1269–1278

38 Acevedo E, Castañeda O, Ugaz M, Beaulieu AD, Pons-Estel B, Caeiro F, Casas N, Garza-Elizondo M, Irazoque F, Hinojosa W et al (2001) Tolerability profiles of rofecoxib (Vioxx) and Arthrotec. A comparison of six weeks treatment in patients with osteoarthritis. *Scand J Rheumatol* 30: 19–24

39 Cannon GW, Caldwell JR, Holt P, McLean B, Seidenberg B, Bolognese J, Ehrich E, Mukhopadhyay S, Daniels B (2000) Rofecoxib, a specific inhibitor of cyclooxygenase 2, with clinical efficacy comparable with that of diclofenac sodium: results of a one-year, randomized, clinical trial in patients with osteoarthritis of the knee and hip. *Arthritis Rheum* 43: 978–987

40 Day R, Morrison B, Luza A, Castaneda O, Strusberg A, Nahir M, Helgetveit KB, Kress B, Daniels B, Bolognese J et al (2000) A randomized trial of the efficacy and tolerability of the COX-2 inhibitor rofecoxib vs ibuprofen in patients with osteoarthritis. *Arch Intern Med* 160: 1781–1787

41 Ehrich EW, Schnitzer TJ, McIlwain H, Levy R, Wolfe F, Weisman M, Zeng Q, Morrison B, Bolognese J, Seidenberg B et al (1999) Effect of specific COX-2 inhibition in osteoarthritis of the knee: a 6 week double blind, placebo controlled pilot study of rofecoxib. *J Rheumatol* 26: 2438–2447

42 Geba GP, Weaver AL, Polis AB, Dixon ME, Schnitzer TJ. Vioxx, Acetaminophen, Celecoxib Trial (VACT) Group (2002) Efficacy of rofecoxib, celecoxib, and acetaminophen in osteoarthritis of the knee: a randomized trial. *JAMA* 287: 64–71

43 Saag K, van der HD, Fisher C, Samara A, DeTora L, Bolognese J, Sperling R, Daniels B (2000) Rofecoxib, a new cyclooxygenase 2 inhibitor, shows sustained efficacy, comparable with other nonsteroidal anti-inflammatory drugs: a 6-week and a 1-year trial in patients with osteoarthritis. *Arch Fam Med* 9: 1124–1134

44 Truitt KE, Sperling RS, Ettinger WH, Jr, Greenwald M, DeTora L, Zeng Q, Bolognese J, Ehrich E (2001) A multicenter, randomized, controlled trial to evaluate the safety profile, tolerability, and efficacy of rofecoxib in advanced elderly patients with osteoarthritis. *Aging Clin Experiment Res* 13: 112–121

45 Li Q (1999) Statistical Review – NDA 21-042 – Rofecoxib. NDA 21-042. *http://www.fda.gov/cder/foi/nda/99/021042_52_Vioxx.htm*. Accessed Dec. 10, 2003

46 Schnitzer TJ, Truitt K, Fleischmann R, Dalgin P, Block J, Zeng Q, Bolognese J, Seidenberg B, Ehrich EW (1999) The safety profile, tolerability, and effective dose range of rofecoxib in the treatment of rheumatoid arthritis. *Clin Ther* 21: 1688–1702

47 Catella-Lawson F, McAdam B, Morrison BW, Kapoor S, Kujubu D, Antes L, Lasseter KC, Quan

H, Gertz BJ, FitzGerald GA (1999) Effects of specific inhibition of cyclooxygenase-2 on sodium balance, hemodynamics, and vasoactive eicosanoids. *J Pharmacol Exp Ther* 289: 735–741

48 Degner F, Richardson B (2001) Review of gastrointestinal tolerability and safety of meloxicam. *Inflammopharmacology* 9: 71–80

49 Hawkey C, Kahan A, Steinbruck K, Alegre C, Baumelou E, Begaud B, Dequeker J, Isomaki H, Littlejohn G, Mau J et al (1998) Gastrointestinal tolerability of meloxicam compared to diclofenac in osteoarthritis patients. Meloxicam Large-Scale International Study Safety Assessment (MELISSA). *Br J Rheumatol* 37: 937–945

50 Dequeker J, Hawkey C, Kahan A, Steinbruck K, Alegre C, Baumelou E, Begaud B, Isomaki H, Littlejohn G, Mau J (1998) Improvement in gastrointestinal tolerability of the selective cyclooxygenase (COX)-2 inhibitor, meloxicam, compared with piroxicam: results of the Safety and Efficacy Large- Scale Evaluation of COX-inhibiting Therapies (SELECT) trial in osteoarthritis. *Br J Rheumatol* 37: 946–951

51 Zeidler H, Kaltwasser JP, Leonard JP, Kohlmann T, Sigmund R, Degner F, Hettich M (2002) Prescription and tolerability of meloxicam in day-to-day practice: postmarketing observational cohort study of 13,307 patients in Germany. *J Clin Rheumatol* 8: 305–315

52 Layton D, Heeley E, Hughes K, Shakir SA (2003) Comparison of the incidence rates of selected gastrointestinal events reported for patients prescribed rofecoxib and meloxicam in general practice in England using prescription-event monitoring data. *Rheumatology* 42: 622–631

53 Rinder HM, Tracey JB, Souhrada M, Wang C, Gagnier RP, Wood CC (2002) Effects of meloxicam on platelet function in healthy adults: a randomized, double-blind, placebo-controlled trial. *J Clin Pharmacol* 42: 881–886

54 Hawkey C, Laine L, Simon T, Beaulieu A, Maldonado-Cocco J, Acevedo E, Shahane A, Quan H, Bolognese J, Mortensen E (2000) Comparison of the effect of rofecoxib (a cyclooxygenase 2 inhibitor), ibuprofen, and placebo on the gastroduodenal mucosa of patients with osteoarthritis: a randomized, double-blind, placebo-controlled trial. The Rofecoxib Osteoarthritis Endoscopy Multinational Study Group. *Arthritis Rheum* 43: 370–377

55 Laine L, Harper S, Simon T, Bath R, Johanson J, Schwartz H, Stern S, Quan H, Bolognese J (1999) A randomized trial comparing the effect of rofecoxib, a cyclooxygenase 2-specific inhibitor, with that of ibuprofen on the gastroduodenal mucosa of patients with osteoarthritis. *Gastroenterology* 117: 776–783

56 Bombardier C, Laine L, Reicin A, Shapiro D, Burgos-Vargas R, Davis B, Day R, Ferraz MB, Hawkey CJ, Hochberg MC et al VIGOR Study Group (2000) Comparison of upper gastrointestinal toxicity of rofecoxib and naproxen in patients with rheumatoid arthritis. *N Engl J Med* 343: 1520–1528

57 Division of Gastrointestinal and Coagulation Drug Products (1998) Medical Officer's Consult Review. *http://www.fda.gov/cder/foi/adcomm/98/adac_12198-consult.pdf.* (Accessed Dec. 10, 2003)

58 Emery P, Zeidler H, Kvien TK, Guslandi M, Naudin R, Stead H, Verburg KM, Isakson PC, Hubbard RC, Geis GS (1999) Celecoxib *versus* diclofenac in long-term management of rheumatoid arthritis: randomised double-blind comparison. *Lancet* 354: 2106–2111

59 Simon LS, Weaver AL, Graham DY, Kivitz AJ, Lipsky PE, Hubbard RC, Isakson PC, Verburg KM, Yu SS, Zhao WW, Geis GS (1999) Anti-inflammatory and upper gastrointestinal effects of celecoxib in rheumatoid arthritis: a randomized controlled trial. *JAMA* 282: 1921–1928

60 Goldstein JL, Correa P, Zhao WW, Burr AM, Hubbard RC, Verburg KM, Geis GS (2001) Reduced incidence of gastroduodenal ulcers with celecoxib, a novel cyclooxygenase-2 inhibitor, compared to naproxen in patients with arthritis. *Am J Gastroenterol* 96: 1019–1027

61 FDA-Study 21 (1998) *http://www.fda.gov/cder/foi/adcomm/98/adac_12198-consult.pdf.* Study 21 (Accessed Dec. 10, 2003)

62 FDA-Study 71 (1998) *http://www.fda.gov/cder/foi/adcomm/98/adac_12198-consult.pdf.* Study 71 (Accessed Dec. 10, 2003)

63 CLASS-12 Month Data (2000) *http://www.fda.gov/ohrms/dockets/ac/01/briefing/367761_05_gi.doc.* (Accessed Dec. 10, 2003)

64 Silverstein FE, Faich G, Goldstein JL, Simon LS, Pincus T, Whelton A, Makuch R, Eisen G, Agrawal NM, Stenson WF et al (2000) Gastrointestinal toxicity with celecoxib vs nonsteroidal anti-inflammatory drugs for osteoarthritis and rheumatoid arthritis: the CLASS study: A randomised controlled trial. Celecoxib Long-term Arthritis Safety Study. *JAMA* 284: 1247–1255

65 Chan FK, Hung LC, Suen BY, Wu JC, Lee KC, Leung VK, Hui AJ, To KF, Leung WK, Wong VW

et al (2002) Celecoxib *versus* diclofenac and omeprazole in reducing the risk of recurrent ulcer bleeding in patients with arthritis. *N Engl J Med* 347: 2104–2110

66 Silverstein FE, Graham DY, Senior J, Davies HW, Struthers BJ, Bittman RM, Geis GS (1995) Misoprostol reduces serious gastrointestinal complications in patients with rheumatoid arthritis receiving nonsteroidal anti-inflammatory drugs. A randomized, doubleblind, placebo-controlled trial. *Ann Intern Med* 123: 241–249

67 Canadian Coordinating Office for Health Technology Assessment (2003) Gastroduodenal ulcers associated with the use of non-steroidal anti-inflammatory drugs: a systematic review of preventive pharmacological interventions. Final Draft January 2003. *http://www.ccohta.ca/publications/pdf/gastroprotective_fr_e.pdf.* (Accessed Dec. 10, 2003)

68 FDA – Study 62 (2000) *http://www.fda.gov/cder/foi/adcomm/98/adac_12198-consult.pdf.* Study 62 (Accessed Dec. 10, 2003)

69 Goldstein JL, Silverstein FE, Agrawal NM, Hubbard RC, Kaiser J, Maurath CJ, Verburg KM, Geis GS (2000) Reduced risk of upper gastrointestinal ulcer complications with celecoxib, a novel COX-2 inhibitor. *Am J Gastroenterol* 95: 1681–1690

70 Singh G, Goldstein J, Bensen W, Agrawal N, Eisen G, Fort J, Bello A, Boots S (2001) SUCCESS-1 in Osteoarthritis (OA) Trial: celecoxib significantly reduces the risk of serious upper GI complications compared to NSAIDs while providing similar efficacy in 13,274 randomized patients. Abstr Perspect 5 (1): *http://www.hopkins-arthritis.som.jhmi.edu/eular2001/oa-treatments-cox2.html* (accessed June 2003)

71 Langman MJ, Jensen DM, Watson DJ, Harper SE, Zhao PL, Quan H, Bolognese JA, Simon TJ (1999) Adverse upper gastrointestinal effects of rofecoxib compared with NSAIDs. *JAMA* 282: 1929–1933

72 Goei Thè HS, Lund B, Distel MR, Bluhmki E (1997) A double-blind, randomized trial to compare meloxicam 15 mg with diclofenac 100 mg in the treatment of osteoarthritis of the knee. *Osteoarthritis Cartilage* 5: 283–288

73 Hosie J, Distel M, Bluhmki E (1997) Efficacy and tolerability of meloxicam *versus* piroxicam in patients with osteoarthritis of the hip or knee. A six-month double-blind study. *Clin Drug Invest* 13: 175–184

74 Hosie J, Distel M, Bluhmki E (1996) Meloxicam in osteoarthritis: a 6-month, double-blind comparison with diclofenac sodium. *Br J Rheumatol* 35 (Suppl 1): 39–43

75 Distel M, Mueller C, Bluhmki E, Fries J (1996) Safety of meloxicam: a global analysis of clinical trials. Br J Rheumatol 35 (Suppl 1): 68–77

76 Distel M, Mueller C, Bluhmki E (1996) Global analysis of gastrointestinal safety of a new NSAID, meloxicam. Inflammopharmacology 4: 71–81

77 National Institute for Clinical Excellence (2001) Guidance on the use of cyclo-oxygenase (COX) II selective inhibitors, celecoxib, rofecoxib, meloxicam and etodolac for osteoarthritis and rheumatoid arthritis. London, UK. *http://www.nice.org.uk/article.asp?a=18030* (accessed June 2003)

78 Wallace JL, McKnight W, Reuter BK, Vergnolle N (2000) NSAID-induced gastric damage in rats: requirement for inhibition of both cyclooxygenase 1 and 2. *Gastroenterology* 119: 706–714

79 McAdam BF, Catella-Lawson F, Mardini IA, Kapoor S, Lawson JA, FitzGerald GA (1999) Systemic biosynthesis of prostacyclin (COX)-2: the human pharmacology of a selective inhibitor of COX-2. *Proc Natl Acad Sci USA* 96: 272–277

COX-2 Inhibitors
Edited by M. Pairet and J. van Ryn

Role of COX-2 in ulcers and ulcer healing

Joanne van Ryn and Michel Pairet

Departments of [1]Drug Discovery Support and [2]Pulmonary Research, Boehringer Ingelheim Pharma GmbH & Co. KG, Birkendorfer Str. 65, D-88397 Biberach, Germany

Introduction

Non steroidal anti-inflammatory drugs (NSAIDs) are among the most frequently prescribed drugs; in the US alone there are over 110 million prescriptions annually [1]. However, their use is limited by their gastrointestinal (GI) side-effects, which is an issue particularly in the elderly where the risk for GI complications is elevated and ~70% take an NSAID (including aspirin) at least once weekly [2–4]. The occurrence of complicated GI events due to NSAIDs in patients with arthritis is 1–1.5% yearly (i.e., only perforation, ulceration and bleeding) [1]. When considering the large number of patients consuming NSAIDs, this translates into a considerable health care problem. Other risk factors associated with NSAID-induced GI complications are history of previous ulcers, concomitant anticoagulation or corticosteroid use and high dose NSAIDs [3]. Use of low-dose aspirin approximately doubles the risk of bleeding in patients taking NSAIDs [3, 4].

It has been demonstrated clinically that COX-2 selective inhibitors result in a reduced number of GI complications as compared to nonselective inhibitors [5–9], presumably by sparing COX-1 inhibition, and thereby maintaining GI mucosal integrity. In addition, the lack of effect of these agents on platelet aggregation should result in less bleeding episodes. Thus, the advantages of COX-2 selective agents are that the anti-inflammatory activity is maintained, while the amount of COX-1 inhibition is reduced. However, with improved pharmacological tools, it is becoming evident that not only COX-1, but also COX-2 is important for gastric mucosal defense.

NSAID-induced gastric injury

Mucosal defense is a term that describes the mechanisms by which the gastric mucosa is protected from injury from noxious agents, either from endogenous secretion of acid, pepsin etc. or from ingested irritants such alcohol or from NSAIDs [10, 11]. A normal epithelial surface without mucosal layer and the epithelial margin of an ulcer with extensive damage are shown in Figure 1.

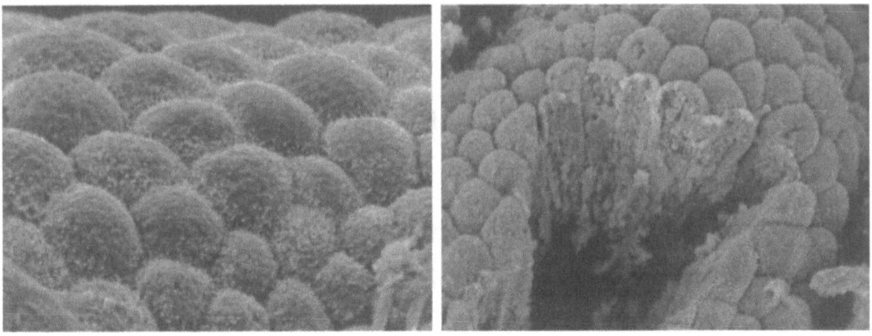

Figure 1. Electron micrograph of healthy human uninjured gastric epithelial cells (A) with the mucus layer removed to allow better visualization of the epithelium. The second panel (B) illustrates the margin of an ulceration. Photographs by Lennart Nielsson, Copyright Boehringer Ingelheim GmbH.

One of the earliest detected forms of damage by particularly aspirin and salicylic acid was the increased gastric permeability and topical injury shortly following administration [11–14]. In addition salicylic acid in concentrations that did not inhibit prostaglandin production could produce gastric injury [12]. The mechanism by which this occurs is still not fully elucidated, whether this involves direct toxic effects on the epithelium, disruption of bicarbonate barrier of the mucosa or disruption of cellular metabolism, such as oxidative phosphorylation remains to be determined [14]. The importance of this effect is also still somewhat controversial, since it is also known that enteric-coated aspirin administration results in similar rates of GI damage as compared to non-coated aspirin [15].

The mucus layer on the luminal side of the gastric epithelial lining provides an important barrier against NSAIDs and other noxious agents and is secreted by protective epithelial cells. Epithelial cells also secrete bicarbonate (Fig. 2), which neutralizes acid in the proximity of the epithelial cells [16] and can undergo rapid responses to mild injury where they proliferate to cover the injured area in an process called restitution [17]. Since the key discovery by Vane that aspirin and NSAIDs inhibit prostaglandins [18], it has become apparent over the years how many gastric protective mechanisms are dependent on prostaglandins and how these mechanisms can be compromised by NSAID intake [10–12, 14]. Prostaglandins are involved in both mucus and bicarbonate production and are both reduced with chronic NSAID administration [19–21].

Another important factor in mucosal integrity is an adequate blood supply. Neuronal input is important in rapid response mechanisms triggering enhanced blood flow that results in removal of toxic products or acid buffering, however, prostaglandins are also important for their vasodilatory properties [10, 11]. NSAID administration has been shown to reduce gastric mucosal blood flow

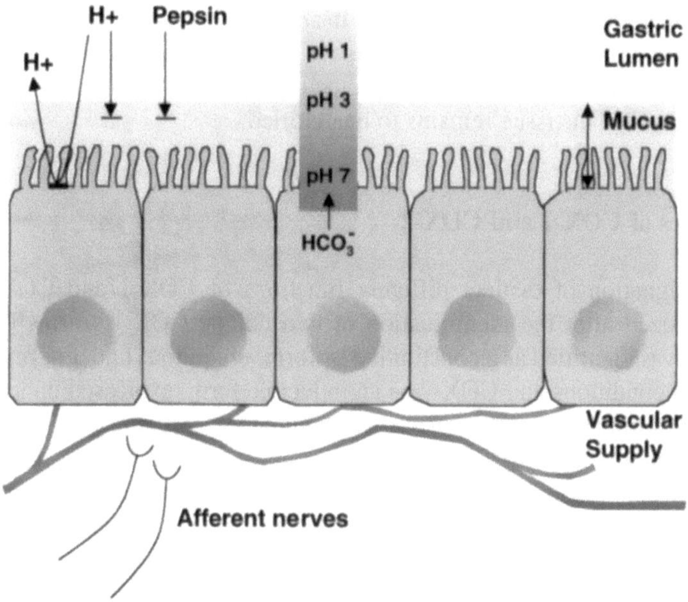

Figure 2. Schematic representation of the mechanisms involved in gastric mucosal defense.

and these effects could be reversed by the administration of prostaglandins [22–24]. This is further exacerbated by neutrophil adhesion to the endothelium in the microcirculation [5, 6, 10]. This results in release of various damaging agents, oxygen-radicals, proteases, cytokines such as tumor necrosis factor α (TNF_{α}) and leukotrienes, which all result in upregulation of adhesion molecules and further neutrophil adhesion [25–28]. This can also result in blockage of blood vessels, leading to a further reduction in blood flow and necrosis. Administration of agents that inhibit neutrophil adherence, prostaglandins, nitric oxide, anti-TNFα antibody or leukotriene inhibitors have all reduced the severity of NSAID-induced damage in animal models [29, 30]. However, unlike animal models, neutrophil and other lymphocyte involvement is not characteristic of GI injury due to NSAID use in humans, thus the relative importance of these findings remains to be determined [6].

It must be stressed that while prostaglandins play a major role in the maintenance of the GI integrity, this is not the only protective mechanism. Nitric oxide duplicates all the protective mechanisms afforded by prostaglandins; it has a role in mucus production [31], gastric mucosal blood flow [32] and inhibition of adhesive molecule expression and thus leukocyte adhesion [33–35].

Though this will not be discussed in detail here, the effect of *Helicobacter pylori* in patients taking NSAIDs is also been extensively studied recently [5, 6]. *H. pylori* infection is the single largest risk factor for ulcer formation in non-NSAID users [6, 36], however, the effect of *H. pylori* in patients taking NSAIDs is controversial and both positive and negative effects on ulcer formation have been reported [37, 38]. It appears that prior infection with H. pylori does not result in an elevated risk of ulcers in NSAID users; in contrast it may even afford some form of protection by elevating gastric adaptation against insult. This issue remains to be clarified.

The roles of COX-1 and COX-2

The designation of distinct different functions of COX-1 and COX-2 was hypothesized after the identification of two unique COX isoforms [39–41]. COX-1 was identified as a constitutive isoform, ubiquitous under normal physiological conditions and COX-2 as an inducible form, expressed under conditions of stimulation, particularly in inflammatory cells [42–44]. This partitioning led to the simple and elegant concept that selective inhibition of COX-2 would result in anti-inflammatory activity, but would not result in the GI and renal side-effects that afflict NSAIDs, because constitutive COX-1 would not be inhibited [45]. While the basic tenets of this concept hold true, selective COX-2 inhibitors do have fewer GI side-effects, they are not devoid of these side-effects.

It has become clear over time that this is not a black and white concept and there are roles for COX-2 in physiological processes just as it appears that COX-1 may also be involved in inflammatory processes. More recent studies identifying COX-1 and -2 in human gastric mucosa [46, 47] have demonstrated the presence of COX-2 in addition to COX-1 in both normal and inflamed gastric tissue, though it appears that basal gastric prostaglandin production is largely dependent on COX-1.

The importance of both COX-1 and COX-2 in maintaining GI mucosa integrity was illustrated by Wallace et al. [48], who used a highly selective COX-1 inhibitor, SC-560 and celecoxib, a selective COX-2 inhibitor, to determine the relative roles of COX-1 and COX-2 inhibition on ulcer formation in healthy rats. SC-560 administration alone resulted in almost complete inhibition of gastric prostaglandin production, consistent with previous reports [49], but did not induce ulceration. Celecoxib did not inhibit gastric prostaglandin production and also did not induce ulceration. These results indicate that COX-2 inhibition alone does not inhibit gastric prostaglandins or induce ulceration; however, surprisingly inhibition of only COX-1 also does not result in ulcer formation, despite inhibition of gastric prostaglandins. When both SC-560 and celecoxib were administered to rats, then ulceration occurred to the same extent as treatment with indomethacin. This clearly demonstrates an additive role for COX-2 inhibition in combination with COX-1 inhibition in ulcer formation. This may also explain why mice lacking the gene for COX-1 do not have spontaneous ulcer formation, despite reduced prostaglandin formation [50].

This phenomenon was explored further by Gretzer et al., who also showed that neither SC-560 nor rofecoxib, a COX-2 inhibitor, induced ulceration when administered to healthy rats. Rats were then exposed to an acid challenge that was shown to elevate COX-2 mRNA levels but not COX-1 mRNA, though this challenge alone was not enough to result in ulceration [51]. When SC-560 was then administered to the acid-challenged rats there was a dose-dependent induction of ulceration. Administration of rofecoxib alone did not induce ulceration, however, the combination of SC-560 and rofecoxib additively induced ulceration as compared to SC-560 alone.

Mechanistically, this may in part be explained by the effect prostaglandins have on gastric blood flow. An important line of defense in the gastric mucosa is the anti-inflammatory reaction, blood flow is increased via afferent nerve stimulation and leukocyte adhesion follows, enhancing gastric defense against toxins and bacteria [52, 53]. Wallace et al. showed that celecoxib or indomethacin increased leukocyte adherence in mesenteric venules in the rat, while SC-560 had no effect [48]. In addition, SC-560 or indomethacin reduced gastric blood flow by ~50%, but celecoxib had no effect. Thus vascularization of the gastric mucosa may be largely dependent on prostaglandins produced via COX-1, while inflammatory processes beginning with leukocyte adherence are regulated by prostaglandins produced by COX-2.

Thus in the normal stomach, both COX-1 and COX-2 inhibition are required to induce ulceration, however, in the presence of a potentially noxious agent, inhibition of COX-1 alone is sufficient to induce these pathological changes, while inhibition of COX-2 under these conditions is not. This illustrates the importance of COX-1 in gastric mucosal protection, but also highlights a role for COX-2. Even mild gastric injury results in the upregulation of COX-2 and the result is a dual resistance against injurious agents.

The importance of COX-1 is illustrated in another study where the ulcerogenic effect of meloxicam, a selective COX-2 inhibitor, was compared to the nonselective NSAID, piroxicam, in equivalent anti-inflammatory doses [54]. Drugs were given subcutaneously to rats 1 h prior to 0.45 N HCl, i.e., doses of acid that induced mild ulcerogenic injury (unpublished data). One hour later, the animals were killed, stomachs removed and analysed for the extent of ulceration. Ulcer area was recorded as a percent of the total glandular area. Meloxicam, at 1.5 and 3 mg/kg s.c., resulted in no additional ulcerogenic damage as compared to placebo animals (14 and 13% compared to 13%, respectively) who received the acid challenge alone. However, pretreatment with the same doses of piroxicam, which inhibits both COX-1 and COX-2 equally, resulted in significantly enhanced ulceration (17 and 22% with 1.5 and 3.0 mg/kg, respectively, Figure 3).

Microvascular rupture associated with ulceration results in blood flow into surrounding tissue with subsequent clot formation and a depletion of nutrients and oxygen in this tissue giving rise to ischemia and necrosis. The subsequent involvement of inflammatory cells is thought to be initiated by activated platelet and thrombin during clotting [55]. Infiltration of monocytes and neutrophils in response to these chemoattractants results in further release of growth factors and cytokines like TNFα, interleukin (IL)-1α and IL-1β [53, 55]. The cytokines

Placebo: 13% ulcer area | Meloxicam 3 mg/kg: 13% ulcer area | Piroxicam 3 mg/kg: 22% ulcer area

Figure 3. Extent of gastric injury in the stomach of rats treated with 0.45 N HCl and either vehicle, meloxicam or piroxicam at the highest dose of 3 mg/kg. The glandular region of the rat stomach is illustrated in the lower portion of each stomach. Areas of ulceration are marked with arrows.

activate endothelial cells, epithelial cells and fibroblasts. This process depends somewhat on the extent of the ulceration, but this inflammatory participation is dependent on COX-2 upregulation in these cells. COX-1 expression remains unchanged after an inflammatory insult, while COX-2 has been localized to monocytes, macrophages, fibroblasts and endothelial cells [46, 47].

There are redundant protective mechanisms in the stomach as described earlier, including nitric oxide and afferent innervation, in addition to COX. It has subsequently been shown that additive inhibition of more protective factors also results in gastric ulceration. Impairment of either COX-2, nitric oxide or afferent denervation in the presence of acid-challenged stomachs did not result in ulceration, however, inhibition of any two mechanisms in stomachs that were also acid-challenged resulted in significant ulceration [31, 56].

The degree of inflammatory burden may also play a role, in rats with adjuvant arthritis the levels of gastric PGE_2 were almost doubled as compared to normal levels [57]. Addition of rofecoxib reduced prostaglandins to levels seen in normal stomachs of non-arthritic rats, implicating COX-2 upregulation in the stomach of arthritic rats [57]. Both rofecoxib and celecoxib induced significant hemorrhagic gastric lesions in these arthritic rats, though the same doses had no injurious effects in normal non-arthritic rats. These data indicate that in the rat in this arthritic model, COX-2 is upregulated and protects against ulceration. When COX-2 is inhibited by selective COX-2 agents, ulceration occurs at doses that would not induce ulceration in the uninjured stomach. Thus COX-2 can play an important role in mucosal defense.

Ulcer healing

An ulcer by definition is a deep defect in the gastric wall resulting from necrosis and involves the entire mucosal thickness, penetrating to the submucosa or deeper [58]. The process of ulcer healing is complex, involving many cell types that are orchestrated by growth factors and cytokines, resulting in cell migration, proliferation, angiogenesis, matrix deposition and eventually scar tissue formation. Ulcers have a distinct ulcer margin composed mostly of epithelial cells. In addition, granulation tissue formed at the base of the ulcer is composed of connective tissue, containing fibroblasts, macrophages and proliferating endothelial cells undergoing angiogenesis [55, 59].

The epithelial cells at the margin of an ulcer dedifferentiate, proliferate and migrate along the surface of either a superficial wound or over the granulation tissue formed at the base a deeper injury, resulting in a continuous layer of epithelium. This is important in protecting the injury site from acid or other irritant exposure. Restitution or re-epithelialization can occur rapidly in response to epidermal growth factor (EGF), transforming growth factor (TGF)-α, basic fibroblast growth factor (bFGF), hepatocyte growth factor (HGF), insulin and platelet-derived growth factor (PDGF) [55, 59]. COX-2 expression is increased at the margins of ulcers and plays an important role in cell proliferation [46, 55, 59], however, this re-epithilization process seems to involve both COX-1 and COX-2. Giap et al. [60] demonstrated *in vitro* that re-epithelialization of a damaged area in a confluent gastric epithelial cell monolayer could be inhibited by indomethacin, but not the COX-2 selective inhibitor NS-398. Stimulation of the epithelial culture by bFGF enhanced repair as compared to basal conditions with no addition of bFGF. This bFGF-enhanced repair could be inhibited by indomethacin and NS-398, this data suggests that both a COX-1- and COX-2-dependent component is required for re-epithelialization.

In addition, epithelial cells from the base of the ulcer margin invade the tissue underlying the ulcer, where they form tubes and branch and form gastric glands. This highly organized event also occurs through growth factor coordination [55, 59]. After several days, adherent inflammatory cells have infiltrated and macrophages become dominant and release a variety of growth factors and cytokines upon activation, including TGFα, EGF, FGF-1, -2 and -4, PDGF, vascular endothelial growth factor (VEGF) and TGFβ [61, 62]. Granulation tissue is present at the base of the ulcer about 2–3 days after ulceration. It consists of proliferating connective tissue cells such as fibroblasts that eventually supply extracellular matrix for the lamina propria, as well as macrophages and endothelial cells [63]. COX-2 expression until day 10 is very high and has been localized to these inflammatory cells, to fibroblasts and to endothelial cells [63, 64]. Inhibition of ulcer healing results in enhanced thickness of the granulation tissue base and resulting scar [63]. COX-1 expression in the area immediately around the ulcer margin was no longer detectable as compared to normal gastric tissue [63]. Inhibition of COX-2 by selective inhibitors or by nonselective inhibitors delayed ulcer healing in experimental models [61–64]. The additional COX-1 inhibitory effect by nonselective inhibitors appears to play no further role. Thus this repair mechanism appears to be dependent on COX-2.

Another important component of wound healing is angiogenesis. Endothelial cell migration, proliferation and microvascular structure formation are triggered by extracellular matrix components (collagens) and growth factors [62, 65]. Two key growth factors are bFGF and VEGF, both of which have been shown to accelerate ulcer healing when administered experimentally [62, 65]. VEGF binds to specific VEGF receptors on endothelial cells, and initiates phosphorylation of numerous cytosolic proteins involved in signal transduction and triggers endothelial cell proliferation [66, 67]. VEGF production is stimulated by a number of growth factors, PDGF, TGF-β1, bFGF, as well as cytokines, nitric oxide, prostaglandins of the E series and importantly, hypoxia. It has also been demonstrated *in vitro* that NSAIDs may interfere with angiogenesis by inhibiting intracellular signaling, suggesting another possible mechanism for delaying ulcer healing [68]. Both NS-398 and indomethacin inhibited *in vitro* angiogenesis in rat aortic cells and human dermal microvascular endothelial cells [68]. This indicates that COX-2 produced prostaglandins are required for angiogenesis and angiogenesis can partially be inhibited by COX-2 inhibition. However, since addition of exogenous prostaglandins did not fully reverse this effect, a further component of NSAID-induced inhibition of angiogenesis may be unrelated to COX inhibition. Higher concentrations of both NS-398 and indomethacin could inhibit endothelial ERK-2, a mitogen-activated protein kinase involved in angiogenesis [68].

Another novel mechanism involving NSAIDs and angiogenesis is the influence of platelet activation on this process. Platelets are involved in wound healing around the ulcer margin and contain a number of growth factors and chemotactic agents which are released with their activation. It has been demonstrated that platelets can modulate angiogenesis by release of angiogenic promoting growth factors such as VEGF [69]. Platelets also contain endostatin, a potent inhibitor of angiogenesis, which is also released when they are activated [69, 70]. This balance of VEGF to endostatin was altered to favor endostatin in rats with delayed ulcer healing after administration of either celecoxib, a selective COX-2 inhibitor or flurbiprofen, a non selective inhibitor. When serum from these rats was added to human umbilical vein cell cultures, cell proliferation was reduced and apoptosis increased. Treatment of rats with a nitric oxide-releasing COX inhibitor did not delay ulcer healing. While increased endostatin levels in the serum of these animals were elevated to a similar extent as with the other drugs, it also increased VEGF levels, thus the ratio of endostatin to VEGF was similar to control levels [70].

Again it must be emphasized that COX is not the only factor involved in promoting ulcer repair and redundant gastric mechanisms such as nitric oxide also play a role in ulcer repair [71, 72]. In addition, the most effective way to negate the damaging effects of NSAIDs on ulcer healing is the addition of a proton pump inhibitor, reducing acid secretion by the gastric mucosa.

Conclusions

It has become apparent in the last few years that COX-2 also plays a role in GI mucosal integrity acting as reinforcement or back up for COX-1 mediated processes and that the original paradigm of "good COX-1" and "bad COX-2" is over simplistic. While it is clear from large, well designed clinical trials that COX-2 inhibitors do result in reduced GI side-effects at least in normal risk arthritic patients, the role of COX-2 inhibitors in patients with existing "silent" ulcers or in patients with an elevated risk of ulceration still needs to be determined. These probably reflect situations of GI inflammation, where COX-2 upregulation is also involved in mucosal integrity, and selective COX-2 inhibition may or may not be of added advantage in these patients.

References

1 Laine L (2003) Gastrointestinal effects of NSAIDs and coxibs. *J Pain Symptom Manage* 25S: S32–S40
2 Talley NJ, Evans JM, Fleming KC, Harmsen WS, Zinsmeister AR, Melton LJ 3rd (1995) Nonsteroidal antiinflammatory drugs and dyspepsia in the elderly. *Dig Dis Sci* 40: 1345–1350
3 Fries JF, Williams CA, Block DA, Michel BA (1991) Non-steroidal anti-inflammatory drug-associated gastropathy: incidence and risk factor models. *Am J Med* 61: 213–222
4 Wolfe MM, Lichtenstein DR, Singh G (1999) Gastrointestinal toxicity of nonsteroidal anti-inflammatory drugs. *N Engl J Med* 340: 1888–1899
5 Hawkey CJ, Langman MJS (2003) Nonsteroidal anti-inflammatory drugs: Overall risks and management. Complementary roles for COX-2 inhibitors and proton pump inhibitors. *Gut* 52: 600–608
6 Chan FKL, Leung WK (2002) Peptic-ulcer disease. *Lancet* 360: 933–941
7 Dequeker J, Hawkey C, Kahan A, Steinbruck K, Alegre C, Baumelou E, Begaud B, Isomaki H, Littlejohn G, Mau J et al (1998) Improvement in gastrointestinal tolerability of the selective cyclooxygenase (COX)-2 inhibitor, meloxicam, compared with piroxicam: results of the Safety and Efficacy Large-scale Evaluation of COX-inhibiting Therapies (SELECT) trial in osteoarthritis. *Brit JRheumatol* 37: 946–951
8 Bombardier C, Laine L, Reicin A, Shapiro D, Burgos-Vargas R, Davis B, Day R, Ferraz MB, Hawkey CJ, Hochberg MC et al VIGOR Study Group (2000) Comparison of upper gastrointestinal toxicity of rofecoxib and naproxen in patients with rheumatoid arthritis. VIGOR Study Group. *N Engl J Med* 343: 1520–1528
9 Silverstein FE, Faich G, Goldstein JL, Simon LS, Pincus T, Whelton A, Makuch R, Eisen G, Agrawal NM, Stenson WF et al (2000) Gastrointestinal toxicity with celecoxib vs nonsteroidal anti-inflammatory drugs for osteoarthritis and rheumatoid arthritis: the CLASS study: A randomized controlled trial. Celecoxib Long-term Arthritis Safety Study. *JAMA* 284: 1247–1255
10 Wallace JL, Bell CJ (1996) Gastromucosal defense. *Curr Opin Gastroenterol* 12: 503–511
11 Halter F, Tarnawski AS, Schmassmann A, Peskar BM (2001) Cyclooxygenase 2- implications on maintenance of gastric mucosal integrity and ulcer healing: controversial issues and perspectives. *Gut* 49: 443–453
12 Hawkey CJ (2000) Nonsteroidal anti-inflammatory drug gastropathy Gastroenterology 119: 521–535
13 Davenport HW (1969) Gastric mucosal hemorrhage in dogs. Effects of acid, aspirin and alcohol. *Gasteroenterology* 56: 439–449
14 Whittle BJR. COX-1 and COX-2 products in the gut: therapeutic impact of COX-2 inhibitors. *Gut* 47: 320–325
15 Kelly JP, Daufman DW, Jurgelon JM, Sheehan J, Koff RS, Shapiro S (1996) Risk of aspirin-associated major upper-gastrointestinal bleeding with enteric-coated or buffered product. *Lancet* 348: 1413–1416
16 Bickel M, Kauffman GL Jr, (1981) Gastric gel mucus thickness: effect of distention, 16, 16-dimethyl prostaglandin E_2 and carbenoxolone. *Gastroenterology* 80: 770–775
17 Silen W, Ito S (1985) Mechanisms for rapid re-epithelialization of the gastric mucsal surface. *Annu Rev Physiol* 47: 217–229

18 Vane JR (1971) Inhibition of prostaglandin synthesis as a mechanism of action for aspirin-like drugs. *Nature* 231: 232–235
19 Rainsford KD, Willis C (1982) Relationship of gastric mucosal damage induced inpigs by antiinflammatory drugs to their effects on prostaglandin production. *Dig Dis Sci* 27: 624–625
20 Whittle BJ (1981) Temporal relationship between cyclooxygenase inhibition, as measured by prostacyclin biosynthesis, and the gastrointestinal damage induced by indomethacin in the rat. *Gastroenterology* 80: 94–98
21 Uribe A, Johansson C, Rubio C (1987) Cell proliferation of the rat gastrointestinal mucosa after treatment with E2 prostaglandins and indomethacin. *Digestion* 36: 238–245
22 Kitahora T, Guth PH (1987) Effect of aspirin plus hydrochloric acid on the gastric mucosal microcirculation. *Gastroenterology* 93: 810–817
23 Gana TJ, Huhlewych R, Koo J (1987) Focal gastric mucosal blood flow in aspirin-induced ulceration. *Ann Surg* 205: 399–403
24 Rainsford KD (1983) Microvascular injury during gastric damage by anti-inflammatory drugs in pigs and rats. *Agents Actions* 13: 457–460
25 Andrews FJ, Malcontenti-Wilson C, O'Brien PE (1994) Effect of nonsteroidal anti-inflammatory drugs on LFA-1 and ICAM-1 expression in gastric mucosa. *Am J Physiol* 266: 657–664
26 Appleyard CB, McCafferty DM, Tigley AW, Swain MG, Wallace JL (1996) Tumor necrosis factor mediation of NSAID-induced gastric damage: role of leukocyte adherence. *Am J Physiol* 270: 42–48
27 Wallace JL, Reuter B, Cicala C, McKnight W, Grisham MB, Cirino G (1994) Novel nonsteroidal anti-inflammatory drug derivatives with markedly reduced ulcerogenic properties in the rat. *Gastroenterology* 107: 173–179
28 Vaananen PM, Keenan CM, Grisham MB, Wallace JL (1992) Pharmacological investigations of the role of leukotrienes in the pathogenesis of experimental NSAID gastropathy. *Inflammation* 16: 227–240
29 Ham EA, Soderman DD, Zanetti ME, Dougherty HW, McCauley E, Kuehl FA (1983) Inhibition by prostaglandins of leukotriene B_4 release from activated neutrophils. *Proc Natl Acad Sci USA* 80: 4349–4353
30 Gryglewski RJ, Szczeklik A, Wandzilak M (1987) The effect of six prostaglandins, prostacyclin and iloprost on generation of superoxide anions by human oplymorphonuclear leukocytes stimulated by zymosan or formyl-methionyl-leucyl-phenylalanine. *Biochem Pharmacol* 36: 4209–4212
31 Whittle BJR, Lopez-Belmonte J, Moncada S (1990) Regulation of gastric mucosal integrity by endogenous nitric oxide: interactions with prostanoids and sensory neuoropeptides in the rat. *Br J Pharmacol* 99: 607–611
32 Elliot SN, McKnight W, Cirino G, Wallace JL (1995) A nitric oxide-releasing nonsteroidal anti-inflammatory drug accelerates gastric ulcer healing in rats. *Gastroenterology* 109: 524–530
33 Kubes P, Suzuki M, Granger DN (1991) Nitric oxide: an endogenous modulator of leukocyte adhesion. *Proc Natl Acad Sci USA* 88: 4651–4655
34 Kosonen O, Kankaanranta H, Malo-Ranta U, Moilanen E (1999) Nitric oxide-releasing compounds inhibit neutrophil adhesion to endothelial cells. *Eur J Pharmacol* 382: 111–117
35 Kim YM, Talanian RV, Li J, Billiar TR (1998) Nitric oxide prevents IL-1beta and IFN-gamma-inducing factor (IL-18) release from macrophages by inhibiting caspase-1(IL-1beta-converting-enzyme). *J Immunol* 161: 4122–4128
36 Pawlik T, Konturek PC, Konturek JW, Konturek SJ, Brzozowski T, Czesnikiewicz M, Plonka M, Dielanski W, Areny H (2002) Impact of *Helicobacter pylori* and nonsteroidal anti-inflammatory drugs on gastric ulcerogenesis in experimental animals and in humans. *Eur J Pharmacol* 229: 1–15
37 Hawkey CJ, Tulassay Z, Szczepanski L, van Rensburg CJ, Filipowicz-Sosnowska A, Lanas A, Wason CM, Peacock RA, Gillon KRW (2002) Randomised controlled trial of *Helicobacter pylori* eradication in patients on non-steroidal anti-inflammatory drugs: HELP NSAIDs study. *Lancet* 352: 1016–1102
38 Chan FKL, To KF, Wu JCY, Yung MY, Leung WK, Kwok T, Hui Y, Chan HLY, Chan CSY, Hui E et al (2002) Eradication of *Helicobacter pylori* and risk of peptic ulcers in patients starting long-term treatment with non-steroidal anti-inflammatory drugs: a randomised trial. *Lancet* 359: 9–13
39 Xie W, Chipman JG, Robertson DL, Erikson RL, Simmons DL (1991) Expression of a mitogen-responsive gene encoding prostaglandin synthase is regulated by mRNA splicing. *Proc Natl Acad Sci USA* 88: 2692–2696

40 Kujubu DA, Fletcher BS, Varnum BC, Lim RW, Herschman HR (1991) TIS10, a phorbol ester tumor promotor-inducible mRNA from Swiss 3 T3 cells, encodes a novel prostaglandin synthase/cyclooxygenase homologue. J Biol Chem 266: 12866–12872
41 O'Banion MK, Sadowski HB, Winn V, Young DA (1991) A serum- and glucocorticoid-regulated 4-kilobase mRNA encodes a cyclooxygenase-related protein. *J Biol Chem* 266: 23261–23267
42 Fu J-Y, Masferrer JL, Seibert K, Raz A, Needlman P (1990) The induction and suppression of prostaglandin H2 synthase (cyclooxygenase) in human monocytes. *J Biol Chem* 265: 16737–16740
43 O'Neill GP, Ford-Hutchinson AW (1993) Expression of mRNA for cyclooxygenase-1 and cyclooxygenase-2 in human tissues. *FEBS Lett* 330: 156–160
44 Kargman S, Charleson S, Cartwright M, Frank J, Riendeau D, Mancini J, Evans J, O'Neill G (1996) Characterization of prostaglandin G/H synthase 1 and 2 in rat, dog, monkey and human gastrointestinal tracts. *Gastroenterology* 111: 445–454
45 Vane J (1994) Towards a better aspirin. *Nature* 367: 215–216
46 Jackson LM, Wu KC, Mahida YR, Jenkins D, Hawkey CJ (2000) Cyclooxygenase (COX) 1 and 2 in normal, inflamed, and ulcerated human gastric mucosa. *Gut* 47: 762–770
47 To KF, Chan FKL, Cheng ASL, Lee TL, Ng YP, Sung JJY (2001) Up-regulation of cyclooxygenase-1 and -2 in human gastric ulcer. *Aliment Pharmacol Ther* 15: 25–34
48 Wallace JL, McKnight W, Reuter BK, Vergnolle N (2000) NSAID-Induced gastric damage in rats: requirement for inhibition of both cyclooxygenase 1 and 2. *Gastroenterology* 119: 706–714
49 Smith CJ, Zhang Y, Koboldt CM, Muhammad J, Zweifel BS, Shaffer A, Talley JJ, Masferrer JL, Seibert K, Isakson PC (1998) Pharmacological analysis of cyclooxygenase-1 in inflammation. *Proc Natl Acad Sci USA* 95: 13313–13318
50 Langenbach R, Morham SG, Tiano HF, Loftin CD, Ghanayem BI, Chulada PC, Mahler JF, Lee CA, Goulding EH, Kluckman KD et al (1995) Prostaglandin synthase 1 gene disruption in mice reduces arachidonic acid-induced inflammation and indomethacin-induced gastric ulceration. *Cell* 83: 483–492
51 Gretzer B, Maricic N, Respondek M, Schuligoi R, Peskar BM (2001) Effects of specific inhibition of cyclo-oxygenase-1 and cyclo-oxygenase-2 in the rat stomach with normal mucosa and after acid challenge. *Brit J Pharmacol* 132: 1565–1573
52 Asako H, Kubes P, Wallace JL, Wolf RE, Granger DN (1992) Modulation of leukocyte adhesion in rat mesenteric venules by aspirin and salicylate. *Gastroenterology* 103: 146–152
53 Wallace JL, McKnight W, Miyasaka M, Tamatani T, Paulson J, Anderson DC, Granger DN, Kubes P (1993) Role of endothelial adhesion molecules in NSAID-induced gastric mucosal injury. *Am J Physiol* 256: G99–G998
54 Pairet M, van Ryn J, Schierok H, Mauz A, Trummlitz G, Engelhardt G (1998) Differential inhibition of cyclooxygenases-1 and -2 by meloxicam and its 4'-isomer. *Inflamm Res* 47: 270–276
55 Tanawski A (2000) Molecular mechanisms of ulcer healing. *Drug News Perspect* 13: 158–168
56 Maricic N, Ehrlich KH, Schuligoi R, Peskar BM (1999) Cyclooxygenase-2-derived prostaglandins interact with nitric oxide and afferent neurons in the maintenance of rat gastric mucosal integrity. Gastroenterology 116: A280–A281
57 Kato S, Ogawa Y, Kanatsu K, Okayama M, Watanabe T, Arakawa T, Takeuchi K (2002) Ulcerogenic influence of selective cyclooxygenase-2 inhibitors in the rat stomach with adjuvant-induced arthritis. *J Phamacol Exp Ther* 303: 503–509
58 McCarthy DM (2001) Prevention and treatment of gastrointestinal symptoms and complications due to NSAIDs. *Best Practice Res Clin Gastroenterol* 15: 755–773
59 Schmassman A (1998) Mechanisms of ulcer healing and effects of nonsteroidal anti-inflammatory drugs. *Am J Med* 104 (Suppl 3): 43S–51S
60 Giap AQ, Tarnawski A, Hoa NT, Akotia V, Ma TY (2002) NSAID inhibition of RGMI gastric monolayer wound re-epithelialization: comparison of selective COX-2 *versus* non-selective COX inhibitors. *Life Sciences* 70: 3029–3037
61 Shigeta J-I, Takahashi S, Okabe S (1998) Role of cyclooxygenase-2 in healing of gastric ulcers in rats. *J Pharmacol Exp Ther* 286: 1383–1390
62 Szabo S, Folkman J, Vattay P, Morales RE, Pinkus GS, Kato K (1994) Accelerated healing of duodenal ulcers by oral administration of a mutein of basic fibroblast growth factor in rats. *Gastroenterology* 106: 1106–1111
63 Schmassmann A, Peskar BM, Stettler C, Netzer P, Stroff T, Flogerzi B, Halter F (1998) Effects of inhibition of prostaglandin endoperoxide synthase-2 in chronic gastro-intestinal ulcer models in

rats. *Brit J Pharmacol* 123: 795–804
64 Berenguer B, de la Lastra CA, Moreno FJ, Martin MJ (2002) Chronic gastric ulcer healing in rats subjected to selective and non-selective cyclooxygenase-2 inhibitors. *Eur J Pharmacol* 442: 125–135
65 Schmassmann A, Tarnawski A, Peskar BM, Varga L, Flogerzi B, Halter F (1995) Influence of acid and angiogenesis on kinetics of gastric ulcer healing in rats: interaction with indomethacin. *Am J Physiol* 268: G276–G285
66 Leung DW, Cachianes G, Kung WJ, Goeddel DV, Ferrera N (1989) Vascular endothelial growth factor is a secreted angiogenic mitogen. *Science* 246: 1306–1309
67 Szabo S, Folkman J, Vincze A, Sandor ZS, Gombos A (1997) Modulation of vascular factors by VEGF/VPF is sufficient for chronic ulcer healing and acute gastroprotection. *Gastroenterology* 122:A303
68 Jones MK, Wang H, Peskar BM, Levin E, Itani RM, Sarfeh IJ, Tarnawski AS (1999) Inhibition of angiogenesis by nonsteroidal anti-inflammatory drugs: Insight into mechanisms and implications for cancer growth and ulcer healing. *Nature Med* 5: 1418–1423
69 Ma L, Elliott SN, Cirino G, Buret A, Ignarro LJ, Wallace JL (2001) Platelets modulate gastric ulcer healing: Role of endostatin and vascular endothelial growth factor release. *Proc Natl Acad Sci USA* 98: 6470–6475
70 Ma L, del Soldato P, Wallace JL (2002) Divergent effects of new cyclooxygenase inhibitors on gastric ulcer healing: Shifting the angiogenic balance. *Proc Natl Acad Sci USA* 99: 13243–13247
71 Konturek SJ, Brzozowski T, Majka J, Pytoko-Polonczyk J, Stachura J (1993) Inhibition of nitric oxide synthase delays healing of chronic gastric ulcers. *Eur J Pharmacol* 239: 215–217
72 Elliot SN, McKnight W, Cirino G, Wallace JL (1995) A nitric oxide releasing non steroidal anti-inflammatory drug accelerates gastric ulcer healing in rats. *Gastroenterology* 109: 524–530

COX-2 Inhibitors
Edited by M. Pairet and J. van Ryn

COX-2 and the cardiovascular system

Rahul Nayak and Brendan F. McAdam

Division of Cardiovascular Medicine, Vanderbilt University Medical Center, Nashville TN 37232-8802, USA

Introduction

COX-1 derived-prostaglandins mediate physiological responses, such as platelet activation, gastrointestinal (GI) and vascular integrity. Conversely, COX-2 is the isoform responsible for the elaboration of inflammatory prostaglandins and its inhibition is anti-inflammatory [1, 2]. The development of agents selective for COX-2 were expected to maintain the anti-inflammatory efficacy of existing NSAIDs, but have a superior GI profile due to their lack of effect on cytoprotective prostaglandins [3, 4]. This has subsequently been largely verified with clinical studies in target populations and lends credence to the Vane hypothesis that these enzymes serve distinct biological functions [5]. However, studies in animals rendered deficient in both iso-enzymes have shown some surprising deviations from the original Vane paradigm and have suggested that the distinction between mutually exclusive "physiological" and "pathological" COX pathways is somewhat artificial [6–9]. Perhaps even more intriguing was the finding that COX-2 is hydraulically expressed in vascular endothelium *in vivo* and this isoform, and not constitutive COX-1, contributes to the formation of prostacyclin (PGI_2) [10]. It is now known that there are overlaps in expression patterns and functions of both iso-enzymes and this is most apparent in pathological conditions. For example, the expression of both isoforms is increased and co-localized in atherosclerotic lesions, and in myocardial injury and dysfunction from a variety of etiologies [11, 12]. While the role of both isoforms in platelet-vessel wall interactions is expanding, the functional contribution and interplay between these COX pathways in mediating pathobiological responses in atherosclerotic plaque is not well understood.

Prostaglandins in vascular homeostasis

The COX products, thromboxane A_2 (TxA_2) and PGI_2 play an important role in vascular homeostasis by their interaction with cognate cell surface G protein coupled receptors [13–15]. They are formed from the endoperoxide intermediate PGH_2 by the action of thromboxane A_2 and PGI_2 synthases [15] (Fig. 1).

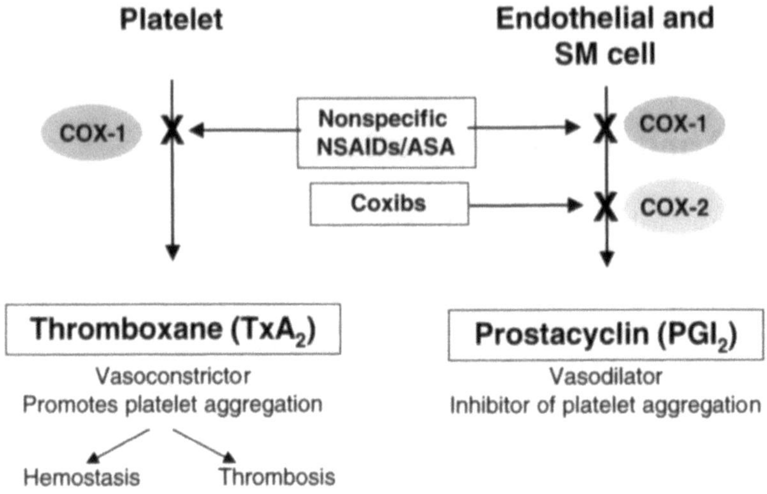

Figure 1. Vascular prostaglandins and their formation by the COX isoforms. The inhibition of endothelial- and smooth muscle (SM) cell-derived PGI_2 and platelet-derived TxA_2 by aspirin, NSAIDs and the COX-2 selective drugs is indicated.

TxA_2 is formed primarily through the activity of platelet COX-1 and is a potent platelet agonist and vasoconstrictor [16]. TxA_2 not only promotes platelet adhesion and aggregation, but acts as an amplifying signal for other agonists, thus playing a central role as a mediator of acute vascular occlusive syndromes [16, 17]. The pro-thrombotic effects of TxA_2 are in part counterbalanced by PGI_2 which has opposite biological effects on platelet function and vascular tone [18]. The interplay between these eicosanoids in the regulation of platelet-vessel wall interactions *in vivo* was confirmed in studies of genetically engineered mice lacking TxA_2 and PGI_2 receptors [19].

PGI_2 can be formed by the activity of both COX-1 and COX-2 in vascular tissue [20]. Although vascular cells under static conditions *in vitro* do not express COX-2, it has been reported that COX-2 can be upregulated in response to hydraulic flow forces of laminar, but not turbulent shear flow conditions [10]. This physiological role for COX-2 *in vivo* has been confirmed by pharmacological studies in healthy subjects, which showed maximal suppression of systemic formation of PGI_2 with selective COX-2 inhibitors, confirming that under normal conditions COX-2 is the main source of PGI_2 *in vivo* [21, 22] (Fig. 2a, 2b).

Increased TxA_2 and PGI_2 generation *in vivo* has been demonstrated in patients with syndromes of platelet activation, not only in severe atherosclerosis, but also in hyperlipidemia and smoking [23, 24]. The increased formation of TxA_2 is probably derived from platelet COX-1, as low-dose aspirin suppresses its formation [25]. Previously, the increased formation of PGI_2 observed in atherosclerosis and in acute coronary syndromes, was thought to reflect a compensatory homeostatic response to increased platelet-vessel wall

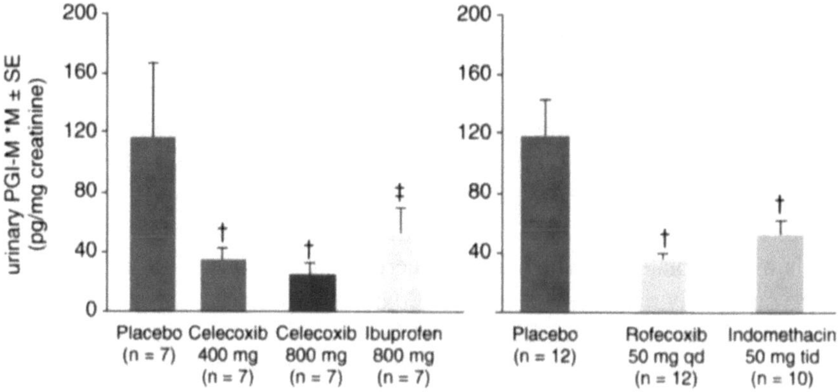

Figure 2 (a) The effect of single dose administration of celecoxib and ibuprofen in healthy volunteers on systemic biosynthesis of PGI_2 as measured by the quantification of its major urinary metabolite PGI-M (2,3 dinor-6-keto $PGF_{1\alpha}$) [21]. (Printed with permission, Copyright National Academy of Sciences, USA). (b) The effect of chronic administration of rofecoxib compared with indomethacin on PGI_2 formation in healthy elderly subjects [22]. (Adapted with permission, Copyright American Society for Pharmacology and Experimental Therapeutics).

interactions [26]. However, recent studies in transgenic mice and in animal models of thrombosis, confirm that this represents a protective response to limit further platelet-dependent thrombosis *in vivo* [19]. Recent reports of thrombosis in patients with lupus with celecoxib and the differential effect of rofecoxib and naproxen on the incidence of myocardial infarction in patients with rheumatoid arthritis has focused attention on the effect of both selective and non-selective agents on vascular eicosanoid balance [27, 28].

Pharmacology of COX inhibition in platelets and in the vasculature

Conventional NSAIDs inhibit COX activity by preventing access of arachidonic acid into the catalytic site of the enzyme and thereby reducing formation of PGH_2, the substrate for TxA_2 and PGI_2 synthases [4, 29]. Traditional NSAIDs do not discriminate between the COX isoforms and by binding reversibly to the active site of COX-1 and COX-2, they concomitantly suppress the formation of both platelet derived TxA_2 and vascular PGI_2 to a similar extent. This occurs in proportion to the plasma concentration and the dosing interval of these drugs [5, 29]. The result of the VIGOR trial showing a difference in the rate of cardiovascular events between naproxen and rofecoxib has prompted renewed interest in whether non-selective inhibitors, particularly naproxen, can exert sustained antithrombotic inhibitory effects that may account for such a difference [28]. There are no prospective controlled studies to specifically address the hypothesis that naproxen is cardioprotective. Unlike NSAIDs, which reversibly inhibit COX, the mode of aspirin is unique as it results in permanent inhibition of the enzyme [30]. Aspirin covalently inhibits COX by irreversibly acetylating a key serine amino acid near the active site of both enzymes (Ser529 in the COX-1 and Ser516 in the COX-2) [31]. In the case of the anucleate mature platelet, which only expresses COX-1 (and cannot synthesize new enzyme), the effect of aspirin on TxA_2 dependent platelet function is permanent for the lifetime of the circulating platelet [32]. The formation of TxA_2 recovers as a function of platelet turnover, requiring the synthesis of new platelets, which are regenerated at a daily rate of 10%. In this way a once daily regimen of aspirin < 100 mg has cumulative effects and completely inhibits TxA_2-dependent platelet function [32].

Thus, the antithrombotic effect of aspirin is primarily derived from its suppression of COX-1 derived TxA_2 dependent platelet aggregation and clinical studies support the marked efficacy of this agent in the secondary prevention of cardiovascular disease [33]. While aspirin is non selective for both isoenzymes, it is more potent in inhibiting COX-1 and low doses can selectively inhibit platelet COX-1 in the portal circulation with a relative sparing of vascular derived PGI_2 [32, 34]. COX-2 is less sensitive to the inhibitory effects of low-dose aspirin and this may account for the substantial residual (COX-2-dependent) PGI_2 biosynthesis *in vivo* at daily doses of aspirin in the range < 100 mg [32, 35]. Moreover, given the short plasma half-life of aspirin, nucleated vascular cells can rapidly re-synthesize the enzyme, thus overcoming the irreversible effect of aspirin on vascular COX [35]. Interestingly, despite development of different formulations of aspirin that result in increased biochemical selectivity for TxA_2 and sparing of PGI_2, it has not been clearly established that more profound suppression of PGI_2 formation by higher doses of aspirin attenuates the beneficial effects of this agent *in vivo* [35]. However, the ACE trial suggested that there might be such an effect, as lower doses of aspirin showed greater efficacy in stroke prevention than higher doses in patients with carotid artery disease [36].

The development of the selective COX-2 inhibitors held the promise of increased anti-inflammatory efficacy and a lower incidence of bleeding (and GI toxicity) related to inhibition of COX-1 [1, 5]. Consistent with their selective effect, clinical studies have confirmed that these agents at therapeutic plasma concentrations have no effect on COX-1 dependent platelet aggregation or on bleeding time [5, 22]. However, they do suppress the formation of systemic PGI_2 to the same extent as traditional NSAIDs in healthy subjects, suggesting that COX-2 is hydraulically induced and is the dominant source of PGI_2 *in vivo* [21, 22]. While this data reflects global vascular synthesis of PGI_2, it is unlikely that this is a uniform pattern of expression. Certain locations in the vasculature are subjected to more turbulent flow conditions where local COX-2 may not be upregulated. This may be relevant to sites such as vascular bifurcations that are predisposed to focal atherogenesis. However, other factors such as inflammatory stimuli are known to increase the expression of COX-2 at these local sites and net local formation of PGI_2 may be dependent on the balance between these and other factors. Although an inflammatory effect for PGI_2 has been reported, the alterations in PGI_2 likely reflect a protective response that limits the consequences of platelet activation [19, 37].

Since selective COX-2 inhibitors have no effect on platelet function, the biological consequences of an isolated, yet incomplete reduction in PGI_2 on platelet-vessel wall interactions *in vivo* has been the subject of much discussion. Mice rendered deficient in both copies of the PGI_2 receptor (IP) have a normal phenotype and do not exhibit spontaneous thrombosis, but do so only in response to a thrombotic stimulus [37]. Pharmacological reduction of systemic biosynthesis of PGI_2 by selective inhibitors is not comparable to mice that are homozygous for IP receptor deletion. It is not known if the heterozygous mouse (which may be equivalent to the 60% reduction in systemic biosynthesis that is observed with selective inhibitors) is protected from risk of thrombosis. However, these mice are protected from the development of vascular lesions when crossed with a mouse model of atherosclerosis [38]. Selective inhibition of PGI_2 in healthy subjects appears to have no clinical sequelae and indeed short-term treatment with celecoxib did not alter endothelial function in volunteers [22, 39]. Moreover, the clinical experience with these agents (at recommended doses with prolonged exposure) in patients particularly with osteoarthritis and otherwise healthy older populations with Alzheimer's disease, have confirmed their cardiovascular safety [5, 40].

Is selective inhibition of COX-2 prothrombotic in "at risk" subjects?

The more important and intensely controversial issue is whether these selective agents, by reducing PGI_2, have the potential to increase the likelihood of arterial thrombotic events in populations who may be at increased risk, or indeed do conventional NSAIDs have cardioprotective effects (Fig. 3). Many of the patients who are candidates for longer-term treatment with these agents are also at increased risk for or have established atherosclerosis (either documented or clinically silent). The biosynthesis of both TxA_2 and PGI_2 is increased not only in syndromes of altered platelet activation such as acute coronary syndromes, but also in patients with extensive atherosclerosis, in diabetics and in smokers [23, 24, 26, 41]. Suppression of PGI_2 without coincident inhibition of platelet COX-1-dependent TxA_2 formation may be undesirable in these clinical conditions. Most of the clinical studies on the longer-term effect of these agents have been performed in relatively healthy subjects and the available data on the cardiovascular effects of both selective and traditional NSAIDs is derived from recent clinical studies with the two new coxibs [5, 28].

Studies in experimental models suggest that there may be thrombotic risk. Transgenic mice that lack PGI_2 receptors have an increased risk of thrombosis when subjected to appropriate stimuli [37]. Importantly, PGI_2 modulated the prothrombotic actions of TxA_2 in these mutant mice, which confirmed a critical vaso-protective role for this autacoid under thrombotic conditions [19]. Hennan et al. showed in a canine model of coronary thrombosis that celecox-

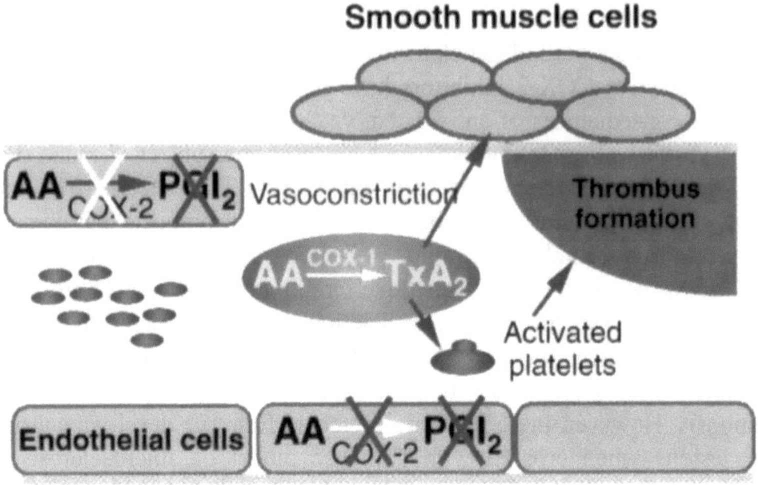

Figure 3. Vascular expression of COX and potential pro-thrombotic effects of selective COX-2 inhibitors.

ib attenuated the beneficial effects of aspirin on time to vessel occlusion and prolongation of bleeding time when it was administered as a high dose followed by a recovery period (permitting regeneration of vascular COX) [42]. Interestingly, administration of celecoxib alone had a similar effect to high-dose aspirin and did not alter time to coronary vessel occlusion despite inhibiting vasodilator responses suggesting that inhibition of COX-2 derived PGI_2 alone did not accelerate the process of thrombosis in this model of vessel injury. These data also demonstrate the complex temporal interplay between platelet-vessel wall interactions in the setting of sequential and selective inhibition of COX activity.

Possible thrombotic hazard with coxibs was reported in four patients with lupus anticoagulant when vascular thrombotic events occurred after the initiation of therapy with celecoxib [27]. The potential importance of this finding was further suggested by the results of the VIGOR trial, which examined the long-term effects of rofecoxib and naproxen in patients with rheumatoid arthritis [28]. This study was designed to explore efficacy and demonstrate superior GI safety of the coxib over naproxen in patients with rheumatoid arthritis. This multi-center study enrolled over 8,000 mostly women (mean age 58 years) with median follow up of 9 months. Aspirin use was not permitted to allow specific evaluation of the effects of this agent on GI toxicity. There was no difference in overall mortality, but the incidence of vascular thrombotic events diverged significantly between the treatment groups, 1.7% in the rofecoxib treatment group compared with 0.7% in the naproxen arm. This represented a five-fold difference in rates of myocardial infarction and this effect appeared within 6–8 weeks after the start of the study (Fig. 4) [28, 43]. Patients with

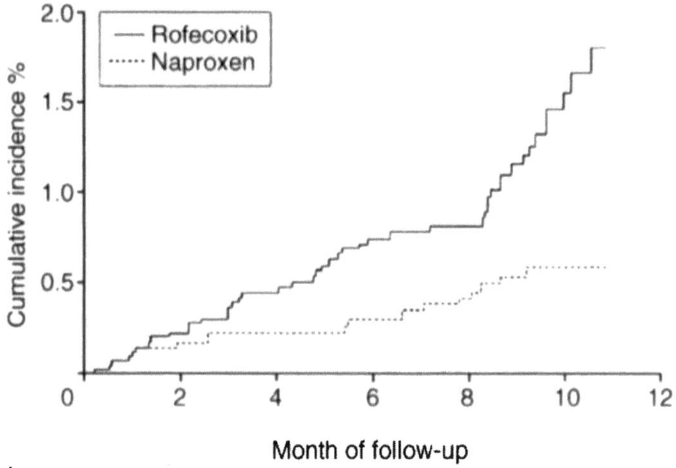

No. at risk							
Rofecoxib	4047	3643	3405	3177	2806	1067	531
Naproxen	4029	3647	3395	3172	2798	1073	514

Figure 4. Kaplan-Meier analysis of the time to cardiovascular events with rofecoxib and naproxen in the VIGOR trial [51].

rheumatoid arthritis have a well-documented increased risk of cardiovascular events, but it is unknown if this is related to increased platelet activation and/or to other inflammatory mechanisms that may impact on vascular dysfunction [44, 45]. Despite exclusion of patients with prior "cardiovascular disease" in this study, a review of the patient demographics showed that 40% had a "history of some form of vascular pathology" and 4% (?) met FDA criteria for the use of aspirin for secondary prevention [28, 43]. *Post hoc* analysis showed that this subset of subjects accounted for the excess of myocardial infarctions (>30%) and after re-analysis excluding this group, there was no significant difference in the event rates between naproxen and rofecoxib.

These adverse vascular effects were not observed in the other large arthritis study (CLASS) which was designed to examine the effect of celecoxib in 8,059 patients (mean age 60 years) with osteoarthritis and rheumatoid arthritis compared with ibuprofen and diclofenac, with follow-up over 6 months. Again most of the patients were females (average age 60 years). Aspirin use was permitted for those with an indication for such treatment (22%) [46]. There were no differences in cardiovascular outcomes at 6 months between the treatment groups and the overall event rates were low (<1% after 6 months of follow-up) [46, 47].

Although these studies were not designed or powered to demonstrate adverse thrombotic events (a pre-specified end point) between these agents and comparator drugs, and while the absolute numbers were small, the differences in the rates of myocardial infarction were significant in VIGOR. However, there was no placebo arm to adjudicate background rates, and the doses used were higher than those currently recommended [28, 46].

Several mechanisms have been proposed to explain the differential rate of myocardial infarction in the VIGOR study between naproxen and rofecoxib and why such an effect was not observed in the CLASS trial with a similarly selective agent.

Naproxen effect or rofecoxib effect on vascular eicosanoids?

Assuming that the difference between naproxen and rofecoxib was not due to chance, two mechanisms have been proposed based on known effects of these inhibitors on vascular eicosanoids: either an anti-platelet effect of naproxen or a possible pro-thrombotic effect of rofecoxib or both.

Unlike aspirin, traditional NSAIDs inhibit both COX-1 derived TxA_2 and COX-2 derived PGI_2 to varying degrees for the duration of the dosing interval. Their effects on platelet-vessel wall function *in vivo* at anti-inflammatory dosing schedules are not well characterized. Knijff-Dutmer et al. reported that treatment with naproxen 500 mg bid for 2 weeks in patients with rheumatoid arthritis resulted in significant inhibition of platelet aggregation *ex vivo* when compared with nabumetone, but there was no effect on bleeding time [48]. The exact timing of the sampling for platelet aggregation studies in relation to the administration of naproxen was not reported. Van Hecken et al. showed that treatment with ibuprofen 800 mg tid and naproxen sodium 500 mg bid for 5 days resulted in >90% inhibition in the capacity of platelet COX-1 to generate TxA_2 and inhibited platelet aggregation for up to 8 h [49]. However, in both studies the comparative effects of these agents on urinary prostaglandins was not reported. Prolonged functional inhibition of TxA_2 dependent platelet aggregation requires >95% sustained suppression of platelet capacity to generate TxA_2. It is not clear whether this reversible pharmacodynamic effect that has been reported with naproxen is similar to the irreversible effect of aspirin. It remains to be determined whether this effect is sufficient to translate into cardiovascular benefit and is currently the subject of significant debate [17].

There is some indirect evidence that other NSAIDs can have protective cardiovascular effects [50]. In the absence of prospective controlled data on possible cardio-protective effects of naproxen, several recent case control and observational studies have yielded either conflicting results or suggested modest benefits with the acknowledged limitations of such analyses [51–54]. However, a much larger cohort study in subjects with an increased incidence of coronary artery disease by Ray et al. demonstrated that there was no protective effect of naproxen and other traditional NSAIDs on the risk of cardiac events in 532,634 person years of follow-up [55]. However, the significant difference in myocardial infarction between groups (despite the small numbers of events) in VIGOR exceeds the modest and variable benefit reported in the case control studies and indeed the maximum benefit of aspirin in placebo-controlled trials, suggesting that other mechanisms may be operant [56].

The difference between the groups may also reflect the expected lack of effect of rofecoxib on platelet COX-1 and/or be linked with the second hypothesis that PGI_2 is important in restraining platelet activation and that its suppression would lead to the unopposed effects of TxA_2. To address this hypothesis, we examined the short-term effect of COX-2 inhibition in smoking, a well-documented paradigm of platelet activation [41]. The increased biosynthesis of TxA_2 in smoking, as measured by its urinary metabolite, 11-dehydro-TxA_2, was reduced to the level of non-smokers with rofecoxib without alteration in serum TxB_2. (McAdam BF, unpublished work) These data suggest that COX-2 and not COX-1 is responsible for the increased TxA_2 biosynthesis in smoking and this may reflect induction of COX-2 in inflammatory cells [57]. Interestingly, Patrono et al. has previously suggested that the aspirin insensitive increase in TxA_2 formation in unstable coronary disease may be derived from COX-2 in activated circulating monoyctes [58]. Another possible explanation for this finding relates to the expression of COX-2 in immature platelets as reported by Rocca et al., although increased platelet turnover has not been reported in smoking [59].

COX inhibition in the "inflamed" atheromatous plaque?

Despite the well-documented effect of selective inhibitors on the vascular formation of eicosanoids, there is very limited data on the effects of therapeutic doses of COX inhibitors on the prostaglandin formation in the vessel wall itself. It is possible that these agents, either selective or non selective, may have effects within the vascular wall depending on degree of tissue penetration. Consistent with the new perspective of inflammatory signaling in atherosclerosis, both COX enzymes are expressed in several cells within the atheromatous plaque and elevated plasma C reactive protein (CRP) levels have been reported to reflect arterial wall inflammation [11, 60, 61]. COX-2 derived PGE_2 in inflammatory cells modulates the expression of matrix metalloproteinases, which regulate structural content of the fibrous cap of atheroma and may play a role in the biology of plaque rupture [57, 62]. In this microenvironment, prostaglandins can signal not only through cell surface receptors, but are also able to regulate the transcription of inflammatory genes, such cytokines (and interestingly COX-2 itself) by binding to nuclear receptors, PPARs. Thus, by modulating upstream elements of several intracellular signaling pathways, prostaglandins may act as an alternative stimulus for cytokine release and subsequent increased formation of CRP [63, 64]. While NSAIDS have been shown to act as ligands for these nuclear receptors, it is unknown if the selective inhibitors can transduce biological effects by this pathway [65]. Moreover, it is unknown if naproxen could exert a cardioprotective effect as a result of its anti-inflammatory activity in the vessel wall. However, the effect of COX inhibition within the plaque is likely to be complex and may vary as a function of time and the degree of inflammation. It remains to be determined how COX derived prostanoids alter plaque-platelet and leukocyte interactions and how this might impact on the progression of this disease or on its acute manifestations.

Can an effect on hypertension lead to vascular events in VIGOR?

The increased incidence of hypertension and fluid retention observed with the 50 mg dose of rofecoxib has been raised as a possible explanation for the increased in myocardial infarction and stroke observed with rofecoxib in VIGOR [28]. More patients developed hypertension in the rofecoxib arm than in the naproxen arm (4.6 mmHg compared with 1.0 mmHg in mean systolic BP). While this may have long-term consequences on risk of vascular events, it is unlikely to explain the differences between these agents after 6–8 weeks of treatment [43]. There were no differences between celecoxib and the comparator NSAIDs in the CLASS study on this parameter [47].

Disparity in cardiovascular event rates between CLASS and VIGOR?

The apparent divergent results on rate of vascular thrombotic events with two selective inhibitors that produce similar suppressive effects on vascular PGI_2 without altering platelet function, is initially surprising. However, there are important differences between the studies that may in part account for the disparate findings. Perhaps, of most relevance were the study populations; in VIGOR, all the subjects had rheumatoid arthritis but only 28% of patients in CLASS had this diagnosis. Although overall thrombotic event rates were small and similar in both studies, patients with rheumatoid arthritis have an increased risk of adverse vascular events compared with those with osteoarthritis and the general population [44, 45]. Aspirin use was not permitted in the VIGOR study, whereas 22% of patients were on this agent in CLASS, however, subsequent analysis in those subjects not taking aspirin showed no increased event rates [28, 46]. Finally, different NSAIDs did not permit useful comparisons between these studies as differences in their vascular pharmacology have now been reported.

A recent widely reported meta-analysis comparing the cardiovascular event rates from four trials (including CLASS and VIGOR) with those from the placebo group of a meta-analysis of data from aspirin prevention trials, heightened the controversy and suggested that both celecoxib and rofecoxib were pro-thrombotic. However, this analysis had significant limitations in methodology and used statistical comparisons that were not valid between the studies with different populations [66]. Konstam et al. performed a meta-analysis of all trials with rofecoxib and did not observe increased cardiovascular events when compared with placebo, however, most of the subjects in these combined studies were at lower risk for events [40].

Other selective agents, such as meloxicam and nimesulide have been in use for several years without apparent adverse CV events based on post market surveillance. Though it must be noted there are significant limitations in accurate assessment of causal relationships of these clinically important adverse events, particularly since this potential effect was not appreciated until recently. These agents were also never formally studied in large prospective trials with long-term follow-up.

Not a selective COX-2 class effect but a single compound effect?

Are there differences between COX-2 selective agents to account for the differential effect on vascular outcomes between the CLASS and VIGOR trials? A trend for adverse vascular events was also observed in two recent smaller studies in the same population with rheumatoid arthritis with a more potent selective inhibitor, etoricoxib, when compared with placebo or naproxen over a short treatment period. The absolute event rate numbers were small (3 in >600 patients), but this occurred in a population that was already carefully screened for cardiovascular disease and aspirin use was permitted [67, 68]. This effect was not seen in trials in the same patient population when celecoxib (200–800 mg daily) was compared to naproxen over a 12-week period [69]. It is possible that there are differences in the pharmacokinetic or pharmacodynamic properties of both coxibs that account for a drug-specific effect rather than a class effect.

While all the coxibs are lipophilic and have a similar volume of distribution, rofecoxib and etoricoxib exhibit higher specificity for COX-2 and have a slower dissociation from the active site (i.e., almost irreversible binding) and have longer half-lives compared to celecoxib and valdecoxib [5, 70]. Whether these characteristics contribute to possible differences in clinical outcome remains to be determined as surprisingly little is known about the effects of these agents on tissue prostaglandin formation in inflamed synovial tissue (or in vascular tissue). Thus, for example, a coxib with a shorter half-life and/or with quicker dissociation from the COX active site might permit recovery of endothelial PGI_2 formation and thereby not predispose to continuous unopposed platelet COX-1 dependent TxA_2 formation. As there are no head to head studies at comparable doses in the same population to assess the impact of these effects on anti-inflammatory end points, renal and hypertensive effects and on potential vascular adverse effects, definitive conclusions cannot be drawn.

The initiation of atherothrombosis involves a process of platelet adhesion, activation and finally aggregation. Some of the selective agents have a modest inhibitory effect on the capacity of platelet COX-1 to generate TxA_2 without altering aggregation responses (due to the non-linear relationship between inhibition of aggregation *ex vivo* and of the capacity to generate TxA_2) [17, 22, 49]. It is not known if a modest effect on TxA_2 reduces the release of soluble mediators that influences platelet-vessel interactions, such as expression of adhesion molecules and activation of the tissue factor pathway with CD40 ligand [71]. These possibilities are somewhat speculative but merit further investigation.

Apart from recent intense interest in the vascular effects, COX inhibitors appear to have effects in other aspects of cardiovascular biology, such as atherogenesis and cardiac function.

Do COX pathways modulate atherosclerosis?

Inflammatory signaling is now considered central to the pathobiology of atherosclerosis and increased expression of both COX enzymes has been reported in atherosclerotic plaque [11, 60]. In addition to their vasoactive properties and effects on hemostasis, both PGI_2 and TxA_2 have been shown *in vitro* to have opposing actions on mitogenic responses in cells critical to the atherosclerotic process. Pharmacological TxA_2 receptor antagonism reduced atherosclerosis in the apolipoprotein E knockout mouse and deletion of both copies of the IP receptor accelerated lesion formation in the low-density lipoprotein receptor knockout mouse [38, 72]. What is the functional contribution of COX to this process and what are the implications for pharmacologically induced sustained 60% reduction in PGI_2 on the progression of atherosclerosis?

Two groups reported divergent results on the effects of pharmacologically defined inhibition of both COX isoenzymes using selective and non-selective inhibitors in the same LDL receptor deficient mouse model [73, 74]. An impressive reduction in the burden of plaque was observed with 6 weeks of treatment with rofecoxib that was not reproduced with the less selective COX-2 inhibitor, nimesulide, when administered over a longer period. In both studies, the non-selective indomethacin also exhibited a marked inhibitory effect [73, 74]. Thus, it remains to be defined if this protective effect is a COX-1 or COX-2 mediated response, although genetic manipulation of macrophage COX-2 in these animals resulted in a significant reduction in the extent of progression in the early stages of atherosclerosis [74]. It is not known whether pharmacological differences between the COX inhibitors in the mouse or whether activation of PPARs are responsible for the disparate data in this model [65].

While useful insights into COX biology can be obtained in these animal models, there are limitations in extrapolating these findings to human disease, as the process of plaque development in the mouse is significantly different. Periods of intermittent plaque rupture and fissuring that are mechanistically important in human disease are not features of this process in these mouse models [75].

Does COX modulate cardiac function?

Although COX-1 is the only isoform expressed in normal cardiomyocytes, recently the expression of COX-2 has been shown in cardiac tissue from patients with advanced cardiac dysfunction and also in myocardial infarction [12, 76]. Inhibition of this isoform was found to improve myocardial function in a rat model of infarction [76]. These beneficial effects in infarction contrast with the role of COX-2 in ischemic preconditioning in rabbits and anthracycline induced cardiac toxicity in rodents, when inhibition of COX-2 derived prostaglandins was deleterious [77, 78].

Recently a small pilot, open-label, randomized, prospective study examined the effect of the selective COX-2 inhibitor, meloxicam, in patients with acute coronary syndromes (who also were given aspirin) and reported that there was a trend towards benefit in these subjects, perhaps reflecting an anti-inflammatory effect on the vasculature [79]. These data suggest that the expression of COX-2 plays an important but variable biological role in cardiac disease and further studies are necessary to elucidate the role of COX across the spectrum of cardiac disease.

Unresolved issues in the use of selective inhibitors in higher risk patients

Many questions remain about therapeutic options in elderly patients who have co-existing rheumatic complaints and possible risk factors for atherosclerosis. Thus, when considering treatment options in patients with rheumatic disorders, the decision to treat with traditional NSAIDs or selective agents will require at least some assessment of both the risk of GI adverse events and risk of cardiovascular events.

If there is pro-thrombotic potential with these agents, the intuitive response would be to administer aspirin, however, the advantage of superior GI tolerability may be lost when aspirin is co-administered with the coxib relative to those patients who received standard NSAIDs as noted in the CLASS study [46]. Whether this combination maintains an advantage over nonselective NSAIDs remains to be determined and the cost effectiveness of this strategy will require further investigation. However, the same issue pertains to those patients at risk who may be administered an NSAID. Co-administration of aspirin with NSAIDs such as ibuprofen may attenuate the antithrombotic effect of aspirin by preventing its access to the catalytic pocket of COX-1 [80]. This effect has not been observed with the rofecoxib or meloxicam [81]. Low doses of aspirin also partially inhibit PGI_2 formation and it is unclear if this effect may accentuate the effect of the selective inhibitor on this eicosanoid [35, 36].

Interestingly administration of celecoxib (and valdecoxib) after aspirin attenuates its prolongation of bleeding time [82]. This is similar to the effect seen in a dog model of coronary thrombosis when the beneficial effects of aspirin on vessel occlusion were lost after celecoxib administration [42]. While this raises some theoretical concerns about co-administration of aspirin with these agents, no adverse effects were noted in the subset of subjects from CLASS who were given aspirin with celecoxib [46, 47].

Conclusions

COX-2 inhibitors were introduced on the premise that they would have an improved safety profile owing to fewer toxic GI effects that had complicated the use of traditional NSAIDs. Recent studies have largely confirmed their superiority in a broad range of patients with rheumatic disorders and their cardiovascular safety profile in normal subjects and those at low risk for events has been reassuring. However, the unexpected findings of a potential for increased cardiovascular thrombosis in certain subsets of higher risk patients raises important concerns over the use of these agents. There is limited clinical experience with the longer-term effect of this class of drugs in higher risk subjects such as those with hyperlipidemia, diabetes and those with unrecognized atherosclerosis, which are characterized by increased TxA_2 generation. Certainly more information is required on the effects of these agents in these higher risk subjects as well as those with hypertension and renal disease. Perhaps the results of the VIGOR trial highlight the potential for adverse interactions in higher risk subsets as COX expression is modified by pathophysiological circumstance as noted in arteriosclerosis. This is perhaps consistent with the findings in IP knockout mice, whose cardiovascular phenotype under physiological conditions is benign but in the setting of adverse vascular stimuli, exhibits deleterious responses.

Presently, it is not clear if there are one or more mechanisms that account for such a possible thrombotic hazard. Is it mediated by an alteration in vascular eicosanoids, through an effect on COX pathways in atherosclerotic plaque or is it related to specific kinetic or pharmacodynamic features of the individual agents or does all of the above play a role. Several investigators have called for studies to specifically examine possible hazards with existing agents, particularly in patients at increased risk for vascular events. However, at risk patients, like those in VIGOR, would not likely be enrolled without concomitant aspirin, which would limit the ability of clinical trials to resolve this cardiovascular controversy. Based on the studies with aspirin and the differences between the rates of events in both CLASS and VIGOR, it is estimated that several thousand subjects with extended follow up would be required to perform a trial that would have the power to ascertain differences. Such a study is underway with lumiracoxib and as aspirin use is permitted, the cardiovascular dilemma will probably not be adequately addressed despite its large size. It will, however, yield information on the effect of aspirin co-therapy on GI outcomes when administered with this coxib.

The original Vane paradigm continues to undergo modification as new clinical and experimental studies reveal novel insights into the biology of COX expression and its functional contribution not only in the vasculature, but also across the spectrum of cardiovascular disease. While the risk of vascular thrombosis with the use of these selective inhibitors in "at risk" elderly populations with sub-clinical atherosclerosis needs to be clarified, by comparison, the role of COX and its inhibition may yet be a novel new target for anti-inflammatory therapy in the earlier stages of this disease.

References

1 Vane J (1994) Towards a better aspirin. *Nature* 367: 215–216
2 Masferrer JL, Zweifel BS, Manning PT, Hauser SD, Leahy KM, Smith WG, Isakson PC, Seibert K (1994) Selective inhibition of inducible cyclooxygenase 2 *in vivo* is anti-inflammatory and nonulcerogenic. *Proc Natl Acad Sci USA* 91: 3228–3232
3 Kurumbail RG, Stevens AM, Gierse JK, McDonald JJ, Stegeman RA, Pak JY, Gildehaus D, Miyashiro JM, Penning TD, Seibert K et al (1996) Structural basis for selective inhibition of cyclooxygenase-2 by anti-inflammatory agents. *Nature* 384: 644–648
4 Vane JR, Botting R (1996) Overview-Mechanisms of action of anti-inflammatory drugs. In: JR Vane, J Botting, R Botting (eds): *Improved Non steriodal anti-inflammatory drugs: COX-2 enzyme Inhibitors*, Kluwer Academic Publishers, Dordrecht, 1–27
5 FitzGerald GA, Patrono C (2001) The coxibs, selective inhibitors of cyclooxygenase-2. *N Engl J Med* 345: 433–442
6 Morham SG, Langenbach R, Loftin CD, Tiano HF, Vouloumanos N, Jennette JC, Mahler JF, Kluckman KD, Ledford A, Lee CA (1995) Prostaglandin synthase 2 gene disruption causes severe renal pathology in the mouse. *Cell* 83: 473–482
7 Langenbach R, Morham SG, Tiano HF, Loftin CD, Ghanayem BI, Chulada PC, Mahler JF, Lee CA, Goulding EH, Kluckman KD (1995) Prostaglandin synthase 1 gene disruption in mice reduces arachidonic acid-induced inflammation and indomethacin-induced gastric ulceration. *Cell* 83: 483–492
8 Smith, WL, Langenbach, R (2001) Why there are two cyclooxygenase isozymes. *J Clin Invest* 107: 1491–1495
9 McAdam BF, Mardini I, Habib A, Burke A, Lawson JA, Kapoor S, FitzGerald, GA (2000) Effect of regulated expression of human cyclooxygenase isoforms on eicosanoid and isoeicosanoid production in inflammation *J Clin Invest* 105: 1473–1482
10 Topper JN, Cai J, Falb D, Gimbrone MA Jr, (1996) Identification of vascular endothelial genes differentially responsive to fluid mechanical stimuli: cyclooxygenase-2, manganese superoxide dismutase, and endothelial cell nitric oxide synthase are selectively up-regulated by steady laminar shear stress. *Proc Natl Acad Sci USA* 93: 10417–10422
11 Schonbeck U, Sukhova GK, Graber P, Coulter S, Libby P (1999) Augmented expression of cyclooxygenase-2 in human atherosclerotic lesions. *Am J Pathol* 155: 1281–1291
12 Wong Y, Fukuchi M, Melnyk P, Rodger IW, Giaid A (1998) Induction of COX-2 and activation of NFκB in myocardium of patients with CHF. *Circulation* 98: 100–103
13 Smith WL (1992) Prostanoid biosynthesis and mechanism of action. *Am J Physiol* 263: F181–F191
14 Dubois RN, Abramson SB, Crofford L, Gupta RA, Simon LS, Van De Putte LB, Lipsky PE (1998) Cyclooxygenase in biology and disease. *FASEB J* 2: 1063–1073
15 FitzGerald GA, Patrono C (1998) Antiplatelet drugs. In: M Verstraete, V Fuster, E Topol (eds): *Cardiovascular thrombosis- thrombocardiology, thromboneurology*. Lippincott-Raven, Philadelphia, 121–139
16 Hamberg M, Svensson J, Samuelsson B (1975) Thromboxanes: a new group of biologically active compounds derived from prostaglandin endoperoxides. *Proc Natl Acad Sci USA* 72: 2294–2298
17 FitzGerald GA (1991) Mechanisms of platelet activation: TxA_2 as amplifying signal for other ago-

nists. *Am J Cardiology* 68: 11B–15B
18 Moncada S, Gryglewski R, Bunting S, Vane JR (1976) An enzyme isolated form arteries transforms prostaglandin endoperoxides to an unstable substance that inhibits platelet aggregation. *Nature* 263: 663–665
19 Cheng Y, Austin SC, Rocca B, Koller BH, Coffman TM, Grosser T, Lawson JA, FitzGerald GA (2002) Role of prostacyclin in the cardiovascular responses to thromboxane A_2. *Science* 296: 539–541
20 Habib A, Creminon C, Frobert Y, Grassi J, Pradelles P, Maclouf J (1994) Demonstration of an inducible COX in human endothelial cells using antibodies raised against the carboxyl-terminal region of COX-2. *J Biol Chem* 268: 23448–23454
21 McAdam BF, Catella-Lawson F, Mardini IA, Kapoor S, Lawson JA, FitzGerald GA (1999) Systemic biosynthesis of prostacyclin by cyclooxygenase (COX)-2: the human pharmacology of a selective inhibitor of COX-2. *Proc Natl Acad Sci USA* 96: 272–277
22 Catella-Lawson F, McAdam B, Morrison BW, Kapoor S, Kujubu D, Antes L, Lasseter KC, Quan H, Gertz BJ, FitzGerald GA (1999) Effects of specific inhibition of COX-2 on sodium balance, hemodynamics and vasoactive eicosanoids. *J Pharm and Exper Therap* 289: 735–741
23 Fitzgerald DJ, Roy L, Catella F, FitzGerald GA (1986) Platelet activation in unstable coronary disease *N Engl J Med* 315: 983–989
24 Davi G, Gresele P, Violi F, Basili S, Catalano M, Giammarresi C, Volpato R, Nenci GC, Ciabattoni G, Patrono C (1997) Diabetes mellitus, hypercholesterolemia and hypertension but not vascular disease *per se* are associated with persistent platelet activation *in vivo*. *Circulation* 96: 69–75
25 Patrignani P, Filabozzi P, Patrono C (1982) Selective cumulative inhibition of platelet thromboxane production by low-dose aspirin in healthy subjects. *J Clin Invest* 69: 1366–1372
26 FitzGerald, GA, Smith B, Pedersen AK, Brash AR (1984) Increased prostacyclin biosynthesis in patients with severe atherosclerosis and platelet activation. *N Engl J Med* 310: 1065–1068
27 Crofford LJ, Oates JC, McCune WJ, Gupta S, Kaplan MJ, Catella-Lawson F, Morrow JD, McDonagh KT, Schmaier AH (2000) Thrombosis in patients with connective tissue diseases treated with specific COX-2 inhibitors. A report of four cases. *Arthritis Rheum* 48: 1891–1896
28 Bombardier, C, Laine L, Reicin A, Shapiro D, Burgos-Vargas R, Davis B, Day R, Ferraz M, Hawkey CJ, Hochberg MC et al (2000) Comparison of upper gastrointestinal toxicity of rofecoxib and naproxen in patients with rheumatoid arthritis: VIGOR study group. *N Engl J Med* 343: 1520–1528
29 Vane JR (1971) Inhibition of prostaglandin synthesis as a mechanism of action for aspirin-like drugs. *Nature New Biol* 231: 232–235
30 Roth GJ, Stanford N, Majerus PW (1975) Acetylation of prostaglandin synthase by aspirin. *Proc Natl Acad Sci USA* 72: 3073–3077
31 Loll PJ, Picot D, Garavito RM (1995) The structural basis of aspirin activity inferred from the crystal structure of inactivated prostaglandin H synthase. *Nat Struct Biol* 2: 637–643
32 Patrono C (1994) Aspirin as an anti-platelet drug. *N Engl J Med* 330: 1287–1294
33 Antithrombotic Trialists' Collaboration (2002) Collaborative meta-analysis of randomised trials of antiplatelet therapy for prevention of death, myocardial infarction, and stroke in high-risk patients. *Brit Med J* 324: 71–86
34 Pedersen AK, FitzGerald GA (1984) Dose related kinetics of aspirin. Pre-systemic acetylation of platelet cyclooxygenase. *N Engl J Med* 311: 1206–1211
35 Clarke RJ, Mayo G, Price P, FitzGerald GA (1991) Suppression of thromboxane A_2 but not systemic prostacyclin by controlled-release aspirin. *N Engl J Med* 325: 1137–1141
36 Ferguson G, Eliasziw M, Barr HW, Clagett GP, Barnes RW, Wallace MC, Taylor DW, Haynes RB, Finan JW, Hachinski VC et al (1999) Low-dose and high-dose acetylsalicylic acid for patients undergoing carotid endarterectomy: a randomized controlled trial. *Lancet* 353: 2179–2184
37 Murata T, Ushikubi F, Matsuoka T, Hirata M, Yamasaki A, Sugimoto Y, Ichikawa A, Aze Y, Tanaka T, Yoshida N et al (1997) Altered pain perception and inflammatory response in mice lacking prostacyclin receptor. *Nature* 388: 678–682
38 Egan KM, Austin SC, Smyth EM, FitzGerald GA (2000) *Circulation* 102, 234 (abstr.)
39 Verma S, Raj SR, Shewchuk L, Mather KJ, Anderson TJ (2001) Cyclooxygenase-2 blockade does not impair endothelial vasodilator function in healthy volunteers: Randomized evaluation of rofecoxib *versus* naproxen on endothelium-dependent vasodilatation. *Circulation* 104: 2879–2882
40 Konstam MA, Weir MR, Reicin A, Shapiro D, Sperling RS, Barr E, Gertz BJ (2002) Cardiovascular thrombotic events in controlled, clinical trials of rofecoxib. *Circulation* 104:

2280–2288
41 Nowak J, Murray JJ, Oates JA, FitzGerald GA (1987) Biochemical evidence of a chronic abnormality in platelet and vascular function in healthy individuals who smoke cigarettes. *Circulation* 76: 6–14
42 Hennan JK, Huang J, Barrett TD, Driscoll EM, Willens DE, Park AM, Crofford LJ, Lucchesi BR (2001) Effects of selective cyclooxygenase-2 inhibition on vascular responses and thrombosis in canine coronary arteries. *Circulation* 104: 820–825
43 Food and Drug Administration (2001) Arthritis Advisory Committee Open Panel Meeting 2/8/01: review of VIGOR, Cardiovascular Safety Review of Rofecoxib. Rockville MD, FDA. URL: *www.fda.gov/ohrms/dockets/ac/01/briefing/3677b2_06/cardio.pdf*
44 Wallberg-Jonsson S, Ohman ML, Dahlqvist SR (1997) Cardiovascular morbidity and mortality in patients with seropositive rheumatoid arthritis in northern Sweden. *J Rheumatol* 24: 445–451
45 De Maria, M (2002) Relative risk of CV events in patients with rheumatoid arthritis *Am J Card* 89S: 33D–38D
46 Silverstein FE, Faich G, Goldstein JL, Simon LS, Pincus T, Whelton A, Makuch R, Eisen G, Agrawal NM, Stenson WF et al (2000) Gastrointestinal toxicity with celecoxib vs nonsteroidal anti-inflammatory drugs for osteoarthritis and rheumatoid arthritis: the Celecoxib Long-Term Arthritis Safety Study (CLASS): a randomized controlled trial. *J Am Med Assoc* 284: 1247–1255
47 White WB, Faich G, Whelton A, Maurath C, Ridge NJ, Verburg KM, Geis GS, Lefkowith JB (2002) Comparison of thromboembolic events in patients treated with celecoxib, a cyclooxygenase-2 specific inhibitor, *versus* ibuprofen or diclofenac. *Am J Cardiol* 89: 425–430
48 Knijff-Dutmer EA, Martens A, vd Laar MA (1999) Effects of nabumetone compared with naproxen on platelet aggregation in patients with rheumatoid arthritis. *Ann Rheum Dis* 58: 257–259
49 Van Hecken A, Schwartz JI, Depre M, De Lepeleire I, Dallob A, Tanaka W, Wynants K, Buntinx A, Arnout J, Wong PH et al (2000) Comparative inhibitory activity of rofecoxib, meloxicam, diclofenac, ibuprofen, and naproxen on COX-2 *versus* COX-1 in healthy volunteers. *J Clin Pharmacol* 40: 1109–1120
50 Brochier ML (1993) Evaluation of flurbiprofen for prevention of re-infarction and reocclusion after successful thrombolysis or angioplasty in acute myocardial infarction: the Flurbiprofen French Trial. *Eur Heart J* 14: 951–957
51 Rahme E, Pilote L, LeLorier J (2002) Association between naproxen use and protection against acute myocardial infarction. *Arch Int Med* 162: 1111–1115
52 Solomon DH, Glynn RJ, Levin R, Avron J (2002) Nonsteroidal anti-inflammatory drug use and acute myocardial infarction. *Arch Intern Med* 162: 1099–1104
53 Watson, DJ, Rhodes T, Cai B, Guess HA (2002) Lower risk of thromboembolic cardiovascular events with naproxen among patients with rheumatoid arthritis. *Arch Intern Med* 162: 1105–1110
54 Garcia Rodriguez LA, Varas C, Patrono C (2000) Differential effects of aspirin and non-aspirin nonsteroidal anti inflammatory drugs in the primary prevention of myocardial infarction in post-menopausal women. *Epidemiology* 11: 382–387
55 Ray WA, Stein CM, Hall K, Daugherty JR, Griffin MR (2002) NSAIDs and risk of serious coronary heart disease: an observational cohort study. *Lancet* 359: 118–123
56 Sanmuganathan PS, Ghahramani P, Jackson PR, Wallis EJ, Ramsay LE (2001) Aspirin for primary prevention of coronary heart disease: safety and absolute benefit related to coronary risk derived from meta-analysis of randomized trials. *Heart* 85: 265–271
57 Mertz PM, DeWitt DL, Steller-Stevenson WG, Wahl LM (1994) IL-10 suppression of monocyte COX-2. Mechanism of inhibition of prostaglandin dependant matrix metalloproteinases production. *J Biol Chem* 269: 21322–21329
58 Cipollone F, Patrignani P, Greco A, Panara MR, Padovano R, Cuccurullo F, Patrono C, Rebuzzi AG, Liuzzo G, Quaranta G et al (1997) Differential suppression of thromboxane biosynthesis by indobufen and aspirin in patients with unstable angina. *Circulation* 96: 1109–1116
59 Rocca B, Secchiero P, Ciabattoni G, Ranelletti FO, Catani L, Guidotti L, Melloni E, Maggiano N, Zauli G, Patrono C (2002) Cyclooxygenase-2 expression is induced during human megakaryopoiesis and characterizes newly formed platelets. *Proc Natl Acad Sci USA* 99: 7634–7639
60 Ross R (1999) Atherosclerosis: an inflammatory disease. *N Engl J Med* 340: 115–126
61 Ridker PM, Hennekens CH, Buring JE, Rifai N (1997) Inflammation, aspirin, and the risk of cardiovascular disease in apparently healthy men. *N Engl J Med* 336: 973–979
62 Cipollone F, Prontera C, Pini B, Marini M, Fazia M, De Cesare D, Iezzi A, Ucchino S, Boccoli G, Saba V et al (2001) Over expression of functionally coupled COX-2 and PGE synthase in symp-

tomatic atherosclerotic plaques as a basis of PGE_2 dependent plaque instability. *Circulation* 104: 921–927
63 Ricote M, Huang J, Fajas L, Li A, Welch J, Najib J, Witztum JL, Auwerx J, Palinski W, Glass CK (1998) Expression of the peroxisome proliferator-activated receptor γ (PPARγ) in human atherosclerosis and regulation in macrophages by colony stimulating factors and oxidized low density lipoprotein. *Proc Natl Acad Sci USA* 95: 7614–7619
64 Inoue H, Tanabe T, Umesono K (2000) Feedback control of cyclooxygenase-2 expression through PPARγ. *J Biol Chem* 275: 28028–28032
65 Lehmann JM, Lenhard JM, Oliver BB, Lenhard JM, Oliver BB, Ringold GM, Kliewer SA (1997) Peroxisome proliferator-activated receptors α and γ are activated by indomethacin and other non-steroidal anti-inflammatory drugs. *J Biol Chem* 272: 3406–3410
66 Mukherjee D, Nissen SE, Topol EJ (2001) Risk of cardiovascular events associated with selective COX-2 inhibitors. *J Am Med Assoc* 286: 954–959
67 Matsumoto AK, Melian A, Mandel DR, McILwain H, Borenstein D, Zhao PL, Lines CR, Gertz BJ, Curtis S (2002) A randomized, controlled trial of etoricoxib in the treatment of rheumatoid arthritis. *J Rheumatol* 29: 1623–1630
68 Collantes E, Curtis SP, Lee KW, Casas N, McCarthy T, Melian A, Zhao PL, Rodgers DB, McCormick CL, Lee M et al (2002) A multinational randomized controlled clinical trial of etoricoxib in the treatment of rheumatoid arthritis. *BMC Fam Pract* 3: 1–10
69 Simon LS, Weaver Al, Graham DY, Kivitz AJ, Lipsky PE, Hubbard RC, Isakson PC, Verburg KM, Yu SS, Zhao WW et al (1999) Anti-inflammatory and upper gastrointestinal effects of celecoxib in rheumatoid arthritis: a randomized controlled trial. *J Am Med Assoc* 282: 1921–1928
70 McMurray RW, Hardy KJ (2002) COX-2 inhibitors: today and tomorrow. *Am J Med Sci* 323: 181–189
71 Libby P (2001) Current concepts of the pathogenesis of the acute coronary syndromes. *Circulation* 104: 365–372
72 Cayette AJ, Du Y, Oliver-Krasinki J, Lavielle G, Verbeuren TJ, Cohen RA (2000) The thromboxane receptor antagonist S1 8886 but not aspirin inhibits atherogenesis in apo E deficient mice: evidence that eicosanoids other than thromboxane contribute to atherosclerosis. *Arterioscler Thromb Vasc Biol* 20: 1724–1728
73 Pratico D, Tillmann C, Zhang ZB, Li H, FitzGerald GA (2001) Acceleration of atherogenesis by COX-1-dependent prostanoid formation in low density lipoprotein receptor knockout mice. *Proc Natl Acad Sci USA* 98: 3358–3363
74 Burleigh ME, Babaev VR, Oates JA, Harris RC, Gautam S, Riendeau D, Marnett LJ, Morrow JD, Fazio S, Linton MF (2002) Cyclooxygenase-2 promotes early atherosclerotic lesion formation in LDL receptor-deficient mice. *Circulation* 105: 1816–1823
75 Fuster V, Badimon L, Badimon JJ, Chesebro JH (1992) The pathogenesis of coronary artery disease and acute coronary syndromes. *N Engl J Med* 326: 310–318
76 Saito T, Rodger IW, Hu F, Shennib H, Giaid A (2000) Inhibition of COX-2 improves cardiac function in myocardial infarction. *Biochem Biophys Res Commun* 273: 772–775
77 Shinmura K, Tang XL, Wang Y, Xuan YT, Liu SQ, Takano H, Bhatnagar A, Bolli R (2000) Cyclooxygenase-2 mediates the cardioprotective effects of the late phase of ischemic preconditioning in conscious rabbits. *Proc Natl Acad Sci USA* 97: 10197–10202
78 Dowd NP, Scully M, Adderley SR, Cunningham AJ, Fitzgerald DJ (2001) Inhibition of cyclooxygenase-2 aggravates doxorubicin-mediated cardiac injury *in vivo*. *J Clin Invest* 108: 585–590
79 Altman R, Luciardi HL, Muntaner J, Del Rio F, Berman SG, Lopez R, Gonzalez C (2002) Efficacy assessment of meloxicam, a preferential COX-2 inhibitor, in acute coronary syndromes without ST segment elevation: the NSAIDs in Unstable Angina Treatment 2 (NUT-2) pilot study. *Circulation* 106: 191–195
80 Catella-Lawson F, Reilly MP, Kapoor SC, Cucchiara AJ, DeMarco S, Tournier B, Vyas SN, FitzGerald GA (2001) Cyclooxygenase inhibitors and the antiplatelet effects of aspirin. *New Engl J Med* 345: 1809–1817
81 van Ryn J, Kink-Eiband M, Kuritsch I, Feifel U, Hanft G, Wallenstein G, Trummlitz G, Pairet M (2003) Meloxicam does not interfere with the antiplatelet effect of aspirin in healthy volunteers. EULAR 2003, Lisbon, Portugal, Abstract THU114
82 Gotta AW (2002) Valdecoxib (Pharmacia). *Curr Opin Investig Drugs* 3: 240–245

COX-2 Inhibitors
Edited by M. Pairet and J. van Ryn

Effects on the kidney: role of COX-2 inhibitors

Dirk O. Stichtenoth

Institute of Clinical Pharmacology, Medizinische Hochschule Hannover, D-30623 Hannover, Germany

Introduction

Renal side-effects of non-steroidal anti-inflammatory drugs (NSAIDs) range from mild to life threatening and a life-long necessity for dialysis. They can be divided into three categories [1–3]: (1) Renal side-effects of NSAIDs due to inhibition of prostanoid synthesis. Most of the unwanted renal effects of NSAIDs are related to this mechanism, including reduction in renal blood flow and glomerular filtration rate, sodium and water retention, and hyperkalemia. (2) Analgesic nephropathy. Habitual use of NSAIDs and non-NSAID analgesics can cause analgesic nephropathy with the characteristic papillary necrosis. Most likely chronic hypoperfusion of the renal medulla caused by NSAIDs leads to ischaemic damage [1]. For the non-NSAID analgesics the mechanism of damage is unknown [4]. (3) Acute interstitial nephritis, which is a cell-mediated immune response to NSAIDs and many other drugs [3].

The incidence of all NSAID renal side-effects can be as high as 18%; severe renal side-effects with clinically apparent symptoms occur in 1% of individuals taking NSAIDs [5]. Though this percentage is small, the absolute numbers are high, due to the large exposure to NSAIDs [3]. More than 10% of the population in developed countries ingest NSAIDs intermittently or routinely per year. As a result, 3% of all cases of acute renal failure are caused by NSAIDs and these drugs are implicated as causal agents in up to 30% of patients with endstage renal failure [6, 7]. Thus, any effort to avoid renal side-effects of NSAIDs is of great epidemiological importance.

The value of selective COX-2 inhibitors in terms of renal safety will be reviewed here, beginning with a discussion of physiological mechanisms in the kidney affected by NSAIDs and the roles of the COX isoenzymes in the kidney.

Prostanoid functions in the kidney and NSAID effects

Prostaglandin E_2 (PGE_2), prostacyclin (PGI_2) and to a minor degree $PGF_{2\alpha}$ are the predominant prostanoids in the kidney [8, 9]. Renomedullary interstitial cells synthesize mainly PGE_2 and renal vasculature and glomeruli produce predominantly PGI_2 [10]. The effects of these prostanoids are mediated via G-protein-coupled receptors. Many receptor subtypes and a specific pattern of distribution are responsible for the multiple and at times apparently opposing functional effects of a single prostaglandin [11].

Regulation of renal hemodynamics

Infusion of PGE_1 into renal arteries of dogs produced an increase in renal blood flow [12]. The same effect was also shown for PGE_2, PGD_2 and PGI_2 [13]. In conscious dogs COX inhibition by indomethacin had no effect on renal blood flow [14]. In contrast, in anesthetized, laparotomized dogs indomethacin led to striking reductions of renal blood flow, pointing to the important role of other mediators which, when unopposed by renal prostanoids, cause pronounced vasoconstriction [15].

Since these initial studies, much has been learned about the crucial role of prostanoids in pathophysiological states for maintenance of renal blood flow and glomerular filtration rate and we are now aware of the risk factors for acute renal failure induced by NSAIDs (Tab. 1) [2]. Accordingly, indomethacin

Table 1. Risk factors for renal side-effects of NSAIDs [2]. Edema, hypertension and hyperkalemia are listed when they may occur as isolated symptoms, however, a reduction of GFR/RBF can also cause edema, hypertension and hyperkalemia.

Risk factor	NSAID renal side-effect
Preexisting renal impairment	reduction of GFR/RBF, acute renal failure, edema, hypertension, hyperkalemia
Congestive heart failure	reduction of GFR/RBF, acute renal failure, edema, hypertension
Sodium depletion by diet or diuretics	reduction of GFR/RBF, acute renal failure
Dehydratation	reduction of GFR/RBF, acute renal failure
Liver cirrhosis with ascites	reduction of GFR/RBF, acute renal failure, edema, hypertension
Lupus nephritis	reduction of GFR/RBF, acute renal failure, edema, hypertension, hyperkalemia
Status post surgery	reduction of GFR/RBF, acute renal failure
Co-medication with nephrotoxic drugs (e.g., aminoglycosides, ciclosporine)	reduction of GFR/RBF, acute renal failure, edema, hypertension
Co-medication with ACE inhibitors, angiotensin receptor antagonists	reduction of GFR/RBF, acute renal failure, hyperkalemia
Co-medication with potassium sparing diuretics	hyperkalemia

GFR = Glomerular filtration rate. RBF = Renal blood flow

caused a striking decrease in glomerular filtration rate in patients with disorders characterized by a low effective plasma volume like liver cirrhosis with ascites, heart failure and nephrotic syndrome [16–18]. In all these diseases vasoconstrictor effects of renal nerve activity, catecholamines, angiotensin II and vasopressin play an important role. Other risk factors for acute renal failure by inhibition of prostanoid synthesis are preexisting chronic renal insufficiency, dietary sodium depletion, therapy with diuretics, and systemic lupus erythematosus with renal involvement [2].

The renal medulla has a very low oxygen tension, because blood flow is provided exclusively by the juxtamedullary nephrons. Here, prostanoids play a critical role in protecting the renal medulla from ischaemic damage [19]. Thus, one important factor in the development of analgesic nephropathy is the redistribution of blood flow away from the renal medulla caused by the habitual use of NSAIDs.

Electrolyte excretion

PGE_2 reduces sodium reabsorption in a segment beyond the proximal tubule, primarily in the juxtamedullary nephrons [20, 21]. Sodium retention is difficult to demonstrate in normal volunteers [22], however, in patients with post malignant hypertension on a 100 mEq sodium diet, inhibition of renal prostanoid production by indomethacin 25 mg three times daily (TID) caused sodium retention and weight gain [22]. In contrast, sodium retention is easily observed after prostanoid inhibition in patients with congestive heart failure, liver cirrhosis or nephrotic syndrome and may lead to significant clinical deterioration (Tab. 1) [16–18]. It must be stressed that the simultaneous decreases of renal blood flow, glomerular filtration rate and sodium excretion make it difficult to localize the point of action of the drugs. However, sodium retention is more severe than can be explained solely by a reduction in the filtered load and an increased tubular reabsorption is most likely also involved.

NSAIDs also have a direct effect on potassium excretion. In the macula densa PGI_2 and PGE_2 mediate renin release [23, 24]. PGI_2 synthesized in the afferent arterioles directly stimulate the juxtaglomerular cells to release renin, whereas PGE_2 from tubular cells stimulate renin release via the macula densa [24]. We showed that indomethacin completely abrogates the increase in plasma renin activity seen within minutes after giving furosemide intravenously to healthy volunteers [22]. In sodium depleted volunteers inhibition of renin release by NSAIDs can be observed only by inhibiting sympathetic stimulation of renin release, i.e., by beta receptor blockade [25]. In patients with normal renal function hyperkalemia is not a common problem. However, hyperkalemia may become a severe threat to patients with renal disease, which may in itself can cause potassium retention or when drugs that cause potassium retention are taken simultaneously (Tab. 1).

Antidiuretic hormone

Antidiuretic hormone (ADH) stimulates intrarenal synthesis of PGE_2 and $PGF_{2\alpha}$, which attenuates the hydroosmotic effect of ADH as part of a negative feedback loop [26]. NSAIDs interrupt this negative feedback mechanism. Indomethacin led to an increase of urine osmolality in response to ADH in hypophysectomized dogs [27]. In humans, infusion of hypertonic saline increases urine osmolality and free water reabsorption; both effects are enhanced by indomethacin, showing that endogenous ADH, released in large amounts by hypertonic saline, will become more effective when renal prostanoid production is suppressed [28].

COX-1 and COX-2 in the kidney

In the late 1970s Smith and coworkers located COX-1 in arterial vascular endothelial cells, collecting tubules and medullary interstitial cells of five mammalian species, including rat, rabbit, guinea pig, sheep and cow [29]. This was performed using polyclonal COX antibodies, which were retrospectively classified as COX-1 antibodies. COX antigenicity was not detectable in the macula densa or in the adjoining juxtaglomerular cells, proximal or distal convoluted tubules, or Henle's loop in any species [29]. After discovery of COX-2, Harris et al. confirmed these results in the rat kidney [30]. In addition, a new surprising finding was that COX-2 mRNA and protein was detected under control conditions at locations where no COX-1 was detectable: in the interstitial cells of the papilla, macula densa and the adjacent cortical thick ascending limb [30]. Furthermore, cortical COX-2 expression under a normal diet was restricted to a few macula densa, whereas chronic salt depletion increased COX-2 expression strikingly, both in terms of frequency and number of cells within the macula densa [30].

Yang et al. corroborated these findings and were able to demonstrate in the rat kidney a divergent regulation of COX-2 in the macula densa and renal medulla by changing concentrations of dietary salt [31]. Medullary COX-2 mRNA and protein was increased by high dietary sodium intake, whereas COX-2 expression in macula densa was decreased by this treatment [31]. In contrast, sodium depletion induced a three-fold increase in COX-2 in the macula densa and tended to reduce COX-2 mRNA in the inner medulla [31]. COX-2 expression in the macula densa could also be stimulated by inhibition of salt transport in the loop of Henle [32]. In this study, treatment of rats with furosemide for 1 week led to a three-fold increase in COX-2 protein in the macula densa, whereas hydrochlorothiazide had no effect on COX-2 mRNA or immunoreactivity [32].

COX-2 also seems to play an important role in mediating prostanoid-ADH antagonism. Addition of a hypertonic medium to collecting duct cell cultures induced COX-2 and increased PGE_2 synthesis. This could be completely blocked by the COX-2 inhibitor NS-398 or by indomethacin [33]. Accordingly, in the medulla of the rat kidney, dehydration led to enhanced COX-2 expression in collecting ducts [33].

In the human kidney both COX-isoforms are expressed constitutively. Low levels of both COX-1 and COX-2 mRNA were observed in the kidneys of victims of accidents [34]. Subsequently, Kömhoff et al. confirmed the constitutive expression of COX-2 in similar amounts as COX-1 [35]. COX-1 mRNA and protein were localized in endothelial and smooth muscle cells of arteries and veins, collecting duct cells, and in interstitial cells of the renal cortex and medulla [35]. For COX-2 the same pattern of vascular expression was shown as with COX-1. COX-2 was abundantly expressed intraglomerularly in podocytes, suggesting that COX-2 may contribute to the regulation of renal perfusion and filtration properties [35]. The same study also investigated the localization of COX-1 and COX-2 in the human fetal kidney, because animal data point to participation of COX-2 in renal development [36–39]. COX-2 was detected only at the endstage of renal development in podocytes, whereas COX-1 was expressed abundantly with a similar localization as in adults and additionally in podocytes [35].

Further studies yielded divergent results, which may be explained by different methodological approaches and functional state of the kidneys investigated. While Khan et al. found the same restriction of COX-2 to blood vessels and podocytes [40], Nantel et al. revealed that COX-2 is also expressed in the human macula densa, thick ascending limb of Henle's loop and medullary interstitial cells [41]. Moreover, in human fetal kidneys marked COX-2 levels in macula densa and thick ascending limb of Henle's loop were demonstrated [42]. This COX-2 expression declines during nephron maturation and COX-2 was no longer detected in the adult kidney [42].

Thus, COX-2 is expressed in the human kidney with a pattern similar to the known intrarenal localization in other mammals: Vascular endothelium and smooth muscle cells, podocytes, macula densa, thick ascending limb of Henle's loop and medullary interstitial cells. In the fetal kidney COX-2 was detected in macula densa, thick ascending limb of Henle's loop and at the endstage of renal development in podocytes. Thus, one might expect that COX-2 is involved in all prostanoid-controlled mechanisms in the human kidney and further in renal development. However, before drawing any definite conclusions, the studies on the renal effects of COX-2 inhibitors in animals and humans *in vivo* must be reviewed.

Animal experimental work with selective COX-2 inhibition *in vivo*

As far as renal hemodynamics are concerned, experiments in sodium and volume depleted conscious dogs, volume depleted rats and rats with liver cirrhosis and ascites indicate that COX-1 is the critical enzyme [43–46]. All of these pathophysiological conditions are risk factors for the deterioration of renal blood flow and glomerular filtration rate by NSAIDs. In these studies, COX-2 inhibitors had no effects on renal hemodynamics, whereas nonselective NSAIDs or the selective COX-1 inhibitor SC-560 in comparable doses reduced renal perfusion markedly [43–46].

In contrast, for sodium and water excretion and the control of renin release it has repeatedly been shown in rats and mice that COX-2 is the responsible isoenzyme. Treatment of rats with the COX-2 inhibitor celecoxib for 3 weeks caused significant weight gain (20%) and an increase of systolic blood pressure of 30 mm Hg relative to vehicle-treated controls [47]. In hypertensive rats, the same treatment schedule with celecoxib led to a mean increase in systolic blood pressure of 33 mm Hg [47].

The COX-2 inhibitor NS-398 blocked the increase in renal renin content in sodium depleted mice [48]. Also in rats on a low sodium diet the furosemide stimulated increase of plasma renin activity was blocked by the COX-2 inhibitor SC-58125 to the same degree as indomethacin [49]. Interestingly, urinary PGE_2 excretion was markedly reduced by indomethacin but unaffected by SC-58125, suggesting, that PGE_2 in the normal kidney is synthesized largely by COX-1, whereas COX-2 appears to contribute qualitatively but not quantitatively [49]. COX-2 in the macula densa and renin release are coordinately upregulated in the rat model of renovascular hypertension [50, 51]. Administration of celecoxib almost normalized plasma renin activity and lowered mean arterial blood pressure, thereby proving the functional linkage between COX-2 and renin release [51]. Mice genetically devoid of COX-1 have normal plasma renin activity, the increase in plasma renin activity induced by captopril could be blocked, again, by celecoxib [52].

A detailed insight into the roles of the COX-isoenzymes in renal development is provided by studies with COX-1 and COX-2 knockout mice. Renal malformations primarily due to a block in the development of glomeruli occurred in COX-2, but not in COX-1 knockout mice. Most of the animals died within 16 weeks of birth [36–38]. In the rat, COX-2 mRNA and immunoreactivity were detected in a subset of the epithelial cells of the developing thick ascending limb of Henle [39]. Both the time course and magnitude of COX-2 expression support the conclusion that COX-2 plays an important role in rat nephron development. More relevant to the therapeutic situation, Kömhoff et al. demonstrated in mice and rats that treatment with celecoxib from postnatal day 0 to day 21, when COX-2 expression peaked in renal cortex, severely reduced glomerular diameter. In contrast, treatment limited to pregnancy duration did not affect glomerular size [53].

In the inflamed kidney COX-2 expression is induced and COX-2 is the responsible enzyme for synthesis of pro-inflammatory prostanoids. Seibert et al. [54] investigated the contribution of COX-1 and COX-2 to prostanoid production in the inflamed kidney using a model of unilateral ureteral obstruction, which leads to hydronephrosis and inflammatory infiltration. The 10-fold higher PGE_2 excretion in the venous effluent of the hydronephrotic kidney was inhibited by the COX-2 inhibitor DuP-697 and by indomethacin, dexamethasone and cycloheximide, an inhibitor of protein biosynthesis [54]. In the normal kidney, only indomethacin was able to reduce PGE_2 production, indicating that the majority of PGE_2 in the normal kidney is produced by COX-1 [54]. Several other studies revealed enhanced COX-2 expression in experimental glomerulonephritis [55–57], nephrotoxic nephritis [58], passive Heymann nephritis [59], subtotal renal ablation [60, 61] and lupus nephritis in mice and man [62]. Treatment with COX-2 inhibitors reduced inflammation, proteinuria and retarded progression of renal injury in these models [59, 61].

Clinical experience with selective COX-2 inhibition

The first experiences with COX-1 sparing drugs were made before the discovery of the COX-isoenzymes. Glucocorticoids potently suppress COX-2 expression but have no inhibitory activity on preformed COX [63, 64]. Dexamethasone at dosages that induce a Cushingoid appearance had no effect on renal and total body prostanoid production in healthy volunteers [65] rabbits [66] and rats [67]. Clinically, glucocorticoids have impressive anti-inflammatory actions but they are devoid of the NSAID typical side-effects. The absence of unwanted renal effects of glucocorticoids shows that prostanoids synthesized by an inducible COX-2 are of no importance for renal physiology, otherwise glucocorticoids should have detrimental effects on renal function. It must be stressed that this holds true only for the inducible COX-2, and that there is also a constitutive pool of COX-2 as described above, however, its expression seems to be insensitive to glucocorticoids [68].

Salicylic acid is the first NSAID. It is the active substance present in herbal remedies that have been used by humans for more than 3500 years. Surprisingly, salicylic acid and other non-acetylated salicylates are selective COX-2 inhibitors and suppress the induction of COX-2 [69, 70]. In healthy volunteers, sodium salicylate in the anti-inflammatory dose of 53 mg/kg body weight did not inhibit renal PGE_2 synthesis, whereas the equivalent dose of acetylsalicylic acid led to a marked decrease [71]. Salsalate, the prodrug of salicylic acid, is a potent NSAID, but does not cause reduction in renal blood flow and glomerular filtration rate, sodium and water retention, or a decrease in lithium clearance [72–74]. Unfortunately, the therapeutic value of non-acetylated salicylates is undeservedly forgotten in many countries.

Preferential COX-2 inhibitors

Etodolac, meloxicam, nimesulide and non-acetylated salicylates are classified as preferential COX-2 inhibitors, i.e., NSAIDs with a moderate COX-2 selectivity [69, 75]. Amongst them, meloxicam is the best documented. In healthy volunteers we demonstrated that meloxicam at the therapeutic dose of 7.5 mg OD had no effects on platelet aggregation and thromboxane formation in contrast to a marked inhibition by indomethacin 25 mg TID [76]. Thus, meloxicam at this dose level is COX-1 sparing in humans *in vivo*. In the same study, renal PGE_2 synthesis remained unaffected by meloxicam but was markedly inhibited by indomethacin [76]. Thus, under physiological conditions the majority of renal PGE_2 is synthesized by COX-1. This is in agreement with the animal experimental findings [49, 54].

This selectivity observed in normal volunteers translates into improved clinical safety as far as renal hemodynamics is concerned. Meloxicam in the maximum recommended dose of 15 mg OD for 28 days did not cause further deterioration of renal function in 25 patients with mild renal impairment, a risk factor for NSAID-induced renal failure [77]. Convincingly, in patients with chronic heart failure, another risk factor for deterioriation of renal function by NSAIDs, treatment with meloxicam 15 mg OD for 7 days caused no renal side-effects nor attenuated the diuretic effect of furosemide [78].

Two prospective, large scale, double-blind randomized trials support the findings of the above mentioned studies. In the MELISSA (n = 9323) and SELECT (n = 8656) trial meloxicam 7.5 mg was compared to diclofenac 100 mg slow release and piroxicam 20 mg, respectively, for 28 days in patients with osteoarthritis [79, 80]. While the clinical efficacy of meloxicam was equivalent to the comparator drugs in both trials, renal toxicity of meloxicam, measured as increases in creatinine and urea in serum, was significantly less as compared to diclofenac and piroxicam [79, 80]. The results of MELISSA and SELECT show a reduced risk of renal side-effects with meloxicam, it is obvious, however, that they were not eliminated.

To address the question, whether renin release is COX-2 dependent in humans, we undertook a randomized cross-over study in healthy volunteers comparing the effects of meloxicam in the COX-1 sparing dosage of 7.5 mg/d to indomethacin 25 mg TID on plasma renin activity [81]. Meloxicam abolished the furosemide stimulated plasma renin as effectively as indomethacin, suggesting that the COX-isoenzyme responsible for renin release in man is COX-2 (Fig. 1) [81]. Thus, inhibition of renin release with subsequent hyperkalemia must be considered as a side-effect of COX-2 inhibition. We also determined renal PGE_2 synthesis, which was, in accordance with our previous finding, not inhibited by meloxicam whereas indomethacin led to a significant reduction [76, 81].

First generation COX-2 inhibitors

Preferential COX-2 inhibitors like meloxicam can exert at the highest therapeutic dosage clinically relevant COX-1 inhibition. Compounds with a higher selectivity for COX-2 are classified as specific or highly selective COX-2 inhibitors [69, 75]. The first specific COX-2 inhibitors were celecoxib and rofecoxib.

As far as renal perfusion is concerned, even two-fold the therapeutic dosage of celecoxib or rofecoxib (Fig. 2) caused no decrease of glomerular filtration

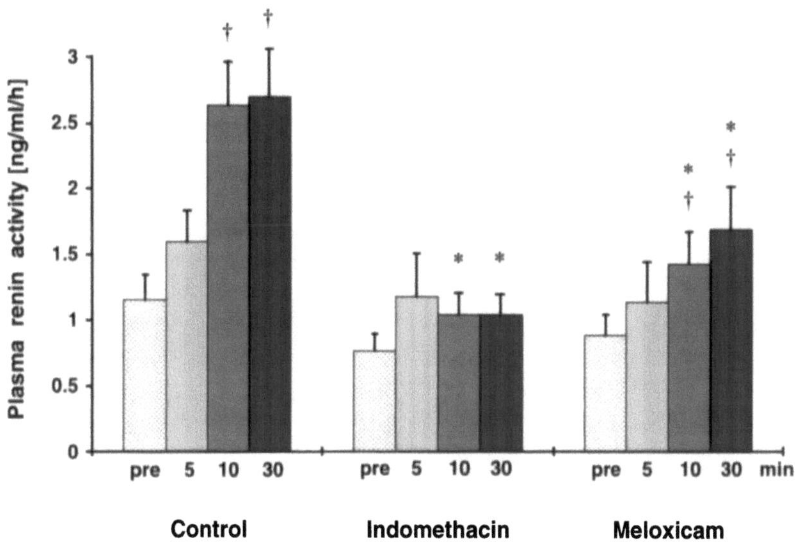

Figure 1. Effects of meloxicam and indomethacin on plasma renin activity before (pre) and 5, 10 and 30 min after furosemide 20 mg i.v. Mean ± SE of plasma renin activity in 15 healthy volunteers on control day or on the last day of treatment with meloxicam 7.5 mg once daily (OD) for 6 days or indomethacin 25 mg TID for 3 days in a randomized, cross-over study design. * = $p < 0.05$ *versus* control. † = $p < 0.05$ *versus* pre furosemide. (Figure reproduced from [81] with the kind permission of the *Journal of Investigative Medicine*.)

rate in healthy elderly volunteers and normotensive salt-depleted young healthy volunteers [82–85]. Both groups are at moderate risk for impairment of renal function by NSAIDs. However, when a second risk factor is present, such as older age plus sodium and volume depletion, rofecoxib at therapeutic dosage led to the same reduction of glomerular filtration rate as indomethacin 50 mg TID (Fig. 3) [86]. In addition, a number of case reports of acute renal failure caused by celecoxib or rofecoxib alert us, that the gain of renal safety in terms of renal perfusion by use of specific COX-2 inhibitors is small [87–89]. Thus, the same precautions as for conventional NSAIDs must be used.

While the clinical data support the experimental results in animals that COX-2 is only partly involved in the regulation of renal hemodynamics, they clearly show that COX-2 is the critical COX-isoenzyme for sodium and water excretion. Several double blind, placebo-controlled and randomized studies provide detailed data on this issue [82–86]. In healthy, elderly volunteers on a controlled normal sodium diet, intake of rofecoxib 50 mg once daily as well as indomethacin 50 mg TID for 14 days, each reduced sodium excretion by 20% during the first 3 days of administration, thereafter returning to normal [82]. No edema occurred, body weight and diastolic blood pressure remained unchanged. Similar effects were found for celecoxib in a study in normotensive salt-depleted young healthy volunteers [83]. Celecoxib was given for 7 days in a dosage of 200 mg BID and 400 mg BID, the latter is twice the maximum recommended dosage in rheumatoid arthritis, the comparators were

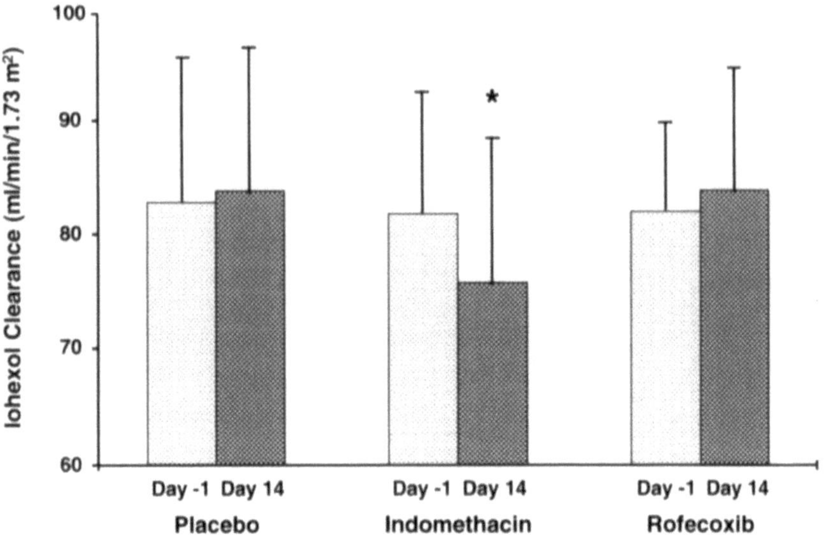

Figure 2. Effects of rofecoxib and indomethacin on glomerular filtration rate. Mean ± SD of iohexol clearance in 36 healthy elderly volunteers treated with either rofecoxib 50 mg OD, indomethacin 50 mg TID or placebo for 14 days in a double-blind, randomized, parallel-group study design. *$p < 0.05$ before (day -1) *versus* treatment day 14. Data from [82].

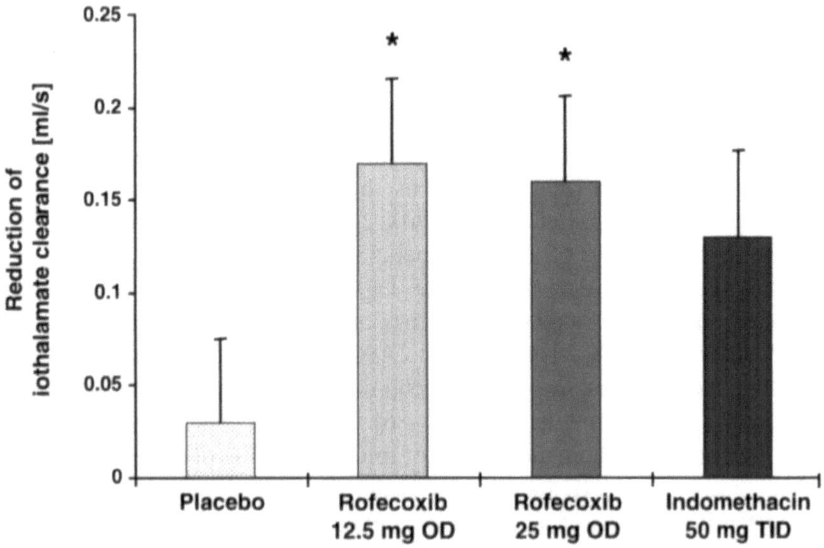

Figure 3. Effects of rofecoxib and indomethacin on glomerular filtration rate. Mean ± SE of iothalamate clearance in 60 healthy elderly volunteers receiving a low-sodium diet on the 6th day of treatment with either rofecoxib 12.5 mg OD, rofecoxib 25 mg OD, indomethacin 50 mg TID or placebo in a double-blind, randomized, parallel-group study design. * $p < 0.05$ *versus* placebo. Data from [86].

naproxen 500 mg BID for 7 days and placebo [83]. Celecoxib caused sodium retention to the same degree as the nonselective naproxen, even under the low sodium diet used in this study. The reduction of urinary sodium excretion was most pronounced 2–3 h after the first dose, but still present on the last day of treatment [83]. Endogenous lithium clearance, reflecting tubular sodium reabsorption, was decreased accordingly (Fig. 4). Remarkably, urine flow was markedly reduced for several hours after intake of celecoxib and naproxen; the effect was attenuated on the last day of treatment but still significant for the two doses of celecoxib [83]. This observation strongly suggests medullary COX-2 as the responsible isoenzyme for ADH antagonism, since reduction in urine output is a well known renal effect of NSAIDs and caused by inhibition of prostanoid-ADH antagonism.

The analysis of the safety data from the large clinical trials with celecoxib and rofecoxib show that as anticipated from the studies in healthy volunteers, the incidence of peripheral edema and hypertension occurring with celecoxib or rofecoxib was significantly higher than with placebo and comparable to traditional NSAIDs [90–92]. Furthermore, as demonstrated for rofecoxib, there is a clear dose-dependency with highest incidence of edema in patients treated with 50 mg rofecoxib once daily, a dosage twice the maximum recommended dose for rheumatoid arthritis but still COX-1 sparing [93]. In a 6 week head-to-head trial rofecoxib 25 mg OD was compared to celecoxib 200 mg OD in 810 older hypertensive osteoarthritis patients. The incidence of peripheral edema and

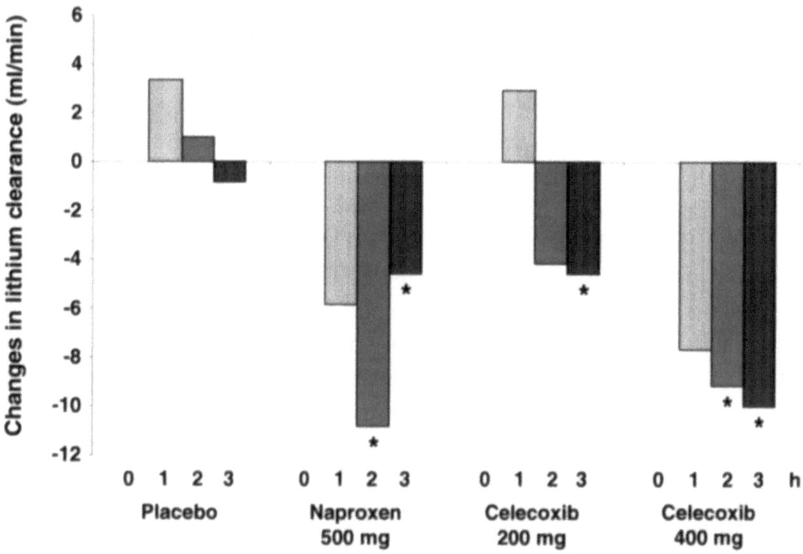

Figure 4. Effects of celecoxib and naproxen on lithium clearance. Mean of 40 healthy volunteers receiving a low-sodium diet after the first administration of either celecoxib 200 mg, celecoxib 400 mg, naproxen 500 mg or placebo in a double-blind, randomized, parallel-group study design. $*p < 0.05$ *versus* placebo. Data from [83].

increase in systolic blood pressure was higher with rofecoxib (9.5% and 17%, respectively) than with celecoxib (4.9% and 11%, respectively) [94]. However, the dose of celecoxib 200 mg OD is less antiphlogistically potent than rofecoxib 25 mg OD; a further bias was the non comparable pharmacokinetics of once daily celecoxib with a half-life of 10 h *versus* rofecoxib with 17 h. Another recently published head-to-head comparison of rofecoxib 12.5 and 25 mg OD to celecoxib 200 mg OD and acetaminophen 1 g QID in 382 osteoarthritis patients without concurrent diseases demonstrated superior efficacy of rofecoxib 25 mg OD over the other treatments and no difference in the incidence of edema and hypertension, which were low (1–4%) in this study population [95].

With regard to renin release, we recently confirmed our previous results obtained with meloxicam [81] in a double-blind, randomized three-way crossover study in healthy volunteers comparing celecoxib 200 mg BID with indomethacin 50 mg TID and placebo. The unpublished results clearly demonstrate a decrease of furosemide stimulated and unstimulated plasma renin activity by celecoxib. Thus, the COX-isoenzyme responsible for renin release in humans is most likely COX-2.

Second generation COX-2 inhibitors

Valdecoxib, its water-soluble prodrug parecoxib, etoricoxib and lumiracoxib represent the second generation of COX-2 inhibitors. The COX-2 selectivity of valdecoxib and parecoxib in the whole blood assay is comparable to rofecoxib, whereas etoricoxib and lumiracoxib possess a several-fold higher COX-2 selectivity than rofecoxib [96]. To date there is only limited experience on the renal effects of these compounds.

For valdecoxib incidence of water retention and edema is evident, 12% with the 80 mg OD dose, 9% with 40 mg OD and 10% with 10 mg OD was observed [97]. Thus, with the therapeutic doses of valdecoxib, 10 and 20 mg OD, the rate of renal side-effects is similar to naproxen, ibuprofen or diclofenac. The available data from clinical trials for lumiracoxib show a frequency of moderate to severe edema of 1%. The corresponding incidence rates were 3% for diclofenac and 0% for placebo [98].

Although experience with the renal safety of the second generation of COX-2 inhibitors is very limited, the speculation is justified that their higher COX-2 selectivity does not translate into renal sparing effects.

References

1 Schlöndorff D (1993) Renal complications of nonsteroidal anti-inflammatory drugs. *Kidney Int* 44: 643–653

2 Stichtenoth DO, Frölich JC (2000) COX-2 and the kidneys. *Curr Pharm Des* 6: 1737–1753

3 Whelton A (1999) Nephrotoxicity of nonsteroidal anti-inflammatory drugs: physiologic foundations and clinical implications. *Am J Med* 106: 13S–24S

4 De Broe ME, Elseviers MM (1998) Analgesic nephropathy. *N Engl J Med* 338: 446–452
5 Murray MD, Brater DC, Tierney WM, Jui SL, McDonald CJ (1990) Ibuprofen-associated renal impairment in a large general internal medicine practice. *Am J Med Sci* 299: 222–229
6 Kleinknecht D (1993) Diseases of the kidney caused by non-steroidal antiinflammatory drugs. In: JH Stewart (ed.): *Analgesic and NSAID-induced kidney disease*. Oxford University Press, Oxford, 160–179
7 Schwarz A, Offermann G, Keller F (1992) Analgesic nephropathy and renal transplantation. *Nephrol Dial Transplant* 7: 427–432
8 Larsson C, Anggard E (1973) Regional differences in the formation and metabolism of prostaglandis in the rabbit kidney. *Eur J Pharmacol* 21: 30–36
9 Whorton AR, Smigel M, Oates JA, Frölich JC (1978) Regional differences in prostaglandin formation by the kidney: Prostacyclin is a major prostaglandin of renal cortex. *Biochim Biophys Acta* 529: 176–180
10 Garcia-Perez A, Smith WL (1984) Apical-basolateral membrane assymetry in canine cortical collecting tubule cells: bradykinin, arginine vasopressin, prostaglandin E_2 interrelationships. *J Clin Invest* 74: 63–74
11 Breyer MD, Jacobson HR, Breyer RM (1996) Functional and molecular aspects of renal prostaglandin receptors. *J Am Soc Nephrol* 7: 8–17
12 Johnston HH, Herzog JP, Lauler DP (1967) Effect of prostaglandin E_1 on renal hemodynamics, sodium and water excretion. *Am J Physiol* 213: 939–946
13 Lifschitz MD (1981) Prostaglandins and renal blood flow. *Kidney Int* 19: 781–785
14 Kirschenbaum MA, Stein JH (1976) Effect of inhibition of prostaglandin synthesis on urinary sodium excretion in conscious dog. *J Clin Invest* 57: 517–521
15 Aiken JW, Vane JR (1973) Intrarenal prostaglandin release attenuates the renal vasoconstrictor activity of angiotensin. *J Pharmacol Ther* 184: 678–687
16 Arisz L, Donker AJM, Brentjens JRH, van der Hem GK (1976) The effect of indomethacin on proteinuria and kidney function in the nephrotic syndrome. *Acta Med Scand* 199: 121–125
17 Walshe JJ, Venuto RC (1979) Acute oliguric renal failure induced by indomethacin: possible mechanism. *Ann Intern Med* 91: 47–49
18 Antillon M, Cominelli F, Lo S Moran M, Somberg K, Reynolds T, Zipser RD (1990) Effects of oral prostaglandins on indomethacin induced renal failure in patients with cirrhosis and ascites. *J Rheumatol Suppl* 20: 46–49
19 Brezis M, Rosen S (1995) Hypoxia of the renal medulla – its implications for disease. *N Engl J Med* 332: 647–655
20 Kauker ML (1977) Prostaglandin E_2 effect from the luminal side on renal tubular ^{22}Na efflux: tracer microinjection studies. *Proc Soc Exp Biol Med* 154: 274–277
21 Stokes JB, Kokko JP (1977) Inhibition of sodium transport by prostaglandin E_2 across the isolated, perfused rabbit collecting tubule. *J Clin Invest* 59: 1099–1104
22 Frölich JC, Hollifield JW, Dormois JC Frölich BL, Seyberth H, Michaelis AM, Oates JA (1976) Suppression of plasma renin activity by indomethacin in man. *Circ Res* 39: 447–452
23 Whorton AR, Misono K, Hollifield J, Frölich JC, Inagami T, Oates JA (1977) Prostaglandins and renin release: I. Stimulation of renin release from rabbit renal cortical slices by PGI_2. *Prostaglandins* 14: 1095–1104
24 Ito S, Carretero OA, Abe K, Beierwaltes WH, Yoshinaga K (1989) Effect of prostanoids on renin release from rabbit isolated afferent arterioles with and without attached macula densa. *Kidney Int* 35: 1138–1144
25 Frölich JC, Hollifield JW, Michelakis AM Vesper BS, Wilson JP, Shand BG, Seyberth HJ, Frölich WH, Oates JA (1979) Reduction of plasma renin activity by inhibition of the fatty acid cyclooxygenase in human subjects: independence of sodium retention. *Circ Res* 44: 781–787
26 Walker LA, Whorton AR, Smigel M, France R, Frölich JC (1978) Antidiuretic hormone increases renal prostaglandin synthesis *in vivo*. *Am J Physiol* 235: F180–F185
27 Anderson RJ, Berl T, McDonald KM, Schrier RW (1975) Evidence for an *in vivo* antagonism between vasopressin and prostaglandin in the mammalian kidney. *J Clin Invest* 56: 420–426
28 Kramer HJ, Backer A, Hinzen S, Dusing R (1978) Effects of inhibition of prostaglandin-synthesis on renal electrolyte excretion and concentrating ability in healthy man. *Prostaglandins Med* 1: 341–349
29 Smith WL, Bell TG (1978) Immunohistochemical localization of the prostaglandin forming cyclooxygenase in renal cortex. *Am J Physiol* 235: F451–F457
30 Harris RC, McKanna JA, Akai Yg, Jacobson HR, Dubois RN, Breyer MD (1994)

Cyclooxygenase-2 is associated with the macula dense of rat kidney and increases with salt restriction. *J Clin Invest* 94: 2504–2510
31 Yang T, Singh I, Pham H, Sun D, Smart A, Schnerman JB, Briggs JP (1998) Regulation of cyclooxygenase expression in the kidney by dietary salt intake. *Am J Physiol* 274: F481–F489
32 Mann B, Hartner A, Jensen BL, Kammerl M, Krämer BK, Kurtz A (2001) Furosemide stimulates macula densa cyclooxygenase-2 expression in rats. *Kidney Int* 59: 62–68
33 Yang T, Schnermann JB, Briggs JP (1999) Regulation of cyclooxygenase-2 expression in renal medulla by tonicity *in vivo* and *in vitro*. *Am J Physiol* 277: F1–F9
34 O'Neill GP, Ford-Hutchinson AW (1993) Expression of mRNA for cyclooxygenase-1 and 2 in human tissues. *FEBS Lett* 330: 156–160
35 Kömhoff M, Gröne H-J, Klein T, Seyberth HW, Nüsing RM (1997) Localization of cyclooxygenase-1 and -2 in adult and fetal human kidney: implication for renal function. *Am J Physiol* 272: F460–F468
36 Dinchuk JE, Car BD, Focht RJ, Johnston JJ, Jaffee BD, Covington MB, Contel NR, Eng VM, Collins RJ, Czerniak PM et al (1995) Renal abnormalities and an altered inflammatory response in mice lacking cyclooxygenase II. *Nature* 378: 406–409
37 Morham SG, Langenbach R, Loftin CD, Tiano HF, Vouloumanos N, Jennette JC, Mahler JF, Kluckman KD, Ledford A, Lee CA (1995) Prostaglandin synthase 2 gene disruption causes severe renal pathology in the mouse. *Cell* 83: 473–482
38 Langenbach R, Morham SG, Tiano HF, Loftin CD, Ghanayem BI, Chulada PC, Mahler JF, Lee CA, Goulding EH, Kluckman KD (1995) Prostaglandin synthase 1 gene disruption in mice reduces arachidonic acid-induced inflammation and indomethacin-induced gastric ulceration. *Cell* 83: 483–492
39 Zhang M-Z, Wang J-L, Cheng H-F, Harris RC, McKanna JA (1997) Cyclooxygenase-2 in rat nephron development. *Am J Physiol* 273: F994–F1002
40 Khan KN, Venturini CM, Bunch RT, Brassard JA, Koki AT, Morris DL, Trump BF, Maziasz TJ, Alden CL (1998) Interspecies differences in renal localization of cyclooxygenase isoforms: implications in nonsteroidal antiinflammatory drug-related nephrotoxicity. *Toxicol Pathol* 26: 612–620
41 Nantel F, Meadows E, Denis D, Connolly B, Metters KM, Giaid A (1999) Immunolocalization of cyclooxygenase-2 in the macula densa of human elderly. *FEBS Letters* 457: 475–477
42 Koki A, Seibert K, Khan NK, Masferrer JL (1999) Cyclooxygenase-2 expression in normal human tissues and pathologic disease. Abstract book of Poster Presentations Third International Workshop on COX-2, Hawaii, Abstr 46
43 Gans KR, Galbraith W, Roman RJ, Haber SB, Kerr JS, Schmidt WK, Smith C, Hewes WE, Ackerman NR (1990) Anti-inflammatory and safety profile of DuP 697, a novel orally effective prostaglandin synthesis inhibitor. *J Pharmacol Exp Ther* 254: 180–187
44 Bosch-Marce M, Claria J, Titos E, Masferrer JL, Altuna R, Poo JL, Jimenez W, Arroyo V, Rivera F, Rodes J (1999) Selective inhibition of cyclooxygenase-2 spares renal function and prostaglandin synthesis in cirrhotic rats with ascites. *Gastroenterology* 116: 1167–1175
45 Black SC, Brideau C, Cirino M, Belley M, Bosquet J, Chan CC, Rodger IW (1998) Differential effect of a selective cyclooxygenase-2 inhibitor *versus* indomethacin on renal blood flow in conscious, volume-depleted dogs. *J Cardiovasc Pharmacol* 32: 686–694
46 Lopez-Parra M, Claria J, Planaguma A, Titos E, Masferrer JL, Woerner BM, Koki AT, Jimenez W, Altuna R, Arroyo V et al (2002) Cyclooxygenase-1 derived prostaglandins are involved in the maintenance of renal function in rats with cirrhosis and ascites. *Br J Pharmacol* 135: 891–900
47 Muscara MN, Vergnolle N, Lovren F, Triggle CR, Elliott SN, Asfaha S, Wallace JL (2000) Selective cyclo-oxygenase-2 inhibition with celecoxib elevates blood pressure and promotes leukocyte adherence. *Br J Pharmacol* 129: 1423–30
48 Harding P, Sigmon DH, Alfie ME, Huang PL, Fishman MC, Beierwaltes WH, Carretero OA (1997) Cyclooxygenase-2 mediates increased renal renin content induced by low sodium diet. *Hypertension* 29: 297–302
49 Pairet M, Churchill L, Engelhardt G (1996) Differential inhibition of cyclooxygenase 1 and 2 by NSAIDs. In: N Bazan, J Botting, J Vane (eds): *New targets in inflammation: Inhibitors of COX-2 or adhesion molecules.* Kluwer Academic Publishers, London, 23–37
50 Hartner A, Goppelt-Struebe M, Hilgers KF (1998) Coordinate expression of cyclooxygenase-2 and renin in the rat kidney in renovascular hypertension. *Hypertension* 31: 201–205
51 Wang J-L, Cheng H-F, Harris RC (1999) Cyclooxygenase-2 inhibition decreases renin content and lowers blood pressure in a model of renovascular hypertension. *Hypertension* 34: 96–101

52 Cheng HF, Wang SW, Zhang MZ, McKanna JA, Breyer R, Harris RC (2002) Prostaglandins that increase renin production in response to ACE inhibition are not derived from cyclo-oxygenase-1. *Am J Physiol Regul Integr Comp Physiol* 283: R638–R646
53 Kömhoff M, Wang JL, Cheng HF, Langenbach R, McKanna JA, Harris RC, Breyer MD (2000) Cyclooxygenase-2-selective inhibitors impair glomerulogenesis and renal cortical development. *Kidney Int* 57: 414–422
54 Seibert K, Masferrer JL, Needleman P, Salvemini D (1996) Pharmacological manipulation of cyclo-oxygenase-2 in the inflamed hydronephrotic kidney. *Br J Pharmacol* 117: 1016–1020
55 Chanmugam P, Feng L, Liou S, Jang BC, Boudreau M, Yu G, Lee JH, Kwon HJ, Beppu T, Yoshida M et al (1995) Radicicol, a protein tyrosine kinase inhibitor, suppresses the expression of mitogen-inducible cyclooxygenase in macrophages stimulated with lipopolysaccharide and in experimental glomerulonephritis. *J Biol Chem* 270: 5418–5426
56 Hirose S, Yamamoto T, Feng L, Yaoita E, Kawasaki K, Goto S, Fujinaka H, Wilson CB, Arakawa M, Kihara I (1998) Expression and localization of cyclooxygenase isoforms and cytosolic phospholipase A2 in anti-Thy-1 glomerulonephritis. *J Am Soc Nephrol* 9: 408–416
57 Schneider A, Harendza S, Zahner G, Jocks T, Wenzel U, Wolf G, Thaiss F, Helmchen U, Stahl RA (1999) Cyclooxygenase metabolites mediate glomerular monocyte chemoattractant protein-1 formation and monocyte recruitment in experimental glomerulonephritis. *Kidney Int* 55: 430–441
58 Khan KN, Alden CL, Gleissner SE, Gessford MK, Maziasz TJ (1998) Effect of papillotoxic agents on expression of cyclooxygenase isoforms in the rat kidney. *Toxicol Pathol* 26: 137–142
59 Heise G, Grabensee B, Schror K, Heering P (1998) Different actions of the cyclooxygenase 2 selective inhibitor flosulide in rats with passive Heymann nephritis. *Nephron* 80: 220–226
60 Wang J-L, Cheng H-F, Zhang M-Z, McKanna JA, Harris RC (1998) Selective increase of cyclooxygenase-2 expression in a model of renal ablation. *Am J Physiol* 275: F613–F622
61 Sanchez PL, Salgado LM, Ferreri NR, Escalante B (1999) Effect of cyclooxygenase-2 inhibition on renal function after renal ablation. *Hypertension* 34: 848–853
62 Tomasoni S, Noris M, Zapella S (1998) Upregulation of renal and systemic cyclooxygenase-2 in patients with active lupus nephritis. *J Am Soc Nephrol* 9: 1202–1212
63 Fu JY, Masferrer JL, Seibert K, Raz A, Needleman P (1990) The induction and suppression of prostaglandin H2 synthase (cyclooxygenase) in human monocytes. *J Biol Chem* 265: 16737–16740
64 Masferrer JL, Seibert K, Zweifel B, Needleman P (1992) Endogenous glucocorticoids regulate an inducible cyclooxygenase enzyme. *Proc Natl Acad Sci USA* 89: 3917–3921
65 Rosenkranz B, Náray-Fejes-Tóth A, Fejes-Tóth G, Fischer C, Sawada M, Frölich JC (1985) Dexamethasone effect on prostanoid formation in healthy man. *Clin Sci* 68: 681–685
66 Naray-Fejes-Toth A, Fejes-Toth G, Fischer C, Frölich JC (1984) Effect of dexamethasone on *in vivo* prostanoid production in the rabbit. *J Clin Invest* 74: 120–123
67 Nasjletti A, Erman A, Cagen LM, Baer PC (1984) Plasma concentrations, renal excretion, and tissue release of prostaglandins in the rat with dexamethasone-induced hypertension. *Endocrinology* 114: 1033–1040
68 Vio CP, Cespedes C, Gallardo P, Masferrer JL (1997) Renal identification of cyclooxygenase-2 in a subset of thick ascending limb cells. *Hypertension* 30: 687–692
69 Frölich JC (1997) A classification of NSAIDs according to the relative inhibition of cyclooxygenase isozymes. *Trends Pharmacol Sci* 18: 30–34
70 Wu KK, Sanduja R, Tsai AL, Ferhanoglu B, Loose-Mitchell DS (1991) Aspirin inhibits interleukin-1-induced prostaglandin H synthase expression in cultured endothelial cells. *Proc Natl Acad Sci USA* 88: 2384–2387
71 Rosenkranz B, Fischer C, Frölich JC (1986) Effects of salicylic and acetylsalicylic acid alone and in combination on platelet aggregation and prostanoid synthesis in man. *Br J Clin Pharmacol* 21: 309–317
72 Bombardier C, Peloso PMJ, Goldsmith CH, and the Salsalate-Diclofenac Study Group (1995) Salsalate, a nonacetylated salicylate, is as efficacious as diclofenac in patients with rheumatoid arthritis. *J Rheumatol* 22: 617–624
73 Clive DM, Stoff JS (1984) Renal syndromes associated with nonsteroidal antiinflammatory drugs. *N Engl J Med* 310: 563–572
74 Carmichael J, Shankel S (1985) Effects of nonsteroidal anti-inflammatory drugs on prostaglandins and renal function. *Am J Med* 78: 992–1000
75 The International COX-2 Study Group (1998) The classification of cyclooxygenase inhibitors. *J Rheumatol* 25: 2298–2303

76 Stichtenoth DO, Wagner B, Frölich JC (1997) Effects of meloxicam and indomethacin on cyclooxygenase pathways in healthy volunteers. *J Investig Med* 45: 44–49
77 Bevis PJR, Bird HA, Lapham G (1996) An open study to assess the safety and tolerability of meloxicam 15 mg in subjects with rheumatic disease and mild renal impairment. *Br J Rheumatol* 35 (Suppl.1): 56–60
78 Müller FO, Middle MV, Schall R, Terblanche J, Hundt HK, Groenewoud G (1997) An evaluation of the interaction of meloxicam with frusemide in patients with compensated chronic cardiac failure. *Br J Clin Pharmacol* 44: 393–398
79 Hawkey C, Kahan A, Steinbrück K, Alegre C, Baumelou E, Begaud B, Dequeker J, Isomaki H, Littlejohn G, Mau J, Papazoglou S (1998) Gastrointestinal tolerability of meloxicam compared to diclofenac in osteoarthritis patients. *Br J Rheumatol* 37: 937–945
80 Dequeker J, Hawkey C, Kahan A, Steinbruck K, Alegre C, Baumelou E, Begaud B, Isomaki H, Littlejohn G, Mau J, Papazoglou S (1998) Improvement in gastrointestinal tolerability of the selective cyclooxygenase (COX)-2 inhibitor, meloxicam, compared with piroxicam: results of the safety and efficacy large-scale evaluation of COX-inhibiting therapies (SELECT) trial in osteoarthritis. *Br J Rheumatol* 37: 946–951
81 Stichtenoth DO, Wagner B, Frölich JC (1998) Effect of selective inhibition of the inducible cyclooxygenase on renin release in healthy volunteers. *J Investig Med* 46: 290–296
82 Catella-Lawson F, McAdam B, Morrison BW, Kapoor S, Kujubu D, Antes L, Lasseter KC, Quan H, Gertz BJ, FitzGerald GA (1999) Effects of specific inhibition of cyclooxygenase-2 on sodium balance, hemodynamics, and vasoactive eicosanoids. *J Pharmacol Exp Ther* 289: 735–741
83 Rossat J, Maillard M, Nussberger J, Brunner HR, Burnier M (1999) Renal effects of selective cyclooxygenase-2 inhibition in normotensive salt depleted subjects. *Clin Pharmacol Ther* 66: 76–84
84 Whelton A, Schulman G, Wallemark C, Drower EJ, Isakson PC, Verburg KM, Geis GS (2000) Effects of celecoxib and naproxen on renal function in the elderly. *Arch Int Med* 160: 1465–1470
85 Schwartz JI, Vandormael K, Malice MP, Kalyani RN, Lasseter KC, Holmes GB, Gertz BJ, Gottesdiener KM, Laurenzi M, Redfern KJ et al (2002) Comparison of rofecoxib, celecoxib, and naproxen on renal function in elderly subjects receiving a normal salt diet. *Clin Pharmacol Ther* 72: 50–61
86 Swan SK, Rudy DW, Lasseter KC, Ryan CF, Buechel KL, Lambrecht LJ, Pinto MB, Dilzer SC, Obrda O, Sundblad KJ et al (2000) Effect of cyclooxygenase-2 inhibition on renal function in elderly persons receiving a low salt diet. *Ann Intern Med* 133: 1–9
87 Perazella MA, Eras J (2000) Are selective COX-2 inhibitors nephrotoxic? *Am J Kidney Dis* 35: 937–940
88 Stafford C, Bestoso JT (2000) Celecoxib-induced acute renal failure. *J Am Soc Nephrol* 11: 134A
89 Gadalean F, Barreto G, Epstein M (2001) Acute renal failure induced by COX-2 specific inhibitors. *J Am Soc Nephrol* 12: 779A
90 Briefing document for Celebrex™ (celecoxib) (2001) FDA Arthritis Advisory Committee, 7.2.2001, NDA 20–998
91 Briefing document for Vioxx™ (rofecoxib) (2001) FDA Arthritis Advisory Committee, 8.2.2001, NDA 21-042
92 Whelton A, Maurath CL, Verburg KM, Geis GS (2000) Renal safety and tolerability of celecoxib, a novel cyclooxygenase-2 inhibitor. *Am J Ther* 7: 159–175
93 Scott LJ, Lamb HM (1999) Rofecoxib. *Drugs* 58: 499–505
94 Whelton A, Fort JG, Puma JA, Normandin D, Bello AE, Verburg KM for the SUCCESS VI Study Group (2001) Cyclooxygenase-2-specific inhibitors and cardiorenal function: A randomized, controlled trial of celecoxib and rofecoxib in older hypertensive osteoarthritis patients. *Am J Ther* 8: 85–95
95 Geba GP, Weaver AL, Polis AB, Dixon ME, Schnitzer TJ for the VACT Group (2002) Efficacy of rofecoxib, celecoxib, and acetaminophen in osteoarthritis of the knee – a randomized trial. *J Am Med Assoc* 287: 64–71
96 Stichtenoth DO, Frölich JC (2003) The second generation of COX-2 inhibitors: What advantages do the newest offer? *Drugs* 63: 33–45
97 Product Information Bextra™, valdecoxib (2001) Pharmacia Corporation, Chicago, IL
98 Schnitzer T, Geusens P, Hasler P, Patel SK, Poor G, Senftleber J, Jayawardene S, Sloan VS, Macerata RS (2000) Efficacy and safety of COX-189 in osteoarthritis: a multinational study. *Arthritis Rheum* 43 (Suppl.): S336

COX-2 Inhibitors
Edited by M. Pairet and J. van Ryn

Effects on the lungs: role of COX-2 inhibitors

R. Stokes Peebles, Jr.[1] and Koichi Hashimoto[2]

[1] *Division of Allergy, Pulmonary and Critical Care Medicine, Vanderbilt University School of Medicine, Nashville, TN 37232-2650, USA*
[2] *Department of Microbiology, School of Medicine, Fukushima Medical University, 1st Hikariga-oka Fukushima, Fukushima, Japan 960-1295*

Introduction

Intervention trials using cyclooxygenase (COX)-2 inhibitors in both humans and animals reveal that this enzyme has important regulatory roles in myriad facets of lung biology. These trials also indicate that COX-2 inhibitors have tremendous potential to treat several diseases for which there are currently suboptimal therapeutic options. In this chapter, we will first outline data regarding COX-2 expression in the lung. We will then focus on various lung disease models in which COX-2 expression is increased and is thought to have a pathogenic role, and then review investigations in which COX-2 inhibition has altered these models (Tab. 1).

COX-2 expression in the lung begins in the fetus [1]. Immunostaining of the lungs of 16 to 32 week gestational age human fetuses that were aborted for

Table 1. Pharmacologic agents detailed in this review

Pharmacologic agent	References
COX-2 inhibitors	
JTE-522	[16, 52, 63]
NS-398	[17, 18, 52, 62, 64, 65, 67, 68, 83]
Nimesulide	[20]
Celecoxib	[21, 24, 25, 45, 59, 81]
Rofecoxib	[53, 60]
SC-58236	[51, 63, 69]
DUP-697	[63]
FK3311	[74, 75]
Nonspecific COX inhibitors]	
Aspirin	[17, 18, 23, 24]
Indomethacin	[27–29, 46–48, 61, 64]
Thromboxane receptor antagonist	
SQ29.548	[67, 68]

fetal anomalies, but that had normal lungs at autopsy, revealed that COX-2 expression was present in the epithelial cells of the alveoli and bronchi with no expression in the interstitium or endothelium [1]. Bronchial staining was localized to the ciliated epithelial cells and alveolar staining occurred in cells resembling type II pneumocytes and cuboidal cells [1]. In pre-term infants (gestational age 23 to 30 weeks) who died of immaturity and respiratory distress syndrome with hyaline membranes, alveolar staining was greatest for COX-2 in cells resembling type II pneumocytes as well as in the areas of the hyaline membranes [1]. Compared to pre-term infants, there was a similar distribution of COX-2 staining in term infants with gestational age of 38 to 42 weeks who died of cardiac abnormalities; however, the staining was less intense [1)]. Thus, there seems to be a negative correlation between the intensity of COX-2 staining and gestational age. These results imply biologic function for COX-2 in the developing human lung that may be regulated by alveolar type II pneumocytes. In adults, COX-2 expression is negligible in lung epithelium from normal controls [2, 3].

Lung cancer

Prostaglandins affect cell proliferation and alter the immune system response to malignant cells, therefore the COX enzymes may directly affect tumorigenesis. For instance, the COX product prostaglandin E_2 (PGE_2) is suspected to promote the development of tumors by augmenting angiogenesis and decreasing immune surveillance [2]. There are several lines of evidence that PGE_2 may have a role in the regulation of lung cancers. First, increased levels of PGE_2 are present in bronchoalveolar lavage (BAL) fluid in patients with either primary lung cancer or with extrathoracic cancer with lung metastases compared to BAL fluid from patients with extrathoracic cancer without lung metastases or patients with benign lung disease [4]. In addition, microsomal PGE_2 synthase is upregulated in non small cell lung cancers [5]. *In vitro* studies examining the regulation of PGE_2 production in tumor cell lines reveal that COX-2, but not COX-1, inhibition decreases tumor related PGE_2 production [6]. Additionally, the unstable COX product intermediary, PGH_2, is able to isomerize to form malondialdehyde, a potent mutagen, by both enzymatic and nonenzymatic reactions [7]. Finally, peroxidative activity of COX-2 can augment free radical damage by oxidating aromatic and heterocyclic amines and dihydrodiol derivatives [8].

While COX-1 expression in neoplastic tissue is reported to be unchanged from its level of constitutive expression in normal lung [2, 9], there is variable expression of COX-2 dependent upon tumor type and histologic grade. For instance, COX-2 expression seems to be greatest in adenocarcinoma of the lung, with much lower COX-2 expression in squamous cell lung carcinomas and almost absent expression in small cell lung cancers [9–11]. COX-2 expression is reported to be increased in epithelial lung tissue immediately adjacent to neoplastic tissue, while not present in lung epithelium distant to the neoplasm, suggesting possible paracrine regulation [2]. There has been conflicting data on the correlation of COX-2 expression in adenocarcinoma of the lung and histologic grade of the neoplasm. While some report that COX-2 mRNA expression by Northern Blot analysis and COX-2 immunostaining of histologic sections was greatest in well-differentiated adenocarcinomas with much lower expression in poorly differentiated adenocarcinoma [11], others have found consistently high level of COX-2 expression within lung adenocarcinomas irrespective of histologic grade [2, 11]. High levels of COX-2 expression have been noted in invasive adenocarcinomas, with even greater expression in lymph node metastases than in the corresponding primary tumor [2, 10].

The prognostic significance of elevated COX-2 expression in lung cancers has also been investigated by several groups. Achiwa et al. found that there was no relationship between increased COX-2 expression in tumors and clinical outcome in a cohort of 130 patients with resected lung adenocarcinoma. However, after analyzing the prognostic significance of increased COX-2 expression after stratification according to disease stage, a significant relationship between elevated COX-2 expression and shortened survival was observed only in a cohort of patients with stage I disease [12]. Similar results were published by Khuri et al. who found that in patients with stage I non-small cell lung cancer, the strength of COX-2 expression correlated with a significantly worsened survival rate. In this series, the median survival times for patients whose tumors exhibited either strong, intermediate or weak, and no COX-2 expression was 1.04, 5.50, and 8.54 years, respectively [13]. Brabender et al. also found that in curatively resected non-small cell lung cancer, high COX-2 expression in the tumor tissue was significantly associated with poor survival with multivariate analysis revealing that high COX-2 expression was an independent predictor of decreased survival in this patient population [14]. However, others have found no relationship between COX-2 expression and survival [15].

There is a wide range of evidence suggesting that COX-2 inhibitors might be useful in the treatment of lung cancers (Tab. 2). The COX-2 selective inhibitor JTE-522 inhibited growth of the human lung adenocarcinoma cell line ACC-LC-319 *in vitro* as a single agent and also improved the antitumor effect of the anti-cancer drugs docetaxel, vinorelbine, and amrubicin [16]. This same COX-2 inhibitor significantly inhibited the growth of ACC-LC-319 cells xenografted into athymic nude mice [16]. In this *in vivo* model, JTE-522 enhanced the tumor inhibitory effects of docetaxel and vinorelbine, but not amrubicin [16]. JTE-522 also had antiangiogenic effects by reducing the microvessel count within tumors *in vivo* [16].

The COX-2 inhibitor NS-398, as well as aspirin, significantly inhibited tumor multiplicity in a murine model of lung tumorigenesis caused by administration of the tobacco-specific nitrosamine 4-(methylnitrosamino)-1-(3-pyridyl)-1-butanone given in the drinking water for 7 weeks [17]. In this

Table 2. Studies examining the effects of COX-2 inhibitors in cancer models

COX-2 inhibitor	Effect	Reference
JTE-522	inhibited growth of lung adenocarcinoma cell line ACC-LC-319 *in vitro*	[16]
JTE-522	inhibited the growth of ACC-LC-319 cells xenografted into athymic nude mice	[16]
JTE-522	reducing the microvessel count within tumors *in vivo*	[16]
NS-398	inhibited tumor multiplicity in a murine model of lung tumorigenesis	[17]
nimesulide	decreased proliferation of non-small cell lung cancer cell lines *in vitro*	[20]
celecoxib	reduced the growth of Lewis lung carcinoma cells *in vivo*	[21]
celecoxib	reduced growth of Lewis lung carcinoma cells *in vivo*	[22]
celecoxib	not effective in preventing the development of lung tumors	[23]
celecoxib	did not inhibit tumor multiplicity caused by urethane	[23]

model, plasma PGE_2 levels were increased four-fold, but were reduced by both NS-398 and aspirin [17]. In addition, a concentration of either aspirin or NS-398 greater than 100 μM inhibited the proliferation of both 82-132 and LM2 murine lung tumor cells by 50% [17]. In this same model, aspirin and NS-398 significantly increased the apoptotic index, upregulated Fas-L and BAD expression, and decreased expression of cyclin B2 [18].

NS-398 also regulates secretion of cytokines by tumor cells. A549 non-small cell lung cancer cells upregulate COX-2 expression and secrete PGE_2 in response to IL-1β, effects that are blocked by NS-398 [19]. NS-398 inhibited the ability of IL-1β-stimulated A549 cells to induce IL-10 in lymphocytes and macrophages and reversed the tumor-derived PGE_2-dependent inhibition of macrophage IL-12 production when whole blood was cultured in tumor supernatant [19]. Another COX-2 inhibitor, nimesulide, decreased proliferation of non-small cell lung cancer cell lines in a dose-dependent fashion *in vitro* by inducing apoptosis [20]. In this study, there was a substantial relationship between the expression of COX-2 within the tumor cell lines and the effectiveness of nimesulide in inhibiting the growth of the individual tumor lines [20].

Celecoxib, another selective COX-2 inhibitor also has significant anti-tumor effects *in vivo* in some murine models, but not all [21]. Celecoxib administration from the time of implantation of Lewis lung carcinoma cells into the paws of C57/Bl6 mice significantly reduced the growth of these tumors in a dose-dependent fashion (Fig. 1) [21]. In this same experiment, at higher doses celecoxib also reduced the number and size of metastatic nodules in the lung [21]. In a different murine model in which Lewis lung carcinoma cells are implanted subcutaneously, celecoxib reduced tumor growth *in vivo* and also inhibited *in vitro* fibroblast production of vascular endothelial growth

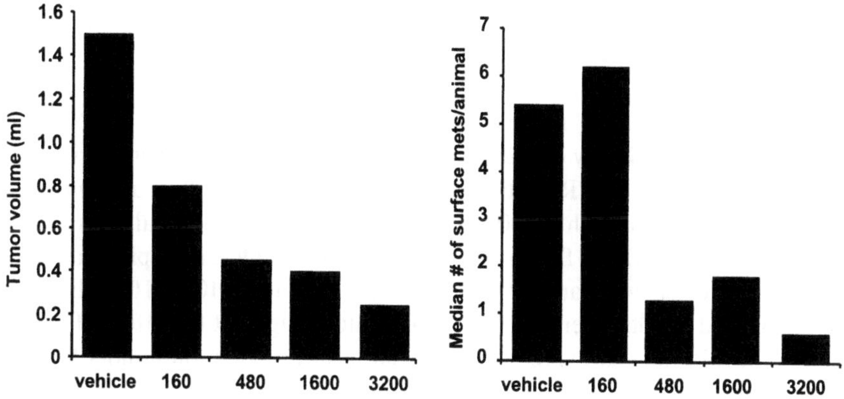

Figure 1. Celecoxib potently inhibited tumor growth and metastasis in the syngeneic Lewis lung model. Lewis lung carcinoma (10^6 cells) was implanted into the paws of C57/Bl6 mice to develop tumors. Celecoxib was administered to mice (n = 20/treatment arm) at 0, 160, 480, 1600, and 3200 ppm in their food from the day of implant. Celecoxib dose-dependently inhibited primary tumor growth (left panel), and number of surface metastatic lesions (right panel).

factor (VEGF) by 92% [22]. However, neither celecoxib nor aspirin were effective in preventing the development of lung tumors induced in a two stage protocol in which dosing of 3-methylcholanthrene precedes chronic butylated hydroxytoluene administration [23]. In another model, celecoxib did not inhibit tumor multiplicity caused by urethane [23]. In fact, the urethane-induced lung tumors were larger in the mice fed chow in which celecoxib was added, even though the tumors in these mice produced 60% less PGE_2 compared to mice fed regular chow [23]. Therefore, COX-2 inhibitors are effective in preventing tumor growth in some, but not all, murine models of lung neoplasia.

Although there have been no studies examining the protective effect of COX-2 inhibitor use in human lung cancer, a prospective study revealed that COX inhibition might be effective in preventing tumor development. For instance, regular users of aspirin had a significantly lower incidence of lung cancer 0.68 (95% CI, 0.49–0.94) [24].

Radiation injury

Given the protective effect of COX-2 inhibitors in some models of lung cancer, it is important to understand the interaction of these agents with forms of cancer therapies that might be used to treat this disease. Some prostaglandins have protective effects against ionizing radiation and inhibit radiation injury [25, 26]. In fact, prostaglandin production by neoplasms is linked to resistance to radiotherapy [25, 27] and inhibition of prostaglandin synthesis by indomethacin increases radiosensitivity [27–29]. Selective COX-2 inhibition also increases the response to radiation. In an *in vivo* model, NS-398 augmented the effect of radiation in delaying growth of human lung cancer cells that expressed COX, NCI-H460, implanted into the left thigh of athymic nude mice [30]. NS-398 did not have this same effect on HCT-116 human colon cancer cells which did not express COX-2, suggesting that the NS-398 enhanced radiosensitivity was dependent upon the cell line's characteristic of expressing COX-2 [30]. Radiation induced apoptosis was also augmented by NS-398 treatment in both H460 cells and rat intestinal epithelial cells stably transfected with COX-2 DNA, but not in either rat intestinal epithelial cells stably transfected with antisense COX-2 DNA nor in HCT-116 cells [30]. These limited results infer that COX-2 inhibitors augment the effect of radiation on neoplasms that express COX-2, but have no effect on tumors that do not express COX-2. Thus, COX-2 inhibitors could have a high potential for increasing the therapeutic ratio of radiotherapy.

Asthma

There have been conflicting reports regarding the expression of COX-2 in the lungs from asthmatic subjects. While a four-fold increase in bronchial epithelial COX-2 immunostaining in asthmatic subjects compared to non-asthmatic controls has been reported [31], another group found no significant differences in the level of immunostaining among asthmatics, chronic bronchitics, or controls with no lung disease [32]. Redington et al. found that COX-2 mRNA expression and immunoreactive protein was increased in the airway epithelium of asthmatics that had not been treated with corticosteroids compared with non-asthmatic controls, while asthmatics receiving corticosteroid therapy had decreased COX-2 expression [33].

COX products have diverse effects on the inflammatory cascade and are presumed to have a pathogenic role in asthma [34, 35]. PGD_2 is increased in the BAL fluid of asthmatic subjects while both PGD_2 and $PGF_{2\alpha}$ contract human airways smooth muscle *in vitro* and are potent bronchoconstrictors *in vivo* [36, 37]. Thromboxane also causes human airway smooth muscle contraction *in vitro* [38]. On the other hand, PGI_2 causes relaxation of isolated precontracted human bronchi, but has minimal effects on airway function *in vivo* [37, 39]. There are several lines of evidence that suggest that PGE_2 has important immunomodulatory effects in human allergic airway inflammation. For instance, PGE_2 inhalation inhibits the pulmonary immediate and late phase responses to inhaled allergen [40, 41]. Compared to vehicle inhalation, inhaled PGE_2 also decreases the change in methacholine airway reactivity and reduces the number of eosinophils after inhaled allergen challenge [40]. There is also evidence that PGE_2 generated during viral respiratory tract infections modulates allergic inflammation in the lung [42, 43]. The results from these studies suggest that PGE_2 has a greater immunomodulatory effect than direct effect on airway caliber. This notion is supported by the fact that PGE_2 inhalation prior to segmental allergen challenge significantly reduced the BAL levels of PGD_2, an important product of mast cell degranulation, and the cysteinyl leukotrienes [44].

To our knowledge, at this time there have been no published studies investigating the effect of selective COX-1 or COX-2 inhibitors on allergic inflammation or asthma in humans; however, in a small study of eight subjects, celecoxib had no effect on pulmonary function, bronchial responsiveness, or cough reflex sensitivity in asthmatics [45]. The role of the COX enzymes in allergic bronchospasm has been examined by treating subjects with indomethacin, a non-selective COX inhibitor that blocks both COX-1 and COX-2, prior to allergen challenge. Indomethacin had no effect on baseline pulmonary function in either allergic asthmatics or nonasthmatic subjects with allergic rhinitis [46]. However, after indomethacin treatment, nonasthmatic subjects with allergic rhinitis had greater sensitivity to allergen challenge as measured by change in the forced expiratory volume in 1 s (FEV_1) and specific airway conductance [46]. Indomethacin treatment prior to allergen challenge caused a small but significant decrease in specific airway conductance in the allergic asthmatic subjects, yet indomethacin had no effect on allergen-induced changes in FEV_1 [46]. Other investigators found that indomethacin treatment had no significant effect on baseline airway responsiveness to histamine or any effect on either the immediate or late phase pulmonary response to allergens in allergic asthmatics [47, 48].

Animal studies suggest that COX-2 is an important modulator of allergic inflammation in the lung. Mice deficient in the COX-1 and COX-2 synthase genes had heightened pulmonary eosinophilia and serum IgE levels compared to wild type mice, while airway hyperresponsiveness was present only in the COX-1 deficient mice [49]. As COX-1 critically affects the transition from $CD4^-CD8^-$ cells to $CD4^+CD8^+$ cells and COX-2 has a very important role in early thymocyte proliferation and differentiation and subsequent maturation of the $CD4^+$ T-lymphocyte lineage [50], experiments were performed using pharmacologic inhibitors of COX-1 and COX-2 to avoid any confounding factors of developmental T-lymphocyte abnormalities in COX-1 and COX-2 deficient mice. Mice treated with a COX-1 inhibitor (SC58560), a COX-2 inhibitor (SC58236), or indomethacin during the development of allergic airway disease all had augmented lung levels of IL-13, increased lung eosinophilia, and airway responsiveness compared to vehicle-treated mice [51]. Since COX inhibition during the development of allergic airway disease resulted in increased allergic inflammation and airway responsiveness, these results suggest that a COX product may restrain allergic inflammation and might be a therapeutic target for the treatment of asthma. It is very important to recognize that in these two murine studies, COX inhibition was present throughout the entire development of allergic disease, from the initial stage of antigen presentation and throughout all airway allergen challenges. In addition, the antigen sensitization procedure consisted of an injection of intraperitoneal ovalbumin formulated with aluminum hydroxide. Different results were obtained in a study performed in guinea pigs that utilized an allergen sensitization protocol of two ovalbumin aerosols administered a week apart [52]. In this study, selective COX-2 inhibitors (NS-398 and JTE-522) administered only during the period of airway challenge each inhibited allergic inflammation in the lung, yet had no effect on airway responsiveness to histamine [52].

Approximately 10% of asthmatics will experience an asthma exacerbation after using aspirin or other NSAIDS [53]. Patients with aspirin intolerant asthma (AIA) have marked overexpression of leukotriene (LT) C_4 synthase, an enzyme that converts LTA_4 to the cysteinyl leukotrienes LTC_4, LTD_4, and LTE_4 [54, 55]. Some investigators theorize that the increase in the cysteinyl leukotrienes, mediators that are known to have a pathogenic role in asthma, may be critical in producing AIA [56]. Inhaled PGE_2 attenuated the decrease in pulmonary function that occurs when subjects with AIA are challenged with inhaled lysine acetylsalicylate and also decreased urinary LTE_4 excretion, a marker of LTC_4 synthase activity [57]. *In vitro* assays confirm this protective effect of PGE_2 in AIA. Although there was no difference in cysteinyl leukotriene production in peripheral blood leukocytes from subjects with AIA and either asthmatics who were tolerant of asthma or non-asthmatic controls, the peripheral leukocytes from the AIA subjects had a significant increase in cysteinyl leukotrienes after aspirin stimulation that was not present in the leukocytes from the other two groups [58]. When PGE_2 was added to the aspirin stimulated cells from the AIA subjects, there was a significant decrease in cysteinyl leukotriene levels [58]. This suggests that there is a significant inhibitory effect of PGE_2 on the production of the cysteinyl leukotrienes in AIA subjects. Recent studies showed that AIA subjects do not develop bronchospasm when they take the selective COX-2 inhibitors celecoxib [59] and rofecoxib [53, 60]. These studies suggest that COX-1 inhibition may be critical for aspirin intolerance [53]. Thus, a COX-1 product, possibly PGE_2, may be critical in restraining the activation of cells responsible for the overproduction of bronchoconstrictor leukotrienes.

Sepsis

Sepsis induces pulmonary inflammation with augmented production of COX products. To our knowledge, there are no clinical trials using specific COX-2 inhibitors in sepsis. However, a non-selective COX inhibitor, ibuprofen, did not reduce the incidence or duration of shock or the acute respiratory distress syndrome and did not significantly improve the rate of survival at 30 days in septic patients compared to placebo treatment [61]. Several animal models have been developed to determine the contribution of COX-2 to sepsis-induced lung injury through the use of selective COX-2 inhibitors (Tab. 3). Most of these models utilize lipopolysaccharide (LPS) administration as the experimental correlate of sepsis-induced injury.

LPS has also been shown to increase COX-2 expression in isolated perfused rat lungs when subtoxic doses of LPS were administered through the pul-

Table 3. Studies examining the effects of COX-2 inhibitors in models of sepsis

COX-2 Inhibitor	Effect	Reference
NS-398	prevented bronchoconstriction induced by LPS and GM-CSF	[62]
NS-398	prevented LPS-induced pulmonary artery pressure, ventilation pressure	[63]
DUP-697	prevented LPS-induced pulmonary artery pressure, ventilation pressure	[63]
SC-236	prevented LPS-induced pulmonary artery pressure, ventilation pressure	[63]
NS-398	attenuated bradykinin-induced vasoconstriction caused by LPS	[64]
NS-398	decreased LPS-induced pyrexia and mortality at 48, but not 72 h	[65]
NS-398	COX-2 KO mice had augmented LPS-induced airway responsiveness	]66]
NS-398	inhibited bronchoconstriction caused by combination of TNF-α, IL-1β, and IFN-γ	[67]
NS-398	abrogated LPS-induced vascular hyperreactivity, only a partial effect on airway hyperreactivity	[68]
SC-236	maintained perfusion distribution after oleic acid and LPS administration	[69]

monary arteries. Wollin and colleagues investigated the effect of GM-CSF in this model to mimic the administration of GM-CSF that is used to treat septic patients when they are neutropenic after chemotherapy for cancer therapy [62]. In this model, the combination of GM-CSF and concomitant LPS administration induced release of thromboxane and TNF into the venous perfusate and also caused airway bronchoconstriction. Both COX-2 inhibition by NS-398 and the administration of a thromboxane receptor antagonist prevented bronchoconstriction but had no effect on TNF release. Anti-TNF antibodies had no effect on bronchoconstriction suggesting that the bronchoconstriction induced by the combined administration of LPS with GM-CSF is likely TNF-independent. In an *in vivo* model, these same investigators found that rats made neutropenic by the intravenous injection of vinblastine were protected against the lung injury caused by the co-administration of GM-CSF and LPS [62].

Isolated rat lungs have been used in other models to study the effects of LPS on lung physiology. For instance, isolated rat lungs were perfused with either buffered-saline or plasma and concentrations of LPS ranging from 1,000 to 10,000 ng/ml during a 2-h priming period. This was associated with LPS-induced production of thromboxane, PGI_2, and TNFα and with increased pulmonary artery pressure, ventilation pressure, and lung weight [63]. These changes were inhibited by the administration of the COX-2 inhibitors NS-398, DUP-697, and SC-236 [63]. With LPS priming, anti-TNFα antibodies had no effect on either the augmented pulmonary artery pressure, ventilation pressure, or lung weight when buffered-saline perfusate was administered; however, anti-TNFα antibodies greatly reduced the heightened ventilation pressure but had no effect on the pulmonary artery pressure response and lung edema formation when plasma perfusate was used [63]. These results suggest that COX-2, probably through thromboxane generation, mediates LPS-induced changes in vascular and smooth muscle physiology. In this system, COX-1 mRNA expression in the lung was unchanged by LPS administration; however, COX-2 lung mRNA expression was increased [63]. Immunohistochemical staining revealed that COX-2 expression was present prior to LPS infusion and subsequently increased in bronchial epithelial cells, bronchial and vascular smooth muscle cells, cells within bronchial associated lymphoid tissue and myocytes of the large hilar veins [63]. LPS administration led to *de novo* expression of COX-2 in endothelial cells and most alveolar macrophages. In contrast, LPS administration led to a decrease of COX-2 expression in perivascular and peribronchial macrophages [63].

In another model examining the effect of LPS on isolated rat lungs, COX-2 blocked sepsis-induced endothelial dysfunction. In this experiment, rats were treated with indomethacin, a nonselective COX inhibitor, or NS-398 by intraperitoneal injection 1 h prior to administration of either LPS or saline [64]. Three hours after either LPS or saline was given, lung isolation was perfomed and pulmonary arterial vasoconstriction was measured in response to either bradykinin or hypoxemia. Both indomethacin and NS-398 attenuated bradykinin-induced vasoconstriction in a dose-dependent manner. However, only indomethacin blocked hypoxic pulmonary vasoconstriction that occurred with LPS administration, while NS-398 had no effect [64].

In a murine model of endotoxemia, LPS administered intraperitoneally induced PGE_2 production in both the blood and lung [65]. This augmented PGE_2 production was decreased by NS-398 treatment administered by intraperitoneal injection [65]. NS-398 significantly decreased pyrexia 1 and 2 h after LPS administration, but had no effect on lung myeloperoxidase activity. Twenty-four hours after LPS administration, NS-398 treatment significantly improved mortality rate from endotoxemia, but there was no effect on survival at 72 h [65]. In a model of bacterial peritonitis induced by cecal ligation and puncture, this same group found PGE_2 was increased in BAL fluid, but not in plasma [65]. Since PGE_2 was increased in BAL fluid, they hypothesized that NS-398 might also have a protective effect; however, the COX-2 inhibitor had no effect on mortality in this model of bacterial peritonitis [65].

A model of local pulmonary endotoxemia has also been developed in which LPS was administered via the aerosol route to determine the role of COX products in airway physiology [66]. In this model, COX-1 and COX-2 deficient mice had augmented airway responsiveness to methacholine compared to wild type mice, although the number of BAL fluid total cells and neutrophils were not different between the groups. In addition, neither the lung histopathology, BAL fluid total protein, nor BAL leukotriene levels were different between the COX-1 deficient, COX-2 deficient, and wild type mice [66].

These results regarding LPS-induced bronchoconstriction are in contrast to a model in which administration of cytokines that are produced as a result of endotoxemia induce bronchoconstriction [67]. In a model utilizing viable lung slices with intact microanatomy harvested from rats, the combination of TNF-α, IL-1β, and IFN-γ induced COX-2 expression, generation of thromboxane, and airway constriction [67]. This bronchoconstriction was hypothesized to be caused by the induction of thromboxane by these cytokines, as administration of NS-398 inhibited bronchoconstriction as well as thromboxane production. The hypothesis that thromboxane was the critical element in the bronchoconstriction was confirmed when the thromboxane receptor antagonist SQ29.548 fully inhibited the bronchoconstriction caused by combined TNF-α, IL-1β, and IFN-γ administration [67].

Isolated, blood-free perfused lungs from mice have also been used to determine the effect of LPS on lung function [68]. LPS administration through the pulmonary vasculature caused airway hyperresponsiveness to methacholine as well as pulmonary vascular hyperreactivity to platelet activating factor, while it had no effect on either airway resistance or pulmonary artery pressure. Thromboxane is a key element in these LPS-induced physiologic effects as witnessed by the complete inhibition of the airway hyperresponsiveness and pulmonary vascular hyperreactivity by the thromboxane receptor antagonist SQ29.548. COX-2 inhibition with NS-398 completed abrogated LPS-induced vascular hyperreactivity but had only a partial effect on airway hyperreactivity caused by LPS administration. This suggests that the thromboxane generated by a COX-2-independent pathway mediates LPS-induced airway hyperresponsiveness [68].

One of the other pulmonary consequences of sepsis is acute lung injury. Administration of oleic acid has been used as a model of acute lung injury because it shares some pathophysiologic similarities to the adult respiratory distress syndrome (ARDS), one of which is the development of lung edema [69]. One mechanism to protect against hypoxemia that can result from lung edema is the redistribution of pulmonary perfusion away from injured edematous lung regions to maintain a relatively normal ventilation/perfusion pattern. In the model of oleic acid injury in which LPS is administered prior to oleic acid, this redistribution of pulmonary perfusion does not occur, resulting in decreased oxygenation. The prevention of redistribution of pulmonary perfusion that occurs with LPS administration is associated with increases in PGI_2 in both plasma and tissue. In a canine model of LPS administration prior to oleic acid-induced lung injury, the COX-2 inhibitor SC-236 maintained perfusion distribution after the lung injury occurred and inhibited the increased levels of plasma and tissue PGI_2 that occurred with placebo administration [69].

As summarized above, COX-2 inhibition has profound effects on LPS-mediated physiology in these animal models. These include protection against LPS-induced bronchoconstriction, lung inflammation, pulmonary artery vasoconstriction, pulmonary vascular hyperreactivity, and maintenance of protective redistribution of pulmonary perfusion in the setting of lung edema. Whether these protective effects of COX-2 inhibition are translatable into human sepsis is unknown until clinical trials using COX-2 inhibitors are performed in this condition.

Ischemia/reperfusion injury

Hemorrhagic shock results in an inflammatory response that can lead to organ injury and potentially multi-organ failure. One of the consequences of hemorrhagic shock is COX-2 induction, possibly through NF-κB activation and a subsequent increase in NF-κB binding to the COX-2 promoter, thus upregulating COX-2 gene transcription [70–72]. For example, a seven-fold increase in COX-2 expression has been reported in a rat model in which lungs were subjected to hemorrhagic shock [73]. In this model, COX-2 mRNA upregulation in the lungs was independent of reperfusion and occurred during ischemic phase of hemorrhagic shock [73]. To test the contribution of COX-2 products in ischemia-reperfusion injury, a selective COX-2 inhibitor, FK-3311 was used in a canine model [74]. In these experiments, FK-3311 was administered 15 min prior to ischemia and 15 min before reperfusion and this resulted in improved left pulmonary vascular resistance, cardiac output, arterial oxygen pressure, and wet-to-dry ratio measurements compared to placebo-treated animals [74]. Treatment with FK-3311 inhibited neutrophil infiltration of the lungs and improved survival. FK-3311 treatment reduced serum thromboxane, while levels of 6-keto-$PGF_{1\alpha}$, a prostacyclin (PGI_2) metabolite, were unchanged. Although the beneficial effects of FK-3311 in this model are not fully elucidated, inhibition of thromboxane is a potential mechanism [74]. In a canine model of ischemia-reperfusion using orthotopic lung transplantation, FK-3311improved pulmonary gas exchange, reduced neutrophilic infiltration, and protected against damage as assessed by histologic analysis as compared to vehicle treated animals. In this model, thromboxane was significantly decreased while the PGI_2 metabolite was not changed [75]. Therefore, in two separate models of ischemia-reperfusion injury in dog models, selective COX-2 inhibition is protective.

Idiopathic pulmonary fibrosis

Idiopathic pulmonary fibrosis (IPF) is a disease characterized by the proliferation of lung fibroblast and subsequent collagen deposition by these cells. A possible role for COX-2 products to regulate IPF was hypothesized after discovery that fibroblasts from patients with pulmonary fibrosis are unable to increase COX-2 expression in response to non-specific stimuli such as PMA, or IL-1 compared to fibroblasts from cancer patients [76]. Several different prostanoid products have been implicated in potentially modulating IPF. For instance, PGE_2 prevents fibroblast proliferation and collagen production. A possible role for PGE_2 in pulmonary fibrosis was speculated when it was discovered that lung fibroblasts from patients with pulmonary fibrosis have a two-fold decrease in basal PGE_2 production compared to fibroblasts from patients having surgery for lung cancer resection [76]. Therefore, this deficiency in the ability to produce PGE_2 as a result from the decreased induction of COX-2 may contribute to the inflammatory and fibrogenic response in pulmonary fibrosis. However, another group has found that PGE_2 production is not different from IL-1β-stimulated fibroblasts from IPF patients or control subjects [77]. In this setting fibroblasts from IPF patients have greater IL-1β stimulated production of thromboxane and a lower PGI_2 to thromboxane ratio compared to fibroblasts from control subjects [77]. Several animal models have been developed to dissect the role of COX-2 products in pulmonary fibrosis. Wild type mice administered bleomycin intratracheally have a mild fibrotic reaction 14 days after bleomycin administration which is patchy and with moderate interstitium thickening [78]. However, bleomycin-treated mice deficient in COX-2 experience an aggressive fibroproliferative response with increased neutrophilic and lymphocytic inflammation and complete loss of alveolar architecture [78]. The role of COX-2 in pulmonary fibrosis has also been investigated in another model in which vanadium pentoxide was used as the fibrotic stimulus [79]. The lungs from COX-2 deficient mice were severely inflamed three days after vanadium pentoxide intratracheal administration with pulmonary fibrosis developing two weeks after vanadium pentoxide administration, while these changes were not present in either COX-1 deficient or wild type mice. After vanadium pentoxide treatment, bronchoalveolar PGE_2 levels were increased in the wild type and COX-1 deficient mice, but not in the COX-2 deficient mice [79]. An *in vitro* model using cells from either COX-1 deficient, COX-2 deficient, or wild type mice also suggest that COX-2 regulates fibrogenesis [80]. Alveolar epithelial cells from wild type and COX-1 deficient mice, but not from COX-2 deficient mice, suppress serum-induced fibroblast proliferation. Alveolar epithelial cells from wild type and COX-1 deficient mice also have increased PGE_2 synthesis after stimulation with either LPS or IL-1, while these same cells from COX-2 deficient mice do not [80]. To our knowledge, there have been no reports examining COX-2 specific inhibitors in models of pulmonary fibrosis; however, these human and animal studies strongly suggest that COX-2 is a critical regulator of this disease.

Pleurisy

Carrageenan administered into the pleural space results in acute tissue inflammation and has been used to model pleurisy [81]. In this model, the pleural cavity inflammatory response is characterized by pleural effusion containing large numbers of neutrophils, in addition to lung tissue infiltration of neutrophils, resultant lipid peroxidation, and increased production of PGE_2, TNF-α, and IL-1β. A role for COX-2 mediation of this inflammatory response was shown when celecoxib inhibited the consequences of carrageenan administration [81]. Celecoxib treatment decreased ICAM-1 and P-selectin expression on bronchial epithelium and resulted in a dose-dependent decrease in pleural exudate volume, accumulation of neutrophils in the pleural space, lung myeloperoxidase activity, and lung malondialdehyde activity. In addition, celecoxib decreased pleural fluid concentrations of TNF-α, IL-1β, PGE_2, and the metabolite of PGI_2. These results suggest that COX-2-derived prostanoids mediate the pro-inflammatory effects in the carrageenan model of pleurisy [81].

Pancreatitis-induced lung injury

The lung is the extra-abdominal organ most frequently damaged in patients with pancreatitis [82]. Approximately one-third of patients with pancreatitis will develop acute lung injury and the acute respiratory distress syndrome. These complications of pancreatitis are responsible for 60% of pancreatitis-related deaths within the first week of onset of illness [82]. A mouse model of pancreatitis induced by the administration of the cholecystokinin analog cerulein was used to determine the role of COX products in pancreatitis and pancreatitits-associated lung injury [83]. Cerulein administration induced pancreatitic COX-2 expression, but had no effect on COX-1 expression [83]. In addition, cerulein caused marked necrosis, vacuolization, and inflammatory cell sequestration within the pancreas and marked edema and neutrophilic inflammation in the lung with augmented myeloperoxidase activity. In this model, COX-2 deficient mice were significantly protected against cerulein-induced pancreatitis with markedly reduced severity of lung injury, while this was not true for either COX-1 deficient mice in comparison to wild type mice. Treatment with NS-398 significantly protected against cerulein-induced pancreatitis but there was no mention of its effect in the associated lung disease [83].

Conclusion

The role of COX-2 products in the genesis and pathophysiology of lung disease is becoming clearer with the ability to selectively inhibit COX-2 pharmacologically in both human and animal studies, and through the use of animal models utilizing COX-2 deficient mice. However, we are just in the infancy of recognizing the diseases for which COX-2 inhibitor therapy might be useful. To date, probably the greatest surprise in this field of investigation is the animal data detailing the role of COX-2 in neoplastic disease and the hope that COX-2 inhibitors will provide therapeutic benefit in the treatment of human lung cancer. Most likely, COX-2 inhibition in other models will yield pleasant surprises and add to the motivation to continued study in this area of research.

Acknowledgments
This work was supported by The American Academy of Allergy, Asthma and Immunology ERT Award, American Lung Association of Tennessee, R01-AI-45512, and R01-HL-069449.

References

1 Lassus P, Wolff H, Andersson S (2000) Cyclooxygenase-2 in human perinatal lung. *Pediatr Res* 47: 602–605
2 Soslow RA, Dannenberg AJ, Rush D, Woerner BM, Khan KN, Masferrer J, Koki AT (2000) COX-2 is expressed in human pulmonary, colonic, and mammary tumors. *Cancer* 89: 2637–2645
3 Hasturk S, Kemp B, Kalapurakal SK, Kurie JM, Hong WK, Lee JS (2002) Expression of cyclooxygenase-1 and cyclooxygenase-2 in bronchial epithelium and nonsmall cell lung carcinoma. *Cancer* 94: 1023–1031
4 Funahashi A, Harland RW, LeFever A (1994) Association of increased prostaglandin E2 content in bronchoalveolar lavage fluid and intrathoracic malignancy. *Chest* 106: 166–172
5 Yoshimatsu K, Altorki NK, Golijanin D, Zhang F, Jakobsson PJ, Dannenberg AJ, Subbaramaiah K (2001) Inducible prostaglandin E synthase is overexpressed in non-small cell lung cancer. *Clin Cancer Res* 7: 2669–2674
6 Rozic JG, Chakraborty C, Lala PK (2001) Cyclooxygenase inhibitors retard murine mammary tumor progression by reducing tumor cell migration, invasiveness and angiogenesis. *Int J Cancer* 93: 497–506
7 Plastaras JP, Guengerich FP, Nebert DW, Marnett LJ (2000) Xenobiotic-metabolizing cytochromes P450 convert prostaglandin endoperoxide to hydroxyheptadecatrienoic acid and the mutagen, malondialdehyde. *J Biol Chem* 275: 11784–11790
8 Koki AT, Masferrer JL (2002) Celecoxib: a specific COX-2 inhibitor with anticancer properties. *Cancer Control* 9(2 Suppl): 28–35
9 Watkins DN, Lenzo JC, Segal A, Garlepp MJ, Thompson PJ (1999) Expression and localization of cyclo-oxygenase isoforms in non-small cell lung cancer. *Eur Respir J* 14: 412–418
10 Hida T, Yatabe Y, Achiwa H, Muramatsu H, Kozaki K, Nakamura S, Ogawa M, Mitsudomi T, Sugiura T, Takahashi T (1998) Increased expression of cyclooxygenase 2 occurs frequently in human lung cancers, specifically in adenocarcinomas. *Cancer Res* 58: 3761–3764
11 Wolff H, Saukkonen K, Anttila S, Karjalainen A, Vainio H, Ristimaki A (1998) Expression of cyclooxygenase-2 in human lung carcinoma. *Cancer Res* 58: 4997–5001
12 Achiwa H, Yatabe Y, Hida T, Kuroishi T, Kozaki K, Nakamura S, Ogawa M, Sugiura T, Mitsudomi T, Takahashi T (1999) Prognostic significance of elevated cyclooxygenase 2 expression in primary, resected lung adenocarcinomas. *Clin Cancer Res* 5: 1001–1005
13 Khuri FR, Wu H, Lee JJ, Kemp BL, Lotan R, Lippman SM, Feng L, Hong WK, Xu XC (2001) Cyclooxygenase-2 overexpression is a marker of poor prognosis in stage I non-small cell lung cancer. *Clin Cancer Res* 7: 861–867
14 Brabender J, Park J, Metzger R, Schneider PM, Lord RV, Holscher AH, Danenberg KD, Danenberg PV (2002) Prognostic significance of cyclooxygenase 2 mRNA expression in non-small cell lung cancer. *Ann Surg* 253: 440–443
15 Hosomi Y, Yokose T, Hirose Y, Nakajima R, Nagai K, Nishiwaki Y, Ochiai A (2000) Increased cyclooxygenase 2 (COX-2) expression occurs frequently in precursor lesions of human adenocarcinoma of the lung. *Lung Cancer* 30: 73–81
16 Hida T, Kozaki K, Ito H, Miyaishi O, Tatematsu Y, Suzuki T, Matsuo K, Sugiura T, Ogawa M, Takahashi T et al (2002) Significant growth inhibition of human lung cancer cells both *in vitro* and *in vivo* by the combined use of a selective cyclooxygenase 2 inhibitor, JTE-522, and conventional anticancer agents. *Clin Cancer Res* 8: 2443–2447
17 Rioux N, Castonguay A (1998) Prevention of NNK-induced lung tumorigenesis in A/J mice by acetylsalicylic acid and NS-398. *Cancer Res* 58: 5354–5360
18 Yao R, Rioux N, Castonguay A, You M (2000) Inhibition of COX-2 and induction of apoptosis: two determinants of nonsteroidal anti-inflammatory drugs' chemopreventive efficacies in mouse lung tumorigenesis. *Exp Lung Res* 26: 731–742
19 Huang M, Stolina M, Sharma S, Mao JT, Zhu L, Miller PW, Wollman J, Herschman H, Dubinett SM (1998) Non-small cell lung cancer cyclooxygenase-2-dependent regulation of cytokine balance in lymphocytes and macrophages: up-regulation of interleukin 10 and down-regulation of interleukin 12 production. *Cancer Res* 58: 1208–1216
20 Hida T, Kozaki K, Muramatsu H, Masuda A, Shimizu S, Mitsudomi T, Sugiura T, Ogawa M, Takahashi T (2000) Cyclooxygenase-2 inhibitor induces apoptosis and enhances cytotoxicity of various anticancer agents in non-small cell lung cancer cell lines. *Clin Cancer Res* 6: 2006–2011
21 Koki AT, Khan NK, WoernerBM, Seibert K, Harmon, JL, Dannenberg AJ, Soslow RA, Masferrer

JL (2002) Characterization of cyclooxygenase-2 (COX-2) during tumorigenesis in human epithelial cancers: evidence for potential clinical utility of COX-2 inhibitors in epithelial cancers. *Prostag Leukotr Ess* 66: 13–18
22 Williams CS, Tsujii M, Reese J, Dey SK, DuBois RN (2000) Host cyclooxygenase-2 modulates carcinoma growth. *J Clin Invest* 105: 1589–1594
23 Kisley LR, Barrett BS, Dwyer-Nield LD, Bauer AK, Thompson DC, Malkinson AM (2002) Celecoxib reduces pulmonary inflammation but not lung tumorigenesis in mice. *Carcinogenesis* 23: 1653–1660
24 Schreinemachers DM, Everson RB (1994) Aspirin use and lung, colon, and breast cancer incidence in a prospective study. *Epidemiology* 5: 138–146
25 Hanson WR, Thomas C (1983) 16, 16-dimethyl prostaglandin E_2 increases survival of murine intestinal stem cells when given before photon radiation. *Radiat Res* 96: 393–398
26 Pillsbury HC, III, Webster WP, Rosenman J (1986) Prostaglandin inhibitor and radiotherapy in advanced head and neck cancers. *Arch Otolaryngol Head Neck Surg* 112: 552–553
27 Milas L, Hanson WR (1995) Eicosanoids and radiation. *Eur J Cancer* 31A: 1580–1585
28 Furuta Y, Hunter N, Barkley T, Jr, Hall E, Milas L (1988) Increase in radioresponse of murine tumors by treatment with indomethacin. *Cancer Res* 48: 3008–3013
29 Milas L, Furuta Y, Hunter N, Nishiguchi I, Runkel S (1990) Dependence of indomethacin-induced potentiation of murine tumor radioresponse on tumor host immunocompetence. *Cancer Res* 50: 4473–4477
30 Pyo H, Choy H, Amorino GP, Kim JS, Cao Q, Hercules SK, DuBois RN (2001) A selective cyclooxygenase-2 inhibitor, NS-398, enhances the effect of radiation *in vitro* and *in vivo* preferentially on the cells that express cyclooxygenase-2. *Clin Cancer Res* 7: 2998–3005
31 Sousa A, Pfister R, Christie PE, Lane SJ, Nasser SM, Schmitz-Schumann M, Lee TH (1997) Enhanced expression of cyclo-oxygenase isoenzyme 2 (COX-2) in asthmatic airways and its cellular distribution in aspirin-sensitive asthma. *Thorax* 52: 940–945
32 Demoly P, Jaffuel D, Lequeux N, Weksler B, Creminon C, Michel FB, Godard P, Bousquet J (1997) Prostaglandin H synthase 1 and 2 immunoreactivities in the bronchial mucosa of asthmatics. *Am J Respir Crit Care Med* 155: 670–675
33 Redington AE, Meng QH, Springall DR, Evans TJ, Creminon C, Maclouf J, Holgate ST, Howarth PH, Polak JM (2001) Increased expression of inducible nitric oxide synthase and cyclo- oxygenase-2 in the airway epithelium of asthmatic subjects and regulation by corticosteroid treatment. *Thorax* 56: 351–357
34 Holtzman MJ (1991) Arachidonic acid metabolism. Implications of biological chemistry for lung function and disease. *Am Rev Respir Dis* 143: 188–203
35 Wenzel SE (1997) Arachidonic acid metabolites: mediators of inflammation in asthma. *Pharmacotherapy* 17(Pt 2): 3S–12S
36 Murray JJ, Tonnel AB, Brash AR, Roberts LJ 2nd, Gosset P, Workman R, Capron A, Oates JA (1986) Release of prostaglandin D2 into human airways during acute antigen challenge. *N Engl J Med* 315: 800–804
37 Gardiner PJ, Collier HO (1980) Specific receptors for prostaglandins in airways. *Prostaglandins* 19: 819–841
38 Armour CL, Johnson PR, Alfredson ML, Black JL (1989) Characterization of contractile prostanoid receptors on human airway smooth muscle. *Eur J Pharmacol* 165: 215–222
39 Hardy C, Robinson C, Lewis RA, Tattersfield AE, Holgate S (1985) Airway and cardiovascular responses to inhaled prostacyclin in normal and asthmatic subjects. *Am Rev Respir Dis* 131: 18–21
40 Gauvreau GM, Watson RM, O'Byrne PM (1999) Protective effects of inhaled PGE2 on allergen-induced airway responses and airway inflammation. *Am J Respir Crit Care Med* 159: 31–36
41 Pavord ID, Wong CS, Williams J Tattersfield AE (1993) Effect of inhaled prostaglandin E2 on allergen-induced asthma. *Am Rev Respir Dis* 148: 87–90
42 Varner AE, Busse WW, Lemanske RF, Jr, (1998) Hypothesis: decreased use of pediatric aspirin has contributed to the increasing prevalence of childhood asthma. *Ann Allergy Asthma Immunol* 81: 347–351
43 Varner AE (1999) The cyclooxygenase-2 theory of atopy and asthma. *Pediatr Asthma Allergy Immunol* 13: 43–50
44 Hartert TV, Dworski RT, Mellen BG, Oates JA, Murray JJ, Sheller JR (2000) Prostaglandin E(2) decreases allergen-stimulated release of prostaglandin D(2) in airways of subjects with asthma. *Am J Respir Crit Care Med* 162: 637–640

45 Dicpinigaitis PV (2001) Effect of the cyclooxygenase-2 inhibitor celecoxib on bronchial responsiveness and cough reflex sensitivity in asthmatics. *Pulm Pharmacol Ther* 14: 93–97
46 Fish JE, Ankin MG, Adkinson NF, Jr, Peterman VI (1981) Indomethacin modification of immediate-type immunologic airway responses in allergic asthmatic and non-asthmatic subjects: evidence for altered arachidonic acid metabolism in asthma. *Am Rev Respir Dis* 123: 609–614
47 Kirby JG, Hargreave FE, Cockcroft DW, O'Byrne PM (1989) Effect of indomethacin on allergen-induced asthmatic responses. *J Appl Physiol* 66: 578–583
48 Sladek K, Dworski R, Fitzgerald GA, Buitkus KL, Block FJ, Marney SR, Jr, Sheller JR (1990) Allergen-stimulated release of thromboxane A2 and leukotriene E4 in humans. Effect of indomethacin. *Am Rev Respir Dis* 141: 1441–1445
49 Gavett SH, Madison SL, Chulada PC, Scarborough PE, Qu W, Boyle JE, Tiano HF, Lee CA, Langenbach R, Roggli VL et al (1999) Allergic lung responses are increased in prostaglandin H synthase-deficient mice. *J Clin Invest* 104: 721–732
50 Rocca B, Spain LM, Pure E, Langenbach R, Patrono C, Fitzgerald GA (1999) Distinct roles of prostaglandin H synthases 1 and 2 in T-cell development. *J Clin Invest* 103: 1469–1477
51 Peebles RS, Jr, Hashimoto K, Morrow JD, Dworski R, Collins RD, Hashimoto Y, Christman JW, Kang KH, Jarzecka K, Furlong J et al (2002) Selective cyclooxygenase-1 and -2 inhibitors each increase allergic inflammation and airway hyperresponsiveness in mice. *Am J Respir Crit Care Med* 165: 1154–1160
52 Oguma T, Asano K, Shiomi T, Fukunaga K, Suzuki Y, Nakamura M, Matsubara H, Sheldon HK, Haley KJ, Lilly CM et al (2002) Cyclooxygenase-2 expression during allergic inflammation in guinea-pig lungs. *Am J Respir Crit Care Med* 165: 382–386
53 Szczeklik A, Nizankowska E, Bochenek G, Nagraba K, Mejza F, Swierczynska M (2001) Safety of a specific COX-2 inhibitor in aspirin-induced asthma. *Clin Exp Allergy* 31: 219–225
54 Cowburn AS, Sladek K, Soja J, Adamek L, Nizankowska E, Szczeklik A, Lam BK, Penrose JF, Austen FK, Holgate ST et al (1998) Overexpression of leukotriene C4 synthase in bronchial biopsies from patients with aspirin-intolerant asthma. *J Clin Invest* 101: 834–846
55 Sampson AP, Cowburn AS, Sladek K, Adamek L, Nizankowska E, Szczeklik A, Lam BK, Penrose JF, Austen KF, Holgate ST (1997) Profound overexpression of leukotriene C4 synthase in bronchial biopsies from aspirin-intolerant asthmatic patients. *Int Arch Allergy Immunol* 113: 355–357
56 Picado C, Valero A (2001) COX-1 sparing drugs in aspirin-sensitive asthma. *Clin Exp Allergy* 31: 179–181
57 Sestini P, Armetti L, Gambaro G, Pieroni MG, Refini RM, Sala A, Vaghi A, Folco GC, Bianco S, Robuschi M (1996) Inhaled PGE_2 prevents aspirin-induced bronchoconstriction and urinary LTE4 excretion in aspirin-sensitive asthma. *Am J Respir Crit Care Med* 153: 572–575
58 Celik G, Bavbek S, Misirligil Z, Melli M (2001) Release of cysteinyl leukotrienes with aspirin stimulation and the effect of prostaglandin E_2 on this release from peripheral blood leucocytes in aspirin-induced asthmatic patients. *Clin Exp Allergy* 31: 1615–1622
59 Woessner KM, Simon RA, Stevenson DD (2002) The safety of celecoxib in patients with aspirin-sensitive asthma. *Arthritis Rheum* 46: 2201–2206
60 Martin-Garcia C, Hinojosa M, Berges P, Camacho E, Garcia-Rodriguez R, Alfaya T, Iscar A (2002) Safety of a cyclooxygenase-2 inhibitor in patients with aspirin- sensitive asthma. *Chest* 121: 1812–1817
61 Bernard GR, Wheeler AP, Russell JA, Schein R, Summer WR, Steinberg KP, Fulkerson WJ, Wright PE, Christman BW, Dupont WD et al (1997) The effects of ibuprofen on the physiology and survival of patients with sepsis. The Ibuprofen in Sepsis Study Group. *N Engl J Med* 336: 912–918
62 Wollin L, Uhlig S, Nusing R, Wendel A (2001) Granulocyte-macrophage colony-stimulating factor amplifies lipopolysaccharide-induced bronchoconstriction by a neutrophil- and cyclooxygenase 2-dependent mechanism. *Am J Respir Crit Care Med* 163: 443–450
63 Ermert M, Merkle M, Mootz R, Grimminger F, Seeger W, Ermert L (2000) Endotoxin priming of the cyclooxygenase-2-thromboxane axis in isolated rat lungs. *Am J Physiol Lung Cell Mol Physiol* 278: L1195-L1203
64 Fischer LG, Hollmann MW, Horstman DJ, Rich GF (2000) Cyclooxygenase inhibitors attenuate bradykinin-induced vasoconstriction in septic isolated rat lungs. *Anesth Analg* 90: 625–631
65 Reddy RC, Chen GH, Tateda K, Tsai WC, Phare SM, Mancuso P, Peters-Golden M, Standiford TJ (2001) Selective inhibition of COX-2 improves early survival in murine endotoxemia but not in

bacterial peritonitis. *Am J Physiol Lung Cell Mol Physiol* 281: L537–L543

66 Zeldin DC, Wohlford-Lenane C, Chulada P, Bradbury JA, Scarborough PE, Roggli V, Langenbach R, Schwartz DA (2001) Airway inflammation and responsiveness in prostaglandin H synthase-deficient mice exposed to bacterial lipopolysaccharide. *Am J Respir Cell Mol Biol* 25: 457–465

67 Martin C, Uhlig S, Ullrich V (2001) Cytokine-induced bronchoconstriction in precision-cut lung slices is dependent upon cyclooxygenase-2 and thromboxane receptor activation. *Am J Respir Cell Mol Biol* 24: 139–145

68 Held HD, Uhlig S (2000) Mechanisms of endotoxin-induced airway and pulmonary vascular hyperreactivity in mice. *Am J Respir Crit Care Med* 162: 1547–1552

69 Gust R, Kozlowski JK, Stephenson AH, Schuster DP (1999) Role of cyclooxygenase-2 in oleic acid-induced acute lung injury. *Am J Respir Crit Care Med* 160: 1165–1170

70 Hierholzer C, Harbrecht B, Menezes JM, Kane J, MacMicking J, Nathan CF, Peitzman AB, Billiar TR, Tweardy DJ (1998) Essential role of induced nitric oxide in the initiation of the inflammatory response after hemorrhagic shock. *J Exp Med* 187: 917–928

71 Meldrum DR, Shenkar R, Sheridan BC, Cain BS, Abraham E, Harken AH (1997) Hemorrhage activates myocardial NFkappaB and increases TNF-alpha in the heart. *J Mol Cell Cardiol* 29: 2849–2854

72 Schmedtje JF, Jr, Ji YS, Liu WL, DuBois RN, Runge MS (1997) Hypoxia induces cyclooxygenase-2 via the NF-kappaB p65 transcription factor in human vascular endothelial cells. *J Biol Chem* 272: 601–608

73 Hierholzer C, Harbrecht BG, Billiar TR, Tweardy DJ (2001) Hypoxia-inducible factor-1 activation and cyclo-oxygenase-2 induction are early reperfusion-independent inflammatory events in hemorrhagic shock. *Arch Orthop Trauma Surg* 121: 219–222

74 Sunose Y, Takeyoshi I, Tsutsumi H, Kawata K, Tokumine M, Iwazaki S, Tomizawa N, Ohwada S, Matsumoto K, Morishita Y (2001) Effects of FK3311 on pulmonary ischemia-reperfusion injury in a canine model. *J Surg Res* 95: 167–173

75 Sunose Y, Takeyoshi I, Tsutsumi H, Ohwada S, Oriuchi N, Matsumoto K, Morishita Y (2001) Effect of a cyclooxygenase-2 inhibitor, FK3311, in a canine lung transplantation model. *Ann Thorac Surg* 72: 1165–1171

76 Wilborn J, Crofford LJ, Burdick MD, Kunkel SL, Strieter RM, Peters-Golden M (1995) Cultured lung fibroblasts isolated from patients with idiopathic pulmonary fibrosis have a diminished capacity to synthesize prostaglandin E2 and to express cyclooxygenase-2. *J Clin Invest* 95: 1861–1868

77 Cruz-Gervis R, Stecenko AA, Dworski R, Lane KB, Loyd JE, Pierson R, King G, Brigham KL (2002) Altered prostanoid production by fibroblasts cultured from the lungs of human subjects with idiopathic pulmonary fibrosis. *Respir Res* 3: 17

78 Keerthisingam CB, Jenkins RG, Harrison NK, Hernandez-Rodriguez NA, Booth H, Laurent GJ, Hart SL, Foster ML, McAnulty RJ (2001) Cyclooxygenase-2 deficiency results in a loss of the anti-proliferative response to transforming growth factor-beta in human fibrotic lung fibroblasts and promotes bleomycin-induced pulmonary fibrosis in mice. *Am J Pathol* 158: 1411–1422

79 Bonner JC, Rice AB, Ingram JL, Moomaw CR, Nyska A, Bradbury A, Sessoms AR, Chulada PC, Morgan DL, Zeldin DC et al (2002) Susceptibility of cyclooxygenase-2-deficient mice to pulmonary fibrogenesis. *Am J Pathol* 161: 459–470

80 Lama V, Moore BB, Christensen P, Toews GB, Peters-Golden M (2002) Prostaglandin E_2 synthesis and suppression of fibroblast proliferation by alveolar epithelial cells is cyclooxygenase-2-dependent. *Am J Respir Cell Mol Biol* 27: 752–758

81 Cuzzocrea S, Mazzon E, Sautebin L, Dugo L, Serraino I, De Sarro A, Caputi AP (2002) Protective effects of Celecoxib on lung injury and red blood cells modification induced by carrageenan in the rat. *Biochem Pharmacol* 63: 785–795

82 Shields CJ, Winter DC, Redmond HP (2002) Lung injury in acute pancreatitis: mechanisms, prevention, and therapy. *Curr Opin Crit Care* 8: 158–163

83 Ethridge RT, Chung DH, Slogoff M, Ehlers RA, Hellmich MR, Rajaraman S, Saito H, Uchida T, Evers BM (2002) Cyclooxygenase-2 gene disruption attenuates the severity of acute pancreatitis and pancreatitis-associated lung injury. *Gastroenterology* 123: 1311–1322

COX-2 Inhibitors
Edited by M. Pairet and J. van Ryn

Reproduction: role of COX-2 and its inhibition

Aarthi R. Mohan and Phillip R. Bennett

Imperial College Parturition Research Group, Institute of Reproductive and Developmental Biology, Hammersmith Hospital Campus, Du Cane Road, London W12 ONN, UK

Introduction

Prostaglandins play a key role in human reproduction. There is increasing evidence to support their involvement, not only in the processes of term and preterm labor, but also in menstruation, ovulation and blastocyst implantation. Our aim is to review the physiological significance of COX-2 and prostaglandins in reproduction, and to discuss the implications of COX-2 inhibition with regard to tocolysis.

COX-2 and male reproduction

Prostaglandins were first discovered as a constituent of human seminal fluid in 1934. They were found to be acidic lipoid factors that modify uterine smooth muscle contractility [1]. Despite a great deal of information on the role of prostaglandins in virtually all mammalian organ systems, very little information is available on their role in male reproduction. McKanna et al. [2] found that COX-2 is the predominant isoform in the adult rat male reproductive system and is heavily localized to the epithelium of the distal vas deferens, where constitutive expression is much greater than in any other organ of the body. COX-2 was not detected in the proximal half of the vas deferens or in the testes, epididymis, seminal vesicles or prostate. It was shown that COX-2 levels were not affected by the elimination of luminal sperm by vasectomy, whereas the levels were severely depleted by castration. However, replacement with androgens post-castration restored the levels, indicating that COX-2 expression in the vas deferens is androgen dependent. A role in penile erection has also been suggested in view of the fact that the distal vas encompasses a large submucosal venous plexus, in the rodent, connected to the penile corpora cavernosa. A similar venous plexus has not been described in primates, however, but the injection of prostaglandins into the corpora cavernosa has been used as an effective remedy for human erectile impotence [3].

It has been widely emphasised that COX-2 expression in most tissues is low in normal adult animals and can be increased in response to injury and various inflammatory stimuli. However, Harris et al. [4] found sustained COX-2 expression in the kidney cortex of normal adult rats. The findings previously described in normal rat vas deferens are, therefore, another example of localized COX-2 expression in the absence of inflammation or dysplasia.

COX-2 and menstruation

Van Voorhis et al. [5] have studied the localization of COX in the female reproductive tract. In the uterus, COX was primarily localized to endometrial epithelium with surface epithelium containing more immunoreactive COX than glandular epithelium. Fallopian tube secretory epithelial cells also contained COX while ciliated epithelial cells did not. It was shown that a higher percentage of glands in endometriosis and adenomyosis, as compared with ectopic endometrium, contained COX suggesting that these tissues are also capable of prostaglandin synthesis.

COX was first thought to be hormone dependent when $PGF_{2\alpha}$ and PGE_2 levels were measured at different stages of the menstrual cycle [6]. There was found to be a marked increase in the ability of the endometrium to synthesise $PGF_{2\alpha}$ in the luteal phase of the cycle. The concentration of prostaglandins is determined not only by their rate of synthesis but also by their rate of metabolism. Progesterone stimulates the metabolism of $PGF_{2\alpha}$ and PGE_2 to the inactive 13-14-dihydro-15-keto form. Therefore, during the luteal phase there is an increased ability of the endometrium to metabolise these prostaglandins due to increased prostaglandin dehydrogenase activity.

Administration of an antiprogestogen in the early luteal phase of the menstrual cycle inhibits the appearance of prostaglandin dehydrogenase and this may be one factor leading to the increase in uterine activity and menstrual bleeding when an antiprogestogen is given in the mid luteal phase of the cycle. However, as stated above, the withdrawal of progesterone alone would be expected to cause release of prostaglandins. In fact, when stromal cells are maintained *in vitro* the addition of the progesterone antagonist mifepristone does induce a dose-related release of $PGF_{2\alpha}$, suggesting that under normal circumstances progesterone is inhibiting the release of prostaglandins.

COX-2 and ovulation

Ovulation is accompanied by induction of prostaglandin synthesis as a consequence of the gonadotrophin LH surge [7, 8]. COX-2 is induced in the ovary and has been found to be necessary for the successful rupture of the ovarian follicle, perhaps by mediating the activation of proteolytic enzymes necessary for the release of the ovum [9]. The processes of ovulation and implantation are considered analogous to 'pro-inflammatory' mechanisms; hence the past speculation of prostaglandin involvement in these processes [10, 11].

Prostaglandins are formed predominantly in the granulosa cells of the pre-ovulatory follicle, and highest concentrations are found at the time of ovulation. Sirois et al. [12] showed that in bovine granulosa cells of preovulatory follicles, COX-2 mRNA and protein is transiently induced by gonadotrophins just prior to ovulation. The number of ovulations is reduced in COX-2 knockout mice, however, follicular development is normal in these mice and administration of exogenous gonadotrophins did not restore the ovulation number [13]. The selective COX-2 inhibitor, meloxicam, was shown to inhibit ovulation in rabbits in a dose- and time-dependent manner [14]. COX-2 inhibition has also been linked to a cause of reversible female infertility. The luteinised unruptured follicle (LUF) syndrome, an anovulatory condition characterized by clinical signs of ovulation but the absence of follicular rupture and ovum release, has been described by Stone et al. [15]. They have identified a possible link between regular NSAID use and reversible female infertility using both animal and clinical studies.

COX-2, implantation and placentation

Prostaglandins play a major role in blastocyst implantation. In the rodent, increased uterine vascular permeability at sites of blastocyst apposition is one of the earliest events in the implantation process. This is preceded by generalized uterine edema and luminal closure, and coincides with the initial attachment reaction between the trophectoderm and luminal epithelium [16]. It is thought that vasoactive prostaglandins are involved in these processes. Chakraborty et al. [17] found that the COX genes are differentially regulated in the peri-implantation mouse uterus. During the pre-implantation period, COX-1 is expressed in the uterine epithelium until the initiation of the attachment reaction, after which expression is down regulated. This COX-1 expression coincides with the generalized uterine oedema required for luminal closure. COX-2, however, is expressed in the luminal epithelium and subepithelial stromal cells at the anti-mesenteric pole exclusively surrounding the blastocyst at the time of the attachment reaction. These results suggest that ovarian steroids influence uterine COX-1 expression, and prostaglandins produced by COX-1 are involved in decidualization and/or continued localized endometrial vascular permeability. However, the COX-2 gene is regulated by the implanting blastocyst during early pregnancy, and COX-2 generated prostaglandins are involved in angiogenesis for the establishment of placentation. There may be a possible role for COX-2 inhibitors as contraceptives in view of their ability to inhibit ovulation, however, their long-term side-effects on other body systems may prove to be a problem.

COX knock-out experiments and fetal development

There have been a number of reports of transgenic mice with knock-outs for COX-1, COX-2 and the prostaglandin E (EP4) and F (FP) receptors [18–21]. Langenbach et al. [18] have shown, surprisingly, that COX-1 knock-out mice show less indomethacin-induced gastrointestinal lesions than wild-type mice, despite the apparent importance of COX-1 mediated prostaglandin synthesis in the stomach. The reason for this could be that the pharmacological inhibition of COX leaves peroxidase activity intact. However, when homozygous COX-1 knock-out mice were crossbred, there was a high mortality of their offspring. Heterozygous COX-1 knock-out mice when mated with homozygotes produced litters of normal size and survival, indicating that COX-1 expression in only 50% of these pups or their placentas is sufficient for survival. These studies showed that COX-1 is an essential requirement for mouse parturition but not for ovulation or spermatogenesis.

Morham et al. [19] first reported a COX-2 knock-out mouse and showed that homozygous COX-2 knock-out mice begin to show signs of nephropathy by six weeks of age, which increases in severity until death at around 8 weeks. All the other tissues in these mice were normal. Examination of COX-2 deleted mice embryos showed that kidney maturation had prematurely ceased after only a small percentage of nephrons had developed and the majority of glomeruli and tubules remained small and immature. During postnatal life, the small number of nephrons was seen to atrophy, with the development of glomerulosclerosis, interstitial inflammation and fibrosis leading, ultimately, to kidney failure. In those mice that did survive to adulthood, kidney lesions ranged from mild to severe, with the more severe lesions being more common in female than male homozygotes. This report clearly illustrates an important role for COX-2 in the embryological development of the kidney.

The dilatation of the ductus arteriosus (DA) *in utero* is an active process maintained primarily by PGE_2. The prostaglandin receptors that may have a role in dilatation of the DA include the PGE_2 receptors, EP2 [22] and EP4 [23], and a receptor for prostacyclin [24]. Nguyen et al. [20] showed abnormal function of the DA in EP4 receptor-deficient mice, with greater than 95% of these mice dying within 48 h of birth. Loftin et al. [25] reported that the absence of either or both COX isoforms in mice did not result in premature closure of the DA *in utero*. However, 35% of COX-2 (–/–) mice died with a patent DA within 48 h of birth. In contrast, the absence of only the COX-1 isoform did not affect ductal closure. The mortality (35%) and patent DA incidence due to absence of COX-2 was, however, significantly increased (79%) when only one copy of the gene encoding COX-1 was also inactivated. All of the mice deficient in both isoforms died with a patent DA within 12 h of birth, indicating that in COX-2 deficient mice, the contribution of COX-1 to DA closure is gene dosage-dependent. Loftin et al. found that indomethacin, a non-steroidal anti-inflammatory drug, induces ductal closure in both wild type and heterozygous COX-2-deficient mice, irrespective of their COX-1 genotype. However, indomethacin did not induce ductal closure in fetuses that lacked COX-2 expression. Therefore, indomethacin-induced closure of the DA requires fetal COX-2 expression, and neither maternal nor fetal COX-1 expression is required for the DA to be sensitive to indomethacin. It has been suggested that COX-2 in the DA may produce constrictor prostaglandins important for DA contraction. Therefore, contraction of the DA following maternal indomethacin exposure may result from the inhibition of dilatory PGE_2 synthesis in the placenta without sufficient inhibition of ductal COX-2 to attenuate DA contraction.

Sugimoto et al. [21] found that mice with deleted FP receptors had normal estrous cycles, normal ovulation and normal implantation. The onset of labor is stimulated by regression of the corpus luteum and withdrawal of ovarian progesterone in mice, allowing upregulation of the uterine oxytocin receptor. However, pregnant homozygous FP knock-out mice failed to initiate labor because of failure of regression of the corpus luteum. Ovariectomy in these mice ceased the progesterone inhibition permitting the onset of labor.

There have been numerous reports of the roles of COX-1 and COX-2 in rodents, with respect to their menstrual cycle, ovulation and implantation, which may be similar to other species and the primate. However, there are fundamental differences between the two groups in the general mechanism of the onset of parturition and in the roles of COX-1 and 2 in labor.

COX-2 in human labor

Prostaglandins play a central role in human parturition. A number of studies have shown that the fetal membranes are a major source of intrauterine prostaglandins in labor [26, 27]. Prostaglandins mediate cervical ripening and directly stimulate uterine contraction [28]. They indirectly increase fundally dominant myometrial contractility by upregulation of gap junctions and oxytocin receptors and synchronization of contractions [29, 30].

As discussed above, prostaglandins are formed from the precursor arachidonic acid, which is itself a substrate for at least three enzyme groups; COX, lipoxygenase and epoxygenase. Prior to labor, endogenous arachidonic acid metabolism in the amnion is principally via the lipoxygenase enzyme pathways. Bennett et al. [31] have shown that with labor there is an increase in arachidonic acid metabolism and a change in the ratio of COX to lipoxygenase metabolism to favor synthesis of PGE_2. It has been suggested that the lipoxygenase metabolites of arachidonic acid may play a role in pre-labor (Braxton-Hicks) contractions [32, 33].

It was previously thought that phospholipase activity was the limiting factor in prostaglandin production in the fetal membranes. However, any stimulus to prostaglandin production must also increase COX activity since it has a short half-life and undergoes destruction after a limited number of reactions [34]. The change in the ratio of COX: lipoxygenase metabolism with labor suggests that the activity of the two enzymes, COX and phospholipase must be independently controlled [31, 35].

In sheep, the placental cotyledon is the principal source of prostaglandins in late gestation [36] and upregulation of COX-2 expression takes place gradually as pregnancy advances and precedes the labor-associated peak of maternal estrogen concentrations. COX-1 expression within the sheep uterus does not change with gestational age [37, 38].

The principle sources of uterine prostaglandins in humans are the fetal membranes and the myometrium. Both COX-1 and COX-2 expression has been demonstrated in fetal membranes using *in situ* hybridization [39]. COX-2 expression is mainly within the amnion epithelial layer, the reticular layer of the chorion and the decidua. COX-2 is not expressed in the trophoblast layer of the chorion. COX-1 expression is more diffuse and is seen in the amnionic epithelium, amnionic mesoderm and also in the chorionic mesoderm and decidua. The levels of expression of the two COX isoforms are very different. In amnion at term, expression of COX-2 mRNA is 100-fold greater than that of COX-1 mRNA [40]. COX-2 expression also doubles in association with labor [39–41]. Slater et al. [40] showed that the mRNA levels of COX-1 do not change significantly through gestation and with labor onset, whereas COX-2 mRNA increases steadily throughout gestation with a significant increase at term. COX-2 protein concentrations in amnion also increase from the first trimester to term with a further increase with the onset of labor. Although COX-1 mRNA concentrations remain low and do not change with labor, COX-1 protein concentrations are similar to those of COX-2, although there are no labor or gestation-associated changes [42]. This apparent discrepancy between mRNA and protein concentrations for COX-1 and COX-2 may be explained in terms of the known functions of the enzymes. COX is thought to be a 'suicide enzyme', undergoing destruction during catalysis [43]. Increased prostaglandin synthesis via COX-2 would therefore mean increased turnover of the enzyme itself. This would mean a large increase in mRNA concentrations to drive protein synthesis, but because the protein itself is also being actively destroyed, the COX-2 protein concentration measured at any one time would be increased to a much lesser extent. As COX-1 is not active at this time in the fetal membranes, the protein may be present at concentrations similar to that of COX-2 but COX-1 mRNA concentrations would be very low. Enzyme kinetic studies suggest that the increase in COX activity with the onset of labor is due to an increase in the synthesis of the COX-2 enzyme.

COX-2 mRNA and protein concentrations within the human fetal membranes increase in the third trimester of pregnancy many weeks before the onset of labor. Slater et al. [44] reported a similar increase in COX activity, suggesting that the fetal membranes, in particular the amnion, have the capacity to synthesize prostaglandins several weeks prior to the onset of labor at term. It is thought that this is opposed by prostaglandin dehydrogenase (PGDH) activity. Sangha et al. [45] have shown a reduction in PGDH expression with the onset of labor, and have also demonstrated that a subgroup of patients presenting in 'idiopathic' pre-term labor have reduced expression of chorionic PGDH both at the mRNA and protein levels. Van Meir et al. [46, 47] have shown a labor-associated reduction in PGDH activity in chorion near to the cervix and chorioamnionitis, leading to pre-term delivery, is associated with reduced chorionic PGDH expression. They have suggested that the opposing effects of progesterone and cortisol may control PGDH activity in the fetal membranes.

The role of progesterone and IL-8 in labor

Progesterone represses myometrial contractility by changing the electrical properties of the plasma membrane and restricting the propagation of the action potential. It increases calcium binding to intracellular sites [48], prevents the formation of gap junctions [49] and oxytocin receptors [50] and inhibits arachidonic acid release [51]. The corpus luteum regresses in the first trimester of human pregnancy and progesterone synthesis is taken over by the placenta and chorion. Unlike the rodent, a corpus luteum-dependent species, there is no detectable progesterone withdrawal in the human to signal the onset of labor. However, there is upregulation of a range of pro-labor genes, including gap junction proteins, oxytocin receptor, prostaglandin receptors and interleukin-8 (IL-8), which are normally repressed by the presence of progesterone. IL-8 is a cytokine whose expression increases and acts as a chemoattractant for neutrophils that invade the cervical tissue, fetal membranes and myometrium. Elliott et al. [52] have shown that IL-8 is produced by the term human placenta, with increased production during labor and by treatment with the antigestagen onapristone. It was also shown that biopsies of first-trimester cervices produced IL-8 [53] and that this production is increased if the woman is pre-treated with the progesterone antagonist mifepristone prior to collection of the biopsy. In both guinea pigs and rabbits, application of IL-8 to the cervix induces cervical ripening [54].

The promoter of the COX-2 gene has been cloned and sequenced and has been shown to have some similarities with the IL-8 promoter. Several putative transcription factor binding sites have been identified, including a cyclic AMP response element (CRE), a nuclear factor IL-6 (NF-IL6) site and two nuclear factor-κB (NF-κB) sites. Each of these has been demonstrated to regulate COX-2 transcription in various cells. It has been shown that the promoter region of the IL-8 gene is regulated in a cell-line specific manner by NF-κB and either an AP-1 or CCAAT enhancer-binding protein site (C/EBP). NF-κB is the transcription factor family classically associated with inflammation. The presence of NF-κB sites in both COX-2 and IL-8 promoters, and the view that the process of labor may be similar to an inflammatory reaction, has led to the suggestion that NF-κB may play a role in labor.

We have used primary cultures of amnion cells to study the roles of the NF-κB sites in COX-2 regulation. Mutation of the downstream NF-κB DNA-binding site showed significantly less reporter expression than the wild type COX-2 promoter only in post-labor amnion cells. However, mutation of the upstream NF-κB DNA-binding site significantly reduced reporter expression in both pre- and post-labor amnion cells. We have also performed similar studies using reporter constructs of the IL-8 gene promoter with site mutations in the NF-κB, C/EBP and AP-1 sites. Mutation of the NF-κB site reduced both basal and IL-1β-stimulated promoter activity in primary amnion cells whilst mutation of the C/EBP or the AP-1 sites did not appear to have a significant effect. NF-κB, therefore, appears to be essential for the transcriptional regulation of both COX-2 and IL-8 in primary amnion cells. In myometrium, NF-κB plays a role in IL-1β stimulated COX-2 expression but is not activated with labor. NF-κB is, therefore, an important regulator of the labor-associated inflammatory mediators and plays a central role in the onset of human parturition.

Preterm labor and COX-2 inhibitors

Preterm birth occurs in 5–10% of all pregnancies but is associated with 70% of all neonatal deaths and up to 75% of neonatal morbidity. Premature infants are at increased risk of cerebral palsy, deafness, blindness and respiratory problems. The social and financial costs are enormous, but most distressing is the impact of long-term handicap on the individuals and their caretakers. Unfortunately, despite research directed to the development of drugs to inhibit myometrial contractions, there has been no reduction in the incidence of preterm birth. Present therapies cannot prevent preterm delivery, but at best provide sufficient delay in order to attempt treatments, which will ameliorate the consequences of prematurity. It is probable that improved therapies for preterm labor will be targeted less directly at myometrial contractility, but at the earlier biochemical events of labor, including the increase in synthesis of prostaglandins and inflammatory cytokines.

Preterm labor is a multi-factorial syndrome. Most theories for the biochemical etiology of preterm labor have, as an important endpoint, increased prostaglandin synthesis mediated via increased COX-2 expression. Upregulation of inflammatory cytokines and prostaglandins that occur over a period of several weeks leads to cervical ripening and membrane rupture and, ultimately, to myometrial contractility and labor [55]. The most common cause of early preterm labor is generally thought to be genital tract infection. Bennett et al. [56] reported that prostaglandins are not synthesized by bacteria themselves, but increase prostaglandin synthesis both by release of phospholipase A [32] and via the endotoxin lipopolysaccharide (LPS) induced upregulation of COX-2 expression. It has been shown that LPS increases prostaglandin synthesis in fetal membranes and increases COX-2 expression.

Preterm labor due to multiple pregnancy may be due to increased placental mass, so that the production of endocrine factors, such as CRH, which upregulate COX-2 synthesis is reached at an earlier gestational age. COX-2 and other contraction-associated proteins are stretch-dependent which may also contribute to early delivery in multiple pregnancy and hydramnios. Where there is fetal growth retardation or placental dysfunction, increased fetal cortisol may upregulate placental CRH synthesis [57] leading to a premature increase in COX-2 expression and the onset of preterm labor.

Studies in the sheep, rat and wallaby, have shown that stretch increases myometrial expression of the oxytocin receptor [58, 59], COX-2 [58] and the gap junction protein connexin-43 [60]. We have shown that, in human myometrium, stretch increases the expression of the oxytocin receptor and COX-2, and that these are associated with activation of AP-1 and C/EBP but not NF-κB. We have also shown that stretch leads to the activation of p38 MAPK and ERK-1 and –2 and that inhibition of ERK blocks stretch induced COX-2 expression (unpublished data).

Beta-sympathomimetics, the most commonly used tocolytics, exhibit tachyphylaxis and cause potentially life-threatening maternal side-effects [61, 62]. Although they are effective in inhibition of preterm contractions and delay of delivery, the use of non-selective COX inhibitors like indomethacin is limited by fetal side-effects, including constriction of the DA, renal tubular dysfunction and oligohydramnios, and neonatal complications including intracranial hemorrhage and necrotising enterocolitis [63]. Since prostaglandin synthesis in the fetus is constant, it is probably mainly mediated by COX-1. Therefore, a COX-2 selective inhibitor may have the efficacy of indomethacin without the fetal side-effects.

Nimesulide is a non-steroidal anti-inflammatory drug, which has been shown not to affect prostaglandin synthesis in tissues in which COX-1 is the predominant isoform [64] but does inhibit COX-2 mediated prostaglandin synthesis [65, 66]. *In vitro*, nimesulide displays 30–100 fold selectivity for COX-2 compared with COX-1 [66, 67]. Studies in animals and in explanted human myometrial cells have demonstrated its efficacy at inhibiting myometrial contractility. Sawdy et al. [68] have described the successful clinical use of nimesulide as a tocolytic agent from 16 to 34 weeks with no effect on amniotic fluid volume or patency of the DA. Locatelli et al. [69] evaluated the efficacy and safety of oral nimesulide for the treatment of preterm labor in a pilot study of five women. They showed that tocolysis with nimesulide was associated with fetal side-effects similar to those described with indomethacin therapy. Oligohydramnios occurred within 9 days of therapy in all treated patients. This could be explained by the fact that nimesulide could have a lower COX-2 selectivity than previously reported. Also, nimesulide could lose its COX-2 selectivity in a dose-time-dependent manner. Studies in sheep have shown that at high doses, nimesulide administration results in inactivation of both COX-1 and –2 [70]. The doses used in clinical practice may result in concentrations where COX-2 selectivity is lost.

Stika et al. [71] have carried out a trial comparing the safety of celecoxib, a selective cyclo-oxygenase-2 inhibitor, with indomethacin administered for the treatment of preterm labor. However, exposure to either drug was only for 48 h, there were only 12 women in each arm of the trial, and it is doubtful whether all of the 24 women in the trial were actually in preterm labor. The trial suggested that the two drugs were equally effective in the maintenance of tocolysis, but the trial was not adequately powered to show this. Both drugs were associated with a transient decrease in amniotic fluid volume, however, celecoxib produced fewer fetal side-effects when compared with indomethacin.

Further randomized, controlled trials using larger numbers of patients focussing on long term safety and drug efficacy are warranted before the widespread introduction of selective COX-2 inhibitors for the treatment of preterm labor.

References

1 Euler US von (1936) On the specific vaso-dilating and plain muscle stimulating substances from accessory genital glands in man and certain animals (prostaglandin and vesiglandin). *J Physiol (Lond)* 88: 213–234

2 McKanna JA, Zhang M-Z, Wang J-L, Cheng H-F, Harris RC (1998) Constitutive expression of cyclooxygenase-2 in rat vas deferens. *Am J Physiol* 275 (*Regulatory Integrative Comp Physiol* 44): R227–R233

3 Wagner G, Kaplan HS (1993) *The New Injection Treatment for Impotence: Medical and Psychological Aspects.* Brunner/Mazel, New York

4 Harris RC, McKanna JA, Akai Y, Jacobson HR, DuBois RN, Breyer MD (1994) Cyclooxygenase-2 is associated with the macula densa of rat kidney and increases with salt restriction. *J Clin Invest* 94: 2504–2510

5 Van Voorhis BJ, Huettner PC, Clark MR, Hill JA (1990) Immunohistological localization of prostaglandin H synthase in the female reproductive tract and endometriosis. *Am J Obstet Gynaecol* 163 (1 Pt 1): 57–62

6 Downie J, Poyser NL, Wunderlich M (1974) Levels of prostaglandins in human endometrium during the normal menstrual cycle. *J Physiol* 236: 465–472

7 LeMaire WJ, Marsh JM (1975) Interrelationships between prostaglandins, cyclic AMP and steroids ovulation. *J Reprod Fertil* 22 (Suppl): 53–74

8 Armstrong DT (1981) Prostaglandins and follicular functions. *J Reprod Fertil* 62: 283–291

9 Tsafriri A (1995) Ovulation as a tissue remodelling process. Proteolysis and cumulus expansion. *Adv Exp Med Biol* 377: 121–140

10 McMaster MT, Dey SK, Andrews GK (1993) Association of monocytes and neutrophils with early events of blastocyst implantation in mice. *J Reprod Fertil* 99: 561–569

11 Espey LL (1994) Current status of the hypothesis that mammalian ovulation is comparable to an inflammatory reaction. *Biol Reprod* 50: 233–238

12 Sirois J (1994) Induction of prostaglandin endoperoxide synthase-2 by human chorionic gonadotropin in bovine preovulatory follicles *in vivo*. *Endocrinology* 135: 841–848

13 Lim H, Paria BC, Das SK, Dinchuk JE, Langenbach R, Trzaskos JM, Dey SK (1997) Multiple Female Reproductive Failures in Cyclooxygenase 2-Deficient Mice. *Cell* 91: 197–208

14 Salhab AS, Gharaibeh MN, Shomaf MS, Amro BI (2001) Meloxicam inhibits rabbit ovulation. *Contraception* 63: 329–333

15 Stone S (2002) Nonsteroidal anti-inflammatory drugs and reversible female infertility: is there a link? *Drug Saf* 25: 545–551

16 Kennedy TG (1983) Embryonic signals and the initiation of blastocyst implantation. *Aust J Biol Sci* 36: 531–543

17 Chakraborty I, Das SK, Wang J, Dey SK (1996) Developmental expression of the cyclooxygenase-1 and cyclooxygenase-2 genes in the peri-implantation mouse uterus and their differential

regulation by the blastocyst and ovarian steroids. *J Mol Endocrinol* 16: 107–122
18 Langenbach R, Morham SG, Tiano HF, Loftin CD, Ghanayem BI, Chulada PC, Mahler JF, Lee CA, Goulding EH, Kluckman KD et al (1995) Prostaglandin synthase 1 gene disruption in mice reduces arachidonic acid-induced inflammation and indomethacin-induced gastric ulceration. *Cell* 83: 483–492
19 Morham SG, Langenbach R, Loftin CD, Tiano HF, Vouloumanos N, Jennette JC, Mahler JF, Kluckman KD, Ledford A, Lee CA et al (1995) Prostaglandin synthase 2 gene disruption causes severe renal pathology in the mouse. *Cell* 83: 473–482
20 Nguyen MT, Camenisch T, Snouwaert JN, Hicks E, Coffman TM, Anderson PAW, Malouf NN, Koller BH (1997) The prostaglandin receptor EP_4 triggers remodelling of the cardiovascular system at birth. *Nature* 390: 78–81
21 Sugimoto Y, Yamasaki A, Segi E, Tsuboi K, Aze Y, Nishimura T, Oida H, Yoshida N, Tanaka T, Katasuyama M et al (1997) Failure of parturition in mice lacking the prostaglandin F receptor. *Science* 277: 681–683
22 Bhattacharya M, Asselin P, Hardy P, Guerguerian AM, Shichi H, Hou X, Varma DR, Bouayad A, Fouron JC, Clyman R et al (1999) Developmental changes in prostaglandin E(2) receptor subtypes in porcine ductus arteriosus. Possible contribution in altered responsiveness to prostaglandin E(2). *Circulation* 100: 1751–1756
23 Smith GC, Coleman RA, McGrath JC (1994) Characterization of dilator prostanoid receptors in the fetal rabbit ductus arteriosus. *Pharmacol Exp Ther* 271: 390–396
24 Smith GC, McGrath JC (1993) Characterisation of the effect of oxygen tension on response of fetal rabbit ductus arteriosus to vasodilators. *Cardiovasc Res* 27: 2205–2211
25 Loftin CD, Trivedi DB, Tiano HF, Clark JA, Lee CA, Epstein JA, Morham SG, Breyer MD, Nguyen MT, Hawkins BM et al (2001) Failure of ductus arteriosus closure and remodelling in neonatal mice deficient in cyclooxygenase-1 and cyclooxygenase-2. *Proc Natl Acad Sci* 98: 1059–1064
26 Kierse MJ, Mitchell MD, Turnbull AC (1977) Changes in prostaglandin F and 13,14-dihydro-15-keto-prostaglandin F concentrations in amniotic fluid at the onset of and during labour. *British Journal of Obstetrics and Gynaecology* 84: 743–746
27 Skinner KA, Challis JRG (1985) Changes in the synthesis and metabolism of prostaglandins by human fetal membranes and decidua at labour. *American Journal of Obstetrics and Gynaecology* 151: 519–523
28 Crankshaw DJ, Dyal R (1994) Effects of some naturally occurring prostanoids and some cyclooxygenase inhibitors on the contractility of the human lower uterine segment *in vitro*. *Can J Physiol Pharmacol* 72: 870–874
29 Liggins G (1989) Initiation of labour. *Biol Neonate* 55: 366–375
30 Garfield RE, Hertzberg EL (1990) Cell-to-cell coupling in the myometrium: Emil Bozler's prediction. *Prog Clin Biol Res* 327: 673–681
31 Bennett PR, Slater D, Sullivan M, Elder MG, Moore GE (1993) Changes in amniotic arachidonic acid metabolism associated with increased cyclooxygenase gene expression. *Br J Obstet Gynaecol* 100: 1037–1042
32 Bennett PR, Elder MG, Myatt L (1987) The effects of lipoxygenase metabolites of arachidonic acid on human myometrial contractility. *Prostaglandins* 33: 837–844
33 Walsh SW (1989) 5-Hydroxyeicosatetraenoic acid, leukotriene C4 and prostaglandin F2α in amniotic fluid before and during term and preterm labor. *Am J Obstet Gynaecol* 161: 1352
34 Marshall P, Kulmacz R, Lands W (1979) Constraints on prostaglandin synthesis in tissues. *J Biol Chem* 262: 3510–3517
35 Bennett PR, Rose MP, Myatt L, Elder MG (1987) Preterm labor: Stimulation of arachidonic acid metabolism in human amnion cells by bacterial products. *Am J Obstet Gynecol* 156: 649–655
36 Rice GE, Payne MJ, Wong MH, Thorburn GD (1992) Immunoreactive prostaglandin G/H synthase content increases in ovine cotyledons during late gestation. *Placenta* 13: 429–437
37 Wimsatt J, Nathanielsz PW, Sirois J (1993) Induction of prostaglandin endoperoxide synthase isoform-2 in ovine cotyledonary tissues during late gestation. *Endocrinology* 133: 1068–1073
38 Rice GE, Freed KA, Aitken MA, Jacobs RA (1995) Gestational- and labour-associated changes in the relative abundance of prostaglandin G/H synthase-1 and –2 mRNA in ovine placenta. *J Mol Endocrinol* 14: 237–245
39 Slater DM, Berger LC, Newton R, Moore GE, Bennett PR (1995) Expression of cyclooxygenase types 1 and 2 in human fetal membranes at term. *Am J Obstet Gynecol* 172: 77–82

40 Slater D, Berger L, Newton R, Moore G, Bennett P (1994) The relative abundance of type 1 to type 2 cyclo-oxygenase mRNA in human amnion at term. *Biochem Biophys Res Commun* 198: 304–308
41 Hirst JJ, Teixeira FJ, Zakar T, Olson DM (1995) Prostaglandin endoperoxide-H synthase-2 expression increases in human gestational tissues with spontaneous labour onset. *Reproduction, Fertility and Development* 7: 633–637
42 Sawdy RJ, Slater DM, Dennes WJB, Sullivan MHF, Bennett PR (2000) The roles of the cyclo-oxygenases types one and two in human fetal membranes at term. *Placenta* 21: 54–57
43 Marshall P, Kulmacz R and Lands W (1979) Constrants on prostaglandin synthesis in tissues. *J Biol Chem* 262: 3510–3517
44 Slater D, Dennes W, Sawdy R, Allport V, Bennett P (1999) Expression of cyclo-oxygenase types-1 and –2 in human fetal membranes throughout pregnancy. *J Mol Endocrinol* 22: 125–130
45 Sangha RK, Walton JC, Ensor CM, Tai HH, Challis JR (1994) Immunohistochemical localization, messenger ribonucleic acid abundance, and activity of 15-hydroxyprostaglandin dehydrogenase in placenta and fetal membranes during term and preterm labor. *J Clin Endocrinol MeTab* 78: 982–989
46 Van Meir CA, Ramirez MM, Matthews SG, Calder AA, Keirse MJ, Challis JR (1997) Chorionic prostaglandin catabolism is decreased in the lower uterine segment with term labour. *Placenta* 18: 109–114
47 Van Meir CA, Sangha RK, Walton JC, Matthews SG, Keirse MJ, Challis JR (1996) Immunoreactive 15-hydroxyprostaglandin dehydrogenase (PGDH) is reduced in fetal membranes from patients at preterm delivery in the presence of infection. *Placenta* 17: 291–297
48 Carsten ME (1979) Calcium accumulation by human uterine microsomal preparations: effects of progesterone and oxytocin. *Am J Obstet Gynecol* 133: 598–601
49 Garfield RE, Merrett D, Grover AK (1980) Gap junction formation and regulation in myometrium. *Am J Physiol* 239: C217–C228
50 Fuchs AR, Periyasamy S, Soloff MS (1983) Systemic and local regulation of oxytocin receptors in the rat uterus, and their functional significance. *Can J Biochem Cell Biol* 61: 615–624
51 Wilson T, Liggins GC, Aimer GP, Watkins EJ (1986) The effect of progesterone on the release of arachidonic acid from human endometrial cells stimulated by histamine. *Prostaglandins* 31: 343–360
52 Elliott CL, Kelly RW, Critchley HO, Riley SC, Calder AA (1998) Regulation of interleukin 8 production in the term human placenta during labor and by antigestagens. *Am J Obstet Gynecol* 179: 215–220
53 Elliott CL, Brennard JE, Calder AA (1998) The effects of mifepristone on cervical ripening and labor induction in primigravidae. *Obstet Gynecol* 92: 804–809
54 Chwalisz K, Garfield RE (1994) Antiprogestins in the induction of labor. *Ann NY Acad Sci* 734: 387–413
55 Bennett PR, Elder MG, Myatt L (1990) Secretion of phospholipases by bacterial pathogens may initiate preterm labor. *Am J Obstet Gynecol* 163 (1 Pt 1): 241–242
56 Bennett PR, Allport V, Loudon J, Elliott C (2001) Prostaglandins, the fetal membranes and the cervix: the endocrinology of parturition: basic science and clinical application. *Front Horm Res* 27: 147–164
57 Challis JR, Matthews SG, Van Meir C, Ramirez MM (1995) Current topic: the placental corticotrophin-releasing hormone-adrenocorticotrophin axis. *Placenta* 16: 481–502
58 Wu WX, Ma XH, Nathanielsz PW (1999) Tissue-specific ontogenic expression of prostaglandin H synthase 2 in the ovine myometrium, endometrium, and placenta during late gestation and at spontaneous term labor. *Am J Obstet Gynecol* 181: 1512–1519
59 Parry LJ, Bathgate RA (2000) The role of oxytocin and regulation of uterine oxytocin receptors in pregnant marsupials. *Exp Physiol* 85: 91S–99S
60 Ou CW, Orsino A, Lye SJ (1997) Expression of connexin-43 and connexin-26 in the rat myometrium during pregnancy and labor is differentially regulated by mechanical and hormonal signals. *Endocrinology* 138: 5398–5407
61 Morrison JJ (1996) Physiology and pharmacology of uterine contraction. In: J Studd (ed.): *The Yearbook of the Royal College of Obstetricians and Gynaecologists*. RCOG Press, London, 45–61
62 Carroll SG, Seibre NJ, Nicolaides KH (1996) *Preterm Prelabour Amniorrhexis*. Parthenon Publishing Group, New York
63 Norton ME, Merrill J Cooper BA (1993) Neonatal complications after administration of

indomethacin for preterm labour. *N Engl J Med* 329: 1602–1607

64 Bianco S, Robuschi M, Gambaro G, Spagnotto S, Petrigni G (1991) Bronchial inflammation and NSAIDs. *Drug Invest* 3 (Suppl 2): 33–38

65 Taniguchi Y, Ikesue A, Yokoyama K, Noda K, Deguchi H, Nakamura T, Toda A, Shimeno H (1995) Selective inhibition by nimesulide, a novel non-steroidal anti-inflammatory drug, with prostaglandin endoperoxide synthase-2 activity *in vitro*. *Pharmacol Sci* 1: 173–175

66 Miralpeix M, Camacho M, Lopez-Belmonte J, Canalias F, Beleta J, Palacios JM, Vila L (1997) Selective induction of cyclo-oxygenase-2 activity in the permanent human endothelial cell line HUV-EC-C: biochemical and pharmacological characterization. *Br J Pharmacol* 121: 171–180

67 Yamada M, Niki H, Yamashita M, Mue S, Ohuchi K (1997) Prostaglandin E_2 production dependent upon cyclo-oxygenase-1 and cyclo-oxygenase-2 and its contradictory modulation by auranofin in rat peritoneal macrophages. *J Pharmacol Exp Ther* 281: 1005–1012

68 Sawdy R, Slater D, Fisk N, Edmonds DK, Bennett P (1997) Use of a cyclo-oxygenase type-2-selective non-steroidal anti-inflammatory agent to prevent preterm delivery. *Lancet* 350: 265–266

69 Locatelli A, Vergani P, Bellini P, Strobelt N, Ghidini A (2001) Can a cyclo-oxygenase type-2 selective tocolytic agent avoid the fetal side-effects of indomethacin? *Br J Obstet Gynecol* 108: 325–326

70 Poore KR, Young IR, Hirst JJ (1999) Efficacy of the selective prostaglandin synthase type 2 inhibitor nimesulide in blocking basal prostaglandin production and delaying glucocorticoid-induced premature labor in sheep. *Am J Obstet Gynecol* 180: 1244–1253

71 Stika CS, Gross GA, Laguizamon G, Gerber S, Levy R, Mathur A, Bernhard LM, Nelson DM, Sadovsky Y (2002) A prospective randomised safety trial of celecoxib for treatment of preterm labor. *Am J Obstet Gynecol* 187: 653–660

COX-2 Inhibitors
Edited by M. Pairet and J. van Ryn

COX-2 in cancer

Kirsi Saukkonen[1, 4], Christianne J. Buskens[5], Anna Sivula[1, 4], Bastiaan P. van Rees[6], Tiina-Liisa Erkinheimo[2, 4], Johanna Rintahaka[1, 4], Leena Pohjola[1, 4], J. Jan B. van Lanschot[5], G. Johan A. Offerhaus[6], Caj Haglund[3] and Ari Ristimäki[1, 4]

[1] *Departments of Pathology,* [2]*Obstetrics and Gynecology and* [3]*Surgery, Helsinki University Central Hospital and* [4]*Molecular and Cancer Biology Research Program, Biomedicum Helsinki, University of Helsinki, P.O. Box 63, FIN-00014 Helsinki, Finland*
[5] *Departments of Surgery and* [6]*Pathology, Academic Medical Center, University of Amsterdam, NL-1105 AZ Amsterdam, The Netherlands*

Role of COX-2 in colorectal carcinogenesis

Multiple epidemiological studies indicate that the use of aspirin and other NSAIDs are associated with reduced risk of malignancies especially in the digestive tract. In addition, another NSAID sulindac causes regression of colorectal adenomatous polyps in patients with familial adenomatous polyposis (FAP) [1, 2]. Recent studies suggest that COX-2 is a rational target of NSAIDs in prevention of colorectal cancer. First, elevated levels of COX-2 mRNA and protein, but not those of COX-1, were found in colorectal adenocarcinomas and in their adenomatous precursors [3–5]. Second, selective COX-2 inhibitors suppress neoplasia formation in rodent models of colorectal cancer [6–10]. Importantly, genetic disruption of COX-2 suppresses the polyp formation in $Apc^{\Delta 716}$-knockout and in Min mice, which are models for FAP [6, 11]. However, it should be noted that disruption of COX-1 gene also reduced the polyp burden in the Min mouse model [11]. Finally, elevated COX-2 expression was shown to associate with poor prognosis in colorectal carcinoma [12].

Several ongoing clinical trials are investigating the chemopreventive or -therapeutic (adjuvant) effects of COX-2 selective inhibitors [1, 2]. The first one of these studies has been published, which indicates that a selective COX-2 inhibitor celecoxib reduced polyp burden in patients who suffer from FAP based on a 6-month, randomized controlled clinical trial [13]. The US Food and Drug Administration granted accelerated approval in December 1999 for celecoxib for adjunctive therapy in patients with FAP. However, it should be kept in mind that the absolute decrease in polyp number in this trial in the celecoxib treatment group was quite modest (28% using 400 mg twice a day) and that no data exist that this treatment would reduce the incidence of colorectal cancer in FAP patients. Thus, at the moment this drug treatment cannot replace standard follow-up and treatment practices.

We have recently characterized COX-2 expression in several different types of human adenocarcinomas. Our data show that COX-2 is primarily expressed in neoplastic epithelial cells in human adenocarcinomas, it is present in both noninvasive precursor lesions and invasive carcinomas, and elevated level of COX-2 can be a marker for poor prognosis in certain types of adenocarcinomas. Our aim is to review data on COX-2 expression in several adenocarcinomas and evaluate the role of COX-2 in human carcinogenesis.

Esophageal adenocarcinoma

Adenocarcinoma of the esophagus is a highly lethal disease and its incidence has markedly increased during the last few decades [14–16]. Surgical resection is currently the preferred curative option, but postoperative mortality and morbidity are significant even after "curative" resection and thus the 5-year survival rates rarely exceed 25%. It is now generally accepted that esophageal adenocarcinoma develops from a premalignant lesion of the esophagus, also referred to as Barrett's esophagus. Barrett's esophagus is a metaplastic change of the normal squamous epithelium of the esophagus to a columnar type epithelium due to longstanding gastro-esophageal reflux [17].

Epidemiological studies suggest that the use of NSAIDs is associated with reduced incidence of esophageal cancer. Of the five published observational studies, four demonstrated a protective effect of NSAIDs (Tab. 1) [18–22]. The role of COX-2 in esophageal carcinogenesis has been emphasized by studies that show that COX-2 expression is elevated in the Barrett's metaplasia-dysplasia-adenocarcinoma sequence. In these studies, the normal squamous epithelium of the esophagus was consistently negative or only weakly positive, while elevated expression of COX-2 was found in 78–100% of the adenocar-

Table 1. Epidemiological studies on the protective effect of NSAIDs for esophageal cancer

Design	Sample size	Endpoint	RR (95% CI)	Reference
Cohort				
	635,031	Mortality	0.59 (0.34–1.03)	Thun et al. 1993 [18]
	14,407	Incidence	0.10 (0.01–0.76)	Funkhouser et al. 1995 [19]
Case control	Cases: 650 Controls: 695	Incidence	0.37 (0.24–0.58)	Farrow et al., 1998 [20]
	Cases: 12,174 Controls: 34,934	Incidence	0.64 (0.41–0.98)	Langman et al. 2000 [21]
	Cases: 1,149 Controls: 5,952	Incidence	0.80 (0.5–1.4)	Coogan et al. 2000 [22]

RR = Risk ratio; CI = Confidence interval.

Table 2. COX-2 expression in normal squamous epithelium of the esophagus and in the metaplasia-dysplasia-adenocarcinoma sequence

Normal	Metaplasia	Dysplasia	Adenocarcinoma	Reference
0%	81%	ND	100%	Wilson et al., 1998 [23]
0%	0%	ND	78%	Zimmerman et al. 1999 [24]
Significant increase in expression of COX-2				Shirvani et al. 2000 [25]
0%	75%	83% (LG) 100% (HG)	100%	Morris et al. 2001 [26]
0%	50%	ND	99%	Buskens et al. 2002 [27]

ND = Not determined; LG = Low-grade; HG = High-grade.

cinomas (Tab. 2) [23–27]. Furthermore, Wilson et al. reported an elevated COX-2 mRNA expression in 81% of the Barrett's metaplasias, and in two studies expression of COX-2 was shown to increase progressively in each stage of esophageal carcinogenesis from Barrett's metaplasia via dysplasia to adenocarcinoma on both mRNA and protein level [25, 26]. Importantly, we demonstrated recently that elevated expression of the COX-2 protein is associated with reduced survival in patients undergoing potentially curative esophageal resection due to Barrett carcinoma (Fig. 1) [27]. Our data show that patients with tumors with high COX-2 expression had a more aggressive course of disease because they were more likely to develop locoregional recurrences and distant metastases. Multivariate analysis indicated that elevated COX-2 expression was an independent prognostic variable for patient survival [27].

Currently several clinical trials are being performed on chemoprevention of esophageal cancer by using COX-2 selective inhibitors [1, 2]. In a phase II multicenter study, Forastiere et al. (Baltimore, Maryland, USA) randomize patients with low- or high-grade columnar dysplasia between the daily selective COX-2 inhibitor celecoxib and no additional treatment, with dysplasia regression as primary endpoint. In addition, randomized trials have been planned and/or recently started using COX-2 selective inhibitors after potentially curative esophageal resection as an adjuvant treatment modality. In the Amsterdam Medical Center, we are currently testing the value of celecoxib as a neo-adjuvant therapy in patients with a Barrett carcinoma by analyzing its effect on various biomarkers.

Gastric adenocarcinoma

Gastric cancer is one of the most frequent and lethal malignancies worldwide and the 5-year survival rate is only about 20% [28]. Etiological background of

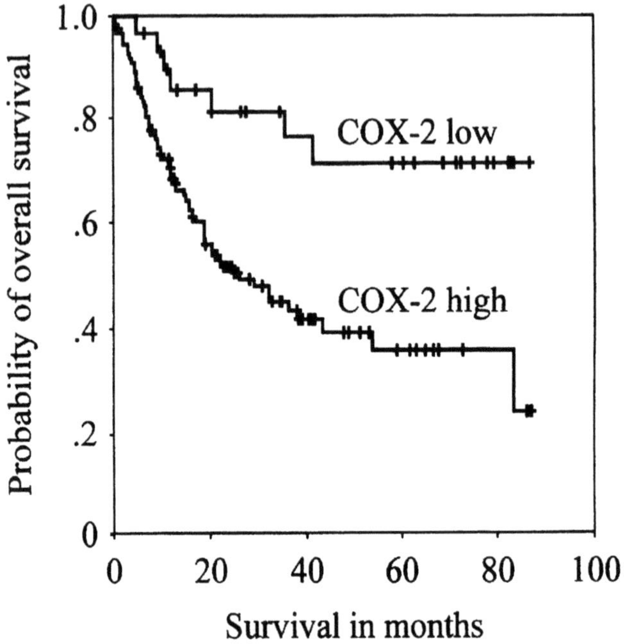

Figure 1. Survival of 145 patients with esophageal adenocarcinoma according to COX-2 protein expression as detected by immunohistochemistry. There were 30 patients with low (scores 0–1) and 115 with high (scores 2–3) COX-2 expression. A statistically significant difference was observed between the 2 groups ($p = 0.002$; log-rank test). Reproduced from Buskens et al. (2002) [27] with permission from the copyright holder.

stomach cancer is complex and combinations of environmental (diet), host factors (including *Helicobacter pylori* infection, partial gastrectomy and gastric adenomas) and genetic factors (hereditary nonpolyposis colorectal cancer syndrome and mutations of E-cadherin) play a role in gastric carcinogenesis. Gastric adenocarcinoma can be divided into histological subtypes, denoted as intestinal and diffuse types by Lauren's classification [29]. The intestinal variant has certain precursor lesions that lead to invasive carcinoma, which include chronic atrophic gastritis, intestinal metaplasia and dysplasia.

Four observational epidemiological studies have been published in which the association between NSAID use and gastric cancer was studied [18, 21, 30, 31]. Three of these studies showed a protective effect of NSAIDs against gastric cancer (Tab. 3). COX-1 was expressed in normal gastric mucosa and COX-2 expression was low or below the detection limit [32]. Expression of COX-2 was elevated in gastric adenocarcinomas when compared to the adjacent non-neoplastic mucosa, suggesting that COX-2 may play a role in gastric carcinogenesis (Tab. 4) [33–44]. Our group has shown that COX-2 protein was expressed in 58% (25/43) of intestinal-type carcinomas, but only in one of 18 diffuse type-tumors (6%) [42]. In agreement with these data, colorectal tumors with a histological pattern of signet ring cells, a typical feature of diffuse type

Table 3. Epidemiological studies on the protective effect of NSAIDs for gastric cancer

Design	Sample size	Endpoint	RR (95% CI)	Reference
Cohort				
	635,031	Mortality	0.53 (0.34–0.81)	Thun et al. 1993 [18]
	12,668	Incidence	0.93 (0.49–1.74)	Schreinemachers et al. 1994 [30]
Case control	Cases: 448 Controls: 610	Incidence	0.60 (0.41–0.90)	Zaridze et al. 1999 [31]
	Cases: 12,174 Controls: 34,934	Incidence	0.51 (0.33–0.79)	Langman et al. 2000 [21]

RR = Risk ratio; CI = Confidence interval.

Table 4. COX-2 expression in gastric adenocarcinoma

Method	COX-2 positive tumors (n)	COX-2 expression	Reference
NB	73% (11)	In contrast to COX-2, expression of COX-1 was not elevated in the tumors.	Ristimäki et al. 1997 [33]
RT-PCR	51% (37)	Correlated with depth of invasion and diameter of the tumor.	Uefuji et al. 2001 [34]
WB	73% (6)	In contrast to COX-2, expression of $cPLA_2$ and COX-1 were not elevated in the tumors.	Soydan et al., 1997 [35]
	83% (23)	Did not correlate with histology of the tumors.	Uefuji et al., 1998 [36]
	70% (100)	Correlated with LN metastasis and advanced tumor stage.	Yamamoto et al. 1999 [37]
	67% (15)	Correlated with LN metastasis.	Murata et al. 1999 [38]
	74% (42)	Correlated with LN metastasis and microvessel density.	Uefuji et al. 2000 [39]
IHC	47% (50)	More frequent in corpus cancer than in cardia cancer.	Ratnasinghe et al. 1999 [40]
	100% (104)	Was also detected in (dysplastic) gastric adenomas.	Lim et al. 2000 [41]
	43% (61)	More frequent in intestinal type than in diffuse type cancers, and was also detected in gastric dysplasias.	Saukkonen et al. 2001 [42]
	61% (33)	More frequent in intestinal type than in diffuse type cancers, and was also detected in intestinal metaplasias.	Yamagata et al. 2001 [43]
	62% (53)	Increased during progression from nonneoplastic epithelium to dysplasia.	van Rees et al. 2002 [44]

NB = Northern blot; RT-PCR = Reverse transcriptase-polymerase chain reaction; WB = Western blot; IHC = Immunohistochemistry; $cPLA_2$ = Cytosolic phospholipase A_2; LN = Lymph node.

gastric cancer, contain low levels of COX-2 immunoreactivity [45]. Recently, Yamagata et al. confirmed our results and reported that COX-2 expression is associated with intestinal type gastric cancer [43]. Since a different set of genetic alterations may take place in the sequence leading to the two histologically distinct types of gastric carcinoma [28], it is possible that a cellular event(s) typical for intestinal type carcinogenesis is responsible for induction of COX-2 expression. Furthermore, overexpression of COX-2 protein is less frequent in gastric and colorectal cancers with microsatellite instability (MSI) than in sporadic ones [37, 45, 46], which suggests that genetic changes typical for sporadic tumors more readily induce COX-2 overexpression than those associated with MSI. Interestingly, COX-2 expression is not restricted to invasive gastric adenocarcinomas, since dysplastic precursor lesions of intestinal type gastric cancer express COX-2, suggesting that COX-2 may contribute to an early event in gastric carcinogenesis [41, 42, 44].

Recent studies have shown that elevated expression of COX-2 protein is associated with invasion into gastric wall lymphatic vessels as well as with metastasis to lymph nodes [37–39]. Thus, similar to esophageal adenocarcinoma (see above) overexpression of COX-2 by gastric tumors may induce invasive and metastatic processes that can lead to a more aggressive behavior of the disease. However, in contrast to adenocarcinomas of the esophagus and the colorectum, expression of COX-2 was not found to correlate with prognosis in gastric cancer patients [38, 46]. This may indicate that the prognostic value of COX-2 is restricted to a certain subgroup of gastric cancer patients (for example early cancers), or that the role of COX-2 is different in the stomach when compared to adenocarcinomas arising in other parts of the gastrointestinal tract.

Lung adenocarcinoma

Lung cancer is one of the most frequent and fatal malignancies, the 5-year survival rate being approximately 14% [47]. The role of external risk factors such as cigarette smoke, exposure to radiation and asbestos are particularly important in the pathogenesis of lung cancer. The great majority of primary lung cancers (90–95%) are carcinomas, which are divided into two groups: small cell lung carcinoma (SCLC) and non-small cell lung carcinoma (NSCLC); adenocarcinoma and squamous cell lung cancer belong to the NSCLC group. The histological distinction between NSCLC and SCLC is important, since there are substantial differences between the two groups in both treatment and prognosis. In general, SCLC is a more aggressive disease and has already metastasized at the time of diagnosis. However, SCLC responds initially well to the treatment with chemotherapy and/or radiation therapy.

The protective role of NSAIDs is less clear for lung cancer than in gastrointestinal cancers [18, 21, 30]. However, a recent case control study showed that daily intake of NSAIDs for at least 2 years prior to interview among 489 lung cancer patients and 978 control subjects was associated with a 68% reduction in the relative risk of lung cancer (RR, 0.32; 95% CI, 0.23–0.44) [48].

Non-neoplastic human lung epithelium expresses only low or nondetectable levels of COX-2 mRNA and protein [49–51]. Elevated levels of COX-2 mRNA and protein expression have been detected in human lung NSCLC by Northern blot [51], RT-PCR [52, 53], *in situ* hybridization [54], immunoblotting [55] and immunohistochemistry [49–51, 56–61] (Tab. 5). Among

Table 5. COX-2 expression in lung adenocarcinoma

Method	COX-2 positive tumors (n)	COX-2 expression	Reference
RT-PCR	100% (19)	AdCa > SCC > SCLC	Ochiai et al. 1999 [52]
(Real-time)	100% (33)	Was an independent predictor of worse survival in patients with NSCLC	Brabender et al. 2002 [53]
ISH	60% (62)	Correlated with shortened survival of patients with stage I disease	Khuri et al. 2001 [54]
WB	82% (17)	No correlation between COX-2 and Bcl-2 expression	Sweeney et al. 2002 [55]
IHC	70% (23)	Was also detected in atypical adenomatous hyperplasia, and was higher in LN metastasis than in primary tumors. AdCa > SCC > SCLC	Hida et al. 1998 [49]
	100% (8)	Lung cancer cell derived PGE_2 mediate immunosuppression *in vitro*	Huang et al. 1998 [50]
(IHC and NB)	90% (21)	Was also detected in atypical alveolar epithelium, and was higher in well- than in poorly differentiated tumors. AdCa > SCC	Wolff et al. 1998 [51]
	72% (130)	Correlated with shortened survival of patients with stage I disease	Achiwa et al. 1999 [56]
(IHC and ISH)	80% (10)	AdCa > SCC	Watkins et al. 1999 [57]
	90% (88)	Was also detected in atypical adenomatous hyperplasia	Hosomi et al. 2000 [58]
	92% (13)	AdCa > SCC	Soslow et al. 2000 [59]
	41% (51)	AdCa > SCC	Hastürk et al. 2002 [60]
	95% (102)	Co-localized with laminin-5 at the invasive front of early-stage tumors	Niki et al. 2002 [61]

RT-PCR = Reverse transcriptase-polymerase chain reaction; ISH = *In situ* hybridization; WB = Western blot; IHC = Immunohistochemistry; NB = Northern blot; AdCa = Adenocarcinoma; SCC = Squamous cell carcinoma; SCLC = Small cell lung cancer; NSCLC = Non-small cell lung cancer; LN = Lymph node.

NSCLCs the level of COX-2 expression has been consistently found to be higher in adenocarcinomas than in squamous cell carcinomas, while SCLCs express only low or undetectable levels of COX-2 [49, 51, 52, 57, 59, 60]. Interestingly, COX-2 overexpression appears to be a marker for poor survival among patients with early stage adenocarcinomas of the lung, but not in more advanced lung tumors [53, 54, 56]. To this end it is interesting to note that we found highest level of COX-2 mRNA and protein in the well-differentiated adenocarcinomas of the lung [51], and that COX-2 is already expressed in precursor lesions of this disease [49, 51, 58]. Thus, COX-2 expression is an early event in the carcinogenic sequence leading to adenocarcinoma of the lung and may play a more significant role at an early stage than in advanced carcinomas. Mechanistically COX-2 may be related to invasion and metastasis, since co-localization of COX-2 and laminin-5 was found at the invasive front of early-stage lung adenocarcinomas [61] and since higher COX-2 levels were found in lymph node metastases than in primary tumors [49]. Currently phase II trials are performed using COX-2 selective inhibitors as a neoadjuvant treatment in conjunction with preoperative chemotherapy or as an adjuvant modality with postoperative chemotherapy of NSCLC.

Breast cancer

Breast cancer is the most frequently diagnosed cancer in women [47]. The protective role of NSAIDs is not clear for breast cancer [18, 21, 30], but a recent meta-analysis suggested that the incidence of this disease might be reduced among NSAID users [62]. Furthermore, experimental work suggests that COX-2 may be linked to mammary gland carcinogenesis, since selective COX-2 inhibitors suppress tumorigenesis in rodent models of breast cancer [63–68]. Moreover, expression of COX-2 as such was sufficient for formation of invasive breast tumors in a transgenic mouse model [69].

COX-2 mRNA and protein are expressed in breast adenocarcinoma specimens, but not in benign epithelium of the breast [59, 70, 71]. Our results suggest that COX-2 is mainly expressed by the neoplastic cells of the tumors when compared to the stromal compartment, and that elevated COX-2 expression is more frequent in ductal type than in lobular type carcinomas (40% *versus* 30% positive, respectively) [72]. Localization of COX-2 to the cytoplasm of the cancer cells is consistent with observations in other studies [59, 70, 73]. In addition to invasive carcinomas, *in situ* carcinomas of ductal type expressed COX-2 [73]. A recent study also showed that COX-2 levels are elevated in breast cancer samples that overexpress HER-2 [74].

Our data show that expression of COX-2 protein is associated with an unfavorable outcome as detected by immunohistochemistry in tissue array of 1576 invasive breast cancer specimens (Fig. 2) [72]. In this material, elevated expression of COX-2 correlated with poor prognostic markers, such as tumor size over 2 cm, axillary lymph node metastasis, high histological grade, high

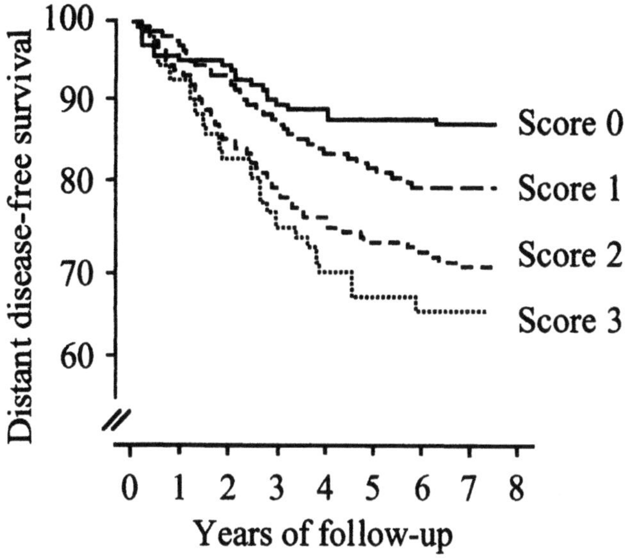

Figure 2. Distant disease-free survival (DDFS) of 1576 breast cancer patients according to COX-2 protein expression as detected by immunohistochemistry. DDFS of patients with no COX-2 expression (score 0; $n = 133$) or with weak (score 1; $n = 854$), moderate (score 2; $n = 511$) or strong (score 3; $n = 78$) COX-2 expression. Elevated expression of COX-2 protein correlated with reduced survival ($p < 0.0001$; log-rank test for trend). Reproduced from Ristimäki et al. (2002) [72] with permission from the copyright holder.

proliferation rate, negative estrogen receptor immunoreactivity, high p53 immunopositivity and amplification of HER-2 oncogene. Interestingly, the prognostic value of COX-2 expression tends to be more marked in certain subgroups of patients, e.g., in cancers with estrogen receptor positivity, low level of p53 expression and no amplification of the HER-2 oncogene (Tab. 6).

Thus, our data suggest that the tumor promoting effect of COX-2 may not be evenly distributed throughout the material and that certain subgroups of patients might benefit more from COX-2 targeted treatment. The fact that elevated expression of COX-2 is associated with poor survival in estrogen receptor positive tumors is of particular interest, since COX-2-derived prostanoids have been implicated in the enhancement of stromal cell aromatase expression [75, 76]. It is thus possible that elevated COX-2 expression in estrogen receptor positive cancers could enhance a growth-promoting microenvironment for the tumor cells by inducing estrogen production via the aromatase pathway. It should be pointed out that induction of aromatase expression is most likely not the only tumor promoting effect of COX-2 (see below), and that our results do not rule out the possibility that COX-2 inhibitors would work as adjuvant drugs in combination with for example HER-2 inhibitors. To this end, several ongoing and/or proposed trials study chemopreventive, neoadjuvant (biomarker studies) and adjuvant (combined with anti-estrogen and/or anti-HER-2) effects of selective COX-2 inhibitors.

Table 6. Five-year distant disease-free survival (DDFS) according to COX-2 expression level

Parameter	COX-2 score	N-value (%)	DDFS (CI 95%)	P-value*
ER positive	0–1	658 (66.6%)	86 (84–89)	<0.0001
	2–3	330 (33.4%)	76 (72–81)	
ER negative	0–1	213 (47.7%)	70 (64–77)	NS; 0.396
	2–3	234 (52.3%)	68 (62–74)	
p53 negative/low	0–1	679 (63.5%)	85 (82–88)	<0.0001
	2–3	391 (36.5%)	74 (70–79)	
p53 positive	0–1	99 (40.9%)	64 (54–74)	NS; 0.842
	2–3	143 (59.1%)	68 (60–76)	
HER-2 negative	0–1	732 (63.3%)	86 (84–89)	<0.0001
	2–3	425 (36.7%)	77 (73–81)	
HER-2 positive	0–1	130 (49.6%)	63 (54–71)	NS; 0.729
	2–3	132 (50.4%)	59 (50–68)	

*Log-rank test
CI = Confidence interval
NS = Not significant
ER = Estrogen receptor
Modified from Ristimäki et al. (2002) [72] with the permission from the copyright holder.

Mechanism of COX-2 -induced carcinogenesis

It is not known whether COX-2 contributes to human carcinogenesis by removing excess arachidonic acid, by producing prostanoids or by metabolizing other compounds. It has, however, been demonstrated that in experimental conditions many effects of COX-2 can be facilitated by PGE_2, which include stimulation of proliferation (in many cases by inhibition of apoptosis), enhanced invasion and metastasis, promotion of angiogenesis and induction of immunosuppression (Fig. 3) [77, 78]. The central role of PGE_2 in tumorogenesis has been further emphasized using homozygous deletion of one of the PGE_2 receptors EP2, which lead to reduced number and size of intestinal polyps in the $Apc^{\Delta 716}$ model [79]. Our clinical data support the role of COX-2 in promoting formation of metastases in both esophageal and breast adenocarcinomas [27, 72]. In case of gastric cancer other groups have found COX-2 expression to correlate with lymph node metastases and angiogenesis [37–39].

The mechanism by which COX-2 is upregulated in human cancers is largely unknown. One possibility is that cancer cells become intrinsically more active in expressing COX-2, since COX-2 gene expression is induced by activation of oncogenes, such as Ras in case of colorectal cancer cells. This leads to subsequent activation of signal transducers (MAPKs and pAkt) that can stimulate COX-2 transcription and increase the stability of the transcript (Fig. 4) [77, 78]. Furthermore, since COX-2 gene has been shown to be

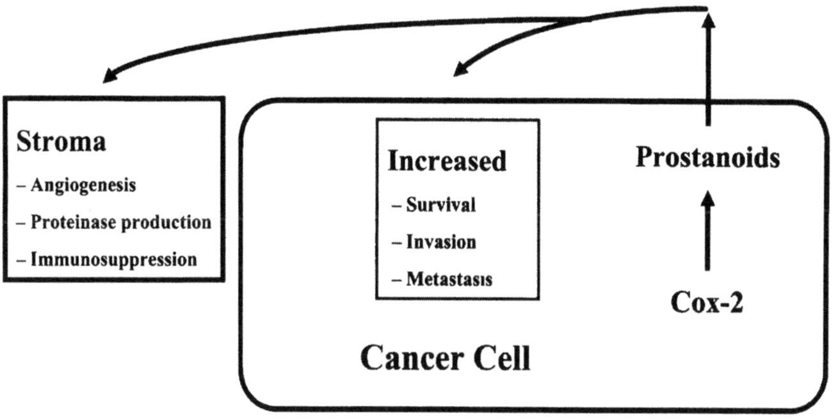

Figure 3. COX-2-derived prostanoids (in most cases PGE_2) inhibit tumor cell apoptosis, enhance production of angiogenic factors, such as vascular endothelial growth factor, increase invasion and metastasis, possibly via activating matrix metalloproteinases and act as immunosuppressive agents. Target cells can be either the cancer cell itself (autocrine) or stromal cells (paracrine).

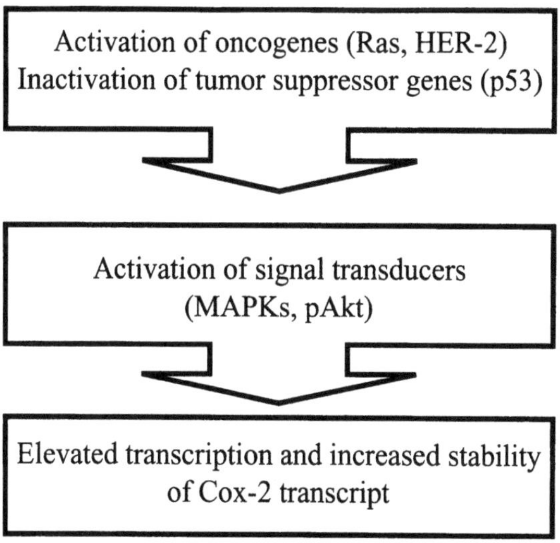

Figure 4. Regulation of COX-2 expression in cancer cells is primarily regulated by intracrine phenomenons such as dysregulation of oncogenes (activating mutation or gene amplification) or tumor suppressor genes (inactivating mutation, gene silencing or deletion). This leads to subsequent activation of signal transducers (MAPKs and pAkt) that can stimulate COX-2 transcription and increase stability of the transcript.

induced in p53 defective cells and downregulated by wild type p53 [80, 81], a direct link between a defective p53 pathway and elevated levels of COX-2 expression in cancer cells may exist. Thus, inactivation of tumor suppressor genes and activation of oncogenes may induce COX-2 expression in human carcinomas. Our results support this hypothesis, because elevated COX-2 expression was significantly more common in invasive breast cancers with high expression of p53 (a marker for inactivating mutations of p53) or with amplification of the HER-2 oncogene [72]. Other researchers have also been able to correlate COX-2 expression to overexpression of HER-2 in case of breast cancer [74] and to mutations of p53 in the case gastric cancer [82]. However, because elevated COX-2 expression is not restricted to p53- and HER-2-positive tumors, other factors and/or mechanism are probably linked to elevated COX-2 expression. Since the stroma of human adenocarcinomas often harbors inflammatory cells, it is possible that for example macrophages release factors that regulate tumor cell COX-2 expression in a paracrine manner.

Conclusions

COX-2 is widely expressed in different types of human adenocarcinomas. However, even within a single organ site the frequency of COX-2 expression may differ considerably depending on histology (intestinal > diffuse gastric cancer; ductal > lobular breast cancer) [42, 43, 72] and genotype (sporadic > MSI positive gastric and colorectal cancers) of the tumor [37, 45, 46]. Even more importantly, the prognostic significance of COX-2 may not be evenly distributed even within one subtype of adenocarcinoma in a single organ site. The prognostic value of COX-2 seems to depend on tumor stage in the case of lung adenocarcinoma [54, 56] and on molecular make up of the tumor in the case of breast adenocarcinoma [72]. Interestingly, it was recently suggested that a COX-2 variant (Val511Ala) might be associated with a reduced risk for colorectal neoplasia [83]. Thus, certain polymorphisms in COX-2 gene may mimic the effect of NSAIDs in reduction of colorectal cancer. While prognostic significance of COX-2 has been most clearly shown for certain types of adenocarcinomas, it was recently published that elevated bone marrow COX-2 levels are associated with reduced survival in chronic myeloid leukemia [84]. Thus, COX-2 selective inhibitors might also have a role in the treatment of non-epithelial malignancies.

Clinically it will be important to determine whether selective COX-2 inhibitors will deliver safer and at least as effective anti-cancer properties as nonselective NSAIDs. In addition, subtypes of premalignant lesions and invasive cancers (based on histology, stage and/or genotype) that are especially sensitive to these drugs should be recognized. Finally, it will be important to perform biomarker studies using selective COX-2 inhibitors in a neoadjuvant setting, integrating them into already existing radio- and/or chemotherapy protocols as adjuvant treatments and at the same time continue research on alternative and hopefully less toxic combination treatments of cancer. One such novel treatment option could be the combination of NSAIDs with inhibitors of Erb/HER family of growth factor receptors [68, 84–86]. All this should help to target the use of COX-2 selective drugs to those cancer patients who most likely will benefit from this treatment. Furthermore, understanding the mechanisms of COX-2 overexpression and the discovery of its role in tumor formation can eventually lead to development of new treatment modalities for cancer.

Acknowledgements
Original work was supported by the Academy of Finland, the Finnish Cancer Foundation and the Helsinki University Central Hospital Research Funds.

References

1 Anderson WF, Umar A, Viner JL, Hawk ET (2002) The role of cyclooxygenase inhibitors in cancer prevention. *Curr Pharm Des* 8: 1035–1062
2 Thun MJ, Henley SJ, Patrono C (2002) Nonsteroidal anti-inflammatory drugs as anticancer agents: mechanistic, pharmacologic, and clinical issues. *J Natl Cancer Inst* 94: 252–266
3 Eberhart CE, Coffey RJ, Radhika A, Giardiello FM, Ferrenbach S, DuBois RN (1994) Up-regulation of cyclooxygenase 2 gene expression in human colorectal adenomas and adenocarcinomas. *Gastroenterology* 107: 1183–1188
4 Kargman SL, O'Neill GP, Vickers PJ, Evans JF, Mancini JA, Jothy S (1995) Expression of prostaglandin G/H synthase-1 and -2 protein in human colon cancer. *Cancer Res* 55: 2556–2559
5 Sano H, Kawahito Y, Wilder RL, Hashiramoto A, Mukai S, Asai K, Kimura S, Kato H, Kondo M, Hla T (1995) Expression of cyclooxygenase-1 and -2 in human colorectal cancer. *Cancer Res* 55: 3785–3789
6 Oshima M, Dinchuk JE, Kargman SL, Oshima H, Hancock B, Kwong E, Trzaskos JM, Evans JF, Taketo MM (1996) Suppression of intestinal polyposis in $Apc^{\Delta 716}$ knockout mice by inhibition of cyclooxygenase 2 (COX-2). *Cell* 87: 803–809
7 Reddy BS, Rao CV, Seibert K (1996) Evaluation of cyclooxygenase-2 inhibitor for potential chemopreventive properties in colon carcinogenesis. *Cancer Res* 56: 4566–4569
8 Kawamori T, Rao CV, Seibert K, Reddy BS (1998) Chemopreventive activity of celecoxib, a specific cyclooxygenase-2 inhibitor, against colon carcinogenesis. *Cancer Res* 58: 409–412
9 Jacoby RF, Seibert K, Cole CE, Kelloff G, Lubet RA (2000) The cyclooxygenase-2 inhibitor celecoxib is a potent preventive and therapeutic agent in the min mouse model of adenomatous polyposis. *Cancer Res* 60: 5040–5044
10 Reddy BS, Hirose Y, Lubet R, Steele V, Kelloff S, Paulson S, Seibert K, Rao C (2000) Chemoprevention of colon cancer by spesific cyclooxygenase-2 inhibitor, celecoxib, administered during different stages of carcinogenesis. *Cancer Res* 60: 293–297
11 Chulada PC, Thompson MB, Mahler JF, Doyle CM, Gaul BW, Lee C, Tiano HF, Morham SG, Smithies O, Langenbach R (2000) Genetic disruption of Ptgs-1, as well as Ptgs-2, reduces intestinal tumorigenesis in Min mice. *Cancer Res* 60: 4705–4708
12 Sheehan KM, Sheahan K, O'Donoghue DP, MacSweeney F, Conroy RM, Fitzgerald DJ, Murray FE (1999) The relationship between cyclooxygenase-2 expression and colorectal cancer. *JAMA* 282: 1254–1257
13 Steinbach G, Lynch PM, Phillips RK, Wallace MH, Hawk E, Gordon GB, Wakabayashi N, Saunders B, Shen Y, Fujimura T et al (2000) The effect of celecoxib, a cyclooxygenase-2 inhibitor, in familial adenomatous polyposis. *N Engl J Med* 342: 1946–1952
14 Blot WJ, Devesa SS, Kneller RW, Fraumeni JF Jr, (1991) Rising incidence of adenocarcinoma of the esophagus and gastric cardia. *JAMA* 265: 1287–1289
15 Pera M, Cameron AJ, Trastek VF, Carpenter HA, Zinsmeister AR (1993) Increasing incidence of adenocarcinoma of the esophagus and esophagogastric junction. *Gastroenterology* 104: 510–513
16 Dolan K, Sutton R, Walker SJ, Morris AI, Campbell F, Williams EM (1999) New classification of esophageal and gastric carcinomas derived from changing patterns in epidemiology. *Br J Cancer* 80: 834–842
17 Spechler SJ, Goyal RK (1986) Barrett's esophagus. *N Engl J Med* 315: 362–371
18 Thun MJ, Namboodiri MM, Calle EE, Flanders WD, Heath CW Jr, (1993) Aspirin use and risk of fatal cancer. *Cancer Res* 53: 1322–1327

19 Funkhouser EM, Sharp GB (1995) Aspirin and reduced risk of esophageal carcinoma. *Cancer* 76: 1116–1119
20 Farrow DC, Vaughan TL, Hansten PD, Stanford JL, Risch HA, Gammon MD, Chow WH, Dubrow R, Ahsan H, Mayne ST et al (1998) Use of aspirin and other nonsteroidal anti-inflammatory drugs and risk of esophageal and gastric cancer. *Cancer Epidemiol Biomarkers Prev* 2: 97–102
21 Langman MJS, Cheng KK, Gilman EA, Lancashire RJ (2000) Effect of anti-inflammatory drugs on overall risk of common cancer: Case control study in general practice research database. *BMJ* 320: 1642–1646
22 Coogan PF, Rosenberg L, Palmer JR, Strom BL, Zauber AG, Stolley PD, Shapiro S (2000) Nonsteroidal anti-inflammatory drugs and risk of digestive cancers at sites other than the large bowel. *Cancer Epidemiol Biomarkers Prev* 1: 119–123
23 Wilson KT, Fu S, Ramanujam KS, Meltzer SJ (1998) Increased expression of inducible nitric oxide synthase and cyclooxygenase-2 in Barrett's esophagus and associated adenocarcinoma. *Cancer Res* 58: 2929–2934
24 Zimmermann KC, Sarbia M, Weber AA, Borchard F, Gabbert HE, Schror K (1999) Cyclooxygenase-2 expression in human esophageal carcinoma. Cancer Res 59: 198–204
25 Shirvani VN, Ouatu-Lascar R, Kaur BS, Omary MB, Triadafilopoulos G (2000) Cyclooxygenase-2 expression in Barrett's esophagus and adenocarcinoma: *Ex vivo* induction by bile salts and acid exposure. *Gastroenterology* 118: 487–496
26 Morris CD, Armstrong GR, Bigley G, Green H, Attwood SEA (2001) Cyclooxygenase-2 expression in the Barrett's metaplasia-dysplasia-adenocarcinoma sequence. *Am J Gastroenterol* 96: 990–996
27 Buskens CJ, van Rees BP, Sivula A, Reitsma JB, Haglund C, Bosma PJ, Offerhaus GJ, van Lanschot JJ, Ristimäki A (2002) Prognostic significance of elevated cyclooxygenase-2 expression in patients with adenocarcinoma of the esophagus. *Gastroenterology* 122: 1800–1807
28 Stadtländer CT, Waterbor JW (1999) Molecular epidemiology, pathogenesis and prevention of gastric cancer. *Carcinogenesis* 20: 2195–2208
29 Laurèn P (1965) The two histological main types of gastric carcinoma: diffuse and so-called intestinal-type carcinoma. *Acta Path Microbiol Scand* 64: 31–49
30 Schreinemachers DM, Everson RB (1994) Aspirin use and lung, colon, and breast cancer incidence in a prospective study. *Epidemiology* 5: 138–146
31 Zaridze D, Borisova E, Maximovitch D, Chkhikvadze V (1999) Aspirin protects against gastric cancer: results of a case-control study from Moscow, Russia. *Int J Cancer* 82: 473–476
32 Kargman SL, O'Neill GP, Vickers PJ, Evans JF, Mancini JA, Jothy S (1995) Expression of prostaglandin G/H synthase-1 and -2 protein in human colon cancer. *Cancer Res* 55: 2556–2559
33 Ristimäki A, Honkanen N, Jänkälä H, Sipponen P, Härkonen M (1997) Expression of cyclooxygenase-2 in human gastric carcinoma. *Cancer Res* 57: 1276–1280
34 Uefuji K, Ichikura T, Mochizuki H (2001) Expression of cyclooxygenase-2 in human gastric adenomas and adenocarcinomas. *J Surg Oncol* 76: 26–30
35 Soydan AS, Gaffen JD, Weech PK, Tremblay NM, Kargman S, O'Neill G, Bennett A, Tavares IA (1997) Cytosolic phospholipase A_2, cyclo-oxygenases and arachidonate in human stomach tumours. *Eur J Cancer* 33: 1508–1512
36 Uefuji K, Ichikura T, Mochizuki H, Shinomiya N (1998) Expression of cyclooxygenase-2 protein in gastric adenocarcinoma. *J Surg Oncol* 69: 168–172
37 Yamamoto H, Itoh F, Fukushima H, Hinoda Y, Imai K (1999) Overexpression of cyclooxygenase-2 protein is less frequent in gastric cancers with microsatellite instability. *Int J Cancer* 84: 400–403
38 Murata H, Kawano S, Tsuji S, Tsuji M, Sawaoka H, Kimura Y, Shiozaki H, Hori M (1999) Cyclooxygenase-2 overexpression enhances lymphatic invasion and metastasis in human gastric carcinoma. *Am J Gastroenterol* 94: 451–455
39 Uefuji K, Ichikura T, Mochizuki H (2000) Cyclooxygenase-2 expression is related to prostaglandin biosynthesis and angiogenesis in human gastric cancer. *Clin Cancer Res* 6: 135–138
40 Ratnasinghe D, Tangrea JA, Roth MJ, Dawsey SM, Anver M, Kasprzak BA, Hu N, Wang QH, Taylor PR (1999) Expression of cyclooxygenase-2 in human adenocarcinomas of the gastric cardia and corpus. *Oncol Rep* 6: 965–968
41 Lim HY, Joo HJ, Choi JH, Yi JW, Yang MS, Cho DY, Kim HS, Nam DK, Lee KB, Kim HC (2000) Increased expression of cyclooxygenase-2 protein in human gastric carcinoma. *Clin Cancer Res* 6: 519–525

42 Saukkonen K, Nieminen O, van Rees B, Vilkki S, Härkönen M, Juhola M, Mecklin JP, Sipponen P, Ristimäki A (2001) Expression of cyclooxygenase-2 in dysplasia of the stomach and in intestinal-type gastric adenocarcinoma. *Clin Cancer Res* 7: 1923–1931

43 Yamagata R, Shimoyama T, Fukuda S, Yoshimura T, Tanaka M, Munakata A (2001) Cyclooxygenase-2 expression is increased in early intestinal-type gastric cancer and gastric mucosa with intestinal metaplasia. *Eur J Gastroenterol Hepatol* 14: 359–363

44 van Rees BP, Saukkonen K, Ristimäki A, Polkowski W, Tytgat GN, Drillenburg P, Offerhaus GJ (2002) Cyclooxygenase-2 expression during carcinogenesis in the human stomach. *J Pathol* 196: 171–179

45 Karnes WE, Jr, Shattuck-Brandt R, Burgart, LJ, DuBois RN, Tester DJ, Cunningham JM, Kim CY, McDonnell SK, Schaid DJ, Thibodeau SN (1998) Reduced COX-2 protein in colorectal cancer with defective mismatch repair. *Cancer Res* 58: 5473–5477

46 Lee TL, Leung WK, Lau JY, Tong JH, Ng EK, Chan FK, Chung SC, Sung JJ, To KF (2001) Inverse association between cyclooxygenase-2 overexpression and microsatellite instability in gastric cancer. *Cancer Lett* 168: 133–140

47 Jemal A, Thomas A, Murray T, Thun M (2002) Cancer statistics. *CA Cancer J Clin* 52: 23–47

48 Harris RE, Beebe-Donk J, Schuller HM (2002) Chemoprevention of lung cancer by non-steroidal anti-inflammatory drugs among cigarette smokers. *Oncol Rep* 9: 693–695

49 Hida T, Yatabe Y, Achiwa H, Muramatsu H, Kozaki K, Nakamura S, Ogawa M, Mitsudomi T, Sugiura T, Takahashi T (1998) Increased expression of cyclooxygenase 2 occurs frequently in human lung cancers, specifically in adenocarcinomas. *Cancer Res* 58: 3761–3764

50 Huang M, Stolina M, Sharma S, Mao JT, Zhu L, Miller PW, Wollman J, Herschman H, Dubinett SM (1998) Non-small cell lung cancer cyclooxygenase-2-dependent regulation of cytokine balance in lymphocytes and macrophages: up-regulation of interleukin 10 and down-regulation of interleukin 12 production. *Cancer Res* 58: 1208–1216

51 Wolff H, Saukkonen K, Anttila S, Karjalainen A, Vainio H, Ristimäki A (1998) Expression of cyclooxygenase-2 in human lung carcinoma. *Cancer Res* 58: 4997–5001

52 Ochiai M, Oguri T, Isobe T, Ishioka S, Yamakido M (1999) Cyclooxygenase-2 (COX-2) mRNA expression levels in normal lung tissues and non-small cell lung cancers. *Jpn J Cancer Res* 90: 1338–1343

53 Brabender J, Park J, Metzger R, Schneider PM, Lord RV, Holscher AH, Danenberg KD, Danenberg PV (2002) Prognostic significance of cyclooxygenase 2 mRNA expression in non-small cell lung cancer. *Ann Surg* 235: 440–443

54 Khuri FR, Wu H, Lee JJ, Kemp BL, Lotan R, Lippman SM, Feng L, Hong WK, Xu XC (2001) Cyclooxygenase-2 overexpression is a marker of poor prognosis in stage I non-small cell lung cancer. *Clin Cancer Res* 7: 861–867

55 Sweeney CJ, Marshall MS, Barnard DS, Heilman DK, Billings SD, Cheng L, Marshall SJ, Yip-Schneider MT (2002) Cyclo-oxygenase-2 expression in primary cancers of the lung and bladder compared to normal adjacent tissue. *Cancer Detect Prev* 26: 238–244

56 Achiwa H, Yatabe Y, Hida T, Kuroishi T, Kozaki K, Nakamura S, Ogawa M, Sugiura T, Mitsudomi T, Takahashi T (1999) Prognostic significance of elevated cyclooxygenase 2 expression in primary, resected lung adenocarcinomas. *Clin Cancer Res* 5: 1001–1005

57 Watkins DN, Lenzo JC, Segal A, Garlepp MJ, Thompson PJ (1999) Expression and localization of cyclo-oxygenase isoforms in non-small cell lung cancer. *Eur Respir J* 14: 412–418

58 Hosomi Y, Yokose T, Hirose Y, Nakajima R, Nagai K, Nishiwaki Y, Ochiai A (2000) Increased cyclooxygenase 2 (COX-2) expression occurs frequently in precursor lesions of human adenocarcinoma of the lung. *Lung Cancer* 30: 73–81

59 Soslow RA, Dannenberg AJ, Rush D, Woerner BM, Khan KN, Masferrer J, Koki AT (2000) COX-2 is expressed in human pulmonary, colonic, and mammary tumors. *Cancer* 89: 2637–2645

60 Hastürk S, Kemp B, Kalapurakal SK, Kurie JM, Hong WK, Lee JS (2002) Expression of cyclooxygenase-1 and cyclooxygenase-2 in bronchial epithelium and nonsmall cell lung carcinoma. *Cancer* 94: 1023–1031

61 Niki T, Kohno T, Iba S, Moriya Y, Takahashi Y, Saito M, Maeshima A, Yamada T, Matsuno Y, Fukayama M et al (2002) Frequent co-localization of Cox-2 and laminin-5 gamma2 chain at the invasive front of early-stage lung adenocarcinomas. *Am J Pathol* 160: 1129–1141

62 Khuder SA, Mutgi AB (2001) Breast cancer and NSAID use: a meta-analysis. *Br J Cancer* 84: 1188–1192

63 Alshafie GA, Abou-Issa HM, Seibert K, Harris RE (2000) Chemotherapeutic evaluation of cele-

coxib, a cyclooxygenase-2 inhibitor, in rat mammary tumor model. *Oncology Reports* 7: 1377–1381
64 Harris RE, Alshafie GA, Abou-Issa H, Seibert K (2000) Chemoprevention of breast cancer in rats by celecoxib, a cyclooxygenase 2 inhibitor. *Cancer Res* 60: 2101–2103
65 Nakatsugi S, Ohta T, Kawamori T, Mutoh M, Tanigawa T, Watanabe K, Sugie S, Sugimura T, Wakabayashi K (2000) Chemoprevention by nimesulide, a selective cyclooxygenase-2 inhibitor, of 2-amino-1-methyl-6-phenylimidazo[4,5-b]pyridine (PhIP)-induced mammary gland carcinogenesis in rats. *Jpn J Cancer Res* 91: 886–892
66 Kundu N, Fulton AM (2002) Selective cyclooxygenase (COX)-1 or COX-2 inhibitors control metastatic disease in a murine model of breast cancer. *Cancer Res* 62: 2343–2346
67 Connolly EM, Harmey JH, O'Grady T, Foley D, Roche-Nagle G, Kay E, Bouchier-Hayes DJ (2002) Cyclo-oxygenase inhibition reduces tumour growth and metastasis in an orthotopic model of breast cancer. *Br J Cancer* 87: 231–237
68 Howe LR, Subbaramaiah K, Patel J, Masferrer JL, Deora A, Hudis C, Thaler HT, Muller WJ, Du B, Brown AM et al (2002) Celecoxib, a selective cyclooxygenase 2 inhibitor, protects against human epidermal growth factor receptor 2 (HER-2)/neu-induced breast cancer. *Cancer Res* 62: 5405–5407
69 Liu CH, Chang SH, Narko K, Trifan OC, Wu MT, Smith E, Haudenschild C, Lane TF, Hla T (2001) Overexpression of cyclooxygenase-2 is sufficient to induce tumorigenesis in transgenic mice. *J Biol Chem* 267: 18563–18569
70 Parrett ML, Harris RE, Joarder FS, Ross MS, Clausen KP, Robertson FM. Cyclooxygenase-2 expression in human breast cancer (1997) *Int J Oncol* 10: 503–507
71 Hwang D, Scollard D, Byrne J, Levine E (1998) Expression of cyclooxygenase-1 and cyclooxygenase-2 in human breast cancer. *J Natl Cancer Inst* 90: 455–460
72 Ristimäki A, Sivula A, Lundin J, Lundin M, Salminen T, Haglund C, Joensuu H, Isola J (2002) Prognostic significance of elevated cyclooxygenase-2 expression in breast cancer. *Cancer Res* 62: 632–635
73 Half E, Tang XM, Gwyn K, Sahin A, Wathen K, Sinicrope FA (2002) Cyclooxygenase-2 expression in human breast cancers and adjacent ductal carcinoma *in situ*. *Cancer Res* 62: 1676–1681
74 Subbaramaiah K, Norton L, Gerald W, Dannenberg AJ (2002) Cyclooxygenase-2 is overexpressed in HER-2/neu-positive breast cancer: evidence for involvement of AP-1 and PEA3. *J Biol Chem* 277: 18649–18657
75 Brueggemeier RW, Quinn AL, Parrett ML, Joarder FS, Harris RE, Robertson FM (1999) Correlation of aromatase and cyclooxygenase gene expression in human breast cancer specimens. *Cancer Lett* 140: 27–35
76 Brodie AM, Lu Q, Long BJ, Fulton A, Chen T, Macpherson N, DeJong PC, Blankenstein MA, Nortier JW, Slee PH et al (2001) Aromatase and COX-2 expression in human breast cancers. *J Steroid Biochem Mol Biol* 79: 41–47
77 Dannenberg AJ, Altorki NK, Boyle JO, Dang C, Howe LR, Weksler BB, Subbaramaiah K (2001) Cyclo-oxygenase 2: a pharmacological target for the prevention of cancer. *Lancet Oncol* 2: 544–551
78 Gupta RA, Dubois RN (2001) Colorectal cancer prevention and treatment by inhibition of cyclooxygenase-2. *Nature Rev Cancer* 1: 11–21
79 Sonoshita M, Takaku K, Sasaki N, Sugimoto Y, Ushikubi F, Narumiya S, Oshima M, Taketo MM (2001) Acceleration of intestinal polyposis through prostaglandin receptor EP2 in Apc(Delta 716) knockout mice. *Nat Med* 7: 1048–1051
80 Subbaramaiah K, Altorki N, Chung WJ, Mestre JR, Sampat A, Dannenberg AJ (1999) Inhibition of cyclooxygenase-2 gene expression by p53. *J Biol Chem* 274: 10911–10915
81 Han JA, Kim JI, Ongusaha PP, Hwang DH, Ballou LR, Mahale A, Aaronson SA, Lee SW (2002) P53-mediated induction of Cox-2 counteracts p53- or genotoxic stress-induced apoptosis. *EMBO J* 21: 5635–5644
82 Leung WK, To KF, Ng YP, Lee TL, Lau JY, Chan FK, Ng EK, Chung SC, Sung JJ (2001) Association between cyclo-oxygenase-2 overexpression and missense p53 mutations in gastric cancer. *Br J Cancer* 84: 335–339
83 Lin HJ, Lakkides KM, Keku TO, Reddy ST, Louie AD, Kau IH, Zhou H, Gim JS, Ma HL, Matthies CF et al (2002) Prostaglandin H synthase 2 variant (Val511Ala) in African Americans may reduce the risk for colorectal neoplasia. *Cancer Epidemiol Biomarkers Prev* 11: 1305–1315
84 Giles FJ, Kantarjian HM, Bekele BN, Cortes JE, Faderl S, Thomas DA, Manshouri T, Rogers A, Keating MJ, Talpaz M et al (2002) Bone marrow cyclooxygenase-2 levels are elevated in chron-

ic-phase chronic myeloid leukaemia and are associated with reduced survival. *Br J Haematol* 119: 38–45

85 Torrance CJ, Jackson PE, Montgomery E, Kinzler KW, Vogelstein B, Wissner A, Nunes M, Frost P, Discafani CM (2000) Combinatorial chemoprevention of intestinal neoplasia. *Nat Med* 6: 1024–1028

86 Mann M, Sheng H, Shao J, Williams CS, Pisacane PI, Sliwkowski MX, DuBois RN (2001) Targeting cyclooxygenase 2 and HER-2/neu pathways inhibits colorectal carcinoma growth. *Gastroenterology* 120: 1713–1719

Index

MIX
Papier aus verantwortungsvollen Quellen
Paper from responsible sources
FSC® C105338

If you have any concerns about our products,
you can contact us on
ProductSafety@springernature.com

In case Publisher is established outside the EU,
the EU authorized representative is:
Springer Nature Customer Service Center GmbH
Europaplatz 3, 69115 Heidelberg, Germany

Printed by Libri Plureos GmbH
in Hamburg, Germany